Plants and Phytomolecules for Immunomodulation

Neelam S. Sangwan • Mohamed A. Farag •
Luzia V. Modolo
Editors

Plants and Phytomolecules for Immunomodulation

Recent Trends and Advances

Springer

Editors
Neelam S. Sangwan
School of Interdisciplinary and Applied Sciences, Department of Biochemistry
Central University of Haryana
Mahendergarh, Haryana, India

Mohamed A. Farag
Pharmacognosy Department
Faculty of Pharmacy
Cairo University
Cairo, Egypt

Luzia V. Modolo
Universidade Federal de Minas Gerais
Instituto de Ciências Biológicas
Belo Horizonte, Brazil

ISBN 978-981-16-8119-6 ISBN 978-981-16-8117-2 (eBook)
https://doi.org/10.1007/978-981-16-8117-2

This Springer imprint is published by the registered company Springer Nature Singapore Pte Ltd.
The registered company address is: 152 Beach Road, #21-01/04 Gateway East, Singapore 189721, Singapore

Preface

Plants have always been part of the human diet and/or survival owing to its myriad of health benefits. Plants have profound effects on the health and well-being of humans attributed to its complex chemical and diverse makeup. Medicinal plants have been known to biosynthesize and accumulate a wide array of metabolites belonging to different chemical categories and classes. These natural phytochemicals have immense diversity and uniqueness of molecules. Molecules belonging to phenylpropanoids, terpenoids, steroids, alkaloids, tannins, etc., possess a variety of pharmacological activities. Besides, many of these bioactives exert effects even when taken as part of diet, supplements, and/or as traditional herbal medicine.

At a time when the whole world has suffered from the COVID-19 pandemic that is caused by novel coronaviruses emergence, it is strongly believed that this and other viruses affect the human body by challenging the immune system. Because of that our body fails to combat the initial attack of viruses or other pathogens leading to a disease and/or even death. Thus, if innate immunity remains strong enough, disease incidences and conception can be controlled to a great extent.

The concept of this book is to capitalize upon plants which possess immunomodulation properties and hence can boost the immune system. There is a great need at present as well as in future since immunomodulation processes will always be relevant to address various emerging as well as existing diseases in humans. Better immune system will itself trigger cellular responses for neutralizing and combating the onset of disease conditions. Studies have reported a myriad of medicinal plants, their extracts, or isolated molecules which have shown potential medicinal immunomodulatory properties. Establishing such plants along with their cultivation, propagation, standardization, and bioactive characterization using modern biochemical technologies, i.e., metabolomics in relation to genomics will pave the way for a better understanding of these immune enhancer biomolecule production and moreover the scientific knowledge about their effects. This book will provide the state of the art on scientific knowledge, activities, and mechanisms and leads in the area of medicinal plants and phytochemicals with immunomodulation properties.

This book is devoted to students, academics, researchers, and industry professionals interested in biology, biotechnology, biomedical science, life science, and pharmacology. This will be a source of knowledge for a long time on terrestrial plants with immunomodulatory properties.

Such compilation as presented in this textbook shall be useful to undergraduate and graduate students besides various central and department libraries as well as to Science and Medical departments of universities.

We, as editors of this book, do hope that the reader enjoys such an exhaustive compilation and find it of great benefit.

Mahendergarh, Haryana, India — Neelam S. Sangwan
Cairo, Egypt — Mohamed A. Farag
Belo Horizonte, Brazil — Luzia Valentina Modolo

Contents

Editors and Contributors

About the Editors

Neelam S. Sangwan PhD, FNA, FNASc, FNAAS has published more than 100 research papers in journals of international repute and implemented externally competitive grant projects. She is recipient of several national awards. Her major area of research is specialized metabolites and metabolism in medicinal and aromatic plants, genomics, molecular understanding of traditional therapies, and phytoactive molecules for pharmacological investigation. Dr. Sangwan has supervised over 50 postdoctoral, PhD, and postgraduate students and mentored young interns and women scientists. She has visited The Samuel Noble Foundation, USA; John Innes Centre, Norwich, UK, for scientific research programs; and as Women Leaders in Crop Science at Clare College and Cambridge University under Newton-Bhabha program.

Mohamed A. Farag PhD Specializing in metabolomics and natural products chemistry, Mohamed A. Farag completed his PhD at Texas Tech University, USA, in 2003. Dr. Farag now works at the American University in Cairo (AUC) where his research work focuses primarily around applying innovative biochemical technologies (metabolomics) to help answer complex biological questions in medicine, herbal drugs analysis, and agriculture.

He has coauthored more than 216 scientific papers and has more than 6000 citations and an H index of 41.

In 2010, Dr. Farag was selected as a top researcher in the field of plant biology in Africa by the American Society of Plant Biology, USA. Dr. Farag is a current TWAS Fellow in Agriculture from Egypt.

Luzia V. Modolo PhD received her PhD degree in Functional and Molecular Biology in 2004 from the State University of Campinas (SP, Brazil). She is faculty member of the Department of Botany at the Federal University of Minas Gerais (MG, Brazil). Prof. Modolo is the coordinator of the Network for the Development of Novel Urease Inhibitors and her research interests include plant nutrition and secondary metabolism and signaling processes in plant tissues triggered by environmental stress. She has published over 70 peer-reviewed articles and is the author of 21 book chapters and 7 patents.

Contributors

Faridah Abas Department of Food Science, Faculty of Food Science and Technology, Universiti Putra Malaysia, Serdang, Selangor, Malaysia

Hossam M. Abdallah Department of Pharmacognosy, Faculty of Pharmacy, Cairo University, Cairo, Egypt
Department of Natural Products and Alternative Medicine, King Abdulaziz University, Jeddah, Saudi Arabia

Noha Fawzy Abdelkader Department of Pharmacology and Toxicology, Faculty of Pharmacy, Cairo University, Cairo, Egypt

Nabil Ali Al-Mekhlafi Atta-ur-Rahman Institute for Natural Product Discovery, Puncak Alam, Selangor, Malaysia
Biochemical Technology Program, Department of Chemistry, Faculty of Applied Science, Thamar University, Dhamar, Yemen

Juliana Mendes Amorim Laboratório de Farmacognosia, Programa de Pós-graduação em Ciências Farmacêuticas, Departamento de Produtos Farmacêuticos, Faculdade de Farmácia, Universidade Federal de Minas Gerais. Faculdade de Farmácia, Belo Horizonte, Minas Gerais, Brazil

Avijit Banik Department of Human Physiology, Vidyasagar University, Midnapore, West Bengal, India
Department of Microbiology, Vidyasagar University, Midnapore, West Bengal, India

Francisco Geraldo Barbosa Department of Organic and Inorganic Chemistry, Science Centre, Federal University of Ceará, Fortaleza, Ceará, Brazil

Subir Kumar Bose Agriculture Department, Himalayan Garhwal University, Pauri, Garhwal, Uttarakhand, India

Rachel Oliveira Castilho Laboratório de Farmacognosia, Programa de Pós-graduação em Ciências Farmacêuticas, Departamento de Produtos Farmacêuticos, Faculdade de Farmácia, Universidade Federal de Minas Gerais. Faculdade de Farmácia, Belo Horizonte, Minas Gerais, Brazil

Nagendra Singh Chauhan Drugs Testing Laboratory Avam Anusandhan Kendra (State Government Lab of AYUSH), Government Ayurvedic College, Raipur, Chhattisgarh, India

Narayan D. Chaurasiya Southern Research, Birmingham, AL, USA

Lei Chen College of Food Science and Technology, Guangdong Ocean University, Zhanjiang, China

Manisha Chownk Department of Biotechnology, Panjab University, Chandigarh, India

Raymond Cooper Department of Applied Biology and Chemical Technology, The Hong Kong Polytechnic University, Hung Hom, Hong Kong

Schirley Costalonga Laboratório de Genética Vegetal e Toxicológica, Departamento de Ciências Biológicas, Universidade Federal do Espírito Santo, Vitória, Espírito Santo, Brazil

Rosana C. Cruz College of Dentistry, Faculdade Padre Arnaldo Janssen, Belo Horizonte, MG, Brazil
Bioengineering Laboratory, Department of Mechanical Engineering, Universidade Federal de Minas Gerais, Belo Horizonte, MG, Brazil

Mainã Mantovanelli da Mota Laboratório de Genética Vegetal e Toxicológica, Departamento de Ciências Biológicas, Universidade Federal do Espírito Santo, Vitória, Espírito Santo, Brazil

Jose de Brito Vieira Neto Department of Physiology and Pharmacology, Drug Research and Development Center, Federal University of Ceará, Fortaleza, Brazil

Marcos Carlos de Mattos Department of Organic and Inorganic Chemistry, Science Centre, Federal University of Ceará, Fortaleza, Ceará, Brazil

Judá BenHur de Oliveira Laboratório de Genética Vegetal e Toxicológica, Departamento de Ciências Biológicas, Universidade Federal do Espírito Santo, Vitória, Espírito Santo, Brazil

Maria do Carmo Pimentel Batitucci Laboratório de Genética Vegetal e Toxicológica, Departamento de Ciências Biológicas, Universidade Federal do Espírito Santo, Vitória, Espírito Santo, Brazil

Vanessa Silva dos Santos Laboratório de Genética Vegetal e Toxicológica, Departamento de Ciências Biológicas, Universidade Federal do Espírito Santo, Vitória, Espírito Santo, Brazil

Jean Carlos Vencioneck Dutra Laboratório de Genética Vegetal e Toxicológica, Departamento de Ciências Biológicas, Universidade Federal do Espírito Santo, Vitória, Espírito Santo, Brazil

Riham Salah El-Dine Department of Pharmacognosy, Faculty of Pharmacy, Cairo University, Cairo, Egypt

Ali M. El-Halawany Department of Pharmacognosy, Faculty of Pharmacy, Cairo University, Cairo, Egypt

Mohamed A. Farag Pharmacognosy Department, Faculty of Pharmacy, Cairo University, Cairo, Egypt

Paulo Michel Pinheiro Ferreira Department of Biophysics and Physiology, Federal University of Piauí, Teresina, Brazil

Suiany Vitorino Gervásio Laboratório de Genética Vegetal e Toxicológica, Departamento de Ciências Biológicas, Universidade Federal do Espírito Santo, Vitória, Espírito Santo, Brazil

Chandradipa Ghosh Department of Human Physiology, Vidyasagar University, Midnapore, West Bengal, India

Kartik Chandra Guchhait Department of Human Physiology, Vidyasagar University, Midnapore, West Bengal, India

Neelima Gupta Department of Life Sciences, Chhatrapati Shahu Ji Maharaj University, Kanpur, UP, India

Sabrin R. M. Ibrahim Department of Chemistry, Preparatory Year Program, Batterjee Medical College, Jeddah, Saudi Arabia
Department of Pharmacognosy, Faculty of Pharmacy, Assiut University, Asyut, Egypt

Pir Mohammad Ishfaq Cancer Biology Laboratory, Department of Zoology, School of Biological Sciences, Dr. Harisingh Gour Central University, Sagar, MP, India

Nor Hadiani Ismail Atta-ur-Rahman Institute for Natural Product Discovery, Puncak Alam, Selangor, Malaysia

Jyoti Singh Jadaun Department of Botany, Dayanand Girls Postgraduate College, Kanpur, Uttar Pradesh, India

Debarati Jana Department of Human Physiology, Vidyasagar University, Midnapore, West Bengal, India

Nurkhalida Kamal Institute of Systems Biology (INBIOSIS), Universiti Kebangsaan Malaysia, Bangi, Selangor, Malaysia

Monalisha Karmakar Department of Human Physiology, Vidyasagar University, Midnapore, West Bengal, India

Swati Kumari School of Applied Sciences and Biotechnology, Shoolini University of Biotechnology and Management Sciences, Solan, HP, India

Paula Mendonça Leite Laboratório de Farmacognosia, Programa de Pós-graduação em Ciências Farmacêuticas, Departamento de Produtos Farmacêuticos, Faculdade de Farmácia, Universidade Federal de Minas Gerais. Faculdade de Farmácia, Belo Horizonte, Minas Gerais, Brazil

Jair Mafezoli Department of Organic and Inorganic Chemistry, Science Centre, Federal University of Ceará, Fortaleza, Ceará, Brazil

Lingjun Ma College of Food Science and Nutritional Engineering, China Agricultural University, Beijing, China

Tuhin Manna Department of Human Physiology, Vidyasagar University, Midnapore, West Bengal, India

Lana Grasiela Alves Marques Postgraduate in Biotechnology and Natural Resources, Federal University of Ceara, Fortaleza, Brazil

Ahmed Mediani Institute of Systems Biology (INBIOSIS), Universiti Kebangsaan Malaysia, Bangi, Selangor, Malaysia

Siddhartha Kumar Mishra Department of Life Sciences, Chhatrapati Shahu Ji Maharaj University, Kanpur, UP, India
Cancer Biology Laboratory, Department of Zoology, School of Biological Sciences, Dr. Harisingh Gour Central University, Sagar, MP, India

Luzia V. Modolo Departamento de Botânica, Instituto de Ciências Biológicas, Universidade Federal de Minas Gerais, Belo Horizonte, MG, Brazil

Gamal A. Mohamed Department of Natural Products and Alternative Medicine, King Abdulaziz University, Jeddah, Saudi Arabia
Department of Natural Products and Alternative Medicine, Faculty of Pharmacy, King Abdulaziz University, Jeddah, Saudi Arabia

Keshab Chandra Mondal Department of Microbiology, Vidyasagar University, Midnapore, West Bengal, India

Passant Elwy Moustafa Department of Pharmacology, Medical Research Division, National Research Centre, Giza, Egypt

Fátima Miranda Nunes Department of Organic and Inorganic Chemistry, Science Centre, Federal University of Ceará, Fortaleza, Ceará, Brazil

Joy Ifunanya Odimegwu Department of Pharmacognosy, Faculty of Pharmacy, College of Medicine Campus, University of Lagos, Lagos, Nigeria

Maria Conceição Ferreira Oliveira Department of Organic and Inorganic Chemistry, Science Centre, Federal University of Ceará, Fortaleza, Ceará, Brazil

Carlos Roberto Koscky Paier Department of Physiology and Pharmacology, Drug Research and Development Center, Federal University of Ceará, Fortaleza, Brazil

Amiya Kumar Panda Department of Chemistry, Vidyasagar University, Midnapore, West Bengal, India
Sadhu Ram Chand Murmu University of Jhargram, Jhargram, West Bengal, India

Neha Pandey Department of Botany, CMP PG College, University of Allahabad, Prayagraj, India

Hamza Ahmed Pantami Department of Chemistry Faculty of Science, Gombe State University, Gombe, Nigeria

Paula Roberta Costalonga Pereira Laboratório de Genética Vegetal e Toxicológica, Departamento de Ciências Biológicas, Universidade Federal do Espírito Santo, Vitória, Espírito Santo, Brazil

Claudia Pessoa Department of Physiology and Pharmacology, Drug Research and Development Center, Federal University of Ceará, Fortaleza, Brazil

Priyanka Raul Department of Human Physiology, Vidyasagar University, Midnapore, West Bengal, India

Farzana Sabir Linking Landscape, Environment, Agriculture and Food (LEAF), Instituto Superior de Agronomia, Universidade de Lisboa, Lisbon, Portugal
Research Institute for Medicines (iMed.ULisboa), Faculty of Pharmacy, Universidade de Lisboa, Lisbon, Portugal

Mohammed S. M. Saleh Department of Pharmacology, Faculty of Medicine, Universiti Kebangsaan Malaysia, Kuala Lumpur, Malaysia

Neelam S. Sangwan School of Interdisciplinary and Applied Sciences, Department of Biochemistry, Central University of Haryana, Mahendergarh, Haryana, India
School of Interdisciplinary and Applied Sciences, Department of Biochemistry, Central University of Haryana, Mahendergarh, Haryana, India

Ajay Sharma Department of Chemistry, Chandigarh University, Mohali, Punjab, India

Vikas Sharma Shri Rawatpura Sarkar Institute of Pharmacy, Datiya, Madhya Pradesh, India

Mohamed Sheashea Medicinal and Aromatic Plants Department, Desert Research Center, Cairo, Egypt

Maria Francilene Souza Silva Department of Physiology and Pharmacology, Drug Research and Development Center, Federal University of Ceará, Fortaleza, Brazil

Deependra Singh University Institute of Pharmacy, Pt. Ravishankar Shukla University, Raipur, Chhattisgarh, India

Manju Rawat Singh University Institute of Pharmacy, Pt. Ravishankar Shukla University, Raipur, Chhattisgarh, India

Anupam Tiwari Department of Botany, Dayalbagh Educational Institute, Agra, Uttar Pradesh, India

Swati Tripathi Amity Institute of Microbial Technology, Amity University, Noida, UP, India

Fang Wang College of Food Science and Engineering, Nanjing University of Finance and Economics, Nanjing, China

Jing Wang College of Biosystems Engineering and Food Science/Ningbo Research Institute, Zhejiang University, Hangzhou, China

Jianbo Xiao Department of Analytical Chemistry and Food Science, University of Vigo, Vigo, Spain

Krishna Yadav University Institute of Pharmacy, Pt. Ravishankar Shukla University, Raipur, Chhattisgarh, India

Nisha Yadav School of Interdisciplinary and Applied Sciences, Department of Biochemistry, Central University of Haryana, Mahendergarh, Haryana, India

Ritesh Kumar Yadav National Institute of Plant Genome Research, New Delhi, India

Fei Zhou College of Standardization, China Jiliang University, Hangzhou, China

Immune System and Mechanism of Immunomodulation

1

Manju Rawat Singh, Krishna Yadav, Narayan D. Chaurasiya, and Deependra Singh

Abstract

The immune system and its constituents establish an interface with the environment suitable for distinguishing "self" from "nonself" and then obliterating those elements identified as nonself. It provides both active and passive defense against anticipated intruders and works as a safety component in critical situations. The immune system is well skilled in fighting a wide array of potential microorganisms. On a physiological level, the immune system's response to an intruder is a complex and facilitated network of connections between many diverse categories of molecules, proteins, and cell types. As a result, the immune system eliminates bacteria, viruses, and other germs and particles that may pose a threat to the living form. The utilitarian immune system shields against pathogens while maintaining homeostasis in a changing environment. The condition depreciates only when the link between the host and the pathogenic intruder causes enough damage to disrupt homeostasis. Therefore, preventing any illness in such conditions requires enough knowledge, innovative interventions or therapeutic expert, and both nonspecific and explicit immunomodulation. The immune system's primary functions include disease management, tissue upkeep, and the elimination of damaged or cancerous cells. It is sometimes unavoidable for a component of the immune system to check prospective invaders, necessitating the need for other experts to manage illness. Immunomodulators are external or inborn chemicals that control or modify the extent, kind, duration, or capacity of the immune response. This book section provides a summary of the immune

M. R. Singh · K. Yadav · D. Singh (✉)
University Institute of Pharmacy, Pt. Ravishankar Shukla University, Raipur, Chhattisgarh, India

N. D. Chaurasiya
Southern Research, Birmingham, AL, USA

N. S. Sangwan et al. (eds.), *Plants and Phytomolecules for Immunomodulation*,
https://doi.org/10.1007/978-981-16-8117-2_1

system's components, including the role of immunomodulators, their sources, and comprehensive information about immunomodulation interaction.

Keywords

Immune system · Immunity · Immunomodulators · Infection · Homeostasis

1.1 Introduction

The immune system (IMS) is a fierce and multidimensional protection system that defends the individual against dangerous foreign and internal substances, such as infectious agents, cancer cells, and heterogeneous particulate matter. Mammalian IMSs essentially have the concern of distinguishing oneself from nonself. If nonself is detected, the IMS's effector mechanisms are involved; they intervene to remove what would be deemed nonself. The IMS's efficacy is negatively affected as it turns against the host organism itself. Immunity overreacts, and host antigens are attacked by the IMS, such as in autoimmune conditions, asthma, and rejection of transplants (Kumar et al. 2011; Kitcharoensakkul and Cooper 2020).

Immunomodulation (IMD) is a method of modifying the IMS, either by activating or suppressing the immune reaction, to achieve the desired response (Fig. 1.1). In some pathologies, in countering pathological agents, the IMS needs upregulation or at least the right stimulation. For both prevention and treatment, regulation of IMS response has been employed (Kitcharoensakkul and Cooper 2020). Vaccinations have been implemented to protect the body from infectious diseases. Immunotherapy for cancer treatment has since been used. In this scenario,

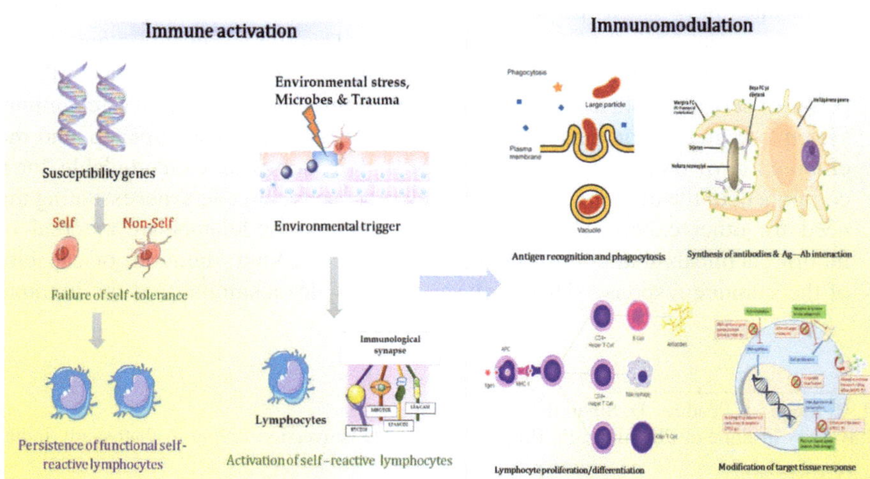

Fig. 1.1 An illustration representing the mode of immune activation and various immunomodulation approaches

the immunity needs to be improved and guided toward carcinogenesis-altered, initially owned antigens (Yadav et al. 2020). The immune response, such as autoimmunity (AI) or inflammation, reaches its initial defensive capacity at the reverse aggressive response of the IMS, and host tissues are impaired or increased production of pro-inflammatory mediators harms the host. Immunosuppressors are commonly used in this case to restore the reaction of the organism. IMD takes into account all treatment procedures intended to alter the immune response. Improved immune responses are beneficial in order to avoid pathogens, fight infections, and battle cancer in the condition of immunodeficiency (Calabrese et al. 2020; Khan and Gerber 2020).

There is a heterogeneous grouping of conditions under which disorders are caused by primary immune dysregulation. Instances of diseases within this class comprise autoimmune conditions, including rheumatoid arthritis and systemic lupus erythematosus (Amaya-Uribe et al. 2019). In such disorders, the typical control of the IMS is disrupted and triggering the IMS effector mechanisms that mistakenly damage their own body tissue to "eradicate" them. Autoimmune conditions entail the manifestation of circulating autoantibodies and immunoglobulins that react with self-antigens (Amaya-Uribe et al. 2019). There are also a large number of diseases without the typical manifestations of AI, aside from classical autoimmune diseases, but characterized by pathological system dysregulation of overactivity. Inflammatory bowel diseases, seronegative arthritis, and chronic inflammatory skin diseases such as psoriasis and eczema are examples of chronic inflammatory disorders. The hyperactive IMS corresponds to systemic inflammation that destroys tissue. In certain allergic disorders, such as asthma, there may be severe chronic inflammation, and immunosuppressive medications can be treated (Ponce 2018).

In the subsequent major clinical group, which comprises two scenarios in which the IMS is correctly roused but harms the patient, medications acting on the IMS are pretty beneficial. The very first is perceived in people undergoing allogeneic organ transplants wherein immunosuppressive medications are necessary for the prevention and treatment of transplanted allograft rejection (Faustman 2013; Socié and Blazar 2013). A predominantly successful way to treat patients with organ failure is to substitute failed organs, including the kidneys, liver, lungs, and heart, with healthy organs acquired from brain-dead donors or, in some circumstances, healthy families (Calabrese et al. 2020). However, since transplanted organs liberate antigens that the IMS identifies as nonself unless immunosuppressive medications are given, they will be immediately rejected by the IMS of transplant recipients. Graft-versus-host disease is the other clinical situation in allogeneic bone marrow transplant recipients. In this case, the immune cells of the donor identify the antigens of the recipient as nonself and develop the immune response of the recipient. Immunosuppressive drugs are very successful in regulating graft-versus-host disease by overpowering the function of the donor immune cells (Calabrese et al. 2020).

In the widest context, the term IMD is used in this chapter, which covers every intervention to alter the clinical response of the IMS. There are different health conditions in which the IMS could be manipulated and the therapeutic methods used in each of them are discussed herein.

1.2 Immune System

The IMS is an extremely specialized well-coordinated and well-programmed multi-faceted machinery of the human body involving various cellular and tissue-level entities in their appropriate functioning. They achieve their functions with the assistance of the physical boundaries separating the body from surroundings, cells inspecting signs of distress or other anomalies, other cells conducting IMS functions, and the organ in which cells work (Calabrese et al. 2020). The IMS cells perform extremely specialized functions mainly through the activities of particular molecules expressed by them.

1.2.1 The Anatomical Barriers

Epithelia shields the body from the outside and may also be regarded as part of the IMS by making this specialized membrane. Epithelia are distinct cell membranes that protect the skin, the mucous membrane of the eye, and the oral cavity, the respiratory epithelium, and the gastrointestinal epithelium that cover varied sections of the body. In order to perform the essential functions needed by that region of the body, each of these various epithelia is structured differently. The skin is also a multicellular membrane that safeguards against physical damage and averts unwarranted water evaporation (Gudjonsson et al. 2020). A single layer of columnar epithelial cells is present in the respiratory epithelium, distributed over a lamina propria that comprises structural elements and several distinct immune cells (Gea-Banacloche 2006). The epithelium in much of the gastrointestinal tract is designed to retain nutrients and fluids. Tight interstices obstruct the particulate and microbial passage between the cells of the single-layer epithelium of the large or small intestines, which is an imperative feature of the physical boundary between the host source of food and the lumen. Epithelial cells provide external information of surroundings to specialized lamina propria immune cells (Gea-Banacloche 2006).

Aside from the physical barrier raised by epithelium cells, other factors also contribute to preserving immunity on epithelial surfaces. Dendritic cells are antigen-presenting cells present under the epithelium surface where they examine the epithelium environment and potentially indeed the exterior epithelium surface. Plasma cells sited on the lamina propria of epithelial secret antibodies, in particular the secretory immunoglobulin A (IgA) subtype, passing through the epithelium barrier. The Secretory IgA performs a vivacious role in the neutralization of potentially hazardous bacteria on the exterior surface of the epithelium (Ponce 2018; Khan and Gerber 2020).

1.2.2 Cellular Warriors of Immunity

There are numerous factions of highly specialized cells in the IMS (Calabrese et al. 2020). A systematic understanding of the existing knowledge of these cells and the

Table 1.1 The different kinds of cells involving various functions in the progression of immunity

Immune cell	Specification	Function in immunity
Dendritic cells	Dexterous cells for antigen processing and presentation	Antigens are taken up, processed, and presented to LYM
Eosinophils	Comprise thick eosinophilic granules	Plays the role in the elimination of antibody-coated pathogens
Lymphocytes (LYM)	T cells, which indeed mature in the thymus, and B-LYM, which mature in the bone marrow, are the two types of LYM	Simply expressing receptors specific to antigens allow invading infections to be recognized
Mast cells	Comprise granules harboring the vasoactive arbitrator histamine, which is a facilitator of allergic responses	Whenever a degranulation emits histamine, it causes plasma and cells to flow out of nearby capillaries
Monocytes and macrophages	Tissue macrophages are migrating in the specific form of monocytes	Bacterial phagocytosis and demise, as well as antigen processing and presentation from foreign pathogens
Natural killer cells	They are similar in nature to LYM but without antigen receptors	Competent of surging lytic granules strong enough to kill virus-infected cells
Neutrophils	The most prevalent sort is leukocyte, a multinucleate cell with a limited half-life that is generated and secreted from the bone marrow	During an infection, it performs phagocytosis of pathogenic microbes, and their elimination increases exponentially

specialized molecules expressed in them is beyond the scope of this chapter, but a brief framework of the fundamental notions is given in Table 1.1.

1.2.3 Organ of Immunity

The body comprises an assortment of lymphoid organs that function as anatomical regions for most of the IMS's action. Cells and IMS molecules act together and coordinate the elusive and multifarious processes involved in immune responses at these sites (Zimmerman et al. 2020).

1.2.3.1 Primary and Secondary Lymphoid Organs

The development site of most cells of the IMS is the bone marrow. The cells are made in the bone marrow from a pluripotent hematopoietic stem cell and follow a process of differentiation which results in the creation of lively mature immune cells. B and T lymphocytes (LYM) derive from a single lymphoid progenitor, while neutrophils, eosinophils, basophils, monocytes, and dendritic cells arise as a single myeloid precursor (Gea-Banacloche 2006; Zimmerman et al. 2020). From the bone marrow, LYM are released and circulating in the blood. During development, B and T LYM acquire the expression of an extremely complex continuum of antigen receptors on their cell exterior (Ponce 2018; Zimmerman et al. 2020). The T-cell receptor is a transmembrane multiprotein receptor which is multifarious, whereas the

B-cell receptor is a type of transmembrane immunoglobulin (Gea-Banacloche 2006). Complex somatic rearrangement contrivance in the gene segments encodes the antigen-binding parts of the receptors and produces the variation of the repertoire of these receptors during lymphocyte development. Moreover, certain fragments of these genes may develop somatic hypermutation, which enhances even more diversity. This mechanism leads to the slightly varying antigen receptors in each receptor being of different antigen affinity on the surface (Botos et al. 2011; Garzorz-Stark et al. 2018).

LYM are freed into circulation after development in the bone marrow. The thymus is home to immature T LYM, while the B LYM returns to the bone marrow. The immature T and B LYM encounter a selection and maturation course there, early in life, culminating in the preservation of LYM that does not identify self-antigens and the removal of those that do (Santori 2015). The theory of clonal selection suggests that LYM expressing receptors having a great affinity for antigens that they experience in the thymus and bone marrow during maturation are chosen for apoptosis removal (Santori 2015). Conversely, LYM whose receptors do not undergo a high-affinity antigen live. The consequence of this filtering scheme is that only LYM that exhibits receptors for non-self-antigens survives, whereas those that exhibit receptors with higher self-affinity are abolished (Santori 2015). This is the foundation on which self and nonself will be marginalized against the IMS. B and T LYM then circulate in small numbers and as a large cell population with a strikingly large antigen receptor repertoire capable of noticing a myriad of non-self-antigens.

The secondary lymphoid organs comprise lymph nodes, mucosal-associated lymphoid structures, and spleen (Santori 2015). Interaction of immune cell is taken place here at these sites and is desired for the proper functioning of the IMS.

1.2.4 Innate and Adaptive Immunity

The IMS's cells, tissues, and molecules interact in several functional paradigms that can be defined generally as innate and adaptive (acquired) immunity. Innate immunity relates to immune responses that are not antigen-specific, while adaptive immunity belongs to reactions that are specific to the antigen (Gea-Banacloche 2006). There is a strong degree of intercellular communication between innate and adaptive IMSs. In Table 1.2, the main aspects of the innate and adaptive IMSs are compared.

Stimulation of innate immune cells, like macrophages and neutrophils, is essential for a typical immune response. A multitude of factors may activate the cells, but they typically involve the binding of a microbial source of the pattern recognition receptor found in these cells. For example, several cells of the innate IMS display a model-recognition receptor named a toll-like receptor on the cell surface (Kumar et al. 2011; Zimmerman et al. 2020). A wide range of microbial ligands, such as lipopolysaccharides or flagellins, can trigger toll-like receptors. The induced phenotype is followed by cytokine secretion, such as interleukin (IL)-1 tumor necrosis

Table 1.2 A comparative depiction of the functioning of innate and adaptive immune system

Function/ specification	Inherent immune response-based system	Acquired immune response-based system
Receptor specificity	Non-explicit	Specific to antigen and pathogenic microbes
Composition	Composed of leukocytes, macrophages, and natural killer cells	Composed of B cells, T cells, and antigens
Time of response	Instantaneous utmost response on initial exposure	The time elapsed between the first stimulation and peak reaction.
Cellular involvement	Cellular and humoral modules	Both cellular and humoral modules
Memory	There is no immunological remembrance.	Initial exposure causing the development of immunological memory
Receptor diversity	Low	High

factor (TNF)-α, IL-6, chemokines such as IL-8, and other immune arbiters such as eicosanoids (Gea-Banacloche 2006; Zimmerman et al. 2020). In sum, an innate immune response is a fast early defense antigen-specific mechanism whose primary purpose is to deter and suppress microbial infections.

The innate immune response is generally adequate to handle small microbial infections and prevent them from spreading. The innate IMS, however, also stimulates the adaptive IMS, which is necessary for many infections to be overcome (Kumar et al. 2011). Invading bacteria are engulfed and destroyed by cells like dendritic cells and macrophages and subsequently converted their proteins into small peptides. Peptides retrieved from viral proteins expressed in certain cells can be processed by most virus-infected cells. In the form of large histocompatibility complex molecules, the microbial-derived peptides are then present on the surface of certain antigen-presenting cells (Kumar et al. 2011).

The adaptive immune response is activated by an antigenic-specific receptor on the surface of B and T LYM that recognizes these peptides. B or T LYM are triggered because of the high affinity between their receptor and an antigen which results in an array of lymphocytic molecular events. Herein, the stimulated expression of IL-2 is evident that is liable for the growth of LYM (Santori 2015). In this fashion, even erratic cells in a diverse lymphocyte populace can greatly extend their clonal system so that they can perform their immune functions. These cells need activation by costimulatory receptor-like CD28, which is most essential for effective activation of T LYM, in addition to attaining an activation signal by the T-cell receptor. Two signals are also essential for the complete activation of B LYM. They first arise from the B-cell receptor, which typically originates from neighboring T LYM under the influence of secreted cytokines after it has been identified with a high-affinity antigen (Santori 2015).

1.3 Immunity-Associated Disorders

The IMS is standout to be the most essential in the human body due to its specificity, adaptive, and inductive nature. A deficit in functioning of our natural defense mechanism leads to deviated action casing different disordered conditions that can be categories into AI, immunodeficiency, and hypersensitivity.

1.3.1 Autoimmunity

Autoimmune diseases are pathophysiological circumstances that is an outcome from loss of self-resistance and the succeeding immune annihilation of host tissues. AI is intervened by a mixture of molecular and cellular events and feedbacks. The advancement of autoimmune diseases is an enormously composite process in which acknowledgment of self-antigens by LYM is crucially engaged with pathologic organ damage. Autoimmune diseases are developed as an intricate trait, with different loci governing different characteristics of disease susceptibility (Scanlin 2014; Antiochos and Rosen 2019). Certain ecological impacts, for example, tobacco smoke, UV light, or infections- and diseases-causing agents, may interplay with this hereditary inclination to start the disease progression (Fig. 1.2). Some IMS reactions prompt by a pathogen, whose protein(s) hold auxiliary structural similarity to locales on proteins of the host. Therefore, antibodies induced against a pathogen may cross-respond with a self-protein and acting as autoantibodies, and the concerned autoantigen at that point gives a source to tenacious incitement (Ray et al. 2013; Hirsch and Ponda 2015; Rosenblum et al. 2015).

The body contains huge quantities of potentially autoreactive B and T cells even though there is an assortment of components working to set up self-resistance amid lymphocyte development. This is especially valid for T cells that are not erased by self-epitopes. The manifestation of autoreactive B cells in the typical inhabitants is all around displayed by the advancement of autoantibodies when autoantigens are infused with adjuvants in normal animals (e.g., antithyroglobulin). The existence of autoreactive T cells in a healthy individual is displayed by the creation of IMS lines of T cells when typical circulating T cells are stimulated by the suitable autoantigen (e.g., myelin essential protein) and IL-2 (Skepner et al. 2014; Zitti and Bryceson 2018).

The idea of AI was first anticipated by Nobel Laureate Paul Ehrlich toward the beginning of the nineteenth century, and he portrayed it as "horror autotoxicus." His investigations drove him to presume that the IMS is ordinarily centered on reacting to foreign materials and has a built-in propensity to abstain from assaulting self-tissues. However, when this procedure turns out erroneously, the IMS can assault self-tissues causing autoimmune disease (Bray et al. 2018). Ehrlich far along redefined his hypothesis to perceive the probability of autoimmune tissue assaults but assumed that certain inborn protection contrivances would keep the immune reaction from getting to be obsessive. In 1904, this hypothesis was tested by the revelation of an element in the patient's serum with paroxysmal cold hemoglobinuria

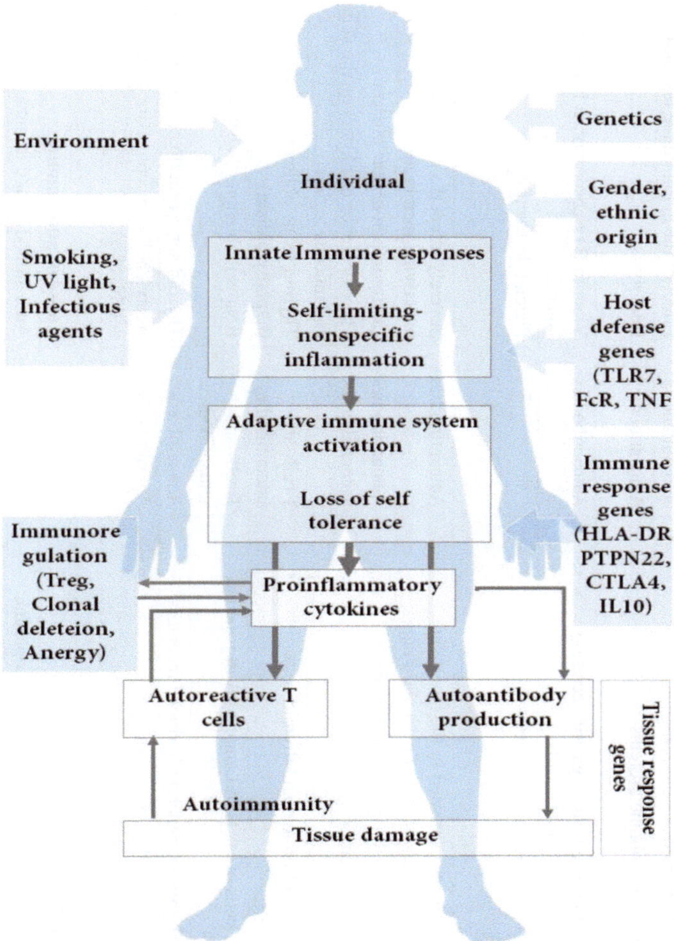

Fig. 1.2 Various factors and their recognized pathways influencing the progression and perpetuation of autoimmune diseases

that responded with RBCs. Amid the next decades, various conditions were interrelated to autoimmune reactions, although, the regulatory status of Ehrlich's hypothesis hindered the comprehension of these discoveries. Immunology turned into a biochemical instead of a clinical discipline (Poletaev et al. 2012; Rose and Mackay 2019a). By the 1950s, the cutting-edge comprehension of autoantibodies and IMS diseases began to spread. Lately, it has turned out to be accredited that autoimmune reactions are an obligatory part of the vertebrate IMS (now and again named "natural AI"), typically preventing from building disease by the mechanism of immunological tolerance to self-antigens (Zheng et al. 2020).

Autoimmune disorders can be widely classified into systemic and organ-specific (Table 1.3), contingent upon the foremost clinicopathologic character of each

Table 1.3 Classification of human autoimmune and their cause of autoimmunity

SN	Autoimmune disorder	Affected organ	Self-antigen	Cause of autoimmune	Effect of the autoimmune response
Human organ-specific autoimmune diseases					
1.	Autoimmune hemolytic	RBCs	RBC membrane proteins	Autoantibodies	Hemolysis
2.	Pernicious anemia	RBCs	a. Intrinsic factor I gastric secretion b. Gastric parietal cells	Autoantibodies	Vitamin B12 absorption is hampered (anemia).
3.	Idiopathic thrombocytopenic purpura	Platelets	Platelet membrane proteins		Platelet destruction
4.	Goodpasture's syndromes	Kidney and lungs	Antigens expressed in the basement membranes of the kidneys and lungs	Autoantibodies	Severely affect the kidney and lungs' underlying membranes
5.	Bullous pemphigoid	Skin	Skin underlying membrane zone	Autoantibodies	Configuration of stiff blisters
6.	Pemphigous vulgaris	Skin	Desmoglein-3	Autoantibodies	The emergence of blisters on the skin and mucous region
7.	Hashimoto's thyroiditis	Thyroid	Thyroid protein and thyroid cells	T cells and autoantibodies	Thyroid tissue obliteration (hypothyroidism)
8.	Insulin-dependent diabetes mellitus	Pancreas	β cells of islets of Langerhans in pancreas	T cells and autoantibodies	Beta-cell ablation (diabetes)
9.	Graves disease	Thyroid	Thyroid-stimulating hormone receptors	Autoantibodies	Thyroid activation (hyperthyroidism)
10	Myasthenia gravis	Muscle	Acetylcholine receptors	Autoantibodies	Acetylcholine receptors are destroyed and blocked.
Human systemic autoimmune diseases					
SN.	**Autoimmune disease**		**Self-antigens**		**Cause of autoimmune**
1.	Systemic lupus erythematosus (SLE)		DNA, nuclear proteins, and cytoplasmic		Autoantibodies and immune complex
2.	Rheumatoid arthritis		Joint		Autoantibodies and immune complex

3.	Scleroderma	Heart, lung, kidneys, and gastrointestinal tract	Autoantibodies
4.	Sjogren's syndrome	Salivary glands, liver, kidney, and thyroids	Autoantibodies
5.	Multiple sclerosis	Brain	Autoantibodies and T_C cells

disorder. It is categorized as organ-specific when the reaction is essentially against antigens confined to specific organs or as non-organ-specific when the reaction is harmonized against widespread antigens. Many autoimmune disorders have been affirmed, and IMS components are associated to supplement the pathogenesis with about the same number of more constant provocative conditions. Some autoimmune disorders influence a solitary organ (e.g., immune-mediated [Hashimoto's] thyroiditis), others destruct numerous sites of a solitary organ system (e.g., immune-intervened vasculitis), but others disrupt various organs spreading over numerous systems (e.g., systemic lupus erythematosus [SLE]) (Zeher and Szegedi 2007; Rose 2014). As a minimum, one autoimmune disorder has been recognized for almost every organ and tissue in the body. While a few organs are assaulted by various autoimmune disorders, the diseases influencing a solitary site frequently have particular clinical and pathological appearances, probably because of dissimilar antigenic boosts rather than anatomical and physiological contrasts among people. Thriving proof recommends that other health circumstances, counting yet not restricted to atherosclerosis, amyotrophic lateral sclerosis, schizophrenia, infertility, and postmenopausal osteoporosis, may have an autoimmune element as well (McGonagle and McDermott 2006; Zeher and Szegedi 2007; Rose 2014).

1.3.2 Immunodeficiencies

Immunodeficiency, also referred to as immunocompromised, is a condition that weakens or completely loses the capacity of the IMS to cope with infectious diseases and cancer (Agarwal and Cunningham-Rundles 2019). Often, circumstances ("secondary") are acquired through environmental conditions influencing the IMS of the patient. Examples such as HIV infection and environmental influences like nutrition are alien factors. Immunodeficiencies ensue when one or more mechanisms of the IMS have got inactive. In both young and elderly persons, the capacity of the IMS to react against pathogens is decreased and immune defenses start to decrease at the age of around 50 years owing to immunosuppression (Agarwal and Cunningham-Rundles 2019; Amaya-Uribe et al. 2019). In developed countries, obesity, smoking, and the consumption of medications are the recurrent reasons for poor immune function, whereas in developing countries, starvation is the most prevalent source of immunodeficiency. Infectious cell-mediated immunity, supplement activity, functional phagocytes, IgA antibody concentrations, and the development of cytokines are interconnected with a diet lacking appropriate proteins. Moreover, a genetic mutation or operative elimination trigger in early thymus failure contributes to extreme immunodeficiency and a strong susceptibility to infection (Kitcharoensakkul and Cooper 2020). Immunodeficiencies can similarly be acquired or inherited. Chronic granulomatous disease in which phagocytes are becoming less and less capable of attacking pathogenic agents is an instance of an inherited immunodeficiency, whereas AIDS and some forms of cancer are due to acquired immunodeficiency (Kitcharoensakkul and Cooper 2020).

1.3.2.1 Primary Immunodeficiencies

Genetically inherited immunodeficiencies with immune and nonimmune abnormalities constitute primary immunodeficiency syndromes. Nonimmune characteristics are quite often better known than immunodeficiency indications. Instances include cartilage-hair hypoplasia, ataxia-telangiectasia, Wiskott-Aldrich syndrome, DiGeorge, and IgE syndrome. Some people also experience autoimmune diseases despite the existence of immunodeficiencies. Typically, immunodeficiency is a recurring disease. The age at which recurring infections have begun indicates the function of the IMS (Chinen and Cowan 2018; Amaya-Uribe et al. 2019).

These disorders are defined genetically; they can happen on their own or as a part of a disorder. In 2017, 354 inborn immune errors and 344 genes were found to be related to primary immunodeficiency disturbances by the International Union of Immunological Sciences (Picard et al. 2018). The molecular base is known for nearly 80%. Primary immunodeficiencies commonly occur as abnormally regular (recurring) or unusual infections during pregnancy and infancy. Around 70% of patients are beginning <20 years; 60% of them are male, so they are mostly X-linked. The average symptomatic disease prevalence is roughly 1/280 individuals.

It can be classified into cellular, humoral, and combined as well as based on phagocytic cells and complement proteins (Picard et al. 2018).

Cellular Immunodeficiencies

Approximately 5–10% of primary immunodeficiencies are caused by cell immune deficiencies and are predisposed to infectious diseases by viruses, Pneumocystis jirovecii, pathogens, other opportunistic organisms, and numerous common diseases. Immunoglobulin defects are also caused by T-cell disorders as the IMSs of B and T cells are interdependent. X-linked lymphoproliferative syndrome, zeta-associated protein 70 (ZAP-70) deficiency, DiGeorge syndrome, and chronic mucocutaneous candidiasis are the few examples of such conditions (Chinen and Cowan 2018; Picard et al. 2018).

Primary and extremely unusual natural killer cell abnormalities may be predisposed to viral infections and tumors. In patients with numerous other single or multiple immunodeficiencies, secondary natural killer cell defects can arise.

Humoral Immunodeficiencies

Humoral immunodeficiencies (B-cell deformities) that inflict antibody deficits account for 50–60% of primary immunodeficiencies. Serum antibody levels decline, allowing infectious diseases to become predisposed. It tends to give the impression that the most prevalent B-cell disorder is due to IgA deficiency (Picard et al. 2018).

Combined Cellular and Humoral Immunodeficiencies

Approximately 20% of primary immunodeficiencies occur in a combined cellular (B and T cell) and humoral immunity deficiency (CCD). Severe combined immunodeficiency (SCID) is among the most important form. The combined immunodeficiency, for instance, phosphorylase failure of purine nucleoside, may have average or higher levels of immunoglobulin, but the production of antibodies is compromised as a result of the weak T-cell activity (Picard et al. 2018).

Phagocytic Cell Abnormalities

The potential of phagocytic cells such as monocytes, macrophages, and granulocytes, such as eosinophils and neutrophils, to destroy pathogens is found to be impaired in such immunodeficiency conditions. Phagocytic cell abnormalities make up 10–15% of primary immunodeficient. Cutaneous infections of staphylococcal and gram negative are indicative of these characteristics. Leukocyte adhesivity disorder (type 1 and 2), Chediak-Higashi syndrome, cyclic neutropenia, chronic disease, and granulomatous disease are the most prevalent (although rare) phagocytic cell abnormalities (Picard et al. 2018).

Complement Deficiencies

Complement defects are rare (about 2%); they comprise discrete complement factor or inhibitor deficiencies and can be inherited or acquired. Inherited defects are autosomal recessive, except for defects of the dominantly inherited inhibitor C1 and X-linked prophedin. The deficits lead to faulty pathogens opsonization, phagocytosis, lysis, and malformed antigen-antibody complex removal. The most severe repercussions are reoccurring infection due to faulty opsonization and autoimmune disorders (e.g., SLE, glomerulonephritis), due to antigen-antibody complex misaligned clearance (Picard et al. 2018).

1.3.2.2 Secondary Immunodeficiencies

Secondary immunodeficiency typically occurs in critically ill, elderly, or hospitalized patients. The prolonged serious ailment will impair immune utilities that are often revocable once the principal circumstance is resolved (Chinen and Shearer 2010). Due to enteropathy, nephrotic conditions, and extreme skin burns or dermatitis, immunodeficiency may be resulting from a serum protein deficiency, particularly IgG and albumin. Therapy depends on the inherited issue and includes the use of a supplement high in medium-chain triglycerides, which may decrease and be significantly effective in the decreased level of immunoglobulins and gastrointestinal LYM. Testing must therefore focus on related conditions, including HIV infection, diabetes, primary ciliary dyskinesia, and cystic fibrosis, where a specific secondary immunodeficiency disease is clinically suspected (Chinen and Shearer 2010).

1.3.3 Hypersensitivity

Hypersensitivity (also termed intolerance or hypersensitivity response) signifies unwanted responses, particularly allergies and AI, generated by the natural IMS. They are typically stated as an IMS overreaction, and these responses can be harmful, unpleasant, or sometimes life-threatening. Hypersensitivity responses necessitate a pre-sensitized (immune) condition of the host. The hypersensitivity categorization of Gell and Coombs is by far the most extensively used and characterizes four types of immune response that result in harm to nearby tissue (Bagirova 2007).

Hypersensitivity is an autoimmune reaction that threatens the self-tissues and organs. They are classified into four clusters (Type I–IV) depending on the mechanisms intricated and the extent of the hypersensitive response (Bagirova 2007). Type I hypersensitivity, frequently connected with an allergy, is an instantaneous or anaphylactic response. Symptoms can range from minor irritation to demise. Form I hypersensitivity is regulated by IgE, which causes degranulation of mast cells and basophils once cross-linked by antigen (Bagirova 2007). Type II hypersensitivity happens as the antibodies on the patient's own cells attach to antigens, labeling them for degradation. It is also termed hypersensitivity, which is antibody-dependent (or cytotoxic) and is controlled by IgG and IgM antibodies. Type III hypersensitivity responses are caused by immune complexes (complement proteins, configurations of antigens, and IgM and IgG antibodies) accumulated in different tissues. It typically takes between second and third days for type IV hypersensitivity (often recognized as delayed-type or cell-mediated hypersensitivity) to establish. In certain autoimmune and infectious diseases, Type IV responses are implicated but can also include contact dermatitis (poison ivy). These responses are facilitated by macrophages, T cells, and monocytes (Bagirova 2007).

1.4 Immunomodulation Through Therapeutic Approaches

The IMS response can be modified to defeat undesired responses ensuing from autoimmune disorders, allergic reactions, and transplant rejection and to boost protective immunity toward infectious agents that mainly elude the immune function or cancer. The immunomodulator used for the manipulation of immune function in various immune disorders has been addressed in Table 1.4.

1.4.1 Immunomodulation for Diminution of Destructive Reaction

Allergic reactions and AI are circumstances marked by a negative immune response. Often, they are regarded under the generalized phrase "hypersensitivity" (Bagirova 2007). The major discrepancy between allergy and AI is the basis of the antigen with which the immune reaction is concerned with. It is exterior in allergic reactions and internal in AI. This differentiation, which is perfectly obvious in most instances, becomes distorted as what begins as a response to any external antigen remains autonomous and continues weeks until the threatening agent is (presumably) distant (Faustman 2013). Indeed, multiple autoimmune disorders are specifically caused by environmental causes (primarily infections).

Table 1.4 Therapeutic immunomodulators and their mechanism of activity utilized in various immune conditions

Therapeutic activities	Immunomodulatory activity on various immune conditions	Mechanism of immunomodulation
Conventional therapeutics for immunomodulation		
Azathioprine	Avoidance of organ transplant rejection	Antimetabolite working through inhibition of purine synthesis
Corticosteroids	Basic drugs are utilized to manage and control solid organ rejection, prohibit GVHD, and deal with a variety of autoimmune disorders	There are many action possibilities. Steroid binds to glucocorticoid response elements in the promoter regions of cytokine genes and engages with cytosolic receptors and nucleus translocations. The impact safeguards T-cell activation and recruitment
Cyclosporine A	Preventing the rejection of stable organs and GVHD. Authorized for psoriasis and rheumatoid arthritis; had been used for some other inflammatory disorders	Works by binding to cyclophilins. Complex cyclosporine-cyclophilins link and prevent calcineurin blocking the triggering of the transcription factor nuclear-activated T-cell (NFAT) factor. This restricts T-cell activation and the development of IL-2
Leflunomide	Beneficial in rheumatoid arthritis and solid organ transplantation	Suppresses the de novo mechanism of pyrimidine formation, constraining the propagation of T cells
Methotrexate	Stimulation and management of stable organ transplantation, avoidance of GVHD, and management of vasculitis and rheumatoid arthritis	Antifolate antimetabolite
Mycophenolate mofetil (MMF)	Prevent the occurrence of failure of organ transplants; avoidance of GVHD	Antimetabolite repressing synthesis of purine; preferential for LYM
Sirolimus (rapamycin)	Preventative medicine against rejection in stable organ transplantation; GVHD prophylaxis; GVHD steroid-refractory therapy	It works by binding the FKBP intracellular component. Complex rapamycin-FKBP attracts MTOR, blocking the cell growth between G1 and S
Tacrolimus (FK506)	Preventing the rejection of firm organs and GVHD. Authorized for atopic dermatitis. Frequently used in a variety of other autoimmune disorders	Tacrolimus works by binding the FKBP domain, and this compound binds and suppresses calcineurin with a similar impact as cyclosporine
Modern therapeutics for immunomodulation		
Alemtuzumab (Campath-1H)	It has been authorized for the treatment of certain autoimmune disorders, which include autoimmune cytopenia, refractory rheumatism, and vasculitis. FDA	Humanized monoclonal antibody against CD52

(continued)

Table 1.4 (continued)

Therapeutic activities	Immunomodulatory activity on various immune conditions	Mechanism of immunomodulation
	approved only as an antineoplastic for chronic lymphocytic leukemia. It was also used in allo-HSCT for GVHD prevention	
Antithymocyte immunoglobulin	Aplastic anemia therapy. Acute kidney transplantation rejection therapy	Polyclonal immunoglobulin acquired from rabbit or horse
Basiliximab	It was used as part of the induction system of firm organ transplantation	Anti-CD25 monoclonal antibody (fragments of the IL-2 high-affinity receptor)
BMS-188667	Phase III studies in psoriasis, rheumatoid arthritis, and transplantation	A fusion molecule that links CD80 and CD86 to the exterior of APCs restricting CD28 association that would send a costimulatory transmission to the T cell
CD154 protein	Animal experimental models such as xenotransplantation	Inhibits the association between CD40 on macrophages and B cells and CD40 membrane protein present on activated T cells
Daclizumab	It is used as part of the induction system of firm organ transplants; it's being used to treat steroid-refractory GVHD of allo-HSCT	Humanized monoclonal antibody against CD25 (fragment of the IL-2 high-affinity receptor)
Efalizumab	Phase I/II trials in rheumatoid arthritis and transplantation; phase III trials in psoriasis	Forms a complex with LFA-1, avoiding interaction between LFA-1 and ICAM, leading to obstruction, trapping, and triggering of T-cell adhesion
Etanercept	Often used for psoriasis, rheumatoid arthritis, and ankylosing spondylitis. Under investigation for various other autoimmune conditions	A recombinant variant of the extracellular dimensions of human p75 receptor tumor necrosis factor coupled to the segment of human immunoglobulin G1. Blocks TNF-α activity
Infliximab	Used in the condition of Crohn (autoimmune colitis) and ankylosing spondylitis. In allo-HSCT, steroid-refractory GVHD has been treated	Chimeric antibody counter to human TNF-α. Inhibits TNF-α activity
Muromonab-CD3	Acute rejection treatment in the transplantation of solid organs. Allo-HSCT treatment with GVHD	Anti-CD3 monoclonal antibody
Natalizumab	Authorized for usage in remitting-relapsing multiple sclerosis and Crohn's disease	Anti-alpha-4-integrin, a humanized monoclonal antibody, inhibits interactions between 5–007-4 integrin on the membrane of inflammatory cells and VCAM-1 on vascular endothelial cells, blocking leukocyte transport through the

(continued)

Table 1.4 (continued)

Therapeutic activities	Immunomodulatory activity on various immune conditions	Mechanism of immunomodulation
		blood-brain barrier in multiple sclerosis

GVHD graft-versus-host disease; *HSCT* hematopoietic stem cell transplantation

1.4.1.1 Immunomodulation in Hypersensitivity

IMD of allergic conditions is often sought in clinical practice. Desensitization is the expression used to designate the incremental delivery of larger doses of the allergen to the individual with a view to changing the reaction to a typical Th1 phenotype (Agarwal and Cunningham-Rundles 2019). Desensitization operates by various pathways in a number of therapeutic environments. When accelerated desensitization (executed within hours) is being used to prescribe a lifesaving medication (usually an antibiotic) to which the individual is allergic, there is little time to adjust the immune response. However, the treatment is largely efficacious, indicating that the incremental incorporation of an increase in the concentration of allergens has contributed to the saturation of all IgE receptors without inducing cross-linking and eventual degranulation (Gea-Banacloche 2006; Nader 2017; Agarwal and Cunningham-Rundles 2019). In this situation, the patient is not "cured" by the allergy, and the treatment would have to be repetitive if the medication were required another time. In the occurrence of an allergy to insect venom, the continual implementation of a desensitization regimen over the years may effectively lead to the generation of a regular immune reaction to the particular antigen, and in certain circumstances, it is possible to avoid the administration of venom and to sustain a desensitized condition (Gea-Banacloche 2006; Stewart and Keselowsky 2017). The great majority of desensitization methods approved for the treatment of allergic rhinitis or asthma are not more efficacious than placebo in randomized clinical trials, even though they are regularly attempted ("allergy shots").

1.4.1.2 Immunomodulation in Autoimmunity

The treatment for autoimmune disease is deeply disappointing. In comparison to allergic conditions, antigen prevention is not an option. It is hormonal replacement if the glandular functioning has been compromised or the suppression of improper hormonal fabrication is the treatment of choice for many autoimmune diseases involving the endocrine system and the defective immune response is not controlled at all (Gea-Banacloche 2006; Yang et al. 2018). In AI, as in transplantation, IMD relies on the utilization of medications that regulate inflammation and dramatically reduce immune tasks. The predicted adverse consequence of immunosuppressive medications is that they find the patient vulnerable to infections, but the use of these drugs is accompanied by numerous other toxicities (Gea-Banacloche 2006).

1.4.1.3 Immunomodulators

The various available immunomodulators can be categorized as per the portion of the immune response in which they are involved more. Immune interactions proceed through multiple phases that can be functionally differentiated as follows:

- Activation of T cell
- Processing of antigen and their presentation to T cells
- T-cell proliferation and successive possessions

This categorization could afford a context for understanding the possibility of standard treatment with a variety of diverse immunosuppressive agents. Ideally, medications that operate on various stages of the immune response may have a synergic activity (Lebish and Moraski 1987; Gea-Banacloche 2006; Mukherjee et al. 2014; Ponce 2018). The most effective immunomodulators/immunosuppressive drugs in clinical practice are addressed in Table 1.4.

Immunomodulators for Inhibition of T-Cell Activation

Calcineurin antagonists, cyclosporine A, and tacrolimus are potentially the most effective medicines in surgical transplantation. Cyclosporine A and tacrolimus (FK506) function by modulating the activity of somatic mutations that are stereotypically expressed during T-cell activation, particularly IL-2, the most significant T-cell growth factor. These medicines are more effective in controlling immunological response and are thus much more essential in transplantation (to avoid rejection) than in autoimmune diseases (in which the immune response is previously in headway) (Gea-Banacloche 2006; Flores et al. 2019). Cyclosporine and tacrolimus attach to various target proteins named immunophilins (cyclosporine to cyclophilin, tacrolimus to the FK-binding protein); however, all results in the binding and repression of calcineurin, a phosphatase that is triggered in T cells after T-cell interaction. Calcineurin stimulates a variety of transcription factors and inhibits successful T-cell activation and clonal expansion. Tacrolimus and cyclosporine function at the very same level of T-cell activation and have significant similarities (Gea-Banacloche 2006; Flores et al. 2019; Khan and Gerber 2020). The combination of the two medications would not be anticipated to be more successful than one of them at the optimal dosage and therefore would not usually be used.

Immunomodulators Targeting Antigen Processing and Their Presentation to T Cells

In this class, most agents are investigational. Communication between APCs and LYM can be controlled at a variety of stages, along with prevention of activity between CD40 on B cells and CD40 binding sites (CD154) on activated T cells with anti-CD154 antibodies (ineffectual in a randomized controlled lupus trial) and the prevention of successful costimulation by trapping CD80 and CD86 (B7.1, B7.2) on APCs with CTLA4-Ig (appealing in a rheumatoid and psoriasis) (Ville et al. 2015).

Immunomodulators Targeting T-Cell Proliferation

Proliferation is one of the results of T-cell activation. In myriad ways, T-cell proliferation can be blocked. IL-2 can be inhibited by monoclonal antigens (daclizumab or basiliximab) against CD25 (the IL-2 receptor α chain). The T-cell growth factor, which has been made by T-cell activation, may be repressed (Sageshima et al. 2009). The association inhibition (that only expresses on activated T cells) between IL-2 and its high-affinity receptor prevents the proliferation of active T cells. These antibodies have been utilized in organ transplantation to avoid rejection and can work with calcineurin blockers synergistically.

Cyclophosphamide is a cytotoxic medication in the nitrogen mustard group of combinations. It attacks LYM and several other dividing cells. Its toxicity has led to reduced usage, even though it is the medication of preference for certain significant autoimmune disorders (Wegener's granulomatosis and lupus nephritis) (Kim and Chan 2016; Baldo 2016).

Antimetabolites, which function by inhibiting purine synthesis, are the azathioprine and mycophenolate mofetil (MMF). Azathioprine effectively inhibits adenine and guanine nucleotides in the de novo cycle and affects all cells that replicate it (usually leukopenia and thrombocytopenia). MMF is metabolized into mycophenolic acid, blocking inosine monophosphate dehydrogenase lymphocyte-specific subtype, which prevents the formation of guanine nucleotides mostly in LYM. During the transplantation phase, MMF substituted azathioprine similar chemical mode of action and effectiveness with decreased toxicity (Allison and Eugui 2005; Dalal et al. 2010).

The de novo sequence of pyrimidine synthesis works by interfering with the leflunomide which restricts T-cell division when triggered. It contributes to G1 arrest as rapamycin. This drug was widely utilized in rheumatoid arthritis but was restricted due to safety issues during transplantation.

Sirolimus (rapamycin) links (like tacrolimus) to the FK-binding protein, but the rapamycin-FK-binding protein interaction does not contend with calcineurin. Rather, it binds to other intracellular proteins, MTOR (a molecular target of rapamycin), which control the cell cycle (Tong and Jiang 2015). The final result is cell arrest in the G1 step of the cell cycle. As it operates at a different level than cyclosporine and tacrolimus, the interaction of rapamycin with any of these agents is anticipated to result in more effective obstruction of T cells working (Gea-Banacloche 2006; Tong and Jiang 2015).

1.4.2 Immunomodulation Against Infections

Vaccination is the best and most effective modulation of the immune response. In vaccination, preventative adaptive immune reactions are developed by the exposure of the host to a harmless portion of the microorganism (e.g., inactivated or attenuated viruses, such as influenza or measles) or portions thereof (e.g., tetanus toxoid, harmless preparation of tetanus toxin, tetanus agent caused by the bacterium Clostridium tetani) (Molina and Shoenfeld 2005; Pulendran and Ahmed 2011; Zhu et al.

2014; Rauch et al. 2018). Captivatingly, most effective vaccines predate the existing thorough biochemical insight of the IMS over the centuries (in the context of smallpox vaccines) or decades. However, a new understanding has also culminated in more efficient vaccines toward bacteria (Hemophilus influenzae) and viruses (hepatitis B) and continues to contribute to better and more operational vaccines in the forthcoming (Gea-Banacloche 2006; Zhu et al. 2014).

There are three main vaccination techniques in clinical usage: attenuated infectious compounds, exterminated whole infectious agents, and selected pieces of infectious agents that are commonly synthesized in the research laboratory (Singh and O'Hagan 1999; Chinen and Shearer 2010; Azizi et al. 2010; Gregory et al. 2013). The newest approach, experimental at this stage, includes the use of DNA from pathogens of interest.

Attenuated pathogens are by far the most powerful type of vaccine since they replicate the virus and manifest in a very close immune reaction to the pathogen-induced virulent type, comprising CD4 cells, CD8 cells, B cells, and, sometimes, mucosal immunity. Vaccines of the attenuated virus are being used against pneumonia, measles, mumps, rubella, and typhoid fever (Molina and Shoenfeld 2005; Pulendran and Ahmed 2011). Attenuated viruses are strains that have evolved in nonhuman cells for decades to come. Evolution and natural selection ensues in viruses that are far less capable of causing disease in humans than their wild counterparts (Pulendran and Ahmed 2011). However, since they do induce inflammation, the same mechanisms since pathogenic variants (such as the major histocompatibility complex [MHC] class I mechanism that produces CD8 cytotoxic reactions) are introduced to the IMS, consequential in perhaps the most comprehensive immune response (Rauch et al. 2018).

If an attenuated variant of the pathogen is not an alternative, an effort can be made to pick the correct antigen to produce a defensive immune reaction. Knowledge of the IMS functioning is vital in this sense. For example, though polysaccharide antigens are usually T-cell-independent (i.e., they produce antibody reaction even without T-cell "assistance"), they do not elicit a sufficient immune reaction in infants (Chaplin 2010; Ray et al. 2013; Nicholson 2016). The bacterium H. influenza and Streptococcus pneumonia have a polysaccharide capsule that can be used as a vaccine but are the most significant infections in children. The capsule by itself does not induce an appropriate immune reaction in the target set of respondents. This dilemma can be solved by the use of polysaccharide antigens covalently attached to the protein carrier (Sadarangani 2018). The inclusion of the protein carrier is activated by the T-cell component of the immune reaction, and a much more powerful antibody response is obtained.

It is understood that dendritic cells, the competent antigen-presenting cells (APCs), present antigens to B and T cells in lymphoid organs. As the conception of the detection and processing of antigens by dendritic cells has progressed, the potential for action at this level has increased. This seems plausible to establish the mechanisms for favoring the desired immune reaction by administering cytokines (e.g., IL-12 while the target is to respond to Th1) synchronously with the administration of the antigen (Hughes et al. 2016).

The lessons can be learned by the use of attenuated viruses that have been understood via the use of viral vectors as transporters of antigens of interest. There have been some viruses, such as fowl pox, which cannot manifest in a viable infection in humans (Gea-Banacloche 2006). However, if the virus is being used as a transporter of the genes of another virus, an abortive infection will transpire in a few cells that will synthesize (and progress through the MHC class I pathway) proteins from the virus of interest (Gea-Banacloche 2006). This is most appealing when seeking to produce immune function to pathogens for which CD8-mediated cytotoxicity is deemed vital (HIV is a prime example, although related techniques are often utilized in cancer immunotherapy).

A DNA vaccine is in the emerging stage of finding the intramuscular administration of DNA from a given virus leading to the production of antibodies and cytotoxic T cells that enable the individual to resist exposure to the entire infectious virus. This means that "naked" DNA is ultimately picked up and transcribed by cells, giving birth to viral proteins that are introduced by class I human leukocyte antigen (HLA) and class II HLA mechanisms (Renneberg et al. 2017; Lee et al. 2018a, b). DNA of interest can be incorporated into a plasmid containing other genes, for example, cytokines such as granulocyte/macrophage colony-stimulating factor (GM-CSF), which may induce the immune reaction to the viral gene. There are no vaccines in therapeutic use that have been using this new technique, although this method is being utilized experimentally to improve vaccines for tumors and vaccines toward chronic infections (Renneberg et al. 2017; Lee et al. 2018a, b).

1.4.3 Immunomodulation Against Cancer

The uncontrolled expansion of a transformed cell phenotype is the product of cancer. Based on what other new antigens they express at the site of the tumor and subsequent necrosis or inflammation, these transformed cells may be identified. In experimental animals, and primarily when tumors are secondary to the action of a virus, immune activity toward tumors is reasonably simple to make evident. For centuries, the intense potency of the cancer-immune reaction in some of these prototypes has driven cancer immunotherapy, with comparatively few successful outcomes (Psimadas et al. 2012; Calabrese et al. 2020).

In order to combat cancer, there are many approaches to employ against the IMS. Improving such a reaction will help with certain tumors that cause an immune reaction. Intravesical bacillus Calmette-Guerin, a mycobacterium-attenuated utilized in the treatment of primary bladder cancer as a tuberculosis vaccine, is known for many years to serve as a potent immune adjuvant. The best description of the immunotherapy-responsive tumor is renal cell carcinoma (Gea-Banacloche 2006; Psimadas et al. 2012; Calabrese et al. 2020). A small proportion of patients with renal cell carcinoma receive significant long-term respond from IL-2. Another IMS modulation is not expected, indicating that a particular immune reaction was evident and that only the supraphysiological dose of cytokine.

IFN-α is utilized for the chemotherapy of renal cell carcinoma, hair cell leukemia, and chronic myelogenous leukemia and in certain CD8 T-cell reactions to leukemia. The usage of natural killer cells in patients, extended and activated in vitro with IL-2 lymphokine-activated killer cells, is a variant of the same strategy. However, the usage of lymphokine-activated killer cells has been correlated with substantial toxic effects and merely limited efficacy and has acted mostly as a prototype than a feasible clinical methodology (Gea-Banacloche 2006; Schön 2019; Serrano-del Valle et al. 2020).

Metastatic melanoma is very often managed with IFN-α, but its effectiveness in this context is uncertain. Kaposi sarcoma, the most prevalent tumor in HIV-infected persons, can be controlled with IFN-5-α, although the exact mechanism rests undefined. Interestingly, successful anti-HIV treatment, accompanied by immune reestablishment, has been linked with growth retardation and even reversal in certain Kaposi sarcoma tumors.

The systemic administration of cytokines that boost the antitumor immune response is functionally comparable to anti-CTLA-4 administration. This is a CTLA-4 monoclonal antibody (CD 152), a lymphocyte-activated molecule that downregulates the immune reaction on the exterior of follicular dendritic cells and perhaps other APCs regarding its interplay with B7.2. The CTLA-4 antimicrobial prevents B7.2-CTLA-4 interaction (Gea-Banacloche 2006; Krown 2007; Cesarman et al. 2019). An improved immune reaction is accomplished by suppressing the inhibitory signal. It is important to remember that the medicinal application of this agent has succeeded in partial responses to melanoma, but also in autoimmune conditions, in particular colitis, hepatitis, and hypophysitis disease (Gea-Banacloche 2006; Krown 2007; Cesarman et al. 2019). This discovery indicates that considering the mechanisms of positive and negative preference in the thymus, the autoreactive variants of T cells are present and that peripheral resistance is far more essential than ever in avoiding AI.

The administration of antibodies primarily targeting malignant cells offers a different approach. It is reasonable to accept this treatment as a form of immunotherapy, but the function of the immune reaction of the patient is uncertain. Monoclonal antibodies toward antigens present in the cancerous cell can be produced (Nelson and Ballow 2003; Rose and Mackay 2019b). As soon as only or exclusively the target molecule exists in malignant cells, this treatment is considered to have excellent precision and very few adverse effects.

A variety of different antibody-based treatment mechanisms can occur. For example, three different pathways of action can exist for rituximab: antibody-dependent cell-mediated cytotoxicity (ACD), cell-mediated complementary cell lysis, and active apoptosis induction in target cells. The effect on several malignancies of different monoclonal antibodies is distinct (Wang et al. 2015). Monoclonal antibodies can improve their efficacy by being used in synergy with other agents, such as rituximab. Usually, all other drugs are chemotherapeutic, but immunomodulatory treatment may be incorporated. When ADCC is activated, which is most commonly regulated by the natural killer cells, the key activities of an antibody may lead to increased efficacy of agents that augment their number or function (such as IL-2) (Wang et al. 2015).

The antibody may pick up toxic molecules and deliver them to malignant cells (immunotoxins) in a particular portion of this methodology. A monoclonal CD22-directed antibody is constructed enduring responses for two-thirds of patients with chemotherapy-refractory disorders in malignant cells with B cells in hairy cell leukemia. A radioactive isotope, local radiation treatment, or even enzymes that can turn a prodrug into an active cytotoxic agent (antibody-directed enzyme drug therapies) may be used to link to antibodies as well (Basheer et al. 2014).

Using tumor vaccinations requires more sophisticated efforts to control the IMS. The very first step is to recognize the unique antigens of the tumor that evoke an effective immune response. This is also the limiting factor since these so-termed tumor rejection antigens are not expressed by most tumors. If the antigens have been detected, the objective is to find the safest way to produce an efficient immune response against them by delivering themselves to the patient with a suitable vector and costimulation (Gonzalez et al. 2018). A series of scientific techniques are being used to improve the immunogenicity of tumor vaccines, which include transfecting tumor cells with genes that encode costimulatory molecules (e.g., B7) or cytokines that activate APCs (e.g., GM-CSF) (Gea-Banacloche 2006; Gonzalez et al. 2018).

1.4.4 Immunomodulation in Organ Transplant

There is a destructive immune reaction that occurs during transplantation due to AI that needs to be suppressed. There are two very major functional distinctions though. First, the duration and severity of the immune reaction are predictive in the case of solid organ rejection (and its counterpart in hematopoietic cell transplantation, or GVHD), so avoidance is a possibility (Taylor et al. 2005; Dalal et al. 2010). Early detection is the best immediate solution as it has been proven that it is much harder to control immune reactions than to avoid them. Second, human circumstances are more equivalent to experimental animals of transplantation than to animal models of AI. This has enabled a seamless transition in transplant immunology between studies and clinical practice that is not evident in autoimmune diseases.

In certain clinical cases, transplantation is the only choice for end-organ injury. Medical-surgical methods are becoming common for kidney, liver, heart, and lung transplantation. Experimental treatments that are getting closer to the quality of treatment are pancreatic islet cells and intestinal transplantation. Other than the supply of donors, the biggest hurdle to successful transplantation is rejection. Either of two mechanisms, called direct and indirect, can carry out an immune response to the transplanted organ. The direct pathway includes the identification by the recipient's T cells of donor HLA antigens on donor APCs. As a large fraction of the T-cell repertoire is guided toward allo-MHC molecules, this direct identification of non-self-MHC molecules on the transplanted organ is anticipated to be the utmost effective mechanism for rejection. The indirect route includes the operation of the

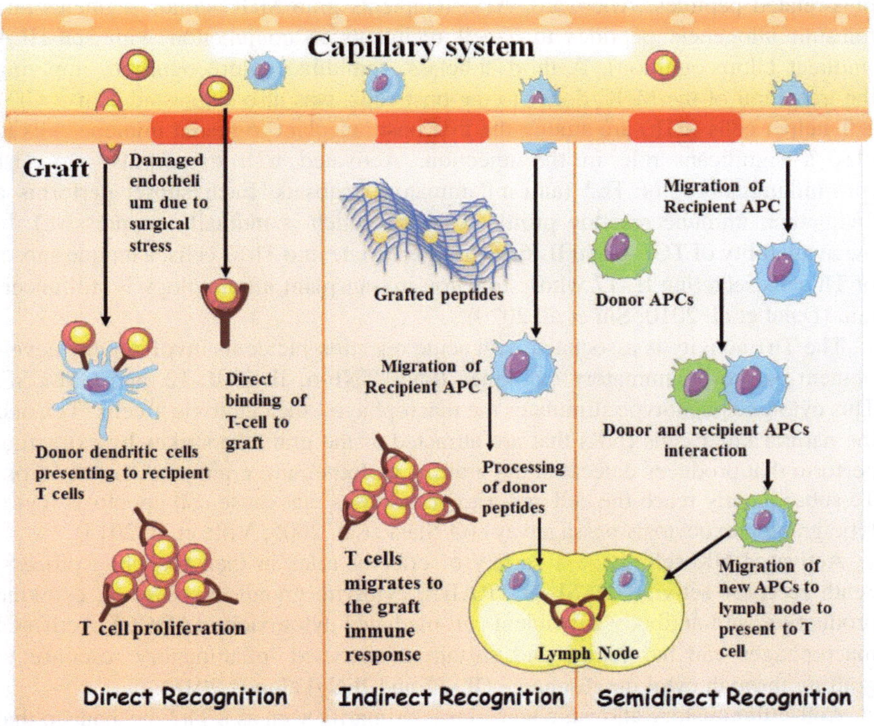

Fig. 1.3 The subset of recognition pattern and expression of the grafting alloantigen

specialist APCs of the recipient that process and display a variability of gratin donor antigens (Taylor et al. 2005; Dalal et al. 2010; Song et al. 2020).

Allografts are referred to as transplants from a genetically unrelated donor with a similar species. The final cure for chronic organ failure shall be called allogeneic transplantation. Denial remains the key challenge in posttransplantation, often with the assistance in organ protection and advancement in immunosuppression. Rejection happens despite pretransplant tissue trying to type/blood analysis and could be seen to differing degrees in nearly all transplant patients. Excluding hyperacute rejection, which arises owing to the combination of preexisting antibodies (tends to consequence from birth, blood transfusions, and/or prior transplants), transplantation rejection may closely categorize into two groupings: acute and chronic. Acute rejection is considered to be the only autoimmune reaction, whereas chronic rejection requires both immunological and non-immunological processes (Gea-Banacloche 2006).

Allorecognition is the way to prepare an expression of the grafting antigen (alloantigen) and is classified into two major subtypes: direct and indirect (Fig. 1.3). Dendritic cells trying to voyage from the graft activate direct allorecognition, where the receptor T cells specifically recognize allogeneic MHC plus-related peptides. Afterward, APC recipients grab MHC donor segments and transmit allogeneic peptides to T-cell recipients in conjunction with self-HLA (indirect allorecognition). A third subclass, semidirect allorecognition, implying the transition of the MHC donor to the host cells, was also suggested. Naïve CD4 + T helper cells (nTh) are among the first posttransplant-triggered immune cells to play a significant role in the rejection. Activated nTh transforms into Th1 (pro-inflammatory) or Th2 (anti-inflammatory) subsets. Each subset performs a distinctive, immune reaction profile (each of which is mutually suppressive). In the availability of TGF-β and IL-6, nTh differentiate into Th17 cells, a unique subset of Th cells secreting IL-17 whose function in transplant immunology is still uncertain (Dalal et al. 2010; Siu et al. 2020).

The Th1 activity is associated with acute rejection incidents involving the development of pro-inflammatory cytokines IFNγ, TNF-α, IL-2, IL-12, and GM-CSF. This cytokine phenotype stimulates the macrophages, the cytotoxic T cells (Tc), and the natural killer cells (NK) that are attracted to the graft. Tc strikes by extracting perforin that produces defects in the grafting endothelium; granzymes emitted from Tc subsequently reach the cell and trigger caspases that cause cell apoptosis (cytolytic granule exocytosis passageway) (O'Shea et al. 2005; Ville et al. 2015).

Activated NK cells have a variety of effector roles in their recycling process: death receptor activity (FASL + TRAIL), cytolytic granule exocytosis, cytokine production, and antibody-dependent cell-mediated cytotoxicity (ADCC). Activated macrophages can harmonize and sustain a restricted inflammatory response to grafting through cytokine discharge (IL-12 and IFNγ) (Jervis 2016).

Anti-inflammatory allogenic behavior is primarily seen as a Th2 mechanism that has a clear link with chronic rejection. Th2 cells are the product of the activation of B cells. Cell-to-cell interaction and the interchange of cytokines between both the Th2 and B cells are obligatory for the development of antibodies to the graft. B cells exhibit MHC class II that involves Th2 cells (indirect) culminating in Th2 activation/proliferation. The resulting Th2 cells are alloantigen specific, at first presented as B cell and IL-2 secrete for B-cell proliferation and IL-4 and IL-5 for antibody class switching. The majority of the activated B cells distinguish between plasma cell secretory antibody (mainly IgM and IgG) and graft selectivity. The bonding of Ab to the graft endothelium ultimately contributes to the triggering of the complement, leading to cell lyses. Conversely, B cells grow into memory cells and migrate to the bone marrow, establishing memory cells for the graft (Papp et al. 2017).

Acute rejection starts after a couple of days and results in T cells (both CD4 and CD8) triggered predominantly by the alloantigens control or recognition pathway, as mentioned above. This is the inevitable response that IMD can avoid (Harper et al. 2015). Acute rejection incidents can arise years after transplantation, and for prevention, continuing immunosuppression is desired. When an acute rejection event follows, more intense immunosuppression is appropriate following prophylaxis. Chronic rejection is only partly immunologic, does not respond to treatment, and involves retransplant or organ removal (Harper et al. 2015; Lin and Gill 2016).

Once an acute rejection is introduced, care must be aimed at already activated T cells in order to avoid an extension of the immune reaction, even though the potential impediments on proliferation and activation remain. The treatment of acute rejection is preferred through strong intravenous glucocorticoids (relating to few other comparable tests that have been conducted), but the precise scheme of preference is unknown (Coutinho and Chapman 2011). The mode of action of the high-dose steroid is unclear, while lymphocytic apoptosis of the peripheral blood is shown.

Triple therapy with corticosteroids, the calcineurin inhibitor (tacrolimus/cyclosporine), and the antilymphoproliferative agent azathioprine, MMF, or rapamycin are comprised in the standard regimen. To reduce the chance of acute rejection, certain teams add "antibody induction," using ATG, anti-CD3, or anti-CD25 monoclonal antibodies (Gea-Banacloche 2006; Dalal et al. 2010; Cortazar et al. 2013). Whether either of these antibody therapies results in a stronger long-term outcome is not understood. As per the so-called Edmonton Regimen, which incorporates tacrolimus, sirolimus, and daclizumab, the probability of corticosteroid-free immunosuppression for organ transplantation has been investigated for islet cell transplantation for diabetes (McCall and Shapiro 2012; Bottino et al. 2018). In all transplants, steroid abolition is a desirable outcome, and more experiments are being performed using this strategy for transplants other than islet cells.

1.5 Future Prospects

IMD has provided substantial therapeutic value across several negative immune conditions, including the therapy of tissue allografts and autoimmune disorders. However, the elevated risk with large active immunomodulators is problematic that can be addressed by more focused immune modulation. Based on the cumulative epidemiologic studies in people with immune deficits, patient observations and biomarkers, and experimental statistics, a close link between immune status and negative outcomes in human populations can be established to overcome such situations. These facts must refine our understanding of the IMS's role in promoting and suppressing through immunomodulators and its parallel impressions on the various immune conditions. The immunomodulators are very efficient in eliciting specific immune responses by modulating the functioning of the IMS, but their efficacy still lacks due to appropriate targeting and delivery viewpoint. The effectiveness of such immunomodulators, such as vaccines, biologics, and other synthetic agents, is impaired by the inaccessibility of an appropriate delivery carrier that can be accomplished by integrating with nanotechnology approaches to biomedicine and can open up numerous opportunities in the future.

Acknowledgments The authors are thankful to the DHR-ICMR, FN. V.25011/286-HRD/2016-HR for funding relating to this work.

Declaration of interest: None.

References

Agarwal S, Cunningham-Rundles C (2019) Autoimmunity in common variable immunodeficiency. Ann Allergy Asthma Immunol 123:454–460. https://doi.org/10.1016/j.anai.2019.07.014

Allison A, Eugui E (2005) Mechanisms of action of mycophenolate mofetil in preventing acute and chronic allograft rejection. Transplantation 80:S181–S190. https://doi.org/10.1097/01.tp.0000186390.10150.66

Amaya-Uribe L, Rojas M, Azizi G et al (2019) Primary immunodeficiency and autoimmunity: a comprehensive review. J Autoimmun 99:52–72. https://doi.org/10.1016/j.jaut.2019.01.011

Antiochos B, Rosen A (2019) Mechanisms of autoimmunity. In: Clinical immunology, pp 677–684.e1. https://doi.org/10.1016/B978-0-7020-6896-6.00050-8

Azizi A, Kumar A, Diaz-Mitoma F, Mestecky J (2010) Enhancing oral vaccine potency by targeting intestinal m cells. PLoS Pathog 6:e1001147. https://doi.org/10.1371/journal.ppat.1001147

Bagirova SF (2007) Hypersensitivity. In: Dyakov YT, Dzhavakhiya VG, Korpela T (eds) Comprehensive and molecular phytopathology. Elsevier, Amsterdam, pp 247–263

Baldo BA (2016) Monoclonal antibodies approved for cancer therapy. In: Safety of biologics therapy. Springer, pp 57–140. https://doi.org/10.1007/978-3-319-30472-4_3

Basheer F, Bloxham DM, Scott MA, Follows GA (2014) Hairy cell leukemia—immunotargets and therapies. ImmunoTargets Ther 3:107–120. https://doi.org/10.2147/ITT.S31425

Botos I, Segal DM, Davies DR (2011) The structural biology of toll-like receptors. Structure 19: 447–459

Bottino R, Knoll MF, Knoll CA et al (2018) The future of islet transplantation is now. Front Med 5: 202. https://doi.org/10.3389/fmed.2018.00202

Bray F, Ferlay J, Soerjomataram I et al (2018) Global cancer statistics 2018: GLOBOCAN estimates of incidence and mortality worldwide for 36 cancers in 185 countries. CA Cancer J Clin 68:394–424. https://doi.org/10.3322/caac.21492

Calabrese LH, Caporali R, Blank CU, Kirk AD (2020) Modulating the wayward T cell: new horizons with immune checkpoint inhibitor treatments in autoimmunity, transplant, and cancer. J Autoimmun 115:102546. https://doi.org/10.1016/j.jaut.2020.102546

Cesarman E, Damania B, Krown SE et al (2019) Kaposi sarcoma. Nat Rev Dis Prim 5:9. https://doi.org/10.1038/s41572-019-0060-9

Chaplin DD (2010) Overview of the immune response. J Allergy Clin Immunol 125:S3–S23. https://doi.org/10.1016/j.jaci.2009.12.980

Chinen J, Cowan MJ (2018) Advances and highlights in primary immunodeficiencies in 2017. J Allergy Clin Immunol 142:1041–1051. https://doi.org/10.1016/j.jaci.2018.08.016

Chinen J, Shearer WT (2010) Secondary immunodeficiencies, including HIV infection. J Allergy Clin Immunol 125:S195–S203. https://doi.org/10.1016/j.jaci.2009.08.040

Cortazar F, Diaz-Wong R, Roth D, Isakova T (2013) Corticosteroid and calcineurin inhibitor sparing regimens in kidney transplantation. Nephrol Dial Transplant 28:2708–2716. https://doi.org/10.1093/ndt/gft231

Coutinho AE, Chapman KE (2011) The anti-inflammatory and immunosuppressive effects of glucocorticoids, recent developments and mechanistic insights. Mol Cell Endocrinol 335:2–13. https://doi.org/10.1016/j.mce.2010.04.005

Dalal P, Shah G, Chhabra D, Gallon L (2010) Role of tacrolimus combination therapy with mycophenolate mofetil in the prevention of organ rejection in kidney transplant patients. Int J Nephrol Renovasc Dis 3:107–115. https://doi.org/10.2147/ijnrd.s7044

Faustman DL (2013) Immunomodulation, 4th edn. Elsevier

Flores C, Fouquet G, Moura IC et al (2019) Lessons to learn from low-dose cyclosporin-A: a new approach for unexpected clinical applications. Front Immunol 10:588. https://doi.org/10.3389/fimmu.2019.00588

Garzorz-Stark N, Lauffer F, Krause L et al (2018) Toll-like receptor 7/8 agonists stimulate plasmacytoid dendritic cells to initiate TH17-deviated acute contact dermatitis in human

subjects. J Allergy Clin Immunol 141:1320–1333.e11. https://doi.org/10.1016/j.jaci.2017.07.045
Gea-Banacloche JC (2006) Immunomodulation. In: Runge MS (ed) Principles of molecular medicine. Humana Press, pp 269–287
Gonzalez H, Hagerling C, Werb Z (2018) Roles of the immune system in cancer: from tumor initiation to metastatic progression. Genes Dev 32:1267–1284. https://doi.org/10.1101/gad.314617.118
Gregory AE, Titball R, Williamson D (2013) Vaccine delivery using nanoparticles. Front Cell Infect Microbiol 4:13
Gudjonsson JE, Kabashima K, Eyerich K (2020) Mechanisms of skin autoimmunity: cellular and soluble immune components of the skin. J Allergy Clin Immunol 146:8–16. https://doi.org/10.1016/j.jaci.2020.05.009
Harper SJF, Ali JM, Wlodek E et al (2015) CD8 T-cell recognition of acquired alloantigen promotes acute allograft rejection. Proc Natl Acad Sci U S A 112:12788–12793. https://doi.org/10.1073/pnas.1513533112
Hirsch DL, Ponda P (2015) Antigen-based immunotherapy for autoimmune disease: current status. ImmunoTargets Ther 4:1–11. https://doi.org/10.2147/ITT.S49656
Hughes CE, Benson RA, Bedaj M, Maffia P (2016) Antigen-presenting cells and antigen presentation in tertiary lymphoid organs. Front Immunol 7:481. https://doi.org/10.3389/fimmu.2016.00481
Jervis S (2016) Transplant rejection: T-helper cell paradigm. Br Soc Immunol. https://www.immunology.org/public-information/bitesized-immunology/organs-and-tissues/transplant-rejection-t-helper-cell. Accessed 16 Oct 2020
Khan S, Gerber DE (2020) Autoimmunity, checkpoint inhibitor therapy and immune-related adverse events: a review. Semin Cancer Biol 64:93–101. https://doi.org/10.1016/j.semcancer.2019.06.012
Kim J, Chan J (2016) Cyclophosphamide in dermatology. Australas J Dermatol 58:5–17. https://doi.org/10.1111/ajd.12406
Kitcharoensakkul M, Cooper MA (2020) The autoimmune diseases. In: Rose NR, Mackay I (eds) The autoimmune diseases, 6th edn. Academic Press, pp 513–532
Krown SE (2007) AIDS-associated Kaposi's sarcoma: is there still a role for interferon alfa? Cytokine Growth Factor Rev 18:395–402. https://doi.org/10.1016/j.cytogfr.2007.06.005
Kumar H, Kawai T, Akira S (2011) Pathogen recognition by the innate immune system. Int Rev Immunol 30:16–34. https://doi.org/10.3109/08830185.2010.529976
Lebish IJ, Moraski RM (1987) Mechanisms of immunomodulation by drugs. Toxicol Pathol 15:338–345. https://doi.org/10.1177/019262338701500312
Lee J, Arun Kumar S, Jhan YY, Bishop CJ (2018a) Engineering DNA vaccines against infectious diseases. Acta Biomater 80:31–47. https://doi.org/10.1016/j.actbio.2018.08.033
Lee LYY, Izzard L, Hurt AC (2018b) A review of DNA vaccines against influenza. Front Immunol 9:1568. https://doi.org/10.3389/fimmu.2018.01568
Lin CM, Gill RG (2016) Direct and indirect allograft recognition: pathways dictating graft rejection mechanisms. Curr Opin Organ Transplant 21:40–44. https://doi.org/10.1097/MOT.0000000000000263
McCall M, Shapiro AMJ (2012) Update on islet transplantation. Cold Spring Harb Perspect Med 2:a007823–a007823. https://doi.org/10.1101/cshperspect.a007823
McGonagle D, McDermott MF (2006) A proposed classification of the immunological diseases. PLoS Med 3:e297. https://doi.org/10.1371/journal.pmed.0030297
Molina V, Shoenfeld Y (2005) Infection, vaccines and other environmental triggers of autoimmunity. Autoimmunity 38:235–245. https://doi.org/10.1080/08916930500050277
Mukherjee PK, Nema NK, Bhadra S et al (2014) Immunomodulatory leads from medicinal plants. Indian J Tradit Knowl 13:235–256
Nader ND (2017) Immunomodulation mechanisms in disease and in the surgical patient. Immunol Investig 46:765–768. https://doi.org/10.1080/08820139.2017.1373906

Nelson RP, Ballow M (2003) Immunomodulation and immunotherapy: drugs, cytokines, cytokine receptors, and antibodies. J Allergy Clin Immunol 111:720–732. https://doi.org/10.1067/mai.2003.146

Nicholson LB (2016) The immune system. Essays Biochem 60:275–301. https://doi.org/10.1042/EBC20160017

O'Shea JJ, Park H, Pesu M et al (2005) New strategies for immunosuppression: interfering with cytokines by targeting the Jak/Stat pathway. Curr Opin Rheumatol 17:305–311. https://doi.org/10.1097/01.bor.0000160781.07174.db

Papp G, Boros P, Nakken B et al (2017) Regulatory immune cells and functions in autoimmunity and transplantation immunology. Autoimmun Rev 16:435–444. https://doi.org/10.1016/J.AUTREV.2017.03.011

Picard C, Bobby Gaspar H, Al-Herz W et al (2018) International union of immunological societies: 2017 primary immunodeficiency diseases committee report on inborn errors of immunity. J Clin Immunol 38:96–128. https://doi.org/10.1007/s10875-017-0464-9

Poletaev AB, Churilov LP, Stroev YI, Agapov MM (2012) Immunophysiology versus immunopathology: natural autoimmunity in human health and disease. Pathophysiology 19:221–231. https://doi.org/10.1016/J.PATHOPHYS.2012.07.003

Ponce R (2018) Immunomodulation and cancer: using mechanistic paradigms to inform risk assessment. Curr Opin Toxicol 10:98–110. https://doi.org/10.1016/j.cotox.2018.06.002

Psimadas D, Georgoulias P, Valotassiou V, Loudos G (2012) Molecular nanomedicine towards cancer. J Pharm Sci 101:2271–2280. https://doi.org/10.1002/jps

Pulendran B, Ahmed R (2011) Immunological mechanisms of vaccination. Nat Immunol 12:509–517. https://doi.org/10.1038/ni.2039

Rauch S, Jasny E, Schmidt KE, Petsch B (2018) New vaccine technologies to combat outbreak situations. Front Immunol 9:1963. https://doi.org/10.3389/fimmu.2018.01963

Ray S, Sonthalia N, Kundu S, Ganguly S (2013) Autoimmune disorders: an overview of molecular and cellular basis in today's perspective. J Clin Cell Immunol 01:1–12. https://doi.org/10.4172/2155-9899.S10-003

Renneberg R, Berkling V, Loroch V (2017) Viruses, antibodies, and vaccines. Biotechnol Beginners:165–200. https://doi.org/10.1016/B978-0-12-801224-6.00005-9

Rose NR (2014) Human organ-specific autoimmune disease. Ref Modul Biomed Sci. https://doi.org/10.1016/B978-0-12-801238-3.00124-0

Rose NR, Mackay IR (2019a) The autoimmune diseases, 6th edn. Academic Press

Rose NR, Mackay IR (2019b) The autoimmune diseases. Elsevier/Academic Press

Rosenblum MD, Remedios KA, Abbas AK (2015) Mechanisms of human autoimmunity. J Clin Invest 125:2228–2233. https://doi.org/10.1172/JCI78088

Sadarangani M (2018) Protection against invasive infections in children caused by encapsulated bacteria. Front Immunol 9:2674. https://doi.org/10.3389/fimmu.2018.02674

Sageshima J, Ciancio G, Chen L, Burke GW 3rd (2009) Anti-interleukin-2 receptor antibodies-basiliximab and daclizumab-for the prevention of acute rejection in renal transplantation. Biologics 3:319–336. https://doi.org/10.2147/btt.2009.3257

Santori FR (2015) The immune system as a self-centered network of lymphocytes. Immunol Lett 166:109–116. https://doi.org/10.1016/j.imlet.2015.06.002

Scanlin A (2014) Autoimmune diseases: the growing impact. BioSupply Trends Quarterl:44–47

Schön MP (2019) Adaptive and innate immunity in psoriasis and other inflammatory disorders. Front Immunol 10:1764. https://doi.org/10.3389/fimmu.2019.01764

Serrano-del Valle A, Naval J, Anel A, Marzo I (2020) Novel forms of immunomodulation for cancer therapy. Trends in Cancer 6:518–532. https://doi.org/10.1016/j.trecan.2020.02.015

Singh M, O'Hagan D (1999) Advances in vaccine adjuvants. Nat Biotechnol 17:1075–1081

Siu JHY, Motallebzadeh R, Pettigrew GJ (2020) Humoral autoimmunity after solid organ transplantation: germinal ideas may not be natural. Cell Immunol 354:104131. https://doi.org/10.1016/j.cellimm.2020.104131

Skepner J, Ramesh R, Trocha M et al (2014) Pharmacologic inhibition of ROR t regulates Th17 signature gene expression and suppresses cutaneous inflammation in vivo. J Immunol 192: 2564–2575. https://doi.org/10.4049/jimmunol.1302190

Socié G, Blazar BR (2013) Immune biology of allogeneic hematopoietic stem cell transplantation: models in discovery and translation. Elsevier/AP

Song N, Scholtemeijer M, Shah K (2020) Mesenchymal stem cell immunomodulation: mechanisms and therapeutic potential. Trends Pharmacol Sci 41:653–664. https://doi.org/10.1016/j.tips.2020.06.009

Stewart JM, Keselowsky BG (2017) Combinatorial drug delivery approaches for immunomodulation. Adv Drug Deliv Rev 114:161–174. https://doi.org/10.1016/j.addr.2017.05.013

Taylor AL, Watson CJE, Bradley JA (2005) Immunosuppressive agents in solid organ transplantation: mechanisms of action and therapeutic efficacy. Crit Rev Oncol Hematol 56:23–46

Tong M, Jiang Y (2015) FK506-binding proteins and their diverse functions. Curr Mol Pharmacol 9:48–65. https://doi.org/10.2174/1874467208666150519113541

Ville S, Poirier N, Blancho G, Vanhove B (2015) Co-stimulatory blockade of the CD28/CD80-86/CTLA-4 balance in transplantation: impact on memory T cells? Front Immunol 6:411. https://doi.org/10.3389/fimmu.2015.00411

Wang W, Erbe AK, Hank JA et al (2015) NK cell-mediated antibody-dependent cellular cytotoxicity in cancer immunotherapy. Front Immunol 6:368. https://doi.org/10.3389/fimmu.2015.00368

Yadav K, Chauhan NS, Saraf S et al (2020) Challenges and need of delivery carriers for bioactives and biological agents: an introduction. In: Advances and avenues in the development of novel carriers for bioactives and biological agents. Elsevier, pp 1–36

Yang SH, Gao C, Li L et al (2018) The molecular basis of immune regulation in autoimmunity. Clin Sci 132:43–67

Zeher M, Szegedi G (2007) Types of autoimmune disorders. Classification. Orv Hetil 148:21–24. https://doi.org/10.1556/OH.2007.28030

Zheng D, Liwinski T, Elinav E (2020) Interaction between microbiota and immunity in health and disease. Cell Res 30:492–506. https://doi.org/10.1038/s41422-020-0332-7

Zhu M, Wang R, Nie G (2014) Applications of nanomaterials as vaccine adjuvants. Hum Vaccin Immunother 10:2761–2774. https://doi.org/10.4161/hv.29589

Zimmerman KA, Hopp K, Mrug M (2020) Role of chemokines, innate and adaptive immunity. Cell Signal 73:109647. https://doi.org/10.1016/j.cellsig.2020.109647

Zitti B, Bryceson YT (2018) Natural killer cells in inflammation and autoimmunity. Cytokine Growth Factor Rev 42:37–46. https://doi.org/10.1016/J.CYTOGFR.2018.08.001

2 Insights into the Modulation of Immune Response, Chemistry, and Mechanisms of Action of Immunomodulatory Phytomolecules

Rosana C. Cruz, Mohamed Sheashea, Mohamed A. Farag, Neelam S. Sangwan, and Luzia V. Modolo

Abstract

Modulators are defined as substances of different chemical natures that act together or in an antagonistic manner to initiate, maintain, reduce, and self-regulate the immune response. To determine a chemical substance as an immunomodulator, one must consider its chemical structure, which is related to its function and its role in the immune response. The immune system of plants contributes to inhibit the growth of the pathogen and to form or maintain a healthy microbiome in the body of the plant, being capable of modulation. Medicinal plants can be a potential source of natural modulators to treat diseases such as cancer, microbial infections, or immunity-related diseases, without many

Rosana C. Cruz and Mohamed Sheashea contributed equally to this work.

R. C. Cruz
College of Dentistry, Faculdade Padre Arnaldo Janssen, Belo Horizonte, MG, Brazil

Bioengineering Laboratory, Department of Mechanical Engineering, Universidade Federal de Minas Gerais, Belo Horizonte, MG, Brazil

M. Sheashea
Medicinal and Aromatic Plants Department, Desert Research Center, Cairo, Egypt

M. A. Farag
Pharmacognosy Department, Faculty of Pharmacy, Cairo University, Cairo, Egypt

N. S. Sangwan
School of Interdisciplinary and Applied Sciences, Department of Biochemistry, Central University of Haryana, Mahendergarh, Haryana, India

L. V. Modolo (✉)
Departamento de Botânica, Instituto de Ciências Biológicas, Universidade Federal de Minas Gerais, Belo Horizonte, MG, Brazil
e-mail: lvmodolo@icb.ufmg.br

N. S. Sangwan et al. (eds.), *Plants and Phytomolecules for Immunomodulation*,
https://doi.org/10.1007/978-981-16-8117-2_2

side effects caused by using synthetic agents. Plants and their natural compounds can be used as stimulants, suppressors, or adjuvants of the immune response. Compounds with immunostimulatory activity exhibit joint effects of enhancing the number of serum immunoglobulins, stimulating the humoral response, increasing phagocyte activity, activating the innate response, and increasing cell-mediated immunity. Polysaccharides as β-glucans and carotenoids as constituents in garlic are examples that stimulate the immune response. Immunosuppressive substances are synthetic or natural molecules capable of modulating the immune response and widely used in the treatment of autoimmune diseases and in the reduction of allergic responses. Polyphenolic compounds, i.e., flavonoids such as genistein and alkaloids such as caffeine, can suppress the immune response. Adjuvants can be defined as either natural or nonnatural substances capable of increasing and modulating the immunogenicity of the antigen. A combination of different plant formulations with stimulating function can be used as a supportive adjuvant therapy. The Plantae kingdom is vast and presents a treasure trove for isolation and characterization of natural chemical compounds with immunomodulatory potential. In this chapter, the chemical structure and immunomodulation mechanisms of some plants and related natural underlying chemicals will be presented to exemplify the unlimited potential of natural agents as modulators of the immune response.

Keywords

Plants · Natural products · Structure activity relationship · Immunomodulation

2.1 Introduction

The human body can resist almost all types of foreign substances or organisms (microorganisms, toxins, macromolecules) that attempt to attack or damage tissues and organs. This ability is conferred by the immune system, which can be defined as a set of organs, lymphoid tissues, cells, and biomolecules that have different mechanisms and defense strategies for maintaining body homeostasis (Chaplin 2010). These strategies are based on the recognition and differentiation between self and non-self-antigens (Catanzaro et al. 2018). The immune response can be divided into an innate and adaptive response. The innate immunity consists of a set of responses that are activated without previous exposure to an antigen and that is already born with individuals (Alberts et al. 2002). In contrast, the adaptive comprises humoral and cell-mediated immunity with a set of responses that occurs only after previous exposure to a specific antigen (Sharma et al. 2011). Thus, the adaptive response provides extreme protection. Immunomodulation of the immune system comprises changes in defense strategies by stimulation, suppression, or via the use of adjuvants.

Immunostimulant agents are referred to as the agents stimulating or inducing the immune response to act on both innate and adaptive immunity in healthy individuals or immunodeficient individuals (Kumar et al. 2012). On the other hand, controlling or suppressing the immune response comprises the use of immunosuppressant agents to treat autoimmune diseases, organ transplant rejection, and hypersensitivity reactions (Amirghofran 2012).

Immunoadjuvants are agents that used vaccines formulations as they induce specific immune response creating cells of memory. The term immunomodulator could also be inferred to be specific response used in adjuvants or therapeutic vaccines against specific antigens (Jantan et al. 2015). The nonspecific response of an immunomodulator agent has been categorized into immunostimulant or immunosuppressant. The urge of searching for new immunomodulating agents is based upon disclosing any constituent influencing immune system in a specific or nonspecific manner affecting innate and adaptive immune responses.

Immunopharmacological research has attributed in developing regimens of incorporating both nonnatural- and natural-derived agents for immunotherapeutic purposes. The regimens are used as both curative and preventive treatments of various diseases such as cancer, inflammatory conditions, hypersensitivity reactions, infectious diseases, and autoimmune diseases (Kayser et al. 2003). Regardless of the significant advance, several challenges such as safety and efficacy are associated with the use of nonnatural substances in immunotherapy. For example, patients under azathioprine and 6-mercaptopurine treatments demonstrated adverse effects including pancreatitis, hepatitis, and myalgia. Regarding the use of cyclosporines, nephrotoxicity is the most common side effect aside from neurotoxicity, hypertension, and liver damage (Kimura et al. 2012; Shukla et al. 2012). In addition, cirrhosis and fibrosis may take place by the use of methotrexate, while reactivation of tuberculosis has been reported by the use of tumor necrosis factor (TNF) antagonists (Bascones-Martinez et al. 2014).

Finally, the continuous administration of corticosteroids can lead to adrenal crisis, hypertension, glucose intolerance, psychotic disorders, and hepatic steatosis (Dhaou et al. 2012). Risk of infections, decreased immunotherapy response, and generalized effects throughout the immune system have been reported with respect to efficacy (Shukla et al. 2012; Bascones-Martinez et al. 2014). Owing to their limitations, the high cost of conventional nonnatural agents became more unaffordable, warranting for the development of suitable candidates from natural origin likely safer in a relatively low-cost manner.

Medicinal plants are considered golden mines of natural products with new derived constituents barring immunomodulation action. The use of complementary alternative medicine (CAM) is estimated at 80% of the world population (Boccolini and Boccolini 2020). The constituents in CAM are extracted from the traditional use of herbs around the world such as the Traditional Chinese Medicine, the Siddha and Ayurvedic medicine for Indian culture, the Kampo (Japanese medicine), and homeopathy (Venkatesha et al. 2011). Apart from some exceptions, natural immunomodulator entities have usually improved safety profile when compared with nonnatural agents. Additionally, the chemical complexity of plant constituents is a key factor to potentially treat different pathophysiologies either by exhibiting additive or synergistic effects toward other constituents (Burns et al. 2010). Also, drug candidates should have high selectivity and potency when considering the design of novel immunomodulators based on CAM (Jantan et al. 2015).

Several reports on medicinal herbs with immunomodulatory activities are found in the literature. The review work published by Wen et al. (2017) is an example of a good compilation about this subject. The authors addressed some aspects of the mammalian immune system and immune-related disorders and the cell types of the immune response that are modulated by herb medicinal plants in addition to phytochemicals with immunomodulatory potential.

This chapter focuses on the discussion of plant-derived immunomodulators and their pharmacological outcomes based on chemical structures and immune responses triggered. Insights into the plant immune system are also provided to help the reader better understand the basis for certain plant natural products being good candidates for use as immunomodulators in mammalians. Overall, the information included in this chapter shall guide researchers engaged in the search for phytochemicals with immunomodulatory properties.

2.2 Plant-Derived Immunomodulators

There is a wide range of medicinal plants used as immunomodulators based upon some traditional information or scientific knowledge. The following section will focus on some classes of plant natural products and the chemical structure activity relationship in the scope of immunomodulation. To better assist the readership, compounds (Fig. 2.1) are grouped according to the chemical classes they belong to.

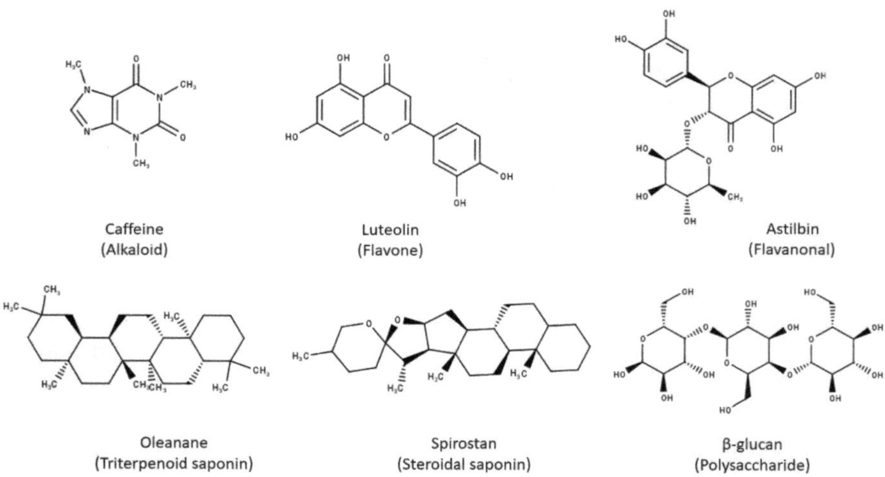

Fig. 2.1 Representative example of chemical classes with immunomodulating activity

2.3 Alkaloids

One of the chemical features of alkaloids is that it comprises nitrogen atom in their structure, usually situated in a cyclic system. Alkaloids are further subclassified based on structure, i.e., indoles, quinolines, pyrrolizidines, pyrrolidines, and pyridines (Kurek 2019). The immunomodulator effect of natural origin alkaloids could be manifested into either immunostimulant or immunosuppressant (Jantan et al. 2015).

2.4 Polyphenols

Polyphenols represent a large class of polyhydroxylated phytochemicals to include more than 8000 compounds isolated from plant kingdom (Dai and Mumper 2010). Carbon skeleton structure (C6-C3 phenylpropanoid unit) is defining wide range of structures within phytophenols, i.e., flavonoids (C6-C3-C6 type), phenolic acid (C6-C1 or C6-C3 structures), coumarins (C6-C3), lignans (C6-C3), stilbenoids (C6-C2-C6 structure), and tannins [(C6-C3-C6)n] (Pereira et al. 2009). The key drift of using plant phenolics as immunomodulator agents lies in the link between antioxidant and anti-inflammatory activities of phenols with nonspecific immune responses (Grigore 2017).

Among plant polyphenols, flavonoids are the most ubiquitous in planta with a wide array of structures of flavonoids exhibiting diverse biological actions including anti-inflammatory, anticancer, and immunomodulator (Durga et al. 2014). Flavonoids exist as free aglycones or glycosidic conjugates in which they are categorized into flavonols, flavanones, isoflavones, flavanols, flavones, and chalcones (Hosseinzade et al. 2019). Flavonoids, asides from the aforementioned effects, exert immunostimulating effect and suppressive action and are acting as vaccine adjuvants (Vajdy 2011; Woods et al. 2017).

2.5 Saponins

Saponins embrace wide range of structure-related compounds containing steroids or polycyclic triterpenes. They are characterized by skeleton derived of 30-carbon precursor oxidosqualene (Vincken et al. 2007). Spirostan is steroidal saponin derived from plants distinguished with 27 carbon atoms from triterpenoidal saponins (Bachran et al. 2014). The latter is mainly comprised of aglycones with C-30 atoms demonstrated in its core structure, i.e., pentacyclic oleanans (Kvasnica et al. 2015). The sugar moieties are usually attached to saponins at position C-3 as glycosides or at position C-28 as esters. D-fructose, D-xylose, D-galactose, L-arabinose, and L-rhamnose are examples of broad monosaccharide sugar moieties in saponins (Bhaskarachary and Joshi 2018). Further, acylation with phenolic or organic acids can occur in saponins and improve its pharmacological effect (Mugford and

Osbourn 2013). Saponins could be short-listed as immunostimulant and immunoadjuvant agents acting on immune system responses (Rajput et al. 2007).

2.6 Polysaccharides

Polysaccharides are the complex carbohydrates comprised of repeating units of monosaccharides or their derivatives linked together with glycosidic bonds forming polymeric structure. They are usually linear and contain various degree of branching classified into homo-polysaccharides and heteropolysaccharides. Another discriminate between two types of polysaccharides is that upon yielding, heteropolysaccharides produce mixture of monosaccharides unlike homo-polysaccharides (i.e., produce one type of monosaccharide) (Delgado and Masuelli 2019). Inulin, starch, β-glucan, and cellulose are typical examples of homo-polysaccharides (Chan et al. 2009), while gums, mucilage, and glycosaminoglycans are types of heteropolysaccharides (Casale and Crane 2019). The therapeutic use of polysaccharides as immunomodulators is underlined in their action to elicit innate immune system (i.e., immunostimulant agents) besides their immunoadjuvant effects (Zhang et al. 2018).

2.7 Structure Activity Relationship (SAR) with Immunomodulation Action

SAR is a type of analysis used to build up relationship between chemical structure of a molecule and its biological activity. The manipulation of chemical structure in silico allows for improving its efficacy and targeting its biological effect. Despite that, the establishment of sufficient SAR study is facing the lack of proper isolation, purification, and structural characterization of chemical classes (Ferreira et al. 2015).

Polyphenols are the major chemical classes that has been examined for SAR features in relation to its immune activity. Such studies have examined either differences between aglycones and its glycosides, sugar moiety type, hydroxylation/methylation of flavonoids, and molecular weight. The latter was demonstrated in the stimulation of innate lymphocytes by procyanidin oligomers, whereas monomers were inactive. With regard to sugar moiety, SAR of astilbin was influenced by the presence of sugar (rhamnose) at position 3 producing immunosuppressive action in rat gut model evaluating liver infiltrating T cells (Guo et al. 2007).

For hydroxylation of flavonoids, luteolin exerted stronger immunomodulator actions than apigenin due to extra hydroxy groups at position 3′ in ring B (Kilani-Jaziri et al. 2016). Likewise, SAR of diosmetin with regard to induction of NO and TNF production was impaired upon 4′-O-methylation, reflecting the critical role of free hydroxy groups in flavones (Wen et al. 2017).

Compared to phenolics, much less is known regarding polysaccharides SAR studies and in context to its immunomodulating activity. In general, type of polysaccharide, conformation, molecular weight, functional groups, and degree of branching are considered critical structural features associated with immunomodulation activity (Ferreira et al. 2015). The conformation is uniquely different between polysaccharides exemplified by a triple helix conformation to exhibit immunostimulating activity for glucans. Nevertheless, random coil of mannans exhibited stronger immunostimulating activity than triple helix (Lee et al. 2010; Deng et al. 2013).

Further acylation in polysaccharides, i.e., acetylation, or in sulfated glucans presented higher immunostimulant effect than glucans with no functional groups regardless of the position of the functional group (Qiao et al. 2010). The more degree of branching in polysaccharides was positively associated with higher immunostimulating activity compared to less branched and linear polysaccharides (Deng et al. 2013; Ferreira et al. 2015). Regarding the molecular weight, the difference in molecular weight between polysaccharides was suggested to be irrelevant unlike other structural features (Ferreira et al. 2015).

2.8 Plants and Plant Natural Products with Immunomodulatory Activities

2.8.1 Insights into the Modulation on Animal Immune Response

By definition, modulators of the inflammatory process are substances that differ in chemical nature and are produced in different types of cells that act in conjunction with or in an antagonistic manner to initiate, maintain, reduce, and self-regulate the immune response to eliminate the antigen and reestablish cell homeostasis (Lebish and Moraski 1987; Fernandez-Villa et al. 2019).

The appearance and development of an immune response usually involve a complex interaction among Pattern Recognition Receptors (PRRs) that recognize Pathogen-Associated Molecular Patterns (PAMPs); different types of cells; chemical mediators such as cytokines, lymphokines, and chemokines; and complement proteins. These interactions promote a variety of defense strategies (either synergistic or antagonistic) that can regulate themselves using mechanisms of modulation (Catanzaro et al. 2018). Lymphokines produced by the cellular immune response Th1 (lymphocyte T helper 1) and Th2 (lymphocyte T helper 2) may work in an antagonistic manner. Interferon gamma (INF-γ) produced by Th1 is involved in the stimulation of innate response mediated by phagocytosis (activation of macrophages and ultimately neutrophils) and natural killer lymphocytes (LNK). This same lymphokine may suppress the activity of Th2. The lymphokine IL-10 produced by Th2 is related to the production of immunoglobulins (IgE) and eosinophils- and mastocytes-mediated immune reactions. The IL-10 can suppress the activities of Th1 (Nicholson 2016). Therefore, a certain mediator may act as a suppressor or as a stimulator depending on the target.

Two features must be considered when defining a substance as an immunomodulator: **(i)** the chemical structure and **(ii)** the role the substance plays in the immune response. The determination of the mechanism of action of a substance toward the immune system is sometimes challenging. The chemical nature of the target receptor is crucial because it frequently determines the type of action of a substance (Abdulkhaleq et al. 2018). Many mediators or chemical substances may participate in the same signal transduction cascade within a cell and therefore behave as anti- or pro-inflammatory agents. Let us take the arachidonic acid cascade as an example. The action of the enzymes cyclooxygenase (COX-1, COX-2) and lipoxygenase (LO) acts on the arachidonic acid leading to the formation of prostaglandins or leukotrienes that may function as activators of the immune response. In contrast, the cleavage of arachidonic acid by the enzyme 5-lipoxygenase (5-LO) results in the formation of lipoxins (LX), which may function as modulating agents of the immune response, inhibiting pro-inflammatory mediators (Hanna and Hafez 2018).

2.8.2 Insights into the Modulation on Plant Immune Response

As it occurs in animals, plants need self-protection against a variety of pathogens and exogenous factors. Plants have a complex immune system that helps surviving to pathogen attacks, specific stress factors, or a combination of stressful factors (Nejat and Mantri 2017). Plants are capable of responding to biotic and antibiotic factors, adjust the signaling pathways accordingly, and balance the immune system in relation to other responses. This versatility is necessary, particularly because different plant tissues and cells are exposed to a variety of microhabitats throughout the day, and different from animals, plant immune cells do not move (Nobori and Tsuda 2019) even though the plant immune system shares similarities with the animal one, especially with respect to the recognition of antigens and respective modulation.

Like in animals, plants have PRRs and leucine-rich-repeat-containing proteins (NLR) which are able to recognize PAMPs and microbe-associated molecular patterns (MAMPs). These receptors are activated when interacting with different types of antigens such as fungi and oomycetes, bacteria and phytoplasmas, herbivorous insects, and viruses or under high salinity, drought, heat, cold, ozone, and high or low osmolarity (Zipfel 2014; Chae et al. 2016). The first and main line of plant immune defense composed of Pattern-Triggered Immunity (PTI) and Effector-Triggered Immunity (ETI) is then activated. The referred events occur in an unspecific manner similar to that of animal innate response. The activation of these responses induces a cascade assisted by intracellular second messengers that ultimately affects the transcription of several genes. Next, an influx of ions through the membrane takes place with an increment of the production of reactive oxygen species (ROS). The Mitogen-Activated Protein Kinases (MAPKs) cascades are subsequently activated, following the activation of heat-shock proteins, hormone-mediated responses (such as jasmonic acid and salicylic acid), and protein phosphorylation in an effort to fight the antigen in which the plant has been exposed to (Nejat and Mantri 2017; Lal et al. 2018). In an event in which the pathogen succeeds in the suppression of plant response, a second line of defense is activated, with the production of a series of pathogen-related proteins that lead to programmed cell death (Chakraborty et al. 2018).

A competent plant immune system inhibits the pathogen spread and efficiently maintain a heathy microbiome in its tissues (Zipfel and Oldroyd 2017). The transcriptional analysis of different MAMPs of Proteobacteria (Pseudomonadales and Rhizobiales) has shown that some interactions between plant-associated bacterial species suppress or modulate the immune response of plants by mechanisms that are still not fully understood (Stringlis et al. 2018). The microbiome seems to exert a direct effect on the plant immune system (suppression or evasion of the immune response) in favor of its own colonization. In fact, the establishment of a microbiome comprises an additional physical protection against pathogens (Teixeira et al. 2019). Therefore, plant responses to pathogens can be modulated according to the extent of the challenge to maintain cell homeostasis. The presence of a balanced microbiome provides plants with adaptive advantages. This is indeed a parallel with the maintenance of indigenous microbiota by the animal immune system. In plants, however, each cell has the potential to defend itself because different types of cells have different functions, and the immunological properties of each cell are quite specific (Nobori and Tsuda 2019).

2.8.3 Plants as Natural Modulators

Medicinal plants are a powerful source of natural modulators for the treatment of immunity-related diseases, cancer, and microbial infections with the advantage of, in most cases, not showing great side effects. For example, the spare use of dietary supplements (herb or extracts) derived from plants of the genera *Rhodiola*, *Echinacea*, *Ginseng,* and *Camellia* has shown to be an interesting alternative to treat bacterial infections because they provided less side effects than those of conventional antibiotics (Lewicka et al. 2019). The modulatory effects of whole plants and related compounds on the animal immune system depend on the plant organ the extract originated from, the purified extract fraction, the concentration of the active compounds, and the markers of innate or adaptive response chosen in the trials for the investigation of potential immunomodulatory agents (Haque et al. 2017). Different concentrations of *Echinacea* spp. extracts may increase the mobility of leukocytes or suppressors by inducing the expression of anti-inflammatory cytokines such as interleukin-10 (IL-10) (Catanzaro et al. 2018). A notable example is the effect of different *Withania somnifera* extracts as adjuvants in the treatment of cancer (Kaur et al. 2017). While the *W. somnifera* root extract stimulates the immune response, the leaf extracts from the same plant species favors anti-inflammatory processes (Kaur et al. 2017).

The Plantae kingdom consists of an unlimited source of specimens for the isolation and characterization of natural chemical compounds with immunomodulatory potential. Some examples of modulatory effects of whole plants and plant natural products on the animal immune response are shown in Table 2.1.

Table 2.1 Some examples of whole plants and plant natural products and corresponding effects on the animal immune system

Plants and biomolecules	Stimulatory action	Suppressive action	Effect on immune system	References
Bioactive metabolites from *Tinospora* sp.	↑ extract concentration-dependent activity of macrophage, IL-1β, and cell proliferation; ↑ ROS and NO production; LB and LT differentiation; cytokines production by Th1 and Th2; ↑ IgA and IgM production	↓ pro-inflammatory cytokine production (IL-17, IL-1β, IL-6, IL-23, IL-8, TNF-α); ↓ activity of COX-1, COX-2, 5-LO, PLA2, and NO	Activation and modulation of humoral and cellular immunity	Haque et al. (2017)
Polysaccharides from plants	Activation of LB, LT, LNK, macrophages, complement proteins, and cytokines (TNF-α, IL-6, IL-1, IL-2)	↑ IL-12 production	Activation and modulation of innate response	Mohamed et al. (2017) and Yin et al. (2019)
Acemannan from *Aloe vera*	Improvement of hematopoiesis; ↑ neutrophil, macrophage, and lymphocyte activity; ↑ expression of NO, IL-1, IL-6, and MHC-II of dendritic cell	–	Activation and modulation of innate response	Liu et al. (2019)
Carotenoids	↑ leukocytes number; ↑ LNK, IFN-γ, and LTCD8⁺ activity and expression of MHC-I molecule	–	Stimulation of cellular immunity; stimulation and modulation of the innate immunity	Mohamed et al. (2017)
Flavonoids	↑ LTCD8⁺ activity and production of IL-12	↓ IL-1 production and COX-2 activity	↓ inflammatory response and modulation of cellular immunity	Mohamed et al. (2017)

(continued)

Table 2.1 (continued)

Plants and biomolecules	Stimulatory action	Suppressive action	Effect on immune system	References
Quercetin	–	↑ Th1; decreases Th2, LO, histamine production and release, IL-4, IL-8, TNF-α, and IgE	Modulation of adaptative response; ↓ innate response and allergic reaction	Mlcek et al. (2016)
Caffeine	–	↓ number of monocytes in the blood, TNF-α, and IL-10	↓ innate response activity	Senchina et al. (2014)
Theophylline and its isomers	–	↓ monocytes and dendritic cells in the blood, IL-3, IL-6, and TNF-α	↓ innate response activity	Senchina et al. (2014)
Gingerols	–	↓ IL-6, TNF-α, and monocytes cells	↓ innate response activity	Senchina et al. (2014)
Rehmannia root	↑ secretion of colony-stimulating factor (G-CSF) by granulocytes	–	↓ leukocytes production	Ota et al. (2019)
Rhodiola spp.	↑ percentage of granulocytes and monocytes in the blood	↓ number of LTCD4$^+$, LTCD19$^+$ cells, and total LNK; ↓ IL-2, Il-12, IFN-γ, and COX 1 and 2	Activation and modulation of innate response and cellular immunity	Kaur et al. (2017) and Lewicka et al. (2019)
Echinacea spp.	↑ mobility of leukocytes, LNK, and macrophages; ↑ pro-inflammatory cytokines production (IL-1, TNF-α, INF-β)	↓ level of antibodies in the plasma and the number of spleens lymphocytes; ↑ anti-inflammatory cytokine IL-10	↓ humoral and cellular immunity and activation and modulation of innate response	Catanzaro et al. (2018) and Lewicka et al. (2019)
Ginseng spp.	↑ IgG total by elevated level of cytokines (IL-2, IL-6, TNF-α, and IFN-γ); stimulation of lymphocytes; ↑ phagocytosis (neutrophils and monocytes)	–	Stimulation of innate and humoral immunity	Lewicka et al. (2019)

(continued)

Table 2.1 (continued)

Plants and biomolecules	Stimulatory action	Suppressive action	Effect on immune system	References
Camellia sp.	↑ pro-inflammatory cytokines; ↓ anti-inflammatory cytokines in serum	↑ ratio of IL-10, TNF-α, and IL-1; inhibition of macrophages penetration	Stimulation and modulation of innate immunity	Lewicka et al. (2019)
Allium sativum	↑ NO production and release; ↑ IFN-γ, IL-2, and TNF-α production; ↑ phagocytosis and proliferation of $LTCD8^+$; ↓ IL-10	Stimulation (at high doses) and activation of LT and LNK, LB, and monocytes	Stimulation of cellular immunity; stimulation and modulation of innate immunity	Akram et al. (2014); Sultan et al. (2014); Arreola et al. (2015); Mohamed et al. (2017); and Rodrigues and Percival (2019)
Curcuma longa	↑ T-cell activation; ↑ IL-12, IgG, and IgM levels	↓ dendritic cell maturation and pro-inflammatory cytokines (IL-12, IL-6, IL-1, TNF-α), NO, COX-2, and 5-LO	Suppression of innate response and modulation of adaptive response	Catanzaro et al. (2018) and Yahfoufi et al. (2018)
Withania somnifera	↑ IFN-γ, IL-2, and granulocyte macrophage colony-stimulating factor (GM-CSF)	Inhibition of TGFβ, TNF-α, and NO production	Stimulation and modulation of innate response	Kaur et al. (2017)
Nigella sativa	↑ IL-4 and IL-10 secretion from lymphocytes; ↑ TNFα production from lymphocytes	↓ COX-1, 5-LO, and NO production; ↓ granuloma formation and edema; ↓ IL-4, Il-5, IL-10, IFN-γ, eosinophils, and basophils	↓ innate response and allergic processes and modulates Th1 and Th2 responses	Majdalawieh and Fayyad (2015)
Moringa oleifera	–	Inhibition of mast cell degranulation and release of IL-4 and TNF-α	↓ allergic process	Abd Rani et al. (2019)
Mahonia aquifolium	–	↑ blood monocytes; ↑ $LTCD4^+$, $LTCD8^+$, LTCD25, and IL-10	Modulation of cellular and adaptive responses against tumor cells	Andreicut et al. (2019)

2.8.4 Natural Immunostimulators

Immunostimulatory substances induce increment in the number of serum immunoglobulins, indicating a stimulation of the humoral response; increase in the activity of phagocytes, indicating activation of the innate response; and increase in cell-mediated immunity (CMI). Immunostimulatory substances may be used in the treatment of cancer, infection, or immunosuppression-related diseases (Akram et al. 2014).

Polysaccharides are a group of carbohydrates with high molecular weight that are ubiquitous in microorganisms, animals, and plants. β-glucans from several plant species are examples of polysaccharides that exhibit stimulatory effects on animal immune system. They show relatively low cytotoxicity and possess no significant side effects. *β*-glucans can interact with Toll-like receptor (TRL-2, TRL-4) depending on MYD-88, dectin-1, and macrophage mannose receptors (MR). These interactions lead to the activation of a cascade of intracellular events that culminate in changes in the transcription of genes, promotion of the biosynthesis of different types of cytokines with pro-inflammatory activity, and stimulation of the innate response and complement proteins (Yin et al. 2019). Acemannan is a polysaccharide isolated from *Aloe vera*, one of the most popular medicinal plants studied. This glycan was shown to stimulate hematopoiesis, stimulate and modulate the innate response, and increase the activity of macrophages, neutrophils, and lymphocytes in addition to increasing the expression of pro-inflammatory cytokines (Liu et al. 2019).

β-carotene and lycopene are outstanding examples of carotenoids that are considered as immunostimulants due to their ability to induce the production of cytokines and consequently the increase in the number of leukocytes, especially natural killer lymphocyte (LNK) (Mohamed et al. 2017). *Allium sativum* (garlic) stimulates an innate and cellular response against intracellular pathogens, such as viruses, by inducing the production of cytokines that increase the activity of LNK and cytotoxic T lymphocyte ($LTCD8^+$) and release of nitric oxide (NO) in addition to increasing phagocytic capacity and presentation of antigens by macrophages (Akram et al. 2014; Sultan et al. 2014). Garlic also stimulates dendritic cells and eosinophils by increased modulation of cytokines secretion and production of immunoglobulin (Arreola et al. 2015). This plant species is considered as a promising candidate for the maintenance of homeostasis of the immune system (Arreola et al. 2015). At high concentrations, substances from fresh garlic stimulated the adaptive response by increasing the expression of genes that encode for the nuclear factor of activated T cells (NFAT) (Rodrigues and Percival 2019).

2.8.5 Natural Immunosuppressors

Immunosuppressive substances (natural or nonnatural ones) by definition are compounds that are able to inhibit and modulate the immune response. Therefore, they are widely used in the treatment of autoimmune diseases. Plant species that produces a variety of natural products with multiple pharmacological actions are usually good candidates for the search of potential immunosuppressors (Catanzaro et al. 2018).

Polyphenolic compounds, widely distributed in fruits and vegetables (over 8000 different examples reported), may be used as anti-inflammatory agents due to their ability to suppress Toll-like receptors (TLR) and the expression of genes related to the biosynthesis of pro-inflammatory cytokines (Yahfoufi et al. 2018). Polyphenols from *Curcuma longa* (turmeric) were shown to suppress the innate response by diminishing the concentration of pro-inflammatory cytokines and reducing T-cell proliferation (Mohamed et al. 2017; Catanzaro et al. 2018). Flavonoids such as genistein, tangeritin, and quercetin exhibit are suppressive due to the ability to reduce the number of pro-inflammatory cytokines, exhibit modulatory effect on the adaptive response, and induce Th1 response in tumor cells and in allergic responses (Mlcek et al. 2016; Mohamed et al. 2017; Catanzaro et al. 2018). Alkaloids such as caffeine, theophylline and related isomers, and ginger alkaloids exhibit suppressive activity on the immune response via inhibition of cells and cytokines production, features typical of innate responses (Senchina et al. 2014).

The suppressive action of *Nigella sativa* (black seed or black cumin) extracts occurs during the innate response by inhibition of the COX-1 and 5-LO pathways. Oral administration of *N. sativa* extract ameliorates inflammatory symptoms caused by allergic rhinitis through the inhibition of basophils and eosinophils activities (Majdalawieh and Fayyad 2015). This plant extract may also reduce the ovalbumin-induced allergic response (Majdalawieh and Fayyad 2015). A decrease in the allergic process has also been observed by the use of a *Moringa oleifera* extract, commonly used in phytotherapy. Such extract inhibits the release of cytokines responsible for histamine degranulation by mastocytes (Abd Rani et al. 2019). The modulatory capacity of bark, leaves, flowers, and fruits extracts from *Mahonia aquifolium* was investigated against three cancer cell lines (DLD-1 colon carcinoma, A2780 ovary adenocarcinoma, and the A375 malignant melanoma). Overall, the *M. aquifolium* extracts induced an increment in the activity of LB cells and the T cells $CD4^+$ and $CD25^+$ and consequently a suppression of tumor cells growth. Increased activity of IL-10 was also reported which, at high concentrations, demonstrated to suppress the tumor cells growth by inhibiting angiogenesis (Andreicut et al. 2019).

2.8.6 Whole Plants and Related Substances as Adjuvants

Adjuvants may be defined as natural or nonnatural substances capable of increasing and modulating the immunogenicity of antigens. These substances may boost response and enhance “immune memory” by stimulating the production of memory lymphocytes (Vetter et al. 2018). Moreover, adjuvants can be combined to obtain an improved response performance, particularly when cell-mediated response activation is necessary (Vetter et al. 2018). The adjuvants may be classified according to their physicochemical properties, origin, and mechanism of action. Their mechanisms of action include the recruitment of immune cells, stimulation of innate response by activation of TLR and NOD-like receptor (NLRs), intensification of antigens importation to favor phagocytosis, and augmentation of the capacity of the molecules histocompatibility complex (MHC) present in cells containing antigens. The latter leads to the activation of cellular and humoral immune responses, with production of antibodies and memory cells. The types of adjuvants include mineral salt, oil-based emulsion, liposomes, saponins and immune-stimulating complexes, polymeric particles, cytokines, bacterial derivatives, and viruslike particles. Stability, safe biological activity, antigenic compatibility, and raw material (natural or nonnatural) are parameters to be considered when choosing or formulating adjuvants (Reed et al. 2013; Apostolico Jde et al. 2016; Bastola et al. 2017).

Plants may offer a plethora of natural compounds that may function as raw material for the synthesis of highly safe and stable adjuvants useful in several clinical applications. For instance, combination of plant species in formulations with stimulatory function may be used as a support adjuvant therapy to improve cell homeostasis during a cancer treatment (Kaur et al. 2017). Allergen-specific immunotherapy (AIT) is a powerful and precise therapy in the inductive treatment of tolerance to specific proteins. The combination of CpG adjuvant with *Arachis hypogaea* (peanut) extract stimulated the recognition by TLRs present in cells hosting antigens. This recognition induced an adaptive response resulting in the production of secretory IgA, IgG, and IL-10 that promoted tolerance to and modulation of response against allergens (Sokolowska et al. 2019). Saponins isolated from the bark of *Quillaja saponaria* trunk were evaluated as potential adjuvants against an antigen from the bacteria *Porphyromonas gingivalis*, one of the etiological agents of periodontal disease (Wang et al. 2019). The saponins stimulated the adaptive response and the production of specific IgG against the bacterial antigen (Wang et al. 2019). Leaves of *Quillaja brasiliensis* have been used to obtain the fraction QB-90 in which, when associated with BVDV (bovine viral diarrhea virus) vaccine, yielded increased production of IFN-γ, LTCD4$^+$, and TCD8$^+$ in mice in comparison with those treated with QB-90-free vaccine (Magedans et al. 2019).

Activation of the immune response is frequently related to the recognition of PAMPs, which are often antigenic (glycans) polysaccharides. TRR receptors present in leukocytes are responsible for this recognition. Plant glycans play a central role in vaccinology: they may serve as adjuvants and/or delivery vehicles or structures for the synthesis of conjugated vaccines. The use of plant polysaccharides as adjuvants has received great attention because of their capacity to enhance capture, processing, and presentation of antigens and subsequently stimulating T cells. The use of liposomes conjugated with polysaccharide isolated from *Lycium barbarum* as adjuvant was found to stimulate innate response by increasing the capacity for antigen presentation by the dendritic cells. It also activates the humoral response by the production of specific IgG against ovalbumin and induces the expression of pro-inflammatory cytokines (Bo et al. 2017). Polysaccharide extract from *Cistanche deserticola* ("ginseng of the deserts") was determined to activate dendritic cell maturation and induce cellular and humoral immunity and the production of IL-12 eTNF-α. No mortality, side effects, or change of behavior was reported in mice tested with these extracts in toxicity trials highlighting their potential for use as adjuvants (Zhang et al. 2018).

2.9 Conclusion

The examples shown in this chapter, together with those emphasized throughout this book, clearly show the unlimited potential of plants and related purified compounds as immunomodulators with a variety of mechanisms of action on immune response cascades. The development and maintenance of a certain immune response depends on a sophisticated balance in the predominance of pro- or anti-inflammatory determinants. One can note that the study of mechanisms of immunomodulation is challenging as the sequence of events that take place rarely occurs straightforward in one and same direction or with a single purpose. The chemical and cellular interactions in living organisms are dynamic and converge toward the maintenance of cell homeostasis, and plants offer an array of substances that can orchestrate the precise modulation of animal immune system.

References

Abd Rani NZ, Kumolosasi E, Jasamai M, Jamal JA, Lam KW, Husain K (2019) In vitro anti-allergic activity of Moringa oleifera Lam. extracts and their isolated compounds. BMC Complement Altern Med 19(1):361

Abdulkhaleq LA, Assi MA, Abdullah R, Zamri-Saad M, Taufiq-Yap YH, Hezmee MNM (2018) The crucial roles of inflammatory mediators in inflammation: a review. Veterinary World 11(5):627–635

Akram M, Hamid A, Khalil A, Ghaffar A, Tayyaba N, Saeed A, Ali M, Naveed A (2014) Review on medicinal uses, pharmacological, phytochemistry and immunomodulatory activity of plants. SAGE Publications, London, England

Alberts B, Johnson A, Lewis J, Raff M, Roberts K, Walter P (2002) Innate immunity. Molecular biology of the cell, 4th edn. Garland Science

Amirghofran Z (2012) Herbal medicines for immunosuppression. Iran J Allergy Asthma Immunol 11:111–119

Andreicut AD, Fischer-Fodor E, Parvu AE, Tigu AB, Cenariu M, Parvu M, Catoi FA, Irimie A (2019) Antitumoral and immunomodulatory effect of mahonia aquifolium extracts. Oxidative Med Cell Longev 2019:6439021

Apostolico Jde S, Lunardelli VA, Coirada FC, Boscardin SB, Rosa DS (2016) Adjuvants: classification, modus operandi, and licensing. J Immunol Res 2016:1459394

Arreola R, Quintero-Fabian S, Lopez-Roa RI, Flores-Gutierrez EO, Reyes-Grajeda JP, Carrera-Quintanar L, Ortuno-Sahagun D (2015) Immunomodulation and anti-inflammatory effects of garlic compounds. J Immunol Res 2015:401630

Bachran C, Bachran S, Sutherland M, Bachran D, Fuchs H (2014) Preclinical studies of saponins for tumor therapy. Recent advances in medicinal chemistry. Elsevier, pp 272–302

Bascones-Martinez A, Mattila R, Gomez-Font R, Meurman JH (2014) Immunomodulatory drugs: oral and systemic adverse effects. Med Oral Patol Oral Cir Bucal 19(1):e24–e31

Bastola R, Noh G, Keum T, Bashyal S, Seo J-E, Choi J, Oh Y, Cho Y, Lee S (2017) Vaccine adjuvants: smart components to boost the immune system. Arch Pharm Res 40(11):1238–1248

Bhaskarachary K, Joshi AK (2018) Natural bioactive molecules with antidiabetic attributes: insights into structure–activity relationships. Studies in Natural Products Chemistry, Elsevier 57:353–388

Bo R, Sun Y, Zhou S, Ou N, Gu P, Liu Z, Hu Y, Liu J, Wang D (2017) Simple nanoliposomes encapsulating Lycium barbarum polysaccharides as adjuvants improve humoral and cellular immunity in mice. Int J Nanomedicine 12:6289–6301

Boccolini P d MM, Boccolini CS (2020) Prevalence of complementary and alternative medicine (CAM) use in Brazil. BMC Complementary Medicine and Therapies 20(1):1–10

Burns JJ, Zhao L, Taylor EW, Spelman K (2010) The influence of traditional herbal formulas on cytokine activity. Toxicology 278(1):140–159

Casale J, Crane JS (2019) Biochemistry, glycosaminoglycans

Catanzaro M, Corsini E, Rosini M, Racchi M, Lanni C (2018) Immunomodulators inspired by nature: a review on curcumin and echinacea. Molecules 23(11)

Chae E, Tran DT, Weigel D (2016) Cooperation and conflict in the plant immune system. PLoS Pathog 12(3):e1005452

Chakraborty J, Ghosh P, Das S (2018) Autoimmunity in plants. Planta 248(4):751–767

Chan GC-F, Chan WK, Sze DM-Y (2009) The effects of β-glucan on human immune and cancer cells. J Hematol Oncol 2(1):1–11

Chaplin DD (2010) Overview of the immune response. J Allergy Clin Immunol 125(2):S3–S23

Dai J, Mumper RJ (2010) Plant phenolics: extraction, analysis and their antioxidant and anticancer properties. Molecules 15(10):7313–7352

Delgado LL, Masuelli M (2019) Polysaccharides: concepts and classification. Evolution Poly Tech J 2(2):180013

Deng C, Fu H, Teng L, Hu Z, Xu X, Chen J, Ren T (2013) Anti-tumor activity of the regenerated triple-helical polysaccharide from Dictyophora indusiata. Int J Biol Macromol 61:453–458

Dhaou BBB, Boussema F, Aydi Z, Baili L, Tira H, Rokbani L (2012) Corticoid-associated complications in elderly. F1000Research 1(37):37

Durga M, Nathiya S, Devasena T (2014) Immunomodulatory and antioxidant actions of dietary flavonoids. Int J Pharm Pharm Sci 6:50–56

Fernandez-Villa D, Aguilar MR, Rojo L (2019) Folic acid antagonists: antimicrobial and immunomodulating mechanisms and applications. Int J Mol Sci 20(20)

Ferreira S, Passos C, Madureira P, Vilanova M, Coimbra M (2015) Structure–function relationships of immunostimulatory polysaccharides: a review. Carbohydr Polym 132:378–396

Grigore A (2017) Plant phenolic compounds as immunomodulatory agents. IntechOpen

Guo J, Qian F, Li J, Xu Q, Chen T (2007) Identification of a new metabolite of astilbin, 3'-O-methylastilbin, and its immunosuppressive activity against contact dermatitis. Clin Chem 53: 465–471

Hanna VS, Hafez EAA (2018) Synopsis of arachidonic acid metabolism: a review. J Adv Res 11: 23–32

Haque MA, Jantan I, Abbas Bukhari SN (2017) Tinospora species: an overview of their modulating effects on the immune system. J Ethnopharmacol 207:67–85

Hosseinzade A, Sadeghi O, Naghdipour Biregani A, Soukhtehzari S, Brandt GS, Esmaillzadeh A (2019) Immunomodulatory effects of flavonoids: possible induction of T CD4+ regulatory cells through suppression of mTOR pathway signaling activity. Front Immunol 10:51

Jantan I, Ahmad W, Bukhari SN (2015) Plant-derived immunomodulators: an insight on their preclinical evaluation and clinical trials. Front Plant Sci 6:655

Kaur P, Robin V, Makanjuola O, Arora R, Singh B, Arora S (2017) Immunopotentiating significance of conventionally used plant adaptogens as modulators in biochemical and molecular signalling pathways in cell mediated processes. Biomed Pharmacother 95:1815–1829

Kayser O, Masihi K, Kiderlen A (2003) Natural products and synthetic compounds as immunomodulators. Expert Rev Anti-Infect Ther 1:319–335

Kilani-Jaziri S, Mustapha N, Mokdad-Bzeouich I, El Gueder D, Ghedira K, Ghedira-Chekir L (2016) Flavones induce immunomodulatory and anti-inflammatory effects by activating cellular anti-oxidant activity: a structure-activity relationship study. Tumor Biol 37(5):6571–6579

Kimura Y, Tsukada J, Tomoda T, Takahashi H, Imai K, Shimamura K, Sunamura M, Yonemitsu Y, Shimodaira S, Koido S (2012) Clinical and immunologic evaluation of dendritic cell–based immunotherapy in combination with gemcitabine and/or S-1 in patients with advanced pancreatic carcinoma. Pancreas 41(2):195–205

Kumar D, Arya V, Kaur R, Bhat ZA, Gupta VK, Kumar V (2012) A review of immunomodulators in the Indian traditional health care system. J Microbiol Immunol Infect 45(3):165–184

Kurek J (2019) Alkaloids: their importance in nature and human life. BoD–Books on Demand

Kvasnica M, Urban M, Dickinson NJ, Sarek J (2015) Pentacyclic triterpenoids with nitrogen-and sulfur-containing heterocycles: synthesis and medicinal significance. Nat Prod Rep 32(9):1303–1330

Lal NK, Nagalakshmi U, Hurlburt NK, Flores R, Bak A, Sone P, Ma X, Song G, Walley J, Shan L, He P, Casteel C, Fisher AJ, Dinesh-Kumar SP (2018) The receptor-like cytoplasmic kinase BIK1 localizes to the nucleus and regulates defense hormone expression during plant innate immunity. Cell Host Microbe 23(4):485–497. e485

Lebish IJ, Moraski RM (1987) Mechanisms of immunomodulation by drugs. Toxicol Pathol 15(3):338–345

Lee JS, Kwon JS, Yun JS, Pahk JW, Shin WC, Lee SY, Hong EK (2010) Structural characterization of immunostimulating polysaccharide from cultured mycelia of Cordyceps militaris. Carbohydr Polym 80(4):1011–1017

Lewicka A, Szymanski L, Rusiecka K, Kucza A, Jakubczyk A, Zdanowski R, Lewicki S (2019) Supplementation of plants with immunomodulatory properties during pregnancy and lactation-maternal and offspring health effects. Nutrients 11(8):1958

Liu C, Cui Y, Pi F, Cheng Y, Guo Y, Qian H (2019) Extraction, purification, structural characteristics, biological activities and pharmacological applications of acemannan, a polysaccharide from aloe vera: a review. Molecules 24(8)

Magedans YV, Yendo AC, Costa F, Gosmann G, Fett-Neto AG (2019) Foamy matters: an update on Quillaja saponins and their use as immunoadjuvants. Future Med Chem 11(12):1485–1499

Majdalawieh AF, Fayyad MW (2015) Immunomodulatory and anti-inflammatory action of Nigella sativa and thymoquinone: a comprehensive review. Int Immunopharmacol 28(1):295–304

Mlcek J, Jurikova T, Skrovankova S, Sochor J (2016) Quercetin and its anti-allergic immune response. Molecules 21(5):623

Mohamed SIA, Jantan I, Haque MA (2017) Naturally occurring immunomodulators with antitumor activity: an insight on their mechanisms of action. Int Immunopharmacol 50:291–304

Mugford S, Osbourn A (2013) Saponin synthesis and function, p 405–424

Nejat N, Mantri N (2017) Plant immune system: crosstalk between responses to biotic and abiotic stresses the missing link in understanding plant defence. Curr Issues Mol Biol 23:1–16

Nicholson LB (2016) The immune system. Essays Biochem 60(3):275–301

Nobori T, Tsuda K (2019) The plant immune system in heterogeneous environments. Curr Opin Plant Biol 50:58–66

Ota M, Nakazaki J, Tabuchi Y, Ono T, Makino T (2019) Historical and pharmacological studies on rehmannia root processing—trends in usage and comparison of the immunostimulatory effects of its products with or without steam processing and pretreatment with liquor. J Ethnopharmacol 242:112059

Pereira D, Valentão P, Pereira J, Andrade P (2009) Phenolics: from chemistry to biology. Molecules 14(6):2202–2211

Qiao D, Liu J, Ke C, Sun Y, Ye H, Zeng X (2010) Structural characterization of polysaccharides from Hyriopsis cumingii. Carbohydr Polym 82(4):1184–1190

Rajput Z, Hu S-h, Xiao C-w, Arijo A (2007) Adjuvant effects of saponins on animal immune responses. J Zhejiang Univ Sci B 8:153–161

Reed SG, Orr MT, Fox CB (2013) Key roles of adjuvants in modern vaccines. Nat Med 19(12):1597–1608
Rodrigues C, Percival SS (2019) Immunomodulatory effects of glutathione, garlic derivatives, and hydrogen sulfide. Nutrients 11(2):295
Senchina DS, Hallam JE, Kohut ML, Nguyen NA, Perera MA (2014) Alkaloids and athlete immune function: caffeine, theophylline, gingerol, ephedrine, and their congeners. Exerc Immunol Rev 20:68–93
Sharma R, Rohilla A, Arya V (2011) A short review on pharmacology of plant immunomodulators. Int J Pharm Sci Rev Res 9:126–131
Shukla S, Bajpai VK, Kim M (2012) Plants as potential sources of natural immunomodulators. Rev Environ Sci Biotechnol 13(1):17–33
Sokolowska M, Boonpiyathad T, Escribese MM, Barber D (2019) Allergen-specific immunotherapy: power of adjuvants and novel predictive biomarkers. Allergy 74(11):2061–2063
Stringlis IA, Proietti S, Hickman R, Van Verk MC, Zamioudis C, Pieterse CMJ (2018) Root transcriptional dynamics induced by beneficial rhizobacteria and microbial immune elicitors reveal signatures of adaptation to mutualists. Plant J 93(1):166–180
Sultan MT, Buttxs MS, Qayyum MMN, Suleria HAR (2014) Immunity: plants as effective mediators. Crit Rev Food Sci Nutr 54(10):1298–1308
Teixeira PJPL, Colaianni NR, Fitzpatrick CR, Dangl JL (2019) Beyond pathogens: microbiota interactions with the plant immune system. Curr Opin Microbiol 49:7–17
Vajdy M (2011) Immunomodulatory properties of vitamins, flavonoids and plant oils and their potential as vaccine adjuvants and delivery systems. Expert Opin Biol Ther 11:1501–1513
Venkatesha SH, Rajaiah R, Berman BM, Moudgil KD (2011) Immunomodulation of autoimmune arthritis by herbal CAM. Evid Based Complement Alternat Med 2011:986797
Vetter V, Denizer G, Friedland LR, Krishnan J, Shapiro M (2018) Understanding modern-day vaccines: what you need to know. Ann Med 50(2):110–120
Vincken JP, Heng L, de Groot A, Gruppen H (2007) Saponins, classification and occurrence in the plant kingdom. Phytochemistry 68(3):275–297
Wang P, Ding X, Kim H, Skalamera D, Michalek SM, Zhang P (2019) Vaccine adjuvants derivatized from Momordica Saponins I and II. J Med Chem 62(21):9976–9982
Wen L, Jiang Y, Yang J, Zhao Y, Tian M, Yang B (2017) Structure, bioactivity, and synthesis of methylated flavonoids. Ann N Y Acad Sci 1398(1):120–129
Woods N, Niwasabutra K, Acevedo R, Igoli J, Altwaijry N, Tusiimire J, Gray A, Watson D, Ferro V (2017) Natural vaccine adjuvants and immunopotentiators derived from plants, fungi, marine organisms, and insects. In: Immunopotentiators in modern vaccines. Elsevier, pp 211–229
Yahfoufi N, Alsadi N, Jambi M, Matar C (2018) The immunomodulatory and anti-inflammatory role of polyphenols. Nutrients 10(11)
Yin M, Zhang Y, Li H (2019) Advances in research on immunoregulation of macrophages by plant polysaccharides. Front Immunol 10:145
Zhang A, Yang X, Li Q, Yang Y, Zhao G, Wang B, Wu D (2018) Immunostimulatory activity of water-extractable polysaccharides from Cistanche deserticola as a plant adjuvant in vitro and in vivo. PLoS One 13(1):e0191356
Zipfel C (2014) Plant pattern-recognition receptors. Trends Immunol 35(7):345–351
Zipfel C, Oldroyd GE (2017) Plant signalling in symbiosis and immunity. Nature 543(7645):328–336

Bioprospection and Clinical Investigations of Immunomodulatory Molecules

3

Maria do Carmo Pimentel Batitucci, Jean Carlos Vencioneck Dutra, Judá BenHur de Oliveira, Mainã Mantovanelli da Mota, Paula Roberta Costalonga Pereira, Schirley Costalonga, Suiany Vitorino Gervásio, and Vanessa Silva dos Santos

Abstract

The immune system, a complex network that guarantees protection against infections, plays an important role in the homeostasis of organs and tissues. Its control and modulation are important factors to be considered for the maintenance of human health. Some natural or synthetic agents are known to stimulate and/or suppress the immune system, being relevant for immunological therapies. The use of plant secondary metabolites by humans has occurred for thousands of years, mainly by traditional knowledge for the treatment of diseases. In this sense, the bioprospection and investigation of new drugs based on plant natural products stand by the search for their direct action on organs or by activation of the immune system. This principle is being addressed by numerous research groups worldwide regarding the isolation and identification of natural substances for the production and/or use as immunomodulators. Therefore, this chapter addresses the importance of bioprospecting phytochemicals, the steps involved in clinical investigation, and some studies dealing with the use of plant-derived substances and their mechanisms in modulating the immune system and their application in immunotherapies, such as tissue repair and cancer.

Keywords

Phytochemicals · Immunomodulation · Bioprospecting · Immune system

M. do Carmo Pimentel Batitucci (✉) · J. C. V. Dutra · J. B. de Oliveira · M. M. da Mota · P. R. C. Pereira · S. Costalonga · S. V. Gervásio · V. S. dos Santos
Laboratório de Genética Vegetal e Toxicológica, Departamento de Ciências Biológicas, Universidade Federal do Espírito Santo, Vitória, Espírito Santo, Brazil

N. S. Sangwan et al. (eds.), *Plants and Phytomolecules for Immunomodulation*, https://doi.org/10.1007/978-981-16-8117-2_3

3.1 Introduction

The Brazilian biodiversity stands out as the largest in species richness on the planet, hosting 15–25% of the total plant species reported up to now (Joly et al. 2011). Due to efforts to inventory these species, plants are among the best studied groups in Brazil. In addition, its high level of endemism is remarkable (55% are endemic to the national territory), with distribution concentrated in two major biomes—Amazon and the Atlantic forests (Simões et al. 2017).

This high biological diversity gives the country an immeasurable source of substances of various classes and chemical structures, with a huge potential for Brazil to be at the forefront of research involving natural compounds and plant metabolism. Therefore, the preservation of Brazilian flora is of fundamental importance given the biological richness and the broad potential for the development of new drugs, cosmetics, agrochemicals, and other bioproducts. The biggest challenge, however, lies in reconciling the management of natural heritage with sustainable practices to prevent biodiversity losses (Simões et al. 2017).

The use of plant substances, known scientifically as phytochemicals or secondary metabolites, by society dates back to millennia, mainly by traditional communities. Traditional knowledge associated with biodiversity is an instrument to indicate proper ways of using natural resources and how to use them wisely. Indeed, one should be aware that the preservation of natural resources is closely related to the individual's own survival. Ethnobotany, or the study of botanical knowledge, and the use of plants by traditional communities are of great value to assist in guiding bioprospecting activity as well as providing support for pharmacological and conservation approaches (Fagundes et al. 2017).

Thus, with the expansion of the human capacity to reveal applications of socioeconomic interests to biodiversity resources, there has been an increase in information and research on how valuable natural resources and knowledge can be. These perspectives, therefore, have gained new dimensions with the advance of biotechnology, more specifically the field of bioprospecting, which is one of the ways to extract economic value from biodiversity (Saccaro Júnior 2011).

In addition, the search for new therapies that combine treatment and patient well-being opens the way for research with plant substances that play an immunomodulatory role in the immune response, which may allow a more assertive treatment, especially in the fight against cancer, with reduction of adverse effects.

3.2 Bioprospecting and Immunomodulation: Concepts and Definitions

The bioprospecting activity in the pharmacological aspect seeks in biodiversity active principles that can generate an innovative product with therapeutic potential. According to Péret (2008), bioprospecting can be defined as the survey of biodiversity to identify a component of genetic heritage and information about traditional knowledge for the development of products of commercial value. The targets of bioprospecting are collectively called genetic resources (Saccaro Júnior 2011).

The idea of ownership on genetic resources and traditional knowledge becomes effective with the Convention on Biological Diversity (CBD), an international agreement launched during the United Nations Conference on Environment and Development in 1992, in which it acknowledges that each country must legislate on the forms of access to genetic resources located in the respective territory (Saccaro Júnior 2011).

In Brazil, in relation to the regulation of bioprospecting in order to adequately protect the genetic resources existing in the country, these guidelines were given through Provisional Measure No. 2.186–16/2001, repealed by Law No. 13.123/2015 (Biodiversity Law) after 15 years of maturing the legal framework (Carvalho et al. 2020). This law establishes new policies for accessing genetic heritage, traditional-associated knowledge and benefit sharing, de-bureaucratizing, and facilitating study procedures. It no longer requires prior authorization for the execution of research or technological development with Brazilian biodiversity. Currently, the regularization of research activity is carried out through electronic registration on the National System of Management of Genetic Heritage and Associated Traditional Knowledge (SisGen). SisGen is an instrument to assist the Genetic Heritage Management Council with regard to the management of genetic heritage and associated traditional knowledge.

In this scenario, bioprospecting and ethnobotany are associated in the investigation of new drugs composed mainly of plant molecules capable of modulating immune responses in an organism. The combination of these sciences, that is, to perform bioprospecting based on ethno knowledge, optimizes and guides the search for biological material, often essential for the selection of research targets.

The World Health Organization (WHO) has observed that the study of traditional medicines through bioprospecting found several pharmaceutical compounds, 11.1% of which were obtained from plants and 8.7% were obtained from animals (Santos et al. 2011). Bioprospecting makes use of biodiversity resources for the discovery and formulation of new sources of chemical compounds, genes, proteins, and other components with potential biotechnological and economic value (Marques et al. 2014).

Given this, the development of new drugs based on bioprospecting of products of plant origin in the treatment of numerous diseases has proved to be an increasingly promising field of research. Some synthetic drugs, even when administered at optimal dosages, are still directly related to toxicity in normal cells and tissues—which makes the area of bioprospecting increasingly challenging, increasing the need to design a therapy that is safer, more effective, selective, and stimulating studies aimed at the use of products of natural origin that have immunomodulatory activities (Diwanaya et al. 2004).

The organisms' defense system is mediated by sequential and coordinated responses called innate and adaptive immunity. Innate immunity is the first line of defense of the body against external pathogens and foreign substances characterized by a rapid response that is independent of external stimuli. Thus, its main function is to immediately prevent the spread and movement of foreign or specific pathogens by the body, regardless of previous contact with these immunogens or aggressor agents. This system consists of physical and chemical barriers and is capable of combating pathogens by recruiting defense cells (macrophages and leukocytes) that are capable of destroying pathogens through inflammatory processes (Abbas et al. 2019). When the innate immune system fails to eliminate pathogens entering the body, the adaptive immune system (or acquired immune response) takes over. This is the second line of defense of the organism and targets the specific antigen of the pathogen that is causing the infection. Adaptive immune response is mediated by cells called lymphocytes (B and T). These cells have a wide variety of receptors that are able to recognize a vast number of antigens. It is through the adaptive immune system that the organism is able to develop an immune memory. When the body is exposed to an antigen for the second time, the response tends to be faster, greater in magnitude, and qualitatively different from the primary response to that antigen (Abbas et al. 2019). Chapter 1 provides with more details on how the mammalian immune system works.

Immunomodulation, by either synthetic or natural substances, can be considered an alternative for the prevention or cure of diseases—consists in controlling the body's response to certain pathogens. This control is performed by an immunomodulatory agent capable of adjusting this response, exacerbating or reducing. For example, cancer treatment using immunotherapy may occur in order to trigger an active immune response to tumor antigens or with the administration of antitumor antibodies, auxiliary T lymphocytes, to establish passive immunity. Immunomodulatory drugs, and the substances produced by cells that are involved in immune responses, can act as immunostimulators, associated with increased innate and adaptive immunity or immunosuppressants responsible for decreased immune system activity (Lima 2007; Nunes-Pinheiro et al. 2003).

The regulatory function attributed to cells that are directly linked to the immune response ensures, and shows, that the immune system itself is capable of self-regulating through the release of immunomodulatory substances in order to restore homeostases (Holland and Vizi 2002). The discovery of immunomodulators allowed interventions in the immune system to restore balance and reduce drug-associated side effects.

Immunomodulators include synthetic, biological (cytokines and antibodies), plant, and microbial products, all capable of influencing immunoregulation cascades to produce their specific stimulator, suppressive or regulatory effects. Substances of plant origin can be sources of synergistic combinations useful in the discovery of drugs. Some classes of plant metabolites with immunomodulatory potential are reported in the literature, such as isoflavonoids, indols, phytosterols, polysaccharides, sesquiterpenes, alkaloids, glucans, and tannins (Patwardhan and Gautam 2005).

Secondary metabolites have a wide spectrum of biological activities, providing herbal medicines or isolated active substances that can be used directly or as prototypes for the synthesis of new potent drugs. In recent years, interest in synergistic effects among these substances has increased due to the development of multicomponent therapeutic strategies, in which drug candidates act simultaneously on several interrelated therapeutic targets (Casanova and Costa 2017).

The combination of substances with synergistic effect is useful to increase the bioavailability of active substances, decrease the required doses, and/or minimize side effects. The constituents of medicinal plants often exhibit ability to interact with proteins and modulate cellular processes. The combination of these agents may provide multi-targeting formulations with higher potency than that of the individual active substances (Casanova and Costa 2017). Therefore, medicinal plants are undoubtedly potential therapeutic agents.

3.3 Biomolecule-Mediated Immunomodulation: A Focus on Bioactive Plant Compounds

The diversity of pathogens has led to the evolutionary development of a system capable of fighting infections and developing protective memory to produce a quick response and prevent the progress of pathologies in reinfections; in this way, a cascade of reactions is required and are responsible for the regulation of key processes in the synthesis, activation, and distribution of components of the immune system machinery.

In view of the interdependence of these reactions, biomolecules capable of intervening in any of these stages will exert an immunomodulatory effect, as either an immunosuppressant, immunoadjuvant, or stimulator of the immune response (Chandana 2016). In this scenario, several natural products, especially plant metabolites, have immunomodulatory capacity (Patwardhan and Gautam 2005) and are promising sources for the bioprospection of new drugs and application in new therapies.

In fact, several studies have already reported the immunomodulatory action of herbal medicines in regulating the secretion of cytokines and immunoglobulins, expression of lymphocytes, and stimulating phagocytosis, among other aspects (Patwardhan and Gautam 2005).

Nunes-Pinheiro et al. (2003) define the immune system as a complex and highly sophisticated system, in which the activation, differentiation, and regulation of defense cells 1 depend on cytokines 2, a group of soluble mediators synthesized by immune cells (Hall and Virella 2007; Pyorala 2002; Abbas et al. 2015). The production pattern of cytokines 2 results in the activation of defense, leading to a pro- or anti-inflammatory response (Sharon 2000; Oliveira and Nunes-Pinheiro 2013).

The first line of defense is triggered by the capture, processing, and presentation of the antigen[1] to the CD4+ T cells (Th0) via APC[2] cells, an event that signals the activation and differentiation of T lymphocytes in different cell subpopulations. The Th1 phenotype controls the proinflammatory and cytotoxic response through the production of IL-2, IL-6, IFN-α, and TNF cytokines involved in the proliferation and activation of T lymphocytes and NK cells as well as acts in the activation of phagocytic macrophages. In turn, the Th2 subtype synthesizes and releases cytokines that will mediate the anti-inflammatory response, meanly IL-4, IL-5, and IL-10 (Chandana 2016; Oliveira and Nunes-Pinheiro 2013).

The second line of defense of the immune system begins with the activation of macrophages by helper T lymphocytes. These cells are responsible for the identification of bacteria captured by T cells, for the elimination of phagocyted microorganisms, and for the secretion of attractive substances from phagocytic neutrophils (Abbas et al. 2015).

Due to the peculiarities and the interdependence of the events necessary for the response to pathogens attack, three main mechanisms are currently acknowledged in the literature: (I) defense cells mediated, (II) cytokines mediated, and (III) microchimerism mediated (Hellings and Blajchman 2009).

3.3.1 Cell-Mediated Immunomodulation Mechanism

Immunomodulation mechanism comprises four main events: (I) antigen recognition, (II) activation of macrophages, (III) cytotoxic response mediated by T lymphocytes, and (IV) production of antibodies via B lymphocytes (Nunes-Pinheiro et al. 2003).

Although it can be mediated by the CD1 protein (Sharon 2000), the recognition of antigenic particles occurs predominantly through MHC molecules compatible with the APC-peptide complex and constitutes an essential step for the activation and proliferation of $CD4^+$ T cells (Crotty 2011), as well as for the beginning of the defense process (Cockburn et al. 2010); the activation of $CD8^+$ T lymphocytes can also occur via IL-12 released from Th1 cells (Nunes-Pinheiro et al. 2003). The proliferation of T cells is suppressed by IFN-γ and nitric oxide (NO) released by macrophages (Fischer et al. 2008), as well as by some classes of secondary metabolites, such as flavonoid glycosides from *Euphorbia microsciadia* Bios. These flavonoids are inhibitors of protein kinase (Modares et al. 2018) together with unsaturated fatty acids and the alkaloid wariftein extracted from *Cissampelos sympodialis* (Oliveira and Nunes-Pinheiro 2013).

The performance of macrophages impacts the entire immune system, being as important as the recognition of antigens. As a result of the phagocytic activity of

[1] Substance foreign to the organism that triggers the production of antibodies.

[2] Antigen-presenting cells that have the Main Histocompatibility Complex (MHC), which is coupled to the antigen (forming the peptide-MHC complex) and presents the antigen to specific receptors (TCR) on the surface of $TCD4^+$ lymphocytes, leading to their activation.

macrophages, it can be observed the generation of free radicals, the mediation of inflammatory processes, and the secretion of a variety of biochemically different substances, such as enzymes, cytokines, and components of the complement system (Fischer et al. 2008). According to the function, these cells can be divided into two subtypes. The M1 phenotype acts in the elimination of infected cells through the production of NO, expression of high levels of IL-12, and low levels of IL-10. The repair phenotype—M2—leads to tissue healing through ornithine synthesis and production of high levels of IL-10 and low levels of IL-12 (Chandana 2016).

The ability to phagocytize pathogenic microorganisms is essential for fulfilling the defense functions of macrophages and depends on the supply of intracellular Ca^{+2} immediately after activation. The Ca^{+2} signals the production of inflammatory cytokines (TNF-α and IFN-γ) characteristics of the Th1. These soluble mediators trigger one of the most important defense strategies of the immune system: the activation of inducible nitric oxide synthase (iNOS) and increase of NO levels (Alves et al. 2017). Relatively high amounts of NO guarantee the elimination of phagocyte pathogens (Melo-Neto et al. 2016).

The NO is highly cytotoxic and interferes with several immunological aspects, such as by inducing apoptosis and the synthesis of inflammatory cytokines (Gonçalves et al. 2015). The stimulation of oxidative metabolism with production of reactive oxygen species (ROS) is also triggered by NO (Ueda-Nakamura et al. 2006). However, the increase in the production of ROS can cause damage to the host organism if this defense response persists for a long period. This damage can be avoided by antioxidant compounds, such as the flavonoid quercetin, which leads to decreased phagocytosis and ultimately the elimination of microorganisms (Côrtes et al. 2013). On the other hand, the benzophenones (flavonoid class) may stimulate the production of ROS (Sen and Chatterjee 2011).

Furanocoumarins stimulate apoptosis via stimulation of NO release (Widelski et al. 2017). Oxylipins, in conjunction with NO, enhance the inflammatory response by stimulating the synthesis of TNF-α in macrophages (Sen and Chatterjee 2011).

Depending on the immune cell in which the secondary metabolite will act, representatives of the same class of compounds can exert antagonistic functions, such as those observed by alkaloids, which induce apoptosis in macrophages by activating NF-κB and increase NO generation in infected neutrophils by stimulating oxidative metabolism (Sen and Chatterjee 2011). However, quinolizidine alkaloids isolated from *Sophora subprostrata* as well as ailantamide extracted from *Zanthoxylum ailanthoides* inhibit the synthesis of superoxide anions in neutrophils (Souto et al. 2011).

The mechanism of cell elimination mediated by $CD8^+$ T lymphocytes involves the release of cytotoxic cytokines toward the target cell. This event destroys the membrane integrity and allows the entry of other toxic substances, which accelerates cell self-destruction (Nunes-Pinheiro et al. 2003). The toxicity of these lymphocytes can be maximized by the association with programmed cell death proteins 1 (PD-1), expressed on the surface of infected cells (Thyagarajan and Sahu 2018), which stimulates the synthesis of IFN-γ and other cytokines by T lymphocytes $CD8^+$. Binding to PD-1 is one of the main checkpoints to ensure the effectiveness of the immune system (Zhu and Li 2018), and the increase in the receptors compatible with this protein in $CD8^+$ T cells is proportional to the increase in infection (Crotty 2011).

In turn, IL-12- and TNF-α-activated NK cells from macrophages (Diwanaya et al. 2004; Fischer et al. 2008) associate with $CD8^+$ T lymphocytes at the beginning of the infection and release high levels of IFN-γ, IL-4, and IL-10 (Cockburn et al. 2010; Seder et al. 2013; Chandana 2016).

Essential oils stand out for inhibiting inflammatory responses. The mechanism is based on the modulation of cytotoxicity, which induces programmed cell death of Th1 lymphocytes and stimulates the differentiation of $CD8^+$ T lymphocytes, by overproduction of ROS and NO. It is worth mentioning that the programmed cell death mechanism may also be modulated by plant fatty acids, which induce lymphocyte apoptosis and necrosis (Oliveira and Nunes-Pinheiro 2013).

The IL-1β is a mediator of necrosis and cell death by apoptosis (Guedes et al. 2004) via activation of NOS that culminates in oxidative stress (Oliveira et al. 2011). Although highly damaging to cellular metabolism, the synthesis of ROS and NO adds to the strategies of $CD8^+$ T lymphocytes and NK cells in the cytotoxic response to infected cells (Abbas et al. 2015; Chandana 2016). The longer the antigen remains in the host, the greater the severity of this response is (Cockburn et al. 2010).

The activation of B lymphocytes is crucial for the synthesis of antibodies, whose class will be determined by the defense response pattern. In the Th1 profile, the activation of these cells leads to the production of immunoglobulin G (IgG) predominantly. In turn, the Th2 profile stimulates the synthesis of IgM, IgG, IgA, and IgE. Unsaturated fatty acids and volatile fatty acids of plant origin exert an immunosuppressive effect on B cell activation and antibody secretion (Oliveira and Nunes-Pinheiro 2013; Chandana 2016).

3.3.2 Immunomodulation Mechanism Mediated by Soluble Factors

The immunomodulation mechanism mediated by soluble factors consists of the release of cytokines, mainly interleukin, interferon, and tumor necrosis factors.

According to Oliveira et al. (2011), these water-soluble molecules are capable of regulating the activity, differentiation, proliferation, and survival of immune cells, as well as regulating the production and activity of other cytokines. In this way, the defense response to a pathogen can be stimulated (profile Th1 and synthesis of IL-1, IL-2, IL-6, IL-7, IFN-γ, and TNF-α) or attenuated (profile Th2 and production of IL-4, IL-10, and IL-13).

Some classes of phytochemicals may enhance the Th1 defense response by stimulating the synthesis of pro-inflammatory cytokines. Phenolic compounds, including benzophenone and tannins, modulate gene expression for the production of pro-inflammatory cytokines, especially TNF-α and IFN-γ (Sen and Chatterjee 2011), and stimulate iNOS activity (Alves et al. 2017). The alkaloid wariftein increases NO production (Souto et al. 2011), while terpenes stimulate the release of NO, TNF-α, and IFN-γ cytokines in host cells (Sen and Chatterjee 2011). The increase in NO levels also occurs in the presence of plant lecithins, making them important agents in the cytotoxic response of the immune system (Afonso-Cardoso et al. 2011; Marim et al. 2010). In contrast, condensed tannins (proanthocyanidins) significantly reduced pro-inflammatory cytokines, favoring the establishment of the anti-inflammatory response—Th2 (Salinas-Sánchez et al. 2017).

Among the pro-inflammatory cytokines, IL-6 promotes maturation and activation of neutrophils and macrophages in addition to differentiation/maintenance of cytotoxic T lymphocytes (Guedes et al. 2004) and NK cells (Oliveira et al. 2011). Flavones can stimulate the production of IL-6, IL-1β, TNF-α, and NF-κB, increasing the pro-inflammatory defense response (Chandana 2016).

Cytokine synthesis begins with the activation and migration of the nuclear transcription factor (NF-κB) to the nucleus to play a fundamental role in establishing the inflammatory response. This is because of its ability to regulate the expression of genes that encode for key components of the immune system, such as protein 1 from monocyte attraction (MCP-1), inflammatory cytokines (interleukin and TNF-α), and the inducible nitric oxide synthase (iNOS) (Fischer et al. 2008). Additionally, NF-κB assists the recruitment of phagocytic cells (macrophages and monocytes) and stimulates lymphocyte activation (Persson-Waller et al. 2003; Guedes et al. 2004).

Phenolic compounds, especially flavonoids such as sakuranetin (extracted from *Baccharis retusa* DC), have stood out as immunosuppressive agents due to their ability to inhibit NF-κB activation, favoring anti-inflammatory immune response due to the predominance of the Th2 profile (Toledo et al. 2013). The inhibition of NF-κB can also occur indirectly, through the inhibition of the proteolysis-inducing factor (PIF), in a mechanism by which alpha-linolenic phenolic acid (ALA), a component of vegetable oils such as olive oil, reduces the inflammatory response in the acute phase of infection, given that this factor participates in the activation of NF-κB. As a result, it stimulates the gene expression of inflammatory cytokines, such as IL-6 and IL-8 (Rosa and Cruz 2016).

The production of inflammatory cytokines such as IFN-γ, TNF-α, IL-1β, and IL-12 (Souto et al. 2011) leads to a cascade of activation of other cytokines via target cells throughout the immune system (Oliveira et al. 2011). A general response then takes place, leading to the Systemic Inflammatory Response Syndrome (SIRS) (Guedes et al. 2004; Souto et al. 2011). Plant fibers can prevent the activation of the NF-κB complex since they can decrease the cell levels of ROS (Oliveira and Nunes-Pinheiro 2013).

Several plants and their metabolites are capable of modulating the production and function of cytokines (Chandana 2016). Plant unsaturated fatty acids, for example, can stimulate the recruitment of leukocytes to inflammatory sites and induce the biosynthesis of anti-inflammatory cytokines in detriment of those that characterize the Th1 profile (Oliveira and Nunes-Pinheiro 2013; Vogt et al. 2013).

Plant lecithins, such as those from soy and sunflower, are important defense metabolites. In the mammal's immune system, plant lecithins may couple to surface receptors of immune cells activating a cascade that culminates in the production of a variety of cytokines depending to the lecithin (Afonso-Cardoso et al. 2011).

The lecithin ScLL from *Synadenium carinatum* Boiss induces the production of IL-12 and consequently the inflammatory defense response. In turn, the lecithin

Artin M (from *Artocarpus* spp.) can modulate responses for the Th1 profile (stimulating the production of IL-12, IFN-γ, and TFN-α) and the Th2 profile as well (via IL-10 biosynthesis) (Araújo et al. 2013; Egito et al. 2017). The Artin M lecithin participates in the activation of macrophages and increment of neutrophils influx to the infection, contributing for the healing (Egito et al. 2017).

Another important action carried out by lecithins concerns the activation of the complement system (Egito et al. 2017), formed by plasma glycoproteins, which induce inflammatory responses by recognizing the surface structures of pathogens (Fischer et al. 2008, Araújo et al. 2013). Aromatic acids and flavonoids, on the other hand, stimulate the macrophage production of C1q protein, which is essential for the activation of the complement system (Fischer et al. 2008).

3.3.2.1 Microchimerism Immunomodulation Mechanism

The microchimerism involves processes between cells that are coordinated by the release of inhibitory cytokines IL-4 and IL-10, characteristics of the Th2 response.

The IL-4, a typical component of the anti-inflammatory defense strategy (Varella et al. 2001; Maghraby and Bahgat 2004), is produced by $CD4^+$ T lymphocytes and inhibits pro-inflammatory cytokines, such as IL-1, IL-6, and TFN-α. In addition, this IL-4 presents antioxidant activity, inhibits the production of ROS (Oliveira et al. 2011), and ameliorates the symptoms and severity of pathogen-induced inflammation.

The anti-inflammatory interleukin-10 (IL-10) is mainly synthesized by activated $CD8^+$ T cells and can interfere at different extents with the immune system. The IL-10 biosynthesis occurs directly from signaling between short-chain $\beta 2 \rightarrow 1$-fructan[3] plant fibers and Toll-like cell receptors (TLR) for the recognition of specific molecular patterns of some pathogens (Vogt et al. 2013). The IL-10 play a key role in neutralizing the pro-inflammatory response by inhibiting both the proliferation of lymphocytes of the Th1 profile (Varella et al. 2001) and the expression of the IFN-γ, IL-2, and IL-12 cytokines and TNF-β (Oliveira et al. 2011) through the prevention of NF-κB complex activation (Guedes et al. 2004).

The alkaloid wariftein induces IL-10 biosynthesis, favoring the anti-inflammatory defense (Costa et al. 2008; Souto et al. 2011). Terpenes, particularly the essential oils from *Helianthus annuus* L. (sunflower), also drive the immune system to the Th2 profile by preventing the cleavage of the protein complex linked to NF-κB, which decreases the number of active molecules (Oliveira and Nunes-Pinheiro 2013) and the synthesis of inflammatory cytokines. Coumarins, like auraptene, also modulate the system leading to the anti-inflammatory response by inhibiting TNF-α biosynthesis and decreasing NF-κB levels in the nucleus. Auraptene was found to show to inhibit the production of superoxide anion and NO (Askari et al. 2018).

The immunomodulatory mechanisms are didactyly classified in three categories since a series of interdependent events that involve the entire immune response take place. An example is the activation of $CD4^+$ T cells, which does not produce a

[3]Long chain $\beta 2 \rightarrow 1$-fructans lead to Th1 response profile with the production of IL-12.

defense response by itself and therefore depends on the production of cytokines such as IL-2, TNF-α, among others (Nunes-Pinheiro et al. 2003).

3.3.3 Clinical Research: From the Basics to the Clinical Trials

Natural products are important sources of therapeutic agents and contributes for the development of new medicines. The interest in natural immunomodulators for the treatment of diseases like cancer, immunodeficiencies, autoimmune, and infectious pathologies and for posttransplant therapies has grown recently (Singh et al. 2016; Mohamed et al. 2017). There are several immunomodulators available in the pharmaceutical industry, but many of them provide with adverse reactions in patients. In this sense, the search for natural products capable of modulating the immune system with low side effects, greater bioavailability, and longer lasting effect has been considerably increasing in recent years (Grigore 2017).

According to Nair et al. (2019), some natural sources of immunomodulatory compounds, namely, biological response modifiers (BRMs) have been described since antiquity (Nair et al. 2019). Example of natural sources includes garlic (*Allium sativum*) for its anti-allergic and antitumor properties, whose isolated constituents can inhibit the transcription of several pro-inflammatory genes, acting hence as immunomodulators. Other plants such as *Artemisia annua* L. (Bai et al. 2019), *Curcuma longa* (Mahana et al. 2018), *Acacia catechu* Willd (Sunil et al. 2019), *Momordica charantia* (Mahamat et al. 2020), and *Brugmansia suaveolens* (Humboldt and Bonpland ex Willdenow) Bercht. and J. Presl (Kumar et al. 2020) are reported to be important modulators of the immune system response.

Several immunomodulators of natural origin or their derivatives are already available by the pharmaceutical industry such as brevenal, an immunomodulator produced by the unicellular alga *Karenia brevis*, used in the treatment of chronic lung disease (COPD), cystic fibrosis, and asthma (Keeler et al. 2019). Fingolimod (FTY720), an immunosuppressant used to treat multiple sclerosis, is a synthetic compound inspired in the structure of myriocin, a secondary metabolite extracted from the fungus *Isaria sinclairii* (Baer et al. 2018; Strader et al. 2011). Paclitaxel, a natural product widely produced by the plant species *Taxus chinensis* (Pilg.) Rehder, is the chemotherapeutic front line for the treatment of various types of cancer (Wang and Du 2018).

The search for new drug candidates from nature is challenging due to the wide range of complex chemical structures present in plants. The advent of new technologies improved the screening of natural products for the development of new medicines (Thomford et al. 2018; Koparde et al. 2019).

There are a series of steps to be considered for a compound of natural origin with potential to modulate the immune response to become a commercialized drug. These steps include basic research, preclinical tests, and clinical trials besides the approval of the new drug by regulatory agencies. Techniques such as High-Performance Liquid Chromatography (HPLC), High-Performance Thin Layer Chromatography (HPTLC), Fourier Transform Infrared Spectroscopy (FTIR), Nuclear Magnetic

Resonance (NMR), and Mass Spectrometry (GC-MS) are essential to determine the chemical identity of biologically active natural product. Thus, the development of a new drug requires a rigorous assessment of the candidate's safety and effectiveness (Corr and Williams 2009; Nugent et al. 2013; Goyal et al. 2016; Koparde et al. 2019).

Clinical investigations are necessary for the discovery of new immunomodulators and the elucidation of the mechanisms of action, which consist of studies carried out with sick and healthy humans, expanding the knowledge about the diseases, techniques, and methodologies for treatment and diagnosis (Brasil, Agência Nacional de Vigilância Sanitária - Anvisa 2020). Prior to testing on human beings, basic research (or preclinical tests) is carried out to expand the knowledge about the drug candidate (Carneiro 2008; Nair et al. 2019). The studies on preclinical stage are of great relevance as they provide initial information on the general and reproductive toxicity of the tested compound and its carcinogenic, mutagenic, or teratogenic potential. Preclinical trials also assist scientists in the strategies devised for clinical trials. For instance, preclinical trials assist in establishing the safe initial dose for administration in humans (Corr and Williams 2009; Nugent et al. 2013).

Preclinical tests require the execution of a series of experiments using in vitro and in vivo models, and more recently, in silico models have been used as a prior or complementary procedure to the aforementioned tests. In vitro studies are relatively quick, simple, and less expensive when compared to in vivo tests. It relies on the use of cell, tissue, or even organ culture to perform assays and evaluate a specific cell component. In general, these tests enable greater control and monitoring of experimental conditions (Honek 2017). In vivo models, in general, are more expensive and require longer time to obtain the results, making large-scale screening difficult. However, the importance of using animals in preclinical tests is to better understand the physiological mechanisms of action of a certain drug in living organisms (Soni et al. 2020). With the advances in bioinformatics, in silico tests have been adopted as strategies to reduce costs and time to market a drug. Such analyses are based on computer simulations using algorithms, which can precede or complement in vitro and in vivo tests (Pappalardo et al. 2018).

Research with immunomodulatory agents should include analyses of their effect on the immune system in preclinical tests. Several tests can be used to evaluate the immunomodulatory capacity of a drug candidate, such as mitogen-induced lymphocyte proliferation assays; evaluation and quantification of cytokinins; inhibition of T-cell proliferation; assessment of phagocytosis capacity in macrophages, antibody-forming cells, or plaque-forming cells (PFC); and assays for the production of specific antibodies, among others (Singh et al. 2016). The immunomodulatory effect of *Momordica charantia* extracts was investigated in mice infected with *Salmonella typhi*, using phagocytosis assays and analysis of cell death mechanisms in macrophages and neutrophils, determination of NO concentration, mobilization of leukocytes, and production of antibodies against *Salmonella typhi* using the hemagglutination technique (Mahamat et al. 2020). Another example is the immunomodulatory capacity of *Schinus terebinthifolia*-derived lectin on spleen cells extracted from mice assessed by cell viability, determination of intracellular Ca^{2+}

concentration, and quantification of NO and interleukin-type (IL) cytokines, tumor necrosis factors (TNF), and interferon gamma (IFN-y) (Dos Santos et al. 2020). The effects of walnut oligopeptides (*Juglans regia* L.) on the mice immune system were determined by the activity of NK cells, phagocytosis capacity of macrophages, proliferation of T lymphocytes, and the quantification of cytokines and immunoglobulins (Mao et al. 2020).

Regardless of its origin (natural or synthetic), when designing a new drug, studies are needed initially to address the safety, viability, and efficacy of the drug so that clinical tests can be started (Teicher and Andrews 2004). A clinical trial consists of a prospective study, carried out in humans to prove efficacy and safety of a new drug. These tests are intended to verify the physiological effects related to the absorption, distribution, metabolism, and excretion of experimental drugs.

Clinical tests are carried out after the approval by the health authorities of the country in which the investigation will take place, and clinical trials must be in accordance with the ethical and scientific standards of regulatory agencies such as American agency (FDA, Food and Drug Administration), the European agency (EMA, European Medicines Agency), the Japanese Pharmaceuticals and Medical Devices Agency (PMDA), and the Brazilian National Health Surveillance Agency (ANVISA) (ANVISA. Agência Nacional de Vigilância Sanitária 1999; Brody 2016; Griffiths et al. 2017).

Clinical studies with therapeutic agents are usually carried out in three distinct stages: Phase I–Phase III. However, there are Phase IV studies carried out after the launch of the new drug, and they consist of large trials or observational studies to assess possible side effects that have not been recorded in Phases I to III. Also, there are studies that do not fit perfectly into these phases, and some clinical trials may overlap two different phases, such as studies involving human behavior, lifestyle, or surgical interventions (Grady et al. 2013; Brody 2016; Wrigth 2017).

Phase I studies typically involve healthy volunteers, but in some therapeutic areas, such as cancer, the drug is tested directly on patients without the need of previous tests in healthy people. The objective in this step is to determine the maximum tolerated dose (MTD), which is commonly estimated by increasing the treatment doses until reaching the dose-limiting toxicity (DLT) of the new drug, being observed the most frequent side effects and aspects related to the pharmacokinetics and pharmacodynamics of the drug and its metabolites (Storer 1989; Friedman et al. 2015; Wrigth 2017). After the safety certification and determination of the correct dosages of the new drug is obtained in Phase I clinical trials, the investigations move to Phase II, where the efficacy and safety of the therapeutic agent in the target population are evaluated, and the tests will involve a higher number of individuals. These studies function as screening stages, enabling the determination of correct dose, the most common short-term side effects, and the best regimen to be used in the next stage. Phase II, therefore, represents a critical point in clinical trials, where many drug candidates do not progress to the next testing phase, either because they have previously unknown toxic side effects or were proven to be ineffective in treating the condition tested (Corr and Williams 2009; Wrigth 2017; Norman 2019). The research carried out in Phase II is usually divided into Phase II-a and Phase II-b. Analyses carried out in Phase II-a provide preliminary data on the safety of the new drug since the side effects may be different in sick and healthy people. In addition, the Phase II-a tests assess the effectiveness of the tested drug, indicating the occurrence of the candidate drug's interaction with its target and the consequent alteration of the disease or condition. Phase II-b trials are larger and use broader dosages, providing more robust data on the effectiveness of the new drug (Corr and Williams 2009; Wrigth 2017). Phase III consists of the final stage of the development of new drugs before registration and aims to confirm the efficacy of the new drug and its dosage, as well as to monitor side effects and to compare the therapeutic agent tested with drugs or treatments frequently used and/or placebo. This stage of the assessment involves hundreds or thousands of patients with the disease or condition of interest. These studies are performed in a randomized manner to ensure that the differences found between the treated groups are actually the result of the treatment and not of genetic or environmental factors within the groups. This phase is essential for the approval of a new drug by the regulatory authorities (Corr and Williams 2009; Swati and Akhilesh 2011; Wrigth 2017). After approval of the new drug, diagnostic or therapeutic procedure will be launched on the market, and Phase IV studies will begin, which consist of tests to monitor its use in the population, enabling the knowledge of effects associated with long-term use that were not detected in clinical research (Wrigth 2017).

A large number of clinical studies have been developed to elucidate the performance of natural products and their applications in the treatment of diseases. Among them, it can be highlighted that the Phase II trial used a formulation of curcuminoid and soy lecithin patented as Meriva®, whose objectives are to test the safety and activity of curcumin as a nutritional supplement for patients diagnosed with pancreatic cancer and undergoing chemotherapy with gemcitabine. In this study, patients were submitted to physical examinations, biomarkers assessments related to inflammation, toxicity, and the quality of life prior to the beginning of each treatment cycle. Each treatment cycle consisted of a daily intake of 2000 mg of the formulation for 28 days and nine treatment cycles. At the end of the evaluation, it was verified that the treatment with Meriva® was safe and effective, helping to fight cancer and bringing benefits to patient's well-being. These finds suggest that the tested curcuminoid may control the inflammation through the activation of the nuclear factor Nrf2, which modulates the expression of genes that act in the inflammatory responses and leads to the inactivation of the NF-κB factor, related to the modulation of genes in response to stress factors (Pastorelli et al. 2018).

The diterpenoid lactone andrographolide, isolated from the stem and leaves of *Andrographis paniculata* (Burm. F.) Nees, has been shown to be effective in modulating the immune response, to have neuroprotective properties. It is currently in Phase II tests for the treatment of progressive multiple sclerosis (PMS). Patients were treated for 24 months with 150–300 mg of andrographolide at 99.5% or with placebo in a double-blind protocol. Every 12 weeks, a trained and unrelated person evaluates the patients who were participating in the study. At the end of the tests, it was concluded that the use of andrographolide was safe and promising for the treatment of progressive multiple sclerosis. The mechanisms of action identified as responsible for immunomodulation are also related to the activation of Nrf2 and the inactivation of NF-κB (Ciampi et al. 2020).

Ingenol mebutate, a substance derived from *Euphorbia peplus*, was tested in Phase III trials for the topical treatment of actinic keratosis (AK), a condition on the skin that manifests itself as a dark spot with a rough appearance and can potentially progress for a squamous cell carcinoma if the patient does not receive proper treatment. These studies aimed to compare the safety of ingenol mebutate gel when applied simultaneously or sequentially in two areas with AK lesions and evaluate the effectiveness and satisfaction with the treatment. Ingenol mebutate has been shown to have an acceptable safety profile and high effectiveness in eliminating AK lesions (Pellacani et al. 2015). The mechanisms of action of the compound are likely related to its immunomodulatory capacity to induce apoptosis and neutrophil-mediated cell cytotoxicity (Kim et al. 2018; Seca and Pinto 2018).

The importance of understanding and well planning clinical studies is evident considering the vast number of diseases and treatments related to the immune system and its modulation.

3.3.4 Plant Metabolites as Medicines

There is a range of compounds from plant metabolism (e.g., peptides, saponins, essential oils, terpenes, flavonoids, and others) that are described as modulators of immune response through complex processes.

Ascorbic acid, popularly known as vitamin C, is an antioxidant used worldwide, whose effects include combating diabetes and assisting fat burning (Cui et al. 2015), besides the role as cofactor in several biosynthetic pathways (Carr et al. 2013) and ROS (reactive oxygen species) (Lira et al. 2012) scavenger. Glycyrrhizic acid, extracted from plant roots (*Glycyrrhiza glabra*), shows hepatoprotective effect by modulating the hepatocyte toxicity levels and decreasing inflammation in liver (Yoshikawa et al. 1997). The role of glycyrrhizic acid in the inhibition of SARS virus replication (Cinatl et al. 2003) and in the increase of resistance to the herpes virus (simplex-1) and reinfection by *Candida albicans* (Sekizawa et al. 2001; Utsunomiya et al. 2000) was reported. *Withania somnifera* Dunal (WS) produces alkaloid and steroids in roots that have immunomodulatory, antitumor, and antioxidant properties (Mishra and Singh 2000).

Much of the compounds derived from plant metabolism used in the treatment of diseases are also investigated as alternatives to immunotherapies to fight cancer. In this way, species of the genus *Morinda*, such as *Morinda elliptica* and *Morinda citrifolia*, traditionally used in the treatment of various diseases, such as headache, cholera, gastric ulcer, and diarrhea (Ong and Norzalina 1999; Wang et al. 2002), are important tools for modern medicine because of the production of damnacanthal (Ali et al. 2000). Damnacanthal is an anthraquinone capable of stimulating the immune system, preventing the formation and proliferation of different types of cancer, including lung cancer (Hirazumi et al. 1996, 1999). Its mechanism of action against cancer cells was elucidated by Hiramatsu et al. (1993) and occurs through the inhibition of RAS proteins (Rat Sarcoma virus), which act as a trigger for various types of cancer related to the imbalance in the mechanisms of cell division control.

Essential oils (monoterpenes, sesquiterpenes, and several oxygenated derivatives), widely used in aromatherapy, phytocosmetic, and food industries, have received great attention with respect to immunomodulatory activity (Hotta et al. 2009; Hart et al. 2000). Its immunomodulatory activity may involve the activation of both immune, pro-inflammatory, and anti-inflammatory response pathways. Essential oils can promote the recruitment of lymphocytes and their phagocytic capacity, intensifying their bactericidal performance (Sonnenberg and Hepworth 2019; Redgrove and McLaughlin 2014). The essential oil extracted from *Melaleuca alternifolia*, a tree native to Australia, improved the immunity of the intestinal mucosa of piglets by increasing the levels of cytokines in both routes, pro-inflammatory (interferon-γ [IFN-γ] and interleukin-12 [IL-2]) and anti-inflammatory (mainly IL-10), in the jejunum and ileum region (Dong et al. 2019). In turn, *ginseng* essential oil (Panax ginseng) increases the pro-inflammatory response via activation of tumor necrosis factor (TNF-α) and stimulates the phagocytic activity of immune cells (Kang and Min 2012; Sandner et al. 2020; Reyes et al. 2017).

The extract produced with leaves of *Baccharis dracunculifolia* DC. is effective in the treatment of bacterial infections in the gastrointestinal tract, acting as a selective modulator of neutrophils, inhibiting enzymatic activity, and eliminating oxidant compounds (Pereira et al. 2016; Figueiredo-Rinhel et al. 2017).

Oleic acid, an essential fatty acid extracted from stem and bark of *Anacardium occidentale*, is used in the treatment of respiratory infections, gastrointestinal disorders, and diabetes. It also has antibacterial, anti-inflammatory, and antiulcerogenic properties (Araújo et al. 2018). Alkylamides are present in the aqueous and ethanolic extracts of the purple *Echinacea* (L.) Moench. Roots are anti-herpes and anti-influenza agents (Hudson et al. 2005). Similar properties were reported for *Phyllanthus urinaria*, which is capable of interacting with reverse transcriptase and inhibits the entry of HIV into cells (Zhang et al. 2017).

3.3.5 Direct Applications in Cancer Therapy: Mechanisms Related to Immunological Modulation

Cancer is a disease that has an ever-increasing mortality rate every year (Bray et al. 2018), and despite extensive studies performed over the past decades, there are no curative treatments available for this disorder. Besides, commonly available therapies, such as chemotherapy and radiotherapy, imply high systemic toxicity, which limits the tolerability and clinical application (Fu et al. 2018).

Plant secondary metabolites, such as alkaloids, terpenoids, organosulfur compounds, and polyphenols, have been extensively investigated for their potential anticancer effects (Nwodo et al. 2016).

Terpenes or isoprenoids are found in almost all life forms. They may present functional actions (Chappell 1995) and interfere with the immunological system.

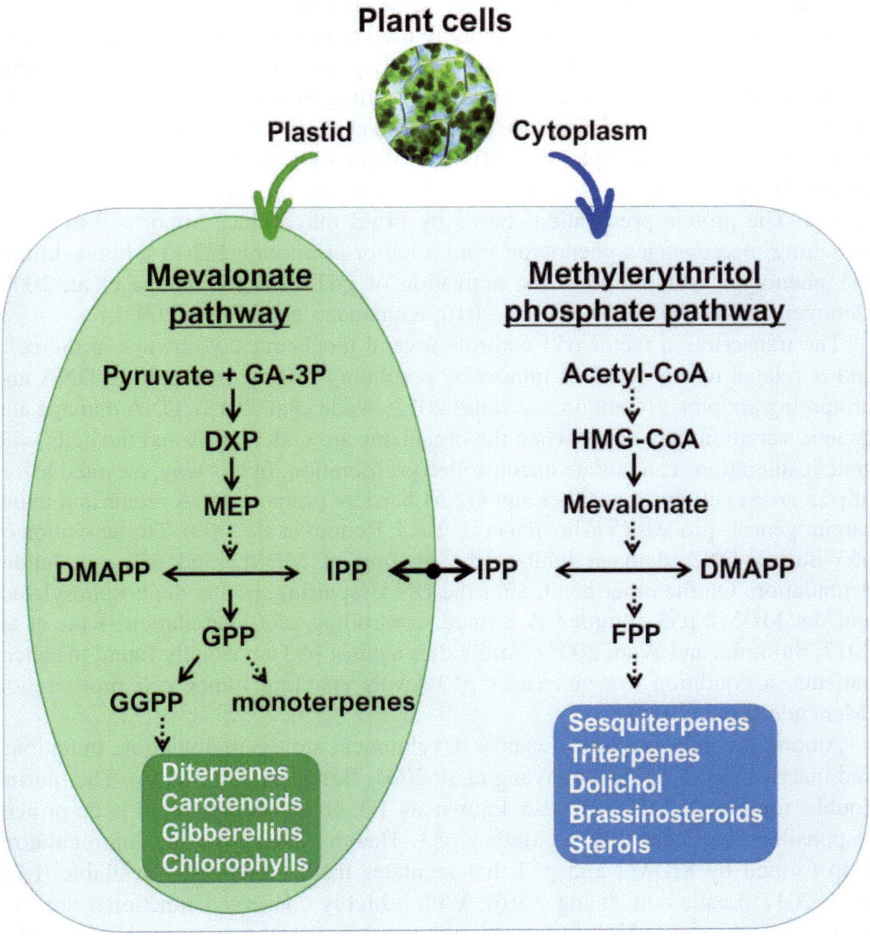

Fig. 3.1 Schematic representation of terpenoid biosynthesis. *GA-3P* glyceraldehyde 3-phosphate; *DXP* 1-deoxy-D-xylulose 5-phosphate; *MEP* 2-C-methyl-d-erythritol-4-phosphate; *DMAPP* dimethylallyl pyrophosphate; *IPP* isopentenyl diphosphate; *GPP* geranyl diphosphate; *GGPP* geranylgeranyl diphosphate; *HMG-CoA* 3hydroxy-3-methylglutaryl coenzyme A reductase; *FPP* farnesyl diphosphate

Terpenes originate from the condensation of dimethylallyl diphosphate (DMAPP) with isopentenyl diphosphate (IPP) (Rohmer 2008; Miziorko 2011; Oldfield and Lin 2012). Such precursors come from methylerythritol phosphate pathway (commonly in plastids) and mevalonate pathway (commonly in cytosol). Figure 3.1 summarizes the biosynthesis of terpenoid by mevalonate and methylerythritol phosphate pathways.

Mevalonate pathway takes place in human cells as well (Lombard and Moreira 2011), and this knowledge has been used to modulate the immune system. Farnesyl diphosphate synthase (FPPS) was shown to be eligible for the development of new cancer drugs. Bisphosphonates, commonly used to treat osteoporosis, has proved to be potent against cancer (Gnant et al. 2009). Bisphosphonates, such as zoledronate, target the allylic site in FPPS, allowing the binding of DMAPP to $[Mg^{2+}]^3$ protein cluster (Aaron and Christianson 2010; Rondeau et al. 2006). This event prevents the biosynthesis of farnesyl diphosphate (FPP) and geranylgeranyl diphosphate (GGPP), accumulating prenyl units in cells, which increases the prenylation of proteins, such as Ras. The protein prenylation started by FPPS may induce tumor cell death by switching macrophages phenotype from a tumor-promoting M2 to a tumor-killing M1 phenotype and promotes the activation of gdT cells (Senaratne et al. 2000; Denoyelle et al. 2003; Coscia et al. 2010; Kunzmann et al. 1997, 2002).

The transcription factor p53 controls several biochemical pathways in the cells and is related to the quash of tumors by regulating cell cycle, repairing DNA and promoting apoptosis (Gannon and Jones 2012; Wade et al. 2013). DNA damage and genetic variation may occur when the organisms are under stress and the cells with genetic alterations can initiate uncontrolled proliferation. In this way, elevated levels of p53 arrest cell cycle at G1-S and G2-M barriers promote DNA repair and avoid carcinogenesis process (Vasilevich et al. 2014; Renouf et al. 2009). The activation of p53 during DNA damage inhibits the binding of MDM2 and p53 and hinder degradation. On the other hand, after the DNA repairing, p53 is dephosphorylated, and the MDM2-p53 complex is formed, controlling p53 degradation (Riaz et al. 2017; Stommel and Wahl 2005). Antibodies against p53 are usually found in cancer patients, a condition that decreases p53 levels enabling tumor cell proliferation (Menendez et al. 2013).

Among the major causes of cancer development are the malfunction, mutations, and inactivation of p53 gene (Wang et al. 2004; Berchuck et al. 1994). The murine double minute 2 (MDM2), also known as E3 ubiquitin ligase, is a co-protein responsible for negatively regulating p53. Healthy cells host an autoregulatory loop formed by MDM2 and p53 that regulates the levels of p53 available (Riaz et al. 2017; Leslie and Zhang 2016). With a highly conserved functional domain, MDM2 can bind to p53 by transactivation and restrains the transcription of p53, forwarding a biochemical signaling that leads to the degradation of the post-transcript in the cell's cytoplasm (Li and Lozano 2013). Since the MDM2-p53 complex decreases the level of free p53, substances capable of preventing the formation of this complex can help in the treatment of cancers. In the search for potent natural MDM2 antagonists, Muhseen and Li (2020) screened a library of 500 terpenes and found out that the 3-trans-*p*-coumaroyl maslinic acid, silvestrol, and betulonic acid are potential inhibitors of MDM2-p53 interaction and could serve as potent antagonists.

As a promising option for the treatment of several cancer types, studies have been performed on the immune checkpoint blockade (Topalian et al. 2012). The monoclonal antibodies that act in immune checkpoint blockers target the pathways of cytotoxic T lymphocyte-associated protein 4 or programmed cell death protein 1. In this way, these inhibitory pathways avoid the uncontrolled immune activation in the host (Keir et al. 2008). However, these mechanisms also can be co-opted by tumors, conferring resistance to the immune attack (Wherry 2011). Anti-PD-1 immune checkpoint blockers may induce clinical responses in cancer, and the effectiveness of antitumor responses requires a subset of tumor-infiltrating dendritic cells, with consequent synthesis of interleukin-12 (IL-12). Dendritic cells are not able to bind to the anti-PD-1. However, interferon gamma (IFN-γ) released by T-cell immunity sensitizes dendritic cells, inducing the production of IL-12. Hence, IL-12 can stimulate antitumor T-cell immunity by dendritic cells cross talk, assisted by IFN-γ and IL-12 (Garris et al. 2018).

Phenolic compounds have proved to be beneficial in the treatment of cardiovascular diseases, diabetes, and cancer, exerting mainly antioxidant and anti-inflammatory effects. Plant phenolic compounds usually influence the nonspecific immune response, enhancing phagocytosis and proliferation of macrophages and neutrophils (Grigore 2017). Immunomodulation by flavonoid glycosides varies according to their corresponding aglycones. Quercetin can concomitantly activate lymphocytes and secretion of IFN-γ, while rutin (quercetin-3-rutinoside) stimulates the secretion of IFN-γ but does not induce the proliferation of human peripheral blood mononuclear cells, suggesting that the glycosides act as the key point for flavonoids immunomodulatory role (Cherng et al. 2008).

In astilbin (taxifolin 3-rhamnoside), the sugar at position 3 is related to a selective immunosuppression and is difficult to be released from astilbin moiety by glucosidases. Thus, the type and position of sugar attached to the aglycone may interfere in the biological action (Vickers 2017; Guo et al. 2007).

The inhibition of proinflammatory cytokine expression seems to be related to hydroxylation of flavonoids at positions 5 and 7, double bond at C2–C3, and at 2′ in B ring (Rice-Evans et al. 1996). Luteolin and apigenin differ from each other by an extra hydroxyl group in B ring of luteolin structure, and the degree of hydroxylation in B ring seems to be critical for the greater immunomodulatory activity of luteolin compared to apigenin (Kilani-Jaziri et al. 2016). Trimethoxy chalcones with fluoro and chloro at the A ring and bromo substitution in B ring have been reported as better inhibitors of nuclear factor-κB (NF-κB) (Yadav et al. 2011). Methoxy groups also have been correlated to the immunomodulatory ability of coumarins. The presence of two methoxy groups in isopimpinellin was correlated to the activation of lymphocytes. On the other hand, one methoxy group in xanthotoxin was correlated to the induction of interferon (IFN) secretion. In comparison to xanthotoxin (8-methoxypsoralen), bergapten (5-methoxypsoralen) exhibited better ability as IFN-γ activators (Cherng et al. 2008).

Alkaloids also have proved to be potent compounds to quash cancer tumors. *Vinca* vinblastine and vincristine alkaloids are characterized as antimicrotubule agents and have been widely used against cancer, such as leukemia, lymphomas, and some solid tumors (Perry 2008). *Vinca* alkaloids interfere with cell cycle by the destabilization of polymerized tubulin, which blocks the attachment of tubulin dimer and prevents microtubules polymerization (Haskell 1995). In this way, it is suggested that the antitumor activity of *Vinca* alkaloids is related to the interference with the normal function of microtubules and blockage of cell cycle progression in the G2-M phase, inducing apoptotic cell death, mainly in solid tumor cells (Haskell 1995; Fan et al. 2001).

Paclitaxel, a taxane alkaloid, presents regulatory effects on IκBα degradation and activation of NF-κB (Huang et al. 2000). NF-κB belongs to the family of Rel transcription factors and has InBa as a specific intracellular inhibitor that mediates or regulates inflammation, immune response, cell proliferation processes, and apoptotic cell death (Baldwin and Albert 1996; Brown et al. 1993). It has also been reported as the role of NF-kB in the coordination of apoptotic cell death control, promoting or inhibiting apoptosis by different apoptotic stimulus (Baeuerle 1991; Grimm et al. 1996; Beg and Baltimore 1996; Wu and Lozano 1994; Lin et al. 1995; Ryan et al. 2000). Huang et al. (2004) demonstrated that *Vinca spp* alkaloids may induce apoptotic cell death through activation of NF-κB/IκB signaling pathway. It is also suggested that the *Vinca spp* alkaloid-induced apoptosis in human breast cancer cells are related to the NF-κB/IκB signaling pathway mediation (Huang et al. 2004).

In general, phytocompounds have been extensively investigated for their proven anticancer action and low toxicity and side effects. However, even with the use of natural compounds in the treatment of diseases, the full action and immunomodulatory mechanisms involved in biological activity of phytocompounds remain unclear.

3.4 Conclusion

It is undeniable the importance that plants exert in the daily life of human being, mainly as a therapeutic alternative for a large part of the population. In recent decades, the potential of plant species to provide a considerable number of compounds with medicinal properties has been recognized, and currently, research involving plant-derived compounds is growing up. In this scenario, bioprospecting is an important tool in the search for new molecules and the elucidation of the mechanisms of action of the promising natural products.

The new therapies seek to provide the patient with more effective medications with fewer side effects, rendering better quality of life, especially in the case of patients affected by cancer. Plant-derived compounds are very promising in this venue. The disclosure of the full mechanism of action of natural products on immune system will contribute for new medicines to become increasingly effective and with little or no unwanted effect.

References

Aaron JA, Christianson DW (2010) Trinuclear metal clusters in catalysis by terpenoid synthases. Pure Appl Chem 82(8):1585–1597

Abbas AK, Lichtman AH, Pillai S (2019) Basic immunology E-book: functions and disorders of the immune system. Elsevier Health Sciences

Abbas AK, Lichtman AH, Pillai S (2015) Imunologia celular e molecular, 8th edn. Elsevier, Rio de Janeiro, RJ

Afonso-Cardoso SR, Silva CV, Ferreira MS, Souza MA (2011) Effect of the Synadenium carinatum latex lectin (ScLL) on Leishmania (Leishmania) amazonensis infection in murine macrophages. Exp Parasitol 128:61–67

Ali AM, Ismail N, Mackeen M, Yazan L, Mohamed S, Ho A, Lajis N (2000) Antiviral, cytotoxic and antimicrobial activities of anthraquinones isolated from the roots of Morinda elliptica. Pharm Biol 38(4):298–301

Alves MMM, Brito LM, Souza AC, Queiroz BCSH, Carvalho TP, Batista JS, Oliveira JSSM, Mendonça IL, Lira SRS, Chaves MH, Gonçalves JCR, Carneiro SMP, Arcanjo DDR, Carvalho FAA (2017) Gallic and ellagic acids: two natural immunomodulatory compounds solve infection of macrophages by Leishmania major. Naunyn Schmiedeberg's Arch Pharmacol 390:893–903

ANVISA. Agência Nacional de Vigilância Sanitária (1999) Resolução N° 391, de 9 de agosto de. Disponível. https://bvsms.saude.gov.br/bvs/saudelegis/anvisa/1999/res0391_09_08_1999.html. Acessado em: 26 Aug 2021

Araújo EMM, Almeida CSC, Godinho Junior JMF, Nascimento FRF, Santos APSA (2013) Ativação in vitro do sistema complemento como mecanismo imunomodulador induzido pelo mesocarpo de babaçu. Rev Ciênc Saúde 15(1):05–10

Araújo S, Sousa IJO, Gonçalves RLG, França ARS, Negreiros PS, Brito AKS, Oliveira AP, Lima EBS (2018) Aplicações Farmacológicas e Tecnológicas da Goma do Cajueiro (Anacardium Occidentale L)—um Produto Obtido da Flora Brasileira. Revistax GEINTEC-Gestão, Inovação e Tecnologias 8(1):4292–4305

Askari VR, Rahimi VB, Rezaee SA, Boskabady MH (2018) Auraptene regulates Th1/Th2/TReg balances, NF-κB nuclear localization and nitric oxide production in normal and Th2 provoked situations in human isolated lymphocytes. Phytomedicine 43:1–10

Baer A, Colon-Moran W, Bhattarai N (2018) Characterization of the effects of immunomodulatory drug fingolimod (FTY720) on human T cell receptor signaling pathways. Sci Rep 8:10910

Baeuerle PA (1991) The inducible transcription activator NF-κB: regulation by distinct protein subunits. Biochim Biophys Acta (BBA) Rev Cancer 1072(1):63–80

Bai L, Li H, Li J, Song J, Zhou Y, Liu B, Lu R, Zhang P, Chen J, Chen D, Pang Y, Liu X, Wu J, Liang C, Zhou J (2019) Immunosuppressive effect of artemisinin and hydroxychloroquine combination therapy on IgA nephropathy via regulating the differentiation of CD4+ T cell subsets in rats. Int Immunopharmacol 70:313–323

Baldwin JR, Albert S (1996) The NF-κB and IκB proteins: new discoveries and insights. Annu Rev Immunol 14(1):649–681

Beg AA, Baltimore D (1996) An essential role for NF-κB in preventing TNF-α-induced cell death. Science 274(5288):782–784

Berchuck A, Kohler MF, Marks JR, Wiseman R, Boyd J, Bast RCJ (1994) The p53 tumor suppressor gene frequently is altered in gynecologic cancers. Am J Obstet Gynecol 170(1): 246–252

Brasil, Agência Nacional de Vigilância Sanitária - Anvisa (2020) Avaliação Clínica de Dispositivos Médicos, Guia n° 31, Brasília

Bray F, Ferlay J, Soerjomataram I, Siegel RS, Torre LA, Jemal A (2018) Global cancer statistics 2018: GLOBOCAN estimates of incidence and mortality worldwide for 36 cancers in 185 countries. CA Cancer J Clin 68(6):394–424

Brody T (2016) Clinical trials: study design, endpoints and biomakers, drug safety, and FDA and ICH guidelines, 2nd edn. Elsevier, London

Brown K, Park S, Kanno T, Franzoso G, Siebenlist U (1993) Mutual regulation of the transcriptional activator NF-kappa B and its inhibitor, I kappa B-alpha. Proc Natl Acad Sci 90(6): 2532–2536

Carneiro AV (2008) Como avaliar a investigação clínica. O exemplo da avaliação crítica de um ensaio clínico 15:30–36

Carr AC, Bozonet SM, Pullar JM, Simcock JW, Vissers MC (2013) Human skeletal muscle ascorbate is highly responsive to changes in vitamin C intake and plasma concentrations. Am J Clin Nutr 97(4):800–807

Carvalho ACB, Kumoto MC, Perfeito JPS (2020) Unraveling the complexities of Brazilian regulations for medicinal plants and herbal medicinal products. In: Modolo LV, Foglio MA (eds) Brazilian medicinal plants. CRC Press Taylor & Francis Group, Boca Raton, pp 1–12

Casanova LM, Costa SS (2017) Interações sinérgicas em produtos naturais: potencial terapêutico e desafios. Rev Virtual Quim 9(2):575–595

Chandana EPS (2016) Immunity and the role of selected plants as immunomodulators: a review. J Univ Ruhuna 4(1–2):1–15

Chappell J (1995) Biochemistry and molecular biology of the isoprenoid biosynthetic pathway in plants. Annu Rev Plant Biol 46(1):521–547

Cherng J-M, Chiang W, Chiang L-C (2008) Immunomodulatory activities of common vegetables and spices of Umbelliferae and its related coumarins and flavonoids. Food Chem 106(3): 944–950

Ciampi E, Uribe-San-Martin R, Cárcamo C, Cruz JP, Reyes A, Reyes D, Pinto C, Vásquez M, Burgos RA, Hancke J (2020) Efficacy of andrographolide in not active progressive multiple sclerosis: a prospective exploratory double-blind, parallel-group, randomized, placebo-controlled trial. BMC Neurol 20(1):173

Cinatl J, Morgenstern B, Bauer G, Chandra P, Rabenau H, Doer HW (2003) Glycyrrhizin, an active component of liquor ice roots, and replication of SARS-associated coronavirus. Lancet 361: 2045–2046

Corr P, Williams DE (2009) The pathway from idea to regulatory approval: examples for drug development. In: Lo B, Field MJ (eds) Conflict of interest in medical research, education and practice. National Academies Press, Washington, DC

Cui Z, Lee Y, Park D (2015) P-synephrine suppresses glucose production but not lipid accumulation in H4IIE liver cells. J Med Food 18(1):76–82

Cockburn IA, Chen YC, Overstreet MG, Lees JR, Rooijen NV, Farbe DL, Zavala F (2010) Prolonged antigen presentation is required for optimal $CD8^+$ T cell responses against Malaria liver stage parasites. PLoS Pathog 6(5):e1000877

Côrtes MA, França EL, Reinaque APB, Scherer EF, Honorio-França AC (2013) Imunomodulação de Fagócitos do Sangue Humano pelo Extrato de Strychnos pseudoquina ST. HILL Adsorvido em Microesferas de Polietilenoglicol. Polímeros 23(3):402–409

Coscia M, Quaglino E, Iezzi M, Curcio C, Pantaleoni F, Riganti C, Holen I, Mönkkönen H, Boccadoro M, Forni G, Musiani P, Bosia A, Cavallo F, Massaia M (2010) Zoledronic acid repolarizes tumour-associated macrophages and inhibits mammary carcinogenesis by targeting the mevalonate pathway. Cell Mol Med 14(12):2803–2815

Costa HF, Bezerra-Santos CR, Filho JMB, Martins MA, Piuvezam MR (2008) Warifteine, a bisbenzylisoquinoline alkaloid, decreases immediate allergic and thermal hyperalgesic reactions in sensitized animals. Int Immunopharmacol 8:519–525

Crotty S (2011) Follicular helper CD4 T cells (TFH). Annu Rev Immunol 29:621–663

Denoyelle C, Hong L, Vannier J-P, Soria J, Soria C (2003) New insights into the actions of bisphosphonate zoledronic acid in breast cancer cells by dual RhoA-dependent and-independent effects. Br J Cancer 88(10):1631–1640

Diwanaya S, Gautamb M, Patwardhan B (2004) Cytoprotection and immunomodulation in cancer therapy. Curr Med Chem Anti-Cancer Agents 4:479–490

Dong L, Liu J, Zhong Z, Wang S, Wang H, Huo Y, Wei Z, Yu L (2019) Dietary tea tree oil supplementation improves the intestinal mucosal immunity of weanling piglets. Anim Feed Sci Technol 255:114209

Dos Santos AJCA et al (2020) *Schinus terebinthifolia* leaf lectin (SteLL) is an immunomodulatory agent by altering cytokine release by mice splenocytes. 3 Biotech 10(4):1–9

Egito J, Daghetti G, Crispim LF, Oliveira CJF, Paulino TP, Chica JEL, Rodrigues WR, Migue CB (2017) Interaction immunomodulatory for control of infectious diseases and noninfectious by activity of lectin Artin M. Revista Saúde Multidisciplinar 4:163–174

Fagundes NC, Oliveira G, Souza B (2017) Etnobotânica de plantas medicinais utilizadas no distrito de Vista Alegre, Claro dos Poções—Minas Gerais. Revista Fitos 11(1):62–80

Fan M, Goodwin ME, Birrer MJ, Chambers TC (2001) The c-Jun NH2-terminal protein kinase/AP-1 pathway is required for efficient apoptosis induced by vinblastine. Cancer Res 61(11): 4450–4458

Figueiredo-Rinhel ASG, Melo LL, Bortot LO, Santos EOL, Andrade MF, Azzolini AECS, Kabeya LM, Caliri A, Bastos JK, Lucisano-Valim YM (2017) Baccharis dracunculifolia DC (Asteraceae) selectively modulates the effector functions of human neutrophils. J Pharm Pharmacol 69(12):1829–1845

Fischer G, Hübner SO, Vargas GD, Vidor T (2008) Imunomodulação pela própolis. Arq Inst Biol 75(2):247–253

Friedman LM, Furberg CD, DeMets DL, Granger RCB (2015) Fundamentals of clinical trials, 5th edn. Springer, New York

Fu B, Wang N, Tan H-Y, Li S, Cheung F, Feng Y (2018) Multi-component herbal products in the prevention and treatment of chemotherapy-associated toxicity and side effects: a review on experimental and clinical evidences. Front Pharmacol 9:1394

Gannon HS, Jones SN (2012) Using mouse models to explore MDM-p53 signaling in development, cell growth, and tumorigenesis. Genes Cancer 3(3–4):209–218

Garris CS, Arlauckas SP, Kohler RH, Trefny MP, Garren S, Piot C, Engblom C, Pfirschke C, Siwicki M, Gungabeesoon J, Freeman GJ, Warren SE, Ong S, Browning E, Twitty CG, Pierce RH, Le MH, Algazi AP, Daud AI, Pai SI, Zippelius A, Weissleder R, Pittet MJ (2018) Successful anti-PD-1 cancer immunotherapy requires T cell-dendritic cell crosstalk involving the cytokines IFN-γ and IL-12. Immunity 49(6):1148–1161

Gnant M, Mlineritsch B, Schippinger W, Luschin-Ebengreuth G, Pöstlberger S, Menzel C, Jakesz R, Seifert M, Hubalek M, Bjelic-Radisic V, Samonigg H, Tausch C, Eidtmann H, Steger G, Kwasny W, Dubsky P, Fridrik M, Fitzal F, Stierer M, Rücklinger E, Greil R (2009) Endocrine therapy plus zoledronic acid in premenopausal breast cancer. N Engl J Med 360(7): 679–691

Gonçalves AD, Maia DCG, Ferreira LS, Monnazzi LGS, Alegranci P, Placeres MCP, Batista-Duharte A, Carlos IZ (2015) Involvement of major components from sporothrix schenckii cell wall in the caspase-1 activation, nitric oxide and cytokines production during experimental sporotrichosis. Mycopathologia 179:21–30

Goyal S, Gupta N, Chatterjee S (2016) Investigating therapeutic potential of trigonella foenum-graecum L. as our defense mechanism against several human diseases. J Toxicol 2016:1–10

Grady D, Cummings SR, Hulley SB (2013) Alternative clinical trial designs and implementation issues. In: Hulley SB, Cummings SR, Browner WS, Grady DG, Newman TB (eds) Designing clinical research, 4th edn. Lippincott Williams & Wilkins, a Wolters Kluwer Business, Philadelphia, pp 151–170

Griffiths AJF, Wessler SR, Carroll SB, Doebley J (2017) Introdução a Genética, 11th edn. Guanabara Koogan, Rio de Janeiro

Grigore A (2017) Plant phenolic compounds as immunomodulatory agents. In: Phenolic compounds–biological activity. IntechOpen, London, UK, pp 75–98

Grimm S, Bauer MKA, Baeuerle PA, Schulze-Osthoff K (1996) Bcl-2 down-regulates the activity of transcription factor NF-kappaB induced upon apoptosis. J Cell Biol 134(1):13–23

Guedes JC, Bendicho MT, Lemaire D (2004) Citocinas e imunomodulação: novos avanços no tratamento das pancreatites. R Ci Méd Biol 3(2):242–251

Guo J, Qian F, Li J, Xu Q, Chen T (2007) Identification of a new metabolite of astilbin, 3′-O-methylastilbin, and its immunosuppressive activity against contact dermatitis. Clin Chem 53(3): 465–471

Hall P, Virella G (2007) Immune system modulators. In: Virella G (ed) Medical immunology, 6th edn. Informa Healthcare, New York, NY, pp 335–356

Hart PH, Brand C, Carson CF, Riley TV, Prager RH, Finlay-Jones JJ (2000) Terpinen-4-ol, the main component of the essential oil of Melaleuca alternifolia (tea tree oil), suppresses inflammatory mediator production by activated human monocytes. Inflamm Res 49(11):619–626

Haskell CM (1995) Antineoplastic agents: cancer treatment. In: Cancer treatment, 4th edn. WB Saunders Co, Philadelphia, pp 78–165

Hellings S, Blajchman MA (2009) Transfusion-related immunosuppression. Anaesthesia Int Care Med 10(5):231–234

Hiramatsu T, Imato M, Koyono T, Umezawa K (1993) Induction of normal phenotypes in Ras transformed cells by damnacanthal from Morinda citrifolia. Cancer Lett 73(2–3):161–166

Hirazumi A, Furusawa E, Chou SC, Hokama Y (1996) Immunomodulation contributes to the anti-cancer activity of Morinda citrifolia (noni) fruit juice. Proc West Pharmacol Soc 39:7–9

Hirazumi A, Furusawa E, Chou SC, Hokama Y (1999) Immunomodulatory polysaccharide-rich substances from the fruit juice of Morinda citrifolia (noni) with anti-tumor activity. Phytother Res 13(5):380–387

Holland SM, Vizi ES (2002) Immunomodulation. Curr Opin Pharmacol 2:425–427

Honek J (2017) Preclinical research in drug development. Medical Writing 26(4):5–8

Hotta M, Nakata R, Katsukawa M, Hori K, Takahashi S, Inoue H (2009) Carvacrol, a component of thyme oil, activates PPARα and γ and suppresses COX-2 expression. J Lipid Res 51(1): 132–139

Huang Y, Fang Y, Wu J, Dziadyk JM, Zhu X, Sui M, Fan W (2004) Regulation of Vinca alkaloid-induced apoptosis by NF-κB/IκB pathway in human tumor cells. Mol Cancer Ther 3(3): 271–277

Huang Y, Johnson KR, Norris JS, Fan W (2000) Nuclear factor-κB/IκB signaling pathway may contribute to the mediation of paclitaxel-induced apoptosis in solid tumor cells. Cancer Res 60(16):4426–4432

Hudson J, Vimalanathan S, Kang L, Amiguet VT, Livesey J, Arnason JT (2005) Characterization of antiviral activities in Echinacea root preparations. Pharm Biol 43(9):790–796

Joly CA, Haddad CFB, Verdade LM, Oliveira MC, Bolzani WS, Berlinck RGS (2011) Diagnóstico da pesquisa em biodiversidade no Brasil. Revista USP 89:114–133

Kang S, Min H (2012) Ginseng, the 'immunity boost': the effects of panax ginseng on immune system. J Ginseng Res 36(4):354–368

Keeler DM, Grandal MK, McCall JR (2019) Brevenal a marine natural product, is anti-inflammatory and an immunomodulator of macrophage and lung epithelial cells. Mar Drugs 17(184):1–14

Keir ME, Butte MJ, Freeman GJ, Sharpe AH (2008) PD-1 and its ligands in tolerance and immunity. Annu Rev Immunol 26:677–704

Kilani-Jaziri S, Mustapha N, Mokdad-Bzeouich I, Gueder DE, Ghedira K, Ghedira-Chekir L (2016) Flavones induce immunomodulatory and anti-inflammatory effects by activating cellular anti-oxidant activity: a structure-activity relationship study. Tumor Biol 37(5):6571–6579

Kim Y, Yang J, Yoon J, Jo S, Ahn H, Song K-H, Lee D-Y, Chung K-Y, Won Y-H, Kim I-H (2018) A multicentre, open, investigator-initiated phase IV clinical trial to evaluate the efficacy and safety of ingenol mebutate gel, 0·015% on the face and scalp, and 0·05% on the trunk and extremities, in Korean patients with actinic keratosis (perfect). Br J Dermatol 179:836–843

Koparde AA, Doijad RC, Magdum CS, Koparde AA (2019) Natural products in drug discovery. In: Doijad RC (ed) Pharmacognosy—medicinal plants. IntechOpen, Rijeka

Kumar S, Gupta A, Saini RV, Kumar A, Dhar KL, Mahindroo N (2020) Immunomodulation-mediated anticancer activity of a novel compound from Brugmansia suaveolens leaves. Bioorg Med Chem 28(12):115552–115552

Kunzmann V, Jomaa H, Feurle J, Bauer E, Herderich M, Wilhelm M (1997) Stimulation of human gamma delta T cells by aminobisphosphonates used in treatment of bone disorders. Blood 90(10):2558–2558

Kunzmann V, Eckstein S, Lindner A, Tony H, Wilhelm M (2002) Induction of anti-lymphoma activity by pamidronate activated gamma delta T cells: results from a phase I/II trial. Blood 100(11):309B–309B

Leslie PL, Zhang Y (2016) MDM2 oligomers: antagonizers of the guardian of the genome. Oncogene 35(48):6157–6165

Li Q, Lozano G (2013) Molecular pathways: targeting Mdm2 and Mdm4 in cancer therapy. Clin Cancer Res 19(1):34–41
Lima HC (2007) Fatos e mitos sobre imunomoduladores. An Bras Dermatol 82(3):207–221
Lin Y-Z, Yao SY, Veach RA, Torgerson TR, Hawiger J (1995) Inhibition of nuclear translocation of transcription factor NF-κB by a synthetic peptide containing a cell membrane-permeable motif and nuclear localization sequence. J Biol Chem 270(24):14255–14258
Lira FAS, Brasileiro-Santos MS, Borba BBL, Costa MJC, Dantas PROF, Santos AC (2012) Influência da vitamina C na modulação autonômica cardíaca no repouso e durante o exercício isométrico em crianças obesas. Revista Brasileira de Saúde Materno Infantil 12(3):259–267
Lombard J, Moreira D (2011) Origins and early evolution of the mevalonate pathway of isoprenoid biosynthesis in the three domains of life. Mol Biol Evol 28(1):87–99
Maghraby A, Bahgat M (2004) Immunostimulatory effect of coumarin derivatives before and after infection of mice with the parasite Schistosoma mansoni. Drug Res 54(9):545–550
Mahamat O, Flora H, Tume C, Kamanyi A (2020) Immunomodulatory activity of Momordica charantia L. (Cucurbitaceae) leaf diethyl ether and methanol extracts on salmonella typhi-infected mice and LPS-induced phagocytic activities of macrophages and neutrophils. Evid Based Complement Alternat Med 2020:1–11
Mahana N, Abou-Eldahab M, Abd AEA, Attia A, Ramadan R (2018) Immunomodulatory activity of moringa oleifera and curcuma longa extracts in cyclophosphamide-immunosuppressed male rats. Egypt J Zool 69(69):89–106
Mao R, Wu L, Zhu N, Liu X, Hao Y, Liu R, Du Q, Li Y (2020) Immunomodulatory effects of walnut (Juglans regia L.) oligopeptides on innate and adaptive immune responses in mice. J Funct Foods 73:1–8
Marim FM, Silveira TM, Lima DS, Zamboni DS (2010) A method for generation of bone marrow-derived macrophages from cryopreserved mouse bone marrow cells. PLoS One 5(12):1–8
Marques LG, Santos MRMC, Raffo J, Pessoa C (2014) Redes de Bioprospecção no Brasil: cooperação para o desenvolvimento tecnológico. RDE-Revista de Desenvolvimento Econ 15(28)
Melo-Neto B, Leitão JMSR, Oliveira LGC, Santos SEM, Carneiro SMP, Rodrigues KAF, Chaves MH, Arcanjo DDR, Carvalho FAA (2016) Inhibitory effects of Zanthoxylum rhoifolium Lam. (Rutaceae) against the infection and infectivity of macrophages by Leishmania amazonensis. An Acad Bras Cienc 88(3 Suppl)
Menendez D, Shatz M, Resnick MA (2013) Interactions between the tumor suppressor p53 and immune responses. Curr Opin Oncol 25(1):85–92
Mishra LC, Singh BB (2000) Scientific basis for the therapeutic use of Withania somnifera (Ashwagandha): a review. Altern Med Rev 5(4):334–346
Miziorko HM (2011) Enzymes of the mevalonate pathway of isoprenoid biosynthesis. Arch Biochem Biophys 505(2):131–143
Modares M, Zarasvand MA, Madani M (2018) The effects of hydro-alcoholic extract of Artemisia dracunculus L. (Tarragon) on hematological parameters in mice. J Bas Res Med Sci 5(1):10–14
Mohamed SIA, Jantan I, Haque MA (2017) Naturally occurring immunomodulators with antitumor activity: an insight on their mechanisms of action. Int Immunopharmacol 50:291–304
Muhseen ZT, Li G (2020) Promising terpenes as natural antagonists of cancer: an in-silico approach. Molecules 25(1):155
Nair A, Chattopadhyay D, Saha B (2019) New look to phytomedicine. Elsevier
Norman GA (2019) Phase II trials in drug development and adaptive trial design. JACC Basic Trans Sci 4(3):428–437
Nugent P, Duncan JN, Colagiovanni DB (2013) Preparation of a preclinical dossier to support an investigational new drug (IND) application and first-in-human clinical trial. In: Faqi AS (ed) A comprehensive guide to toxicology in nonclinical drug development, 2nd edn. Academic Press
Nunes-Pinheiro DCS, Melo-Leite AKRM, Farias VM, Braga LT, Lopes CAP (2003) Immunomodulatory activity of the medicinal plants: perspectives for veterinary medicine. Ciência Animal 13(1):23–32

Nwodo JN, Ibezim A, Simoben CV, Ntie-Kang F (2016) Exploring cancer therapeutics with natural products from African medicinal plants, part II: alkaloids, terpenoids and flavonoids. Anti-Cancer Agents Med Chem 16(1):108–127
Oldfield E, Lin F-Y (2012) Terpene biosynthesis: modularity rules. Angew Chem Int Ed 51(5): 1124–1137
Oliveira CMB, Sakata RK, Issy AM, Gerola LR, Salomão R (2011) Citocinas e Dor. Rev Bras Anestesiol 61(2):255–265
Oliveira MLM, Nunes-Pinheiro DCS (2013) Cellular and molecular biomarkers involved in immune-inflammatory response modulated by unsaturated fatty acids. Acta Veterinaria Brasilica 7(2):113–124
Ong HC, Norzalina J (1999) Malay herbal medicine in Gemencheh, Negeri Sembilan, Malaysia. Fitoterapia 70(1):10–14
Pappalardo F, Russo G, Tshinanu FM, Viceconti M (2018) In silico clinical trials: concepts and early adoptions. Brief Bioinform 20(5):1699–1708
Pastorelli D, Fabricio ASC, Giovanis P, D'Ippolito S, Fiduccia P, Soldà C, Buda A, Sperti C, Bardini R, Dalt GD, Rainato G, Gion M, Ursini F (2018) Phytosome complex of curcumin as complementary therapy of advanced pancreatic cancer improves safety and efficacy of gemcitabine: results of a prospective phase II trial. Pharmacol Res 132:72–79
Patwardhan B, Gautam M (2005) Botanical immunodrugs: scope and opportunities. Drug Discov Today 10(7):495–502
Pellacani G, Peris K, Guillen C, Clonier F, Larsson T, Venkata R, Puig S (2015) A randomized trial comparing simultaneous vs. sequential field treatment of actinic keratosis with ingenol mebutate on two separate areas of the head and body. J Eur Acad Dermatol Venereol 29:2192–2198
Pereira CA, Costa AC, Liporoni PC, Rego MA, Jorge AO (2016) Antibacterial activity of Baccharis dracunculifolia in planktonic cultures and biofilms of Streptococcus mutans. J Infect Public Health 9(3):324–330
Péret PJS (2008) A bioprospecção no Brasil: contribuições para uma gestão ética. In: The chemotherapy source book. Lippincott Williams & Wilkins, Brasília
Perry MC (ed) (2008) The chemotherapy source book. Lippincott Williams & Wilkins
Persson-Waller K, Colditz IG, Lun S, O'stensson K (2003) Cytokines in mammary lymph and milk during endotoxin-induced bovine mastitis. Res Vet Sci 74(1):31–36
Pyorala S (2002) New strategies to prevent mastitis. Reprod Domest Anim 37(4):211–216
Redgrove KA, McLaughlin EA (2014) The role of the immune response in chlamydia trachomatis infection of the male genital tract: a double-edged sword. Front Immunol 5:534
Renouf B, Hollville E, Pujals A, Tétaud C, Garibal J, Wiels J (2009) Activation of p53 by MDM2 antagonists has differential apoptotic effects on Epstein–Barr virus (EBV)-positive and EBV-negative Burkitt's lymphoma cells. Leukemia 23(9):1557–1563
Reyes AWB, Hop HT, Arayan LT, Huy TXN, Park SJ, Kim KD, Min W, Lee HJ, Rhee MH, Kwak Y-S, Kim S (2017) The host immune enhancing agent Korean red ginseng oil successfully attenuates Brucella abortus infection in a murine model. J Ethnopharmacol 198:5–14
Riaz M, Ashfaq UA, Qasim M, Yasmeen E, Qamar MTU, Anwar F (2017) Screening of medicinal plant phytochemicals as natural antagonists of p53–MDM2 interaction to reactivate p53 functioning. Anti-Cancer Drugs 28(9):1032–1038
Rice-Evans CA, Miller NJ, Paganga G (1996) Structure-antioxidant activity relationships of flavonoids and phenolic acids. Free Radic Biol Med 20(7):933–956
Rohmer M (2008) From molecular fossils of bacterial hopanoids to the formation of isoprene units: discovery and elucidation of the methylerythritol phosphate pathway. Lipids 43(12):1095–1107
Rondeau J-M, Bitsch F, Bourgier E, Geiser M, Hemmig R, Kroemer M, Lehmann S, Ramage P, Rieffel S, Strauss A, Green JR, Jahnke W (2006) Structural basis for the exceptional in vivo efficacy of bisphosphonate drugs. ChemMedChem 1(2):267–273
Rosa LPS, Cruz DJ (2016) Effects of nutritional imunomodulator in oncology: review of scientific evidence. Rev Saúde Com 12(2):561–565
Ryan KM, Ernst MK, Rice NR, Vousden KH (2000) Role of NF-κB in p53-mediated programmed cell death. Nature 404(6780):892–897

Saccaro Júnior NL (2011) A regulamentação de acesso a recursos genéticos e repartições de benefícios: disputas dentro e fora do Brasil. Ambiente Sociedade XIV(1):229–244

Salinas-Sánchez DO, Jiménez-Ferrer E, Sánchez-Sánchez V, Zamilpa A, González-Cortazar M, Tortoriello J, Herrera-Ruiz M (2017) Anti-inflammatory activity of a polymeric proanthocyanidin from Serjania schiedeana. Molecules 22(6):2–19

Sandner G, Mueller AS, Zhou X, Stadlbauer V, Schwarzinger B, Schwarzinger C, Wenzel U, Maenner K, Van der Klis JD, Hirtenlehner S, Aumiller T, Weghuber J (2020) Ginseng extract ameliorates the negative physiological effects of heat stress by supporting heat shock response and improving intestinal barrier integrity: evidence from studies with heat-stressed caco-2 cells, C. elegans and growing broilers. Molecules 25(4):835

Santos LD, Pieroni M, Menegasso ARS, Pinto JRAS, Palma MS (2011) A new scenario of bioprospecting of Hymenoptera venoms through proteomic approach. J Venom Anim Toxins Incl Trop Dis 17(4):364–377

Seca AML, Pinto DCGA (2018) Plant secondary metabolites as anticancer agents: successes in clinical trials and therapeutic application. Int J Mol Sci 19(1):263

Seder RA, Chang LJ, Enama ME, Zephir KL, Sarwar UN, Gordon IG, Holman LA, Billingsley JPF, Gunasekera A, Richman A, Chakravarty S, Manoj A, Velmurugan S, Li M, Ruben AJ, Li T, Eappen AG, Stafford RE, Plummer SH, Hendel CS, Novic L, Costner PJM, Mendoza FH, Saunders JG, Nason MC, Richardson JH, Murphy J, Davidson SA, Richie TL, Sedegah M, Sutamihardja A, Fahle GA, Lyke KE, Laurens MB, Roederer M, Tewari K, Epstein JE, Sim KL, Ledgerwood JE, Graham BS, Hoffman SL (2013) Protection against malaria by intravenous immunization with a nonreplicating sporozoite vaccine. Science 341(6152):1359–1365

Sekizawa T, Yanagi K, Itoyama Y (2001) Glycyrrhizin increases survival of mice with herpes simplex encephalitis. Acta Virol 45(1):51–54

Sen R, Chatterjee M (2011) Plant derived therapeutics for the treatment of Leishmaniasis. Phytomedicine 18:1056–1069

Senaratne SG, Pirianov G, Mansi JL, Arnett TR, Colston KW (2000) Bisphosphonates induce apoptosis in human breast cancer cell lines. Br J Cancer 82(8):1459–1468

Sharon J (2000) Imunologia básica. Guanabara Koogan, Rio de Janeiro, RJ

Simões CMO et al (2017) Farmacognosia: do produto natural ao medicamento. 1.ed. Artmed, Porto Alegre

Soni S, Bandyopadhayaya S, Mandal CC (2020) Animal models systems of cancer for preclinical trials. In: Kumar M, Sharma A, Kumar P (eds) Pharmacotherapeutic botanicals for cancer chemoprevention. Singapore, Springer Nature

Singh N, Tailang M, Mehta SC (2016) A review on herbal plants as immunomodulators. Int J Pharm Sci Res 7(9):3602–3610

Sonnenberg GF, Hepworth MR (2019) Functional interactions between innate lymphoid cells and adaptive immunity. Nat Rev Immunol 19(10):599–613

Souto AL, Tavares JF, Da Silva MS, Diniz MFFM, De Athayde-Filho PF, Barbosa Filho JM (2011) Anti-inflammatory activity of alkaloids: an update from 2000 to 2010. Molecules 16:8515–8534

Stommel JM, Wahl GM (2005) A new twist in the feedback loop: stress-activated MDM2 destabilization is required for p53 activation. Cell Cycle 4(3):411–417

Storer BE (1989) Design and analysis of phase I clinical trials. Biometrics 45(3):925–937

Strader CR, Pearce CJ, Oberlies NH (2011) Fingolimod (FTY720): a recently approved multiple sclerosis drug based on a fungal secondary metabolite. J Nat Prod 74(4):900–907

Sunil MA, Sunitha VS, Ashitha A, Neethu S, Midhun SJ, Radhakrishnan EK, Jyothis M (2019) Catechin rich butanol fraction extracted from Acacia catechu L. (a thirst quencher) exhibits immunostimulatory potential. J Food Drug Anal 27(1):195–207

Swati R, Akhilesh G (2011) Regulatory requirements for drug development and approval in United States: a review. Asian J Pharm Res 1(1):1–6

Thyagarajan A, Sahu RP (2018) Potential contributions of antioxidants to cancer therapy: immunomodulation and radiosensitization. Integr Cancer Ther 17(2):210–216

Teicher BA, Andrews PA (2004) Anticancer drug development guide. Springer, New York

Toledo AC, Sakoda CPP, Perini A, Pinheiro NM, Magalhães RM, Grecco S, Tibério IFLC, Câmara NO, Martins MA, Lago JHG, Prado CM (2013) Flavonone treatment reverses airway inflammation and remodelling in an asthma murine model. Br J Pharmacol 168:1736–1749. www.brjpharmacol.org. Accessed 28 July 2020

Thomford NE, Senthebane DA, Rowe A, Munro D, Seele P, Maroyi A, Dzobo K (2018) Natural products for drug discovery in the 21st century: innovations for novel drug discovery. Int J Mol Sci 19:1578

Topalian SL, Hodi FS, Brahmer JR, Gettinger SN, Smith DC, McDermott DF, Powderly JD, Carvajal RD, Sosman JA, Atkins MB, Leming PD, Spigel DR, Antonia SJ, Horn L, Drake CG, Pardoll DM, Chen L, Sharfman WH, Anders RA, Taube JM, McMiller TL, Xu H, Korman AJ, Jure-Kunkel M, Agrawal S, McDonald D, Kollia GD, Gupta A, Wigginton JM, Sznol M (2012) Safety, activity, and immune correlates of anti–PD-1 antibody in cancer. N Engl J Med 366(26): 2443–2454

Ueda-Nakamura T, Mendonca-Filho RR, Morgado-Díaz JA, Maza PK, Filho BPD, Cortez DAG, Alviano DS, Rosa MSS, Lopes AHCS, Alviano CS, Nakamura CV (2006) Antileishmanial activity of Eugenol-rich essential oil from Ocimum gratissimum. Parasitol Int 55(2):99–105

Utsunomiya T, Kobayashi M, Ito M, Pollard RB, Suzuki F (2000) Glycyrrhizin improves the resistance of MAIDS mice to opportunistic infection of Candida albicans through the modulation of MAIDS-associated type 2 T cell responses. Clin Immunol 95(2):145–155

Varella PPV, Neves Forte WC (2001) Citocinas: revisão. Rev bras alergia imunopatol:146–154

Vasilevich NI, Afanasyev II, Kovalskiy DA, Genis DV, Kochubey VS (2014) A re-examination of the MDM2/p53 interaction leads to revised design criteria for novel inhibitors. Chem Biol Drug Des 84(5):585–592

Vickers NJ (2017) Animal communication: when i'm calling you, will you answer too? Curr Biol 27(14):R713–R715

Vogt L, Ramasamy U, Meyer D, Pullens G, Venema K, Faas MM, Schols HA, Vos P (2013) Immune modulation by different types of β2→1-fructans is toll-like receptor dependent. PLoS One 8(7):1–12

Wade M, Li Y-C, Wahl GM (2013) MDM2, MDMX and p53 in oncogenesis and cancer therapy. Nat Rev Cancer 13(2):83–96

Wang J-L, Zheng B-Y, Li X-D, Angström T, Lindström MS, Wallin K-L (2004) Predictive significance of the alterations of p16INK4A, p14ARF, p53, and proliferating cell nuclear antigen expression in the progression of cervical cancer. Clin Cancer Res 10(7):2407–2414

Wang L, Du GH (2018) Paclitaxel. In: Natural small molecule drugs from plants. Springer, Singapore

Wang MY, West B, Jensen CJ, Nowicki D, Su C, Palu AK, Anderson G (2002) Morinda citrifolia (noni): a literature review and recent advances in noni research. Acta Pharmacol Sin 23(12): 1127–1141

Wherry EJ (2011) T cell exhaustion. Nat Immunol 12(6):492–499

Widelski J, Kukula-Koch W, Baj T, Kedzierski B, Fokialakis N, Magiatis P, Pozarowski P, Rolinski J, Graikou K, Chinou I, Skalicka-Wozniak K (2017) Rare coumarins induce apoptosis, G1 cell block and reduce RNA content in HL60 cells. Open Chem 15:1–6

Wrigth B (2017) Clinical trial phases. In: Delva S, Brenda W (eds) A comprehensive and practical guide to clinical trials. Academic Press

Wu H, Lozano G (1994) NF-kappa B activation of p53. A potential mechanism for suppressing cell growth in response to stress. J Biol Chem 269(31):20067–20074

Yadav VR, Prasad S, Sung B, Aggarwal B (2011) The role of chalcones in suppression of NF-κB-mediated inflammation and cancer. Int Immunopharmacol 11(3):295–309

Yoshikawa M, Kawamoto H, Umemoto N, Oku K, Koizumi M, Yamao J, Kuriyama S, Nakano H, Hozumi N, Ishizaka S, Fukui H (1997) Effects of glycyrrhizin on immune-mediated cytotoxicity. J Gastroenterol Hepatol 12(3):243–248

Zhang J, Cui Y, Wang R, Zhang L (2017) The anti-HIV-1 activity of polyphenols from Phyllanthus urinaria and the pharmacokinetics and tissue distribution of its marker compound, gallic acid. J Tradit Chin Med Sci 4(2):158–166

Zhu HF, Li Y (2018) Small-molecule targets in tumor immunotherapy. Nat Prod Bioprospect 8: 297–301

Pathway and Genomics of Immunomodulator Natural Products

4

Jing Wang, Lingjun Ma, Fei Zhou, Fang Wang, Lei Chen, and Jianbo Xiao

Abstract

Immune dysfunction has become a major public health problem, so the search for natural products with efficacy, safety, and therapeutic effects on the host immune functions has attracted extensive attention. Several in vitro and in vivo models have been used to outline the immunomodulation effects of natural products due to their anti-inflammatory property, induction of phagocytosis, and immune cells stimulation activity. The updated research progress of immunomodulatory mechanisms of natural products was reviewed. Furthermore, several immune-related pathways, including NF-κB, IL-6/JAK/STAT, MAPK, and PI3K pathways, were elucidated. It is hoped that by analyzing and summarizing the research progress of the immunomodulatory activity of phytochemicals (mainly

J. Wang
College of Biosystems Engineering and Food Science/Ningbo Research Institute, Zhejiang University, Hangzhou, China

L. Ma
College of Food Science and Nutritional Engineering, China Agricultural University, Beijing, China

F. Zhou
College of Standardization, China Jiliang University, Hangzhou, China

F. Wang
College of Food Science and Engineering, Nanjing University of Finance and Economics, Nanjing, China

L. Chen
College of Food Science and Technology, Guangdong Ocean University, Zhanjiang, China

J. Xiao (✉)
Department of Analytical Chemistry and Food Science, University of Vigo, Vigo, Spain
e-mail: jianboxiao@uvigo.es

N. S. Sangwan et al. (eds.), *Plants and Phytomolecules for Immunomodulation*,
https://doi.org/10.1007/978-981-16-8117-2_4

flavonoids, isoflavones, saponins, catechins, anthocyanins, and alkaloids), a more comprehensive understanding of the immunomodulatory activity of NPs can be provided.

Keywords

Immunomodulatory · Natural products · Phytochemicals · Inflammation · Antioxidative

4.1 Introduction

The immune system plays a key important role in maintaining the physiological functions of the host to protect the body against the onset of disease. Disruption of the immune system can lead to autoimmune diseases, inflammatory diseases, and even cancer. When the immune system is deficient, it can lead to tumors and infections. Conversely, when the immune system is overactive, it can lead to autoimmune diseases like rheumatoid arthritis, type I diabetes, and systemic lupus erythematosus. Immune cells, an important part of the immune system, participate in regulating innate and adaptive immunity, mainly including lymphocytes, dendritic cells (DC), monocytes/macrophage cells, NK cells, CD^{8+} cytotoxic T cells (CTLs), myeloid-derived suppressor cells (MDSCs), and so on. These immune cells play critical roles in various immunomodulatory processes due to their different structures and functions. Except for such immune cells, other inflammatory cytokines and chemokines including TNF-α, TGF-β, ILs, and IFN-γ can be triggered and regulated (Galon and Bruni 2020; Ivashkiv 2018; Wenjun and O'Garra 2019). Furthermore, signaling pathways such as NF-κB, PD-1, CTLA-4, MAPKs, ERK, JNK, and STAT are involved in immunomodulation (Bilanges et al. 2019; Clara et al. 2019; Peluso et al. 2017).

Natural products and their derivatives have attracted extensive attention worldwide because of their diverse structures, biological activities, and pharmacological activities, as well as the low toxicity and side effects (Harvey et al. 2015). Notably, natural products have been developed as biological modulators of immunotherapy due to the potential to maintain homeostasis by regulating the immune cells or producing antibodies (Ali et al. 2021). In fact, several well-known natural products including alkaloids (e.g., leonurine, berberine), flavonoids (e.g., liquiritigenin, silymarin), saponins (e.g., astragaloside and glycyrrhizin), catechins (e.g., (-)-epicatechin gallate, (-)-epicatechin, (-)-epigallocatechin, and (-)-epigallocatechin gallate), anthocyanins (e.g., cyanidin-3-glucoside), and capsaicin have been reported to have potential immunomodulatory effects (Deng et al. 2020; Grudzien and Rapak 2018; Pan et al. 2019). The immunoregulation mechanisms of these natural products are complex and involve multiple signal transduction pathways.

4.2 Toll-Like Receptor Signaling

The immune system is very complex and evolved in protecting against various pathogens in the surrounding environment, populated in both achordate and chordates. There are several pattern recognition receptors (PRRs) expressed in the immune system responsible for detecting danger by recognizing specific pathogen-associated molecular patterns. The toll-like receptors (TLRs), a kind of type I protein, were first discovered in Drosophila melanogaster (Lemaitre et al. 1996). TLRs have also been observed in humans, and this series of proteins are highly conserved, with multiple TLRs sequences homology similar to toll sequences observed in Drosophila (Rock et al. 1998). TLRs provided a critical role in the immune system due to triggered pro-inflammatory cytokines and chemokine (Chen et al. 2017).

Ten functional TLRs in humans and 13 active TLRs in mice have been determined and can be further divided into subfamilies depending on the ligand type. Some TLRs identify lipopeptides and glycolipids (e.g., TLR1, TLR2, TLR4, and TLR6), ssRNA, and unmethylated CpG DNA. TLR3 recognizes dsRNA associated with the viral infection. Furthermore, TLR1, in association with TLR2 (TLR1/TLR2), identifies diacyl triacylaliopeptides. TLR2 and TLR6 (TLR2/TLR6) will form heterodimers and recognize diacyl lipopeptides. TLR2 can further concert with TLR1 or TLR6 to recognize many lipoproteins, peptidoglycan, lipopeptides, and zymosan from bacteria and fungi. TLR5 identifies bacterial flagellin, and TLR11 and TLR12 recognize profilin (Takeda and Akira 2015). TLR7 and TLR8 can interact with single-stranded RNA during virus replication. TLR3 identifies double-stranded RNA, and TLR9 recognizes the nonmethylated deoxycytidyl-deoxyguanosine (CpG) motif ubiquitous (Chen et al. 2017). TLR4 identifies heat-shock proteins, fibronectin, and lipopolysaccharides. TLR10 can be homodimerized or heterodimerized with TLR1 and TLR2. In contrast, TLR11 was found to be non-functional due to termination codon's presence in the gene (Yu et al. 2010).

TLR is mainly expressed on macrophages and DCs, T and B lymphocytes, epithelial cells, endothelial cells, and fibroblasts (Kawasaki and Kawai 2014). However, the subsets are expressed in different immune cells. Depending on the cell's location, TLRs 1, 2, 4, 5, and 6 are located on the cell surface and TLRs 3, 7, 8, and 9 in endosomes. As the primary immune mediators, most TLRs are expressed in monocytes/macrophages, and some of them (e.g., TLR2, TLR4, TLR6, and TLR8) are observed in expression of mast cells, and TLRs 1, 2, 4, and 5 are found in the immature dendritic cells (Takeda et al. 2003). In addition to innate immune cells, TLRs were found to be expressed in several other cells. TLR4 was expressed in various types of epithelial cells widely distributed in the intestine, kidney, and cornea, which are always regarded as the first defense against microbial invasion (Yang et al. 2015).

The dysfunction of TLR signaling pathways could lead to many diseases, such as inflammation, autoimmune diseases, metabolic diseases, neurological diseases, and cancers (Wang et al. 2013). Additionally, many key transcription factors responsible for inflammation and immunity, such as NF-κB, JNK, MAPK, p38, and ERK, can be regarded as downstream signaling pathways after TLR activation (Mcgettrick and O'Neill 2010). Generally, by activating TLRs and their related signaling like NF-κB and MAPKs, the cells can produce cytokines or pro-inflammatory mediators to achieve anti-inflammatory effects (Allan et al. 2001). MyD88 is identified downstream of the mammalian TLR and IL-1R families, which binds TLRs to IRAKs and activates NF-κB and MAPKs (Deguine and Barton 2014). MyD88 can further interact with TRAF6 to activate and translocate NF-κB. MyD88 participates inflammation signaling pathway (Takeuchi et al. 2000) and intracellular protozoan *Toxoplasma gondii* (Scanga et al. 2002). In general, TLRs-mediated signaling pathways can be divided into MyD88-dependent and MyD88-dependent pathways, which yield pro-inflammatory cytokines by different signaling molecules. Myd88-dependent signaling pathway, on the one hand, depends on activation of NF-κB by regulating some related genes, such as COX-2, IL-6, and TNF-α (Keating et al. 2007). On the other hand, it depends on interaction with TRAF6, which can further activate MAPKs, p38, and ERK signal pathways (Doyle and O'Neill 2006). Alternatively, the MyD88-independent pathway is also regarded as a TRIF-dependent pathway. TLR recruits TIR-domain containing adaptor, inducing TRIF and TRAM (Rahimifard et al. 2017). The dimerization of both transcription factors further stimulates TBK, IKKε, and IRFs. Additionally, IRF3 and IRF7 can stimulate IFN-α and IFN-β transcription. NK cells produce IFN-γ and TNF-α, which indirectly promotes T-cell response initiation by activating macrophage-associated cytokines (Patra et al. 2020). In a word, both pathways contribute to the immune response (Takeda and Akira 2004).

In recent years, many phytochemicals have shown in vitro and in vivo anti-inflammation activity studies (Chung-Yi et al. 2018; Pérez-Jiménez et al. 2013). As mentioned above, it is possible to believe that phytochemicals exert their anti-inflammatory action by targeting TLRs. Various studies have demonstrated the effect of different phytochemicals on TLRs gene and protein expression. In in vitro experiments, the plant lignin phyllanthusmin C showed anti-inflammatory effects by recognizing TLR1 and/or TLR6 to induce NF-κB activation and binding IFN-γ promoter in the human natural killer cell model (Deng et al. 2014). Isoglycyrrhizin significantly downregulated TLR4 expression in RAW 264.7 macrophage line (Park and Youn 2010).

Similarly, engelsin and lactinin have been shown to inhibit LPS-stimulated J774 macrophage TLR4 gene expression (Huang et al. 2011). However, the TLR signaling-related downstream gene expression induced by P. gingivalis LPS was significantly downregulated after adding baicalin into human oral keratinocytes culture, but the TLR1-9 gene expression remained unchanged by PCR array (Luo et al. 2012). In addition, quercetin suppressed TLR4 and TLR2 expression induced by ox-LDL. It amended the TLR-NF-κB signaling pathway and further downregulated the activity of inflammatory enzymes (Bhaskar et al. 2011). Kaempferol 3-*O*-sophoroside is similar to downregulated TLR2 and TLR4 expression in several cells (Kim et al. 2012; Li et al. 2012). Naringin can inhibit TLR2 gene expression in 3T3-L1 cells but had no effect on TLR4 expression (Yoshida et al. 2013). Curcumin has shown inhibitory effects on TLR2 in various cells (Shuto et al.

2010). Furthermore, curcumin can suppress LPS-induced expression of inflammatory mediators through TLR4-MAPK/NF-κB pathway (Zhe et al. 2013). 6-Shogaol can modulate TLR4-/NF-κB-mediated inflammation (Ahn et al. 2009).

Moreover, flavopiridol, which is a quercetin analogue, can inhibit TNF-α level by suppressing the activation of NF-κB and MAPKs pathway (Haque et al. 2011). EGCG inhibited IKKβ expression in the MyD88-dependent pathway of TLRs in RAW264.7 cells (Youn et al. 2006). Berberine is an isoquinoline with anti-inflammatory activity via TLR4 signaling pathway and suppressed the activation of Src in macrophages and v-Src-transformed cells (Cheng et al. 2015).

The applicability of phytochemicals, especially as a ligand for TLRs, requires in vivo data. Till now, various in vivo studies have confirmed the effects of different phytochemicals on TLRs signaling pathways. The expression of TLR in immune system of mice fed with cocoa rich in procyanidins was found to change (Pérez-Berezo et al. 2011). For example, TLR4 and TLR9 genes were upregulated, and TLR2 and TLR7 genes were downregulated in mesenteric lymph nodes. In contrast, TLR4 and TLR9 mRNA levels were lower, and TLR2 and TLR7 mRNA levels were higher in the small intestine of the cocoa-fed mice (Pérez-Berezo et al. 2012). In the streptozotocin-induced diabetic rat model, total glucosides of peony from the root of *Paeonia lactiflora* pall reduced the levels of TNF-α and IL-1β by inhibiting the activation of TLR2 and TLR4 in kidneys (Wu et al. 2009b). Luteolin protected rat brains against focal ischemia through downregulating the expressions of TLR4, TLR5, NF-κB, and p38MAPK and upregulating ERK expression in the cerebral cortex (Qiao et al. 2012). The anti-inflammation potential of curcumin was verified in LPS-induced mastitis mice and traumatic brain injury mice model by inhibiting TLR4-mediated NF-κB signaling pathways (Fu et al. 2014; Zhu et al. 2014). Resveratrol can be helpful in TLR-mediated inflammation responses in different mice models (Lee et al. 2011; Wang et al. 2020a, b; Xiong et al. 2020). In humans, two interventional studies indicated that orange juice could prevent diet-induced oxidative stress and inflammation by increasing endotoxin and TLR expression in volunteers (Deopurkar et al. 2010; Ghanim et al. 2010). In addition to the phytochemicals mentioned above, various natural compounds and extracts also were able to show anti-inflammation through regulating TLRs family members. Achyrocline satureioides extract rich in quercetin and luteolin suppressed neutrophil functions by reducing TLR4 expression on neutrophils (Barioni et al. 2013). Furthermore, various fruit extracts from strawberry, blackberry, and feijoa can mediate anti-inflammatory responses through the combined action of TLR2 and TLR4 pathways (Noha et al. 2014). Besides, a Korean fermented soybean extract showed anti-inflammatory effects by regulating TLR ligands, mainly by inhibiting NF-κB activation against TLR2-, TLR3-, TLR4-, and TLR9-specific ligands (Lee et al. 2014).

In summary, the TLRs family and their downstream signaling pathways play an important role in the inflammatory resistance and immune response. Activating these signaling pathways promotes the expression of various anti-inflammatory or/and pro-inflammatory molecules. Therefore, members of the TLRs family have become important targets for the regulation of inflammation and immunity. The regulatory role of phytochemicals in TLRs and their downstream pathways has been and will continue to be used as TLR agonists or antagonists to prevent and treat inflammatory and immune-related diseases.

4.3 Jak/Stat: IL-6 Receptor Signaling

4.3.1 JAK/STAT

Janus kinases (JAKs) as well as signal transducer transcriptional activators (STATs) were discovered in the 1990s when studying interferon signaling (Darnell Jr et al. 1994; Leonard and O'Shea 1998). JAK/STAT are involved in conveying information from extracellular polypeptide signals to target gene promoters in the nucleus, regulating immune, cell proliferation, differentiation, apoptosis, etc. (Mertens and Darnell Jr. 2007). JAK/STAT are essential components of many cytokine receptor systems.

In mammals, JAKs, namely, JAK1, JAK2, JAK3, and Tyk2, are mainly composed of FERM domain, SH2-like domain, pseudokinase domain, and kinase domain (Villarino et al. 2020). JAKs played a critical role in cytokine signaling. JAKs bind to a subset of cytokine receptors, and upon cytokine activation, JAKs use ATP to phosphorylate themselves and the intracellular tail of the receptor subunits. This creates docking sites for the recruitment of downstream signaling molecules (Gadina et al. 2020).

In 1994, members of Jim Darnell's laboratory identified an essential class of JAK substrates, the tyrosine-phosphorylated STAT family of DNA binding proteins (Darnell Jr et al. 1994). STAT proteins are a family of potential cytoplasmic transcription factors that share seven forms (STAT1, STAT2, STAT3, STAT4, STAT5A, STAT5B, STAT6) with five domains: N-terminal domain, coiled-coil domain, DNA binding domain, SH2 domain, and a transcription activation domain at carboxy terminus (Mertens and Darnell Jr. 2007). A hallmark of activated STAT proteins is the phosphorylation of tyrosine residues in the C terminus (Y ~ 700), and inactivated stats exist as inverted dimers in the cytoplasm. Activated JAKs can activate STAT proteins to form dimers and execute rapid transcription (Mao et al. 2005). The binding of cytokine ligands initiates the JAK-STAT signaling pathway to cell surface receptors, four JAKs are linked to specific cytokine receptors, and the activation of JAKs bound to the receptors can be facilitated by different cytokine ligands (IFNs, IL-6, IL-10 family, TSLP, IL-12, IL-13, and IL-17) binding to specific cytokine receptors (Gadina et al. 2020).

4.3.2 IL-6

In 1973, researchers at Osaka University discovered IL-6 secreted by T cells, which can co-regulate the production of antibodies by B cells. The IL-6 has become a key pathway in immune response in many diseases (Kishimoto 1973; Tanaka et al.

2018). IL-6 signal transduction system is composed of two receptor chains and downstream signal molecules. Its signal transduction is realized by the combination of IL-6 and IL-6 receptor (IL-6R) (Mihara et al. 2012). Once IL-6 was bound to IL-6R, the complex will induce the dimerization of gp130, thus triggering a downstream signal cascade (Choy et al. 2020). The functional receptor is a hexamer with two IL-6 molecules, two IL-6R molecules, and two gp130 molecules. The pleiotropy of IL-6 can be explained by the extensive expression of gp130 in various cells (Garbers et al. 2015).

4.3.3 IL-6/JAK/STAT

The homogeneity of gp130 can promote the activation of downstream JAKs, and the activated JAKs are close to each other, leading to the phosphorylation of key tyrosine residues of gp130 in the cytoplasm, thus recruiting specific STATs through its SH2 domain (Tanaka et al. 2018). Activated JAKs can activate the phosphorylation of tyrosine residues (Y ~ 700) at the C-terminal of STAT proteins to form a dimer, which is transported into the nucleus through the nuclear transporter importin-α (Imp-α) in dependent manner. STATs are classical transcription factors combined with DNA regulatory elements (DRE) with fixed sequences to control the transcription of downstream genes (Villarino et al. 2020).

STATs are widely integrated into the whole genome in promoters. The genes encoding STATs have superstrong enhancers, which can help STATs to control transcription. JAK/STAT is negatively regulated in many ways. In the nucleus, nuclear protein tyrosine phosphatases (NPTPs) can dissociate STAT from DNA and inactivate it. The combination of STAT dimer and DNA is also inhibited by protein inhibitor of activated STAT (PIAS). Besides, members of the suppressor of cytokine signaling proteins (SOCS) family can inhibit JAK/STAT receptor signal transduction through homologous or heterologous feedback regulation (Fig. 4.1) (Mertens and Darnell Jr. 2007; Morris et al. 2018).

4.3.4 Immunomodulation

IL-6/JAK/STAT signaling pathway is abnormally activated in patients with chronic inflammation or solid tumors. IL-6 is yielded during infection and tissue injury (Kumari et al. 2016). IL-6 is a specific cytokine to maintain homeostasis. When the homeostasis is destroyed by infection or tissue injury, IL-6 will be produced immediately. It promotes host defense by stimulating acute phase response and hematopoietic and immune response. In the microenvironment with disrupted homeostasis, various cell types can produce IL-6, which leads to the activation of the JAK/STAT signal in immune cells. STATs are over-activated in immune cells and regulate dendritic cells (DC), neutrophils, effector T cells, and natural killer cells (Fig. 4.2) (Johnson et al. 2018). The IL-6/JAK/STAT signaling pathway's targeted components can inhibit tumor cells growth and alleviate the tumor

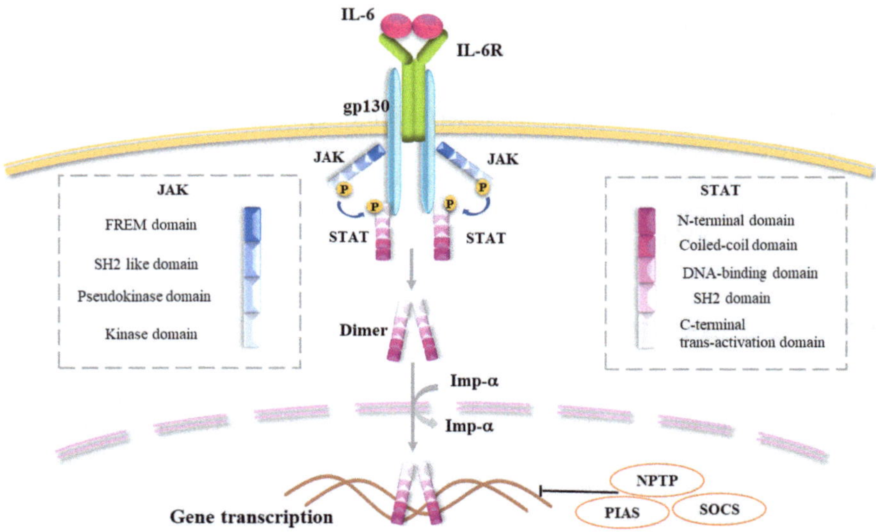

Fig. 4.1 IL-6/JAK/STAT signaling pathways: IL-6 combines with IL-6R and forms a hexamer with gp130 to activate JAK downstream. Activated JAK promotes tyrosine phosphorylation of gp130, and phosphorylated gp130 collects stat. Activated JAK promotes the phosphorylation of stat tyrosine to form a dimer, transported into the nucleus through imp-α to participate in gene transcription. NPTP, PIAS, and SOCS negatively regulated transcription

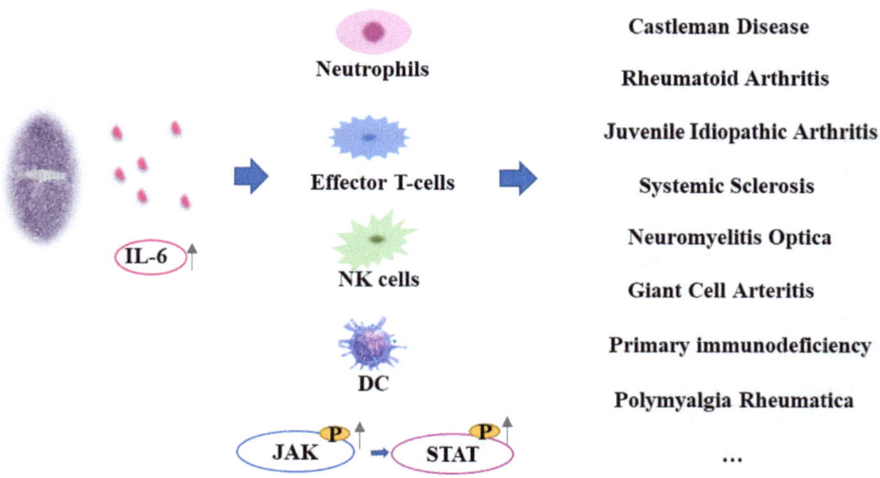

Fig. 4.2 IL-6 in immunoregulation. When tissues are infected or damaged, IL-6 is produced, leading to JAK/STAT signal activation in immune cells (neutrophils, natural killer cells, effector T cells, and dendritic cells), and regulates Castleman disease, rheumatoid arthritis, and other immune diseases

microenvironment's immune control. IL-6/JAK/STAT signaling pathway is closely related to regulate many diseases (Choy et al. 2020).

FDA has approved IL-6, JAK inhibitors, and IL-6 receptors for treating various malignant tumors (Gadina et al. 2020). JAK inhibitors have been applied to treat inflammatory bowel disease, arthritis, and myeloproliferative tumors.

4.3.5 Plant Active Ingredients

1. **Alkaloids**
   Alkaloids are a kind of nitrogen-containing essential organic compounds with significant biological activities and are essential effective components in Chinese herbal medicine. *Leonurus japonicus* Houtt. is a traditional Chinese herbal medicine widely used in various diseases such as menstrual disorders, dysmenorrhea, etc. Leonurine is an active substance extracted from *L. japonicus*. It has been proved that leonurine can resist oxidation, apoptosis, anti-inflammation, and antibacterial (Yang et al. 2019). Recently, through the establishment of mouse osteoarthritis (OA) model, a study found that leonurine could affect IL-6 through IL-1β, which effectively inhibited the degradation of cartilage and joint and focal inflammatory reaction, and gavage of leonurine in OA mice delayed the development of the disease (Yin and Lei 2018). Besides, Ma's laboratory studies have found that berberine reduces inflammation in human bronchial epithelial cells by reducing the production of IL-6 (Ma et al. 2020). Other alkaloids, such as ergometrine and ephedrine, have also shown to alleviate inflammation by reducing the production of IL-6 (Guillermo García-Laínez et al. 2017; Zheng et al. 2013).
2. **Flavonoids**
   Flavonoids are widely-distributed polyphenolic secondary metabolites with diverse biological activities in plants in nature. Flavonoid extracts can reduce the expression of pro-inflammatory cytokine IL-6, regulate inflammatory markers, and prevent nerve injury by regulating MAPK and NF-κB signaling pathways in microglia (Spagnuolo et al. 2018). Michaud Leveque's laboratory provides new insight into quercetin's role as a blocker of IL-6 stimulating STAT3 activation pathway, which has a potential role in preventing and treating glioblastoma (Michaud-Levesque et al. 2012). It has been reported that luteolin can alleviate the inflammation of HT-29 colon epithelial cells stimulated by cytokines by inhibiting the JAK/STAT pathway, which may have a potential therapeutic effect on inflammatory bowel disease (IBD) (Nunes et al. 2017). Zhou et al. found that formononetin (a flavonoid extracted from *Astragalus membranaceus*) can improve high-glucose-diet rats' endothelial function through JAK/STAT pathway, which can be used to prevent and treat diabetic complications (Zhou et al. 2019).
3. **Saponins**
   Saponins are glycosidic compounds having aglycones as triterpenoids or spirostane compounds widely available in Chinese herbal medicines.

Ginsenoside, a predominantly available saponin in Chinese medicine, has antioxidant, anti-inflammatory, and anticancer activities. Ginsenosides regulate the immune response during bacterial and viral infection via inflammatory factor IL-6 (Nguyen and Nguyen 2019). A simulation study on the binding and docking of six saponins with IL-6 showed that IL-6 was successfully combined with six analyzed saponin ligands, indicating that the pharmacological target of saponin's anti-inflammatory and anti-obesity may be IL-6 (Cui et al. 2018). Teng et al. used different concentrations of total saponins of *Rhizoma Paridis* (RPTS) to treat SW480 colorectal cancer (CRC) cells and explored its effect on apoptosis CRC cells. They found that RPTs inhibited the secretion of IL-6 and IL-6/JAK/STAT protein signaling pathway, significantly promoting the apoptosis of CRC cells, which has reference value for clinical treatment of colon cancer (Teng et al. 2015).

4. **Isoflavones**
   Isoflavones have a similar estrogen structure. A study has shown that the level of IL-6 in human serum is significantly reduced in the diet of high-dose soybean isoflavone and soybean protein, which indicates that soybean isoflavone may improve the inflammatory response through the IL-6 pathway (Gholami et al. 2020). Isoflavone extract can increase skin surface hydration, reduce the activation of JAK/STAT in normal human epidermal keratinocytes, and reduce epidermal proliferation and inflammatory cell infiltration, which can treat inflammatory skin diseases (Li et al. 2018). Isoflavones can improve the phosphorylation of p38MAPK and reduce the phosphorylation of JAK/STAT in myoblasts, thus activating myogenic differentiation. Besides, isoflavones showed various bioactivities, such as anticancer effect on lung, reducing insulin resistance, affecting growth and immune activity of broilers (Soundharrajan et al. 2019).
5. **Catechins**
   Catechin is a kind of phenolic active substances extracted from natural plants such as green tea. Among catechins, EGCG has the most prominent bioactivity. EGCG can downregulate the gene expression level of IL-6 in prostate cancer cells and the cytoplasmic level of IL-6 in the prostate cancer rats (Zhang et al. 2017c). EGCG treatment may induce phenoloxidase, prophenoloxidase, and JAK-STAT pathways (Wu et al. 2014). EGCG may effectively improve innate immunity (Li et al. 2017).
6. **Anthocyanins**
   Anthocyanins as water-soluble natural pigments are obtained by hydrolysis of anthocyanins. It is of great significance to reduce intestinal inflammation symptoms. Anthocyanins can inhibit the expression levels of NF-κB and IL-6 by scavenging the excessive accumulation of oxidative free radicals and affect the endogenous cell signaling pathway and intestinal flora (Fallah et al. 2020). Anthocyanins can inhibit the migration and proliferation of FLS cells in arthritic rats, activate PIAS3 protein, and inhibit STAT3-specific transcriptional activation, which can be used as a therapeutic drug for rheumatoid arthritis (RA) (Samarpita et al. 2020).

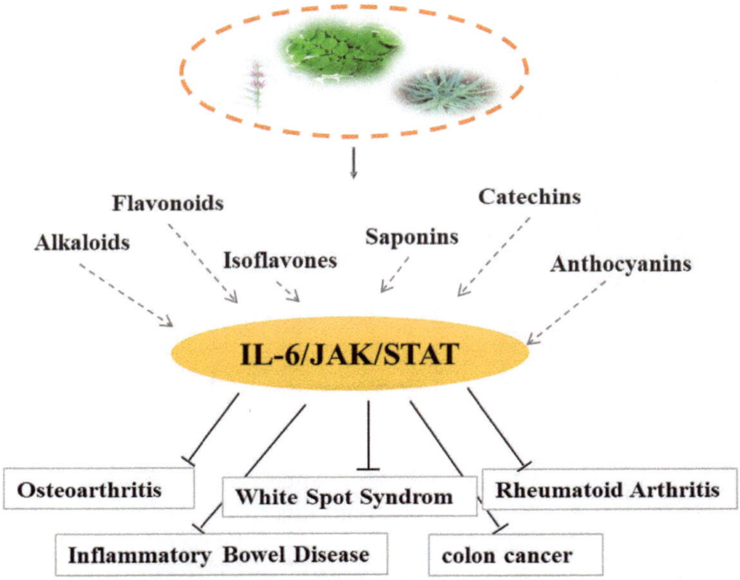

Fig. 4.3 Scheme showing effects of plant extracts (alkaloids, flavonoids, saponins, isoflavones, catechins, anthocyanins) on immunomodulatory diseases through IL-6/JAK/STAT signaling pathway

In short, the natural active components extracted from plants (alkaloids, flavonoids, saponins, catechins, etc.) may regulate the immune system through the IL-6/JAK/STAT signaling pathway, which may be the direction of future research on the immune regulation of active plant components (Fig. 4.3).

4.4 T-Cell Receptor Signaling

T-cell receptor (TCR) is the surface receptor of T cells, which is mainly in charge of recognizing the specific antigen peptide on the surface of antigen-presenting cell (APC) to form TCR-peptide-MHC complex (TCR-pMHC complex) (Garcia et al. 2009; Rudolph et al. 2006). TCR is a heterodimer mainly composed of α and β chains. However, TCR is composed of γ and δ chains in some T cells. The specificity of TCR is determined by the binding sites formed by α/β chains or γ/δ chains (Germain and Stefanova 1999; Samelson 2002; Call and Wucherpfermig 2005).

TCR plays a vital role in the formation of immune synapses. Activation of TCR promotes lots of signal transduction cascades by regulating cytokine products and cell differentiation and proliferation. Typical intracellular signals of TCR activation include MAPK, PKC, and calcium signaling pathways (Fig. 4.4). The activation of these signals will eventually activate specific gene expression of T cells, cause cell proliferation, and make T-cells differentiation (Smith-Garvin et al. 2009).

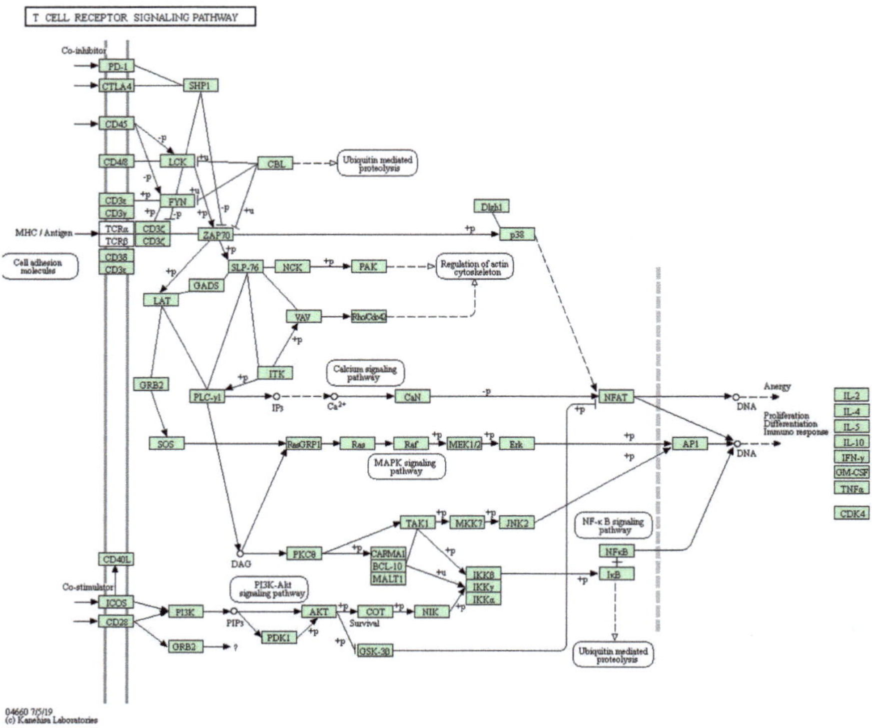

Fig. 4.4 T-cell receptor signaling pathway. (From https://www.genome.jp/kegg-bin/show_pathway?hsa04660)

TCR-induced intracellular signaling starts from tyrosine phosphorylation in the sequence of immunoreceptor tyrosine-based activation motif (ITAM) in the CD3 molecule. After the formation of the TCR-pMHC complex, protein tyrosine kinases (PTKs) such as lymphocyte protein tyrosine kinase (Lck), proto-oncogene tyrosine protein kinase (Fyn), ζ-chain-associated protein kinase of 70 kDa (ZAP-70), and IL2-inducible T-cell kinase (ITK) were activated. Lck is the first kinase activated. Lck is bound to the cytoplasmic regions of CD4 and CD8 molecules in T cells and phosphorylated (Nika et al. 2010). CD4 or CD8 are recruited into TCR-pMHC complex, and tyrosine in ITAM of CD3 molecule in the cytoplasmic region is phosphorylated. The phosphorylated CD3 ζ ITAM site then recruited ZAP-70 and activated it by Lck phosphorylation. ZAP-70 can phosphorylate linker to activate T cells (LAT) (van der Merwe and Dushek 2011).

The phosphorylated ZAP-70 can bind to many downstream signal proteins, such as Grb2, Grb2-related adapter protein 2 (GADS), phospholipase C gamma 1 (PLC γ1), etc. These proteins can be collected together to form LAT signalosome (Balagopalan et al. 2010). GADS in LAT signalosome can further recruit SLP-76 and ITK. Phosphorylation of ITK activates phospholipase PLC γ1. Activated PLC γ1 can decompose phosphatidylinositol 4,5-bisphosphate (PIP2) on the cell membrane and produce second-messenger inositol trisphosphate (IP3) and diglyceride (DAG). IP3 binds to inositol trisphosphate receptors (IP3Rs) to activate endoplasmic reticulum calcium channels and release endoplasmic reticulum calcium ions into cells. The decrease of endoplasmic reticulum calcium concentration further promotes the opening of calcium channels on the cell membrane and leads to extracellular calcium influx. The increase of intracellular Ca^{2+} concentration can activate calcineurin, catalyze the dephosphorylation of nuclear factor of activated T cell (NFAT), and mediate the expression of downstream genes, such as IL-2; DAG can activate MAPK signal pathway and activate activator AP-1 and activate NF-κB through activating PKC to regulate the expression of target genes. The initiation and transmission of TCR signal are finely regulated, including positive and negative regulation (Smith-Garvin et al. 2009). Once the balance is destroyed, it will affect the immune response's effect or induce autoimmunity (Starr et al. 2003).

Natural products exhibit various pharmacological and physiological functions. Some of them may be the potential modulators of the TCR signaling to prevent or treat immune diseases. Here, we focus on the classifications of alkaloids, flavonoids, saponins, isoflavones, catechins, and anthocyanins regarding their effects on TCR signaling. These phytochemicals mainly affect the downstream signaling pathways of TCR and related cytokines to achieve immune regulation.

4.4.1 Alkaloids

Berberine has the therapeutic potential for autoimmune diseases in animal models. The mechanisms via immunomodulatory and anti-inflammation of berberine on autoreactive inflammatory responses were recently summarized by Ehteshamfar et al. (2020). Sinomenine suppresses lymphocytes' activities by mainly inhibiting their proliferation, inducing T lymphocytes' apoptosis, and attenuating the Th1/Th2 imbalance. Wang and Li (2011) summarize the immunosuppressive and anti-inflammatory activities, mechanisms, and applications of sinomenine.

Caffeine caused alterations in the immune system (Orban et al. 2018). Caffeine hardly shows immunosuppressive effects on neonatal T cells. Caffeine improves calcium influx in activated T cells to enhance short-term activation in neonatal T lymphocytes to a more considerable extent via increased cAMP levels. However, the cytokine production of neonatal CD4 cells remains relatively unaffected.

4.4.2 Flavonoids

Flavonoids include flavonols and flavones, flavanols, flavanones, etc. and have the potential to treat autoimmune diseases. Rengasamy et al. summarized flavonoids' role in autoimmune diseases, including intracellular signalling pathways targeted by flavonoids (Rengasamy et al. 2019). The imbalance between regulatory T cell (Treg) and T helper 17 cell (Th17) is critical to many immune diseases. Phloretin, a dihydrochalcone structural flavonoid, could decrease Th17 cell production, and phospho-Stat3 expression increases Treg generation and phospho-Stat5 expression during Th17/Treg differentiation (Jiao et al. 2020). Phloretin can mediate the Th17/ Treg balance by regulating metabolism via the AMP-activated protein kinase signal pathway. Liquiritigenin shows an anti-atopic effect by controlling T-cell activation and exhibits therapeutic potential for T-cell-mediated disorder (Lee et al. 2020a, b). Liquiritigenin blocks the expression of IL-2 and CD69 from activated T cells by PMA/A23187 or anti-CD3/CD28 antibodies. The expression levels of CD40L and CD25 were reduced in T cells pretreated with liquiritigenin by repressing NF-κB and MAPK pathways. Silymarin has immunoregulatory effects on Tregs (Shariati et al. 2019). Silymarin promoted Tregs proliferation and enhanced IL-10 and TGF-beta levels by increasing FOXP3, JAK3, transducer, and activator of STAT5 gene expression (Keyhanmehr et al. 2019).

4.4.3 Saponins

The main bioactive components in *Panax ginseng* are ginsenosides such as ginsenoside Rg1, Rg3, Rg5, Rb1, Rh2, F2, etc. Ginsenoside Rg1 enhanced $CD4^+$ T-cell proliferation in a dose-dependent manner (Lee et al. 2004). Ginsenoside Rg1 affects Th-cell responses from a Th1-dominant to a Th2-dominant pattern by increasing IL-4 and IL-2 mRNA expression while decreasing IFN-γ expression. Ginsenoside Rg3 inhibited IFN-γ and T-bet expression in T cells under Th1-skewing condition (Cho et al. 2019). Ginsenoside Rh2 exhibits immunoregulatory and anti-inflammatory properties. Ginsenoside Rh2 triggered cytotoxicity in spleen lymphocytes and increased T-lymphocyte infiltration in tumor (Wang et al. 2017a, b). Ginsenoside Rh2-B1, a sulfated derivative of ginsenoside Rh2, could stimulate cell proliferation and IFN-γ production by activating p38 MAPK and ERK signaling pathways in T cells (Lv et al. 2014). Ginsenoside F2 weakened alcoholic liver injury by increasing mRNA expression levels of IL-10 and Foxp3 and reducing IL-17 mRNA expression. Ginsenoside F2 inhibited differentiation from naïve T cells to Th17 cells by suppressing RORγt expression and facilitating Foxp3 expression (Kim et al. 2020).

Saikosaponin-d from *Bupleurum falcatum* L. (Umbelliferae) showed an immunoregulatory effect on splenic T lymphocytes of C57BL/6 mice by promoting IL-2 production and IL-2 receptor expression (Kato et al. 1994). The target of Saikosaponin-d is located at the transcription or pretranscription of the c-fos gene and after the activation of T-cell receptor/CD3-mediated protein tyrosine kinase. Saikosaponin-d suppresses T-cell activation by inhibiting CD69 and CD71 expressions and IL-2 production (Leung et al. 2005). Astragaloside, a major active component of *Astragalus membranaceus* (Heine et al. 2018) Bge (Huang-Qi) induced T-cell activation, regulated effector/regulatory T-cell balance, and enhanced CD45 phosphatase activity. Qi et al. reviewed the recent progress of ASIs in immunoregulatory activity and its mechanism (Qi et al. 2017). Glycyrrhizin has immunomodulatory effects on $CD4^+$ T-cell responses during liver fibrogenesis

(Tu et al. 2012). Glycyrrhizin suppressed ConA-induced proliferation of splenic $CD4^+$ T cells and improved the IFN-γ and IL-10 mRNA expressions in these cells by JNK, ERK, and PI3K/AKT pathways.

4.4.4 Catechins

Catechins are well known for their immunosuppressive properties. EC, EGC, ECG, and EGCG induced significant cytotoxicity on leukemia cells (Tang et al. 2014). ECG synergistically improved EGCG-induced phosphorylation of JNK and the expression level of IFN-γ in Jurkat cells. EGCG displayed a high binding affinity with ZAP-70 and formed a stable complex, ZAP-70-EGCG. The complex effectively suppressed ZAP-70, LAT, MAPK, ERK, and PLC γ1 activities in CD3-activated T-cell leukemia (Shim et al. 2008). EGCG dose-dependently inhibited the activation of AP-1 and IL-2 induced by CD3 and induced caspase-mediated apoptosis in P116.cl39 ZAP-70-expressing leukemia cells. EGCG suppressed the CD3-mediated T-cell-induced pathways in leukemia cells. EGCG could induce Foxp3 and IL-10 expression in $CD4^+$ Jurkat T cells in vitro. EGCG significantly improved Treg frequencies and numbers in spleen and lymph nodes in mice (Wong et al. 2011). EGCG supplementation may benefit patients with abnormal T-cell function (Wu et al. 2009a).

4.4.5 Anthocyanins

Cyanidin 3-*O*-glucoside (C3G) is a major component of anthocyanins and can prevent or treat bone-related immune diseases. C3G inhibited NF-κB receptor activator-induced osteoclast differentiation (Park et al. 2015). C3G significantly reduced ERK activation, JNK, and p38 MAPK in osteoclast precursor cells. Besides, C3G suppressed the expression of c-Fos and NFATc1. Protocatechuic acid (PCA), a metabolite of anthocyanins, repressed RANKL-induced osteoclast differentiation dose-dependently in mouse bone marrow macrophages (Park et al. 2016). PCA inhibited RANKL-stimulated c-Fos and NFATc1 at the mRNA and protein levels.

Delphinidin- and gallic acid-boost T cells differentiate into Tregs (Hyun et al. 2019). The secretion of various immunosuppressive proteins increased, and the function of induced Tregs was enhanced. Both compounds promoted the differentiation of regulatory T cells and inhibited the function of memory T cells. Delphinidin stimulates immunostimulatory effects by promoting cytokine production through Ca^{2+} channel and NFAT activation (Jara et al. 2014). Delphinidin improved cytosolic-free Ca^{2+} level by releasing intracellular Ca^{2+} and enhancing Ca^{2+} entry. BTP2 and gadolinium can reduce calcium entry induced by delphinidin. Delphinidin improved the production of IL-2 in Jurkat cells, which was inhibited by BTP2.

4.4.6 Isoflavones

Puerarin can treat immune diseases attributed to Treg deficiency (Yamamoto et al. 2019). It is a natural enhancer of retinoic acid required to induce Tregs and suppresses the development of food allergy in a mouse model.

4.5 NF-κB Signaling

4.5.1 NF-κB Signaling Pathway to Immunomodulation

NF-κB, a nuclear protein, was initially extracted from B lymphocytes nucleus in 1986 by Sen and Baltimore (1986). NF-κB widely exists in eukaryotic cells and has a vital role in regulating cell differentiation and growth, immune response, and inflammatory cytokines (Lingappan. 2018). NF-κB complex includes three protein families (Fig. 4.5): NF-κB/Rel, Iκb, and IKK. There are five members of NF-κB/Rel, which contain an amino terminal of 300 amino acids named Rel homology domain (RHD). Meanwhile, this region presents N-terminal DNA/IκB binding sites and also contains a C-terminal dimerization site. Generally, NF-κB exists in the form of the heterodimer, which combined P50 and p65. Under a resting state, NF-κB combines with IκB to form NF-κB-IκB complex. When stimulated by LPS or TNF-α, IκB will be phosphorylated and degraded to form free NF-κB. Subsequently, activated NF-κB exposed the nuclear localization sequence and then entered the nucleus combined with the specific DNA site, started gene transcription, and released IL-1β, IL-6, IL-12, and IL-17. The released inflammatory factors could also activate NF-κB in reverse to form a positive feedback loop, which makes the inflammatory

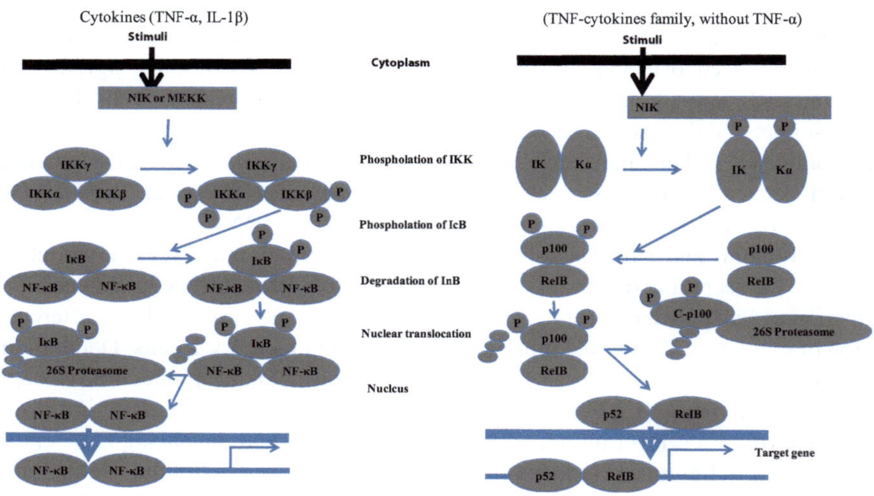

Fig. 4.5 NF-κB-regulated processes and factors involved in immunomodulation

reaction cascade amplification. Pro-inflammatory cytokines and infections activate this pathway by activating IKKb, which will degrade IκB and liberate NF-κB dimers (Diomede et al. 2017).

4.5.2 Natural Polyphenols and NF-κB Signaling Pathway

Many dietary polyphenols have been proved to be potential modulators of NF-κB (Khan et al. 2020; Miao et al. 2019; Zhao et al. 2019). In this section, we would like to focus on the inhibitory effects of polyphenols on NF-κB. As for targeting NF-κB signaling pathway, polyphenols can inhibit the phosphorylation of kinases, which will inhibit NF-κB translocation. Furthermore, polyphenols also suppress the NF-κB-DNA interaction (Ruiz and Haller 2006) (Fig. 4.6).

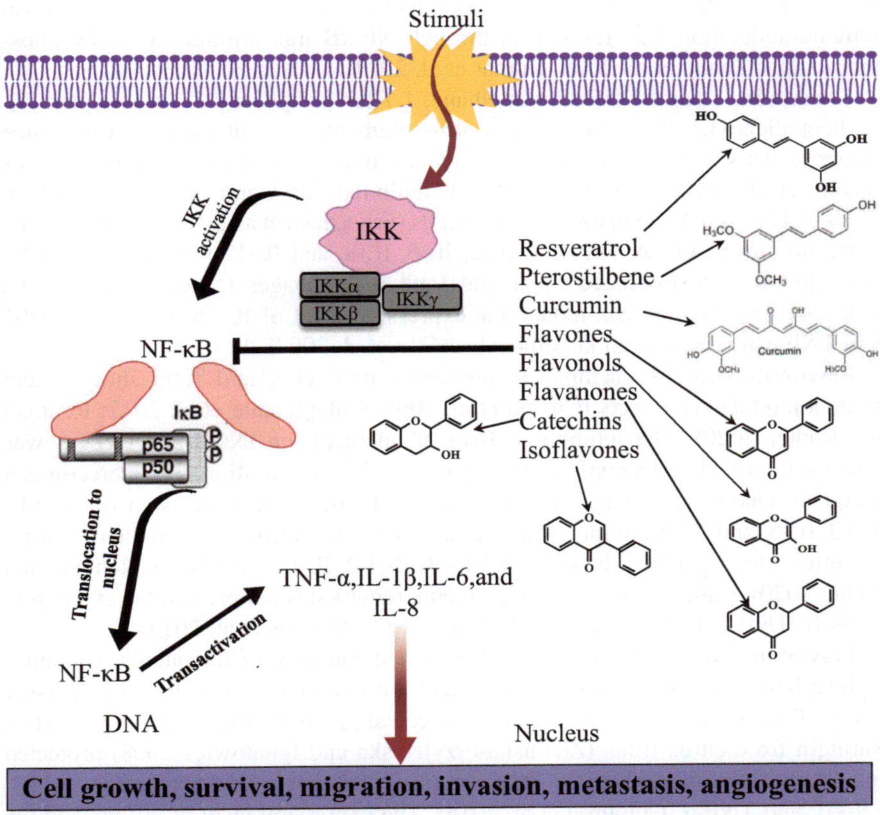

Fig. 4.6 Inactivation of NF-κB and downstream pro-inflammatory cascades by phytochemicals

Resveratrol, a stilbene mainly from grapes and peanuts, has shown many bioactivities (Pan et al. 2016; Lai et al. 2017). Resveratrol could lower the expression of iNOS, PGE2, and COX2 in Aβ-activated glial cells (Seo et al. 2018). Through suppressing the NF-κB activity, resveratrol could mitigate LPS-induced microglial inflammation (Zhang et al. 2015). By attenuating the NF-κB signaling pathway, resveratrol inhibited LPS-induced mice mastitis (Zhang et al. 2017a, b) and suppressed inflammatory cytokines production in RAW264.7 cells (Liu et al. 2019). Resveratrol not only can inhibit NF-κB-mediated cytokine expression in adipocytes (Haghighatdoost and Hariri 2019) but also suppresses iNOS expression and nitric oxide generation (Truong et al. 2018).

Pterostilbene (*trans*-3,5-dimethoxy-4′-hydroxystilbene) is a natural dimethyl ether analog of resveratrol (Tsai et al. 2017). Pterostilbene has shown significant anti-inflammation effects (Li et al. 2016; Chen et al. 2017; Wang et al. 2020a, b). Interestingly, pterostilbene usually exhibits a more substantial immunomodulator effect when compared with the analog of resveratrol (Gómez-Zorita et al. 2020; Suh et al. 2007). Researchers found pterostilbene with a more potent effect on immunomodulation than resveratrol through NF-κB transcription and JNK phosphorylation (Zhang et al. 2020; Kumar et al. 2017).

Curcumin is named diferuloylmethane, 1,7-bis [4-hydroxy-3-methoxyphenyl]-1,6-heptadiene-3,5-dione, occurring yellow pigment, and obtained from the spice turmeric. This compound exhibited several immunomodulatory characteristics (Yadav et al. 2005; Gao et al. 2004). It inhibited AP-1 and NF-κB induced by IL-1α or TNF-α in bone marrow stromal cells (Srivastava et al. 2011). Moreover, the expression levels of TNF-α, IL-1, IL-2, IL-6, IL-8, and IL-12 were suppressed by curcumin in LPS-stimulated monocytes and macrophages (Machova Urdzikova et al. 2016). Curcumin can reduce the expression level of IL-1β, IL-6, and cyclin E in TNF-α-treated human keratinocytes (Cho et al. 2007).

Flavonols such as kaempferol, quercetin, myricetin, and rutin show potent immunomodulatory effects (Crespo et al. 2008; Galleggiante et al. 2017; Rehman and Rather 2020). The inhibitory effect of quercetin on iNOS and COX-2 was stronger than that of kaempferol (Crespo et al. 2008). Additionally, quercetin can attenuate pancreatic damage and significantly reduced the expression of NF-κB, IL-1β, IL-6, and TNFα in rats (Zheng et al. 2016). Myricetin reduced the deteriorative effects by regulating the level of NF-κB, Nrf-2, IL-6, and TNF-α (Rehman and Rather 2020). Rutin, a glycoside of quercetin, remarkably suppressed the expression levels of TNF-α, IL-6, NF-κB, COX-2, and iNOS (Nafees et al. 2015).

Flavanones belong to the 2-benzo-γ-pyrone category of flavonoids, contain a 2-phenyl-1-benzopyran-4-one skeleton, and are most commonly found in parsley celery. They are also widely distributed in cereal grains (Duodu and Awika 2019). Naringin from citrus fruits (Zielińska-Przyjemska and Ignatowicz 2008) protected against cisplatin-induced striatal injury in rats by enhancing the expression of P53, NF-κB, and TNF-α (Chtourou et al. 2015). The expression level of NF-κB and the NF-κB-DNA binding activity were inhibited by naringin in rats (Sahu et al. 2014). Naringenin is a potential therapeutic agent for sepsis, fulminant hepatitis, fibrosis, and cancer (Zeng et al. 2018). For example, Xie and coauthors found that naringenin significantly inhibited oxidative stress and inflammation by suppressing NF-κB and MAPK pathways (Xie et al. 2019). Naringenin could reduce the expression levels of IL-6 and TNF-α in the colon mucosa by inactivating NF-κB p65 in macrophages (Dou et al. 2013). Similarly, naringenin inhibited the expression level of TNF-α-induced TLR2 in differentiated adipocytes (Hasnat et al. 2015), as well as the levels of IL-33, TNF-α, and IL-1β in the paw skin (Pinho-Ribeiro et al. 2016). Diosmin inhibited LPS-induced NF-κB activation in cells (Qiao et al. 2016) and animal models (Chung et al. 2020). Nevertheless, diosmetin downregulated the protein expression level of inducible NO synthase in cocultured macrophages and adipocytes and suppressed the phosphorylation of MAPK translocation of p65 and p50 to the nucleus (Lee et al. 2020a, b).

Eriodictyol translocated NF-κB to the nucleus and could be considered an immune response regulator during immune responses (Huang et al. 2010). Eriodictyol also showed a potent inhibitory effect on neuroinflammatory response to experimental stroke by reducing the expression level of TNF-α and iNOS (de Oliveira Ferreira et al. 2016). Hesperetin, one of the other essential flavanones, is the primary therapeutic ingredient of the genus citrus fruit (Krishnan et al. 2017). Administration of hesperetin has been reported to inhibit NF-κB signaling pathway in diabetic myocardium and the reduced expression levels of TNF-α, IL-1β, and ICAM-1 (Yin et al. 2017).

Catechins in green tea has displayed anti-inflammatory activity (Wang et al. 2018; Lakshmi et al. 2020). Catechin treatment inhibited the expression level of NF-κB p65 protein in the mouse nasal mucosa (Pan et al. 2018). EGCG inhibited NF-κB-p65 transcriptional activity by preventing NF-κB-p65 binding to κBs (Lakshmi et al. 2020). Catechins reduced LPS-stimulated inflammation by suppressing the activation of NF-κB pathway (Wang et al. 2020a, b).

Isoflavones are a kind of phytochemical present in the soybean. It is well recognized that inflammation has been modulated by dietary soybean product. Genistein, the primary active isoflavones in soybean, have shown the anti-inflammatory effects (Ji et al. 2012; Zhou et al. 2014). Genistein could suppress NF-κB activation in LPS-treated macrophages and stressed animals (Ji et al. 2012; Palanisamy et al. 2011; Zhang et al. 2017a, b; Kim et al. 2011). Genistein inhibited LPS-induced TNF-α and IL-6 release in part by inhibiting NF-κB activation. Daidzein has the potential in inhibiting NF-κB signaling pathway and was applied in breast cancer therapy (Guo et al. 2020).

4.6 B-Cell Receptor Signaling

The B-cell reactor (BCR) is the most numerous surface receptor on the B cells (Wen et al. 2019; Kurosaki and Wienands 2016). As shown in Fig. 4.7, PLC-γ2, PI3K, and MAPK pathways are the major BCR-mediated signaling pathways.

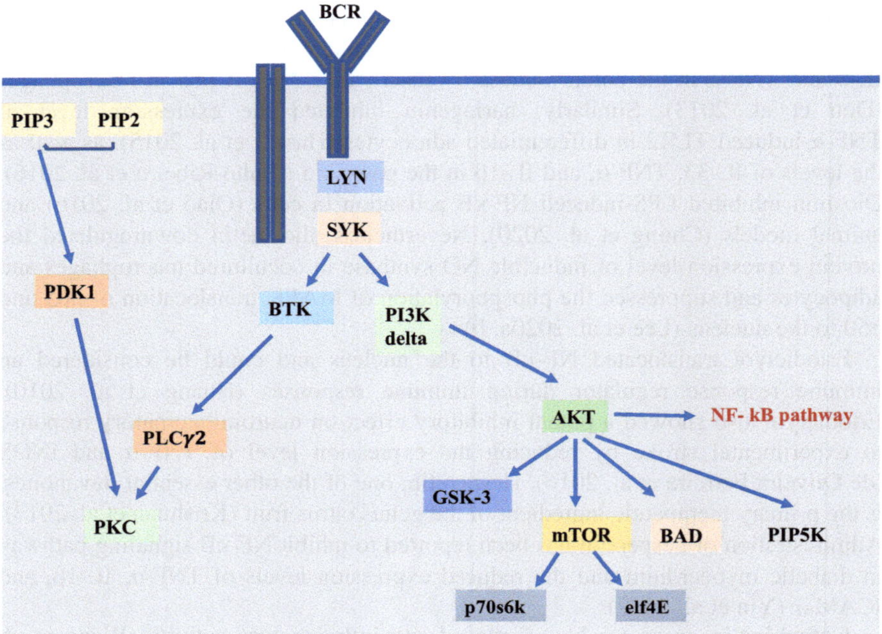

Fig. 4.7 BCR signaling pathway

4.6.1 Flavones and Flavonols

Apigenin, luteolin, and quercetin have shown anti-inflammatory effects by targeting various molecules in cellular pathways to decrease the level of TNF-α and reduce the production of NF-κB, p38, and MAPK (Hostetler et al. 2012). By inhibition of p65 phosphorylation, these flavones can interfere MAPK pathway mediated by BCR. Particularly, inhibition of p38, ERK, and CK2 activation is linked to reduce TNF-a release from macrophages (Xagorari et al. 2002). Apigenin inhibited the expression levels of HIF-1α and VEGF by blocking PI3K/Akt signaling pathway, as well as the expression levels of LPS-induced pro-inflammatory cytokines by inactivating NF-kB (Pan et al. 2009). Quercetin displayed anti-inflammatory activity by downregulating NF-κB, ERK, JNK, and ROS signaling pathway.

4.6.2 Flavanones and Flavanonols

Eriodictyol and naringenin could reduce mRNA expression levels of pro-inflammatory cytokines. Naringenin suppressed the gene expression levels of VCAM-1, ICAM-1, and MCP-1 in endothelial cells (Chanet et al. 2013). Meanwhile, hesperidin and hesperetin inhibited ERK/Nrf2 signaling pathway to reduce inflammation (Parhiz et al. 2015). Hesperetin inhibited platelet aggregation induced by collagen and arachidonic acid by inhibiting PLC-2 phosphorylation and cyclooxygenase-1 activity. Naringin inhibited TNF-a-induced expression of MMP-9 in vascular smooth muscle cells (VSMC) by modulating PI3K/AKT/mTOR/p70S6K pathway and suppressing transcriptional activity of AP-1 and NF-κB.

4.6.3 Catechins

Galloyl moiety on the C ring and the hydroxyl group on the B ring of catechins play a significant role in binding catechins to proteins. Catechol-type catechins possess a stronger binding affinity with proteins than pyrogallol-type catechins. EGCG significantly inhibited UVB-mediated activation of IKKα, degradation and phosphorylation of IκBα, and nuclear translocation of the NF-κB/p65 transcription factor (Afaq et al. 2003). In H-ras-transformed JB6 cells, EGCG also caused potent cell growth inhibition by suppressing MAPK signaling pathway and AP-1 activity (Sah et al. 2004). EGCG inhibits HER2 receptor tyrosine phosphorylation in a mouse mammary tumor cell line, and this was associated with the inhibition of PI3K/Akt kinase and NF-κB signaling pathways (Pianetti et al. 2002). EGCG suppressed the activation of the EGFR. This is associated with the inhibition of Akt and ERK activation (Sah et al. 2004). In H-ras-transformed mouse epidermal cells, treatment with EGCG caused a decrease in activation of ERK and MEK1 and decreased the association of Raf-1 with MEK1 and inhibition of AP-1 activity (Chung et al. 2001). EGCG inhibits activation of EGFR and ERK1/2, as well as AKT activity.

4.6.4 Anthocyanins

The most important anthocyanidins (cyanidin, delphinidin, petunidin, peonidin, malvidin, and pelargonidin) differ by number and position of the hydroxyl groups on the flavan nucleus. Anthocyanidins effectively decreased the expression levels of mRNA and COX-2 by suppressing C/EBP, AP-1, and NF-κB (Hou et al. 2005). Besides, C3G and anthocyanin fraction of blackberry extract reduced the expression level of iNOS by attenuating NF-kB and/or MAPK activation (Pergola et al. 2006).

The structure-activity relationship studies indicated that ortho-dihydroxy phenyl structure on the B ring of anthocyanidins is, at least, required to suppress COX-2 expression (Triebel et al. 2012). Several studies suggest that the number of hydroxyl groups on the B ring of anthocyanidins is associated with their biological activities' potency. In general, ortho-dihydroxy phenyl anthocyanidins such as delphinidin and cyanidin showed potent anti-inflammatory activity (Cespedes et al. 2010; Shih et al. 2007), while pelargonidin, peonidin, and malvidin lacking ortho-dihydroxy phenyl structure failed to show the activities as mentioned above (Hou et al. 2005). Indeed, a previous study reported by Hou et al. (2004) indicated that delphinidin, but not peonidin, could inhibit the activation of MAPKK (SEK and MEK) and MAPK (ERK and JNK) and consequently suppress AP-1 activation and cell transformation.

A further study (Hou et al. 2005) confirmed that anthocyanidins' ortho-dihydroxy structure is essential for suppressing COX-2 expression and also critical for the inhibition of tyrosine kinase and protein kinase C.

4.7 Conclusion

Phytochemicals are common in human diets and have been demonstrated with a broad spectrum of biological activities, including an anti-inflammatory property. It can attenuate inflammation by targeting different intracellular signaling pathways mediated by B-cell receptors. The anti-inflammatory mechanisms of flavonoids involve the modulation of inflammatory signaling, reduction of inflammatory molecule production, diminishing recruitment and activation of inflammatory cells, and regulation of cellular function and antioxidative property, a potential role in preventive and therapeutic roles in chronic inflammatory conditions. Extensive research of phytochemicals in strengthening the network of inflammatory response needs to be studied in the future.

References

Afaq F, Adhami VM, Ahmad N, Mukhtar H (2003) Inhibition of ultraviolet B-mediated activation of nuclear factor κ B in normal human epidermal keratinocytes by green tea Constituent (-)-epigallocatechin-3-gallate. Oncogene 22(7):1035–1044

Ahn SI, Lee JK, Youn HS (2009) Inhibition of homodimerization of toll-like receptor 4 by 6-shogaol. Mol Cells 27(2):211–215

Ali SA, Singh G, Datusalia AK (2021) Potential therapeutic applications of phytoconstituents as immunomodulators: preclinical and clinical evidences. Phytother Res 35(7):3702–3731

Allan SM et al (2001) Cytokines and acute neurodegeneration. Nat Rev Neurosci 2(10):734–744

Balagopalan L, Coussens NP, Sherman E et al (2010) The LAT story: a tale of cooperativity, coordination, and choreography. Cold Spring Harb Perspect Biol 2(8):a005512

Barioni ED et al (2013) Achyrocline satureioides (Lam.) D.C. hydroalcoholic extract inhibits neutrophil functions related to innate host defense. Evid Based Complement Alternat Med 2013:787916

Bhaskar S, Shalini V, Helen A (2011) Quercetin regulates oxidized LDL induced inflammatory changes in human PBMCs by modulating the TLR-NF-κB signaling pathway. Immunobiology 216(3):367–373

Bilanges B, Posor Y, Vanhaesebroeck B (2019) PI3K isoforms in cell signalling and vesicle trafficking. Nat Rev Mol Cell Biol 20(9):515–534

Call ME, Wucherpfermig KW (2005) The T cell receptor: critical role of the membrane environment in receptor assembly and function. Annu Rev Immunol 23:101–125

Cespedes CL, Alarcon J, Avila JG, Nieto A (2010) Anti-inflammatory Activity of Aristotelia chilensis Mol. (Stuntz)(Elaeocarpaceae). Boletin Latinoamericano y del Caribe de Plantas Medicinales y Aromaticas 9(2):127–135

Chanet A, Milenkovic D, Claude S, Maier JA, Khan MK, Rakotomanomana N, Shinkaruk S, Bérard AM, Bennetau-Pelissero C, Mazur A (2013) Flavanone metabolites decrease monocyte adhesion to TNF-α-activated endothelial cells by modulating expression of atherosclerosis-related genes. Br J Nutr 110(4):587–598

Chen RJ, Lee YH, Yeh YL, Wu WS, Ho CT, Li CY, Wang YJ (2017) Autophagy-inducing effect of pterostilbene: a prospective therapeutic/preventive option for skin diseases. J Food Drug Anal 25(1):125–133

Cheng W-E et al (2015) Berberine reduces Toll-like receptor-mediated macrophage migration by suppression of Src enhancement. Eur J Pharmacol 757:1–10

Cho JW, Lee KS, Kim CW (2007) Curcumin attenuates the expression of IL-1β, IL-6, and TNF-α as well as cyclin E in TNF-α-treated HaCaT cells; NF-κB and MAPKs as potential upstream targets. Int J Mol Med 19(3):469–474

Cho M, Choi G, Shim I et al (2019) Enhanced Rg3 negatively regulates Th1 cell responses. J Ginseng Res 43(1):49–57

Choy EH, De Benedetti F, Takeuchi T, Hashizume M, John MR, Kishimoto T (2020) Translating IL-6 biology into effective treatments. Nat Rev Rheumatol 16(6):335–345

Chtourou Y, Aouey B, Kebieche M, Fetoui H (2015) Protective role of naringin against cisplatin induced oxidative stress, inflammatory response and apoptosis in rat striatum via suppressing ROS-mediated NF-κB and P53 signaling pathways. Chem Biol Interact 239:76–86

Chung JY, Park JO, Phyu H, Dong Z, Yang CS (2001) Mechanisms of inhibition of the Ras-MAP kinase signaling pathway in 30.7 b Ras 12 cells by tea polyphenols (−)-epigallocatechin-3-gallate and theaflavin-3, 3′-digallate1. FASEB J 15(11):2022–2024

Chung S, Kim HJ, Choi HK, Park JH, Hwang JT (2020) Comparative study of the effects of diosmin and diosmetin on fat accumulation, dyslipidemia, and glucose intolerance in mice fed a high-fat high-sucrose diet. Food Sci Nutr 8(11):5976–5984

Chung-Yi C, Chiu-Li K, Chi-Ming L (2018) The cancer prevention, anti-inflammatory and anti-oxidation of bioactive phytochemicals targeting the TLR4 signaling pathway. Int J Mol Sci 19(9):2729

Clara JA, Monge C, Yang Y, Takebe N (2019) Targeting signalling pathways and the immune microenvironment of cancer stem cells—a clinical update. Nat Rev Clin Oncol 17(Suppl. 5): 204–232

Crespo I, García-Mediavilla MV, Gutiérrez B, Sánchez-Campos S, Tuñón MJ, González-Gallego J (2008) A comparison of the effects of kaempferol and quercetin on cytokine-induced pro-inflammatory status of cultured human endothelial cells. Br J Nutr 100(5):968–976

Cui C, Zong J, Sun Y, Zhang L, Ho CT, Wan X, Hou R (2018) Triterpenoid saponins from the genus Camellia: structures, biological activities, and molecular simulation for structure-activity relationship. Food Funct 9(6):3069–3091

Darnell JE Jr, Kerr IM, Stark GR (1994) Jak-STAT Pathways and Transcriptional Activation in response to IFNs and other extracellular signaling proteins. Science 264(5164):1415–1421

de Oliveira Ferreira E, Fernandes MYSD, de Lima NMR, Neves KRT, do Carmo MRS, Lima FAV, de Andrade GM (2016) Neuroinflammatory response to experimental stroke is inhibited by eriodictyol. Behav Brain Res 312:321–332

Deguine J, Barton GM (2014) MyD88: a central player in innate immune signaling. F1000prime Rep 6:97

Deng Y et al (2014) The natural product phyllanthusmin C enhances IFN-γ production by human NK cells through upregulation of TLR-mediated NF-κB signaling. J Immunol 193(6):2994

Deng L, Qi M, Li N, Lei Y, Zhang D, Chen J (2020) Natural products and their derivatives: promising modulators of tumor immunotherapy. J Leukoc Biol 108(2):493–508

Deopurkar R et al (2010) Differential effects of cream, glucose, and orange juice on inflammation, endotoxin, and the expression of toll-like receptor-4 and suppressor of cytokine signaling-3. Diabetes Care 33(5):991–997

Diomede F, Zingariello M, Cavalcanti MF, Merciaro I, Pizzicannella J, de Isla N, Trubiani O (2017) MyD88/ERK/NFkB pathways and pro-inflammatory cytokines release in periodontal ligament stem cells stimulated by Porphyromonas gingivalis. Eur J Histochem EJH 61(2)

Dou W, Zhang J, Sun A, Zhang E, Ding L, Mukherjee S et al (2013) Protective effect of naringenin against experimental colitis via suppression of Toll-like receptor 4/NF-κB signalling. Br J Nutr 110(4):599–608

Doyle SL, O'Neill LAJ (2006) Toll-like receptors: from the discovery of NFkB to new insights into transcriptional regulations in innate immunity. Biochem Pharmacol 72(9):1102–1113

Duodu KG, Awika JM (2019) Phytochemical-related health-promoting attributes of sorghum and millets. In: Sorghum and millets. AACC International Press, pp 225–258

Ehteshamfar SM, Akhbari M, Afshari JT et al (2020) Anti-inflammatory and immune-modulatory impacts of berberine on activation of autoreactive T cells in autoimmune inflammation. J Cell Mol Med 24(23):13573–13588

Fallah AA, Sarmast E, Fatehi P, Jafari T (2020) Impact of dietary anthocyanins on systemic and vascular inflammation: systematic review and meta-analysis on randomised clinical trials. Food Chem Toxicol 135:110922

Fu Y et al (2014) Curcumin attenuates inflammatory responses by suppressing TLR4-mediated NF-κB signaling pathway in lipopolysaccharide-induced mastitis in mice. Int Immunopharmacol 20(1):54–58

Gadina M, Chisolm DA, Philips RL, McInness IB, Changelian PS, O'Shea JJ (2020) Translating JAKs to Jakinibs. J Immunol 204(8):2011–2020

Galleggiante V, De Santis S, Cavalcanti E, Scarano A, De Benedictis M, Serino G, Kunde D (2017) Dendritic cells modulate iron homeostasis and inflammatory abilities following quercetin exposure. Curr Pharm Des 23(14):2139–2146

Galon J, Bruni D (2020) Tumor immunology and tumor evolution: intertwined histories. Immunity 52(1):55–81

Gao X, Kuo J, Jiang H, Deeb D, Liu Y, Divine G, Gautam SC (2004) Immunomodulatory activity of curcumin: suppression of lymphocyte proliferation, development of cell-mediated cytotoxicity, and cytokine production in vitro. Biochem Pharmacol 68(1):51–61

Garbers C, Aparicio-Siegmund S, Rose-John S (2015) The IL-6/gp130/STAT3 signaling axis: recent advances towards specific inhibition. Curr Opin Immunol 34:75–82

Garcia KC, Adams JJ, Feng D et al (2009) The molecular basis of TCR germline bias for MHC is surprisingly simple. Nat Immunol 10(2):143–147

Germain RN, Stefanova I (1999) The dynamics of T cell receptor signaling: complex orchestration and the key roles of tempo and cooperation. Annu Rev Immunol 17:467–522

Ghanim H et al (2010) Orange juice neutralizes the proinflammatory effect of a high-fat, high-carbohydrate meal and prevents endotoxin increase and Toll-like receptor expression. Am J Clin Nutr 91(4):940–949

Gholami A, Baradaran HR, Hariri M (2020) Can soy isoflavones plus soy protein change serum levels of interlukin-6? A systematic review and meta-analysis of randomized controlled trials. Phytother Res 35(3):1147–1162

Gómez-Zorita S, González-Arceo M, Trepiana J, Aguirre L, Crujeiras AB, Irles E, Portillo MP (2020) Comparative effects of pterostilbene and its parent compound resveratrol on oxidative stress and inflammation in steatohepatitis induced by high-fat high-fructose feeding. Antioxidants 9(11):1042

Grudzien M, Rapak A (2018) Effect of natural compounds on NK cell activation. J Immunol Res 2018:1–11

Guillermo García-Laínez MS, García-Bayarri V, Orzáez M (2017) Identification and validation of uterine stimulant methylergometrine as a potential inhibitor of caspase-1 activation. Apoptosis 22(10):1310–1318

Guo S, Wang Y, Li Y, Li Y, Feng C, Li Z (2020) Daidzein-rich isoflavones aglycone inhibits lung cancer growth through inhibition of NF-κB signaling pathway. Immunol Lett 222:67–72

Haghighatdoost F, Hariri M (2019) Can resveratrol supplement change inflammatory mediators? A systematic review and meta-analysis on randomized clinical trials. Eur J Clin Nutr 73(3): 345–355

Haque A et al (2011) Flavopiridol inhibits lipopolysaccharide-induced TNF-α production through inactivation of nuclear factor-κB and mitogen-activated protein kinases in the MyD88-dependent pathway. Microbiol Immunol 55(3):160–167

Harvey AL, Edrada-Ebel RA, Quinn RJ (2015) The re-emergence of natural products for drug discovery in the genomics era. Nat Rev Drug Discov 14(2):111–129

Hasnat MA, Pervin M, Cha KM, Kim SK, Lim BO (2015) Anti-inflammatory activity on mice of extract of Ganoderma lucidum grown on rice via modulation of MAPK and NF-κB pathways. Phytochemistry 114:125–136

Heine M, Fischer AW, Schlein C, Jung C, Straub LG, Gottschling K et al (2018) Lipolysis triggers a systemic insulin response essential for efficient energy replenishment of activated brown adipose tissue in mice. Cell Metab 28(4):644–655.e4

Hostetler G, Riedl K, Cardenas H, Diosa-Toro M, Arango D, Schwartz S, Doseff AI (2012) Flavone deglycosylation increases their anti-inflammatory activity and absorption. Mol Nutr Food Res 56(4):558–569

Hou D-X, Kai K, Li J-J, Lin S, Terahara N, Wakamatsu M, Fujii M, Young MR, Colburn N (2004) Anthocyanidins inhibit activator protein 1 activity and cell transformation: structure–activity relationship and molecular mechanisms. Carcinogenesis 25(1):29–36

Hou D-X, Yanagita T, Uto T, Masuzaki S, Fujii M (2005) Anthocyanidins inhibit cyclooxygenase-2 expression in LPS-evoked macrophages: structure–activity relationship and molecular mechanisms involved. Biochem Pharmacol 70(3):417–425

Huang TC, Tseng KY, Tsai SS, Liu HJ, Ho CT, Lin HY et al (2010) Eriodictyol decreases very late antigen-4 (VLA-4) expression, cellular adhesion, and migration through an NFκB-dependent pathway in monocytes. J Funct Foods 2(4):263–270

Huang H et al (2011) Isolation and characterization of two flavonoids, engeletin and astilbin, from the leaves of engelhardia roxburghiana and their potential anti-inflammatory properties. J Agri Food Chem 59(9):4562–4569

Hyun KH, Gil KC, Kim SG et al (2019) Delphinidin chloride and its hydrolytic metabolite gallic acid promote differentiation of regulatory T cells and have an anti-inflammatory effect on the allograft model. J Food Sci 84(4):920–930

Ivashkiv LB (2018) IFNγ: signalling, epigenetics and roles in immunity, metabolism, disease and cancer immunotherapy. Nat Rev Immunol 18(9):545–558

Jara E, Hidalgo MA, Hancke JL et al (2014) Delphinidin activates NFAT and induces IL-2 production through SOCE in T cells. Cell Biochem Biophys 68(3):497–509

Ji G, Zhang Y, Yang Q, Cheng S, Hao J, Zhao X, Jiang Z (2012) Genistein suppresses LPS-induced inflammatory response through inhibiting NF-κB following AMP kinase activation in RAW 264.7 macrophages. PLoS One 7(12):e53101

Jiao A, Yang Z, Fu X et al (2020) Phloretin modulates human Th17/Treg cell differentiation in vitro via AMPK signaling. Biomed Res Int 2020:1–12

Johnson DE, O'Keefe RA, Grandis JR (2018) Targeting the IL-6/JAK/STAT3 signalling axis in cancer. Nat Rev Clin Oncol 15(4):234–248

Kato M, Pu MY, Isobe KI et al (1994) Characterization of the immunoregulatory action of saikosaponin-d. Cell Immunol 159(1):15–25

Kawasaki T, Kawai T (2014) Toll-like receptor signaling pathways. Front Immunol 5:461

Keating SE et al (2007) IRAK-2 participates in multiple toll-like receptor signaling pathways to NFκB via activation of TRAF6 ubiquitination. J Biol Chem 282(46):33435–33443

Keyhanmehr N, Motedayyen H, Eskandari N (2019) The effects of silymarin and cyclosporine A on the proliferation and cytokine production of regulatory T cells. Immunol Investig 48(5): 533–548

Khan H, Ullah H, Castilho PCMF, Gomila AS, D'Onofrio G, Filosa R et al (2020) Targeting NF-κB signaling pathway in cancer by dietary polyphenols. Crit Rev Food Sci Nutr 60(16):2790–2800

Kim JM, Uehara Y, Choi YJ, Ha YM, Ye BH, Yu BP, Chung HY (2011) Mechanism of attenuation of pro-inflammatory Ang II-induced NF-κB activation by genistein in the kidneys of male rats during aging. Biogerontology 12(6):537–550

Kim TH, Ku SK, Bae JS (2012) Inhibitory effects of kaempferol-3-O-sophoroside on HMGB1-mediated proinflammatory responses. Food Chem Toxicol 50(3–4):1118–1123

Kim MH, Kim HH, Jeong JM et al (2020) Ginsenoside F2 attenuates chronic binge ethanol-induced liver injury by increasing regulatory T cells and decreasing Th17 cells. J Ginseng Res 44: 815–822

Kishimoto TI (1973) Regulation of antibody response in vitro VII. Enhancing soluble factors for IgG and IgE antibody response. J Immunol 111:1194–1205

Krishnan G, Subramaniyan J, Subramani PC, Muralidharan B, Thiruvengadam D (2017) Hesperetin conjugated PEGylated gold nanoparticles exploring the potential role in anti-inflammation and anti-proliferation during diethylnitrosamine-induced hepatocarcinogenesis in rats. Asian J Pharmaceut Sci 12(5):442–455

Kumar A, Rimando AM, Levenson AS (2017) Resveratrol and pterostilbene as a microRNA-mediated chemopreventive and therapeutic strategy in prostate cancer. Ann N Y Acad Sci 1403(1):15–26

Kumari N, Dwarakanath BS, Das A, Bhatt AN (2016) Role of interleukin-6 in cancer progression and therapeutic resistance. Tumour Biol 37(9):11553–11572

Kurosaki T, Wienands J (2016) B cell receptor signaling. In: Current topics in microbiology and immunology, 1st edn. Springer International Publishing: Imprint: Springer, Cham. (pp. 1 online resource (X, 231 pages))

Lai X, Cao M, Song X, Jia R, Zou Y, Li L, Ye G (2017) Resveratrol promotes recovery of immune function of immunosuppressive mice by activating JNK/NF-κB pathway in splenic lymphocytes. Can J Physiol Pharmacol 95(6):763–767

Lakshmi SP, Reddy AT, Kodidhela LD, Varadacharyulu NC (2020) The tea catechin epigallocatechin gallate inhibits NF-κB-mediated transcriptional activation by covalent modification. Arch Biochem Biophys 695:108620

Lee EJ, Ko E, Lee J et al (2004) Ginsenoside Rg1 enhances CD4+ T-cell activities and modulates Th1/Th2 differentiation. Int Immunopharmacol 4(2):235–244

Lee KW, Bode AM, Dong Z (2011) Molecular targets of phytochemicals for cancer prevention. Nat Rev Cancer 11(3):211–218

Lee WH et al (2014) Specific oligopeptides in fermented soybean extract inhibit NF-κB-dependent iNOS and cytokine induction by toll-like receptor ligands. J Med Food 17(11):1239–1246

Lee HS, Kim EN, Jeong GS (2020a) Oral administration of liquiritigenin confers protection from atopic dermatitis through the inhibition of T cell activation. Biomolecules 10(5)

Lee H, Sung J, Kim Y, Jeong HS, Lee J (2020b) Inhibitory effect of diosmetin on inflammation and lipolysis in coculture of adipocytes and macrophages. J Food Biochem 2020:e13261

Lemaitre B et al (1996) The dorsoventral regulatory gene cassette spatzle/toll/cactus controls the potent antifungal response in drosophila adults. Cell 86(6):973–983

Leonard WJ, O'Shea JJ (1998) Jaks and STATs: biological implications. Annu Rev Immunol 16: 293–322

Leung CY, Liu L, Wong RNS et al (2005) Saikosaponin-d inhibits T cell activation through the modulation of PKC0, JNK, and NF-κB transcription factor. Biochem Biophys Res Commun 338(4):1920–1927

Li HY et al (2012) Role of baicalin in regulating Toll-like receptor 2/4 after ischemic neuronal injury. Chin Med J 9:1586–1593

Li J, Deng R, Hua X, Zhang L, Lu F, Coursey TG, Li DQ (2016) Blueberry component pterostilbene protects corneal epithelial cells from inflammation via anti-oxidative pathway. Sci Rep 6:19408

Li J, Zhao F, Wang Y, Chen J, Tao J, Tian G et al (2017) Gut microbiota dysbiosis contributes to the development of hypertension. Microbiome 5:14.

Li HJ, Wu NL, Lee GA, Hung CF (2018) The therapeutic potential and molecular mechanism of isoflavone extract against psoriasis. Sci Rep 8(1):6335

Lingappan K (2018) NF-κB in oxidative stress. Curr Opin Toxicol 7:81–86

Liu J, Huang H, Huang Z, Ma Y, Zhang L, He Y, Zheng X (2019) Eriocitrin in combination with resveratrol ameliorates LPS-induced inflammation in RAW264. 7 cells and relieves TPA-induced mouse ear edema. J Funct Foods 56:321–332

Luo W et al (2012) Baicalin downregulates porphyromonas gingivalis lipopolysaccharide-upregulated IL-6 and IL-8 expression in human oral keratinocytes by negative regulation of TLR signaling. PLoS One 7(12):e51008

Lv S, Yi PF, Shen HQ et al (2014) Ginsenoside Rh2-B1 stimulates cell proliferation and IFN-γ production by activating the p38 MAPK and ERK-dependent signaling pathways in CTLL-2 cells. Immunopharmacol Immunotoxicol 36(1):43–51

Ma J, Chan CC, Huang WC, Kuo ML (2020) Berberine Inhibits Pro-inflammatory Cytokine-induced IL-6 and CCL11 Production via Modulation of STAT6 Pathway in Human Bronchial Epithelial Cells. Int J Med Sci 17(10):1464–1473

Machova Urdzikova L, Karova K, Ruzicka J, Kloudova A, Shannon C, Dubisova J, Jendelova P (2016) The anti-inflammatory compound curcumin enhances locomotor and sensory recovery after spinal cord injury in rats by immunomodulation. Int J Mol Sci 17(1):49

Mao X, Ren Z, Parker GN, Sondermann H, Pastorello MA, Wang W, McMurray JS, Demeler B, Darnell JE Jr, Chen X (2005) Structural bases of unphosphorylated STAT1 association and receptor binding. Mol Cell 17(6):761–771

Mcgettrick AF, O'Neill LA (2010) Localisation and trafficking of Toll-like receptors: an important mode of regulation. Curr Opin Immunol 22(1):20–27

Mertens C, Darnell JE Jr (2007) SnapShot: JAK-STAT signaling. Cell 131(3):612

Michaud-Levesque J, Bousquet-Gagnon N, Beliveau R (2012) Quercetin abrogates IL-6/STAT3 signaling and inhibits glioblastoma cell line growth and migration. Exp Cell Res 318(8): 925–935

Mihara M, Hashizume M, Yoshida H, Suzuki M, Shiina M (2012) IL-6/IL-6 receptor system and its role in physiological and pathological conditions. Clin Sci (Lond) 122(4):143–159

Miao LC, Tao HX, Peng Y et al (2019) The anti-inflammatory effects of purslane extract by partial suppression on NF-κB and MAPK activation. Food Chem 290:239–245

Morris R, Kershaw NJ, Babon JJ (2018) The molecular details of cytokine signaling via the JAK/STAT pathway. Protein Sci 27(12):1984–2009

Nafees S, Rashid S, Ali N, Hasan SK, Sultana S (2015) Rutin ameliorates cyclophosphamide induced oxidative stress and inflammation in Wistar rats: role of NFκB/MAPK pathway. Chem Biol Interact 231:98–107

Nguyen NH, Nguyen CT (2019) Pharmacological effects of ginseng on infectious diseases. Inflammopharmacology 27(5):871–883

Nika K, Soldani C, Salek M et al (2010) Constitutively active lck kinase in T cells drives antigen receptor signal transduction. Immunity 32(6):766–777

Noha N et al (2014) Anti-inflammatory activity of fruit fractions in vitro, mediated through toll-like receptor 4 and 2 in the context of inflammatory bowel disease. Nutrients 6(11)

Nunes C, Almeida L, Barbosa RM, Laranjinha J (2017) Luteolin suppresses the JAK/STAT pathway in a cellular model of intestinal inflammation. Food Funct 8(1):387–396

Orban C, Vasarhelyi Z, Bajnok A et al (2018) Effects of caffeine and phosphodiesterase inhibitors on activation of neonatal T lymphocytes. Immunobiology 223(11):627–633

Palanisamy N, Kannappan S, Anuradha CV (2011) Genistein modulates NF-κB-associated renal inflammation, fibrosis and podocyte abnormalities in fructose-fed rats. Eur J Pharmacol 667(1–3):355–364

Pan M-H, Lai C-S, Dushenkov S, Ho C-T (2009) Modulation of inflammatory genes by natural dietary bioactive compounds. J Agric Food Chem 57(11):4467–4477

Pan W, Yu H, Huang S, Zhu P (2016) Resveratrol protects against TNF-α-induced injury in human umbilical endothelial cells through promoting sirtuin-1-induced repression of NF-KB and p38 MAPK. PLoS One 11(1):e0147034

Pan Z, Zhou Y, Luo X, Ruan Y, Zhou L, Wang Q, Chen J (2018) Against NF-κB/thymic stromal lymphopoietin signaling pathway, catechin alleviates the inflammation in allergic rhinitis. Int Immunopharmacol 61:241–248

Pan P, Huang YW, Oshima K, Yearsley M, Zhang J, Arnold M et al (2019) The immunomodulatory potential of natural compounds in tumor-bearing mice and humans. Crit Rev Food Sci Nutr 59(6):992–1007

Parhiz H, Roohbakhsh A, Soltani F, Rezaee R, Iranshahi M (2015) Antioxidant and anti-inflammatory properties of the citrus flavonoids hesperidin and hesperetin: an updated review of their molecular mechanisms and experimental models. Phytother Res 29(3):323–331

Park SJ, Youn HS (2010) Suppression of homodimerization of toll-like receptor 4 by isoliquiritigenin. Phytochemistry 71(14–15):1736–1740

Park KH, Gu DR, So HS et al (2015) Dual role of cyanidin-3-glucoside on the differentiation of bone cells. J Dent Res 994(12):1676–1683

Park SH, Kim JY, Cheon YH et al (2016) Protocatechuic acid attenuates osteoclastogenesis by downregulating JNK/c-Fos/NFATc1 signaling and prevents inflammatory bone loss in mice. Phytother Res 30(4):604–612

Patra MC, Shah M, Choi S (2020) Toll-like receptor-induced cytokines as immunotherapeutic targets in cancers and autoimmune diseases. Semin Cancer Biol 64:61–82

Pedro A, Ruiz Dirk H (2006) Functional diversity of flavonoids in the inhibition of the proinflammatory NF-κB IRF and Akt signaling pathways in murine intestinal epithelial cells. J Nutr 136(3):664–671. https://doi.org/10.1093/jn/136.3.664

Peluso I, Yarla NS, Ambra R, Pastore G, Perry G (2017) MAPK signalling pathway in cancers: olive products as cancer preventive and therapeutic agents. Semin Cancer Biol 56:185–195

Pérez-Berezo T et al (2011) Cocoa-enriched diets modulate intestinal and systemic humoral immune response in young adult rats. Mol Nutr Food Res 55(Supplement 1):S56–S66

Pérez-Berezo T et al (2012) Mechanisms involved in down-regulation of intestinal IgA in rats by high cocoa intake. J Nutrit Biochem 23(7):838–844

Pérez-Jiménez J, Díaz-Rubio ME, Saura-Calixto F (2013) Non-extractable polyphenols, a major dietary antioxidant: occurrence, metabolic fate and health effects. Nutr Res Rev 26(02):118–129

Pergola C, Rossi A, Dugo P, Cuzzocrea S, Sautebin L (2006) Inhibition of nitric oxide biosynthesis by anthocyanin fraction of blackberry extract. Nitric Oxide 15(1):30–39

Pianetti S, Guo S, Kavanagh KT, Sonenshein GEJCR (2002) Green tea polyphenol epigallocatechin-3 gallate inhibits Her-2/neu signaling, proliferation, and transformed phenotype of breast cancer cells. Cancer Res 62(3):652–655

Pinho-Ribeiro FA, Zarpelon AC, Mizokami SS, Borghi SM, Bordignon J, Silva RL, Verri WA Jr (2016) The citrus flavonone naringenin reduces lipopolysaccharide-induced inflammatory pain and leukocyte recruitment by inhibiting NF-κB activation. J Nutr Biochem 33:8–14

Qi Y, Gao F, Hou L et al (2017) Anti-inflammatory and immunostimulatory activities of astragalosides. Am J Chin Med 45(6):1157–1167

Qiao H et al (2012) Luteolin downregulates TLR4, TLR5, NF-κB and p-p38MAPK expression, upregulates the p-ERK expression, and protects rat brains against focal ischemia. Brain Res 1448:71–81

Qiao J, Liu J, Jia K, Li N, Liu B, Zhang Q, Zhu R (2016) Diosmetin triggers cell apoptosis by activation of the p53/Bcl-2 pathway and inactivation of the Notch3/NF-κB pathway in HepG2 cells. Oncol Lett 12(6):5122–5128

Rahimifard M et al (2017) Targeting the TLR4 signaling pathway by polyphenols: a novel therapeutic strategy for neuroinflammation. Ageing Res Rev 36:11–19

Rehman MU, Rather IA (2020) Myricetin abrogates cisplatin-induced oxidative stress, inflammatory response, and goblet cell disintegration in colon of wistar rats. Plan Theory 9(1):28

Rengasamy KRR, Khan H, Gowrishankar S et al (2019) The role of flavonoids in autoimmune diseases: therapeutic updates. Pharmacol Ther 194:107–131

Rock FL et al (1998) A family of human receptors structurally related to Drosophila Toll. Proc Natl Acad Sci U S A 95(2):588–593

Rudolph MG, Stanfield RL, Wilson IA (2006) How TCRs bind MHCs, peptides, and coreceptors. Annu Rev Immunol 24:419–466

Ruiz PA, Haller D (2006) Functional diversity of flavonoids in the inhibition of the proinflammatory NFkappaB, IRF, and Akt signaling pathways in murine intestinal epithelial cells. J Nutr 136(3):664–71
Sah JF, Balasubramanian S, Eckert RL, Rorke EA (2004) Epigallocatechin-3-gallate inhibits epidermal growth factor receptor signaling pathway evidence for direct inhibition of ERK1/2 and AKT kinases. J Biol Chem 279(13):12755–12762
Sahu BD, Tatireddy S, Koneru M, Borkar RM, Kumar JM, Kuncha M, Sistla R (2014) Naringin ameliorates gentamicin-induced nephrotoxicity and associated mitochondrial dysfunction, apoptosis and inflammation in rats: possible mechanism of nephroprotection. Toxicol Appl Pharmacol 277(1):8–20
Samarpita S, Ganesan R, Rasool M (2020) Cyanidin prevents the hyperproliferative potential of fibroblast-like synoviocytes and disease progression via targeting IL-17A cytokine signalling in rheumatoid arthritis. Toxicol Appl Pharmacol 391:114917
Samelson LE (2002) Signal transduction mediated by the T cell antigen receptor: the role of adapter proteins. Annu Rev Immunol 20:371–394
Scanga CA et al (2002) Cutting edge: MyD88 is required for resistance to toxoplasma gondii infection and regulates parasite-induced IL-12 production by dendritic cells. J Immunol 168(12):5997–6001
Sen R, Baltimore D (1986) Inducibility of κ immunoglobulin enhancer-binding protein NF-κB by a posttranslational mechanism. Cell 47(6):921–928
Seo EJ, Fischer N, Efferth T (2018) Phytochemicals as inhibitors of NF-κB for treatment of Alzheimer's disease. Pharmacol Res 129:262–273
Shariati M, Shaygannejad V, Abbasirad F et al (2019) Silymarin restores regulatory T cells (tregs) function in multiple sclerosis (MS) patients in vitro. Inflammation 42(4):1203–1214
Shih P-H, Yeh C-T, Yen G-CJJ, o. a., & chemistry, f. (2007) Anthocyanins induce the activation of phase II enzymes through the antioxidant response element pathway against oxidative stress-induced apoptosis. J Agric Food Chem 55(23):9427–9435
Shim JH, Choi HS, Pugliese A et al (2008) (−)-Epigallocatechin gallate regulates CD3-mediated T cell receptor signaling in leukemia through the inhibition of ZAP-70 kinase. J Biol Chem 283(42):28370–28379
Shuto T et al (2010) Curcumin decreases toll-like receptor-2 gene expression and function in human monocytes and neutrophils. Biochem Biophys Res Commun 398(4):647–652
Smith-Garvin JE, Koretzky GA, Jordan MS (2009) T cell activation. Annu Rev Immunol 27: 591–619
Soundharrajan I, Kim DH, Kuppusamy P, Choi KC (2019) Modulation of osteogenic and myogenic differentiation by a phytoestrogen formononetin via p38MAPK-dependent JAK-STAT and Smad-1/5/8 signaling pathways in mouse myogenic progenitor cells. Sci Rep 9(1):9307
Spagnuolo C, Moccia S, Russo GL (2018) Anti-inflammatory effects of flavonoids in neurodegenerative disorders. Eur J Med Chem 153:105–115
Srivastava RM, Singh S, Dubey SK, Misra K, Khar A (2011) Immunomodulatory and therapeutic activity of curcumin. Int Immunopharmacol 11(3):331–341
Starr TK, Jameson SC, Hogquist KA (2003) Positive and negative selection of T cells. Annu Rev Immunol 21:139–176
Suh N, Paul S, Hao X, Simi B, Xiao H, Rimando AM, Reddy BS (2007) Pterostilbene, an active constituent of blueberries, suppresses aberrant crypt foci formation in the azoxymethane-induced colon carcinogenesis model in rats. Clin Cancer Res 13(1):350–355
Takeda K, Akira S (2004) Microbial recognition by Toll-like receptors. J Dermatol Sci 34(2):73–82
Takeda K, Akira S (2015) Toll-like receptors. Curr Protoc Immunol 109:14.12.1–14.12.10
Takeda K, Kaisho T, Akira S (2003) Toll-like receptors. Curr Protoc Immunol 21:335–376
Takeuchi O, Hoshino K, Akira S (2000) Cutting edge: TLR2-deficient and MyD88-deficient mice are highly susceptible to Staphylococcus aureus infection. J Immunol 165(10):5392–5396
Tanaka T, Narazaki M, Kishimoto T (2018) Interleukin (IL-6) immunotherapy. Cold Spring Harb Perspect Biol 10(8):a028456

Tang Y, Abe N, Qi H et al (2014) Tea catechins inhibit cell proliferation through hydrogen peroxide-dependent and -independent pathways in human T lymphocytic leukemia jurkat cells. Food Sci Technol Res 20(6):1245–1249

Teng WJ, Chen P, Zhu FY, Di K, Zhou C, Zhuang J, Cao XJ, Yang J, Deng LJ, Sun CG (2015) Effect of Rhizoma paridis total saponins on apoptosis of colorectal cancer cells and imbalance of the JAK/STAT3 molecular pathway induced by IL-6 suppression. Genet Mol Res 14(2): 5793–5803

Triebel S, Trieu H-L, Richling E (2012) Modulation of inflammatory gene expression by a bilberry (Vaccinium myrtillus L.) extract and single anthocyanins considering their limited stability under cell culture conditions. J Agric Food Chem 60(36):8902–8910

Truong VL, Jun M, Jeong WS (2018) Role of resveratrol in regulation of cellular defense systems against oxidative stress. Biofactors 44(1):36–49

Tsai HY, Ho CT, Chen YK (2017) Biological actions and molecular effects of resveratrol, pterostilbene, and 3′-hydroxypterostilbene. J Food Drug Anal 25(1):134–147

Tu CT, Li J, Wang FP et al (2012) Glycyrrhizin regulates CD4+ T cell response during liver fibrogenesis via JNK, ERK and PI3K/AKT pathway. Int Immunopharmacol 14(4):410–421

van der Merwe PA, Dushek O (2011) Mechanisms for T cell receptor triggering. Nat Rev Immunol 1 1(1):47–55

Villarino AV, Gadina M, O'Shea JJ, Kanno Y (2020) SnapShot: Jak-STAT signaling II. Cell 181(7):1696–1696.e1691

Wang Q, Li XK (2011) Immunosuppressive and anti-inflammatory activities of sinomenine. Int Immunopharmacol 11(3):373–376

Wang X, Smith C, Yin H (2013) Targeting Toll-like receptors with small molecule agents. Chem Soc Rev 42:4859–4866

Wang M, Yan SJ, Zhang HT et al (2017a) Ginsenoside Rh2 enhances the antitumor immunological response of a melanoma mice model. Oncol Lett 13(2):681–685

Wang Z, Sun B, Zhu F (2017b) Epigallocatechin-3-gallate inhibit replication of white spot syndrome virus in Scylla paramamosain. Fish Shellfish Immunol 67:612–619

Wang W, Zhang Y, Jin W, Xing Y, Yang A (2018) Catechin weakens diabetic retinopathy by inhibiting the expression of NF-κB signaling pathway-mediated inflammatory factors. Ann Clin Lab Sci 48(5):594–600

Wang B et al (2020a) Resveratrol attenuates TLR-4 mediated inflammation and elicits therapeutic potential in models of sepsis. Sci Rep 10(1):18837

Wang YJ, Chen YY, Hsiao CM, Pan MH, Wang BJ, Chen YC, Chen RJ (2020b) Induction of autophagy by pterostilbene contributes to the prevention of renal fibrosis via attenuating NLRP3 inflammasome activation and epithelial-mesenchymal transition. Front Cell Dev Biol 8:436

Wen Y, Jing Y, Yang L, Kang D, Jiang P, Li N, Cheng J, Li J, Li X, Peng Z, Sun X, Miller H, Sui Z, Gong Q, Ren B, Yin W, Liu C (2019) The regulators of BCR signaling during B cell activation. Blood Sci 1(2):119–129

Wenjun O, O'Garra A (2019) IL-10 family cytokines IL-10 and IL-22: from basic science to clinical translation. Immunity 50(4):871–891

Wong CP, Nguyen LP, Noh SK et al (2011) Induction of regulatory T cells by green tea polyphenol EGCG. Immunol Lett 139(1–2):7–13

Wu D et al (2009a) Green tea EGCG suppresses T cell proliferation through impairment of IL-2/IL-2 receptor signaling. Free Radic Biol Med 47(5):636–643

Wu Y et al (2009b) Renoprotective effect of total glucosides of paeony (TGP) and its mechanism in experimental diabetes. J Pharmacol Ences 109(1):78–87

Wu X, Shao F, Yang Y, Gu L, Zheng W, Wu X, Gu Y, Shu Y, Sun Y, Xu Q (2014) Epigallocatechin-3-gallate sensitizes IFN-gamma-stimulated CD4+ T cells to apoptosis via alternative activation of STAT1. Int Immunopharmacol 23(2):434–441

Xagorari A, Roussos C, Papapetropoulos A (2002) Inhibition of LPS-stimulated pathways in macrophages by the flavonoid luteolin. Br J Pharmacol 136(7):1058–1064

Xie C, Li X, Zhu J, Wu J, Geng S, Zhong C (2019) Magnesium isoglycyrrhizinate suppresses LPS-induced inflammation and oxidative stress through inhibiting NF-κB and MAPK pathways in RAW264. 7 cells. Bioorg Med Chem 27(3):516–524

Xiong X et al (2020) Pterostilbene reduces endothelial cell apoptosis by regulation of the Nrf2mediated TLR4/MyD88/NFκB pathway in a rat model of atherosclerosis. Exp Ther Med 20(3):2090–2098

Yadav VS, Mishra KP, Singh DP, Mehrotra S, Singh VK (2005) Immunomodulatory effects of curcumin. Immunopharmacol Immunotoxicol 27(3):485–497

Yamamoto T, Matsunami E, Komori K et al (2019) The isoflavone puerarin induces Foxp3(+) regulatory T cells by augmenting retinoic acid production, thereby inducing mucosal immune tolerance in a murine food allergy model. Biochem Biophys Res Commun 516(3):626–631

Yang Z et al (2015) Bovine TLR2 and TLR4 mediate Cryptosporidium parvum recognition in bovine intestinal epithelial cells. Microb Pathog 85:29–34

Yang L, Liu G, Zhu X, Luo Y, Shang Y, Gu XL (2019) The anti-inflammatory and antioxidant effects of leonurine hydrochloride after lipopolysaccharide challenge in broiler chicks. Poult Sci 98(4):1648–1657

Yin W, Lei Y (2018) Leonurine inhibits IL-1beta induced inflammation in murine chondrocytes and ameliorates murine osteoarthritis. Int Immunopharmacol 65:50–59

Yu L, Wang L, Chen S (2010) Endogenous toll-like receptor ligands and their biological significance. J Cell Mol Med 14(11):2592–2603

Zeng W, Jin L, Zhang F, Zhang C, Liang W (2018) Naringenin as a potential immunomodulator in therapeutics. Pharmacol Res 135:122–126

Zhang Q, Yuan L, Zhang Q, Gao Y, Liu G, Xiu M, Liu D (2015) Resveratrol attenuates hypoxia-induced neurotoxicity through inhibiting microglial activation. Int Immunopharmacol 28(1): 578–587

Zhang R, Xu J, Zhao J, Chen Y (2017a) Genistein improves inflammatory response and colonic function through NF-κB signal in DSS-induced colonic injury. Oncotarget 8(37):61385

Zhang X, Wang Y, Xiao C, Wei Z, Wang J, Yang Z, Fu Y (2017b) Resveratrol inhibits LPS-induced mice mastitis through attenuating the MAPK and NF-κB signaling pathway. Microb Pathog 107:462–467

Zhang Y, Kong W, Jiang J (2017c) Prevention and treatment of cancer targeting chronic inflammation: research progress, potential agents, clinical studies and mechanisms. Sci China Life Sci 60(6):601–616

Zhang H, Chen Y, Chen Y, Ji S, Wang T (2020) Comparison of the protective effects of resveratrol and pterostilbene against intestinal damage and redox imbalance in weanling piglets. J Anim Sci Biotechnol 11(1):1–16

Zhao W, Ma L, Cai C, Gong X (2019) Caffeine inhibits nlrp3 inflammasome activation by suppressing mapk/nf-κb and a2ar signaling in lps-induced thp-1 macrophages. Int J Biol Sci 15(8):1571–1581

Zhe M et al (2013) Curcumin inhibits LPS-induced inflammation in rat vascular smooth muscle cells in vitro via ROS-relative TLR4-MAPK/NF-κB pathways. Acta Pharmacol Sin 99(7):E13–E13

Zheng Y, Yang Y, Li Y, Xu L, Wang Y, Guo Z, Song H, Yang M, Luo B, Zheng A, Li P, Zhang Y, Ji G, Yu Y (2013) Ephedrine hydrochloride inhibits PGN-induced inflammatory responses by promoting IL-10 production and decreasing proinflammatory cytokine secretion via the PI3K/Akt/GSK3beta pathway. Cell Mol Immunol 10(4):330–337

Zheng J, Wu J, Chen J, Liu J, Lu Y, Huang C, Zeng Y (2016) Therapeutic effects of quercetin on early inflammation in hypertriglyceridemia-related acute pancreatitis and its mechanism. Pancreatology 16(2):200–210

Zhou X, Yuan L, Zhao X, Hou C, Ma W, Yu H, Xiao R (2014) Genistein antagonizes inflammatory damage induced by β-amyloid peptide in microglia through TLR4 and NF-κB. Nutrition 30(1): 90–95

Zhou Z, Zhou X, Dong Y, Li M, Xu Y (2019) Formononetin ameliorates high glucose induced endothelial dysfunction by inhibiting the JAK/STAT signaling pathway. Mol Med Rep 20(3): 2893–2901

Zhu HT et al (2014) Curcumin attenuates acute inflammatory injury by inhibiting the TLR4/ MyD88/NF-κB signaling pathway in experimental traumatic brain injury. J Neuroinflammation 11(1):59

Zielińska-Przyjemska M, Ignatowicz E (2008) Citrus fruit flavonoids influence on neutrophil apoptosis and oxidative metabolism. Phytother Res 22(12):1557–1562

Phytomolecules and Metabolomics of Immunomodulation: Recent Trends and Advances

5

Ahmed Mediani, Nurkhalida Kamal, Hamza Ahmed Pantami, Mohammed S. M. Saleh, Nabil Ali Al-Mekhlafi, Nor Hadiani Ismail, and Faridah Abas

Abstract

Plants have a vital role in human health, including various aspects, such as nutrition and medicine. The huge progress in engineering and biology leads to acceleration in research about plant medicines. Several phytochemicals of plant extracts attract significant concern for their potent uses as pharmacological tools in treating and preventing many diseases. Thus, the identification of novel and/or new therapeutically active metabolites is highly emphasized. Indeed, many holistic approaches as omics tools are applied in systems biology, aiming in the explanation and management of the biological complexity. The omics include genomics, transcriptomics, proteomics, and metabolomics. Metabolomics is the

A. Mediani (✉) · N. Kamal
Institute of Systems Biology (INBIOSIS), Universiti Kebangsaan Malaysia, Bangi, Selangor, Malaysia

H. A. Pantami
Department of Chemistry Faculty of Science, Gombe State University, Gombe, Nigeria

M. S. M. Saleh
Department of Pharmacology, Faculty of Medicine, Universiti Kebangsaan Malaysia, Kuala Lumpur, Malaysia

N. A. Al-Mekhlafi
Atta-ur-Rahman Institute for Natural Product Discovery, Puncak Alam, Selangor, Malaysia

Biochemical Technology Program, Department of Chemistry, Faculty of Applied Science, Thamar University, Thamar, Yemen

N. H. Ismail
Atta-ur-Rahman Institute for Natural Product Discovery, Puncak Alam, Selangor, Malaysia

F. Abas
Department of Food Science, Faculty of Food Science and Technology, Universiti Putra Malaysia, Serdang, Selangor, Malaysia

N. S. Sangwan et al. (eds.), *Plants and Phytomolecules for Immunomodulation*,
https://doi.org/10.1007/978-981-16-8117-2_5

most applied approach in the systems biology field because of its focus on the downstream product (metabolites), which represents phenotype of organisms and directly reflects the cells functionality. It showed significant role in identifying and evaluating biomarkers and study the variation of phytomolecules of biological system. The metabolomics can provide insights and ideas about phytomolecules for studying many immune system deficiency diseases and explaining their mechanisms of action. The stimulation of phytomolecules to the immune system is previously studied. Metabolomics reveals the important metabolites and metabolic pathways contributed to macrophage responses to certain phytomolecules, signifying novel perceptions into their immunomodulatory potentialities. However, additional evidence-based research is required about the current traditional medicine systems for both pure phytomolecules and crude extracts. Thus, this chapter explored the medicinally promising phytomolecules effect on the metabolism of human macrophages for better understanding of their biological targets and mechanisms of immunomodulation. This review may show the prominence of metabolomics approach in enhancing the understanding of immune cell responses.

Keywords

Immunomodulation · Metabolites · Metabolomics · Phytomolecules

5.1 Introduction

In order to defend the body from invading agents, the immune system is a surprisingly a complex defensive system for vertebrates. It is capable of developing cell and molecular varieties that can identify and remove infinite varieties of foreign and unwanted agents. Immune system modulation denotes any improvement in the immune response, which may include any part or step of the immune response of being triggered, expressed, amplified, or inhibited (Albers et al. 2013). The immunomodulator is therefore a material that is used for its impact on the immune system. Based on their effects, there are usually two forms of immunomodulators: immunosuppressants and immunostimulators. They have the capacity to install or protect against pathogens or tumors in an immune response. Immunopharmacology is a quite novel and emerging pharmacology division that tries for the identification of immunomodulators. Alteration of immunodeficiency, such as treating AIDS and suppressing natural or excessive immune function (treatment of graft rejection or autoimmune disease), is the possible application of immunomodulators in clinical medicine (Wilson and Sereti 2013).

Unique immunomodulators, along with antigens identified as immunological adjuvants, are administered to improve the immune response towards the ingredients of the vaccine. For example, plant-based saponin is used in veterinary medicine. In comparison, a generalized state of tolerance to pathogens or tumors is given by nonspecific immunostimulators. Cyclosporine A, a fungal compound, selectively blocks T-lymphocyte function and is used for the prevention of graft rejection (Aiyer-Harini et al. 2013). Metabolomics is an omics strategy that seeks to comprehensively examine the metabolome of the biological system, recognizing the large range of chemically transformed metabolites during metabolism (Lankadurai et al. 2013). In addition, it is an inclusive analysis of low molecular weight molecules in a biological sample, which applied various spectroscopic techniques, such as mass spectrometry (MS) and nuclear magnetic resonance (NMR). Furthermore, the correct approach to endo-metabolome is the phenotyping of metabolomic cells, the detection of metabolic events, and the analysis of cell responses to various signals (Emwas et al. 2015). This method offers the ability to observe changes in biological samples or clinical states when they are compared.

Medicinal plants and their active components are significant and prominent source of immunomodulators. Thus, the production of drugs based on natural compounds for immunomodulation activity has become an enticing mission. We cannot only find some promising immunomodulators through these studies but also explain the mechanism of some conventional medicines' clinical acts. A number of plant and fungal immunomodulatory agents have been reviewed previously (Alamgir and Uddin 2010). The current analysis focused on recent advances in plant immunomodulatory activity using metabolomics approach.

5.2 Immunomodulation Definition and Fundamentals

Immune system is a surprisingly complex defensive system inside vertebrates to defend them from infectious agents. It can produce a range of cells and molecules which are proficient in identifying and removing infinite varieties of foreign and undesirable agents. The modulation of immune system implies some improvement in the immune response that may include the activation, amplification, inhibition, or expression of any element or process of the immune response. Immunomodulation includes all clinical procedures intended to amend the immune response. Increased immune response is desirable in the prevention of immunodeficiency infections, in the fight against existing infections, and in the fight against cancer.

The diagnostic of the cause is most important aspect in immunodeficiency (e.g., malnutrition, HIV). Relevant immune defects can rarely be repaired, and consideration should be given to the development of novel immune system by allogeneic stem cell transplantation. So far, vaccination is the most efficient immunomodulatory strategy to avoid infection. New methods to modify vaccine response include the use of cytokines, viral vectors, and even "naked DNA." Efforts to move the immune response to a Th1-type phenotype are beneficial for treating known infections. In the treatment of cancer, the focus was on making cancer the preferred target of the patient's immune system. To this end, number of methods have been executed, including anti-CTL-4, tumor-specific antibodies, cytokines, and cell therapy using lymphocyte-infiltrating tumors and even stem cell transplantation. The goal of autoimmunity, allergy, and organ transplantation is to deteriorate the immune response. Selected allergies can be handled with special desensitization.

Many drugs are frequently given in autoimmunity and transplantation that can blunt all immune responses. Numerous medications with various targets are prescribed on their own and in combination to cause immunosuppression. Indeed, they interfere with the presentation of antigens (anti-CD 154, CTLA4-Ig), T-cell activation (calcineurin inhibitors, including cyclosporine A and tacrolimus), or T-cell proliferation (sirolimus, mycophenolate mofetil, leflunomide). Corticosteroids function at a variety of stages of immune response and also have strong anti-inflammatory properties. These drugs are required in a solid organ transplantation to avoid or treat rejection if it occurs. When rejection continues despite the use of immunosuppressive agents, the graft is lost, and a new organ is needed. In stem cell transplantation, the equivalent of rejection is graft-versus-host disease, a pathological disorder in which the newly acquired immune system identifies the recipient as foreign and retains an immune response against the recipient. Treatment options are very limited, and the diagnosis is reduced when grafting versus host disease does not respond to steroids.

5.3 Phytomolecules as Immunomodulatory Promising Agents

Conventional medicines have a vital and significant effect in improving, modulating, and suppressing the immune response of the host. Herbal drugs have immunomodulatory characteristics and usually function by suppressing or activating both specific and nonspecific immunities. Several medicinal plants were documented and believed to be immunomodulating (Table 5.1). These plants are used in traditional medicines and have been revealed to possess immunomodulating activities, including *Actaea racemosa*, *Asparagus racemosus*, *Barleria prionitis*, *Tinospora cordifolia*, *Chenopodium ambrosioides*, *Withania somnifera*, and *Cyrtomium*. Complementary medicines and complementary therapies have recently attracted interest in the treatment of many immune disorders. Many more of these are plant-based extracts with promising activity. An extended area of study is the evaluation of the immunomodulatory function of plant extracts. In addition, medicinal plants used for immunomodulation may offer potent alternatives to conventional chemotherapy for a variety of immunity disorders, particularly when the host defense mechanism must be triggered under conditions of impaired immune response. Phytomedicines as immunomodulatory agents for the modulation of immune-related diseases could be a potential approach to the production of new targeted drugs. In depth, plant-derived compounds that demonstrate a promising alternatives as immunomodulatory agents are explained.

5.3.1 Alkaloids

Tetrandrine is an example of alkaloid used as immunomodulatory agent. It is isolated from the tuberous root of *Stephania tetrandra*. This plant is used in traditional Chinese medicines to treat rheumatic disorders, silicosis, and hypertension. It was

Table 5.1 Medicinal plants possessing immunomodulatory properties

No.	Plant name	Part used	Active constituents	Analytical analysis	References
1	*Agave americana L., Agave americana var. marginata Trel, and Agave angustifolia Haw. cv.* (Agavaceae)	Leaves	Steroidal saponins and sapogenins, flavonoids, homoisoflavonoids, phenolic acids, fatty acids, and fatty acid amides	UPLC-MS	El-Hawary et al. (2020)
2	*Protium spruceanum* (Burseraceae)	Leaves	Procyanidin, catechin, rutin, quercitrin, isoquercitrin, and kaempferol-3-*O*-rhamnoside	LC-DAD-MS	Amparo et al. (2019)
3	*Psidium guajava Linn* (Myrtaceae)	Seeds	Proteopolysaccharide (galacturonic acid, glucuronic acid, galactose, glucose, ribose, xylose, arabinose, mannose, fucose, and rhamnose)	Pre-column derivatization HPLC	Lin and Lin (2020)
4	*Ilex latifolia Thunb* (Aquifoliaceae)	Leaves	Heteropolysaccharide	HPLC-UV, FT-IR, methylation, GC-MS analysis, and NMR	Shi et al. (2020)
5	*Carica papaya Linn* (Caricaceae)	Leaves	–	Extract	Otsuki et al. (2010)
6	*Nelumbo nucifera Gaertn* (Nymphaeaceae)	Rhizomes and seeds	–	Extract	Mukherjee et al. (2010)
7	*Terminalia catappa L.* (Combretaceae)	Stem barks	Phenolic compounds	HPLC-DAD-QTOF-MS	Abiodun et al. (2016)
8	*Acanthosicyos naudinianus* (Cucurbitaceae) *Gomphocarpus fruticosus* (Asclepiadaceae) *Cryptolepis decidua* (Apocynaceae)	Leaves Leaves Roots	Triterpenes (cucurbitacins) Glycosylated cardenolides, uscharin, and flavonol diglycoside Indole alkaloids	HPLC-PDA-ELSD-ESIMS	Du Preez et al. (2020)
9	*Oliveria decumbens vent* (Apiaceae)	Aerial parts	Thymol, carvacrol, *p*-cymene, and γ-terpinene	GC-MS	Jamali et al. (2020)
10	*Zataria multiflora* (Labiatae)	Aerial parts	Carvacrol, γ-terpinene, carvacrol, methyl ether, *p*-cymen and thymol	GC-MS	Azadi et al. (2020)

(continued)

Table 5.1 (continued)

No.	Plant name	Part used	Active constituents	Analytical analysis	References
11	*Euphorbia tirucalli L.* (Euphorbiaceae)	Latex	Triterpenes	LC-MS	Martins et al. (2020)
12	*Olea europaea* (Oleaceae)	Olive oil	Polyphenols	^{1}H NMR	Alvarez-Laderas et al. (2020)
13	*Indigofera tinctoria* (Fabaceae)	Leaves	Flavonoid chrysin	–	Boothapandi and Ramanibai (2019)
14	*Ziziphora tenuior* (Lamiaceae)	Aerial parts	Polyphenols		Shahnazi et al. (2016)
15	*Psorospermum febrifugum* (Hypericaceae)	Leaves and stem barks	Fatty acids	TLC and GC-MS	Asogwa et al. (2020)
16	*Panax quinquefolius* (Araliaceae)	Roots	Polysaccharides	HPSEC analysis and GC-MS	Ghosh et al. (2020)
17	*Tinospora crispa* (Menispermaceae)	Stem barks	Cordioside, quercetin, paullinic acid, and boldine	LC-MS	Abood et al. (2014)
18	*Aloe vera* (Liliaceae)	Leaves	Polysaccharides (Acemannan)	HPLC	Aranda-Cuevas et al. (2020)
19	*Momordica charantia L.* (Cucurbitaceae)	Leaves			Mahamat et al. (2020)
20	*Areca catechu* (Arecaceae)	Areca nut	Alkaloid, tannin, and flavonoid (catechin and quercetin)	Preliminary phytochemical screening and HPLC	Sari et al. (2020)
21	*Eurycoma longifolia* (Simaroubaceae)	Roots	Polysaccharide (Ali-1)	NMR, GC-MS, and EI-MS	He et al. (2019)

22	*Salvia officinalis* (Lamiaceae) *Satureja hortensis L.* (Lamiaceae) *Anethum graveolens* (Umbelliferae)	Aerial parts Leaves Leaves	Essential oils (majority α-thujone and camphor) Volatile oil (majority, carvacrol, γ-terpinene, and m-cymene) Volatile oil (majority, α-phellandrene and β-phellandrene)	GC-MS	Popa et al. (2020)
23	*Hypericum perforatum L.* (Hypericaceae)	Flowers and leaves	Essential oils	GC-MS	Schepetkin et al. (2020)
24	*Nitraria retusa* (Nitrariaceae)	Leaves	β-sitosterols and palmitic acid	GC-MS	Boubaker et al. (2018)
25	*Argemone mexicana L.* (Papaveraceae) *Sarcocephalus latifolius* (Rubiaceae) *Vitex doniana* (Verbenaceae)	Aerial parts Roots Leaves	Polysaccharides and monosaccharide	GS-MS	Dénou et al. (2019)
26	*Uraria crinite* (Leguminosae)	Roots	Genistein, lupinalbin A, phenolic acids, fatty acids, and steroids	^{1}H NMR	Tu et al. (2019)
27	*Dracaena angustifolia* (Asparagaceae)	Leaves	Terpenoids and flavonoids	TLC	Handayani et al. (2019)
28	*Cyclea peltate* (Menispermaceae)	Leaves	Tannin, proteins, resins, carbohydrates, alkaloids, and terpenoids	Phytochemical screening	Sukanya and Bhat (2019)
29	*Thymus vulgaris L.*, *Thymus daenensis*, and *Zataria multiflora* (Labiatae)	Aerial parts	Thymol	TLC GC-MS	Amirghofran et al. (2011)
30	*Elephantopus scaber Linn.* (Asteraceae)	Leaves	Deoxyelephantopin	TLC	Aldi et al. (2019)
31	*Galium aparine L.* (Rubiaceae)	Aerial parts	Polyphenols and iridoids	UHPLC-DAD-MS/MS	Ilina et al. (2019)
32	*Leea macrophylla* (Leeaceae)	Leaves	Phenols, flavonoid, 1,2,3-propanetriyl ester, n-hexadecanoic acid, and octadecanoic acid	Spectrophotometrically calculated at 765 and 415 nm GC-MS	Raiyan et al. (2020)

(continued)

Table 5.1 (continued)

No.	Plant name	Part used	Active constituents	Analytical analysis	References
33	*Abrus precatorius L.* (Fabaceae)	Roots	Flavonoids, isoflavones, triterpenes, aromatic carboxylic acids, and aromatic alcohols	FTIR and GC-MS	Okoro et al. (2019)
34	*Aronia melanocarpa* (Rosaceae)	Fruits	Polyphenols; cyanidin, procyanidin B2, B5, and C1 and proanthocyanidin	^{1}H-NMR and ^{13}C-NMR	Ho et al. (2014)
35	*Terminalia catappa (Combretaceae)*	Stem bark	Phenols; ellagic acid, catalagin, and gallic acid	HPLC-DAD-qTOF-MS	Abiodun et al. (2016)
36	*Phaleria nisidai* (Thymelaeaceae)	Leaves	Daphnane diterpenes; simplexin, huratoxin, acetoxyhuratoxin, stelleramacrin B, genkwanine D, and genkwanine H	LC-TOF-MS and chemometric statistical tools	Kulakowski et al. (2014)
37	*Stachytarpheta angustifolia* (Verbenaceae)	Leaves	Phenols	Phytochemical analysis	Awah et al. (2010)
38	*Limoniastrum guyonianum gall* (Plumbaginaceae)	Aerial parts	Flavonoids, tannins, and polyphenols	Quantitative analysis of extract	Krifa et al. (2013)
39	*Portulaca oleracea Linn.* (Portulaceae)	Leaves	Steroid, alkaloid, and terpenoids	Phytochemical analysis	Catap et al. (2018)
40	*Justicia spicigera Schltdl* (Acanthaceae)	Leaves	Flavonoid kaempferitrin	HPLC	Alonso-Castro et al. (2012)
41	*Vaccinium angustifolium* (Ericaceae)	Fruits	Anthocyanins and phenols	HPLC and UPLC	Taverniti et al. (2014)
42	*Allium sativum* (Liliaceae)		Garlic fructans	HPLC and NMR	Chandrashekar et al. (2011)
43	*Terminalia macroptera* (Combretaceae)	Leaves, root bark, and stem bark	Phenols and carbohydrates		Zou et al. (2014a)
44	*Stevia rebaudiana* (Asteraceae)	Leaves	Sterols, alkaloids, saponins, phenols, flavonoids, glycosides, tannins, and resins	Preliminary phytochemical screening	Shukla and Mehta (2015)

45	*Acacia nilotica L.* (Leguminosae)	Leaves	Carbohydrates, glycosides, phytosterols, phenols, saponins, and flavonoids	Phytochemical screening	Sharma et al. (2014)
46	*Petroselinum crispum* (Apiaceae)	Leaves	Essential oil	–	Yousofi et al. (2012)
47	*Punica granatum* (Punicaceae)	Fruits	Polysaccharides	^{1}HNMR, FABMS, and GC-MS	Joseph et al. (2012)
48	*Enicostema axillare* (Gentianaceae)	Whole plants	Swertiamarin	HPLC	Saravanan et al. (2012)
49	*Parkia biglobosa* (Fabaceae)	Barks	Polysaccharides	GC-MS	Zou et al. (2014b)
50	*Ligustrum purpurascens* (Oleaceae)	Leaves	Phenylethanoid glycosides	HPLC and LC-MS	Song et al. (2012)
51	*Phyllanthus amarus and Phyllanthus urinaria* (Euphorbiaceae)	Whole plants	Phyllanthin and hypophyllanthin	HPLC, NMR, and ESI-MS	Ilangkovan et al. (2013)
52	*Tinospora crispa* (Menispermaceae)	Stems	Syringin and magnoflorine	HPLC	Ahmad et al. (2015)
53	*Phyllanthus amarus* (Euphorbiaceae)	Whole plants	Gallic acid, ellagic acid, phyllanthin, hypophyllanthin, corilagin, and geraniin	HPLC	Ilangkovan et al. (2016)
54	*Zingiber zerumbet* (Zingiberaceae)	Rhizomes	Zerumbone	HRESI-MS and NMR	Jantan et al. (2019)

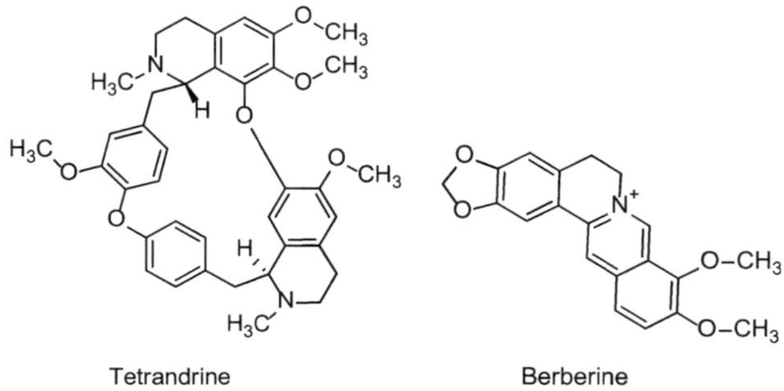

Fig. 5.1 Some alkaloid immunomodulators

proposed that tetrandrine downregulated IκBα kinases-IκBα-NF-κB signaling pathway in human peripheral blood T cell (Ho et al. 2004). Other studies have shown that tetrandrine might modulate microglial activation caused by lipopolysaccharide by inhibiting the NF-κB gun-mediated release of inflammatory factors (Xue et al. 2008). Recent study established tetrandrine as a possible hit that stimulates mesenchymal stem cells to secrete prostaglandin E2 (PGE2), a potent immunosuppressive agent via NF-κB/COX-2 signaling pathway (Yang et al. 2016).

Berberine is a natural alkaloid derived from various berry fruits. Study on combination of fingolimod, an immunomodulator approved for clinical use of multiple sclerosis with berberine on preclinical models of relapsing ulcerative colitis at reduced doses, suppressed the cytokine expression and activation of STAT3 and binding of both drugs to the sphingosine 1-phosphate receptor (Han et al. 2020). Figure 5.1 shows the chemical structure of tetrandrine and berberine as potent immunomodulator compounds.

5.3.2 Flavonoids

Figure 5.2 presents the chemical structure of several flavonoids as potent immunomodulator compounds. Naringenin is an abundant flavonoid found in various citrus fruits, such as *Citrus junos*, *Citrus paradisi,* and *Citrus sinensis* (Heo et al. 2004). Recent review by W. Zeng et al. compiled that naringenin may act as an immunomodulating agent for treating various inflammation-related disease in animal model, such as atherosclerosis, obesity, diabetes, cancer, sepsis and endotoxic shock, hepatitis, and pulmonary fibrosis (Zeng et al. 2018). For example, previous study showed the ability of naringenin as a promising immunomodulator for treating *Chlamydia trachomatis* inflammation. It was the suggested immunomodulation pathway by the modulation of TLR4, TLR2, and CD86 receptors on infected macrophages and downstream via the p38 MAPK pathway (Yilma et al. 2013). For the treatment of tissue fibrosis, Wei et al. observed a substantial reduction in atrial fibrosis and expressions of several primary fibrotic factors, such as TGF-β1,

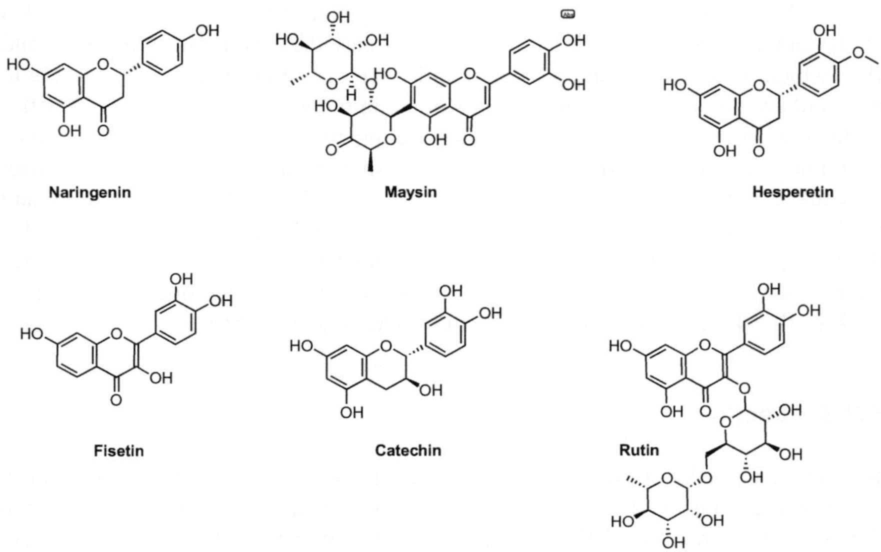

Fig. 5.2 Some flavonoid immunomodulators

Col-1A1/-3A1, and MMP-2 in spontaneous hypertensive rats. In cancer study, naringenin prevents prostate cancer metastases by blocking sodium-gated voltage channels by suppressing the mRNA expression of SCN9A gene (Aktas and Akgun 2018).

Maysin is a major constituent in the Korean tradition herb, corn silk. Previous study suggested maysin as a potential immunomodulator by stimulating macrophages to secrete TNF-5-007 in murine RAW 264.7. It is also shown that maysin induces iNOS expression by triggering the Akt, NF-κB, and MAPK signaling pathways (Lee et al. 2014). Another flavonoid, namely, salvigenin, is isolated from *Tanacetum canescens,* and this compound have shown antitumor immunomodulatory properties by showing substantial decrease in the level of IL-4 and an increase in IFN-γ in vivo and a significant decrease in splenic CD4+CD25+Foxp3+T regulatory cells (Noori et al. 2013).

Hesperetin is a natural flavonoid and one of active ingredients in *Citrus* L. However, hesperetin has poor bioavailability and is easily metabolized in the body which hindered it potential as immunomodulator. Derivative of hesperetin known as HES was synthesized and was tested in immunomodulatory effects against THP-1 cells. The findings suggested that HES contributes in the immune response by improving macrophage phagocytosis to help in the release of NO, IL-6, and IL-1β. Also, it could increase immunity by modulating the expression of Bcl-XL and Bcl-2 proteins (Ma and Li 2019).

Fisetin is a flavonol generally found in vegetables and fruits (Kimira et al. 1998). This research has successfully shown that fisetin prevents rats with adjuvant-induced arthritis by inhibiting inflammation and avoiding bone and cartilage damage. This study demonstrates the potential of fisetin to decrease the development of TNF-α and IL-6 which indicates that the protective influence of fisetin on adjuvant arthritis in rats could be impaired by improvements in the immune system (Guo et al. 2019).

In other study, application of molecular docking was done on catechin to evaluate its immunomodulatory effect. In silico study showed that catechin interacts with several chemokines and inflammatory targets such as TNF-α, IL-1β, IL-6, iNOS, and COX-2, respectively (Ganeshpurkar and Saluja 2018a). Same research group also did molecular docking study on different flavonoid known as rutin. Their findings show that rutin binds with multiple inflammatory and immunomodulatory targets TNF-α, IL-1β, IL-6, and Nos (Ganeshpurkar and Saluja 2018b).

5.3.3 Lectin

Korean mistletoe lectin (KML-IIU) and its subchains displayed immunomodulatory effect by inducing NO development in murine macrophage cells via activation of the iNOS gene expression. This finding indicated that the KML-IIU subchains are effective immunomodulators to improve the effector roles of innate immune cells (Kang et al. 2008). Banana lectin from ripe Musa acuminata was fed orally to the BALB/c male mice. The modulation of pro- and anti-inflammatory cytokines and T cells in the peripheral blood and thymus of mice has shown the immunomodulatory properties of BanLec. Cytokine analysis of peripheral mouse blood showed elevated levels of TNF-alpha, IL-17, and IL-10 and reduced levels of IFN-gamma and IL-6, while increased CD4+ and decreased CD8+ T cells were detected in the thymus following oral administration of BanLec (Sansone et al. 2016).

Lectin isolated from leaves of *Schinus terebinthifolia* Raddi or known as SteLL showed the immunomodulatory properties on mice splenocytes. The involvement of SteLL activated the cells to produce pro-inflammatory cytokines (IL-17A, TNF-α, IFN-γ, and IL-2) as well as IL-4, an anti-inflammation cytokine capable of preventing exacerbated inflammation (dos Santos et al. 2020).

5.3.4 Terpenes

Carnosic acid is the main active component of *Rosmarinus officinalis* (Grace et al. 2017). In male BALB/c mice, carnosic acid decreased the expression of NLRP3 and CASP1, along with myeloperoxidase (MPO) levels. In addition, it raised the level of erythroid 2 nuclear factor (Nrf2) and prevented Nrf2 degradation via a process that impedes the association between Cullin3 and Keap1. In addition, the reduced levels of glutathione (GSH) and superoxide dismutase (SOD) have also increased, and the levels of malondialdehyde (MDA) and iNOS in laboratory animals have decreased by carnosic acid (Yang et al. 2017).

Oridonin is the main active compound isolated from the traditional Chinese medicinal herb *Rabdosia rubescens*. By interfering with cysteine 279 of NLRP3 in

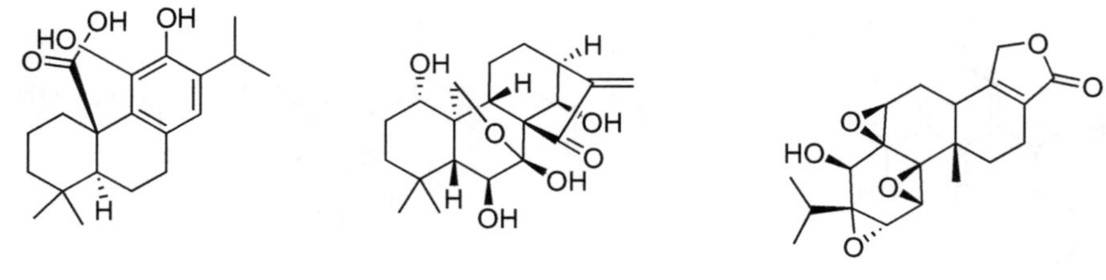

Carnosic acid **Oridonin** **Triptolide**

Fig. 5.3 Some terpenes as immunomodulator agents

the NACHT domain, oridonin demonstrated an anti-inflammatory activity in mice and HEK-293T cells, blocking the association between NLRP3 and NEK7 and inhibiting the development and activation of NLRP3 inflammasomes. It also decreased peritonitis, gouty arthritis, and type 2 diabetes in vivo by preventing NLRP3 activation (He et al. 2018).

Triptolide is a terpene isolated from *Tripterygium wilfordii* Hook F. The NLRP3-TGF1β-Smad pathway was found to be blocked by triptolide in a recent study, indicating that this compound may be an alternative choice for cardiac fibrosis by targeting the NLRP3 inflammasome (Pan et al. 2019). In other study, it was found that this compound downregulated NLRP3 by targeting hsa-miR-20b in male C57BL/6 mice and in THP-1 cells, respectively (Qian et al. 2019). Figure 5.3 shows the chemical structures of some terpene immunomodulators.

5.3.5 Phenolic and Polyphenolic Compounds

Shikonin is a major metabolite of the *Lithospermum erythrorhizon* Sieb. et Zucc. which is used in traditional Chinese medicines. The study found that shikonin suppressed the function of NF-kβ and triggered Th2 cytokines and the expression of Th1 cytokines (Andújar et al. 2010).

Thymoquinone is a quinone derivative found in medicinal *Nigella sativa*. Thymoquinone demonstrated the immunomodulatory effect by inhibiting LPS-induced proliferation of fibroblast and H_2O_2-induced 4-hydroxynonenal generation. It is also capable in inhibiting TNF-α, COX-2, IL-1β, MMP-13, and PGE2 while hindering phosphorylation of ERK1/2, NF-kβp65, and MAPK p38 (Vaillancourt et al. 2011).

Phloretin (PT), a natural dihydrochalcone present in many fruits, was shown to inhibit the activation and activity of dendritic mouse cells (DCs). Phloretin can interact with several intracellular signaling pathways in DCs caused by the lipopolysaccharide (LPS) agonist Toll-like receptor 4 (TLR4), including MAPKs (ERK, JNK, p38 MAPK), ROS, and NF-kβ, thus reducing the development of inflammatory chemokines and cytokines (Lin et al. 2014).

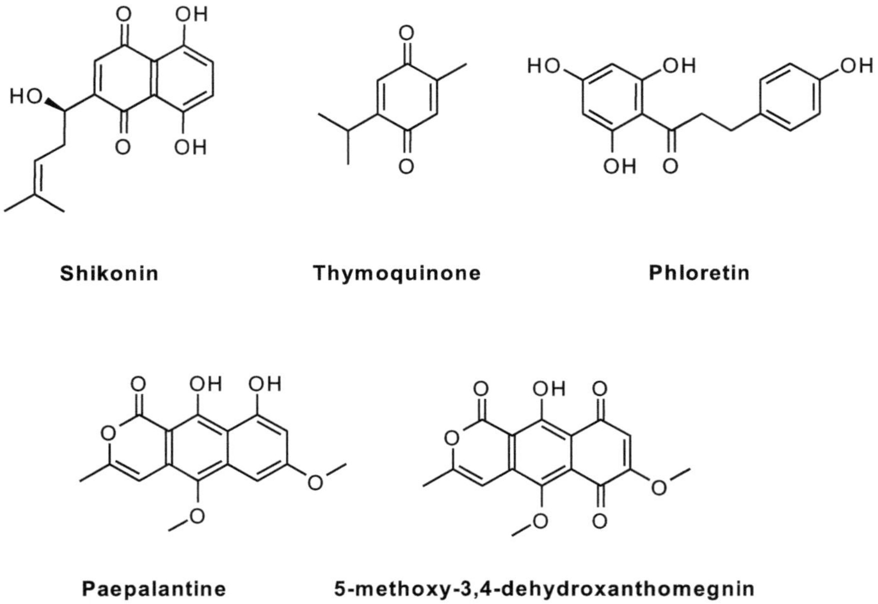

Fig. 5.4 Several phenolics as immunomodulator agents

Two naphthopyranones known as paepalantine and 5-methoxy-3,4-dehydroxanthomegnin were isolated from *Paepalanthus* sp. These compounds displayed immunomodulatory properties by synergistically suppressed pro-inflammatory cytokines (TNF-α and IL-6) and NO through iNOS inhibition and direct capture of NO (Santa Ardisson et al. 2018). Figure 5.4 presents the chemical structures of some phenolics as potent immunomodulators.

5.3.6 Polysaccharides

A group of polysaccharides known as FCPS were detected in *Ficus carica*, a medicinal plant commonly utilized in the Asian region for anticancer purposes. FCPS, as evidenced by the upregulation of CD86, CD80, CD40, and major histocompatibility complex II, might efficiently stimulate DCs and facilitate their maturation, in part via the dectin-1/Syk pathway (MHCII). It was also revealed that FCPS increased the production of dendritic cell cytokines such as IL-12, IFN-γ, IL-6, and IL-23 (Tian et al. 2014).

Inulin-type fructan was isolated from traditional Chinese medicinal plant *Platycodon grandiflorus*. Its immunomodulation properties are shown by a significant increase of anti-inflammatory factors, including IL-4 and IL-10 mRNA levels, but small changes in pro-inflammatory cytokine level, which are IL-1β and TNF-α (Pang et al. 2019). Two acidic polysaccharides, HP1 and HP2, were purified from *Ziziphus jujuba* cv. Huizao, a medicinal plant used in China. It was found that HP1 and HP2 could dramatically raise spleen and thymus indices, facilitate the development of serum hemolysin, increase macrophage phagocytic activity, and prevent mice's footpad edema, respectively (Zou et al. 2018).

A hydrophilic arabinogalactan, namely, KMCP, was isolated from plant *Ixeris polycephala*. KMCP displayed the macrophage-mediated innate immunomodulatory activity by increasing macrophage phagocytosis and NO development, triggering the NF-kappa B signaling pathway, and stimulating dose-dependent proliferation of mice spleen cells, respectively (Luo et al. 2018).

Plant polysaccharides' immunomodulatory function has piqued the interest of many scientists. Polysaccharides from Aloe arborescens leaves were isolated and elucidated in a study. Two acetylated glucomannans (AANP4 and AAAP6), one deoxy-glucogalactan (AANP5), and one deoxy-N-acetyl-[1-4]-galactosamine (AANP2) were isolated and tested for immunomodulation against cytokine modulation in vitro, including IFN-γ, IL-2, IL-12, and TNF-α. AAAP6 was the most potent polysaccharide, increasing IL-12 production by more than tenfold when compared to phytohemagglutinin as a positive control (Nazeam et al. 2020).

5.3.7 Glycoprotein

A glycoprotein was isolated from *Rhus verniciflua* Stokes and demonstrated immunomodulatory effect. According to the findings, RVS glycoprotein reduces lymphocyte proliferation, intracellular ROS production, p38 MAPK activity, and the expression of cytokine-related transcription factors (GATA-3 and T-bet), as well as IL-4 and IL-10 expression in BPA-stimulated primary cultured mouse lymphocytes.

Another glycoprotein known as DOT was isolated from *Dioscorea opposita* Thunb which commonly used as traditional Chinese medicine and is a well-known edible food. The DOT was found to increase TNF-α, IL-6, and nitric oxide development and improve macrophage pinocytosis function. In addition, DOT enhanced the expression of phosphor-p38, JNK, ERK1/2, and NF-kβ p65 protein in peritoneal macrophages (Niu et al. 2017).

5.4 Biological and Pharmaceutical Immunomodulation Assays

Assessing immunomodulation after various methods of boosting immune system is usually and comparatively better accomplished via in vitro studies as few studies involved the in vivo methods. Such methods of improving and analyzing immune parameters are commonly conducted using prophylactic treatment, which involved both the assessment of innate and adaptive immunity levels improved. Nevertheless, the prophylactic treatment is always performed in vivo, while the immunomodulation is assessed in vitro using biological products, such as tissue, urine, fecal, and blood extracts. The in vitro immunomodulation assays include lysozyme activity (LA) that measures the innate immunity level in which blood serum from sample is used in assessing the lysozyme activity in destroying bacterial pathogen. The commonly used bacterial pathogen in such assay is *Micrococcus lysodeikticus*.

Another assay on the serum bactericidal activity (SBA) assesses the extent of the serum from different sample treatment and control in inhibiting the colonization and growth of a selected bacteria. A considerably virulent bacteria *Streptococcus agalactiae* is commonly used in this biological assay. The phagocytosis activity (PA) can be also determined, which can be determined using cell suspension from lymphoid organ, preferably spleen, and can be detected by the property of uptake of particulates (supposedly foreign particles like pathogen) by the lymphoid cells (Anderson and Siwicki 1995). Moreover, the respiratory burst activity (RBA) can assess the ability of the lymphoid organ cell suspension in destroying engulfed pathogen. The lymphoproliferation assay (LPA) is conducted on lymphoid organ cell suspension (spleen) to assess the enhancement by the immunomodulator in the procreation of the lymphoid organ cells, as compared to the control group.

5.5 Application of Omics Approach in Immunomodulation Studies

Omics approach nowadays is the order of the day in addressing and analyzing numerous variations between treated group of observations comprising many variables. The idea behind these approaches is to discriminate and explore the effects of rendered treatment on organisms which are referred to as the animal model in omics studies. The "omics" technologies comprise of genomics, transcriptomics, proteomics, and metabolomics. These recent developmental scientific approaches are widely distributed across many scientific disciplines. The latter mentioned omics approach, metabolomics, is of more relevance in relation to immunomodulation because it directly studies the metabolites modulation in response to immunomodulators and because metabolites are the closest entities to phenotypes. A small variation in immune system can largely result in metabolite perturbation, which can simply be determined using available spectroscopic techniques, namely, NMR and MS. Metabolomics is a powerful tool in studies related to immunomodulation and is chosen to be well discussed in the present book edition, and other omics technologies may be given special consideration in later editions.

5.5.1 Metabolomics in Immunomodulation Studies

Metabolomics is a holistic approach to analyze the overall metabolites in various organisms of a biological system in response to external stimuli, which is applicable in studies involving cells, tissues, and biofluids of an organism, such as environmental stressors, diseases, contaminants, nutritional imbalance, and immunostimulants (Lankadurai et al. 2013). Analytical techniques commonly used in metabolomics studies are NMR and mass spectrometry (GC-MS, HPLC-HRMS, and UPLC-HRMS), and each brings its own advantages and limitations. Because of the great reproducibility of NMR-based techniques and the high sensitivity and selectivity of MS-based techniques, these technologies outperform other analytical techniques. And rather than use one technique independently, some of the studies have been using both techniques to complement each other.

The ultimate purpose of the metabolomics technique is to thoroughly examine a collection of metabolites, which are the end products of biochemical activities, as well as the changes that occur as a result of environmental and genetic interactions in a specific biological system (Fiehn 2002). The continuous progress in metabolomic research makes possible its application in different fields, such as in the fields of nutrition (Hall et al. 2008; Liu et al. 2015), environment (Soule et al. 2015), chemotaxonomy (Kim et al. 2016), and pharmacology (Pontes et al. 2017), in discovering potential biomarkers for a given disease and/or treatment (Chan et al. 2016). The search and identification of biomarkers, as well as the regular association between their concentrations and anomalies, allow not only for treatment and diagnosis but also for a better understanding of complicated biological and biochemical phenomena. Obviously, metabolites are the molecules which are closest to phenotypes, and they provide information on environmental impact on the organism because they are more sensitive to changes (Pontes et al. 2017). Several various metabolomics applications can then be applied in the search and identification of biomarkers from plant extracts (Maulidiani et al. 2015; Sumner et al. 2015) as well as in animal (Sun et al. 2015) and human (Giskeødegård et al. 2015) biofluids and tissues (Allen et al. 2015). Biomarkers can be found using different analyses via metabolomics (Aksenov et al. 2014). NMR (Gowda and Raftery 2015) and MS (Jorge et al. 2016) are the analytical tools often used in large-scale and in complex analyses because the metabolome is constantly changing and it is dependent on the environment and whether the organism is healthy or not. This characteristic allows for possible tracing of treatments, observing the unhealthy organisms retreating to their healthy state.

5.5.2 Metabolomics in Finding Chemical Markers of Immunomodulation

In immunomodulatory field, metabolomics has been embedded in various stages in drug discovery, drug actions, and disease-related study, i.e., from finding the drug candidates (chemical markers) from natural sources until identifying biomarkers in animal or human (Fig. 5.5). General workflow was proposed in finding chemical markers which is by embedding MS and/or NMR-based metabolomics in parallel with bioactivity tests and multivariate data analysis (MVDA) to annotate compounds responsible for the anti-inflammatory or immunomodulation properties. Some of the studies also incorporated *in-house* database, molecular networking, and machine learning to support the workflow. For example, Kulakowski et al. (2014) reported on using HPLC-TOF-MS/MS-based metabolomics, MVDA together with IFN-γ ELISA assays, to identify markers for immunostimulant activity, and they annotated

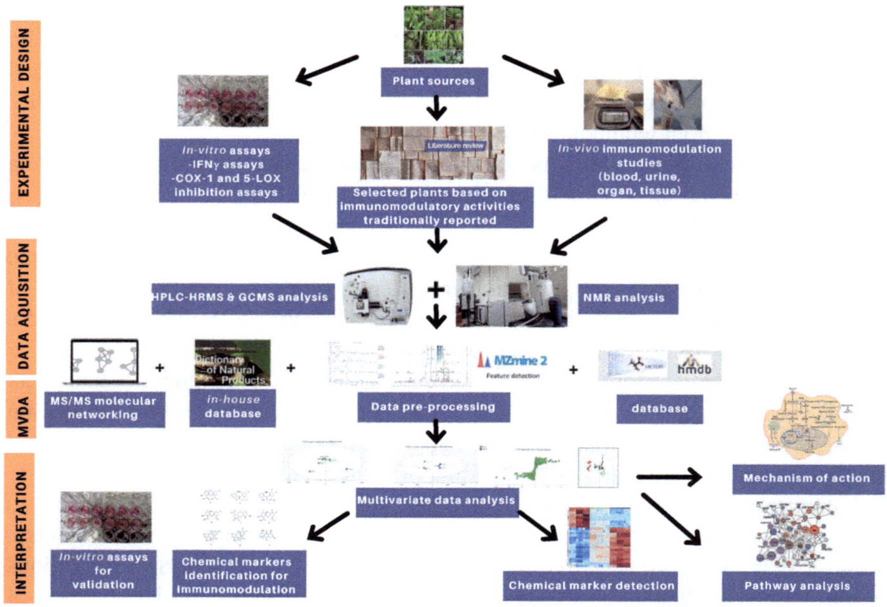

Fig. 5.5 Metabolomics workflow for finding chemical markers and biomarkers of immunomodulation from plant extracts

simplexin and other daphnane diterpenes as potent immunostimulant compounds in the fractions of *Phaleria nisidai*. To corroborate their finding, they tested immunostimulant activity of simplexin on IFN-γ ELISA assays and found that the simplexin showed strong activity without undergoing laborious and time-consuming bioassay-guided isolation work. In other study, detection of various chemical markers which reportedly act as dual inhibitors of COX-1 and 5-LOX was directly detected from 57 leaf extracts from various species of the Asteraceae family without commencing a huge isolation project and bioassays by using HPLC-ESI-HRMS-based metabolomics, incorporated with *in-house* database dedicated for Asteraceae and suitable multivariate data analysis. This study also emphasized the quality of their metabolomics-based approach by detecting tagitinine F, a minor compound in leaves of *T. diversifolia* as one of main markers responsible for the COX-1 and 5-LOX inhibition activities. However, the authors also mentioned that the limitation on this metabolomics approach is not able to differentiate between isomers and suggested using co-elution and/or MS/MS experiments, respectively (Chagas-Paula et al. 2015). Other study by Prinsloo and Vervoort (2018) employed ^{1}H-NMR and LC-MS metabolomic analysis of the plants that were reported previously with antiviral activities against herpes simplex virus (HSV), cytomegalovirus (CMV), and human immunodeficiency virus (HIV). In addition, they also examined some non-active plants with no recorded antiviral activity as control samples. Based on OPLSDA data analysis of both analytical techniques ^{1}H-NMR and LC-MS, they discovered chlorogenic acid and its building blocks as common and valuable chemical markers responsible for the clustering of the anti-HSV samples, respectively.

5.5.3 Metabolomics in Finding Biomarkers of Immunomodulation

Biomarkers are usually indicative of the state of a disease or treatment in a patient with confirmed health problem or recovering, respectively. In some instances, it can indicate the level of biological transformation as with immunomodulation. They can be analytically measured, and they differ from symptoms that can just be felt by the patient because they are subjective (Strimbu and Tavel 2010). However, biomarkers may have a more specific meaning depending on the area of research. For example, biomarkers in nutrition refer to indications about intake of nutrients such as alterations in biochemical and/or physiological profiles (Combs Jr et al. 2013). This is in contrast to plant metabolomics where the study of biosynthesis of secondary metabolites provides an idea about the expected metabolites that may occur in the biological system. The location of secondary metabolites in the plant organism depends on many factors such as specie, age (Achakzai et al. 2009), cultivation area and time (Bourgaud et al. 2001), genetic variation (Pichersky and Gang 2000), etc. In the case of biofluid samples from animals, more concern is given to studies about metabolic networks (Duarte et al. 2007); metabolic subnetworks like kidney, spleen, brain, etc.; and vital biochemical processes that help in the understanding of metabolic profiles. The key point here is that many metabolites are expected and supposedly present in healthy organisms, hence the use of control. Therefore, a slight alteration in the amount or property of specific metabolite may be determinant characteristic for a given disease, vaccine, stress, medicine, or disorder.

For disorder in urea cycle and citric acid cycle, up- or downregulated metabolites can be detected in the urine or blood samples. Example is bipolar disorder that resulted in glycine and alanine as biomarkers (Yoshimi et al. 2016). Another example is the type 2 diabetic coronary heart diseases that results in higher level of glucose compared to the healthy control profile (Liu et al. 2016). For cases of infection, certain biochemical changes that result from the interaction of the pathogen and the host cells are guides to biomarker identification. For example, in the detection of pneumonia in pleural fluid samples, glucose and lactic acid are the biomarkers because they are involved in the bacterial and cellular metabolism via glucose consumption and lactate excretion (Chiu et al. 2016). Thus, the clear understanding of related and relevant biochemical processes aids in the identification of biomarkers. The new and unique approach used in recent research to detect biomarkers of immunogenesis in animals which contrast to the classical methods of determining immune improvements includes metabolite association and pathway analysis of immunogenesis via metabolomics approach, prophylactic treatment, and post-challenge metabolomics studies.

An UPLC-Q/TOF-MS-based metabolomics study showed the latent mechanism of Platycodins Folium (PF) by diminishing depression. The identification of

21 metabolites linked to major depression disorder progress were identified in serum and hippocampus as well as 11 metabolites correlated with PF treatment. In addition, the study of metabolic pathways found that in a mice model with major depression disorder and PF treatment, there were changes in lipid metabolism, amino acid metabolism, energy metabolism, arachidonic acid metabolism, glutathione metabolism, and inositol phosphate metabolism (Wang et al. 2019). Metabolomics analysis of ginseng in vivo was widely studied for various inflammatory diseases, such as Alzheimer's disease, stress, TNBS-induced colitis, cognitive impairment, and cisplatin nephrotoxicity, and it was found that ginseng causes various metabolites to be changed such as bile acids, amino acids, lipids, and phytosphingosine (Yang et al. 2020). Recent study was reported on the identification of biomarkers for immunomodulatory mechanisms of Chinese herbal formulas, yupingfeng granules (YPFG), in cyclophosphamide-induced immunocompromised rats which was accomplished by integration of metabolomics and network pharmacology methods. The potential regulatory mechanism was found to be related to bile acid and glycerophospholipid metabolism, involving the regulation of 11 metabolites, including 8 bile acids, 1 phosphatidylserine, and 2 phosphatidylethanolamines, as determined by UPLC-Q-TOF-MS and MVDA analyses of plasma metabolites. The compound-target network relating to estrogen receptor, PPAR, MAPK, PI3K-Akt, JNK signaling pathways, and ubiquitin-mediated protein degradation was constructed between probable active components of the YPFG and cytokines (Li et al. 2021). In addition, HPLC-HRMS and GCMS metabolomics-based study of two dihydrochalcones, phlorizin and its aglycone phloretin, on human THP-1 macrophage cells discovered that the anti-inflammatory mechanisms activated by the dihydrochalcones mainly affected glycerophospholipid and sphingolipid metabolism and promoted amino acid biosynthesis (Cambeiro-Pérez et al. 2021).

5.6 Chemical Markers and Biomarkers of Immunity Boost

Numerous metabolites from different classes of compounds play an important role in the immunomodulation and immunoregulation within the biological system of an organism. These metabolites are in constant association with each other in between related pathways because of which their proper regulation contributes immensely to the development and proliferation of immune cells. Most metabolites contribute positively to the immune cells when they are upregulated, while few, like glucose and other form of simple sugars, contribute better when in minimal regulatory level or even when downregulated.

An increase in the level of choline in the immunomodulated model organism is a possible indication of mediation of different immune pathways. Its upregulation could be due to an increased demand for immunoregulation of some signaling molecules with immunomodulatory effects as a result of their tendency to activate related nuclear receptor.

Intracellular levels of many amino acids were reported to sharply increase, indicative of increased biosynthesis required for T-cell growth and proliferation. In addition, intracellular pools of fatty acids are elevated, presumably acting as a pool to generate cell membranes suitable for immune cells proliferation (Everts 2018). Thus, the upregulation of many amino acids and fatty acids such as alanine, proline, glycine, choline, linoleic acid, stearic acid, and linolenic acid in an analyzed organism is an indication of improved immune level.

Most member compounds from Krebs cycle (TCA cycle) especially the amino acids, their abundance, is critically involved in inflammatory processes and immune responses of the cells (Alqarni et al. 2019). The most important function of taurine in the immune system is to protect tissues from oxidative stress, which is linked to the pathogenesis of inflammatory disorders. The fact that a treated group had a high level of taurine shows that it has a role in innate immunity. Furthermore, the peroxidase system in activated phagocytes produces a variety of microbial and harmful oxidants. As a result of the high content of taurine in leukocytes, taurine depletion may have an effect on immune cell functioning. In fact, persistent taurine deprivation causes major immune system defects, such as considerable leukopenia, reduced neutrophil respiratory burst, and B-cell depletion from lymph nodes and spleen (Marcinkiewicz and Kontny 2014). Furthermore, to the above information, it was also reported that taurine upregulation is responsible for the smooth antioxidant activity, osmoregulation, membrane stability, and maintenance of Ca^{2+} stability. Hence, the sharp increase of taurine level in treated groups is an indication of an improved innate immunity.

Glutamine and glutamic acid are critical fuel sources for immune cells, including white blood cells (De Oliveira et al. 2016); however, its levels can be decreased due to major injuries that led to excessive blood loss (van Zanten et al. 2015). If a body's need for glutamic acid is greater than its ability to produce it, the body may break down protein stores, such as organ muscles, to release more of this amino acid (Mittendorfer et al. 1999).

The deficiency of some selected vitamins could affect the induction of proper innate as well as adaptive immune responses. For instance, as a result of riboflavin deficiency, typical symptoms called oro-oculo-genital syndrome are observed. In an LPS-stimulated lymphoproliferation model reported, riboflavin-deprived cells released significantly fewer anti-inflammatory agents, while reduction of the release of pro-inflammatory factors as well as enhancement of anti-inflammatory mediators were both recorded with riboflavin-enriched model (Mazur-Bialy et al. 2015).

Histamine is an inflammatory mediator and impact the immune system, usually as pro-inflammatory factors. It acts as a regulatory component to establish homeostasis after injury or prevent the inflammatory process (Branco et al. 2018). Histamine is a chemical messenger that promotes vasodilation and enhanced vascular permeability, as well as possibly contributing to anaphylactic reactions. It also affects cell differentiation, proliferation, hematopoiesis, and cell regeneration, among other physiological activities (O'Mahony et al. 2011). The sharp increase of histamine in treated animal model is an indication of improved innate immunity level.

Innate immune cells require glucose to develop efficient antifungal responses (Tucey et al. 2018). Infected macrophages' reliance on glucose becomes their downfall, according to Tucey et al. (2018). The fungal disease Candida albicans rapidly consumes glucose, leading macrophages to perish. Maintaining host glucose homeostasis is also critical for avoiding life-threatening fungal infection (Tucey et al. 2018). It is well known that excess glucose is also a non-healthy metabolic profile. Therefore, the substantial upregulation in glucose level in the system of a biological sample indicates swift innate immune improvement.

5.7 Metabolites Association and Pathway Analysis

Untargeted metabolomics is a technology that can be used to find new pathways for immunogenesis. This approach's driven insight and biomarkers can be used to obtain the desired information regarding immunogenesis produced by immunomodulators and the difficulties associated with it. The Metabolomics Pathway Analysis (MetPA) coupled with Kyoto Encyclopedia of Genes and Genomes (KEGG) can be employed for such pathway analysis. This combined method can identify the associated metabolites, and their hits can be deduced in relation to the pathways involved. Many pathways such as linoleic acid metabolism; riboflavin metabolism; D-glutamine and D-glutamate metabolism; starch and sucrose metabolism; α-linolenic acid metabolism; glycine, serine, and methionine metabolism; alanine, aspartate and glutamate metabolism; histidine metabolism; and taurine and hypotaurine metabolism included the most possible pathways usually showing high hits with different hit values. The pathway impact factor can be calculated by considering the matched metabolite's importance in the network. In addition, a detailed pathway analysis can be designed and proposed to show the possible association among the metabolic pathways of the most significant affected metabolites due to treatment with immunomodulators.

In MVDA, a loading plot or loading column to a corresponding score plot of a created model is generated to identify variable importance of projection (VIP) metabolites. These metabolites are selected according to a chosen VIP value. In most research, a VIP value of greater than 1 is considered appropriate. Metabolites have met the criteria of being significant metabolites in the discrimination between treatment and control groups because of their high VIP values are subjected to heatmap and box plot analysis for relative quantification. For the heatmap and box plot analysis, based on the observations from the VIP plots, the major contributing metabolites to the chemical classification of the two different groups (treatment and control) are then subjected to the pathway analysis to identify novel perturbed metabolic pathways. The data generated from VIP plot and subjected to statistical analysis in MetaboAnalyst online database should be pareto scaled prior to being subjected to Hierarchical Cluster Analysis (HCA) with Euclidean and clustering algorithm using the Ward method. Using this method, the metabolomics pattern of the key metabolites can be expressed as a square in a heatmap. These squares represent the metabolites and their corresponding contents indicated by colors based on a pareto normalized scale from −1.5 (lowest) to 1.5 (highest). For the box plots of significant individual metabolite, the black dots represent the concentration of the selected feature from all replicates. The notch indicates the 95%

confidence interval around the median of each group, defined as +/− 1.58 × IQR/sqrt(n). The notch is used to evaluate differences between groups; if the notches do not overlap, the medians are likely different. Meanwhile, the mean concentration of each group is indicated with a yellow diamond.

Metabolomics offers an opportunity to discuss the holistic therapeutic process of traditional Chinese medicine. A variety of experiments have been used to elucidate the potential mechanisms of anti-hyperuricemia, antidiabetic, inflammatory, and toxicology. Using metabolic agents, a study showed the potential luteolin and luteolin-4′-*O*-glucoside anti-hyperuricemia mechanism (Lin et al. 2018). There is also interesting research on the Shuanglong Formula mechanism for myocardial infarction (Liang et al. 2011). Other studies indicated a mechanism for the antibiotic effect of *Persicaria capitata* (Buch.-Ham. ex D. Don) H. Gross using GC-MS-based metabolic agents. The LC-ESI-QTOF-MS metabolomics method was used to increase the impact of proteinaceous molecules derived from intestinal bacteria on human peripheral blood mononuclear cells and to propose a mechanism for intervention.

5.8 Conclusions

Results from recent analysis reveal that metabolomics is a potential analytical tool for the identification of chemical markers and biomarkers in which there was induced immune stimulation via various treatment with different immunomodulators. Metabolomics provides an opportunity to discuss the holistic therapeutic process of medicinal plants and gives potent insights on the potential mechanisms. It is believed that the present chapter affords information, which paves a way for future development in related research for mitigating disease outbreak for the development of human health and aquaculture practice. This chapter also suggests an alternative research approach to the classical methods of determining immunomodulation by exploring the protective effect of immunomodulators via tandem metabolomics and metabolism pathway analysis approach.

References

Abiodun OO, Rodríguez-Nogales A, Algieri F, Gomez-Caravaca AM, Segura-Carretero A, Utrilla MP, Rodriguez-Cabezas ME, Galvez J (2016) Antiinflammatory and immunomodulatory activity of an ethanolic extract from the stem bark of *Terminalia catappa L.* (Combretaceae): in vitro and in vivo evidences. J Ethnopharmacol 192:309–319

Abood WN, Fahmi I, Abdulla MA, Ismail S (2014) Immunomodulatory effect of an isolated fraction from *Tinospora crispa* on intracellular expression of INF-γ, IL-6 and IL-8. BMC Complement Altern Med 14(1):205

Achakzai AKK, Achakzai P, Masood A, Kayani SA, Tareen RB (2009) Response of plant parts and age on the distribution of secondary metabolites on plants found in Quetta. Pak J Bot 41(5): 2129–2135

Ahmad W, Jantan I, Kumolosasi E, Bukhari SNA (2015) Immunostimulatory effects of the standardized extract of *Tinospora crispa* on innate immune responses in Wistar Kyoto rats. Drug Des Devel Ther 9:2961

Aiyer-Harini P, Ashok-Kumar H, Kumar GP, Shivakumar N (2013) An overview of immunologic adjuvants—a review. J Vaccines Vaccin 4(1):1000167

Aksenov AA, Pasamontes A, Peirano DJ, Zhao W, Dandekar AM, Fiehn O, Ehsani R, Davis CE (2014) Detection of Huanglongbing disease using differential mobility spectrometry. Anal Chem 86(5):2481–2488

Aktas HG, Akgun T (2018) Naringenin inhibits prostate cancer metastasis by blocking voltage-gated sodium channels. Biomed Pharmacother 106:770–775

Alamgir M, Uddin SJ (2010) Recent advances on the ethnomedicinal plants as immunomodulatory agents. Ethnomedicine 37(661):2

Albers R, Bourdet-Sicard R, Braun D, Calder PC, Herz U, Lambert C, Lenoir-Wijnkoop I, Meheust A, Ouwehand A, Phothirath P (2013) Monitoring immune modulation by nutrition in the general population: identifying and substantiating effects on human health. Br J Nutr 110 (S2):S1–S30

Aldi Y, Dillasamola D, Yanti GR (2019) Immunomodulator activity of ethanol extract of *Tapak liman* leaves (Elephantopus scaber Linn.). Pharm J 11(6s)

Allen PJ, Wise D, Greenway T, Khoo L, Griffin MJ, Jablonsky M (2015) Using 1-D 1 H and 2-D 1 H J-resolved NMR metabolomics to understand the effects of anemia in channel catfish (*Ictalurus punctatus*). Metabolomics 11(5):1131–1143

Alonso-Castro AJ, Ortiz-Sánchez E, Domínguez F, Arana-Argáez V, del Carmen Juárez-Vázquez M, Chávez M, Carranza-Álvarez C, Gaspar-Ramírez O, Espinosa-Reyes G, López-Toledo G (2012) Antitumor and immunomodulatory effects of *Justicia spicigera Schltdl* (Acanthaceae). J Ethnopharmacol 141(3):888–894

Alqarni AM, Dissanayake T, Nelson DJ, Parkinson JA, Dufton MJ, Ferro VA, Watson DG (2019) Metabolomic profiling of the immune stimulatory effect of eicosenoids on PMA-differentiated THP-1 cells. Vaccine 7(4):142

Alvarez-Laderas I, Ramos TL, Medrano M, Caracuel-García R, Barbado MV, Sánchez-Hidalgo M, Zamora R, Alarcón-de-la-Lastra C, Hidalgo FJ, Piruat JI (2020) Polyphenolic extract (PE) from olive oil exerts a potent immunomodulatory effect and prevents graft-versus-host disease in a mouse model. Biol Blood Marrow Transplant 26(4):615–624

Amirghofran Z, Hashemzadeh R, Javidnia K, Golmoghaddam H, Esmaeilbeig A (2011) In vitro immunomodulatory effects of extracts from three plants of the Labiatae family and isolation of the active compound (s). J Immunotoxicol 8(4):265–273

Amparo TR, Seibert JB, Mathias FAS, Vieira JFP, Soares RDOA, Freitas KM, Cabral VAR, Brandão GC, dos Santos ODH, de Souza GHB (2019) Anti-inflammatory activity of *Protium spruceanum* (Benth.) Engler is associated to immunomodulation and enzymes inhibition. J Ethnopharmacol 241(112024)

Anderson D, Siwicki A (1995) Basic hematology and serology for fish health programs. Fish Health Section, Asian Fisheries Society, Manila, Philippines

Andújar I, Recio M, Bacelli T, Giner R, Ríos J (2010) Shikonin reduces oedema induced by phorbol ester by interfering with IκBα degradation thus inhibiting translocation of NF-κB to the nucleus. Br J Pharmacol 160(2):376–388

Aranda-Cuevas B, Tamayo-Cortez J, Vargas LV, Islas-Flores I, Arana-Argáez V, Solís-Pereira S, Cuevas-Glory L, Méndez CHH (2020) Assessment of the immunomodulatory effect of Aloe vera polysaccharides extracts on macrophages functions. Emir J Food Agric 32:408–416

Asogwa FC, Ibezim A, Ntie-Kang F, Asogwa CJ, Okoye CO (2020) Anti-psoriatic and immunomodulatory evaluation of psorospermum febrifugum spach and its phytochemicals. Sci Afr 7: e00229

Awah FM, Uzoegwu PN, Oyugi JO, Rutherford J, Ifeonu P, Yao X-J, Fowke KR, Eze MO (2010) Free radical scavenging activity and immunomodulatory effect of *Stachytarpheta angustifolia* leaf extract. Food Chem 119(4):1409–1416

Azadi M, Jamali T, Kianmehr Z, Kavoosi G, Ardestani SK (2020) In-vitro (2D and 3D cultures) and in-vivo cytotoxic properties of *Zataria multiflora* essential oil (ZEO) emulsion in breast and cervical cancer cells along with the investigation of immunomodulatory potential. J Ethnopharmacol 257:112865

Boothapandi M, Ramanibai R (2019) Immunomodulatory effect of natural flavonoid chrysin (5, 7-dihydroxyflavone) on LPS stimulated RAW 264.7 macrophages via inhibition of NF-κB activation. Process Biochem 84:186–195

Boubaker J, Ben Toumia I, Sassi A, Bzouich-Mokded I, Ghoul Mazgar S, Sioud F, Bedoui A, Safta Skhiri S, Ghedira K, Chekir-Ghedira L (2018) Antitumoral potency by immunomodulation of chloroform extract from leaves of *Nitraria retusa*, Tunisian medicinal plant, via its major compounds β-sitosterol and palmitic acid in BALB/c mice bearing induced tumor. Nutr Cancer 70(4):650–662

Bourgaud F, Gravot A, Milesi S, Gontier E (2001) Production of plant secondary metabolites: a historical perspective. Plant Sci 161(5):839–851

Branco ACCC, Yoshikawa FSY, Pietrobon AJ, Sato MN (2018) Role of histamine in modulating the immune response and inflammation. Mediat Inflamm 2018:9524075

Cambeiro-Pérez N, González-Gómez X, González-Barreiro C, Pérez-Gregorio MR, Fernandes I, Mateus N, de Freitas V, Sánchez B, Martínez-Carballo E (2021) Metabolomics insights of the immunomodulatory activities of phlorizin and phloretin on human THP-1 macrophages. Molecules 26:787

Catap ES, Kho MJL, Jimenez MRR (2018) In vivo nonspecific immunomodulatory and antispasmodic effects of common purslane *(Portulaca oleracea Linn.*) leaf extracts in ICR mice. J Ethnopharmacol 215:191–198

Chagas-Paula DA, Zhang T, da Costa FB, Edrada-Ebel RA (2015) A metabolomic approach to target compounds from the asteraceae family for dual COX and LOX inhibition. Meta 5:404–430

Chan AW, Mercier P, Schiller D, Bailey R, Robbins S, Eurich DT, Sawyer MB, Broadhurst D (2016) 1 H-NMR urinary metabolomic profiling for diagnosis of gastric cancer. Br J Cancer 114(1):59–62

Chandrashekar PM, Prashanth KVH, Venkatesh YP (2011) Isolation, structural elucidation and immunomodulatory activity of fructans from aged garlic extract. Phytochemistry 72(2-3): 255–264

Chiu C-Y, Lin G, Cheng M-L, Chiang M-H, Tsai M-H, Lai S-H, Wong K-S, Hsieh S-Y (2016) Metabolomic Profiling of infectious parapneumonic effusions reveals biomarkers for guiding management of children with Streptococcus pneumoniae Pneumonia. Sci Rep 6(1):1–8

Combs GF Jr, Trumbo PR, McKinley MC, Milner J, Studenski S, Kimura T, Watkins SM, Raiten DJ (2013) Biomarkers in nutrition: new frontiers in research and application. Ann N Y Acad Sci 1278(1):1

De Oliveira DC, da Silva LF, Sartori T, Santos ACA, Rogero MM, Fock RA (2016) Glutamine metabolism and its effects on immune response: molecular mechanism and gene expression. Forum Nutr 41(1):14

Dénou A, Togola A, Inngjerdingen KT, Zhang B-Z, Ahmed A, Dafam DG, Aguiyi JC, Sanogo R, Diallo D, Paulsen BS (2019) Immunomodulatory activities of polysaccharides isolated from plants used as antimalarial in Mali. J Pharmacogn Phytother 11(2):35–42

dos Santos AJCA, da Silva Barros BR, de Souza Aguiar LM, de Siqueira Patriota LL, de Albuquerque LT, Zingali RB, Paiva PMG, Napoleão TH, de Melo CML, Pontual EV (2020) Schinus terebinthifolia leaf lectin (SteLL) is an immunomodulatory agent by altering cytokine release by mice splenocytes. 3. Biotech 10(4):1–9

Du Preez C, Gründemann C, Reinhardt J, Mumbengegwi D, Huber R (2020) Immunomodulatory effects of some Namibian plants traditionally used for treating inflammatory diseases. J Ethnopharmacol 254:112683

Duarte NC, Becker SA, Jamshidi N, Thiele I, Mo ML, Vo TD, Srivas R, Palsson BØ (2007) Global reconstruction of the human metabolic network based on genomic and bibliomic data. Proc Natl Acad Sci 104(6):1777–1782

El-Hawary SS, El-Kammar HA, Farag MA, Saleh DO, El-Dine RS (2020) Metabolomic profiling of five Agave leaf taxa via UHPLC/PDA/ESI-MS in relation to their anti-inflammatory, immunomodulatory and ulceroprotective activities. Steroids 160:108648

Emwas A-H, Luchinat C, Turano P, Tenori L, Roy R, Salek RM, Ryan D, Merzaban JS, Kaddurah-Daouk R, Zeri AC (2015) Standardizing the experimental conditions for using urine in NMR-based metabolomic studies with a particular focus on diagnostic studies: a review. Metabolomics 11(4):872–894

Everts B (2018) Metabolomics in immunology research. In: Clinical metabolomics, vol 1730. Springer, Humana Press, New York, NY, pp 29–42

Fiehn O (2002) Metabolomics—the link between genotypes and phenotypes. In: Functional genomics. Springer, pp 155–171

Ganeshpurkar A, Saluja A (2018a) In silico interaction of catechin with some immunomodulatory targets: a docking analysis. Indian J Biotechnol 17:626–631

Ganeshpurkar A, Saluja A (2018b) In silico interaction of rutin with some immunomodulatory targets: a docking analysis. Indian J Biochem Biophys 55:88–94

Ghosh R, Bryant DL, Arivett BA, Smith SA, Altman E, Kline PC, Farone AL (2020) An acidic polysaccharide (AGC3) isolated from North American ginseng (*Panax quinquefolius*) suspension culture as a potential immunomodulatory nutraceutical. Curr Res Food Sci 3:207–216

Giskeødegård GF, Hansen AF, Bertilsson H, Gonzalez SV, Kristiansen KA, Bruheim P, Mjøs SA, Angelsen A, Bathen TF, Tessem M-B (2015) Metabolic markers in blood can separate prostate cancer from benign prostatic hyperplasia. Br J Cancer 113(12):1712–1719

Gowda GN, Raftery D (2015) Can NMR solve some significant challenges in metabolomics? J Magn Reson 260:144–160

Grace MH, Qiang Y, Sang S, Lila MA (2017) One-step isolation of carnosic acid and carnosol from rosemary by centrifugal partition chromatography. J Sep Sci 40(5):1057–1062

Guo D, Zhao J, Wang H, Zhao Y (2019) Fisetin attenuates cartilage destruction in adjuvant-induced arthritis by modulating cartilage cytokine expression correlated with oxidative status in the early phase in experimental animals. Folia Biol 67(4):177–189

Hall RD, Brouwer ID, Fitzgerald MA (2008) Plant metabolomics and its potential application for human nutrition. Physiol Plant 132(2):162–175

Han Q, Tang H-z, Zou M, Zhao J, Wang L, Bian Z-x, Li Y-h (2020) Anti-inflammatory efficacy of combined natural alkaloid berberine and S1PR modulator fingolimod at low doses in ulcerative colitis preclinical models. J Nat Prod 83(6):1939–1949

Handayani N, Wahyuono S, Hertiani T, Murwanti R (2019) Immunomodulatory activity and phytochemical content determination of fractions of suji leaves (*Dracaena angustifolia* (Medik.) Roxb.). Food Res 4(1):85–90

He H, Jiang H, Chen Y, Ye J, Wang A, Wang C, Liu Q, Liang G, Deng X, Jiang W (2018) Oridonin is a covalent NLRP3 inhibitor with strong anti-inflammasome activity. Nat Commun 9(1):1–12

He P, Dong Z, Wang Q, Zhan Q-P, Zhang M-M, Wu H (2019) Structural characterization and immunomodulatory activity of a polysaccharide from *Eurycoma longifolia*. J Nat Prod 82(2): 169–176

Heo HJ, Kim D-O, Shin SC, Kim MJ, Kim BG, Shin D-H (2004) Effect of antioxidant flavanone, naringenin, from *Citrus junos* on neuroprotection. J Agric Food Chem 52(6):1520–1525

Ho GT, Bräunlich M, Austarheim I, Wangensteen H, Malterud KE, Slimestad R, Barsett H (2014) Immunomodulating activity of *Aronia melanocarpa* polyphenols. Int J Mol Sci 15(7): 11626–11636

Ho LJ, Juan TY, Chao P, Wu WL, Chang DM, Chang SY, Lai JH (2004) Plant alkaloid tetrandrine downregulates IκBα kinases-IκBα-NF-κB signaling pathway in human peripheral blood T cell. Br J Pharmacol 143(7):919–927

Ilangkovan M, Jantan I, Mesaik MA, Bukhari SNA (2016) Inhibitory effects of the standardized extract of *Phyllanthus amarus* on cellular and humoral immune responses in Balb/C mice. Phytother Res 30(8):1330–1338

Ilangkovan M, Jantan I, Mohamad HF, Husain K, Razak A, Faiz A (2013) Inhibitory effects of standardized extracts of *Phyllanthus amarus* and *Phyllanthus urinaria* and their marker compounds on phagocytic activity of human neutrophils. Evid Based Complement Alternat Med 2013:603634

Ilina T, Kashpur N, Granica S, Bazylko A, Shinkovenko I, Kovalyova A, Goryacha O, Koshovyi O (2019) Phytochemical profiles and In vitro immunomodulatory activity of ethanolic extracts from *Galium aparine L.* Plan Theory 8(12):541

Jamali T, Kavoosi G, Ardestani SK (2020) In-vitro and in-vivo anti-breast cancer activity of OEO *(Oliveria decumbens vent* essential oil) through promoting the apoptosis and immunomodulatory effects. J Ethnopharmacol 248:112313

Jantan I, Haque MA, Ilangkovan M, Arshad L (2019) Zerumbone from *Zingiber zerumbet* inhibits innate and adaptive immune responses in Balb/C mice. Int Immunopharmacol 73:552–559

Jorge TF, Rodrigues JA, Caldana C, Schmidt R, van Dongen JT, Thomas-Oates J, António C (2016) Mass spectrometry-based plant metabolomics: metabolite responses to abiotic stress. Mass Spectrom Rev 35(5):620–649

Joseph MM, Aravind S, Varghese S, Mini S, Sreelekha T (2012) Evaluation of antioxidant, antitumor and immunomodulatory properties of polysaccharide isolated from fruit rind of *Punica granatum*. Mol Med Rep 5(2):489–496

Kang TB, Yoo YC, Lee KH, Yoon HS, Her E, Song SK (2008) Korean mistletoe lectin (KML-IIU) and its subchains induce nitric oxide (NO) production in murine macrophage cells. J Biomed Sci 15(2):197–204

Kim W, Peever TL, Park J-J, Park C-M, Gang DR, Xian M, Davidson JA, Infantino A, Kaiser WJ, Chen W (2016) Use of metabolomics for the chemotaxonomy of legume-associated Ascochyta and allied genera. Sci Rep 6(1):20192

Kimira M, Arai Y, Shimoi K, Watanabe S (1998) Japanese intake of flavonoids and isoflavonoids from foods. J Epidemiol 8(3):168–175

Krifa M, Bouhlel I, Ghedira-Chekir L, Ghedira K (2013) Immunomodulatory and cellular antioxidant activities of an aqueous extract of *Limoniastrum guyonianum gall.* J Ethnopharmacol 146(1):243–249

Kulakowski DM, Wu S-B, Balick MJ, Kennelly EJ (2014) Merging bioactivity with liquid chromatography-mass spectrometry-based chemometrics to identify minor immunomodulatory compounds from a Micronesian adaptogen, *Phaleria nisidai*. J Chromatogr A 1364:74–82

Lankadurai BP, Nagato EG, Simpson MJ (2013) Environmental metabolomics: an emerging approach to study organism responses to environmental stressors. Environ Rev 21(3):180–205

Lee J, Kim S-L, Lee S, Chung MJ, Park YI (2014) Immunostimulating activity of maysin isolated from corn silk in murine RAW 264.7 macrophages. BMB Rep 47(7):382

Li M, Gao Y, Yue X, Zhang B, Zhou H, Yuan C, Wu T (2021) Integrated metabolomics and network pharmacology approach to reveal immunomodulatory mechanisms of Yupingfeng granules. J Pharm Biomed Anal 194:113660

Liang X, Chen X, Liang Q, Zhang H, Hu P, Wang Y, Luo G (2011) Metabonomic study of Chinese medicine Shuanglong formula as an effective treatment for myocardial infarction in rats. J Proteome Res 10(2):790–799

Lin C-C, Chu C-L, Ng C-S, Lin C-Y, Chen D-Y, Pan I-H, Huang K-J (2014) Immunomodulation of phloretin by impairing dendritic cell activation and function. Food Funct 5(5):997–1006

Lin H-C, Lin J-Y (2020) Characterization of guava (*Psidium guajava Linn)* seed polysaccharides with an immunomodulatory activity. Int J Biol Macromol 154:511–520

Lin Y, Liu P-G, Liang W-Q, Hu Y-J, Xu P, Zhou J, Pu J-B, Zhang H-J (2018) Luteolin-4′-O-glucoside and its aglycone, two major flavones of *Gnaphalium affine* D. Don, resist hyperuricemia and acute gouty arthritis activity in animal models. Phytomedicine 41:54–61

Liu H, Tayyari F, Khoo C, Gu L (2015) A 1H NMR-based approach to investigate metabolomic differences in the plasma and urine of young women after cranberry juice or apple juice consumption. J Funct Foods 14:76–86
Liu X, Gao J, Chen J, Wang Z, Shi Q, Man H, Guo S, Wang Y, Li Z, Wang W (2016) Identification of metabolic biomarkers in patients with type 2 diabetic coronary heart diseases based on metabolomic approach. Sci Rep 6:30785
Luo B, Dong L-M, Xu Q-L, Zhang Q, Liu W-B, Wei X-Y, Zhang X, Tan J-W (2018) Characterization and immunological activity of polysaccharides from *Ixeris polycephala*. Int J Biol Macromol 113:804–812
Ma J-l, Li C (2019) A hesperetin derivative plays a role in immunoregulatory effect on human macrophages. Cell Mol Biol 65(4):43–47
Mahamat O, Flora H, Tume C, Kamanyi A (2020) Immunomodulatory activity of *Momordica charantia L.* (Cucurbitaceae) leaf diethyl ether and methanol extracts on *Salmonella* typhi-infected mice and LPS-induced phagocytic activities of macrophages and neutrophils. Evid Based Complement Alternat Med 2020:5248346
Marcinkiewicz J, Kontny E (2014) Taurine and inflammatory diseases. Amino Acids 46(1):7–20
Martins CG, Appel MH, Coutinho DS, Soares IP, Fischer S, de Oliveira BC, Fachi MM, Pontarolo R, Bonatto SJ, Fernandes LC (2020) Consumption of latex from *Euphorbia tirucalli L.* promotes a reduction of tumor growth and cachexia, and immunomodulation in Walker 256 tumor-bearing rats. J Ethnopharmacol 255:112722
Maulidiani M, Sheikh BY, Mediani A, Wei LS, Ismail IS, Abas F, Lajis NH (2015) Differentiation of *Nigella sativa* seeds from four different origins and their bioactivity correlations based on NMR-metabolomics approach. Phytochem Lett 13:308–318
Mazur-Bialy A, Pochec E, Plytycz B (2015) Immunomodulatory effect of riboflavin deficiency and enrichment-reversible pathological response versus silencing of inflammatory activation. J Physiol Pharmacol 66(6):793–802
Mittendorfer B, Gore DC, Herndon DN, Wolfe RR (1999) Accelerated glutamine synthesis in critically Ill patients cannot maintain normal intramuscular free glutamine concentration. J Parenter Enter Nutr 23(5):243–252
Mukherjee D, Khatua TN, Venkatesh P, Saha B, Mukherjee PK (2010) Immunomodulatory potential of rhizome and seed extracts of *Nelumbo nucifera Gaertn.* J Ethnopharmacol 128(2):490–494
Nazeam JA, El-Hefnawy HM, Singab A-NB (2020) Structural elucidation of immunomodulators, acetylated heteroglycan and galactosamine, isolated from *Aloe arborescens* leaves. J Med Food 23(8)
Niu X, He Z, Li W, Wang X, Zhi W, Liu F, Qi L (2017) Immunomodulatory activity of the glycoprotein isolated from the Chinese yam (*Dioscorea opposita* Thunb). Phytother Res 31(10): 1557–1563
Noori S, Mohammad Hassan Z, Salehian O (2013) Sclareol reduces CD4+ CD25+ FoxP3+ T-reg cells in a breast cancer model in vivo. Iran J Immunol 10(1):10–21
O'Mahony L, Akdis M, Akdis CA (2011) Regulation of the immune response and inflammation by histamine and histamine receptors. J Allergy Clin Immunol 128(6):1153–1162
Okoro EE, Osoniyi OR, Jabeen A, Shams S, Choudhary M, Onajobi FD (2019) Anti-proliferative and immunomodulatory activities of fractions from methanol root extract of *Abrus precatorius L.* Clinical phytoscience 5(1):1–9
Otsuki N, Dang NH, Kumagai E, Kondo A, Iwata S, Morimoto C (2010) Aqueous extract of *Carica papaya* leaves exhibits anti-tumor activity and immunomodulatory effects. J Ethnopharmacol 127(3):760–767
Pan X-C, Liu Y, Cen Y-Y, Xiong Y-L, Li J-M, Ding Y-Y, Tong Y-F, Liu T, Chen X-H, Zhang H-G (2019) Dual role of triptolide in interrupting the NLRP3 inflammasome pathway to attenuate cardiac fibrosis. Int J Mol Sci 20(2):360

Pang D-J, Huang C, Chen M-L, Chen Y-L, Fu Y-P, Paulsen BS, Rise F, Zhang B-Z, Chen Z-L, Jia R-Y (2019) Characterization of inulin-type fructan from platycodon grandiflorus and study on its prebiotic and immunomodulating activity. Molecules 24(7):1199

Pichersky E, Gang DR (2000) Genetics and biochemistry of secondary metabolites in plants: an evolutionary perspective. Trends Plant Sci 5(10):439–445

Pontes JGM, Brasil AJM, Cruz GC, de Souza RN, Tasic L (2017) NMR-based metabolomics strategies: plants, animals and humans. Anal Methods 9(7):1078–1096

Popa M, Măruţescu L, Oprea E, Bleotu C, Kamerzan C, Chifiriuc MC, Grădişteanu Pircalabioru G (2020) In vitro evaluation of the antimicrobial and immunomodulatory activity of culinary herb essential oils as potential perioceutics. Antibiotics 9(7):428

Prinsloo G, Vervoort J (2018) Identifying anti-HSV compounds from unrelated plants using NMR and LC–MS metabolomic analysis. Metabolomics 14:1–7

Qian K, Zhang L, Shi K (2019) Triptolide prevents osteoarthritis via inhibiting hsa-miR-20b. Inflammopharmacology 27(1):109–119

Raiyan S, Rahman MA, Al Mamun MA, Asim MMH, Makki A, Hajjar D, Alelwani W, Tangpong J, Mathew B (2020) Natural compounds from *Leea macrophylla* enhance phagocytosis and promote osteoblasts differentiation by alkaline phosphatase, type 1 collagen, and osteocalcin gene expression. J Biomed Mater Res A 109:1113–1124

Sansone ACMB, Sansone M, dos Santos Dias CT, do Nascimento JRO (2016) Oral administration of banana lectin modulates cytokine profile and abundance of T-cell populations in mice. Int J Biol Macromol 89:19–24

Santa Ardisson J, Gonçalves RCR, Rodrigues RP, Kitagawa RR (2018) Antitumour, immunomodulatory activity and in silico studies of naphthopyranones targeting iNOS, a relevant target for the treatment of Helicobacter pylori infection. Biomed Pharmacother 107:1160–1165

Saravanan S, Babu NP, Pandikumar P, Raj MK, Paulraj MG, Ignacimuthu S (2012) Immunomodulatory potential of *Enicostema axillare* (Lam.) A. Raynal, a traditional medicinal plant. J Ethnopharmacol 140(2):239–246

Sari LM, Hakim RF, Mubarak Z, Andriyanto A (2020) Analysis of phenolic compounds and immunomodulatory activity of areca nut extract from Aceh, Indonesia, against *Staphylococcus aureus* infection in Sprague-Dawley rats. Veterinary World 13(1):134

Schepetkin IA, Özek G, Özek T, Kirpotina LN, Khlebnikov AI, Quinn MT (2020) Chemical composition and immunomodulatory activity of *Hypericum perforatum* essential oils. Biomol Ther 10(6):916

Shahnazi M, Azadmehr A, Andalibian A, Hajiaghaee R, Saraei M, Alipour M (2016) Protoscolicidal and immunomodulatory activity of *Ziziphora tenuior* extract and its fractions. Asian Pac J Trop Med 9(11):1062–1068

Sharma AK, Kumar A, Yadav SK, Rahal A (2014) Studies on antimicrobial and immunomodulatory effects of hot aqueous extract of *Acacia nilotica* L. leaves against common veterinary pathogens. Vet Med Int 2014:747042

Shi Z, An L, Zhang S, Li Z, Li Y, Cui J, Zhang J, Jin D-Q, Tuerhong M, Abudukeremu M (2020) A heteropolysaccharide purified from leaves of *Ilex latifolia* displaying immunomodulatory activity in vitro and in vivo. Carbohydr Polym 245:116469

Shukla S, Mehta A (2015) Comparative phytochemical analysis and in vivo immunomodulatory activity of various extracts of *Stevia rebaudiana* leaves in experimental animal model. Front Life Sci 8(1):55–63

Song X, Li C-y, Zeng Y, Wu H-q, Huang Z, Zhang J, Hong R-s, Chen X-x, Wang L-y, Hu X-p (2012) Immunomodulatory effects of crude phenylethanoid glycosides from *Ligustrum purpurascens*. J Ethnopharmacol 144(3):584–591

Soule MCK, Longnecker K, Johnson WM, Kujawinski EB (2015) Environmental metabolomics: analytical strategies. Mar Chem 177:374–387

Strimbu K, Tavel JA (2010) What are biomarkers? Curr Opin HIV AIDS 5(6):463

Sukanya B, Bhat P (2019) Phytochemical, antimicrobial, antioxidant and immunomodulatory studies of leaf extracts of *Cyclea peltata* (Lam.) Hook. f. & Thomson. GSC Biol Pharm Sci 9(3):052–063

Sumner LW, Lei Z, Nikolau BJ, Saito K (2015) Modern plant metabolomics: advanced natural product gene discoveries, improved technologies, and future prospects. Nat Prod Rep 32(2): 212–229

Sun H-Z, Wang D-M, Wang B, Wang J-K, Liu H-Y, Guan LL, Liu J-X (2015) Metabolomics of four biofluids from dairy cows: potential biomarkers for milk production and quality. J Proteome Res 14(2):1287–1298

Taverniti V, Fracassetti D, Del Bo' C, Lanti C, Minuzzo M, Klimis-Zacas D, Riso P, Guglielmetti S (2014) Immunomodulatory effect of a wild blueberry anthocyanin-rich extract in human Caco-2 intestinal cells. J Agric Food Chem 62(33):8346–8351

Tian J, Zhang Y, Yang X, Rui K, Tang X, Ma J, Chen J, Xu H, Lu L, Wang S (2014) Ficus carica polysaccharides promote the maturation and function of dendritic cells. Int J Mol Sci 15(7): 12469–12479

Tu P-C, Chan C-J, Liu Y-C, Kuo Y-H, Lin M-K, Lee M-S (2019) Bioactivity-guided fractionation and NMR-based identification of the immunomodulatory isoflavone from the roots of *Uraria crinita* (L.) Desv. ex DC. Foods 8(11):543

Tucey TM, Verma J, Harrison PF, Snelgrove SL, Lo TL, Scherer AK, Barugahare AA, Powell DR, Wheeler RT, Hickey MJ (2018) Glucose homeostasis is important for immune cell viability during Candida challenge and host survival of systemic fungal infection. Cell Metab 27(5): 988–1006.e1007

Vaillancourt F, Silva P, Shi Q, Fahmi H, Fernandes JC, Benderdour M (2011) Elucidation of molecular mechanisms underlying the protective effects of thymoquinone against rheumatoid arthritis. J Cell Biochem 112(1):107–117

van Zanten AR, Dhaliwal R, Garrel D, Heyland DK (2015) Enteral glutamine supplementation in critically ill patients: a systematic review and meta-analysis. Crit Care 19(1):294

Wang C, Lin H, Yang N, Wang H, Zhao Y, Li P, Liu J, Wang F (2019) Effects of *Platycodins folium* on depression in mice based on a UPLC-Q/TOF-MS serum assay and hippocampus metabolomics. Molecules 24:1712

Wilson EM, Sereti I (2013) Immune restoration after antiretroviral therapy: the pitfalls of hasty or incomplete repairs. Immunol Rev 254(1):343–354

Xue Y, Wang Y, Feng D-c, Xiao B-g, Xu L-y (2008) Tetrandrine suppresses lipopolysaccharide-induced microglial activation by inhibiting NF-κB pathway. Acta Pharmacol Sin 29(2):245–251

Yang M, Yan T, Yu M, Kang J, Gao R, Wang P, Zhang Y, Zhang H, Shi L (2020) Advances in understanding of health-promoting benefits of medicine and food homology using analysis of gut microbiota and metabolomics. Food frontiers 1:398–419

Yang N, Xia Z, Shao N, Li B, Xue L, Peng Y, Zhi F, Yang Y (2017) Carnosic acid prevents dextran sulfate sodium-induced acute colitis associated with the regulation of the Keap1/Nrf2 pathway. Sci Rep 7(1):1–12

Yang Z, Concannon J, Ng KS, Seyb K, Mortensen LJ, Ranganath S, Gu F, Levy O, Tong Z, Martyn K (2016) Tetrandrine identified in a small molecule screen to activate mesenchymal stem cells for enhanced immunomodulation. Sci Rep 6:30263

Yilma AN, Singh SR, Morici L, Dennis VA (2013) Flavonoid naringenin: a potential immunomodulator for *Chlamydia trachomatis* inflammation. Mediat Inflamm 2013:102457

Yoshimi N, Futamura T, Kakumoto K, Salehi AM, Sellgren CM, Holmén-Larsson J, Jakobsson J, Pålsson E, Landén M, Hashimoto K (2016) Blood metabolomics analysis identifies abnormalities in the citric acid cycle, urea cycle, and amino acid metabolism in bipolar disorder. BBA clinical 5:151–158

Yousofi A, Daneshmandi S, Soleimani N, Bagheri K, Karimi MH (2012) Immunomodulatory effect of Parsley (*Petroselinum crispum*) essential oil on immune cells: mitogen-activated splenocytes and peritoneal macrophages. Immunopharmacol Immunotoxicol 34(2):303–308

Zeng W, Jin L, Zhang F, Zhang C, Liang W (2018) Naringenin as a potential immunomodulator in therapeutics. Pharmacol Res 135:122–126

Zou M, Chen Y, Sun-Waterhouse D, Zhang Y, Li F (2018) Immunomodulatory acidic polysaccharides from Zizyphus jujuba cv. Huizao: insights into their chemical characteristics and modes of action. Food Chem 258:35–42

Zou Y-F, Ho GTT, Malterud KE, Le NHT, Inngjerdingen KT, Barsett H, Diallo D, Michaelsen TE, Paulsen BS (2014a) Enzyme inhibition, antioxidant and immunomodulatory activities, and brine shrimp toxicity of extracts from the root bark, stem bark and leaves of *Terminalia macroptera*. J Ethnopharmacol 155(2):1219–1226

Zou Y-F, Zhang B-Z, Inngjerdingen KT, Barsett H, Diallo D, Michaelsen TE, El-zoubair E, Paulsen BS (2014b) Polysaccharides with immunomodulating properties from the bark of *Parkia biglobosa*. Carbohydr Polym 101:457–463

6 Knowledge of African Traditional Medicine System: Immunomodulation

Joy Ifunanya Odimegwu

Abstract

African traditional medicine is one of the oldest forms of health-care system in the continent that has continued to be relevant. It is usually holistic, treating mind and body, and includes aromatherapy, bone setting, circumcision, herbs, homeopathy massaging, spiritual therapies, maternity care, psychiatric care, music therapy, and many more. It is a very old and culturally informed method of health management that humans have used against diseases that have threatened existence.

More than 60,000 of the world's higher plant species can be found in sub-Saharan Africa and the Indian Ocean Islands. These are about one-fourth of the global total and less than the 8% of the medicinal plants sold internationally from Africa. This scarcity could be due to lack of data on the traditional uses of many African plants as the knowledge is transferred orally by storytellers and traditional healers though in recent times, there are some information in print.

Immunomodulation is seen as an essential feature of immunotherapy whereby immune responses are provoked, heightened, decreased, or avoided. Immune responses have been observed to be either cellular co-receptor expression, class switching, cytokine secretion, histamine release, immunoglobulin secretion, lymphocyte expression, or phagocytosis. Immune system dysfunction is responsible for various diseases like allergies, asthmas, arthritis, cancers, and infectious diseases. So modulation of immune responses is required in controlling diseases. This is requisite nowadays because of the upsurge of infectious diseases like superbugs caused by Multi-Resistant *Staphylococcus aureus* (MRSA) and the

J. I. Odimegwu (✉)
Department of Pharmacognosy, Faculty of Pharmacy, College of Medicine Campus, University of Lagos, Lagos, Nigeria
e-mail: Jodimegwu@unilag.edu.ng

N. S. Sangwan et al. (eds.), *Plants and Phytomolecules for Immunomodulation*, https://doi.org/10.1007/978-981-16-8117-2_6

coronavirus (COVID-19) plus other emerging diseases. The historical view of African Traditional Medicine (ATMs) will be discussed from the point of view of specific plants used for immunomodulation in the ATMs and their efficacies following the trend of use and development of herbal medicines from crude formulations to refined dosage forms and procedures over time. Most of the ATMs are prepared as tonics and bitters to heighten and keep up immune defenses.

Keywords

Adaptive immunity · African Traditional Medicines · COVID-19 · Herbal medicines · Innate immunity · MRSA · Sub-Saharan Africa

6.1 The Historical Perspective of African Traditional Medicine [ATMs] Systems

In Africa, indigenous peoples typically adopt herbal medicines, a very convenient form of health management and treatments, and it is widely traditionally accepted and a relevant part of the different local health-care systems (Brendler et al. 2010). African Traditional Medicines (ATMs) are one of the oldest, if not the oldest, forms of health-care system in the continent that has stood the test of time, and it is usually holistic, treating mind and body, and involves health management with herbs, orthopedics (bones), mother care (maternity and child delivery), mental health, essential oils therapy, songs therapy, homeopathy, and more (Abdullahi 2011). About 85% of the populace depend on ATMs for general well-being. It is cheaper, more accessible, and with fewer and more noticeable positive side effects compared to orthodox medicines. The trend usually follows knowledge of specific ills that can be discovered through metaphysical methods. Then solution is suggested, typically of herbs or minerals like earth clay that are understood to possess not just remedial abilities but are full of symbols and folklore. ATMs have an underlying belief system that holds that sickness is not an accidental occurrence but is caused by some imbalance in nature. This differs from orthodox medicine in essence and aims (Ubani 2011). Very essentially, the belief of African traditional healing has at its core the rebuilding of the body's natural defense systems, making strong, in and out, mind and body. While orthodox medical scientists may regard ATMs style as nonevidence based, the adherents of ATMs understand its unique rationality most especially in the cultural context where it is usually applied (Okpako 1999).

Africans formulated their wellness traditions over time, and these involve means of diagnosing, naming, and handling diseases (Mokaila 2001; Ngcobo and Gqaleni 2015; Emeagwali 2016), but it is no less effective, and most importantly, it is trusted by the people. Medicinal plants make up an exceedingly essential source for the improvement of the world's drugs and food industries. There are anecdotal narratives of the earlier generations identifying poisonous plants through observation of their domestic animals that never ate certain plants, and those that did usually died. Case in point is the cassava, *Manihot esculenta,* plant that has high cyanide contents in the leaves.

This practice of using herbs as medicines was developed from hundreds of years of oral knowledge transmitted by word of mouth (teachings, songs, and stories). African Traditional Medicine is at a critical point now because of the increased importance of herbal medicines, a going back to nature that has made great swathes of people even those in developed economies to look toward herbal medicines for health care. African Traditional Medicine in the olden days and also in modern times for some locales is the major healing system accessible to multitudes of citizenry in Africa, whether residing in the rural or urban communities (Romero-Daza 2002).

For years, ATMs were the sole provider of health care for most members of the communities (Romero-Daza 2002), before the arrival of western/orthodox medicine. The herbal practitioners were called *Babalawo* by the Yoruba speakers, *Abia ibok* by the Ibibios, *Dibia* by the Igbos, and *Boka* by the Hausa speakers, all in Nigeria, while they are *Sangoma* or *Nyanga* among South Africans (Borokini and Lawal 2014).

In a manner similar to orthodox medicinal practice, the practitioners of African Traditional Medicine specialize in particular areas of their profession. For instance, there are specialized diviners, birth attendants, and bone setters. Each one is an expert in the chosen area having been trained (apprenticed) for years and practicing over time, gaining relevant experiences. The task of the western-trained doctor is to repair the constituent parts of a machine that malfunctioned, but for traditional medicine practitioners, such repairs might merit appeasement of spiritual entities in addition to medical intervention. The former is dwindling in importance, and people become better educated and ceasing to believe that these so-called deities hold their lives. The practitioners are more intentional and looking for evidence for their proffered solutions leading to more evidence-based work in the area. African Traditional Medicine seems to have incorporated that tradition of dualism that some may define as holistic in that it usually embodies both mind and body (Ojiewo et al. 2013). In order for these African Traditional Medicines to be developed over time similar to orthodox medicines, it is imperative to screen them analytically and understand their modes of action, safety status, and effectiveness (Ngcobo et al. 2017) and also prepare them in patient-friendly dosage forms following a standardized way.

6.2 Traditional Medicines and the Immune System

Immunomodulation used in the broadest sense encompasses all therapeutic interventions aimed at changing the immunity reaction. These immune actions/reactions could be in the form of cellular co-receptor expression, class switching, cytokine secretion, histamine release, immunoglobulin secretion, lymphocyte expression, phagocytosis, etc. (Ezekwesili-Ofili and Okaka 2019). Immune system dysfunction is responsible for various diseases like allergies, asthmas, arthritis, cancers, and infectious diseases. Immunomodulation is a requisite nowadays because of the upsurge of infectious diseases like superbugs caused by Multi-Resistant *Staphylococcus aureus* (MRSA) and the coronavirus (COVID-19) plus other emerging diseases (Sofowora et al. 2013; Mothibe et al. 2019). So modulation of immune reactions is required in controlling diseases, and increase of the immune reaction is desirable to prevent infection in conditions of immune deficiency to combat ingrained infections and to manage certain diseases like; cancers (Rajeev 2019).

6.3 Plants/Herbs Used for Immunomodulation in the African Traditional Medicines

Immunity reactions are harnessed by the collective work of molecular means that curbs or triggers immune reactions (Tables 6.1 and 6.2). A change in this delicate balance can lead to increased susceptibility to infections, constant inflammation, or autoimmune diseases (Elbagory et al. 2019).

Table 6.1 Plants with potentiating effects on the immune system

Plants	Effect	References
Andrographis paniculata	Increases lymphocyte proliferation and increases the production of IL-2 and TNF-α	Rajagopal et al. (2003); Kumar et al. (2004); and Sharma et al. (2006)
Echinacea purpurea, E. angustifolia, and *E. pallida*	This triggers macrophages and facilitates the production of cytokine. Triggers NK cells and builds humoral reactions	Yakoot and Salem (2011), and Pannacci et al. (2006)
Ginseng panax quinquefolius	NK cells and lymphocytes are triggered, and antibodies production is elevated	Scaglione et al. (1990), and Sunila and Kuttan (2004)
Piper longum	Elevates leukocyte counts and multiplies amount of plasma cells and circulating antibodies	Pradeep and Kuttan (2004)
Pelargonium sidoides	Indirect activity, possibly through activation of macrophage functions	Mahomoodally (2013), and Kolodziej (2000, 2007)
Vernonia amygdalina	Hydro extracts of *V. amygdalina* (Fig. 6.1b) fresh leaves impart immunity boosting functions on HIV patients by increasing the CD4 count	Momoh et al. (2012)
Telfairia occidentalis	The hydro extract of *T. occidentalis* (Fig. 6.1a) leaves increases iron content in the blood, thereby contributing to immunomodulation. Seed extracts actively reduced oxidative burst action in whole blood, isolated polymorphonuclear cells (PMNs), and mononuclear cells (MNCs)	Okokon et al. (2012)

IL interleukin, *TNF* tumor necrosis factor; *NK* natural killer cell; CD4 T cells: lymphocytes

Table 6.2 Plants with immunosuppressive effects

Plants	Effect	References
Bryophyllum pinnatum	Reduces inflammation	Huang et al. (2010)
Allium cepa (onion) *Malus domestica* (apple)	Can reduce endocytosis and production of IL-1β, TNF-α, IL-6, and IL-12	Carrasco et al. (2009) Tan et al. (2005)
Citrus X *sinensis*	Reduces the activation of NF-*k*B	Rodríguez et al. (2005); Wintergerst et al. (2006)
Morinda citrifolia L.	Reduces inflammation	Márquez et al. (2008)
Pinus leiophylla	Reduces inflammation associated with decreased neutrophil migration	Cocks and Moller (2002)

IL interleukin; *TNF* tumor necrosis factor; *NK* natural killer cell

Fig. 6.1 (**a**) *Telfairia occidentalis* leaves, (**b**) *Vernonia amygdalina* leaves, (**c**) *Garcinia kola* seeds, and (**d**) unripe fingers of *Musa paradisiaca*

Typically, edible vegetables and fruits, e.g., avocados, pawpaw, unripe plantain (*Musa paradisiaca,* Fig. 6.1d) are used to boost immunity in African Traditional Medicines, and over time, the scientific basis for this practice is confirmed. Few are noted in Tables 6.1 and 6.2). Over 400 well-researched plant species that are part of about 53 botanical families are used as vegetables in Africa (Akinola et al. 2020; Ojiewo et al. 2013; Reyes et al. 2016). A few of the popular ones are shown in Table 6.3.

In recent times, essential oils from food spices also have shown significant immunomodulatory effects (Sandner et al. 2020, and Redgrove and McLaughlin 2014). Some of the plants whose immune-boosting potentials have been scientifically studied and proven are listed below:

1. *Astragalus gummifer* Lab (Leguminosae). **Geography**: These grow on steep and dry highlands up to an altitude of 1800 m in the Sahel region. The plants can be found at Sudan, Niger, Chad, and the Kano Emirate of Nigeria (Ezekwesili-Ofili and Okaka 2019). **Pharmacological actions**: Hydro extract of *A. membranaceus* roots exerts total immune restoration of local xenogeneic graft-versus-host reaction (XGVHR) in cancer patients (Sun et al. 1983). In vitro immunity actions of the fractions were first determined from their effects on mononuclear cells (MNCs) gotten from healthy normal donors using the local XGVHR test system (Yamada and Azuma 1977). It was demonstrated that the reversal of the cyclophosphamide-induced immunosuppression by the administration of the active fraction of the plant drug was total since the volume of the abrogated local XGVHR (39.78 ± 8.3 mm^3) was comparable to the value of 34.79 ± 5.69 mm^3 ($p > 0.1$).
2. *Costus afer* Ker (Costaceae). **Geography:** Secondary forest wild growth. Thrives in the wetter areas. Plant is mostly in the rain forest. Widely distributed in the West Coast, Nile Valley, and in parts of Zimbabwe. **Pharmacological actions**: Fixed root oil from *Costus* causes contact dermatitis, and skin sensitivity to the oil is used as a diagnostic tool for studying delayed cell-mediated immune system reactions (Ikewuchi and Ikewuchi 2013).
3. *Garcinia kola.* **Geography** (Fig. 6.1c): Grows well in moist forests and cultivated in family compounds. Well distributed in West and Central Africa and has been located in Sierra Leone, Ghana, Nigeria, Cameroon, and Congo (Ezekwesili-Ofili and Okaka 2019). **Pharmacological actions**: Immunity booster against flu and respiratory diseases (Ezekwesili-Ofili and Okaka 2019).
4. *Ricinus communis* L. (Euphorbiaceae). **Geography**: Indigenous to Africa and grows wild in East Africa. Grown for its oil-bearing seeds. The seeds processed by fermentation is treasured as a probiotic by Igbo people (Nigeria) and is added to soup as flavor. Propagated as an annual (Ezekwesili-Ofili and Okaka 2019). **Pharmacological actions**: Ricin, a toxic phytochemical found in the seeds, is implicated as an antitumor and immune stimulator in very low dosages (Iwu 2014; Ezekwesili-Ofili and Okaka 2019).
5. . *Sansevieria liberica* (Agavaceae) (Gérôme & Labroy). **Geography**: This usually grows in shady places near water bodies. The plant occurs frequently in the

Table 6.3 Popular African vegetables used as immune boosters

Serial number	Common name	Scientific name	Origin	Medicinal use	References
1	Parsley	*Petroselinum crispum*	Tropical Africa and Northern Africa	Powerful antifungal, antidiabetic, analgesic, diuretic, hepatoprotective, and spasmolytic, useful for gastrointestinal disorders and as an immunosuppressant with gastroprotective properties	Agyare et al. (2017)
2	African elemi	*Canarium schweinfurthii* Engl.	Tropical Africa	Anti-inflammatory and antiparasitic. antiparasitic activities	Kuete (2017)
3	Basil	*Ocimum basilicum* L.	Tropics	Immunity stimulant and antiviral	Charvat et al. (2006); Uma (2001)
4.	Black nightshade	*Solanum nigrum*	South Africa	Curbs growth of cervical carcinoma (U14) by modulating immune response of tumor-bearing mice and by introducing apoptosis	Li et al. (2008)
5.	Soapberry	*Quillaja saponaria* Molina	North Africa	Immunity stimulant, evoking both IgG1 and IgG2 antibody and cell-mediated immunity	Iwu (2014)
6.	Baobab	*Adansonia digitata*	Madagascar and mainland Africa	Antioxidant, antimicrobial, antiviral, and anti-inflammatory	Iwu (2014)
7.	African spiderflower	*Gynandropsis gynandra*	Tropical Africa	Immunity development and general healthy functioning of the body	Heuzé et al. (2020), and Tumwet et al. (2014)
8.	Green amaranth	*Amaranthus hybridus*	Tropical Africa and Asia	Blood builder and immune boosting	Tumwet et al. (2014)
9.	Nettle	*Urtica massaica*	Burundi, Congo, Kenya, Rwanda, Tanzania, and Uganda	Stomachache, malaria, and blood builder	Grubben (2004)
10.	Bush okra	*Corchorus olitorius*	Tropical Africa and Asia	Great amount of potassium, antioxidant, and cytotoxic	Oladiran (1986), and Handoussa et al. (2013)

(continued)

Table 6.3 (continued)

Serial number	Common name	Scientific name	Origin	Medicinal use	References
11.	Yellow commelina	*Commelina africana*	Senegal to Ethiopia and south to South Africa	Treatment of a "weak heart" and nervousness. Fertility	Burkill (1985)
12.	Garden cress	*Lepidium sativum* L	Ethiopia	Amoebic dysentery, to treat sore throat, cough, asthma, and headache	Adam (1999), and Ashebir and Ashenafi (1999)
13.	Soap plant	*Helinus integrifolius* (Lam.)	Botswana, Mozambique, and Zambia	Infections/infestations, mental disorders, and skin/subcutaneous cellular tissue disorders)	Gelfand et al. (1985)
14.	Roselle	*Hibiscus sabdariffa* L	Sudan	For treating urinary tract infections. The leaf juice is used to treat conjunctivitis.	Boonkerd et al. (1993)
15.	Egusi melon	*Citrullus lanatus*	Western Kalahari region of Namibia and Botswana	Tar is extracted from the seeds and used for the treatment of scabies and for skin tanning.	Burkill (1985)
16.	Pumpkin	*Cucurbita maxima*	Tropical Africa	Great healing qualities for skin problems such as sores and ulcers	Holland et al. (1991)
17.	Finger tree	*Euphorbia tirucalli* L	Eastern tropical Africa	Is used as potent purgative and emetic. Can manage headaches, asthma, epilepsy, etc.	Bani et al. (2007)
18.	African rock fig	*Ficus glumosa* Delile	Tropical Africa and Western Saudi Arabia	Latex is for sore eyes. Powdered bark is used in Nigeria to plug carious teeth.	Arbonnier (2004)
19.	Wild fig	*Ficus sur*	Tropical Africa, Cape Verde, Somalia, Angola, and South Africa	Latex is used for wounds, toothache, and eye problems.	Chifundera (2001)
20.	Kenaf	*Hibiscus cannabinus* L	African countries south of the Sahara	Leaves are for stomach disorders. Leaf infusion is administered for treating cough.	Akpan (2000), and Burkill (1997)
21.	Wild spikenard	*Hyptis suaveolens*	Senegal to Southern Nigeria	Leaves are applied externally to cuts and wounds and internally to treat fever and catarrh.	Inngjerdongen et al. (2004)

22.	Smooth loofah	*Luffa cylindrica L*	Tropics and subtropics, Africa, or Asia	In Traditional African Medicine, pulp of the whole plant is used as a suppository against constipation.	Bal et al. (2004)
23.	Garden purslane	*Portulaca oleracea* L.	Tropical Africa	Used as a diuretic and to treat rheumatism and gynecological diseases	Phillips (2002); Rubatzky and Yamaguchi (1997)
24.	African padauk	*Pterocarpus soyauxii* Taub.	Southeastern Nigeria, DR Congo, and Northern Angola	As an enema to treat dysentery and against toothache, gonorrhea, and excessive menstruation	Brémaud et al. (2004)
25.	Congo jute	*Urena lobata* L.	Tropical Africa, Cape Verde, and Senegal	Expectorant and emollient. In Nigeria, a preparation of the root is applied externally against rheumatism.	Baert and Raemaekers (2001)
26.	Burweed	*Triumfetta cordifolia* A. Rich.	Moist parts of Tropical Africa	Treatment for sterility in women	Burkill (2000), and Adjanohoun et al. (1991)
27.	Agati sesbania	*Sesbania grandiflora* L. Poir.	West Africa and East Africa	Root is a well-known medicine for malaria. Leaves and flowers are used as poultices.	Neuwinger (2000)
28.	Devil's horsewhip	*Achyranthes aspera*	Tropical Africa	Used as a digestive and stomachic in India, Egypt, and Australia	Yineger et al. (2007)
29.	Tropical primrose	*Asystasia gangetica*	Tropical Africa, Arabia, and Asia	For easing pain during childbirth and in embrocations to treat stiff neck and enlarged spleen in children	Adetula (1987), and Akah et al. (2003)
30.	White goosefoot	*Chenopodium album* L.	All African countries	For use as and an anthelmintic and aphrodisiac	Brenan (1988)

savanna regions and grasslands. Found typically in Chad, Ethiopia, Kenya, Morocco, and Tanzania (Ezekwesili-Ofili and Okaka 2019) **Pharmacological actions:** Leaf extracts were positive for antidiabetic actions. Hypolipidemic, immune-modulating, hepatorenal, and cardioprotective potentials are also attributed to the plants (Ezekwesili-Ofili and Okaka 2019).

6. *Sorghum bicolor* (L.) (Poaceae) Moench. **Geography**: The plant originated from Africa and is grown mostly in the tropical areas of the continent (Ezekwesili-Ofili and Okaka 2019). **Pharmacological actions**: Immuno-modulating properties of Jobelyn® popular in Nigeria and Ghana. The aqueous extract of the leaf sheaths activates immediate upregulation of NK cells, which leads to manyfold increment in the chemokines (Bone 2007; Benson et al. 2013).
7. *Vernonia amygdalina* Del. (Compositae). **Geography** (Fig. 6.1b): Often cultivated widely throughout the African continent as food. Spreads from Sudan to southern Africa. **Pharmacological actions**: Analgesic, antimalarial, antidiabetic, antihypercholesterolemic, and anthelmintic actions and uterine contractility and immunity inducing in HIV patients (Ezekwesili-Ofili and Okaka 2019).

6.4 Formulations of African Traditional Medicines for Immunomodulation

Immunomodulation usually starts with the cleansing of the internal organs to allow them to be receptive to medicines. Colon cleansers produced from herbs act on the immune and ductless systems to supply phyto-compounds that reduce stress (Leclerc-Madlala 1994). The use of traditional medicines to detoxify the blood is a usual practice in Africa and is usually associated to powering body energy where it is lacking. A belief system exists in the continent that states that *a cure or relief is only to be found through purging the blood and cleansing the body* (Ndhlala et al. 2010; Chu et al. 1988). Giving the immune system a boost through naturally derived phytochemicals is very popular and frequently applied by citizens of these areas. Medicines for energy typically produced as teas (infusions), tonics, and bitters, almost all in liquid forms, are used to stimulate the innate immune system (Alamgir and Uddin 2010).

Herbal Tonics: This is a solution or formulation created from selected plant/s and plant parts (Hubbard 2017), and these are usually locally prepared in the following standardized way:

One tablespoon of dried and ground herbs and one cup of cold water (about 200 mL). Add herbs and water into an iron pot and then bring to boil. Lower the heat, cover, and let it simmer from 20 to 45 min, depending on how hardy the plant is. Roots may take longer than leaves. Strain the liquid, and then drink usually twice a day, first in the morning and last after dinner.

Herbal Bitters: Historically originated from ancient Egypt and North Africa. They are usually an alcoholic formulations flavored with herbal materials sometimes fragrant so that the end result is characterized by a bitter, sour or bittersweet flavour.

There exists different brands of bitters originally formulated as patent medicines by alchemists of old. In modern times, they are formulated as polyherbals and used as digestives. The cocktails usually contain sour and sweet flavors and are highly potent. The bitters, in a culinary way, engage the primary taste and so balance out the drink with a more complete flavor profile (Mahomoodally 2013).

6.5 Future Prospects

Africa is rich in plant biodiversity, and due to the peculiar circumstances of the continent and poor infrastructure and resources, the people have not abandoned the ancient ways of medicines which are affordable and available to them and more importantly very effective. There are a lot of orphan medicinal plants (Table 6.3) that with proper research can prove to be great leads for new and powerful pharmaceutical drugs.

The aim of this review is to point to the richness of medicinal plants used in African Traditional Medicines to boost the immune system. Further, focused research in developing these high-value plants will be of global interest because as the COVID-19 pandemic has shown, there is need for collaboration and persistent research in the area of supporting innate immunity of the human system preferably through natural means. An interesting direction for the future could be in developing medicinal plants specifically to yield higher bioactive ingredients in a natural way to reduce side effects. Standardization of medicinal plants in such a way that certifiable dosage forms are generated is a necessary paradigm for growth in the area of promoting and improving African Traditional Medicines for immunomodulation moving forward.

References

Abdullahi AA (2011) Trends and challenges of traditional medicine in Africa. Afr J Tradit Complement Altern Med 8(S):115–123

Adam SEI (1999) Effects of various levels of dietary *Lepidium sativum* L. seeds in rats. Am J Chin Med 27:397–405

Adetula OA (1987) Distribution and cytomorphology of *Asytasia calycina* Bentham and A. gangetica (L.) T. Anderson (Acanthaceae) in Nigeria. MSc Dissertation, University of Lagos, Akoka, Lagos, Nigeria

Adjanohoun EJ, Ahiyi MRA, Aké Assi L, Dramane K, Elewude JA, Fadoju SU, Gbile ZO, Goudote E, Johnson CLA, Keita A, Morakinyo O, Ojewole JAO, Olatunji AO, Sofowora EA (1991) Traditional medicine and pharmacopoeia: contribution to ethnobotanical and floristic studies in western Nigeria. OUA/ST & RC, Lagos, Nigeria, p 420

Agyare C, Appiah T, Boakye YD, Apenteng JA (2017) *Petroselinum crispum*: a review. In: Kuete V (ed) Medicinal spices and vegetables from Africa. Therapeutic potential against metabolic, inflammatory, infectious and systemic diseases. Academic Press, pp 527–547

Akah PA, Ezike AC, Nwafor SV, Okoli CO, Enwerem NM (2003) Evaluation of the anti-asthmatic property of *Asystasia gangetica* leaf extracts. J Ethnopharmacol 89(1):25–36

Akinola R, Pereira LM, Mabhaudhi T, de Bruin F-M, Rusch L (2020) A review of indigenous food crops in Africa and the implications for more sustainable and healthy food systems. Sustainability 12(8):3493
Akpan GA (2000) Cytogenetic characteristics and the breeding system in six Hibiscus species. Theor Appl Genet 100(2):315–318
Alamgir M, Uddin SJ (2010) Recent advances on the ethnomedicinal plants as immunomodulatory agents. In: Chattopadhyay D (ed) Ethnomedicine: a source of complementary therapeutics, pp 227–244
Arbonnier M (2004) Trees, shrubs and lianas of West African dry zones. CIRAD, Margraf Publishers Gmbh, MNHN, Paris, France, p 573
Ashebir M, Ashenafi M (1999) Assessment of the antibacterial activity of some traditional medicinal plants on some food-borne pathogens. Ethiop J Health Dev 13(3):211–216
Baert J, Raemaekers RH (2001) Urena. In: Raemaekers RH (ed) Crop production in tropical Africa. DGIC (Directorate General for International Co-operation), Ministry of Foreign Affairs, External Trade and International Co-operation, Brussels, Belgium, pp 1083–1086
Bal KE, Bal Y, Lallam A (2004) Gross morphology and absorption capacity of cell-fibers from the fibrous vascular system of loofah (Luffa cylindrica). Text Res J 74(3):241–247
Bani S, Kaul A, Khan B, Gupta VK, Satti NK, Suri KA, Qazi GN (2007) Anti-arthritic activity of a biopolymeric fraction from *Euphorbia tirucalli*. J Ethnopharmacol 110(1):92–98
Benson KF, Beaman JL, Ou B, Okubena A, Okubena O, Jensen GS (2013) West African *Sorghum bicolor* leaf sheaths have anti-inflammatory and immune-modulating properties *in vitro*. J Med Food 16(3):230–238
Bone K (2007) The ultimate herbal compendium. Phytotherapy Press
Boonkerd T, Na Songkhla B, Thephuttee W (1993) *Hibiscus sabdariffa* L. In: Siemonsma JS, Piluek K (eds) Plant resources of South-East Asia No 8. vegetables. Pudoc Scientific Publishers, Wageningen, Netherlands, pp 178–180
Borokini TI, Lawal IO (2014) Traditional medicine practices among the Yoruba people of Nigeria: a historical perspective. J Med Plants Stud 6:20–33
Brémaud I, Minato K, Gérard J, Thibaut B (2004) Effect of extractives on vibrational properties and shrinkage of African padauk (*Pterocarpus soyauxii* Taub.). In: Morlier P, Morais J (eds) Proceedings of the 3rd international conference of the European society for wood mechanics, 5–8 September 2004, Vila Real, Portugal. Universidade de Tras-os-Montes e Alto Douro, Portugal. pp 17–24
Brenan JPM (1988) Chenopodiaceae. In: Launert E (ed) Flora Zambesiaca. Volume 9, part 1. Flora Zambesiaca Managing Committee, London, UK, pp 133–161
Brendler T, Eloff JN, Gurib-Fakim A, Phillips LD (eds) (2010) African herbal pharmacopoeia
Burkill HM (1985) The useful plants of West Tropical Africa, vol 1, 2nd edn. Families A–D. Royal Botanic Gardens, Kew, Richmond, UK, p 960
Burkill HM (1997) The useful plants of West Tropical Africa, vol 4, 2nd edn. Families M–R. Royal Botanic Gardens, Kew, Richmond, UK, p 969
Burkill HM (2000) The useful plants of West Tropical Africa, vol 5, 2nd edn. Families S–Z, Addenda. Royal Botanic Gardens, Kew, Richmond, UK, p 686
Carrasco FR, Schmidt G, Romero AL, Sartoretto JL, Caparroz-Assef SM, Bersani-Amado CA, Cuman RK (2009) Immunomodulatory activity of *Zingiber officinale* (Roscoe), *Salvia officinalis* L. and *Syzygium aromaticum* L. essential oils: evidence for humor- and cell-mediated responses. J Pharm Pharmacol 61(7):961–967
Charvat TT, Lee DJ, Robinson WE, Chamberlin AR (2006) Design, synthesis, and biological evaluation of chicoric acid analogs as inhibitors of HIV-1 integrase. Bioorg Med Chem 14(13): 4552–4567
Chifundera K (2001) Contribution to the inventory of medicinal plants from the Bushi area, South Kivu Province, Democratic Republic of Congo. Fitoterapia 72:351–368
Chu DT, Wong W, Mavlight GM (1988) Immunotherapy with Chinese medicinal herbs. J Clin Lab Immunol 25:119–129

Cocks M, Moller V (2002) Use of indigenous and indigenised medicines to enhance personal well-being: a South African case study. Soc Sci Med 54(3):387–397

Elbagory AM, Hussein AA, Meye M (2019) The *in vitro* immunomodulatory effects of gold nanoparticles synthesized from *Hypoxis hemerocallidea* aqueous extract and hypoxoside on macrophage and natural killer cells. Int J Nanomedicine 2019(14):9007–9018

Emeagwali G (2016) African Traditional Medicine revisited. In: Emeagwali G, Shizha E (eds) African indigenous knowledge and the sciences. Anti-colonial educational perspectives for transformative change. Sense Publishers, Rotterdam

Ezekwesili-Ofili JO, Okaka ANC (2019) Herbal medicines in African Traditional Medicine. Herbal Medicine Ed. Philip F. Builders, Kenya

Gelfand M, Mavi S, Drummond RB, Ndemera B (1985) The traditional medical practitioner in Zimbabwe: his principles of practice and pharmacopoeia. Mambo Press, Gweru, Zimbabwe, p 411

Grubben GJH (ed) (2004) Plant resources of tropical Africa: vegetables. PROTA 2004 p 540

Handoussa, Heba; Hanafi, Rasha; Eddiasty, Islam; El-Gendy, Mohamed; Khatib, Ahmed El; Linscheid, Micheal; Mahran, Laila; Ayoub, Nahla (2013). "Anti-inflammatory and cytotoxic activities of dietary phenolics isolated from *C. olitorius* and *Vitis vinifera*". J Funct Foods 5 (3): 1204–1216

Heuzé V, Tran G, Lebas F (2020) African spiderflower (*Gynandropsis gynandra*). Feedipedia, a programme by INRAE, CIRAD, AFZ and FAO. https://www.feedipedia.org/node/144. Accessed 11 Sept 2020

Holland B, Unwin ID, Buss DH (1991) Vegetables, herbs and spices. The fifth supplement to McCance & Widdowson's The Composition of Foods. 4th Edition. Royal Society of Chemistry, Cambridge, UK. 163pp

Huang RY, Yu YL, Cheng WC, Yang CN, Fu E, Chu CL (2010) Immunosuppressive effect of quercetin on dendritic cell activation and function. J Immunol 184:6815–6821

Hubbard L (2017) Everything you need to know about bitters. Town & Country

Ikewuchi JC, Ikewuchi CC (2013) Positive moderation of the hematology, plasma biochemistry and ocular indices of oxidative stress in alloxan-induced diabetic rats, by an aqueous extract of the leaves of *Sansevieria liberica* Gerome and Labroy. Asian Pac J Trop Med 6:1

Inngjerdongen K, Nergård CS, Diallo D, Mounkoro PP, Paulsen BS (2004) An ethnopharmacological survey of plants used for wound healing in Dogonland, Mali, West Africa. J Ethnopharmacol 92:233–244

Iwu MM (2014) Pharmacognostical profile of selected medicinal plants from: handbook of African Medicinal Plants. CRC Press

Kolodziej H (2000) Traditionally used *Pelargonium species*: chemistry and biological activity of umckaloabo extracts and their constituents. Curr Topics Phytochem 3:77–93

Kolodziej H (2007) Fascinating metabolic pools of Pelargonium sidoides and Pelargonium reniforme, traditional and phytomedicinal sources of the herbal medicine Umckaloabo. Phytomedicine 14(1):9–17

Kuete V (2017) Canarium schweinfurthii. In: Kuete V (ed) Medicinal spices and vegetables from Africa. Therapeutic potential against metabolic, inflammatory, infectious and systemic diseases, pp 379–384

Kumar RA, Sridevi K, Kumar NV, Nanduri S, Rajagopal S (2004) Anticancer and immunostimulatory compounds from *Andrographis paniculata*. J Ethnopharmacol 92:291–295

Leclerc-Madlala S (1994) Zulu health, cultural meanings and the reinterpretation of western pharmaceuticals. In: Proceedings of the association of anthropology in South Africa conference, University of Durban-Westville, Durban, South Africa

Li J, Li Q, Feng T, Li K (2008) Aqueous extract of *Solanum nigrum* inhibit growth of cervical carcinoma (U14) via modulating immune response of tumor bearing mice and inducing apoptosis of tumor cells. Fitoterapia 79(7–8):548–556

Mahomoodally MF (2013) Traditional medicines in Africa: an appraisal of ten potent African Medicinal Plants. Evid Based Complement Alternat Med 2013:617459

Márquez F, Alvear R, Montellano R, Meléndez C (2008) Efecto antiinflamatorio de *Pinus leiophylla* Schlechtendal and Cham en la rata. Rev Mex Cienc Farmac 39:22–27
Mokaila A (2001) Traditional vs. Western Medicine-African Context. Drury University, Springfield, Missouri
Momoh MA, Muhamed U, Agboke AA, Akpabio EI, Osonwa UE (2012) Immunological effect of aqueous extract of *Vernonia amygdalina* and a known immune booster called immunace and their admixtures on HIV/AIDS clients: a comparative study. Asian Pac J Trop Biomed 2(3): 181–184
Mothibe ME, Kahler-Venter CP, Osuch E (2019) Evaluation of the in vitro effects of commercial herbal preparations significant in African traditional medicine on platelets. BMC Complement Altern Med 19:224
Ndhlala AR, Finnie JF, Van Staden J (2010) *In vitro* antioxidant properties, HIV-1 reverse transcriptase and acetylcholinesterase inhibitory effects of traditional herbal preparations sold in south africa. Molecules 15(10):6888–6904
Neuwinger HD (2000) African traditional medicine: a dictionary of plant use and applications. Medpharm Scientific, Stuttgart, Germany, p 589
Ngcobo M, Gqaleni N (2015) African Traditional Medicine based immune boosters and infectious diseases: a short commentary. J Mol Biomark Diagn 7:265
Ngcobo M, Gqaleni N, Naidoo V, Cele P (2017) The immune effects of an african traditional energy tonic in *in vitro* and *in vivo* models. Evid Based Complement Alternat Med 2017:6310967
Ojiewo C, Tenkouano A, Hughes J, Keatinge JD (2013) Diversifying food and diets: using indigenous vegetables to improve profitability, nutrition and health in Africa. In: Fanzo J, Hunter D, Borelli T, Mattei F (eds) Diversifying food and diets: using agricultural biodiversity to improver nutrition and health. Routhledge, London, UK
Okokon J, Farooq A, Choudhary M, Antia B (2012) Immunomodulatory, anticancer and anti-inflammatory activities of *Telfairia occidentalis* seed extract and fractions. Int J Food Nutr Safety 2:72–85
Okpako DT (1999) Traditional African medicine: theory and pharmacology viewpoint. TiPS 20: 482–485
Oladiran JA (1986) Effect of stage of harvesting and seed treatment on germination, seedling emergence and growth in Corchorus olitorius 'Oniyaya'. Sci Hortic 28(3):227–233
Pannacci M, Lucini V, Colleoni F, Martucci C, Grosso S, Sacerdote P, Scaglione F (2006) *Panax ginseng* C.A. Mayer G115 modulates pro-inflammatory cytokine production in mice throughout the increase of macrophage toll-like receptor 4 expression during physical stress. Brain Behav Immun 20:546–551
Phillips SM (2002) Portulacaceae. In: Beentje HJ (ed) Flora of Tropical East Africa. A.A. Balkema, Rotterdam, Netherlands, p 40
Ubani LU (2011) Preventive therapy in complimentary medicine. Xlibris Corporation
Romero-Daza N (2002) Traditional medicine in Africa. Ann Am Acad Polit Social Sci 583:173–176
Rajeev KT (2019) Introductory chapter: immunity and immunomodulation. In: Tyagi RK, Bisen PS (eds) Immune response activation and immunomodulation. IntechOpen
Reyes-Munguía A, Carrillo-Inungaray ML, Carranza-Álvarez C, Pimentel-González DJ, Alvarado-Sánchez B (2016) Antioxidant activity, antimicrobial and effects in the immune system of plants and fruits extracts. Front Life Sci 9(2):90–98
Sandner G, Heckmann M, Weghuber J (2020) Immunomodulatory activities of selected essential oils. Biomol Ther 10(8):1139
Redgrove KA, McLaughlin EA (2014) The role of the immune response in *Chlamydia trachomatis* infection of the male genital tract: a double-edged sword. Front Immunol 5:534
Rajagopal S, Kumar RA, Deevi DS, Satyanarayana C, Rajagopalan R (2003) Andrographolide, a potential cancer therapeutic agent isolated from *Andrographis paniculata*. J Exp Ther Oncol 3: 147–158

Sharma M, Arnason JT, Burt A, Hudson JB (2006) *Echinacea* extracts modulate the pattern of chemokine and cytokine secretion in rhinovirus-infected and uninfected epithelial cells. Phytother Res 20:147–152

Yakoot M, Salem A (2011) Efficacy and safety of a multiherbal formula with vitamin C and zinc (Immumax) in the management of the common cold. Int J Gen Med 4:45–51

Scaglione F, Ferrara F, Dugnani S, Falchi M, Santoro G, Fraschini F (1990) Immunomodulatory effects of two extracts of *Panax ginseng*. Drugs Exp Clin Res 16:537–542

Sunila ES, Kuttan G (2004) Immunomodulatory and antitumor activity of *Piper longum* Linn. and piperine. J Ethnopharmacol 90:339–346

Pradeep CR, Kuttan G (2004) Piperine is a potent inhibitor of nuclear factor-kappaB (NF-kappaB), c-Fos, CREB, ATF-2 and proinflammatory cytokine gene expression in B16F-10 melanoma cells. Int Immunopharmacol 4:1795–1803

Wintergerst ES, Maggini S, Hornig DH (2006) Immune enhancing role of vitamin C and zinc and effect on clinical conditions. Ann Nut Metab 50:85–94

Tan PH, Sagoo P, Chan C, Yates JB, Campbell J, Beutelspacher SC, Foxwell BM, Lombardi G, George AJ (2005) Inhibition of NF-kappa B and oxidative pathways in human dendritic cells by antioxidative vitamins generates regulatory T cells. J Immunol 174:7633–7644

Rodríguez RM, Boffil CM, Lorenzo MG, Sanchez FP, López GR, Verdecía MB, Díaz CL (2005) Evaluación preclínica del efecto antiinflamatorio del jugo de *Moringa citrifolia* L. Rev Cubana Plant Med 10:3–4

Sofowora A, Ogunbodede E, Onayade A (2013) The role and place of medicinal plants in the strategies for disease prevention. Afr J Tradit Complement Altern Med 10(5):210–229

Sun Y et al (1983) Immune restoration and/or augmentation of local graft versus host reaction by traditional Chinese medicinal herbs. Cancer 52:70

Yamada Y, Azuma K (1977) Evaluation of the in vitro antifungal activity of allicin. Antimicrob Agents Chemother 2:743

Uma DP (2001) Radioprotective, anti carcinogenic and antioxidant properties of the Indian holy basil, *Ocimum sanctum* (Tulasi). Indian J Expl Biol 39:185–190

Tumwet TN, Kang'ethe EK, Kogi-Makau W, Mwangi AM (2014) Diversity and Immune boosting claims of some African indigenous leafy vegetables in Western Kenya. Afr J Food Agric Nutr Dev 14(1):8529–8544

Rubatzky VE, Yamaguchi M (1997) World vegetables: principles, production and nutritive values, 2nd edn. Chapman & Hall, New York, p 843

Yineger H, Kelbessa E, Bekele T, Lulekal E (2007) Ethnoveterinary medicinal plants at Bale Mountains national park, Ethiopia. J Ethnopharmacol 112:55–70

Middle Eastern Diets as a Potential Source of Immunomodulators

7

Sabrin R. M. Ibrahim, Ali M. El-Halawany, Riham Salah El-Dine, Gamal A. Mohamed, and Hossam M. Abdallah

Abstract

The immune system is a highly developed and complex system. Its optimal functioning is critical to human health, being responsible for safeguarding the human body toward the invading of various pathogens or cancers, and therefore plays a remarkable role in maintaining health. Immunomodulators are agents that change the immunologic function of human, and they include stimulatory and suppressive agents. Diet is one of the main factors that modulate different aspects of the immune functions. The consumption of diets with immunomodulating capacities is known as an efficient tool for preventing the come down of the immune functions and decreasing the risks of infections or cancers, as well as boosting the physiological functions. Recently, an interest has been shifted to the Middle Eastern diet (MED) recognized as one of the healthiest diets, with

S. R. M. Ibrahim
Department of Chemistry, Preparatory Year Program, Batterjee Medical College, Jeddah, Saudi Arabia

Department of Pharmacognosy, Faculty of Pharmacy, Assiut University, Asyut, Egypt

A. M. El-Halawany · R. S. El-Dine
Department of Pharmacognosy, Faculty of Pharmacy, Cairo University, Cairo, Egypt

G. A. Mohamed
Department of Natural Products and Alternative Medicine, King Abdulaziz University, Jeddah, Saudi Arabia

Department of Natural Products and Alternative Medicine, Faculty of Pharmacy, King Abdulaziz University, Jeddah, Saudi Arabia

H. M. Abdallah (✉)
Department of Pharmacognosy, Faculty of Pharmacy, Cairo University, Cairo, Egypt

Department of Natural Products and Alternative Medicine, King Abdulaziz University, Jeddah, Saudi Arabia

N. S. Sangwan et al. (eds.), *Plants and Phytomolecules for Immunomodulation*,
https://doi.org/10.1007/978-981-16-8117-2_7

substantiation of healing and preventing diverse human disorders and increasing longevity. This was attributed to the fact that MED is a wealthy pool of antioxidants, minerals, dietary fibers, essential fatty acids, and vitamins. In this chapter, state-of--the-art knowledge about the effectiveness of the MED on the immune function is reviewed. It is noteworthy that many evidences encourage the consumption of MED aimed at long-term healthy life with improved quality. More future attention should be paid to the consumption of MED with immunomodulatory potential to prohibit the declining of the immune functions and minimize the risk of various disorders such as infections, autoimmune diseases, allergies, or cancer. Moreover, further clinical and mechanistic studies are required to establish the role of MED as immunomodulators.

Keywords

Immune system · Immunomodulation · Middle Eastern diet · Antioxidants · Human disorders

7.1 Introduction

The immune system is an exceptional intricate biological system that integrated into all physiological functions to protect the human body from invading agents by generating varieties of molecules and cells, which are able to recognize and eliminate unlimited sets of undesirable and foreign agents (Saroj et al. 2012). The immune system is comprised of adaptive and innate systems. The innate immune system includes physical barriers that assist to prohibit pathogens entry, in addition to complement system, a diversity of phagocytic cells, and antimicrobial peptides (e.g., neutrophils, mononuclear phagocytes, polymorphonuclear (PMN) and natural killer (NK) cells, macrophages), that represent the host defense first line and operate nonselectively against abnormal and foreign materials/antigens via expressing non-specific pattern recognition receptors (Murphy and Weaver 2016). The innate system acts speedily to identify and devastate non-self-threatening via inflammatory procedures and then solve the inflammation and redress the harm induced by these agents (Castelo-Branco and Soveral 2014). However, this does not raise the speed or efficacy of responses with reiterated exposing to pathogens. The adaptive immune system comprises both cell- and antibody-mediated responses. It is distinguished by its memory and specificity. These responses are instant and of great strength. Antibody-induced immunity is obtained by producing specific antibodies by mature B cells. T cells are the main element of the cell-mediated immunity. CD8 cells are cytotoxic/suppressor cells, while $CD4^+$ phenotype of T cells are inducer/helper cells. These T-cell subtypes work in harmony to regulate and modulate the immune responses. On the basis of the cytokine profile, $CD4^+$ cells can be distinguished into Th1 and Th2 cells (Mosmann and Coffman 1989; De Martinis et al. 1999). Th1 cells produce IFN-γ and IL-2, IL-12, and IL-18, which are pivotal for cell-mediated immunity, while Th2 cells promote the production of antibodies including

IgM-IgG1 and IgE. Also, they produce IL-4, IL-5, IL-6, IL-10, and IL-13. Activating Th2 responses is accompanied by atopy. Developing an effective immune response needs an equilibrium between Th2 and Th1 subsets, and inappropriate deviated responses are linked with pathology (diabetes, Crohn's disease, food allergies, etc.). The activated mechanism relies on the nature of the agents that cause diseases. Intracellular parasites, bacteria, and fungi are monitored by T-cell activation and secretion of cytokines as IFN-γ that stimulate macrophages/monocytes. NK cells and cytotoxic T cells remove the intracellular viruses and cancerous cells. Extracellular viruses and bacteria are controlled or eliminated by complement, phagocytic cells, and antibodies (Murphy and Weaver 2016; Pandya et al. 2016). The adaptive response is engaged subsequently to innate responses. Thus, the effective host's responses depend on the interaction between the different adaptive and innate immune systems components. While adaptive system is accountable for immunological memory generation, whereby the same pathogens reiterated infection, it will produce a fast, vigorous antigen-specific response (Murphy and Weaver 2016; Pandya et al. 2016). Modulation and regulation of the immune system are essential in an over-activation of the immune system as in allergies and autoimmune diseases to overcome these conditions.

Immune system's modulation refers to any change in the immune response, includinginduction, expression, prohibition, or amplification of any phase or part of the immune response. Immunomodulators are substances that affect the immune system responses to any threat upon it. Thus, they are classified into immune stimulators and immune suppressants (Saroj et al. 2012). Clinically, their potential uses include suppressing excessive or normal immune function and the reconstitution of the immune deficiency for treating various health disorders. It is well established that nutrition is one of the leading exogenous factors that has a remarkable role in the optimal functioning and maintenance of the immune system, exactly by aiding to fix the inflammatory responses (Gombart et al. 2020). Nutritional factors and nutrients can assist to maintain good health and influence all human biology aspects by joining the nutrients metabolism and immune system. Nutrient deficiencies and even their suboptimal status lead to impairment of the immune functions, e.g., decrease in the lymphocytes numbers, impairment of killing microbes, and phagocytosis by the innate immune cells, alter the cytokine production, and reduce responses to antibody, as well as impairments in wound healing (Gombart et al. 2020). Sufficient nutrition is essential to assure good supplies of energy sources, micronutrients, and macronutrients that are needed for maintaining, developing, and expressing the immune responses (Maggini et al. 2017). Micronutrients such as vitamins, selenium, beta-carotene, folic acid, zinc, and iron play pivotal roles for the immune system and are needed to sustain immune responses (Alpert 2017).

Among nutrition, infections, and immunity, there is a bidirectional interaction. When nutrition is poor, the immune response is compromised, making individuals liable to infections, and a mal-nutritional condition may be triggered to an infection by the immune response (Maggini et al. 2018).

The Middle East region is an extremely diversified region in the world in terms of political structures (monarchies, republics), ecology (dry yellow deserts and green valleys), stability (civil wars conflicts, and unrest), and economic divergence (including countries that are classified among the world's poorest and richest), all these factors determine nutritional status and health of this region (Galal 2003). MED is a nutritional model inspired by the traditional dietary pattern of the peoples living in Middle Eastern (ME) countries, reflecting the heterogeneous environmental and historical context of this region. Diets vary between these countries and also between regions within the same country. In this work, different nutrients included in MED will be reviewed with focus on their effects on the immune system and action mechanisms.

7.1.1 Milk and Dairy Products

Dairy products and milk are main constituents of the daily diet and represent a wealthy source of bioactive molecules that are able to affect a wide range of physiological functions. In ME countries, milk was available from different animals such as sheep, cows, goats, buffalo, and camel.

Camel milk has been recognized with several health benefits in diseases, including jaundice, dropsy, cancer, diabetes, infant diarrhea, hepatitis, lactose intolerance, allergy, asthma, tuberculosis, piles, anemia, food allergies, autism, and liver damage that can be assigned to the existence of many immunologically significant molecules such as lactoferrin, lysozymes, lactoglobulins, lactalbumin, immunoglobulins, and lactoperoxidase that stimulate both innate and adaptive immunity as well as insulin-like molecules' high levels (Singh et al. 2017). Also, it was reported that camel whey protein (CWP) exhibits strong antioxidant properties via reduction of oxidative stress, increases glutathione levels, and enhances immune system function (Badr et al. 2017). Additionally, camel milk weakened the fibrovascular tissue's main components, macrophage recruitment, vascularization (Hb content), collagen deposition and ILs (IL-1β, IL-6, and IL-17), VEGF (vascular endothelial growth factor), TGF-β (transforming growth factor), and TNF-α levels (Saad 2015).

Goat milk is one of a nutraceutical health drink with better digestibility, buffer capacity, alkalinity, and therapeutic values compared to that of cow milk. Goat milk and its products act as immunity booster and are able to protect individuals from various illnesses such as atherosclerosis, hypertension, metabolic and malabsorption disorders, anemia, osteoporosis, colitis, obesity, diabetes, and inflammatory bowel diseases. Goat milk was able to activate release of NO from blood cells and trigger production of cytokines (IL-10, TNF-α, and IL-6) (Lad et al. 2017).

In ME countries, the main traditional fermented dairy products are Raib, Leben, Jben, Aoules, Klila, Zabadi, Karish (Arish, Kariesh) cheese, Tallaga cheese, Domiati cheese, Mish cheese, Zebda beldia (zebda baladi/zebda beldi), Rigouta, Laban, Laban Zeer, Kishk, Labaneh, Shenglish, Shenineh, Keshkeh, kefir, Akawieh,

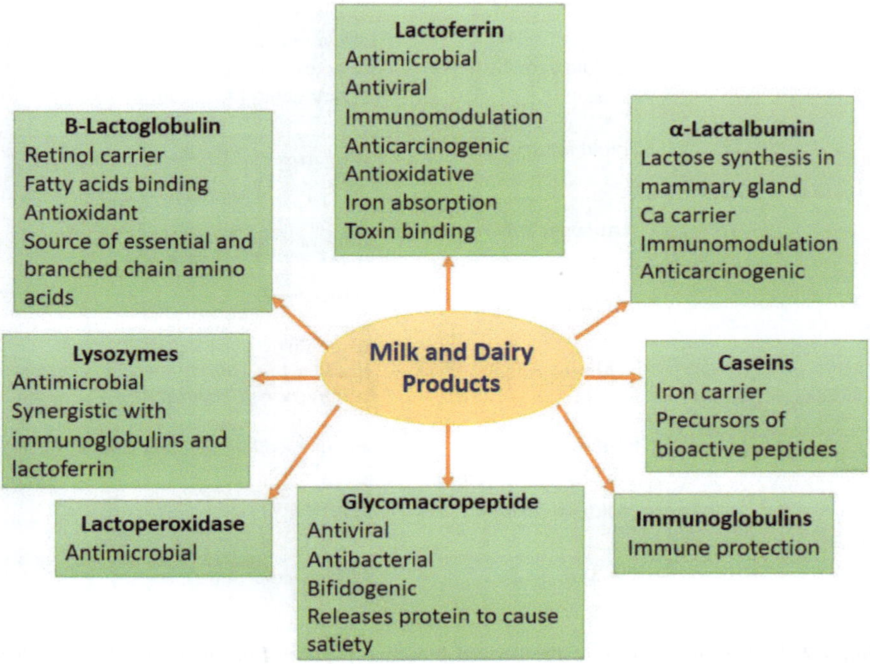

Fig. 7.1 List of milk components and fermented dairy products and their biological activities

Kashta, Akkawi, Kashkaval, and Chelal (Benkerroum 2013; Deoula et al. 2018). They are capable to affect both nonspecific and specific host immune responses aspects (Beaulieu et al. 2006; Cross and Gill 2000; Korhonen et al. 1998). Figures 7.1 and 7.2 illustrate the components of milk and fermented dairy products and their biological activities. It was reported that many fermented dairy products contain active bacterial cultures such as probiotic bacteria that belong to the lactic acid bacteria (LAB) family. LAB supported healthy gut microbiota that has a significant role in maintaining and building the immune system, both systemically and at the intestinal barrier (Gil and Ortega 2019). Also, it enhances the oxidative defense mechanisms, including secretion of cytokines by phagocytic cells and reactive oxygen species production. They boost the phagocytic activity of blood leucocytes (mononuclear and PMN cells) and NK-cell activity (Gill 2003). Also, LAB contribute to human gut healthy microflora, detoxification of xenobiotics and microbial toxins, biosynthesis of vitamins (folic acid, K1, vitamin B12, and biotin), increase of the minerals absorption (e.g., iron, calcium, and magnesium), modulation of intestinal gas production, fermentation of lactose, and short-chain fatty acids (SCFAs) production (Dargahi et al. 2019).

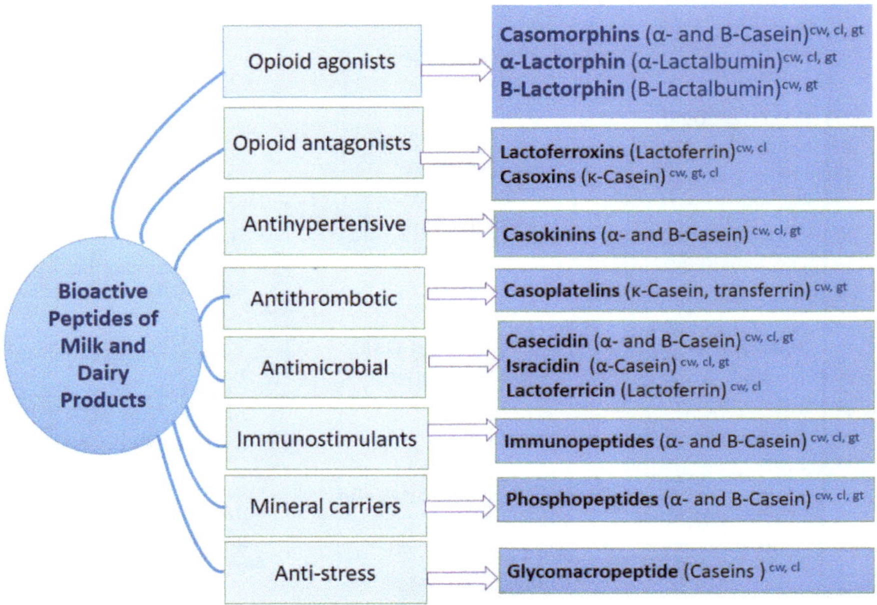

Fig. 7.2 Biological activities of the derived bioactive peptides from milk and dairy product proteins. *cw* cow milk, *cl* camel milk, *gt* goat milk

7.1.2 Hot and Cold Beverages

Many beverages in ME countries are plant-derived. They are used as complementary medicine, and most people use them for their health care to manage diabetes, reduce high cholesterol level and body weight, treat microbial infections, protect against cancer, and augment the function of the immune system. Their beneficial properties for humans rely on their bioactive metabolites such as flavonoids, alkaloids, polysaccharides, terpenoids, lactones or sesquiterpene, and essential oils. They influence the immune system either directly on the immune cell functions in both innate and adaptive immunity and by direct antiviral, antimicrobial, or antifungal properties or indirectly by reduction of inflammation, modulation of nonimmune cell function, modulation of angiogenesis, or affecting the cytokines secretion (Lewicka et al. 1958; Leyva-López et al. 2016; Zhai et al. 2007). In the current part, the most commonly used beverages in ME countries and their immunomodulatory activity are discussed.

Karkade or roselle (*Hibiscus sabdariffa*, Malvaceae) calyx is used all over ME either as cold or hot beverage for management of hypertension, liver diseases, and fever. The water extract contains mainly cyanidin-3-*O*-sambubioside, delphinidin-3-*O*-sambubioside, rutin, quercetin, and chlorogenic acid (Kraus and Franz 1992). The dried calyx water and alcohol extracts stimulated the immune system in mice through production of IL-10 and inhibition of TNF-α in addition to its effect on B cells which is responsible for antibody production (Kraus and Franz 1992; Fakeye et al. 2008). Residual H_2O- and EtOAc-soluble fractions of *H. sabdariffa* showed significant immunomodulatory effect in the experimental animals (Shen et al. 2017). Moreover, heteropolysaccharides, HSP-II, isolated from its flowers showed immune enhancement potential through stimulating macrophages via NF-kB and MAPK signaling pathway (Shen et al. 2017). Recently, its leaves that are used as salad in Sudan also showed immunomodulatory activity that needs further investigation to explore its mechanism of action (El Senousy 2019).

Tea (*Camellia sinensis*, Theaceae) decoction is used as a hot drink in the Arabian region which has potential immunomodulatory and anti-inflammatory action (Chattopadhyay et al. 2012). Tea is considered as a wealthy fountainhead of phenolics, including flavones, catechins, thearubigins, and theaflavins. Its aqueous extract showed immunomodulatory effect through enhancing production of neopterin (cell-mediated immunity sensitive marker) and significant increase in T lymphocytes in vitro in unstimulated HPMC (human peripheral mononuclear cells). Moreover, (−)-epigallocatechin-3-gallate (EGCG) can protect against UVB-induced immunosuppression (Parnham et al. 2019). Epigallocatechin (EGC) and EGCG act as immunomodulator by affecting the T-lymphocyte proliferation and cytokine production and stimulate the production of monocytes, IL-1α and IL-1β, and lymphocytes (Rahayu et al. 2018). In the Middle East especially in Saudi Arabia, regular tea is mixed with cinnamon and cardamom with special ratio to prepare what is known as Karak chai or masala chai. Karak tea is an Indian tea that has a strong flavor and beautiful caramel color. Immunomodulatory activities of cinnamon and cardamom are described in other part of this chapter.

Coffee bean (*Coffea arabica*, Rubiaceae) is originated in Yemen and Ethiopia and considered one of the economically important plants. Roasted coffee beans are common drink in the ME region as well as all over the world. Its popularity referred mainly to the presence of stimulant alkaloid, caffeine, and its characteristic aroma that are produced during roasting process. The seeds are rich in alkaloids, flavonoids, and tannins. Moreover, it contains carbohydrates such as galactomannan, cellulose, and arabinogalactan protein. Arabinogalactan protein (AGP) separated from the instant coffee beans powder influenced immunocompetent cell mediators as TNF-α, IL-2, and IFN-γ cytokines. It is noteworthy that AGP is a good inductor of IFN-gamma and TNF-α, however is less powerful in TNF induction comparing with that of β-D-glucan (biological response modifier) (Nosáľová et al. 2011). Coffee seed alcoholic extract possessed also stimulatory effect on cyclophosphamide-induced immunosuppression and cellular immune function in mice. This effect was mediated through increase DTH (delayed-type hypersensitivity) response and WBC count (Haque et al. 2013). In Arabian region, coffee is a popular traditional drink locally named Qahwa that represents a sign of hospitality. It is blended with clove in a small ratio to intensify its taste. In addition, green coffee is a famous traditional hot beverage in Arabian region especially Saudi Arabia. Green coffee is grind and mixed with ginger, saffron, cardamom, and clove in specific ratios.

Peppermint (*Mentha piperita*, Lamiaceae) leaf infusion is a common hot beverage for digestive disorders due to its volatile constituents (mainly menthol and menthyl acetate). Peppermint oil exhibited a moderate inhibitory potential on T-cell proliferation and potent TNF-α inhibitory effect. Menthofuran and menthol could contribute to the elevated TNF-α suppressing potential of *M. piperita* oil. In addition, menthone, one of the bioactive compounds of the oil, can repress TNF-α and IL-1B (Orhan et al. 2016). Moreover, the alcohol extract of peppermint leaves attenuated oxidative stress and improved survival of macrophages (Ortuño-Sahagún et al. 2017).

Wild mint or habak (Mentha longifolia, Lamiaceae*)* is commonly known as habak, hasawy, or wild mint in western Saudi Arabia region (Almadinah Almunawwarah). It is widely utilized in traditional medicine due to its volatile oil contents for treating several gastrointestinal disorders such as abdominal pain, diarrhea, gut spasm, and ulcers (Murad et al. 2016; Al-Ali et al. 2013; Azhar et al. 2019). Dichloromethane, ethyl acetate, hexane, and butanol fractions of *M. longifolia* were tested for immunomodulatory effects on human peripheral blood lymphocytes (PBLs). *M. longifolia* was reported to reduce IFN-γ levels while boosting IL-4 secretion indicating its capacity to prohibit TH1 inflammatory response toward Th-dominant response. Meanwhile, the aqueous extract showed a weak response (Asemani et al. 2019). Moreover, in broiler chicken diets, *M. longifolia* raised antibody titers toward Newcastle disease virus, indicating that essential oil stimulated the immune system (Talazadeh et al. 2016). Moreover, it was found that feeding of rainbow trout with 0.2% hydroalcoholic extract of *M. longifolia* resulted in improvement of mucosal immunity indices due to the positive effect of its major constituents on immunogenicity and increased rainbow trout resistance to bacterial disease (Heydari et al. 2020).

Liquorice (roots and rhizomes of *Glycyrrhiza glabra*, Leguminosae) is consumed as cold beverage in some regions especially Egypt. It contains varieties of constituents including glycyrrhizin, a triterpenoid saponin, and flavonoids, viz., liquiritin, isoliquiritin, liquiritigenin, and isoliquiritigenin (Pastorino et al. 2018). The grind root and rhizomes are soaked in cold water in presence of sodium carbonate to increase glycyrrhizin in the prepared beverage. *G. glabra* root aqueous extract in vitro was reported to be related to the existence of glycyrrhizin that causes increase lymphocyte and macrophage production from human granulocytes (Pastorino et al. 2018). It showed also antiviral activity against herpes simplex, varicella zoster, and hepatitis. Glycyrrhizin has excellent immunostimulant properties by activating T-lymphocyte proliferation (Pastorino et al. 2018).

7.1.3 Vegetables

Tomato (*Lycopersicon esculentum*, Solanaceae) contains lycopene, lutein, *β*-carotene, zeaxanthin, hydroxycinnamic acid, flavonoids, and glycosides. The leaves exhibited cytotoxic potential toward carcinoma and hematological cell lines, including multidrug-resistant and drug-sensitive phenotypes. It increased the levels of lymphocytes, WBCs, and platelets due to the immunostimulatory potential of the chemical constituents in the extract (Nguenang et al. 2020).

Garlic (*Allium sativum*, Amaryllidaceae) contains diversified bio-compounds (diallyl sulfide, alliin, allicin, diallyl trisulfide, diallyl disulfide, S-allylcysteine, and ajoene). Garlic and its biometabolites exhibited anti-inflammatory, antioxidant, antifungal, antibacterial, immunomodulatory; anticancer; cardiovascular-, renal, neuro-, hepato-, and digestive system protective; anti-obesity; and antidiabetic properties. Garlic polysaccharides had an immune-modulatory potential and regulated interferon-γ, IL-10, TNF-α, and IL-6 expressions in RAW 264.7 because of fructan constituents' degradation during processing. The combination of garlic oil and levamisole can greatly balance the Th1/Th2 response in Wistar rats. Furthermore, the AGE (aged garlic extract) consumption was established to diminish the severity and occurrence of the flu and cold and ameliorate the human immune system functions. In general, polysaccharides represented the main garlic immunomodulating components (Shang et al. 2019). Moreover, *A. sativum* modulated secretion of cytokines which may allocate the mechanism of action for many of its therapeutic potentials (Arreola et al. 2015).

Onion (*Allium cepa*, Amaryllidaceae) and red onion scales (ROS) have flavonoids that are accountable for the established antioxidant, hepato-protective, anticancer, and immune-stimulant capacities. ROS have large amounts of phenolics, mainly gallic, kaempferol, ferulic, quercetin dimer and trimer, quercetin, and quercetin glycosides, which are accountable for its antioxidant potential in comparison to α-tocopherol. The ROS methanolic extract had the capability to reduce the weight of prostate in APH-induced rats because of the immune-modulatory and anti-inflammatory potentials of the extract (Elberry et al. 2014).

Pumpkin (*Cucurbita pepo*, Cucurbitaceae) contains several compounds, including polysaccharides, antifungal proteins, such as α- and β-moschins and myeloid antimicrobial peptide. In addition, it contains β-ryonolic acid, a naturally occurring triterpenoid, identified in the Cucurbitaceae family. It is characterized by its anti-allergic, cytotoxic and antitumor activities due to β-ryonolic acid. Its methanolic and ethyl acetate extracts were more potent than the chloroform extract concerning the innate immunity model. Mainly, the cucurbitacins in addition to proteins and vitamin E are responsible for immune-modulating activities of *C. pepo*. The polysaccharides and polyphenols present in the fruit of pumpkin possessed immune-modulatory effects (Jafarian et al. 2012).

Brassica vegetables (cruciferous vegetables) are another example of immunomodulator vegetables. A high consumption of these vegetables has been linked to a low prevalence of chronic diseases, including various cancer types. *Brassica oleracea* has about seven cultivars which are closely related as rocket (arugula), broccoli, or cabbage which have many beneficial effects due their sulfur-containing glucosinolates (GLSs). Their high consumption is correlating with advantageous effects on cancer development. A cruciferous vegetable-rich diet could promote the health by the activity of isothiocyanates (ITCs) that produced as breakdown products of glucosinolates (GLSs) (Sturm and Wagner 1890). They exhibited anti-inflammatory, immunomodulatory, and cancer preventive properties. They are consumed cooked or rarely raw, and the processing might influence its health-promoting efficacy. *Brassica* vegetable consumption may regulate telomerase effect of T-cell subsets, but processing appears to have a decisive modulating factor. The immune system strongly relies on cell division and clonal expansion, therefore has promoted regulation of telomerase as a strategy for maintenance of telomere (Tran et al. 2019).

Beetroot (*Beta vulgaris*, Chenopodiaceae) is considered as food, food coloring, or medicine. Beetroot pomace and its juice act as anti-inflammatory and antioxidants, and they lowered DNA damage and plasma protein carbonyls in blood leukocytes. It is rich in pigments as betacyanin and betaxanthins of betalain family. Betalains are linked to anti-inflammation, antitumor, and anti-oxidative stress effects. Betaxanthins and betacyanins obtained from beetroot have antioxidant effect. Betanin (the major betacyanin pigment) exhibited a powerful anti-inflammatory potential by scavenging hypochlorous acid and by prohibiting cyclooxygenases (COXs) generated by neutrophils in inflammation. Beetroot remarkably minimized the DNA strand breaks in splenocytes that exposed to irradiation and activated the proliferation of radiosensitive immune cells which reflects its immunostimulatory potential (Cho et al. 2017).

Sweet potato (*Ipomoea batatas*, Convolvulaceae) contains botanical polysaccharides which have immunomodulatory effects. Also, it contains glucan [α-(1→6)-D-glucan] which stimulates lymphocyte proliferation and activates macrophages and NK cells (Yin et al. 2019).

Okra (*Abelmoschus esculentus*, Malvaceae) is a vegetable crop, used traditionally to treat dysentery and diarrhea. Its fruits exhibited antibacterial effect due to tannins. Okra contains vitamin C and flavonoids as antioxidants. Polysaccharides from okra plant are immunomodulators by enhancing the spleen index, phagocytic activity, TNF-α level, and splenocyte proliferation but decreasing level of IL-17 as a response to ban proinflammatory cytokine overexpression, thus considered as effective compound to improve immune response (Wahyuningsih et al. 2018).

Purslane (reglah in Arabic) (*Portulaca oleracea*, Portulacaceae) is an annual grassy plant. It is acting as a febrifuge, diuretic, antispasmodic, vermifuge, and antiseptic, and it has analgesic, skeletal muscle relaxant, antibacterial, anti-inflammatory, wound healing, anticonvulsant, and radical scavenger capacities. It has been reported that its ethanolic extract (10%) had analgesic and anti-inflammatory potentials comparable to synthetic drugs. Its seeds and leaves can be applied topically or used orally to alleviate skin allergy. *P. oleracea* exerted its effects through balancing the innate and adaptive immune system depending on the situation and anti-inflammatory potentials. Moreover, it exhibited immune-modulator and antioxidant effects in the inflammatory cases of asthma, atopic dermatitis, and cancer by the dominance of Th2 response and evoked Th1 disorders, including hepatitis and multiple sclerosis (Rahimi et al. 2019).

Molokhia (*Corchorus olitorius*, Tiliaceae) is a medicinal herb that is used as vegetable in different countries especially Egypt. The leaves of the plant are used in preparing a famous Egyptian dish. The leaves are rich in mucilage and vitamins (D, C, and E), in addition to antioxidant compounds as phenolics and β-carotene.

Medicinally, it showed a potent anti-inflammatory as well as immunomodulatory activities (Ismail et al. 2018). Water-soluble extract of Molokhia (the famous Egyptian dish) exerts immunostimulant activity in immunosuppressed mice through enhancement of NO production as well as immune-related cytokines. Moreover, it augmented NK-cell cytotoxicity toward YAC-1 cells by stimulating the granzyme B release (Park et al. 2018).

7.1.4 Fruits

Entire fruits (e.g., frozen, fresh, dried, or canned) are renowned for their very low to moderate energy density and fiber content and as being substantial sources of phytochemicals (e.g., carotenoids and polyphenols) and healthy nutrients (e.g., vitamin C and potassium), which assist a vast range of health benefits by working synergistically. The Dietary Guidelines for Americans (2015–2020) designated fiber as a main shortfall nutrient of remarkable public health interest. Fiber is known as resistant starch, oligosaccharides, and carbohydrates from plant cell walls that have resistance to gastric acidity, mammalian enzymes hydrolysis, and absorption in the upper GIT. The rich fiber diets improved human GIT and immune and cardio-metabolic systems (Dreher 1833).

Celery (*Apium graveolens*), parsley (*Petroselinum crispum*), fennel (*Foeniculum vulgare*), and coriander (*Coriandrum sativum*) are popular vegetables and spices of the family Umbelliferae. They are characterized by the presence of flavonoids (e.g., rutin, quercetin) and coumarins (e.g., isopimpinellin, bergapten, xanthotoxin). It was reported that these bioactive constituents (mainly xanthotoxin) at cytotoxic concentration showed immunosuppressive effects. Meanwhile nontoxic doses of these plants or their constituents might influence health as immunostimulant, i.e., directly reinforcing multipotent cytokine IFN-gamma secretion and/or lymphocyte activation (Cherng et al. 2008).

Fennel is a wealthy pool of essential oils, which have antioxidant, hepato-protective, antibacterial, emmenagogue, antithrombotic, and antifungal properties. The essential oil of fennel showed a significant immunomodulatory effect through stimulation of the proliferation of PBMC as well as gamma-interferon secretion, which is closely connected to activation of lymphocyte, following mitogen or antigen stimulation (Orhan et al. 2016). Addition of fennel extract to chicken diet significantly improved Newcastle vaccination efficiency and immunoglobulin production and maintained great immunity toward viral infections, bacteria, and new infections. It also had low resistance to infectious bursal disease but higher resistance to infectious bronchitis virus (Safaei-Cherehh et al. 2020; Valdivieso-Ugarte et al. 2019).

Anise is used historically due to its volatile constituents in treatment of colic and aiding in digestion, also promoting milk production in lactating women and protecting against osteoporosis due to its selective estrogen-modulating effect. Recently, it was proved its ability to activate cell-mediated immune mechanisms in mice (Al-Omari et al. 2018).

Pomegranate (*Punica granatum*, Punicaceae) different parts showed specific phytochemical constituents in addition to some bioactivities. The juice possessed antihypertensive, anti-atherosclerotic, potent anti-oxidative, and antiaging properties. It contains anthocyanins, iron, amino acids, glucose, and ascorbic acid, in addition to some phenolics as ellagic, caffeic, and gallic acids, epigallocatechin, catechin, rutin, and quercetin. The seed oil contains mainly sterols and punicic acid, which have nephro-protective capacities. The pericarp contains punicalagins in addition to some flavonoids that possessed antifungal and anti-inflammatory activities. Tannins (e.g., punicafolin and punicalin) and flavones glycosides (e.g., apigenin and luteolin) are the major pomegranate leaves constituents. The flowers contain triterpenoids (e.g., ursolic, asiatic, and maslinic acids) showing antioxidant and hepato-protective capacities and are utilized as a medicament for diabetes mellitus. While the bark and roots contain ellagitannins and piperidine alkaloids, acting as anthelmintic and vermifuge (Prasad and Kunnaiah 2014). *P. granatum* fruits have been utilized for years in traditional medicine due to its antimicrobial and immunomodulatory potentials. Dietary *P. granatum* by-products (PGB) may affect the performance and immunity. Serum IgG and IgA linearly raised in response to PGB (Ahmed and Yang 2017). The aqueous leaf extract of *P. granatum* (PG) exhibited alleviated levels against ovalbumin antibody responses, while its ethanolic fraction has been found to have highest activity (Vadlamani et al. 2016). Immunomodulatory activity of its fruit juice was evaluated through in vivo studies on Wistar albino rats (Sibi and Varghese 2014). The fruit juice exhibited notable effect on humoral and cell-mediated immunity through a significant effect on the percentage of neutrophil adhesion, myelosuppression, and HA titer, when compared to the standard (Levamisole) group (Sibi and Varghese 2014).

Date palm (*Phoenix dactylifera*, Arecaceae) has been developed over no <6000 years ago in the Middle East. The therapeutic effects of *P. dactylifera* are linked to its content of polyphenols, phenolic acids (gallic, hydroxybenzoic, syringic, protocatechuic, and vanillic acids), and flavonoids (apigenin, quercetin, luteolin, proanthocyanidins, and anthocyanins), and different methylated and sulfated flavonoid glycosides also exist in dates. Sulfates are linked to date flavonol glycosides; thus dates are only fruits that contain flavonoid sulfates. Anthocyanins present in fresh fruit only and carotenoids lowered with fruit ripening (Yasin et al. 2015). The effect of mannan-oligosaccharides and β-glucan of *P. dactylifera* seeds was examined on the immunity (El-Far et al. 2016). Supplementation of the broilers with ration having dates at levels of 2% and 4% had a high valuable improving in broiler health with activation of IFN-γ, antibody titer, IL-2, and antioxidant state comparing to β-glucan and mannan-oligosaccharides. Including the dates aqueous extract at a 2 mg concentration in a liquid paraffin oil adjuvanted infectious coryza vaccine could be utilized to enhance immune responses. The aqueous extracts of *P. dactylifera* fruit showed immunomodulatory potential and elevated serum antibodies (Osman et al. 2020).

Mango (*Mangifera indica*, Anacardiaceae) fruit peel and flesh are characterized as a rich source of essential amino acids, fiber, vitamins (C and A), and polyphenols. The phytochemical composition of mango seed contains polyphenols as mangiferin, quercetin, catechins, kaempferol, rhamnetin, phenolic acids (e.g., protocatechuic, ellagic, and gallic acids), anthocyanins, propyl and methyl gallate, and resorcinolic lipids in addition to carotenoids which are accountable for the bright yellow color of the fruit flesh and peel (Ediriweera et al. 2017). The methanol extract of mango leaves in suitable doses exerted immunostimulant effect in mice by reinforcing both adaptive and innate immune systems via increasing spleen index, white blood cells, hemagglutination titer, and DTH (Kumolosasi et al. 2018). The immunomodulatory potential of the ethanolic extract of *M. indica* cultivar Neelam fruit pulp was proved by increasing in humoral antibody (HA) titer and DTH in mice (Naved et al. 2005).

Guava (*Psidium guajava*, Myrtaceae) leave and fruit metabolites and extracts are considered to have antispasmodic, antimicrobial, and anti-diarrhea properties. Moreover, it exhibited hypoglycemic, antitussive, antioxidant, and anti-inflammatory activities (Fernandes et al. 2014). The leaves' ethanol and methanol extracts also inhibited hypotonicity-induced lysis of erythrocyte membrane, and their leaves are considered as immune stimulants as they modified the proliferation response of lymphocyte (Díaz-de-Cerio et al. 2017).

Cantaloupe melon (*Cucumis melo* L. var. *reticulatus,* Cucurbitaceae) includes seeds that are rich in proteins, fibers, and lipids, mainly PUFAs, especially linolenic, in addition to some minerals as potassium, phosphorus, sodium, and magnesium. Their flavonoid and phenolic contents show antiproliferative and antioxidant properties. They contain essential amino acids, such as methionine, isoleucine, tyrosine, valine, and phenylalanine (da Cunha et al. 2020). They had high levels of glycolipids, phospholipids, cucurbitacins, and trypsin inhibitors. The combination of superoxide dismutase (SOD)-rich melon extract and wheat gliadin (glisodin) raised the production of Th1 T lymphocytes and the expression of IL-4 and INF-gamma which reinforced its immunomodulatory action through the activating APC (antigen presenting cells) by that combination (Parle and Kulwant 2011).

Olive fruits (*Olea europaea*, Oleaceae) give olive oil that is used as main source of fat in some ME countries. The olive oil health benefits are specially linked to the intake of extra virgin olive oil (EVOO) with its multiple health positive effects and high nutritional quality. It was reported that the high content of MUFAs, vitamins, and minerals as well as phenolic compounds (phenolic acids, lignans, flavones, flavone glycosides, phenolic alcohols, and secoiridoids), tocopherols, phytosterols, and pigments are accountable for the protective effects of EVOO (Mazzocchi et al. 2019). Several studies suggested that VOO and especially its phenolic compounds possessed preventative effects on various immuno-inflammatory diseases. Its phenolic compounds reduced the secretion of inflammatory cytokines and nitrite and expression of PPARγ, Toll-like receptor 4, and inducible nitric oxide in LPS-treated human monocytes. Also, they prohibited the deregulation of human monocyte subset distribution by LPS and obstructed the M1 macrophages genetic signature while favoring the M2 macrophages phenotype upon canonical polarization of naïve human macrophages (Aparicio-Soto et al. 2018). These phenolics also prevented oxidative damage and beneficially modify immune and inflammatory responses (Carluccio et al. 2003).

7.1.5 Nuts and Seeds

Nuts and seeds have been considered as a part of the human diet since ancient times. Seed is a mature ovule, while nuts are simple fruits containing edible kernels inside hard shells (De 2020). Famous ME nuts include mainly almonds, hazelnuts, pine nuts, pistachios, peanuts, walnuts, and chestnuts (Salas-Salvadó et al. 2011). Meanwhile, flaxseed, nigella, and fenugreek are common seeds in MED. Seeds and nuts are prosperous in essential or polyunsaturated fatty acids (EFAs, PUFAs) (the body cannot synthesize some fatty acids and must be obtained from food, e.g., omega-3 (ω-3) and omega-6 (ω-6)) (Kumar et al. 2019). Linoleic acid (LA, *n*-6 PUFA) is the precursor of ω-6 family and dominates in plant oils as sunflower and soya. Linoleic acid is metabolized to arachidonic acid (AA) which is the precursor of PGs (prostaglandins) and LTs (leukotrienes) that possess potent proinflammatory and immunoregulatory properties.

Meanwhile, α-linolenic (ALA, *n*-3 PUFA) is the precursor of ω-3 family that dominates in green plant tissue and gave eicosapentaenoic (EPA) and docosahexaenoic (DHA) acids on metabolism. Consumption of ω-3 PUFAs is associated by low incidence of autoimmune and inflammatory disorders. It causes lowering in arachidonic acid levels and increasing levels of EPA and DHA that inhibit conversion of arachidonic acid to PG (Calder et al. 2020). Moreover, dietary ω-3 PUFAs influence functions of lymphocytes. Feeding animals with fish oil resulted in decreasing NK-cell activity, lymphocyte prefoliation, and IL-2 production (Calder 1998). Moreover, ω-3 PUFAs can minimize PBMC proliferation and reduce production of IL-6, IL-1, and IL-2 and TNF-α from PBMC (Calder et al. 2020).

Walnuts (*Juglans regia*, Juglandaceae) are prosperous in unsaturated fatty acids, proteins, carbohydrates, and minerals. Walnut protein (glutelin, albumin, and globulin) has a great nutritional value and health benefits as the peptides derived from them showed antifungal, antioxidant, antitumor, hypotensive, anti-inflammatory, and hypoglycemic activities. Oligopeptide derived from its protein enhanced adaptive and innate immunity through increasing macrophages phagocytic capacity, activating NK cells, and improving humoral and cell-mediated immunities (Mao et al. 2020). A novel peptide, CTLEW, with a 651.2795 Da molecular weight was obtained from hydrolysates of walnut residual protein. This peptide showed immunomodulatory activity through enhancing proliferation and secretion of IL-2 from spleen lymphocytes, macrophage NO production, and promoting phagocytosis (Ma et al. 2015). Walnuts contain a unique balance of *n*-6 and *n*-3 PUFAs that influence greatly the immune system. Its deficiency leads to depression of humoral and cellular tolerance responses. Low-fat diet containing little ω-3 causes increase in production of proinflammatory IL-1 and TNF-α cytokines. Meanwhile, diet rich in ω-3 significantly reduced the percentage of T-CD4 cells, increased percentage of T-CD8 cells, and decreased secretion of IL-1, IL-6, and TNF-α (Jedrychowski et al. 2009).

Almond nut (*Prunus dulcis*, Rosaceae) skin is reported to promote the PBMC immune surveillance toward viral infection, both by triggering the Th2 and Th1 subsets (Arena et al. 2010). Almond skins can regulate immunological responses and act positively as novel antiviral agents (Masihuzzaman et al. 2020).

Hazelnut (*Corylus heterophylla*, Betulaceae) is rich in different nutrients, including protein, fat, carbohydrates, fibers, and vitamins. Dregs is the by-product of hazelnut oil production and contains about 58% protein where it considered as fodder and fertilizer. Hazelnut hydrolyzed peptides exerted potential immunomodulatory effects (humoral and cell-mediated immunity) (Ren et al. 2016). The peptide Pro-Glu-Asp-Glu-Phe-Arg (PEDEFR) isolated from protein hydrolysate of hazelnut showed immunomodulatory effect on RAW264.7 cells (Wang et al. 2018).

In pine nuts (*Pinus koraiensis*, Pinaceae), a novel peptide RGAVLH (Alg-Gly-Ala-Val-Leu-His) with a MW of 661.84 Da was purified from its protein and showed immunomodulatory activity (Zhang et al. 2019).

Chestnut (*Aesculus hippocastanum*, Sapindaceae) fruit extract could be considered as immunomodulatory medicine as it recruited lymphocytes to mitotic cycle and, simultaneously, lowered the rate of their proliferation. They also elevated the frequencies of B cells and NK cells and influenced on the induction/suppression-balance of immune system by decreasing the frequency of suppressor/cytotoxic CD8 lymphocytes (Brokos et al. 1999).

Linseed (*Linum usitatissimum*, Linaceae) grows in Middle Eastern region. The fixed oil obtained from linseed is rich in ALA and its metabolic product EPA which are accountable for anti-inflammatory activity of the oil. The oil also has antiarthritic and immunomodulatory activity as proved by decreasing joint swelling, IL-6, and circulating TNF-α levels in complete Freund's adjuvant model of arthritis (Calder 1998; Singh et al. 2012).

Black cumin or black seed (*Nigella sativa*, Ranunculaceae) is utilized traditionally in the Arab region based on Prophet Medicine and also known in different culture. The seed contains mainly volatile oil with thymoquinone (TQ) as a major constituent (30–48%). The seed and its major constituent TQ are used as anti-inflammatory, antihistaminic, antihypertensive, anticancer, hypoglycemic, and immunity-boosting capacities. The reported data denoted that *N. sativa* extracts and oil can potentially suppress humoral immune responses (reduction of antibody titer, B-lymphocyte proliferation, and serum IgA and IgM) while enhancing cellular immune responses (enhancing proliferative capacity of splenocytes and T lymphocytes, elevating IL-3, stimulating $CD4^+$ T lymphocytes, and reducing splenocyte, leukocyte, platelet, and neutrophil counts). Particularly, the immunomodulatory effects of TQ on Th1/Th2 differentiation, humoral immunity, and NK cytotoxic activity require further assessment and validation. While some signaling pathways (e.g., MAPK, PI3K/AKT, and NF-κB) have been targeted by *N. sativa* and TQ, but the exact cellular and molecular mechanisms elaborating TQ and *N. sativa* immunomodulatory effects stay largely questionable (Majdalawieh and Fayyad 2015).

Fenugreek seeds (*Trigonella foenum graecum*, Leguminosae) is used traditionally as condiment and used also as hypoglycemic, anti-ulcerogenic, hypocholesterolemic, and antihypertensive agent. The seed contains mainly diosgenin saponins, trigonelline alkaloid, flavonoids, fixed oil, and tannic acid.

The aqueous extract showed immunostimulatory effect in mice as it caused increase in DTH response and increase humoral immunity through elevation of plaque-forming cell (PFC) assay. Also, plant extract produced a considerable elevation in phagocytic capacity and phagocytic index of macrophages (Bin-Hafeez et al. 2003)

7.1.6 Spices

Spices are collections of nonvolatile and volatile dietary additives that make food enjoyable and add variety of color, flavor, and aroma to the diets. They have a diverse type of phytochemicals that have overlapping and complementary actions, including detoxification enzyme modulation, immune system stimulation, lessening of inflammation, and steroid metabolism modulation as well as antioxidant, antiviral, and antibacterial effects. Commonly, they are used as an integral part of the daily diet in the ME cuisine.

Thyme (*Thymus vulgaris*, Lamiaceae) is an aromatic plant native to ME region and used as a spice, expectorant, anti-inflammatory, and antibacterial, and it has beneficial effect in GIT disturbances. Its effect is referred mainly to its volatile constituents that include mainly thymol. Aqueous extract showed immunomodulatory effect through inhibition of human lymphocyte proliferation and increasing CD-40 expression on DCs indicating its capacity to produce activation to these cells. Moreover, it also prohibited both the allogenic and mitogenic T-cell responses (Amirghofran et al. 2012). Moreover, it showed immunosuppressive effects against mitogen (PHA)-stimulated peripheral blood lymphocytes, and it was indicated that thymol was responsible for this effect (Amirghofran et al. 2011).

Nutmeg (*Myristica fragrans*, Myristicaceae) is one of the famous spices all over the world that are used traditionally for stomach and kidney disorders; it has antioxidant, antimicrobial, and CNS stimulant effect. It is recognized by existence of a lot of secondary metabolites, including fixed and volatile oils, triterpenes, and phenolic compounds (Abourashed and El-Alfy 2016). Fresh mace aqueous extract lignans showed immunomodulatory and radio-modifying properties in mammalian splenocytes (Checker et al. 2008).

Cinnamon bark (*Cinnamomum zeylanicum*, *C. cassia*, Lauraceae) contains volatile oil mainly *trans*-cinnamaldehyde, eugenol, and linalool. The bark is reported to be useful as anti-inflammatory, antidiabetic, antimicrobial, blood pressure-lowering, antioxidant, and anticancer effects (Rao and Gan 2014). Cinnamon extract showed promising anticancer activity as it modulated the activity of the cytotoxic immune cells: T cells. Aqueous extract of *C. cassia* increased the antitumor activities of CD8+ T cells in tumor-draining lymph nodes. It also raised the cytolytic molecules (IFN-ϒ and TNF-α) expressed in the surface of CD8+ T cells. Furthermore, it increased the killing activity of CD8+ T cells toward cancer cells (Larasati and Meiyanto 2018). In addition, it was reported that inclusion of cinnamon nanoparticles (optimum level is 3.0 g CNP/kg diet) to diet of Nile tilapia enhanced the innate immunity (Abdel-Tawwab et al. 2018).

Pepper (*Piper nigrum*, Piperaceae) is one of the most important spices that grow in tropical region and characterized by presence of piperine and its isomers isopiperine, chavicine, and isochavicine. It has different biological activities, including antihypertensive, antioxidant, antitumor, antipyretic, analgesic, and anti-inflammatory (Damanhouri and Ahmad 2014). Aqueous extract of pepper enhanced splenocyte proliferation and T-helper Th1 cytokine release by splenocytes but suppressed Th2 cytokine release by splenocytes (Majdalawieh and Carr 2010). Its ethanolic extract boosted immunity in vitro through increasing production of IgM and B-cell differentiation to plasma cells rather than proliferation of splenocytes (Sarker 2012). Moreover, its aqueous extract played an important role in the serial adaptive immune response (Abbas 2016). Moreover, piperine showed immunomodulatory properties as its administration caused increase bone marrow cellularity, total WBC counts in Bal b/c mice, and alpha-esterase positive cells. In vitro immunomodulatory activity of piperine was estimated to boost rifampicin efficacy in a *Mycobacterium tuberculosis* infection murine model. Mouse splenocytes were utilized to assess in vitro immunomodulation of piperine for macrophage activation, cytokine production, and lymphocyte proliferation. Increasing secretion of Th1 cytokines (IFN-γ and IL-2), T- and B-cell proliferation, and macrophage activation were reported in piperine-treated mouse splenocytes. Protective capacity of rifampicin and piperine (1 mg/kg) combination against *M. tuberculosis* was stated due to immunomodulatory activity (Damanhouri and Ahmad 2014).

Cardamom (*Elettaria cardamomum*, Zingiberaceae) seed is a famous traditionally used spice due to its volatile oil content. The oil has antimicrobial, antioxidant, and anti-inflammatory capacities. The activity could be attributed to eucalyptol, the major constituent in the volatile oil (Han and Parker 2017). It was reported that aqueous extract of cardamom exerted immunomodulatory effects through enhancing of splenocyte proliferation, suppressing Th1 cytokine release by splenocytes, enhancing Th2 cytokine release, reducing NO production by macrophage, enhancing the cytotoxic activity of NK cells, and designating its potential anticancer potential (Majdalawieh and Carr 2010). Induction of *E. cardamomum* distillate to rats in combination with doxorubicin helped in reducing its immunosuppression effect and proved its role as immunostimulant agent for chemotherapy. The distillate increased the amount of white blood, lymphocyte, CD8+, and CD4+ cells in a dose-dependent manner in rats treated with doxorubicin(Raksamiharja et al. 2012).

Capsicum, chili (*Capsicum annuum*, Solanaceae), fruit is globally used as pungent stomachic due to its constituents of the pungent principle capsaicin. Capsicum extract was tested for its immunomodulating activity by examining its effect ex vivo on immune cells of gut-associated lymphoid tissue: Peyer's patch. Capsicum extract remarkably raised the number of CD4+ and CD3+ T cells as well as CD 19+ B cells in comparison to the control but not CD11b + cells. Moreover, the percentage of IFN-γ+/CD4+ and IL-2+/CD4+ cells was highly increased. These data indicated that oral administration of capsicum extract might activate the CD4+ T cells, resulting cytokine production as well as CD19+ B cells in Peyer's patches (Park et al. 2010). A standard diet supplanted with capsicum and curcuma given to broiler chickens infected with *Eimeria acervulina* resulted in reducing fecal oocysts and producing higher serum antibody titers. It significantly enhanced local innate immunity and provided high protective immunity toward *E. acervulina* infection (Lee et al. 2009). In addition, mixture of capsicum oleoresin and curcuma oleoresin in diet of chickens orally challenged with virulent *Eimeria maxima* oocysts and *Clostridium perfringens*. They caused reduction in gut lesion scores, reduced serum alpha-toxin levels, and intestinal IL-8, LITAF (lipopolysaccharide-induced TNF-a factor), and IL-17A and IL-17F mRNA levels, while they caused increase in cytokine/chemokine levels in splenocytes (Lee et al. 2013).

Mastic (*Pistacia lentiscus*, Anacardiaceae) is the oleogum-resin obtained from the trunk of the tree. The produced resin showed antioxidant, antiatherogenic, antitumor, antimicrobial, antiarthritic, anticancer, antigout, hypotensive, antifungal, and anti-inflammatory activities. The oleogum resin can delay the progression and onset of irritable bowel disease due to its anti-inflammatory activity mediated by prohibition of NO and PGE2 production and increasing macrophage migration inhibitory factor (inhibit chemotaxis and migration of monocytes/macrophages). In addition, no considerable changes were observed in concentrations of MCP-1 (monocyte chemotactic protein-1), IL-6, and intracellular antioxidant glutathione (GSH), demonstrating that its oleogum resin acts as an immunomodulator on PBMCs by MIF stimulatory and TNF-α inhibitory activity (Rahimi et al. 2010; Bouriche et al. 2016). In addition, previous studies showed that mastic gum-extracted arabinogalactan proteins can prohibit neutrophil activation in the existence of *Helicobacter pylori* neutrophil-activating protein which indicated its ability to play a critical role in *H. pylori*-associated pathologies in gastric mucosa (Kottakis et al. 2009).

Saffron (*Crocus sativus*, Iridaceae) is the stigma obtained from the flower of *C. sativus*. It is a well-recognized spice that has been traditionally utilized in treating diabetes, cardiovascular disease, menstruation disorders, and depression. These activities are referred to some biometabolites including safranal, crocetin, and crocins (Ghaffari and Roshanravan 2019). *C. sativus* at low doses possessed cell-mediated and humoral immunity, thus establishing its traditional claim as a potential immunostimulant and anticancer agent (Vijayabhargava and Asad 2011). The published data revealed that saffron and crocin enhanced the humoral immune responses production and modulated differential and total leukocyte counts, which elucidated the immunomodulatory capacities of saffron. Saffron extract decreased proinflammatory cytokine production and increased anti-inflammatory cytokines as well as shifted Th1/Th2 balance toward Th2-dominant immunity by increasing IL-4 cytokine expression levels rather than Th1 response in different disorders such as constriction injury, diabetes, asthma, and PHA-stimulated human peripheral blood mononuclear cells. Crocetin, crocin, and safranal also reduced the levels of IL-1B, TNF-α, and IL-4 but raised the levels of IFN-γ and IL-10 in serum, plasma, and tissue in several inflammatory disorders, indicating the potential therapeutic efficacy of the plant constituents on inflammatory diseases by shifting the Th1/Th2 ratio toward Th1 activity (Qadir et al. 2020; Zeinali et al. 2019).

Ginger (*Zingiber officinale*, Zingiberaceae) is characterized by presence of pharmacological active compounds related to phenolics as gingerols and shogaols that showed antiemetic, antifever, anti-cough, antidiabetic, anti-inflammatory, anticancer, and anti-hyperlipidemic capacities (Aryaeian et al. 2019). Essential oil of ginger retrieved the humoral immune response in immunosuppressed mice, but did not have any immunomodulatory activity (Carrasco et al. 2009). In addition, the aqueous ginger extract was tested on nonsmoking volunteers to study its effect on their immunity. Ginger extract caused increased the IgM levels mean, which may result in a stronger humoral immunity or antibody response against infections (Mahassni and Bukhari 2019). Moreover, supplementation of ginger in food of rheumatoid arthritis patients resulted in reduction of the disease manifestation and improving immune system via rising FoxP3 gene expression and by reducing RORγt, NF-kB, and T-bet gene expression (Aryaeian et al. 2019). Meanwhile, inclusion of ginger powder at 3% in zebrafish meal resulted in significant increase in Ig level as well as upregulation of gene expression related to the immune system (Ahmadifar et al. 2019).

Clove (*Syzygium aromaticum*, Myrtaceae) is an important spice all over the world and considered as a wealthy source of phenolics such as eugenol acetate and eugenol and included in cosmetic, agricultural, food, and pharmaceutical applications (Kaur and Chandrul 2017). Clove could enhance the immune system due to its different phytoconstituents, through boosting humoral immunity (Bhowmik et al. 2012). Its essential oil boosted the DTH response and raised the total WBC count in mice. Further, it reinstated humoral and cellular immune responses in a dose-dependent manner in cyclophosphamide-immunosuppressed mice (Carrasco et al. 2009). Administration of its extract increased the lymphocyte proliferation activity, lymphoblasts, and ROI secretion of macrophages in Balb/c mice infected with *S. typhimurium* (Wael et al. 2018).

Rosemary (*Rosmarinus officinalis*, Lamiaceae) contains variety of phytoconstituents, including volatile (cineol, alpha-pinene) and nonvolatile (carnosic acid, rosmanol, carnosol, and rosmarinic acid) compounds (Ahmed and Babakir-Mina 2020). Based on animal studies only, it still debated whether or not rosemary can stimulate immune system. There is a complete shortage of human studies illustrating this effect, whereas all implemented animal studies may have restriction (Ahmed and Babakir-Mina 2020). Meanwhile, it was reported that rosmarinic acid inhibited C-3 convertase, resulting in inhibition of in vitro immunohemolysis of antibody-coated sheep erythrocytes by serum of guinea pig. Also, rosemary boosts immunomodulatory effect through increasing PGE2 production and reduce leukotriene B4 production in human polymorphonuclear leucocytes. Moreover, it was reported also that combination of rosemary extract with diet enriched in casein induced cell proliferation in response to phytohemagglutinin and concanavalin A (Ulbricht et al. 2010).

Turmeric (*Curcuma longa*, Zingiberaceae) is widely used all over the world in nutrition and health-care applications due to its major phenolic coloring matter curcumin and its derivatives. In addition, it contains water soluble polysaccharide ukonans A–D. It was reported that aqueous extract of turmeric had immunomodulatory effects through induction of NO production in RAW264.7 macrophages. Moreover, ukonan polysaccharides enhanced immune response through upregulation of telomere function (Pan et al. 2017). *C. longa* also is reported to increase phagocytosis activity of macrophage, resulting in improving Alzheimer disease via increasing phagocytosis and decreasing accumulation of amyloid. Moreover, it can inhibit production of proinflammatory cytokines through macrophages and lymphocytes (Rezvanirad et al. 2016). Curcumin exhibited powerful immunomodulatory effect that can modify the B-cell activation, T-cells, neutrophils, macrophages, dendritic, and NK cells. It downregulated the expression of proinflammatory cytokines including IL-1, IL-2, IL-6, IL-8, and IL-12 and TNF and chemokines through inactivation of the transcription factor NF-κB and enhancing antibody responses (Jagetia and Aggarwal 2007).

7.1.7 Cereals

Naturally, cereals have a great diversity of polyphenols, which possess a wide array of in vivo and in vitro physiological effects. Their protective effects, including improving redox state and function of immune cells in aged or unhealthy subjects resulted from their powerful antioxidant properties. The cereals, buckwheat flour, wheat germ, wheat middling, and fine rice bran contained variable amounts of phenolic acids (gallic, vanillic, sinapic, *p*-hydroxybenzoic, *p*-coumaric, ferulic, acids), rutin, quercetin, oryzanol, and catechin as main polyphenols, vitamins, minerals or fiber. All cereal fractions improved the studied leukocyte parameters such as microbicidal activity, chemotaxis capacity, IL-2 and TNF-α release, lymphoproliferative response to mitogens, as well as GSSG/GSH ratio, oxidized glutathione (GSSG), lipid oxidative damage, and CAT (catalase) activity. Some of these effects may be attributed to the polyphenol antioxidant capacity. Flavonoids are found in small amounts, even though their multiple bio-effects and implications for chronic diseases and inflammation have been commonly described. Cereal phytochemicals show their benefits for health through multifactorial physiological mechanisms, which include boosting of the hormones and immune system, facilitating crossing of substances through the digestive tract, antioxidant and modulation, metabolism, and antiviral and antibacterial activities. The used cereal fractions improve multiple aspects of the immune function. Consequently, innate immune responses, i.e., intracellular superoxide anion levels (microbicidal activity) and chemotaxis (migration capacity), are highly influenced by the cereals wealthy in polyphenols supplementations. Some of them enhanced the chemotaxis of macrophage, while most of them showed powerful stimulatory effect on this function of lymphocytes, since the cause of this effect could be *p*-hydroxybenzoic acid. The cereals' protective effects could be because of the polyphenol metabolism carried out by the microflora of gut. Polyphenols are naturally found in cereals which may have roles for health safeguarding by acting, at least partly as immune function and redox state modulators (Alvarez et al. 2006).

PUFAs play a major role in various bio-functions. Rice bran oil (RBO) is loaded with linoleic acid, an essential n-6 fatty acid. *n-6* FAs are found to have proinflammatory effects as a consequence of increasing in *n-6* FA-derived eicosanoids. RBO is rich in γ-oryzanol, an unsaponifiable fraction compound, with antioxidant properties. Even though γ-oryzanol acts as immune system modulator, it is not accountable for the immunostimulation effect of RBO. Diets enriched with RBO could be helpful in a situation that requires immune response potentiation. The FA composition might be responsible for this effect more than the unsaponifiable fraction (Sierra et al. 2005).

7.2 Conclusion and Future Prospective

The immune system experiences many change over the course of life and protect human body against different pathogens or cancer. Several studies highlighted that diets differentially affect the immune function. Balanced diet is an important factor in modulating immune function as it decreases the risk of infection and cancer. In addition, an inadequate nutrient intake will affect various immune functions. Thus, the selection of a healthy diet is essential in the prevention of diet-related immune diseases. MED is known to be healthy and balanced diet as it is rich in minerals, vitamins, dietary fibers, and essential fatty acids. A scientific challenge must progress on the basis of reliable scientific evidence which studies the possible immunomodulation that MED exerts over the immune functions that have been outlined throughout this review, which brought to light the MEDs and their influences on the immune system. However, current research still lacks many pieces of the complete picture puzzle that explain the mechanisms by which MEDs can generally improve immune disorders. Additionally, there is a lack of adequate research on the effects of MED on the immune system, and additional in vitro and in vivo research is required to provide detailed information and sound evidence for the immune-modulatory effects of MED. There is a need for further chemical investigation of vegetables, fruits, nuts, seeds, and spices to isolate the phytoconstituents that are responsible for the immune-modulator potential and to highlight their possible mechanisms of immune-modulator effect. Moreover, further clinical and mechanistic studies are required to establish the role of MED as immunomodulators. The reported immunomodulatory effects of some MED components require further assessment and validation.

References

Abbas WH (2016) Immunomodulatory effect of aqueous extract of *Piper nigrum* L. in mice model. Int J Sci Technol 143:1–5

Abdel-Tawwab M, Samir F, Abd El-Naby AS, Monier MN (2018) Antioxidative and immunostimulatory effect of dietary cinnamon nanoparticles on the performance of *Nile tilapia*, *Oreochromis niloticus* (L.) and its susceptibility to hypoxia stress and *Aeromonas hydrophila* infection. Fish Shellfish Immunol 74:19–25

Abourashed EA, El-Alfy AT (2016) Chemical diversity and pharmacological significance of the secondary metabolites of nutmeg (*Myristica fragrans* Houtt.). Phytochem Rev 15:1035–1056

Ahmadifar E, Sheikhzadeh N, Roshanaei K, Dargahi N, Faggio C (2019) Can dietary ginger (*Zingiber officinale*) alter biochemical and immunological parameters and gene expression related to growth, immunity and antioxidant system in zebrafish (*Danio rerio*)? Aquaculture 507:341–348

Ahmed HM, Babakir-Mina M (2020) Investigation of rosemary herbal extracts (*Rosmarinus officinalis*) and their potential effects on immunity. Phytother Res 34(8):1829–1837

Ahmed ST, Yang CJ (2017) Effects of dietary *Punica granatum* L. by-products on performance, immunity, intestinal and fecal microbiology, and odorous gas emissions from excreta in broilers. J Poult Sci 54:157–166. https://doi.org/10.2141/jpsa.0160116

Al-Ali KH, El-Beshbishy HA, Alghaithy AA, Abdallah H, El-Badry AA, Abdel-Sattar E (2013) In vitro antioxidant potential and antiprotozoal activity of methanolic extract of *Mentha longifolia* and *Origanum syriacum*. J Biol Sci 13:207

Al-Omari MM, Qaqish AM, Al-Qaoud KM (2018) Immunomodulatory effect of anise (*Pimpinella anisum*) in BALB/c mice. Trop J Pharm Res 17:1515–1521

Alpert PT (2017) The role of vitamins and minerals on the immune system. Home Health Care Manag Pract 29:199–202

Alvarez P, Alvarado C, Mathieu F, Jiménez L, De la Fuente M (2006) Diet supplementation for 5 weeks with polyphenol-rich cereals improves several functions and the redox state of mouse leucocytes. Eur J Nutr 45:428–438. https://doi.org/10.1007/s00394-006-0616-9

Amirghofran Z, Hashemzadeh R, Javidnia K, Golmoghaddam H, Esmaeilbeig A (2011) In vitro immunomodulatory effects of extracts from three plants of the Labiatae family and isolation of the active compound (s). J Immunotoxicol 8:265–273

Amirghofran Z, Ahmadi H, Karimi MH (2012) Immunomodulatory activity of the water extract of *Thymus vulgaris*, *Thymus daenensis*, and *Zataria multiflora* on dendritic cells and T cells responses. J Immunoassay Immunochem 33:388–402

Aparicio-Soto M, Montserrat-de la Paz S, Sanchez-Hidalgo M, Cardeno A, Bermudez B, Muriana FJ, Alarcon-de-la-Lastra C (2018) Virgin olive oil and its phenol fraction modulate monocyte/macrophage functionality: a potential therapeutic strategy in the treatment of systemic lupus erythematosus. Br J Nutr 120:681–692

Arena A, Bisignano C, Stassi G, Mandalari G, Wickham MS, Bisignano G (2010) Immunomodulatory and antiviral activity of almond skins. Immunol Lett 132:18–23

Arreola R, Quintero-Fabián S, López-Roa RI, Flores-Gutiérrez EO, Reyes-Grajeda JP, Carrera-Quintanar L, Ortuño-Sahagún D (2015) Immunomodulation and anti-inflammatory effects of garlic compounds. J Immunol Res 2015:401630–401630. https://doi.org/10.1155/2015/401630

Aryaeian N, Shahram F, Mahmoudi M, Tavakoli H, Yousefi B, Arablou T, Karegar SJ (2019) The effect of ginger supplementation on some immunity and inflammation intermediate genes expression in patients with active rheumatoid arthritis. Gene 698:179–185

Asemani Y, Bayat M, Malek-Hosseini S, Amirghofran Z (2019) Modulation of in vitro proliferation and cytokine secretion of human lymphocytes by *Mentha longifolia* extracts. Avicenna J Phytomed 9:34

Azhar AS, El-Bassossy HM, Abdallah HM (2019) *Mentha longifolia* alleviates experimentally induced angina via decreasing cardiac load. J Food Biochem 43:e12702

Badr G, Ramadan NK, Sayed LH, Badr BM, Omar HM, Selamoglu Z (2017) Why whey? Camel whey protein as a new dietary approach to the management of free radicals and for the treatment of different health disorders. Iran J Basic Med Sci 20:338

Beaulieu J, Dupont C, Lemieux P (2006) Whey proteins and peptides: beneficial effects on immune health. Clin Pract 3:69

Benkerroum N (2013) Traditional fermented foods of North African countries: technology and food safety challenges with regard to microbiological risks. Compr Rev Food Sci Food Saf 12:54–89

Bhowmik D, Kumar KS, Yadav A, Srivastava S, Paswan S, Dutta AS (2012) Recent trends in Indian traditional herbs *Syzygium aromaticum* and its health benefits. J Pharmacogn Phytochem 1:13–22

Bin-Hafeez B, Haque R, Parvez S, Pandey S, Sayeed I, Raisuddin S (2003) Immunomodulatory effects of fenugreek (*Trigonella foenum graecum* L.) extract in mice. Int Immunopharmacol 3:257–265

Bouriche H, Saidi A, Ferradji A, Belambri SA, Senator A (2016) Anti-inflammatory and immunomodulatory properties of *Pistacia lentiscus* extracts. J Appl Pharm Sci 6:140–146

Brokos B, Gasiorowski K, Noculak-Palczewska A (1999) In vitro antiinflammatory and immunomodulatory activities of extract and intract from Chestnut fruits (*Aesculus hippocastanum* L.). Herba Pol 4:338–344

Calder PC (1998) Fat chance of immunomodulation. Immunol Today 19:244–247

Calder PC, Carr AC, Gombart AF, Eggersdorfer M (2020) Optimal nutritional status for a well-functioning immune system is an important factor to protect against viral infections. Nutrients 12:1181

Carluccio MA, Siculella L, Ancora MA, Massaro M, Scoditti E, Storelli C, Visioli F, Distante A, De Caterina R (2003) Olive oil and red wine antioxidant polyphenols inhibit endothelial activation: antiatherogenic properties of Mediterranean diet phytochemicals. Arterioscler Thromb Vasc Biol 23:622–629

Carrasco FR, Schmidt G, Romero AL, Sartoretto JL, Caparroz-Assef SM, Bersani-Amado CA, Cuman RKN (2009) Immunomodulatory activity of *Zingiber officinale* Roscoe, *Salvia officinalis* L. and *Syzygium aromaticum* L. essential oils: evidence for humor-and cell-mediated responses. J Pharm Pharmacol 61:961–967

Castelo-Branco C, Soveral I (2014) The immune system and aging: a review. Gynecol Endocrinol 30:16–22

Chattopadhyay C, Chakrabarti N, Chatterjee M, Mukherjee S, Sarkar K, Chaudhuri AR (2012) Black tea (*Camellia sinensis*) decoction shows immunomodulatory properties on an experimental animal model and in human peripheral mononuclear cells. Pharm Res 4:15

Checker R, Chatterjee S, Sharma D, Gupta S, Variyar P, Sharma A, Poduval T (2008) Immunomodulatory and radioprotective effects of lignans derived from fresh nutmeg mace (*Myristica fragrans*) in mammalian splenocytes. Int Immunopharmacol 8:661–669

Cherng J-M, Chiang W, Chiang L-C (2008) Immunomodulatory activities of common vegetables and spices of Umbelliferae and its related coumarins and flavonoids. Food Chem 106:944–950

Cho J, Bing SJ, Kim A, Lee NH, Byeon S-H, Kim G-O, Jee Y (2017) Beetroot (*Beta vulgaris*) rescues mice from γ-ray irradiation by accelerating hematopoiesis and curtailing immunosuppression. Pharm Biol 55:306–319. https://doi.org/10.1080/13880209.2016.1237976

Cross ML, Gill H (2000) Immunomodulatory properties of milk. Br J Nutr 84:81–89

da Cunha JA, Rolim PM, Damasceno KS, de Sousa Júnior FC, Nabas RC, Seabra LMAJ (2020) From seed to flour: sowing sustainability in the use of cantaloupe melon residue (*Cucumis melo* L. var. *reticulatus*). PLoS One 15:e0219229. https://doi.org/10.1371/journal.pone.0219229

Damanhouri ZA, Ahmad A (2014) A review on therapeutic potential of *Piper nigrum* L. black pepper: the king of spices. Med Aromat Plants 3:161

Dargahi N, Johnson J, Donkor O, Vasiljevic T, Apostolopoulos V (2019) Immunomodulatory effects of probiotics: can they be used to treat allergies and autoimmune diseases? Maturitas 119:25–38

De L (2020) Edible seeds and nuts in human diet for immunity development. Int J Rec Sci Res 11: 38877–38881

De Martinis GL, D'Ostilio A, Marini L, Loreto M, Quaglino D (1999) The immune system in the elderly: II. Specific cellular immunity. Immunol Res 20:109–115

Deoula M, Hatime Z, Bennani B, El Rhazi K (2018) Dairy products and colorectal cancer in middle eastern and north African countries: a systematic review. BMC Cancer 18:233

Díaz-de-Cerio E, Verardo V, Gómez-Caravaca AM, Fernández-Gutiérrez A, Segura-Carretero A (2017) Health effects of *Psidium guajava* L. leaves: an overview of the last decade. Int J Mol Sci 18:897. https://doi.org/10.3390/ijms18040897

Dreher ML (1833) Whole fruits and fruit fiber emerging health effects. Nutrients 2018:10. https://doi.org/10.3390/nu10121833

Ediriweera MK, Tennekoon KH, Samarakoon SR (2017) A review on ethnopharmacological applications, pharmacological activities, and bioactive compounds of *Mangifera indica* (Mango). Evid Based Complement Alternat Med 2017:6949835. https://doi.org/10.1155/2017/6949835

El Senousy A (2019) Immunomodulatory and anti-inflammatory activities of the defatted alcoholic extract and mucilage of *Hibiscus sabdariffa* L. leaves, and their chemical characterization. J Pharmacogn Phytochem 8:982–990

Elberry AA, Mufti S, Al-Maghrabi J, Abdel Sattar E, Ghareib SA, Mosli HA, Gabr SA (2014) Immunomodulatory effect of red onion (*Allium cepa* Linn) scale extract on experimentally induced atypical prostatic hyperplasia in Wistar rats. Mediat Inflamm 2014:640746. https://doi.org/10.1155/2014/640746

El-Far AH, Ahmed HA, Shaheen HM (2016) Dietary supplementation of Phoenix dactylifera seeds enhances performance, immune response, and antioxidant status in broilers. Oxidative Med Cell Longev 2016. https://doi.org/10.1155/2016/5454963

Fakeye TO, Pal A, Bawankule D, Khanuja S (2008) Immunomodulatory effect of extracts of *Hibiscus sabdariffa* L. (family Malvaceae) in a mouse model. Phytother Res 22:664–668

Fernandes MRV, Azzolini AECS, Martinez MLL, Souza CRF, Lucisano-Valim YM, Oliveira WP (2014) Assessment of antioxidant activity of spray dried extracts of *Psidium guajava* leaves by DPPH and chemiluminescence inhibition in human neutrophils. Biomed Res Int 2014:382891. https://doi.org/10.1155/2014/382891

Galal O (2003) Nutrition-related health patterns in the Middle East. Asia Pac J Clin Nutr 12

Ghaffari S, Roshanravan N (2019) Saffron; An updated review on biological properties with special focus on cardiovascular effects. Biomed Pharmacother 109:21–27

Gil Á, Ortega RM (2019) Introduction and executive summary of the supplement, role of milk and dairy products in health and prevention of noncommunicable chronic diseases: a series of systematic reviews. Adv Nutr 10:S67–S73

Gill H (2003) Dairy products and the immune function in the elderly. Funct Dairy Prod 133:168

Gombart AF, Pierre A, Maggini S (2020) A review of micronutrients and the immune System–Working in harmony to reduce the risk of infection. Nutrients 12:236

Han X, Parker TL (2017) Cardamom (*Elettaria cardamomum*) essential oil significantly inhibits vascular cell adhesion molecule 1 and impacts genome-wide gene expression in human dermal fibroblasts. Cogent Med 4:1308066

Haque MR, Ansari SH, Rashikh A (2013) *Coffea arabica* seed extract stimulate the cellular immune function and cyclophosphamide-induced immunosuppression in mice. Iran J Pharm Res 12:101

Heydari M, Firouzbakhsh F, Paknejad H (2020) Effects of *Mentha longifolia* extract on some blood and immune parameters, and disease resistance against yersiniosis in rainbow trout. Aquaculture 515:734586

Ismail EH, Saqer AMA, Assirey E, Naqvi A, Okasha RM (2018) Successful green synthesis of gold nanoparticles using a Corchorus olitorius extract and their antiproliferative effect in cancer cells. Int J Mol Sci 19:2612. https://doi.org/10.3390/ijms19092612

Jafarian A, Zolfaghari B, Parnianifard M (2012) The effects of methanolic, chloroform, and ethyl acetate extracts of the *Cucurbita pepo* L. on the delay type hypersensitivity and antibody production. Res Pharm Sci 7:217–224

Jagetia GC, Aggarwal BB (2007) "Spicing up" of the immune system by curcumin. J Clin Immunol 27:19–35

Jedrychowski L, Halász A, Németh E, Nagy A (2009) Immunomodulating properties of food components. In: Jedrychowski L, Wichers HJ (eds) Chemical and biological properties of food allergens. CRC Press, Boca Raton, FL, pp 43–82

Kaur D, Chandrul KK (2017) *Syzygium aromaticum* L. (Clove): a vital herbal drug used in periodontal disease. Indian J Pharm Biol 5:45–51

Korhonen H, Pihlanto-Leppäla A, Rantamäki P, Tupasela T (1998) Impact of processing on bioactive proteins and peptides. Trends Food Sci Technol 9:307–319

Kottakis F, Kouzi-Koliakou K, Pendas S, Kountouras J, Choli-Papadopoulou T (2009) Effects of mastic gum *Pistacia lentiscus* var. Chia on innate cellular immune effectors. Eur J Gastroenterol Hepatol 21:143–149

Kraus J, Franz G (1992) Immunomodulating effects of polysaccharides from medicinal plants. In Microbial infections, Springer. pp. 299–308

Kumar NG, Contaifer D, Madurantakam P, Carbone S, Price ET, Van Tassell B, Brophy DF, Wijesinghe DS (2019) Dietary bioactive fatty acids as modulators of immune function: implications on human health. Nutrients 11:2974

Kumolosasi E, Ibrahim SNA, Shukri SMA, Ahmad W (2018) Immunostimulant activity of standardised extracts of *Mangifera indica* leaf and *Curcuma domestica* rhizome in mice. Trop J Pharm Res 17:77–84. https://doi.org/10.4314/tjpr.v17i1.12

Lad SS, Aparnathi KD, Mehta B, Velpula S (2017) Goat milk in human nutrition and health–a review. Int J Curr Microbiol Appl Sci 6:1781–1792

Larasati YA, Meiyanto E (2018) Revealing the potency of cinnamon as an anti-cancer and chemopreventive agent. Indonesian J. Cancer Chemoprevention 9:47–62

Lee SH, Jang SI, Kim DK, Ionescu C, Bravo D, Lillehoj HS (2009) Effect of dietary Curcuma, Capsicum, and Lentinus on enhancing local immunity against *Eimeria acervulina* infection. J Poult Sci 0912030019

Lee SH, Lillehoj HS, Jang SI, Lillehoj EP, Min W, Bravo DM (2013) Dietary supplementation of young broiler chickens with capsicum and turmeric oleoresins increases resistance to necrotic enteritis. Br J Nutr 110:840–847

Lewicka A, Szymański Ł, Rusiecka K, Kucza A, Jakubczyk A, Zdanowski R, Lewicki S (1958) Supplementation of plants with immunomodulatory properties during pregnancy and lactation—maternal and offspring health effects. Nutrients 2019:11

Leyva-López N, Gutierrez-Grijalva EP, Ambriz-Perez DL, Heredia JB (2016) Flavonoids as cytokine modulators: a possible therapy for inflammation-related diseases. Int J Mol Sci 17:921

Ma S, Huang D, Zhai M, Yang L, Peng S, Chen C, Feng X, Weng Q, Zhang B, Xu M (2015) Isolation of a novel bio-peptide from walnut residual protein inducing apoptosis and autophagy on cancer cells. BMC Complement Altern Med 15:413

Maggini S, Maldonado P, Cardim P, Fernandez Newball C, Sota Latino E (2017) Vitamins C, D and zinc: synergistic roles in immune function and infections. Vitam Miner 6:2376–1318. https://doi.org/10.3389/fnut.2020.606398

Maggini S, Pierre A, Calder PC (2018) Immune function and micronutrient requirements change over the life course. Nutrients 10:1531

Mahassni SH, Bukhari OA (2019) Beneficial effects of an aqueous ginger extract on the immune system cells and antibodies, hematology, and thyroid hormones in male smokers and non-smokers. J Nutr Intermed Metab 15:10–17

Majdalawieh AF, Carr RI (2010) In vitro investigation of the potential immunomodulatory and anti-cancer activities of black pepper (*Piper nigrum*) and cardamom (*Elettaria cardamomum*). J Med Food 13:371–381

Majdalawieh AF, Fayyad MW (2015) Immunomodulatory and anti-inflammatory action of *Nigella sativa* and thymoquinone: a comprehensive review. Int Immunopharmacol 28:295–304

Mao R, Wu L, Zhu N, Liu X, Hao Y, Liu R, Du Q, Li Y (2020) Immunomodulatory effects of walnut (*Juglans regia* L.) oligopeptides on innate and adaptive immune responses in mice. J Funct Foods 73:104068

Masihuzzaman AM, Iftikhar MY, Farheen K (2020) Badam shireen (*Prunus dulcis* Mill): a ghiza-e-dawaee and immunomodulator: a latest review. Int J Unani Integ 4:9–14

Mazzocchi A, Leone L, Agostoni C, Pali-Schöll I (2019) The secrets of the mediterranean diet. Does [Only] olive oil matter?. Nutrients 11(12):2941

Mosmann TR, Coffman RL (1989) Heterogeneity of cytokine secretion patterns and functions of helper T cells. Adv Immunol 46:111–147

Murad HA, Abdallah HM, Ali SS (2016) *Mentha longifolia* protects against acetic-acid induced colitis in rats. J Ethnopharmacol 190:354–361

Murphy K, Weaver C (2016) Janeway's immunobiology, 9th edn. Garland Science, New York

Naved T, Siddiqui JI, Ansari SH, Ansari AA, Mukhtar HM (2005) Immunomodulatory activity of *Mangifera indica* L. fruits (cv Neelam). J Nat Remedies 5:137–140

Nguenang GS, Ntyam ASM, Kuete V (2020) Acute and subacute toxicity profiles of the methanol extract of *Lycopersicon esculentum* L. leaves (tomato), a botanical with promising in vitro anticancer potential. Evid Based Complement Altern Med 2020:8935897

Nosáľová G, Prisenžňáková L, Paulovičová E, Capek P, Matulová M, Navarini L, Liverani FS (2011) Antitussive and immunomodulating activities of instant coffee arabinogalactan-protein. Int J Biol Macromol 49:493–497

Orhan IE, Mesaik MA, Jabeen A, Kan Y (2016) Immunomodulatory properties of various natural compounds and essential oils through modulation of human cellular immune response. Ind Crop Prod 81:117–122

Ortuño-Sahagún D, Zänker K, Rawat AKS, Kaveri SV, Hegde P (2017) Natural immunomodulators. J Immunol Res 2017: 7529408. https://doi.org/10.1155/2017/7529408

Osman KM, Kamal OE, Deif HN, Ahmed MM (2020) *Phoenix dactylifera, Mentha piperita* and montanide™ ISA-201 as immunological adjuvants in a chicken model. Acta Trop 202:105281. https://doi.org/10.1016/j.actatropica.2019.105281

Pan MH, Wu JC, Ho CT, Badmaev V (2017) Effects of water extract of *Curcuma longa* (L.) roots on immunity and telomerase function. J Compl Integr Med 14(3):2015-0107

Pandya PH, Murray ME, Pollok KE, Renbarger JL (2016) The immune system in cancer pathogenesis: potential therapeutic approaches. J Immunol Res 2016

Park M-Y, Kim D-H, Jin M-R (2010) Immunomodulatory effects of orally administrated capsicum extract on Peyer's patches. J Physiol Pathol Korean Med 24:446–451

Park HY, Oh MJ, Kim Y, Choi I (2018) Immunomodulatory activities of *Corchorus olitorius* leaf extract: beneficial effects in macrophage and NK cell activation immunosuppressed mice. J Funct Foods 46:220–226. https://doi.org/10.1016/j.jff.2018.05.005

Parle M, Kulwant S (2011) Musk melon is eat-must melon. Int Res J Pharm 2(8):52–57

Parnham MJ, Stepanić V, Tafferner N, Panek M, Verbanac D (2019) Mild plant and dietary immunomodulators. In: Nijkamp and Parnham's principles of immunopharmacology. Springer, Cham, pp 561–587

Pastorino G, Cornara L, Soares S, Rodrigues F, Oliveira MBP (2018) Liquorice (*Glycyrrhiza glabra*): a phytochemical and pharmacological review. Phytother Res 32:2323–2339

Prasad D, Kunnaiah R (2014) *Punica granatum*: a review on its potential role in treating periodontal disease. J Indian Soc Periodontol 18:428–432. https://doi.org/10.4103/0972-124X.138678

Qadir S, Bashir S, John R (2020) Saffron—Immunity system. Elsevier, In Saffron, pp 177–192

Rahayu RP, Prasetyo RA, Purwanto DA, Kresnoadi U, Iskandar RP, Rubianto M (2018) The immunomodulatory effect of green tea (*Camellia sinensis*) leaves extract on immunocompromised Wistar rats infected by Candida albicans. Vet World 11:765

Rahimi R, Shams-Ardekani MR, Abdollahi M (2010) A review of the efficacy of traditional Iranian medicine for inflammatory bowel disease. World J Gastroenterol: WJG 16:4504

Rahimi VB, Ajam F, Rakhshandeh H, Askari VR (2019) A pharmacological review on *Portulaca oleracea* L.: focusing on antiinflammatory, antioxidant, immunomodulatory and antitumor activities. J Pharmacopuncture 22:7–15. https://doi.org/10.3831/KPI.2019.22.001

Raksamiharja R, Sy K, Novarina A, Sasmito E (2012) *Elettaria cardamomum* distillate increases cellular immunity in doxorubicin treated rats. Indonesian J Cancer Chemoprevent 3:437–443

Rao PV, Gan SH (2014) Cinnamon: a multifaceted medicinal plant. Evid Based Complement Alternat Med 2014:642942

Ren D, Wang M, Shen M, Liu C, Liu W, Min W, Liu J (2016) In vivo assessment of immunomodulatory activity of hydrolysed peptides from *Corylus heterophylla* Fisch. J Sci Food Agric 96:3508–3514

Rezvanirad A, Mardani M, Ahmadzadeh SM, Asgary S, Naimi A, Mahmoudi G (2016) *Curcuma longa*: a review of therapeutic effects in traditional and modern medical references. J Chem Pharm Sci 9:3438–3448

Saad B (2015) Greco-Arab and Islamic diet therapy: tradition, research and practice. Arab J Med Aromat Plants 1(1):2–3

Safaei-Cherehh A, Rasouli B, Alaba PA, Seidavi A, Hernández SR, Salem AZ (2020) Effect of dietary *Foeniculum vulgare* Mill. extract on growth performance, blood metabolites, immunity and ileal microflora in male broilers. Agrofor Syst 94:1269–1278

Salas-Salvadó J, Casas-Agustench P, Salas-Huetos A (2011) Cultural and historical aspects of Mediterranean nuts with emphasis on their attributed healthy and nutritional properties. Nutr Metab Cardiovasc Dis 21:S1–S6

Sarker M (2012) Induction of humoral immunity through the enhancement of IgM production in murine splenic cells by ethanolic extract of seed of *Piper nigrum* L. J Sci Res 4:751–756

Saroj P, Verma M, Jha K, Pal M (2012) An overview on immunomodulation. J Adv Sci Res 3:7–12

Shang A, Cao S-Y, Xu X-Y, Gan R-Y, Tang G-Y, Corke H, Mavumengwana V, Li H-B (2019) Bioactive compounds and biological functions of garlic (*Allium sativum* L.). Foods 8:246. https://doi.org/10.3390/foods8070246

Shen C-Y, Zhang W-L, Jiang J-G (2017) Immune-enhancing activity of polysaccharides from *Hibiscus sabdariffa* Linn. via MAPK and NF-kB signaling pathways in RAW264. 7 cells. J Funct Foods 34:118–129

Sibi PI, Varghese P (2014) Evaluation of in vivo immunomodulatory activity of *Punica granatum* Linn. Int J Res Ayurveda Pharm 5:175–178. https://doi.org/10.7897/2277-4343.05235

Sierra S, Lara-Villoslada F, Olivares M, Jiménez J, Boza J, Xaus J (2005) Increased immune response in mice consuming rice bran oil. Eur J Nutr 44:509–516. https://doi.org/10.1007/s00394-005-0554-y

Singh S, Nair V, Gupta Y (2012) Linseed oil: an investigation of its antiarthritic activity in experimental models. Phytother Res 26:246–252

Singh R, Mal G, Kumar D, Patil N, Pathak K (2017) Camel milk: an important natural adjuvant. Agric Res 6:327–340

Sturm C, Wagner AE (1890) Brassica-derived plant bioactives as modulators of chemopreventive and inflammatory signaling pathways. Int J Mol Sci 2017:18. https://doi.org/10.3390/ijms18091890

Talazadeh F, Mayahi M, Naghavi M (2016) The effect of Antibiofin® on the immune response against avian influenza subtype H9N2 vaccine in broiler chickens. Int J Enteric Pathog 4:4–39396

Tran HTT, Schreiner M, Schlotz N, Lamy E (2019) Short-term dietary intervention with cooked but not raw brassica leafy vegetables increases telomerase activity in CD8+ lymphocytes in a randomized human trial. Nutrients 11:786. https://doi.org/10.3390/nu11040786

Ulbricht C, Abrams TR, Brigham A, Ceurvels J, Clubb J, Curtiss W, Kirkwood CD, Giese N, Hoehn K, Iovin R (2010) An evidence-based systematic review of rosemary (*Rosmarinus officinalis*) by the Natural Standard Research Collaboration. J Diet Suppl 7:351–413

Vadlamani S, Durga Devi KB, Poosarala A, Bapatla VK (2016) Identification of plant sources from north Andhra Pradesh exhibiting immunomodulatory activity using Balb/c models. Int J Pharma Bio Sci 7:295–300

Valdivieso-Ugarte M, Gomez-Llorente C, Plaza-Díaz J, Gil Á (2019) Antimicrobial, antioxidant, and immunomodulatory properties of essential oils: a systematic review. Nutrients 11:2786

Vijayabhargava K, Asad M (2011) Effect of stigmas of *Crocus sativus* L. (saffron) on cell mediated and humoral immunity. Nat Prod J 1:151–155

Wael S, Watuguly TW, Arini I, Smit A, Matdoan N, Prihati DR, Sari AB, Wahyudi D, Nuringtyas TR, Wijayanti N (2018) Potential of *Syzygium aromaticum* (clove) leaf extract on immune proliferation response in Balb/c mice infected with *Salmonella typhimurium*. Case Rep Clin Med 7:613–627

Wahyuningsih SPA, Pramudya M, Putri IP, Winarni D, Savira NII, Darmanto W (2018) Crude polysaccharides from okra pods (*Abelmoschus esculentus*) grown in Indonesia enhance the immune response due to bacterial infection. Adv Pharmacol Sci 2018:8505383. https://doi.org/10.1155/2018/8505383

Wang P, Wang M, Liu C, Fang L, Wang J, Liu X, Min W (2018) Isolation, purification and structural identification of immunoactive peptides derived from hazelnut (*Corylus heterophylla* Fisch.) protein. Shipin Kexue/Food. Science 39:200–205

Yasin BR, El-Fawal HAN, Mousa SA (2015) Date (*Phoenix dactylifera*) polyphenolics and other bioactive compounds: a traditional Islamic Remedy's potential in prevention of cell damage, cancer therapeutics and beyond. Int J Mol Sci 16:30075–30090. https://doi.org/10.3390/ijms161226210

Yin M, Zhang Y, Li H (2019) Advances in research on immunoregulation of macrophages by plant polysaccharides. Front Immunol 10:145–145. https://doi.org/10.3389/fimmu.2019.00145

Zeinali M, Zirak MR, Rezaee SA, Karimi G, Hosseinzadeh H (2019) Immunoregulatory and anti-inflammatory properties of *Crocus sativus* (saffron) and its main active constituents: a review. Iran J Basic Med Sci 22:334

Zhai Z, Liu Y, Wu L, Senchina DS, Wurtele ES, Murphy PA, Kohut ML, Cunnick JE (2007) Enhancement of innate and adaptive immune functions by multiple *Echinacea* species. J Med Food 10:423–434

Zhang S, Liang R, Zhao Y, Zhang S, Lin S (2019) Immunomodulatory activity improvement of pine nut peptides by a pulsed electric field and their structure–activity relationships. J Agric Food Chem 67:3796–3810

Immunostimulant Properties of Some Commonly Used Indian Spices and Herbs with Special Reference to Region-Specific Cuisines

8

Monalisha Karmakar, Debarati Jana, Tuhin Manna, Avijit Banik, Priyanka Raul, Kartik Chandra Guchhait, Keshab Chandra Mondal, Amiya Kumar Panda, and Chandradipa Ghosh

Abstract

Indian subcontinent, unique for its *Unity in Diversity*, is made up of 28 provinces and 8 union territories and is known to have amazing diversity of approximately 1402 million people with a wide variety of races and ethnicity staying together harmonically since the ancient time. It is not surprising that India with 15 major languages and about 100 dialects would have a wide variety of cuisines, with each province having its own cooking tradition and taste. Indian ways of food preparation are not the cuisine of a single nationality, but a collective combination of different cuisines from a number of countries with cultural identities that have been heavily influenced by religious and regional particularities. Above the region-specific nature of Indian cuisine, there exists certain common features among the diverse culinary practices. India's history, rulers, trade partners, and moreover, the religious and cultural traditions have great influence on its cuisines. From the ancient ages, *Indian Ayurveda* (the medicinal practices for the well-being of humanity) is considered as a method of *science of life* in a holistic way.

M. Karmakar · D. Jana · T. Manna · P. Raul · K. C. Guchhait · C. Ghosh (✉)
Department of Human Physiology, Vidyasagar University, Midnapore, West Bengal, India
e-mail: ch_ghosh@mail.vidyasagar.ac.in

A. Banik
Department of Human Physiology, Vidyasagar University, Midnapore, West Bengal, India

Department of Microbiology, Vidyasagar University, Midnapore, West Bengal, India

K. C. Mondal
Department of Microbiology, Vidyasagar University, Midnapore, West Bengal, India

A. K. Panda
Department of Chemistry, Vidyasagar University, Midnapore, West Bengal, India

Sadhu Ram Chand Murmu University of Jhargram, Jhargram, West Bengal, India
e-mail: akpanda@mail.vidyasagar.ac.in

N. S. Sangwan et al. (eds.), *Plants and Phytomolecules for Immunomodulation*,
https://doi.org/10.1007/978-981-16-8117-2_8

Ayurvedic Science signifies the importance of natural medicines using herbals that also includes spices. Spices and herbs are found not only to attribute flavour to bland meals, but also influence human metabolic processes and defence mechanisms. Spices have a diverse array of natural phytochemicals that have complementary and overlapping actions that include antioxidant effects, modulation of detoxifying enzymes, stimulation of immune system, reduction of inflammation, modulation of steroid metabolism, and antibacterial as well as antiviral effects. In this review, efforts have been made to take a cursory glance on the traditional Indian spices and herbs being used since the prehistoric times in the preparation of foods. The medicinal, nutritional, and especially immunostimulant properties of different spices and herbs have been reviewed. The great biodiversity observed in India by virtue of its enormous variety of flora and fauna owe to its wide range of climatic conditions and topographical characteristics. While in the colder northern states, dishes are prepared commonly with the warming aromatic spices, and in contrast, to combat the hot climate in the southern Indian states, the foods prepared are generally lighter to make them easy for the digestive system. Natural anti-inflammatory compounds are abundant in different Indian spices that not only add flavour, but also impart different immuno-boosting effects. Anti-inflammatory compounds are plentifully present in Indian spices and herbs, and their additive or synergistic actions protect the human body against a variety of threats. Some of the important bioactive compounds possessing nutritional/immunostimulant values include piperine from black pepper, curcumin from turmeric, allyl sulphides from garlic, eugenol from cloves, capsaicin from red pepper, etc. Moreover, natural polyphenols found in some common herbs used in different Indian cuisines, i.e. coriander, bay, mint, curry leaves, etc., have been found with immunostimulant properties in combating a multitude of disorders.

Keywords

Indian cuisine · Spices · Herbs · Fermented food · Bioactive · Immunostimulant properties · Anti-inflammatory

Abbreviations

AD	Atopic dermatitis
AEC	Absolute eosinophil count
AgPs	Arabinogalactan proteins
AHL	*N*-acyl homoserine lactone
AIDS	Acquired immunodeficiency syndrome
BaP	Benzo-α-pyrene
BIS	Bureau of Indian Standards
CaMK II kinase	Calcium-calmodulin-dependent protein kinase II
CAT	Catalase

CBE	*Cinnamomum burmannii* bark extract
CD4+	Cluster of differentiation 4+
CEO	Clove essential oil
CMP	Cow's milk protein
COX	Cyclooxygenase
CP	Cyclophosphamide
CRP	C-reactive protein
CYP450	Cytochrome P450
DAMPs	Damage-associated molecular patterns
DTH	Delayed-type hypersensitivity
ECD	*Elettaria cardamomum* distillate
ERK	Extracellular signal-regulated kinases
EROD	Ethoxyresorufin-O-deethylase
FOS	Fructo-oligosaccharides
FOXP3	Forkhead box P3
FSP	Fenugreek seed powder
GI	Gastrointestinal
GOS	Galacto-oligosaccharides
GPx	Glutathione peroxidase
ICAM-1	Intercellular adhesion molecule 1
IFN-γ	Interferon-γ
Ig	Immunoglobulin
IL	Interleukin
ImC	Immunomodulatory component
IN	Inulin
iNOS	Pathogen-inducible nitric oxide synthase
ISO	International Organization for Standardization
ITAM	Immune receptor tyrosine based activation motif
JNK	c-Jun N-terminal kinases
LAB	Lactic acid bacteria
LPS	Lipopolysaccharide
MAPKs	Mitogen-activated protein kinases
MEK1/2	Map/Erk kinase I/II
MHC	Major histocompatibility complex
NF-κB	Nuclear factor kappa-B
NK cells	Natural killer cells
NO	Nitric oxide
NSO	*Nigella sativa* oil
PAH	Polycyclic aromatic hydrocarbon
PBMC	Peripheral blood mononuclear cell
PCA	Passive cutaneous anaphylaxis
PFC	Plaque-forming cells
PGD2	Prostaglandin D2
PHA	Phytohemagglutinin
PLC γ	Phospholipase C γ

PWM	Pokeweed mitogen
qRT-PCR	Quantitative real time polymerase chain reaction
RES	Reticuloendothelial system
RGF	Raw garlic fructans
SARS	Severe acute respiratory syndrome
ROS	Reactive oxygen species
SIE	*Sesamum indicum* extract
SIRS	Systemic inflammatory response syndrome
SOD	Superoxide dismutase
SPE	Saffron petal extract
SRBC	Sheep red blood cell
Syk kinase	Spleen tyrosine kinase
TCR	T-cell receptor
TGF-β	Transforming growth factor beta
THQ	Thymohydroquinone
TIg	Antibody titre
TLRs	Toll-like receptors
TNF	Tumour necrosis factor
TQ	Thymoquinone
T_{reg} cells	Regulatory T cells
TRPA1	Transient receptor potential ankyrin 1
TRPV1	Transient receptor potential vanilloid 1
WBC	White blood cells
VCAM1	Vascular cell adhesion molecule 1
VR1	Vanilloid receptor 1

8.1 Introduction

In India, *Unity in Diversity* is commonly used to express the unity between unrelated persons or groups. It is a concept of 'unity without uniformity and diversity without fragmentation' (Lalonde 1994).

The diverse nature of India is incredible. Despite all variations in different corners of the nation, there has been an accord. The diverse geographical locations with its distinctive features and distinguishing cultures could be overcome by the nationals all together with the string of harmony and unity (Mastana 2014). There are 28 states and 8 union territories with an entire population of approximately 1402 million, who communicate through 15 major languages and more than a hundred dialects. Most importantly the aspect that ties up all Indians together is the food and, more explicitly, the wide variety of Indian cuisines (Appadurai 1988).

The development of a high cuisine does not take identical forms or orders in each locale. But with the exceptions of China and Italy, there have been inclination to mark the difference between 'high' and 'low' cuisines, for example, between court food and peasant food, between the urban centres and rural diets, etc. Imperial cuisines ever been pulled out upon regional, provincial, and folk stuffs and recipes. Pre-industrial elites often used to exhibit their political power, wealth, and multinational tastes by depicting raw materials and styles and even cook from far off places. However, these high cuisines, with their direction towards display, attempt for distancing themselves from local sources (Appadurai 1988). In India, different kinds of patterns in high cuisine have been practiced since the time unknown which is, to some extent, unique. In this mode of development of a national cuisine appears to be effectively a post-industrial, postcolonial approach. Cooking in India has deeply been rooted in ethical and medical viewpoints, and therapeutic practices (Sharar 1975). In Indian subcontinent, the national cuisine has emerged with all the regional cuisines playing significant roles not stashing their regional or ethnic roots.

In India foods are as region-specific and diverse as its inhabitants. The history of India in terms of the ruling dynasties, business partners, and in addition to that the religious and cultural conventions of people has great influence on its cuisines. India's spices have greatly been influenced by the European and Arabian merchants, as evidenced throughout its history. Coffee and asafoetida powder were brought in India by the Arabians, and Portuguese traders brought tomatoes, potatoes, and chillies (Mangalassary 2016). India's culinary tradition got the taste of Persian flavours and practices between the early 1500s and late 1600s, courtesy, the Mughals (Collingham 2006).

The notions of disease and health promotion are strongly ingrained in Indian dietary traditions. India is the most recognized country for the traditional uses of spices and herbs that possess a broad array of physiological and pharmacological properties. All the features of traditionally used herbal remedies, capable of establishing modern medicines, have been taken into account in recent studies. Survey by World Health Organization suggested that 70–80% of the world population relies primarily on herbal sources as their foremost healthcare than on modern medicine (Chan 2003). In addition, 80% people from developing countries and around 60% people from whole world trust on plants and herbs for their healthcare gains. Throughout history herbs and spices are known to fortify foods as preservatives, flavour, and therapeutic agents. Herbs and spices have also been utilized worldwide as food additives, not only to increase its organoleptic properties but also to enhance the shelf life by reducing or removing food-borne pathogens. Phenolic compounds from herbs and spices are considered to be valuable substitutes of artificial antimicrobial agents required for food processing. For preventing the growth of certain pathogenic bacteria (*Salmonella enteritidis*, *Staphylococcus aureus*, and *Listeria monocytogenes*, etc.) and fungi, phenolic compounds, such as ellagic acid, oleuropein, tea catechins, coumaric acid, and ferulic acid, have come into use (Shan et al. 2011). Several reports have suggested the use of dietary herbs and spices due to their anti-mutagenic, anti-inflammatory, anti-oxidative, and immunomodulatory properties. Moreover, adding herbs and spices or its extracts to different daily products renders them as useful carriers of nutraceuticals (Conn 1995).

The final profile of every food product is attained principally through two aspects, namely, ingredients and processing methods. Traditional Indian delicacies have developed the trend of using varieties of spices to confer an appropriate flavour profile. The quality of ethnic cuisine depends not only on the regional and ethnic practices but also on different subjective factors, such as skill and style of preparation (Sugasini et al. 2018).

India is essentially the *home sweet home* to different kinds of spices. The spices have become entwined with Indian culture and tradition and made it further rich. Since time immemorial India has gained its prominence as land of quality spices throughout the world. Spices contribute flavour and aroma to foods as natural resources and also act as food seasoning agents. The diverse nature of Indian agroclimatic ecoregions generally facilitates innumerable spices with geospatial distinctiveness to grow. Hence India holds immense potential to cater her huge internal demands and also to strike the export trades.

8.1.1 India: The 'The Home of Spices'

Herbs and spices are widely used in Indian cuisines in order to add aroma, flavour, and visual appeal to foods. For centuries, these botanicals, also known as 'pharmacy of nature', are used as traditional herbal medicines in many parts of India (Lai and Roy 2004). Geneva International Organization has defined spices as 'vegetable products or mixtures thereof, free from extraneous matter, used for flavouring, seasoning, and imparting aroma to foods' (ISO 1995). In general, herbs (Latin *herba*, meaning grass, green stalks, or blades) are leaves of non-woody plants (low-growing shrubs) and have a pleasant taste, while spices are brown, black, or red in colour, derived from the root (ginger, onion, and garlic), bark (cinnamon), seeds (yellow mustard, poppy, and sesame), buds (cloves, saffron), berry (black pepper), or tropical fruit (allspice, paprika) and are generally used as dried products with pungent smell. Some herbs are spices but not all spices come from herbs (Peter and Shylaja 2012, Opara and Chohan 2014).

In rest part of the world India is recognized as the *land of spices* (Peter and Shylaja 2012). The history of spices is similarly old as that of human civilization. All over the globe, Indian spices are well-known for their tastes and strong aromatic flavours. India is the biggest producer, user, and exporter of spices in the world and produces about 86% of the total spice in the globe (Peter and Shylaja 2012). India, mainly due to its varying climate and soil conditions, produces nearly all the spices. Besides, spices are being consumed in Indian households as colouring and flavouring agents, and also render potential benefits through promotion of health. The spices that India produces in large quantity are coriander, cinnamon, clove, pepper, turmeric, cumin, ginger, chilli, celery, nutmeg, mace, fennel, fenugreek, ajwain (bishop's weed), and cassia. Each spice has its own striking flavour, medicinal value, and additional interesting facts.

For more than 7000 years, the legacy of Indian spices is continuing. Based on the data collected from the Bureau of Indian Standards (BIS), 63 spices grow in India (Duke 2002). However, the Spices Board (Government of India) has listed only 52 spices (Nybe 2007). Again, the International Organization for Standardization

(ISO) has approved 75 spices and condiments out of 109 listed by the ISO (Divakaran et al. 2018). In India the state of Kerala is leader in quality spice production. Some of the spices like chillies, cumin, coriander seeds, fennel, and fenugreek come from different other regions of India (Rathore and Shekhawat 2008). At present, around 2.75 million tons of different spices are produced in India costing approximately US$ 4.2 billion, causing it to remain in leading position in the world spices market. The varying climates in India, including the tropical to subtropical conditions that most significantly include temperature range from 45 °C to 0 °C besides other factors, are mainly responsible for the wide variety of spices grown in this country (Bhardwaj et al. 2011). The annual worldwide trade amount of spices being 4,000,000 tons equivalent to U$1.5 billion.

8.1.2 Geographical Origins of Spices and Herbs Growing in India

Spices originating from tree crops are known to be *tree spices*. In India, Kerala, Tamil Nadu, Maharashtra, Andhra Pradesh, and Karnataka are the foremost states to cultivate tree spices in bigger areas. India commonly grows 17 tree spices. Among them cinnamon, clove, nutmeg, garcinia, tamarind, kokam, curry leaf are the important ones. Kanyakumari and Nilgiris districts of Tamil Nadu and few small areas in the states of Kerala and Karnataka are leaders in clove cultivation. Nutmeg is grown predominantly in Kerala and also to some extent in Tamil Nadu and Karnataka (Bhardwaj et al. 2011). Kerala is very much suitable for spice cultivation especially black pepper, cardamom, ginger, turmeric, clove, garcinia, nutmeg, and mace and 95 percent of the total production comes from pepper, cardamom, ginger and turmeric (Sharangi and Pandit 2018).

8.2 India and Its Cuisines

India has a distinctive concoction of regional cuisines that are culturally and climatically diverse. Indian cuisine is a wide array of spicy delicacies. Spices and herbs are considered as one of the most excellent ingredients of Indian cuisine. Indian spices and cuisines go hand in hand as the ethnic food products of India are seasoned with a wide variety of spices. Spices, the dried seeds, fruits, roots, barks, or flowers of plants, and herbs are used in small quantities for flavour and colour or as preservatives. All types of spices and herbs have traditionally been used from the ancient times daily in Indian kitchens that not only stimulate appetite and create visual appeals to food but also fulfil the physiological requirements on routine basis. Many of these substances are also used in traditional medicines for their certain medicinal activities like serving as purgative, laxative, expectorant, carminative, diuretic, etc. (Sachan et al. 2018). In addition to imparting good taste to food, culinary spices have been used as food preservatives and for their health-enhancing properties for centuries in the lifestyle of people.

Indian culinary system has a long history of health-centric dietary practices focused on disease prevention and promotion of health, and these factors influence food preferences and recipe composition by altering the fabric of cuisine. Recipe composition pattern in a cuisine provides a means for investigating its gastronomic history and molecular constitution. Molecular composition of recipes in a cuisine also reveals patterns in food preferences (Jain et al. 2015).

Indian cuisines are enormous arrays of spicy delicacies with its own traditional, region-specific, and unique tastes, based on the spice blends (Sugasini et al. 2018). All the spices have wide variety of biofunctions where their additive and synergistic actions protect the human body from different ailments. Phytochemicals in spices are the secondary metabolites that give protection from herbivorous insects, vertebrates, and pathogens (Srinivasan 2005). With its utmost probability, no single compound in a spice creates flavours, but a combination of different compounds such as phenols, alcohols, esters, alkaloids, organic acids, terpenes, resins, and sulphur-containing compounds contributes to flavours (Yashin et al. 2017). Herbs are the plants that can be grown in own neighbourhood, even if it is not native to the area, whereas spices are the plant parts that cannot be grown at home. Spice includes dried roots, barks, and berries, nearly all parts of plants. When used in larger quantities, spices contribute significant amounts of micronutrients, including calcium, iron, magnesium, and many others in diet (Sachan et al. 2018). Indian cuisine is obsessed with curries, i.e. gravy-like sauce or stew-like dishes with meat, vegetables, or cheese, although the particular ingredients, degree of consistency, and particular use of selected spices are determined by regional preference (Fuhrman and Ferreri 2010). Indian spice mixes often use five or more different spices, sometimes going up to more than ten varieties. Chilli pepper, black mustard seed, cumin, turmeric, fenugreek, ginger, garlic, cardamom, cloves, coriander, cinnamon, nutmeg, saffron, rose petal essence, and asafoetida powder are all used frequently in various combinations (Krishnaswamy 2008). Garam masala is a popular spice mix, a combination of cardamom-cinnamon-clove with the additional spices varying according to regions, personal recipes, and taste preferences. Mint, coriander, and fenugreek leaves are popularly used throughout entire India as they offer their pungent, herby flavours as well as freshness to food.

8.2.1 India as a Land with Diverse Population and Distinctive Gastronomy

The diverse cultural identities seen in Indian population are greatly influenced by regional particularities and religious beliefs. In ancient India, *Ayurveda* – the medicinal practices for the well-being of humanity – was considered as a scheme of *science of life*. Ayurveda signifies the uses of natural medicines using herbs and spices. Ayurvedic teachings are believed to exert its influences over Indian cuisines, dictating cooking rituals. The Ayurvedic food rules differ according to the regional and religious culture. Also, religious beliefs affect other dietary restrictions that dictate as well as shape India's cuisines. Approximately one-third of India's population is religiously vegetarian (Fuhrman and Ferreri 2010).

8.2.1.1 Northern Indian Cuisine

This is the most prevalent form popular as Indian cuisine outside the country, and this has been strongly influenced by Mughal culinary practices. Use of tandoors (clay ovens) is widespread in North India, giving dishes like *tandoori chicken* and *naan bread* with their distinctive charcoal flavour (Dubey 2010). *Korma*, another very popular menu found across all Northern Indian restaurants, is a creamy curry of coconut milk or yogurt, cumin, coriander, and small amounts of cashews or almonds (Sanmugam et al. 2011).

Generally, North Indian food is low to moderate spicy, rich, and creamy. North Indians rely heavily on tomato, onion, ginger, and garlic to prepare their curry. Cinnamon, black pepper, star anise, chillies and other hot peppers, specifically garam masala spice blend are used for increasing flavours. These are also important for increasing metabolic rate and for keeping the body warm.

8.2.1.2 Western Indian Cuisine

The three main regions Maharashtra, Gujarat, and Goa in Western India by their geographical and historical backgrounds bestow prime distinguishing features in this cuisine. Some spices like goda masala, kokum, tamarind, and coconut are unique and essential ingredients in the Maharashtrian kitchen (Nande et al. 2008). The fish- and coconut milk-dominant cuisine of Maharashtra is due to its coastal acquaintance.

Another distinctive feature of Gujarati food is its combination of sweet and sour flavours. To keep the body hydrated in mostly hot and dry Gujarat's coastal climate, sugar, lemon, and tomatoes are commonly used in foods.

Cinnamon, nutmeg, cloves, garlic, and ginger are mainly found to dominate in this regional cuisine, added to impart a unique flavour to the dishes. Use of herb mixture (coriander leaves, fenugreek leaves, mint, etc.) to season food items like paneer, seafood, poultry, and meat in the cases of frying, stewing, baking, or roasting is another attraction of West Indian meals. Saffron is added to increase the delicacy of different dessert preparations.

8.2.1.3 Eastern Indian Cuisine

The Eastern Indian region comprises the states of West Bengal, Odisha, Bihar, Sikkim, Assam, Tripura, Meghalaya, Manipur, Nagaland, Mizoram, and Arunachal Pradesh. Eastern Indian cuisine is recognized all over the country for its desserts. Their admirable light sugariness gives an exotic touch to a meal. Eastern dishes favour mustard oil, poppy seeds, and mustard seeds, providing dishes a light pungency.

In general a trend of using a variety of different spices is found in Eastern Indian dishes. The mixture of five spices—fenugreek seeds, onion seeds, fennel seeds, mustard seeds, and white cumin seeds—is known as *Paanch Phoron* which is the main game changer of this cuisine.

8.2.1.4 Southern Indian Cuisine

Dosa and *Idli* are the main preparations in fascinating and colourful South Indian cuisine that are adored all over India as breakfast and snack items. The Indian states of Kerala, Tamil Nadu, Karnataka, Telangana, and Andhra Pradesh with their own rocky plateau, river valleys, and coastal plains have made the southern part of India extremely different from its Northern counterpart. The famous *sambhar dal* is made with lentil, tamarind, and buttermilk and is found in all South Indian states, though it varies from region to region (Prasad 2005). The food of Kerala needs no introduction. It is simple and flavourful and offers an intelligent combination of potent spices. However, the use of spices differs from that of North Indian culinary.

Tamilian cuisine boasts of local spices like star anise, maratti mokku (dried flower pods), kalpasi (stone flower), and freshly ground spices including bay leaf, fenugreek, tamarind, fennel, clove, cumin, and turmeric (Devasahayam 2001). South Indian food is tangy, hot, and spicy with a handful of curry leaves in most of the preparations. Mustard seed is used as main spice for tempering in any south Indian dish.

8.3 Spices and Herbs in Fermented Foods

The ancestral past of the Indian ethnic groups is foretold in the dietary tradition of ethnicity. Food delicacy invokes culinary creativity and cuisine diversification to compliment the flavour, aroma, taste, and palatability of the final edible products. Ancestors of Indians might have evolved the traditional technology for food fermentation to obtain healthy and nutritionally rich fermented foods and alcoholic beverages besides conserving the perishable plant and animal resources. The other important reasons behind the consumption and popularity of fermented foods and beverages since ancient times are their shorter preparation times, extended shelf life, lower volume, and higher nutritional values in comparison to the non-fermented foods. Moreover, it also aimed in improving new preferred tastes and flavours with the intention to increase delicacy in the diets. With reference to the ancient Indian civilizations and literal works, namely, *Bhagavadgita*, *Ramayana*, and *Manusmriti*, there were clear and distinct food beliefs and practices in every community living in India.

As defined, 'fermentation' is a process in which organic substances are slowly decomposed into simpler substances from complex ones by mediation of microorganisms or enzymes, for example, formation of alcohols or organic acids from carbohydrates. Fermentation is mainly biochemical conversion of carbohydrates by anaerobic or partially anaerobic oxidation (FAO 1998). The word *ferment* has been originated from Latin verb *fervere*, meaning 'to boil'. But fermentation can be carried out without heat. During preparation of fermented foods, the process of fermentation is optimally controlled to achieve acidity and aroma to a desired level. It remains to be constitutive in terms of nutrition as the digestibility, flavour and aroma of foods get improved through this process. It also causes benefits to health as it enriches the foods with different nutritional elements like, essential amino acids, proteins, essential fatty acids, and vitamins as well as destroys several toxicants coming from raw foods. Most importantly, fermented foods became popular into communities since ancient times because of their aids in preservation of foods and development of specific foods and beverages with increased bioavailability of nutrients (Granier et al. 2003).

Fermented foods are essentially interlinked with the age-old acumen of ancient Indian culture. For many years, ethnic fermented foods are being prepared with immense efficiency in different parts of India, and the in-depth preparations differ across the country. Indian fermented foods remain critical in terms of food and nutritional security in a region and are generally developed with the indigenous skills of ethnic people who practise the basic microbiology-centred knowledge, called *ethno-microbiology*. For many generations, traditional astuteness on food processing, techniques for preservation, and therapeutic effectiveness have been proficiently convened in India. Indian traditional fermented foods have also been renowned as *functional foods* due to the presence of 'functional' components, viz. micronutrients, chemicals for body-healing, antioxidants, probiotics, and dietary fibres. These active components play important roles in controlling blood sugar level and body weight, and support immunity.

In general, for preparing fermented foods, a starter culture comprising desirable microorganisms is added to the substrate food materials to carry out the fermentation process in controlled manner to acquire the fermented functional food products. In traditional Indian practices, fermented food products generally originate from milk, vegetables, and fruits. Lactic acid bacteria (LAB) are predominant microflora in these food items, as majority of the LAB have been isolated from them. It is worthy to mention as an example that in dairy industries LAB are of high demand for milk processing through fermentation that not only enhance food safety by production of lactic acid and bacteriocins but also contribute to flavour, aroma, maturation, and texture of the fermented food (e.g. cheese) through proteolytic, lipolytic activities and also polysaccharide products, respectively. Indian traditional fermented foods have been assigned as sources of LAB such as *Lactococcus* spp., *Lactobacillus brevis*, *Lactobacillus casei*, *Lactobacillus fermentum*, *Lactobacillus plantarum*, *Lactobacillus pentosus*, *Leuconostoc fallax*, *Leuconostoc kimchi*, *Leuconostoc mesenteroides*, *Weissella cibaria*, *Weissella confusa*, *Weissella koreensis*, *Pediococcus pentosaceus*, etc., as very important probiotics (Sathe and Mandal 2016). The final profile of each fermented food product is principally based on two components: ingredients and processing methods. In order to obtain an acceptable flavour profile, traditional Indian fermented foods often comprise a variety of spices and herbs. Regional and ethnic traditions and subjective factors, viz. skill and styles of preparations, are responsible for the consistency of ethnic cuisines. There should be certain characteristics in ethnic cuisines, such as a well-defined product profile, a scheduled production process, standardized quality, and stated shelf life.

Bacteria living in the gut have significant influence on the immune system. The fermented foods can facilitate the activities of immune system by virtue of their high probiotic content and can also reduce the risk of infections, for example, the upper respiratory tract infections (Hao et al. 2011). For development of gut immunity and execution of tolerance against food antigens, live commensal microbiota is essential (Brandtzaeg 2002, Rhee et al. 2004). The specific metabolites, generated as a result of fermentation, induce immune responses. Dendritic cells in the gut mucosa play leading role in food antigen presentation and thereby generate immune responses through activating T cells. They also trigger negative regulation by activating T regulatory cells in gut mucosa. Thus there appears two kinds of responses: one is immunogenic and the other, tolerogenic. Balance between these two types depends on the dendritic cell types, differential cytokine generation, co-stimulatory actions and also the survival aptitude of the dendritic cells. Mainly pathogen-associated molecular patterns from distinct bacterial pathogens generate microenvironments that are critical for integration of signalling actions to control the functional abilities of the dendritic cells. For this purpose, the Toll-like receptors (TLRs) expressed on dendritic cells play key role in sensing the microenvironments and further activate several signalling pathways, such as mitogen-activated protein kinases or phosphatidylinositol 3-kinase to regulate immune responses, more specifically induction of 'immunogenic' or 'tolerogenic' dendritic cells. The signalling action triggered by TLRs expressed on dendritic cells varies with the pathogens. In this way the fermented food products exert their immunomodulatory actions by targeting the dendritic cells in gut mucosa (Benacerraf 1978, Bin-Hafeez et al. 2003, Beltran et al. 2007, Lake et al. 2012, Granier et al. 2013, Prasanna and Venkatesh 2015, Safaei-Cherehh et al. 2018, Farsani et al. 2019, Frick et al. 2019).

India is considered to be the hub of diverse food cultures that includes fermented ethnic foods and alcoholic beverages. Spices and herbs used to enhance the flavour of a dish are important part of traditional Indian fermented foods (Table 8.1). In every Indian cuisine, a right blend of aromatic spices is crucial. A multitude of spices and herbs are used to prepare different types of fermented foods and beverages in different regions/states of India (Table 8.2).

8.4 Herbs and Spices as Immunostimulants

As per definition, immunity is the capability to detect and defend infectious agents and allergens that permit the body to resist tissue damage initiated with the infectious diseases and/or allergic disorders and have been developed with enormous potential to settle in various challenges. Immune system is empowered with complex defence mechanisms that act in coordination with the nervous and endocrine systems. In modern biological research, the immune system has been ascribed as sense organ of the body for being able to detect the existence of pathogens, allergens, and tumours, otherwise undetectable in physiological system. Moreover, the attribution as a special sense also gets qualified because of the presence of highly sensitive and explicit intra- and inter-system chemical signalling circuitries working through receptor-ligand repertoire mechanisms against exogenous/endogenous intimidations (Blalock 2007). The immune system is not only involved in protection against pathogens but also is evolved to discriminate self from non-self. Its main function is to maintain the body's internal homeostasis by generating defence barrier against

Table 8.1 Traditional (ethnic) fermented foods and alcoholic beverages from different states of India where spices and/or herbs are used as common ingredients

Fermented foods and beverages	Nature of food	Constituents	Originating Indian state
Dangalbari	Pulse-based	Black gram, dangal, spices like fenugreek (methi), coriander, mustard, mustard green, cumin, turmeric powder, red chilli powder, etc.	Himachal Pradesh
Mashbari	Pulse-based	Black gram, spices like mustard, mustard green, cumin, turmeric powder, red chilli powder, etc.	Himachal Pradesh
Gulgule	Cereal-based	Wheat/rice, salt, spices like fennel seeds, water, ghee	Himachal Pradesh
Balam	Wheat-based	Roasted wheat flour, spices, and herbs like coriander, mustard, mustard green, cumin, etc.	Uttarakhand
Marcha	Cereal-based	Rice, wild herbs, and spices like coriander, mustard, mustard green, and cumin	Darjeeling Hills, Sikkim
Opop	Cereal-based	Rice, wild herbs like *ekkam* leaves (*Phrynium pubinerve*) and spices	Arunachal Pradesh
Paa	Cereal-based	Wild herbs, spices mixed to rice	Arunachal Pradesh
Pee	Cereal-based	Wild herbs, spices mixed to rice	Arunachal Pradesh
Phut	Cereal-based	Wild herbs, spices mixed to rice	Arunachal Pradesh
Kadhi badi	Curd- and besan (gram flour)-based	Curd, besan-based stew, besan and aromatic spices like asafoetida (hing), carom seed, cumin, coriander, turmeric, garam masala, bay leaves, red chilli, etc.	Bihar, and Jharkhand
Dhokla	Cereal-based	Cereals and spices like green chilli, ginger, curry leaves, cumin seeds, sesame seeds, hing	Gujarat
Raita	Milk (cow/buffalo)-based	Several vegetables such as diced onions, tomato, shredded carrots and cucumber; boiled and diced potato, beets, and spinach; herbs like coriander leaves, etc. and also fruits, like pieces of grapes and banana are added to dahi	Gujarat, and Rajasthan
dvi	Cereal-based	buttermilk added to gram flour and, further mixed with sesame seeds and other spices like ginger, green chilli, turmeric, hing, coriander leaves	Gujarat
Idli	Cereal-based	Rice, black gram lentils, curd, and spices like methi seeds	Karnataka

(continued)

Table 8.1 (continued)

Fermented foods and beverages	Nature of food	Constituents	Originating Indian state
Dosa	Cereal-based	Rice, black gram lentils, curd, and aromatic spices like methi seeds, etc.	Karnataka
Chaach	Milk-based	Butter prepared from cream or *dahi*; added with salt and spices like mint leaves, coriander, green chillies, cumin, etc.	Karnataka, Kerala
Lassi	Milk-based	Dahi, sugar, salt, spices like mint, cardamom, cumin, etc.	All parts of Central India (Uttarakhand, Punjab, Haryana, and UP)
Loochi	Flour-based	Fine flour containing mild spices	Jammu and Kashmir
Mattha	Milk-based	Curd with added spices like black pepper, cumin, mint powder, etc.	Maharashtra
Ambil	Cereal-based	Rice/sorghum/ragi flour added with spices like sesame seeds, cumin seeds, chilli, garlic, etc.	Maharashtra
Bibdya	Sorghum-based	Sorghum, spices like chilli powder, and garlic	Northern Maharashtra
Anishi	Leaf-based	Spices such as ginger, chilli powder, and salt are added to leaves of *Colocasia* sp.	Nagaland
Marcha Khesung Phab Buth	Cereal-based	Rice, wild herbs, and spices	Sikkim and West Bengal (Darjeeling)
Moru	Milk-based	Curd with salt and spices/herbs	Tamil Nadu, Kerala, Karnataka, Andhra Pradesh
Taravani or kali	Cereal-based	Boiled rice and spices	Telangana and Andhra Pradesh
Doli ki roti	Cereal-based	Whole wheat flour and spices like fennel seeds, poppy seeds, turmeric, ginger, nutmeg, cloves, black elaichi, cinnamon	Punjab and Haryana
Bari/goyna bari/ rakhiya bari	Legume-based	Pulses (*matar dal, chana dal, urad dal*), spices like cumin seeds, coriander, chilli, etc. along with salts	West Bengal, Odisha, and Chhattisgarh
Papad	Legume-based	Lentil (*Lens culinaris*), *Bengal gram, red* or *green gram*, and *black gram*, added with a small quantity of spices, peanut oil, and common salt	All

(continued)

Table 8.1 (continued)

Fermented foods and beverages	Nature of food	Constituents	Originating Indian state
Achar/ chadnee/ chutney	Fruits/ vegetable-based	Chopped fruits and vegetables, turmeric, chilli, some other spices, and salt	All
Angkur	Cereal-based	Rice mixed with herbs	Assam
Apoppitha	Cereal-based	Rice mixed with herbs	Assam
Bhekur pitha	Cereal-based	Rice mixed with herbs	Assam
Chowan	Cereal-based	Rice mixed with herbs	Tripura
Damdim	Cereal-based	Rice mixed with herbs	Mizoram
Emao	Cereal-based	Rice mixed with herbs	Assam
Hamei	Cereal-based	Rice mixed with wild herbs	Manipur
Humao	Cereal-based	Rice mixed with barks of wild plants	Assam
Perok kushi	Cereal-based	Rice mixed with herbs	Assam
Phab	Wheat-based	Wheat mixed with wild herbs	Sikkim, Ladakh, Arunachal Pradesh
Ranu dabai	Cereal-based	Rice mixed with herbs	West Bengal, Odisha, Jharkhand
Ranu goti	Cereal-based	Rice mixed with herbs	Bihar, Jharkhand, Madhya Pradesh
Apong	Cereal-based	Rice mixed with wild herbs	Arunachal Pradesh
Dhehli	Cereal-based	Mixture of 36 herbs added to roasted flour from barley	Himachal Pradesh
Dawidim	Cereal-based	Rice mixed with wild herbs	Mizoram
Namsing, sukoti	Fish-based	Fishes (small- and medium-sized) and spices like chilli, coriander, cumin seeds, etc.	Assam

external invasions. The two forms of immunity, viz. innate and adaptive immunity, cooperatively work through synergistic interactions. Non-specifically triggered innate immune responses are most active and come into action instantaneously as the infection sets in, while the adaptive immune responses become progressively dominant over time (Krensky et al. 2006). The primary effectors in innate immunity are granulocytes, monocytes/macrophages, natural killer (NK) cells, complements, and mast cells, whereas B and T lymphocytes are the leaders in generation of adaptive immunity. Cellular immune response is mediated by T cells, and $CD4^+$ cells play a central role in cellular and humoral immune interplay. Upon encountering an antigen, the naïve T cells turn into active lymphoblast that is proliferated and differentiated to become effector cells. The effector cells in turn directly kill pathogen or secrete cytokines, i.e. IL-2, IL-4, and IFN-γ to combat pathogens (Yu et al. 2018).

Table 8.2 Fermented foods and beverages from different regions and states of India

Region	State(s)	Fermented foods and beverages	Reference(s)
North-eastern Region	Arunachal Pradesh	Generally include fermented foods made from bamboo, soybean and leafy vegetables, milk, and meat/fish. Common fermented foods are *eup*, *ikhing*, *pikey pila*, *bamboo tenga*, *churapi*, *tapya*, *anpo*, *ziang-sang*, etc. and alcoholic beverages used by different tribes are *apong*, *opo*, *sira-o*, *madua*, etc.	Shrivastava et al. (2012), Anupama et al. (2018), Rawat et al. (2018), Shrivastava et al. (2020)
	Assam	Popular fermented foods and beverages that are unique to Assam are *kahudi*, *khorisa*, *kharoli*, *sukati/namsing*, and *poita bhat*. Fermented fish items are prepared according as the availability of fish species, plant materials, and spices like chilli, garlic, turmeric, and ginger. In cereal-based fermented foods like *Apop pitha*, *Bhekur pitha*, *Emao*, *Humao*, etc., spices and herbs are thoroughly mixed with rice to prevent the growth of unwanted microbes and to get additional organoleptic properties	Das et al. (2012), Bhuyan et al. (2014), Bora et al. (2016), Chowdhury et al. (2019), Barooah et al. (2020)
	Tripura	Here fermented foods are prepared from a variety of natural resources like bamboo shoots, wild leaves, vegetables, fruits, and fishes. A salt-free preserved fish *Sheedal* is widely consumed here; the dominant genus has been found as a rich source of coagulase-negative *Staphylococcus* and LAB. Salt-fermented fish, i.e. *lona fish*, and several other fermented and semi-fermented foods prepared by age-old practices of the rural women are *Melye Amiley*, *Moiya Koshak*, *Moiya Pangsung*, *Midukeye*, etc., using different vegetables like bamboo shoot, amla (Indian	Majumdar (2020)

(continued)

Table 8.2 (continued)

Region	State(s)	Fermented foods and beverages	Reference(s)
		gooseberry), elephant foot yam, etc., which are of wide uses and enormous popularity. Rice is used as fermenting base with the addition of herbs grown locally for preparing alcoholic beverages in traditional methods. Berma is another famous dried and fermented fish item	
	Meghalaya	Foods developed in tribal societies are *tungrymbai*, prepared by spontaneously fermenting locally grown soybean [*Glycine max* (L.)], and *tungtap,* fermented fish preparation using fresh fish (*Puntius* spp. or *Danio* spp.) with their incredible aroma and flavour that have made them tempting above the regular food habits. In the case of fish fermentation, both pre-fermentation salting and post-fermentation salting are adopted	Joshi et al. (2020)
	Manipur	Widely used ethnic fermented foods include fermented products of bamboo shoots (*soidon, soibum, soidonmahi/soijon/soijim, thunbin, thunkheng, thunbinkang, thunkhenakang*), soybean (*khuichang, hawaijar, bekanthu, bethu, theisui*), sesame seed (*sithu*), mustard leaves (*ziangdui, ziangsang, inziangsang, ganang tamdui, ankamthu*), mustard/rapeseed (*hangam maru hawaichar*), kenaf seed (*gankhiangkhui*), milk (*sangom aphamba*), fish (*hentak, ngari, ithiitongka*), and meat (*saayung, sathu, sahro, bongkarot, guaighi kang, aithu*) and fermented alcoholic rice-based beverages, *hamei, khai, chamri*; non-distilled beverages,	Jeyaram et al. (2009), Tamang (2010), Devi and Kumar (2012), Singh et al. (2018), Wahengbam et al. (2020)

(continued)

Table 8.2 (continued)

Region	State(s)	Fermented foods and beverages	Reference(s)
		atingba, *zoungao*, *khor*, *yuangouba*, *pheijou*, *patso*, *timpui*, *waiyu*; distilled liquors, *acham*, *yu*, *zouju/zouzu*. Regular constituents of the traditional fermented Manipuri cuisine are various aromatic spices (e.g. green chillies, ginger, etc.) and herbs (e.g. *Centella asiatica*, bamboo shoots, etc.)	
	Mizoram	Different traditional fermented foods are *tuai-um* (from bamboo shoots), *chhi-um* (from whole seeds), *bekang-um* (from soybean), *ai-um* (from mustard leaves), and *sa-um* (from pork fat), and fermented alcoholic beverages are *zupui*, *zufang* (sweet rice beer) etc. Dawdim/chawl, a dried starter is used for preparation of alcoholic beverages	Dokhuma (1992), Tungoe (2016), Anupma et al. (2018), Thanzami and Lalhlenmawia (2020)
	Nagaland	Some popular ethnic fermented foods and beverages commonly using wild herbs and forest products are *anishi*, *axone/akhonii*, *ashikumna/thevochie*, *bell dzu/dzii*, *bastenga*, *cutocie*, *khulushidzii*, *kothal dzu*, *tsiingen ngashi/tsugu*, *zutho*, *khrei*. Commonly used spices and herbs in cuisine of Nagaland are black pepper, green chillies, turmeric, sesame seeds, cayenne pepper, black salt, ginger, ground cumin, and banana leaves	Teramoto et al. (2002), Tamang et al. (2012), Jamir and Deb (2018), Ajungla et al. (2020)
Eastern Region	West Bengal and Odisha	A variety of fermented foods are prepared that are dairy-made, plant-made, cereal- and pulse-made, and fish-made. Spices like cumin seeds, chilli, coriander, etc. are commonly used during the preparation of *bari* and *papad*, two pulse-made traditional fermented	Sha et al. (2013), Ghosh et al. (2014), Renu and Waghray (2016), Ray et al. (2016)

(continued)

Table 8.2 (continued)

Region	State(s)	Fermented foods and beverages	Reference(s)
		foods to add to the taste and flavour. During making *bakhar*, an amylolytic starter culture, and haria, a common fermented rice-based beverage used in tribal societies, herbs are used	
	Bihar and Jharkhand	Fermented foods like *jalebi, pua, baskarel, dhuska, dahi vada, kadhi badi*, etc. are prepared at different occasions. Most popular beverage among the tribal communities of Jharkhand is *Hadia*, prepared from rice or *mahua*. *Kadhi badi* is also a popular dish. *Kadhi* is made as a stew with curd and besan by mixing curd in gram flour (besan) and enriched with aromatic spices and herbs like ginger, curry leaves, green chilli, etc.	Barik (2006), Geetha et al. (2015), Kumar et al. (2016), Singh et al. (2020)
Northern Region	Jammu and Kashmir	*Loochi* is a traditionally fermented snack prepared with flour containing a little spices. Pickles are very important part of meal that enhance the taste. During the preparation of several pickles (*zimikand, lasooda/gunda, mounji, nadur*), spices like red chilli powder, mustard seeds, turmeric powder, asafoetida, fenugreek, nigella seeds, fennel seeds, coriander seeds, kalonji seeds, carom seeds, garlic, and ginger are used as ingredients	Akhter et al. (2020)
	Himachal Pradesh	Commonly prepared fermented foods of ethnic usage are *bedvin roti, bhaturu, chilra (lwar), dosha, manna, borhe, siddu, marchu, seera, pakk, gulgule, pinni/bagpinni, sepubari, bari, thuktal, churpa*, and *aska*. On the other hand, *sura, angoori/kinnauri, daru/chakti, chhang/lugri, chulli, behmi*, and *arak/ara* are the	Thakur et al. (2004), Kanwar et al. (2007), Savitri and Bhalla (2007), Kanwar and Bhushan (2020)

(continued)

Table 8.2 (continued)

Region	State(s)	Fermented foods and beverages	Reference(s)
		traditional fermented alcoholic beverages. Commonly used spices and herbs in the preparation of fermented foods are fenugreek, coriander, mustard, mustard green, cumin, turmeric and red chilli powder etc.	
	Uttarakhand, Punjab, Haryana, and Uttar Pradesh	The authentic cuisines of Punjab, Haryana, and Uttar Pradesh offer a similar delicacy category of fermented foods such as *lassi*, curd, and *rabadi*. Milk and milk products are of major uses in most of the dishes with good number of traditional Indian spices accompanied by fresh vegetables. *Balam*, made of wheat with spices and herbs, is a starter that prepares fermented alcoholic beverage. *Doli ki roti*, is an ethnic Punjabi fermented cereal product with spices	Bhatia and Khetarpaul (2012), Kumari et al. (2016), Beniwal et al. (2020)
Western Region	Gujarat and Rajasthan	Fermented foods, for example, *dahi*, *raita*, *chhash*, *dhokla*, *khandvi*, *khaman*, *handvo*, and *shrikhand* are essentially consumed on regular basis and are linked with the religious values and customs of people. Commonly used spices and herbs in cuisine of Gujarat and Rajasthan are black pepper, green chillies, basil, turmeric, cilantro, mint, cayenne pepper, black salt, ginger, and ground cumin	Prajapati and Sreeja (2013), Singh and Singh (2014), Sathe and Mandal (2016)
	Maharashtra	Frequently consumed fermented foods and beverages are generally milk products, namely, *mattha*, and *shrikhand*. Besides, *ambil*, *amboli*, *olyafenya*, *kurdai*, *kharodya*, *sandan*, *anarase*, *bibdya*, *dhapode*, *salpapdya*, and *sandge* are cereal- and legume-based foods and	Lonkar et al. (2011), Shinde (2011), Annapure et al. (2020)

(continued)

Table 8.2 (continued)

Region	State(s)	Fermented foods and beverages	Reference(s)
		beverages. Dried fishes, i.e. *Tengli*, *Sode*, *Jawla*, *Karndi*, *Sukke Mandeli*, *Sukke Makul*, *Vaktya*, and *Kharamasa* as well as fruits, vegetables, and meat-made pickles (*lonche*) are used	
Central Region	Chhattisgarh	Wheat, barley, and various lentils used in traditional dishes and fermented foods and beverages are of prime fondness in tribal societies. Some significant fermented foods from Chhattisgarh are *aeersa*, *bara*, *bafaur*, *bijori*, *boreebasi*, *dehori*, *pidiya*, *rakhiya bari*, etc. The premier ethnic beverages from Chhattisgarh are *handia* (made up of cooked rice), *salfi* (from trunk sap of salfi, *Caryota urens*), and *mohua* (made up of mahua flowers, *Madhuca longifolia*)	Kumar and Rao (2007), Gavankar and Chemburkar (2016), Das et al. (2017), Tiwari et al. (2020)
	Madhya Pradesh	Fermented foods and beverages in tribal with ethnic usage i.e. *haria*, *rabadi*, *mahua*, *handia*, *papad*, *dahi*, *ghee*, *lassi*, *panner*, *ranugoti*, *pendum*, *basi*, and *pej/peja* are consumed routinely by people of Madhya Pradesh state. Commonly used spices and herbs in cuisine of Madhya Pradesh are black pepper, basil, green chillies, turmeric, mint, black salt, cayenne pepper, ginger, and ground cumin	Gupta et al. (1992), Arjun (2015), Sharma et al. (2020)
Southern Region	Andhra Pradesh and Telangana	Traditionally fermented foods in these states with certain variations in the preparations lead to several recipes, such as *uttapam*, *dosa*, *idli*, *vada*, *perugu*, and *majjigga/salla* are popularly used fermented milk products, and *toddy* (*kallu*), a fermented alcoholic beverage, is of common use among rural	Mohanty et al. 2018, Palika et al. (2020)

(continued)

Table 8.2 (continued)

Region	State(s)	Fermented foods and beverages	Reference(s)
		folks. Chilli, turmeric, garlic, ginger, onion, and mustard are commonly used spices and herbs in the preparation of fermented foods in these two states. *Ambali (java)* and *Taravan/ kali* (*Ganji*), two traditionally fermented beverages, made with ragi flour and rice seasoned with salt and aromatic spices	
	Karnataka and Tamil Nadu	In these states staple foods are the fermented foods prepared with cereals like rice and pulses such as green gram, red gram, black gram dal, and flours from ragi, wheat, and barley. Most popular and widely consumed are *dosa*, *idli*, *dhokla*, *nan*, *ambali*, *koozhu*, *parotta*, and *pazhaiya soru*. *Idli*/*idly* is a rice-based fermented breakfast item, and fenugreek seeds are used in *idli* for flavour. Besides, other spices, such as ginger, chilli peppers, coriander, cumin, mustard seeds, etc. are used	Sekar and Mariappan (2007), Sarkar et al. (2015), Baruah et al. (2020)
	Kerala	A porridge named as *pazham kanji* with uses of aromatic spices and leafy vegetables is very popular. *Vada* is another traditional fermented breakfast dish. Frequently used spices are fresh peppercorns, coconut pieces, cumin, red chillies, green chillies, finely chopped ginger, fresh coriander leaves, etc.	Tamang (2012), Nampoothiri et al. (2020)

Immunological defence activities are evoked by complicated reactions of innate immune system, generally non-specific in nature in their actions, where the cellular and humoral immune reactions constituting adaptive immunity involve stimulation and inhibition of diverse kinds of immune cells. The cellular network included in innate and adaptive components acts as highly sensitive regulator to maintain host homeostasis through carrying out and restoring tissue functions in the occasion of encountering environmental and microbial factors.

Furthermore, nowadays it is an established fact that survival of multicellular living beings takes place as metaorganisms which focus on the coexistence of host organisms with commensal microbiota through symbiotic association. This milieu of plentiful microbial communities that includes bacteria, fungi, viruses, etc. plays primary role in controlling major part of the host physiology through their remarkable enzymatic and metabolic capabilities. Recent studies have also projected the most significant role played by microbiota in stimulation, education, and tasks of the mammalian immune system (Belkaid and Hand 2014). Hence there exists an alliance between the immune system and microbiota, interweaving the innate and adaptive immunity arms for execution of immune behaviour. Human meta-genome sequencing in recent times has altered the concept about microbiome and has thrown light on the impact of variations in these populations on pathogenesis. In this respect, commensal organisms are highly decisive as they play critical role in inducing immune regulatory responses. Although commensalism is mentioned here, the mode of association between distinct members of microbiota with their host happens to be highly situation-specific even with the same microorganism, depending on the mutualist or parasitic relationship as per nutritional, co-infection, or genetic landscape of host. Gut flora-mediated signalling action determines the immune tolerance, i.e. suppression of inflammatory responses originating from food or any other antigens consumed orally. Active research in this area has shown FOXP3 regulatory T (T_{reg}) cells are dominant in this action, and all through the lifetime of the host, they continue to retain peripheral and mucosal homeostasis. Both thymus and GI-induced T_{reg} cells come into action by activity of antigen-presenting $CD103^{+}CD11b^{+}$ dendritic cells through TGF-β, a cytokine and retinoic acid, metabolite of vitamin A (Coombes et al. 2007, Mucida et al. 2007, Cebula et al. 2013, Belkaid and Hand 2014). Switching in the composition of gut microbiota and their bulk not only is able to influence the local immune responses but also can alter certainly the immune behavioural trends and inflammation in distal organs (Belkaid and Naik 2013).

Immunomodulators are the biological or synthetic elements that can stimulate or inhibit every facet of immune response generation mechanism. In clinical viewpoint, immunomodulators are of three kinds (Agarwal 1999). Immuno-adjuvants are specific immunostimulants that can enhance vaccine effectiveness and are considered as true modulators. On the other hand, immunosuppressants have structural and functional heterogenicity. They are usually used in combinations for treating different autoimmune diseases and preventing rejection of transplants. Immunostimulants are in general non-specific in nature and add to body's opposition towards infection (Sethi and Singh 2015). They are natural or synthetic and biologically active substances with variable chemical characteristics and can influence the host immune system to augment resistance against a variety of infectious agents. They interact with specific receptors and cellular components of innate and adaptive responses to heighten the immune activities (Lake et al. 2012). These bioactive components may cause receptor-mediated direct activation of innate immunity triggering intracellular gene expression leading to the synthesis of effector biomolecules against invaders.

The active components of certain herbs and spices that are antibacterial in nature may destroy cell wall, inhibit synthesis of proteins and replication of DNA, inhibit secretion of enzymes, and hamper the quorum sensing signalling action. Active herbal compounds generally function through interfering viral transcription in the host cells and thus improve non-specific immune tolerance. Most herb and spice extracts are generally active against broad-spectrum pathogens. As they can be easily administered orally, they can act constructively to trigger immunostimulation. They may play very promising role under certain conditions of immunosuppression, like SARS, AIDS, cancer, etc., to empower the immune competence of the host (Petrunov et al. 2014). In immunocompromised patients, they are expected to act as immuno-therapeutic agents for treating severe infections, cancer, and immune deficiency (Lake et al. 2009).

Alternative therapeutic measures by using medicinal plant products have already been established in health management protocols. Medicinal plants are considered to be intensely entrenched components of the cultural heritage of various communities belonging to diverse cultures and countries, and are closely linked to the protection as well as maintenance of good health (Galina et al. 2009). A wide range of medicinal plant parts considered as herbs and spices include roots, leaves, seeds, and flowers of the plants having beneficial compounds are designated as functional foods (Devasagayam et al. 2001). Examples of widely used such functional foods are ginger, garlic, red chilli, mustard, fenugreek, etc. (Kulkarni et al. 2006). The plant-based natural products are two types: the *primary metabolites*, in general proteins, fats, and sugars, etc., are commonly found in every living system, whereas the *secondary metabolites* are complex compounds, such as flavonoids, alkaloids, terpenes, saponins, phenolic acids, tannins, volatile oils, etc., which contribute significantly in compliance and survival means. Among these secondary metabolites, those exhibiting immuno-stimulating and therapeutic aptitudes are recognized as bioactive compounds. Herbs and spices are rich in bioactive compounds, found with the potential of boosting immune system. Here an attempt has been made to highlight possible molecular mechanisms underlying the immunostimulatory effects exerted by different herbs and spices in various animal models, as presented in Table 8.3.

8.5 Conclusions

Indian spices and herbs have taken foremost position in consonance with the rich cultural heritage of India because of their vast acceptability and uses in every corner of the country and matchless quality that has entrenched them with towering demand all over the globe. This also has enabled Indian spices and herbs to attain a foremost position in global spice trade that is unquestionably imperative to invigorate Indian economy. Since ancient times Indian spices and herbs have become integral part of the lives of people not only in everyday culinary practices due to well-heeled flavour, aroma, and taste in creating a pleasant and delightful gastronomy but also given incredible touch in Indian Cuisines to make them unique in rest part of the world.

Table 8.3 Immunostimulant properties of spices and herbs commonly used in Indian cuisines

Spice/herb name in English (Common names in different Indian languages)	Scientific name with representative photographs	Useful part	Active components	Immunostimulant properties	Mode of action
Ajowan (Yavaanika and ugragandha; ajmodika in Sanskrit; ajowan, ajvain in Hindi; ajwaan in Urdu; jowan in Bengali; jonee guti in Assamese; jamain in Maithili; javind, jeven in Kashmiri; ajavana in Punjabi and Marathi; ajavaina in Gujarati, Ajvain; omamu, vamu in Telugu; omam in Tamil; ajamoda; oma in Kannada; and ayamodakam in Malayalam)	***Trachyspermum ammi*** 	Seed	Essential oil of ajowan mainly composed of monoterpene hydrocarbons. Among them, phenol and thymol are the predominant components; others are p-cymene, carvacrol, c-terpinene (Vitali et al. 2016)	Carvacrol plays important role in treatment of inflamed lung airway in asthma by activating Th2 expression in BALB/c mice (Shruthi et al. 2016)	• Shifting of immune response from Th2-type to Th1-type for asthma treatment • Decreases IL-17, IL-4, IL-1c, and TNF-α, TGF-β expression • Rising of FOXP3 and IFN-γ level (Shruthi et al. 2016)
				Thymol in ajowan oil induces lymphocyte proliferation in pig (Vitali et al. 2016)	• Enhances lymphocyte proliferation in peripheral blood mononuclear cell (PBMC) (Shruthi et al. 2016) • Induces circulation of $CD4^+$, $CD8^+$ subsets both and $CD4^-CD8^+$ (CTL) subsets • Activates expression of ϒδT cells after period of mitogen activation (Vitali et al. 2016)
				Arabinogalactan protein (Agp, a 30.7 kDa glycoprotein) in ajowan possesses effective mitogenic activity migrating to the splenocytes. This	• Involve in molecular exchanges and cellular signalling at the cell surface • Participates in somatic and multiplicative cellular growth (Sruthi et al. 2017)

(continued)

Table 8.3 (continued)

Spice/herb name in English (Common names in different Indian languages)	Scientific name with representative photographs	Useful part	Active components	Immunostimulant properties	Mode of action
				glycoprotein functions as significant immunostimulators inducing immunomodulation activity in BALB/c mice (Sruthi et al. 2017)	
				Ajowan seed used for boosting immune system by modulating T-cell responses in BALB/c mice (Sruthi et al. 2016)	Ajowan have five components. Among them D4 exhibits immune-stimulatory activity labelled as immunomodulatory component (ImC) • ImC plays important roles in innate immune responses and production of antibodies by activating B cells and macrophages • ImC induces morphological changes by increasing the production of B lymphocytes (Sruthi et al. 2016)
				Ajowan plays an important role for boosting immune system by increasing memory/activated CD8 α + T cell in pig (Vitali et al. 2016)	Ajowan oil shows proliferation activity initiated by mitogens such as phytohemagglutinin (PHA) and pokeweed mitogen (PWM) • PHA increases both

					percentages of $CD8a^+$ cells and $CD79a^+$ cells • PWM increases the proportion of $CD8^+$ cells and retained fraction of $CD79a^+$cells (Vitali et al. 2016)
Asafoetida (Hingu and raamathan in Sanskrit, heeng in Hindi, hing in Urdu and Bengali, hengu in Oriya, yangu, eng in Kashmiri, higa in Punjabi, hinga in Gujarati and Marathi, inguva in Telugu, perunkayam in Tamil, ingu in Kannada, and kayam in Malayalam)	***Ferula assa-foetida*** 	Rhizome and root	Important phyto-constituents of asafoetida are coumarins, rhamnose, diterpenes, phenolics, sesquiterpenes, falcarinolone, arabinose, oleic acid, α-pinene, b-sitosterol, galactose, luteolin 7b-D-glucopyranoside, ferulic acid, glucuronic acid (Amalraj and Gopi 2017)	α-Pinene and ferulic acid present in the asafoetida exerts potent antiviral and antibacterial activity (Mahendra and Bisht 2012)	Exact mechanism could not be obtained
Bay leaf (Tejapatra in Sanskrit, tej patta in Hindi and Punjabi, kaiyfal ka pata in Urdu, tej pata in Bengali and Kashmiri, tamalapatr,		Dried leaves	Its principal phytochemicals are eucalyptol, terpenes, sesquiterpenes, methyl eugenol, phellandrene, linalool, α- and β-pinenes,	It increases the cellular T-cell count and antibody level in our body (Rengaraj 2014), thereby energizes our immune defence mechanism	Exact mechanism could not be obtained

(continued)

Table 8.3 (continued)

Spice/herb name in English (Common names in different Indian languages)	Scientific name with representative photographs	Useful part	Active components	Immunostimulant properties	Mode of action
tejapatr in Gujarati, tamalpatra in Marathi, patta akulu, talisha in Telugu, talishapattiri, ilavangapattiri in Tamil, pallav elle in Kannada and karuva ella in Malayalam)	***Laurus nobilis*** 		geraniol, terpineol, terpinyl acetate, citric acid, and lauric acid (Batool et al. 2020)		
Black cumin Krishnajira in Sanskrit, kalaunji in Hindi and Urdu, kalo jeera in Bengali, koshur zur in Kashmiri, kala jeera in Odia and Punjabi, kalo jirum in Gujarati, kali jire in Marathi, nalla jirakarra in Telugu, karuppu cirakam in Tamil, kappu jirige in Kannada, karutta jirakam in Malayalam)	***Nigella sativa* Linn.** 	Seed	The constituents of *Nigella sativa* seed contain fixed oil, alkaloid, saponin, proteins, and essential oil. The fixed oil comprises unsaturated fatty acids, viz. arachidonic, linoleic, eicosadienoic, linolenic, palmitoleic, oleic, palmitic, myristic and stearic acids as well as cycloeucalenol, beta-sitosterol, sterol esters cycloartenol, and sterol glucosides. The volatile oil covers saturated fatty acids which include nigellone, only element of the carbonyl fraction of the oil, thymoquinone (TQ),	The feed containing *Nigella sativa* is suitable to enhance the growth and immunity parameters of fish (*Labeo rohita*), used as a natural immunostimulant (Ali et al. 2020)	• Enhancement of white blood cells count and higher lysozyme activity (Ali et al. 2020)
				Nigella sativa oil (NSO) may is capable to stimulate cellular immunity against *Toxoplasma gondii* infected Swiss albino mice (Mady et al. 2016)	• Rises in serum interferon-γ (IFN-γ) (Mady et al. 2016) • Eliciting the lysosomal activity and controlling the metabolic activity of antigen-presenting cells, viz. dendritic cells and macrophages (Sturge et al. 2013)

			dithymoquinone, thymohydroquinone (THQ), carvacrol, d-limonene, thymol, α- and β-pinene, d-citronellol, *p*-cymene (Tembhurne et al. 2014a,b)	Thymoquinone (TQ), the key element of *Nigella sativa* seeds shows an effect on lung inflammation in ovalbumin-sensitized guinea pigs by stimulating immune responses (Keyhanmanesh et al. 2010)	• Suppressive activity of macrophages directly or indirectly by inhibiting Th2 activity and thus regulates the anti-inflammatory responses (Romagnani 1999) • Influencing effect on Th1 and downregulating the effect on Th2 cells (Keyhanmanesh et al. 2010)
Black mustard (Asuri and bimbata in Sanskrit, rai, Banarasi rai, kale sarson in Hindi and Urdu, kalo sorse, sorsa in Bengali, kong in Kashmiri, rai, Banarasi rai, kale sarson in Punjabi, rai in Gujarati, avalu in Telugu, kadugu in Tamil, sasive in Kannada, and kadugu in Malayalam)	***Brassica nigra*** 	Seed	The mustard seeds encompass many active compounds, i.e. alkaloid, inosite, sinapine, albumins, myrosin, sinigrin, gums (Upwar et al. 2011)	Addition of a medicinal plant, *Brassica nigra*, to *Oreochromis niloticus* (*Nile tilapia*) food facilitates both the immune and biotransformation systems after exposure to a polycyclic aromatic hydrocarbon, benzo-α-pyrene (BaP). (Abbas et al. 2019)	Antiproteases show potent effects in the defence mechanism by modifying the activities of destructive proteases • Prevents the action of proteases (Laskowski and Kato 1980) to hamper bacterial growth in fish (Ellis 2001)
				To protect the effects against the lethal impact of benzo-α-pyrene, diet supplement with black mustard seeds extract resulted in increasing of lysozyme activity, which improves immune	• Substantial rises in lysozyme activity in mustard extract-treated group (Abbas et al. 2019) Such observation allowed to consider the potential role of black mustard seed extract to act as an immune-stimulant through

(continued)

Table 8.3 (continued)

Spice/herb name in English (Common names in different Indian languages)	Scientific name with representative photographs	Useful part	Active components	Immunostimulant properties	Mode of action
				functions (Abbas et al. 2019)	strengthening the immune status in resisting the destructive actions of the pollutants (Abbas et al. 2019)
				Acts by lowering biotransformation of cytochrome P (CYP) enzyme and its ethoxyresorufin-o-deethylase (EROD) activity which may be decreased the adverse effects of BaP (Abbas et al. 2019)	• CYP450 was estimated with EROD activity and fortunately, an effective diminution in CYP content, and EROD activity in fishes fed on mustard extract (alcoholic) had been found
Black pepper (Marich in Sanskrit, kali mirch in Hindi, kali mirch, syah mirch in Urdu, kalo morich, golmorich in Bengali, golmaricha in Odiya, thingmarcha in Mizo, marutis in Kashmiri, kali mirch in Punjabi, kalomirich in Gujarati, kali miri in Marathi,	***Piper nigrum*** 	Fruit	Piperine is the first pharmacologically active compound isolated from *Piper nigrum*. Other compounds found in pepper are sabinene, α-pinene, limonene, β-pinene, β-caryophyllene, citral phellandrene, linalool, etc. (Stojanović-Radić et al. 2019)	Piperine augments murine splenocyte proliferation in cadmium (Cd)-induced immunocompromised splenic cells (Pathak and Khandelwal 2007)	• Diminishes the adverse effects of cadmium on IL-2, IFN-γ, and cell proliferative mitogenic response • Reestablishment of the B- and T-cell populations (Pathak and Khandelwal 2007)
				Nitric oxide (NO) production by	• Stimulation of IL-6 and TNF-α by macrophages is

kurumilagu in Tamil, miryalatige in Telugu, karimenasu in Kannada, and kurumulaku in Malayalam)				macrophages gets significantly augmented by black pepper extract, establishes its pro-inflammatory role on cells (Majdalawieh and Carr 2010)	increased • NO production is significantly enhanced by macrophages only in the presence of IFN-γ (Majdalawieh and Carr 2010)
				Piperine shows immune stimulatory properties by suppressing the growth of tumour cells (Hadeel et al. 2010)	• Increases the white blood cells (WBCs) count, thus supporting the body to boost up an influential defence mechanism against invading microbes and cancer cells (Hadeel et al. 2010)
Cardamom (Trutih in Sanskrit, elaichi in Hindi and Urdu, chhoto elach in Bengali, elachi in Assamese and Marathi, aleicha in Odiya, chota alaichi in Kashmiri, ilaici in Punjabi, elaci in Gujarati, elaki in Telugu, elam ancha in Tamil, elakki in Kannada, and elatarri in Malayalam)	***Elettaria cardamomum*** 	Fruit	Cardamom is well-known as 'the queen of spices'. The seeds contain 96.9% oil. Sixty seven compounds have identified, among them, the major components were 1,8-cineole, linalool, α-terpinyl acetate (Savan and Küçükbay 2013)	*Elettaria cardamomum* distillate (ECD) has action on improvement of the immune system by stimulating the specific immune components. Therefore, ECD has potential to be developed as immune-stimulant in chemotherapy (Raksamiharja et al. 2012)	• Inhibition of IL-1β, IL-6, and TNFα secretion, COX-2 and activation of NF-κβ (Cho 2012, Patel et al. 2007, Zhang et al. 2005) • Improves the state of the immune cells through increase in the number of leukocytes, lymphocytes, and neutrophils in female Sprague Dawley rats after doxorubicin chemotherapy (Raksamiharja et al. 2012)
				The cardamom aqueous extracts can significantly enhance splenocyte	• Enhances Th2 cytokines, i.e. IL-4 and IL-10 release (Majdalawieh and Carr

(continued)

Table 8.3 (continued)

Spice/herb name in English (Common names in different Indian languages)	Scientific name with representative photographs	Useful part	Active components	Immunostimulant properties	Mode of action
				proliferation and Th2 cytokine release in mice (Majdalawieh and Carr 2010)	2010) • Induces type-I hypersensitivity reactions (i.e. allergic reactions), which, in turn, established its role in Th2 mediated immune responses
Chilli (Katuvira in Sanskrit, gach-mirichi in Hindi, gol mirch ka powda in Urdu, lonka in Bengali, lonka-jolokiya in Assamese, punjab lal in Punjabi, mirsinga, mirchi in Marathi, molagay or milagai in Tamil, mirapakaya in Telugu, menasinakai in Kannada, and upperiparanki or perangimuluk in Malayalam)	***Capsicum annuum*** 	Fruit	Capsaicin {(E)-N-[(4-hydroxy-3-methoxyphenyl) methyl]-8-methylnon-6-enamide} is the primary bioactive compound of chilli. This hydrophobic alkaloid is responsible for their spicy or pungent flavour (Deng et al. 2016)	Capsaicin has shown its emerging role in autoimmune diseases (Deng et al. 2016)	• Selectively activate the Ca^{2+} permeable ion channel called transient receptor potential vanilloid subfamily 1(TRPV1) via phosphorylation by calcium-calmodulin-dependent protein kinase II (CaMK II kinase) (Deng et al. 2016)
				Capsaicin induces maturation and direct activation of dendritic cells (Beltran et al. 2007)	• Expresses capsaicin receptor VR1 (vanilloid receptor 1) that leads to development of dendritic cells (Beltran et al. 2007)
Cinnamon (Varanga, tanutvak, and darusita in Sanskrit, dalchini in Hindi, darcheeni in Urdu,	***Cinnamomum burmannii verum***	Bark	The main components of *Cinnamomum burmannii* bark essential oil are trans-cinnamaldehyde, eugenol	*Cinnamomum burmannii* bark extract (CBE) administration influences the composition of cellular immune system	• Influences CRP expression (Utomo et al. 2020)

daruchini in Bengali, dalacheni in Assamese, dalcyn in Kashmiri, dalacini in Punjabi and Marathi, taja in Gujarati, lavanga patta in Telugu, ilayangam in Tamil, dalcinni in Kannada and kaṟuvappaṭṭa in Malayalam)			and coumarin (Faishal et al. 2017)	(neutrophil- and lymphocyte-type counts) in pathogen infection (*Staphylococcus aureus* infection) (Nassan et al. 2015)	
				CBE may serve as an influencer of cytokine induced leukocyte proliferation and differentiation (Lee et al. 2011)	• Triggers the IL-2 transcription process (Takeda et al. 2008) Impact on multiplication and leukocyte differentiation such as IL-2 and IL-4 (Lee et al. 2011)
				Cinnamaldehyde activates transient receptor potential ankyrin 1 (TPRA1) to modulate leukocytes during inflammation (Mendes et al. 2016)	• Promotes significant changes in the *leukocyte* population • Triggers an ongoing migration of polymorphonuclears (PMNs) (Mendes et al. 2016)
Clove (Labangam in Sanskrit, laung in Hindi, long in Urdu, labanga in Bengali, lwong in Assamese, roung in Kashmiri, kali in Punjabi, lavinga in Gujarati, lavanga in Marathi, lavangam in Telugu and	***Syzygium aromaticum*** 	Flower buds	The most abundant compounds of clove are eugenol, eugenyl acetate, isocaryophyllene, α-caryophyllene, geranyl-n-butyrate and β-caryophyllene (El-Ghorab and El-Massry 2003). The phenolic compound eugenol (4-allyl-2-	Clove essential oil (CEO) stimulates the immune responses by increasing the number of circulating lymphocytes (Carrasco et al. 2009)	• Proliferation of total WBC count • Stimulates the functioning of T cells and maturation of granulocytes (Carrasco et al. 2009)
				Clove essential oil has shown amplified delayed-type hypersensitivity (DTH) response in	• Activates the sensitized T cells (Th1 subsets) that leads to IFN-γ secretion by antigens

(continued)

Table 8.3 (continued)

Spice/herb name in English (Common names in different Indian languages)	Scientific name with representative photographs	Useful part	Active components	Immunostimulant properties	Mode of action
Kannada, kirampu in Tamil, and grampu in Malayalam)			methoxyphenol) is the major bioactive compound of clove essential oil as it is found in higher concentrations (Carrasco et al. 2009)	experimentally selected non-immunosuppressed mice (Carrasco et al. 2009)	• Initiates macrophage activation (Carrasco et al. 2009)
				CEO improves the humoral and secondary immune responses in rat (Halder et al. 2011)	• Potentiates the primary and secondary immune responses in rats immunized with sheep red blood cells (SRBC) by an increase in anti-SRBC titre (Halder et al. 2011)
Coriander (Dhaneyam in Sanskrit, dhaniya in Hindi and Urdu, dhone in Bengali, dhaniya guti in Assamese, danival in Kashmiri, punjab lal in Punjabi, dhana in Gujarati, kothimbira in Marathi, kottamalli in Tamil, kottimira in Telugu, kottambari in Kannada, and malli in Malayalam)	***Coriandrum sativum*** 	Dried seed Fresh leaves	Linalool is the main bioactive component in *Coriandrum sativum* seeds. Other phytochemicals are γ-terpinene, terpinyl acetate, camphene, E-verbenol, verbenone sabinene, citronellol, β-pinene, geraniol, eugenol, m-cymene carveol, limonene etc. (Farsani et al. 2019)	Coriander has a crucial role in stimulating the initiation of T cells and B cells responses in adaptive immune system (Ishida et al. 2017)	• Triggering of macrophage activation • Secretion of TNF-α and IL-6 • Encourages antibody production by plasma cells • Expressions of cell adhesion molecules • Leads to synthesis of other inflammatory mediators • Differentiation of Th17 cells (IL-17-producing helper T cells) (Ishida et al. 2017)

			Decanal, 2-decen-1-ol, *trans*-2-decenal, cyclodecane, *cis*-2-dodecena, dodecanal, and dodecan-1-ol are the important constituents of *Coriandrum sativum* leaves' essential oil (Mandal and Mandal 2015)	Coriander leaves extract stimulates lymphocyte proliferation (Gomez-Flores et al. 2010)	• Stimulates spleen lymphocyte proliferation • Induces thymus lympho-proliferation (Gomez-Flores et al. 2010)
Cumin (Jeerika in Sanskrit, jeera in Hindi, Assamese, Oriya and Punjabi, zerha in Urdu, jire in Bengali and Marathi, jira in Manipuri, zeur in Kashmiri, jirum in Gujarati, jilakarra in Telugu, cirakam in Tamil, jirige in Kannada, and jirakam in Malayalam)	***Cuminum cyminum*** 	Seed	Cumin seeds contain Cuminaldehyde, terpenoids and cymene. The flavonoid glycoside, 3,5-dihydroxyflavone 7-O-d-galacturonide-4-O-d-glucopyranoside is also an active component of cumin seed (Bettaieb et al. 2011)	*Cuminum cyminum* has the potential to stimulate the cellular immunity in normal and immunosuppressed Swiss albino mice (Chauhan et al. 2010)	• Significantly improved CD4+ and CD8+ T lymphocytes (Chauhan et al. 2010) • Selectively excites the cell-mediated immunity by rising the expression of IFN- γ (Wakil et al. 1998)
				Flavonoid glycoside, an active compound of *Cuminum cyminum* exerting the immune-stimulating properties through modulation of T cells (Chauhan et al. 2010)	• Proliferation of the CD4 and CD8 count and IL-2 expression • The weight of thymus and spleen enlarged, which can be associated with the rise in T-cell count (Frick et al. 2019)
Curry leaves (Mahanimb and girinimb, suravi, and alakavhaya in Sanskrit; kathnim, mitha neem and karee patte in Hindi; kadi patta in Urdu;	***Murraya koenigii (Linn.) Spreng.***	Fresh leaves	Leaves of *Murraya koenigii Linn.* contain flavonoids, alkaloids, and tannins with low levels of saponins and carotenoids	The aqueous extract of *Murraya koenigii* leaves (MKA) stimulates both the specific and non-specific immune mechanisms (Shah and Juvekar 2010)	• Increases sensitivity of T- and B-lymphocyte subsets (Benacerraf 1978) • Enhances in delayed-type hypersensitivity reactions (DTH) in mice in response

(continued)

Table 8.3 (continued)

Spice/herb name in English (Common names in different Indian languages)	Scientific name with representative photographs	Useful part	Active components	Immunostimulant properties	Mode of action
kari pata, barsanga, and kartaphulli in Bengali; noro-singho in Assamese; bhursanga in Odiya, kadipatta in Kashmiri; kari pate in Punjabi; gandhela, gandla, and gani in Kumaon; gorenimb and mitho limbado in Gujarati; kadi patta in Marathi; karivepaku in Telugu; karivempu and kariveppillai in Tamil; karibeva in Kannada; and karuveppilai in Malayalam)			(Uraku and Nwankwo 2015)		to ovalbumin (Shah and Juvekar 2010)
				The methanolic extract of *Murraya koenigii* leaves hold properties that provoke humoral immunity and phagocytic function (Shah et al. 2008)	• Significant rise in the NO production by mouse peritoneal macrophages • Amplify the phagocytic activity of macrophages • Considerably upgraded the total WBC count which in turn restore the myelosuppressive effects made by cyclophosphamide (Shah et al. 2008)
Fennel/Saunf/finocchio (Misreya and madhurika in Sanskrit, moti saunf in Hindi, sonf in Urdu, mouri in Bengali, guvamari in Assamese, hop in Manipuri, peddajilakarra in Telugu, sompu in Tamil, dodda sompu in Kannada, and preumjirakam in Malayalam)	***Foeniculum vulgare*** 	Seed	Main compounds identified from essential oil of fennel are (E)-anethole, *trans*-anethole, limonene, methyl chavicol, fenchone, α-pinene (Al Snafi 2018)	Anethole in fennel oil exhibits anti-inflammatory, anti-carcinogenic properties through modulating TNF- induced cellular responses in rabbits (Chainy et al. 2000)	• Modulates TNF that induce cellular responses in ML1-α cells of human (Chainy et al. 2000)
				Essential oil of fennel mediates innate immune response using various inflammatory materials and	• Proliferation of peripheral blood mononuclear cells (PBMC) • Stimulates antigens and

				activates adaptive immunity in human (Orhan et al. 2016)	mitogens leading to lymphocyte activation through γ-interferon secretion • Induces human cellular immune responses that release reactive oxygen species (ROS), resulting in proliferation of T cells, TNF-α, and IL-2 cytokine production (Orhan et al. 2016)
				Fennel seed extract in broiler chicks exhibits antibacterial activity and boost resistance against Newcastle disease (Safaei-Cherehh et al. 2018)	• Induces IgG and IgM antibody production • Improves Newcastle disease vaccination efficiency in broiler chicks throughout the vaccination period • Presence of fennel extract can reduces *Escherichia coli* infection due to fennel's antibacterial activity in broiler chicks (Safaei-Cherehh et al. 2018)

(continued)

Table 8.3 (continued)

Spice/herb name in English (Common names in different Indian languages)	Scientific name with representative photographs	Useful part	Active components	Immunostimulant properties	Mode of action
Fenugreek (Methika in Sanskrit, methi in Hindi, Urdu, Bangali, Odia, Nepali, Punjabi, Gujarati, and Marathi; methiguti in Assamese, mithhi biyoul in Kashmiri, mentulu in Telugu, ventayam in Tamil, mentya in Kannada, and uluva in Malayalam)	***Trigonella foenum-graecum*** 	Seed	They contain carbohydrates (mainly a water-soluble galactomannan), protein, fat (mainly PUFA), saponins, and alkaloids such as trigonelline (Belguith-Hadriche et al. 2010)	It acts as an immunostimulant in immunocompromised patients (Ramadan et al. 2011)	• Fenugreek seed powder (FSP, especially the high dose) has been found to completely modulate the immunosuppressive activity of CP including leucopenia (Ramadan et al. 2011)
				Fenugreek seed supplemented diets on gilt-head seabream sp. have been able to boost the *humoral immune responses* (Bahi et al. 2017)	• Upregulates the *igm* gene expression (Bahi et al. 2017)
				Fenugreek-activated the immune-related gene expressions in Nile tilapia infected with *Aeromonas hydrophila* (Moustafa et al. 2020)	• Develops immune parameters (lysozyme, immunoglobulin, and respiratory burst activity) • Upregulates the gene expressions in the liver and kidney (IL-1β and TNF-α) (Moustafa et al. 2020)
				The flavonoid and alkaloids of this seed may help to improve the growth and immunity in Nile tilapia (*Oreochromis*	• Stimulates the expression of IL-6 and IL-8 • Boosts interleukins better (Abbas et al. 2019)

				niloticus) (Abbas et al. 2019)	
		Dried leaves	Trigonella leaves have combinations like quercetin, catechin, cinnamic acid, and coumaric acid, along with high content of soluble fibres (Aylanc et al. 2020)	Fenugreek leaves have the ability to strengthen the immune functions of mice (Bin-Hafeez et al. 2003)	• Rising the bone marrow cell counts denotes the stimulatory effect on haematopoietic stem cells of the bone marrow • The increase in thymus weight was got together with rise in cell counts of lymphocytes and bone marrow haematopoietic cells (Bin-Hafeez et al. 2003)
				Ethanolic extract of fenugreek leaves shown stimulatory effects on activation of both humoral immunity and cell-mediated immunity in mice after immunized with sheep red blood cell (SRBC), used as an antigen (Tripathi et al. 2012)	• The high values of haemagglutinating antibody titre indicates that immune-stimulation can be achieved through humoral immunity • Interestingly, proliferation in rosette formation and lymphocyte formation in T-cell population test denotes the effect on cell-mediated immunity • Exhibits significantly

(continued)

Table 8.3 (continued)

Spice/herb name in English (Common names in different Indian languages)	Scientific name with representative photographs	Useful part	Active components	Immunostimulant properties	Mode of action
					high phagocytic index (Tripathi et al. 2012)
Garlic (Lahsuna in Sanskrit, lasun and lahsan in Hindi, lahssan in Urdu, rasun in Bengali, naharu in Assamese, purun-var and purunvar in Mizo, chanam in Manipuri, ruhun in Kashmiri, velluli in Telugu, acanam in Tamil, belluli in Kannada, and vellulli in Malayalam)	***Allium sativum*** 	Bulb	Main component of garlic is Allicin [S-(2- propenyl)-2-propene-1-sulphinothioate. It also contains S-methylcysteine sulphoxide (MCSO) and S-propylcysteine sulphoxide (PCSO) (Zieger and Stitcher 1989)	Garlic extract contains high amount of vitamin C; hence it acts as significant and potent immune system booster (Mirabéau and Samson 2012)	• This extract improves acquired immunodeficiency syndrome (AIDS) patients by activating T lymphocytes and NK cells • Significantly increases WBC count and CD4 cells in garlic extract-treated group in dose-dependent manner (Mirabeau and Samson 2012)
				Aged garlic extract has been shown to boost immune response in BALB/c mice and Wistar rats by immunomodulatory protein, mainly lectins (Chandrashekar and Venkatesh 2009)	Garlic extract contains immunomodulatory proteins; two agglutinins ASAI and ASAII have hemagglutination and mitogenic activities through mitogenic lectins • High-molecular-weight agglutinin (ASA110) has been found to stimulate human and guinea pig peripheral blood lymphocytes

					(Chandrashekar and Venkatesh 2009)
				Fructans from raw garlic extract demonstrate immunostimulatory properties by activating macrophages (Chandrashekar and Venkatesh 2016)	• Induces proliferation of murine splenocytes leading to increase lymphocyte proliferation • Enhances yeast phagocytosis (Chandrashekar and Venkatesh 2016)
				Garlic extract enhances immunity, promotes growth and controls bacterial and viral infections in fish (Shakya and Labh 2014)	• Facilitates phagocytic cells functions and increases bactericidal activities • Stimulates antibody responses, NK cells, lysozyme, and complements • Enhances phagocytosis by macrophages • Exhibits strong innate immunity development through altering levels of albumin, globulin, and serum total proteins (Shakya and Labh 2014)
				Garlic's antiviral efficiency may have preventive measures against SARS-CoV-2 infections (COVID-	In case of COVID-19 infection, caused by SARS-CoV-2, primarily there have been decreases

(continued)

Table 8.3 (continued)

Spice/herb name in English (Common names in different Indian languages)	Scientific name with representative photographs	Useful part	Active components	Immunostimulant properties	Mode of action
				19) to boost immunity (Donma and Donma 2020)	in CD4+ T and CD8+ T cells leading to reducing interferon-Υ production • Short-term garlic extract-supplemented diet increases in CD4+ and CD8+ T cells • Boosts immunity by stimulating lymphocytes, macrophages, NK cells, dendritic cells • Increases total WBC count and CD4+ T cells in treated rats (Donma and Donma 2020)
Ginger (Adrakam in Sanskrit, haldi and adrakh in Hindi, adrak in Urdu, ada in Bengali and Odia, aada in Assamese, shing in Manipuri, shaunth in Kashmiri, adaraka in Punjabi, adu in Gujarati, ale in Marathi, allam in Telugu, inci in Tamil and	***Zingiber officinale*** 	Rhizome	Zingiberol, zingiberene, phellandrene, linalool, gingerols, shogaols, gingerenone, etc. are the most important bioactive compounds of Ginger (Shakya 2015)	For boosting up the immune system, ginger can be applied as an alternative dietary supplement (Shakya 2015) Ginger showed an effective role in the phagocyte activation (Magnadottir 2006)	• Potentiates the humoral and cell-mediated immune responses in immune-suppressed mice (Carrasco et al. 2009) • Improves erythrocyte, WBC, and neutrophil percentages (Haghighi and Rohan 2013) • Powdered ginger rhizome has been seen with a special ability to enhance

Malayalam, sunthi in Kannada)					the lysozyme activity in fish (Magnadottir 2006)
Nutmeg (Jatiphalam in Sanskrit; jaayaphal in Hindi; joz in Urdu; jayaphala in Bengali, Gujarati, and Marathi; jaaiphal in Assamese; jaiphal in Oriya; zaaphal in Kashmiri; japha in Punjabi; jajikaya in Telugu; jatikkay in Tamil; jayikayi in Kannada; and jatikka in Malayalam)	***Myristica fragrans*** 	Seed	Nutmeg contains sabinene, camphene, myristin, elemicin, isoelemicin, eugenol, isoeugenol, methoxyeugenol, safrole, diametric phenylpropanoids, lignans, neolignans, etc. The essential oil isolated from the seeds contain sabinene, α-pinene, α-phellandrene, and γ-terpinol as major constituents (Kuete 2017)	Internal administration of *Myristica fragrans* reduced the allergic inflammatory markers such as circulating serum IgE, absolute eosinophils count (AEC), and infiltration of eosinophils at peribronchial and perivascular region in bronchial tissue (and BAL fluid) (Menthe 2020)	• Significantly reduces IgE a subtle marker of allergic inflammation and asthma when compared to asthma control group • Downregulates Th2 lymphocyte activation and its downstream signalling • Inhibits IL-5 expression of Th2 lymphocyte activation and its downstream signalling (Menthe 2020)
Onion (Palāṇḍuḥ in Sanskrit, pyaj in Hindi, pyaaz in Urdu, pinyaj in Assamese and Bengali, piyaja in Punjabi, kanda in Marathi, dungali in Gujarati, ullipaya in Telugu, venkayam in Tamil, irulli in Kannada, and ulli in Malayalam)	***Allium cepa*** 	Bulb	Main components of onion are fructose, flavonoids,, flavenols, quercetin-3-glucoside, isorhamnetin-4-glucoside, S-alk(en)yl cysteine sulphoxides (Hubbard et al. 2006)	Onion, rich in fructo-oligosaccharides (FOS)/ inulin (IN) known as fructans. Fructans have immunomodulatory activities that help in activation of macrophages and proliferation of immuno-responder cells (Kumar et al. 2015)	• Increases NO production in murine PECs • FOS has lympho-proliferative activity that activates murine peritoneal macrophages to induce intracellular free radical generation (Kumar et al. 2015)
				Fructans in onion exhibit prebiotic functions in elderly people (Watzl et al. 2005)	• Increasing the production of $CD3^+$, $CD4^+$, and $CD8^+$ cells • Decreasing the production of inflammatory markers like IL-6 and phagocytosis (Guigoz et al. 2002)

(continued)

Table 8.3 (continued)

Spice/herb name in English (Common names in different Indian languages)	Scientific name with representative photographs	Useful part	Active components	Immunostimulant properties	Mode of action
				Fructo-oligosaccharides (FOS) and short-chain oligosaccharides in onion modulate immune response in cow milk protein by inducing beneficial antibody profile (Van Hoffen et al. 2009)	• Diminishes Ig levels of the various isotypes, mainly IgE • FOS/GOS supplementation has been found to reduce significantly CMP-specific IgG1 levels (Van Hoffen et al. 2009)
				Fructans can also reduce pathogenicity of pathogenic bacteria in elderly people (Kumar et al. 2015)	• Provides a stronger, continuous immune stimulation to prevent atopic diseases by stimulating turnover of appropriate bacteria (van Hoffen et al. 2009)
				Lectins present in onions exert immunomodulatory activities against infections (Prasanna and Venkatesh 2015)	• Activates and proliferates antigen-specific lymphocyte by entering the pathway of activation-induced cell death (Prasanna and Venkatesh 2015)
				Th1 immune response in murine splenocytes gets induced by onion lectin (Muraillie et al. 1999)	• Increases the level of IL-2 and IFN-γ • Induces the production of NO to activate macrophage • Facilitates the release of

					RAW264, IL-12 and TNF-α (Noroozi et al. 2000)
				Aqueous extracts of onion increase total WBC count and $CD4^+$ cells in a dose-dependent manner that boosts immunity in rats (Mirabeau and Samson 2012)	• Activates NK cells and T lymphocytes particularly in AIDS • Enhances cell-mediated immune response of T cells and promotes proliferation of NK cells in responses to mitogens and antigens • Differential count of WBC (lymphocytes, neutrophils, and monocytes) has been found to increase significantly revealing immunity boosting abilities of onion (Mirabeau and Samson 2012)
Peppermint (Paparaminta in Sanskrit, pudeena in Hindi, podeenay ka sat in Urdu, pudina in Bengali, foutnihh in Kashmiri, miraca in Punjabi, marina dana in Gujarati, peparaminta in Marathi, pipparameṇṭu in Telugu, miḷakukkīrai in Tamil, pudina in Kannada and kurumuḷak in Malayalam)	***Mentha piperita L.*** 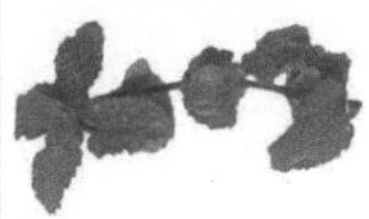	Fresh leaves	It possesses different bioactive compounds such as menthol, menthone, eucalyptol, neo-menthol, menthofuran, menthol acetate, isomenthone, pulegone, limonene, germacrene, terpinen-4-ol, beta-pinene, alpha-pinene, trans-sabinene hydrate, sabinene, piperitone, gamma-terpinene, linalool, etc. (Brahmi et al. 2017)	Peppermint extract strengthens the immune system by increasing total antibody titre, IgM, and IgG (Arab-Ameri et al. 2016)	Peppermint powder (1%) supplemented with diet to experimentally selected birds • Increases total antibody titre (TIg), IgM, and IgG on day 42 of the experiment • Boosts the activity of macrophages, lymphocytes, and natural killer cells by encouraging the phagocytic activity and

(continued)

Table 8.3 (continued)

Spice/herb name in English (Common names in different Indian languages)	Scientific name with representative photographs	Useful part	Active components	Immunostimulant properties	Mode of action
					also production of interferons, which in turn strengthens humoral and cellular immune responses (Arab-Ameri et al. 2016)
				Peppermint essential oils have been found to enhance lymphocyte count in chicken (Awaad et al. 2010)	• Improves phagocytic activity of the macrophages • Amplifies the lymphocyte count by causing lymphocytic hyperplasia and activating Bursa of Fabricius, thymus, spleen, and caecal tonsils, thereby triggering the potency of the immune system (Awaad et al. 2010).
Saffron (Kashmirajanman, keshara, kunkuma, aruna, asra, and asrika in Sanskrit, kesar in Hindi, zafran in Urdu, jafran in Bengali, kong in Kashmiri, kesar, zafran in Punjabi, keshar in Gujarati, keshar, kesara in Marathi, kumkum puvvu in Telugu, kungumapoo in Tamil,	***Crocus sativus*** 	Stigma	Saffron oil is extracted mainly from the flowers and stigma. The major constituents of saffron essential oil are safranal (for aroma), picrocrocin (bitter taste), and crocin (responsible for colour) along with other *carotenoids* and terpenes (Babaei et al. 2014)	The saffron petal may be favourable to rise the secondary immune responses (IgG) against pathogens (Babaei et al. 2014)	• Major surge in IgG level upon treatment with SPE • Encourages the secretion of Th cell cytokines, i.e. IFN-γ, main cytokine for B-lymphocyte activation to produce IgG (Babaei et al. 2014)
				The stigmas of saffron possess humoral and cell-mediated immunity	• Substantial growth in the neutrophil adhesion to nylon fibres

kesari in Kannada, and kashmiram and kunkumapoove in Malayalam)				(Vijayabhargava and Asad 2012)	• *Crocus sativus* at low doses possesses potential for inducing humoral and cell-mediated immunity (Vijayabhargava and Asad 2012)
Sesame (Til in Sanskrit, Hindi, Bengali, and Assamese; thoiding in Manipuri; tiloun ka powda in Urdu; khasa, tila, and rashi in Oriya; tael in Kashmiri; tila in Punjabi and Marathi; tala in Gujarati; nuvvulu in Telugu; el in Tamil; ellu in Kannada; and ell in Malayalam)	***Sesamum indicum* Linn.** 	Seed	*Sesamum indicum* comprises sesamin, sesaminol glucosides, and sesamolin, (Do et al. 2019)	Ethanol extract of *Sesamum indicum* (SIE) possess anti-allergic property (Do et al. 2019)	• Prevents the degranulation process and production of TNF-α, IL-4, IL-6, histamine, and PGD2, on FcεRI-mediated allergic reactions in RBL-2H3 mast cells (Do et al. 2019)
Star anise (Chinese star anise/badian) (thakkolam in Sanskrit, chakra phool in Hindi, biryani ka phool in Urdu, taraka anise in Bengali, star mauri in Assamese, tara anise in Punjabi, stara variyali in Gujarati, stara badicepa in Marathi, naṭcattira compu in Tamil, star sompu in Telugu and Kannada, and takkolam in Malayalam)	***Illicium verum*** 	Seed and pod	Star anise fruits consist greater a number of essential oil, alkaloid, tannin, cis- and trans-anethole, shikimic acid, limone, α-pinene, α-terpineol, farnesol, safrole, and β-phellandrene (Chouksey et al. 2010)	Bioactive molecules of star anise inhibit quorum sensing and biofilm formation of some food-borne bacteria in milk (Rahman et al. 2017)	• Prevents bacterial resistance development ability, and also reduces bacterial virulence properties (Rahman et al. 2017)
				Anethole exhibits anti-inflammatory and anti-carcinogenic properties (Chainy et al. 2000)	• Represses NF-κβ-dependent gene expression by inducing TNF (Chainy et al. 2000)
				Star anise fruits are used in the treatment of chronic inflammatory skin diseases and atopic dermatitis (AD) in the lesional skin of AD	• Stimulates with IFN-ϒ and TNF-α have been seen to cause expression of adhesion molecules, such as ICAM-1, inflammatory

(continued)

Table 8.3 (continued)

Spice/herb name in English (Common names in different Indian languages)	Scientific name with representative photographs	Useful part	Active components	Immunostimulant properties	Mode of action
				model NC/ Nga mice (Sung et al. 2012)	cytokines, and chemokines (Sung et al. 2012)
Turmeric (Haridra, gauri, kanchani, aushadhi, vitkanta, peetvaaluka in Sanskrit, haldi and hardhar in Hindi and Gujarati, haladi in Urdu, halood in Bengali, haladi in Odiya, halodhi in Assamese, ai-eng in Mizo, yaingang in Manipuri, lader in Kashmiri, haldar, haldi in Punjabi, halad and halada in Marathi, haladpito in Konkani, pasupu, haridra in Telugu, mancal in Tamil, arishina in Kannada, and mannal in Malayalam)		Rhizome	The major bioactive compounds of turmeric are curcumin [diferuloylmethane or 1,7-bis (4-hydroxy-3 methoxy-phenyl) hepta-1,6-diene-3,5-dione)], bisdemethoxycurcumin, demethoxycurcumin, and cyclocurcumin (Afolayan et al. 2018)	The therapeutic potential for cerebral malaria management and the influence of curcumin on the immune system have been recommended (Mimche et al. 2011)	• Reduces the production of TNF, IL-12p40 and IL-6 in peripheral blood mononuclear cells (PBMC) primed with trophozoites/ schizont stages of *Plasmodium falciparum* • The expression of adhesion molecules such as ICAM1 (intercellular adhesion molecule 1), VCAM1 (Vascular cell adhesion protein 1), and E-selectin was also downregulated in TNF-activated human endothelial cells following PBMC exposure to curcumin (Mimche et al. 2011)
				Curcumin remarkably stimulates the immune system by potentiating the proliferative capacity of immune cells (Alambra et al. 2012)	• Modulates the stimulation of B cells, T cells, macrophages, natural killer cells, neutrophils, and dendritic cells (Alambra et al. 2012)

				It has been seen that curcumin from turmeric extract increases macrophage phagocytic activity in curcumin-treated animals (Lukita-atmadja et al. 2002)	• Significantly improves NK-cell cytotoxicity and IFN-γ-stimulated NK-cell cytotoxicity • Curcumin has been found to increases the plaque-forming cells (PFC) and also boosts up the macrophage phagocytic activity (Yadav et al. 2005)
				Curcumin has been shown to possess tumouricidal activity by the increased activation of macrophages and NK cells (Bhumik et al. 2000)	• Induces apoptosis in rat histiocytic cells (AK-5) that leads to the inhibition of tumour growth • Initially Th1 cytokine response and NO production by macrophages were downregulated, but their upregulation in NK cells could be pulled up upon prolonged treatment with curcumin that exhibits a strong tumouricidal effect (Bhumik et al. 2000)

Uses of region-specific differential and exclusive blends of assorted spices and herbs have helped in generating exotic cuisines with exemplary traditions and ethnicity. Moreover their proven efficacy to contribute to human health as nutritional and anti-infectious agents by virtue of possessing bioactive compounds in the form of metabolic end-products has added an additional approbation in their reliability and demand. Here an attempt has been made to document systematically the immunostimulant properties of common spices and herbs used in Indian cuisines principally conferring protection against causation of a variety of diseases by strengthening body defence mechanisms. This also remains highly significant in the perspectives of COVID-19 pandemic situation.

Acknowledgement Authors sincerely acknowledge the generous support from Ms. Jhunu Karmakar for arranging and generating the photographs of representative spices and herbs.

References

Abbas WT, Abumourad IM, Mohamed LA et al (2019) The role of the dietary supplementation of fenugreek seeds in growth and immunity in Nile tilapia with or without cadmium contamination. Jordan J Biol Sci 12(5):649–655

Afolayan FID, Erinwusi B, Oyeyemi OT (2018) Immunomodulatory activity of curcumin-entrapped poly d, l-lactic-co-glycolic acid nanoparticles in mice. Integr Med Res 7:168–175

Agarwal SS (1999) Immunomodulation: a review of studies on Indian medicinal plants and synthetic peptides part 1: medicinal plants. Proc Ind Natl Sci Acad 65(3):79–204

Ajungla T, Yeptho L, Kichu A et al (2020) Some ethnic fermented foods and beverages of Nagaland. In: Ethnic fermented foods and beverages of india: science history and culture. Springer, Singapore, pp 459–477

Akhter R, Masoodi FA, Wani TA et al (2020) Ethnic fermented foods and beverages of Jammu and Kashmir. In: Ethnic fermented foods and beverages of India: science history and culture. Springer, Singapore, pp 231–259

Alambra JR, Alenton RRR, Gulpeo PCR et al (2012) Immunomodulatory effects of turmeric, Curcuma longa *(Magnoliophyta, Zingiberaceae*) on *Macrobrachium rosenbergii* (Crustacea, Palaemonidae) against *Vibrio alginolyticus* (Proteobacteria, Vibrionaceae). AACL Bioflux 5: 13–17

Ali A, Aslam N, Naveed JBM et al (2020) Growth performance and immune response of *Labeo rohita* under the dietary supplementation of black seed (*Nigella sativa*). Pure Appl Biol 9(4): 2339–2346

Al-Snafi AE (2018) The chemical constituents and pharmacological effects of *Foeniculum vulgare* - a review. IOSR J Pharm 8:81–96

Amalraj A, Gopi S (2017) Biological activities and medicinal properties of Asafoetida: a review. J Tradit Complement Med 7:347–359

Annapure US, Ghanate AS, Halde PS et al (2020) Ethnic fermented foods and beverages of Maharashtra. In: Ethnic fermented foods and beverages of India: science history and culture. Springer, Singapore, pp 305–348

Anupma A, Pradhan P, Sha SP et al (2018) Traditional skill of ethnic people of the Eastern Himalayas and North East India in preserving microbiota as dry amylolytic starters. Ind J Tradit Knowl 17(1):184–190

Appadurai A (1988) How to make a national cuisine: cookbooks in contemporary India. Comp Stud Soc Hist 30(1):3–24

Arab Ameri S, Samadi F, Dastar B et al (2016) Efficiency of peppermint (*Mentha piperita*) powder on performance, body temperature and carcass characteristics of broiler chickens in heat stress condition. Iran J Appl Anim Sci 6(2):435–445

Arjun J (2015) Comparative biochemical analysis of certain indigenous rice beverages of tribes of Assam with some foreign liquor. J Biosci Biotechnol Res Commun 8:138–144

Awaad MHH, Abdel-Alim GA, Sayed KSS et al (2010) Immunostimulant effects of essential oils of peppermint and eucalyptus in chickens. Pak Vet J 30:61–66

Aylanc V, Eskin B, Zengin G et al (2020) In vitro studies on different extracts of fenugreek (*Trigonella spruneriana* BOISS.): phytochemical profile, antioxidant activity, and enzyme inhibition potential. J Food Biochem 44(11):e13463

Babaei A, Arshami J, Haghparast A et al (2014) Effects of saffron *(Crocus sativus)* petal ethanolic extract on hematology, antibody response, and spleen histology in rats. Avicenna J Phytomed 4(2):103–109

Bahi A, Guardiola FA, Messina CM et al (2017) Effects of dietary administration of fenugreek seeds, alone or in combination with probiotics, on growth performance parameters, humoral immune response and gene expression of gilthead seabream (*Sparus aurata* L.). Fish Shellfish Immunol 60:50–58

Barik R (2006) Land and caste politics in Bihar. Shipra Publications, New Delhi

Barooah M, Bora SS, Goswami G et al (2020) Ethnic fermented foods and beverages of Assam. In: Ethnic fermented foods and beverages of India: science history and culture. Springer, Singapore, pp 85–104

Baruah R, Appaiah KA, Halami PM et al (2020) Ethnic fermented foods and beverages of Karnataka. In: Ethnic fermented foods and beverages of India: science history and culture. Springer, Singapore, pp 209–230

Batool S, Khera RA, Hanif MA et al (2020) Bay leaf. In: Medicinal plants of South Asia. Elsevier, pp 63–74

Belguith-Hadriche O, Bouaziz M, Jamoussi K et al (2010) Lipid-lowering and antioxidant effects of an ethyl acetate extract of fenugreek seeds in high-cholesterol-fed rats. J Agric Food Chem 58(4):2116–2122

Belkaid Y, Hand T (2014) Role of the microbiota in immunity and inflammation. Cell 157(1): 121–141

Belkaid Y, Naik S (2013) Compartmentalized and systemic control of tissue immunity by commensals. Nat Immunol 14:646–653

Beltran J, Ghosh AK, Basu S (2007) Immunotherapy of tumors with neuroimmune ligand capsaicin. J Immunol 178(5):3260–3264

Benacerraf B (1978) Opinion: a hypothesis to relate the specificity of T lymphocytes and the activity of I region-specific Ir genes in macrophages and B lymphocytes. J Immunol 120 (6):1809–1812

Beniwal A, Ghosh T, Bhardwaj KN et al (2020) Ethnic fermented foods and beverages of Uttarakhand, Uttar Pradesh, Haryana, and Punjab. In: Ethnic fermented foods and beverages of India: science history and culture. Springer, Singapore, pp 621–645

Bettaieb I, Bourgou S, Sriti J et al (2011) Essential oils and fatty acids composition of Tunisian and Indian cumin (*Cuminum cyminum* L.) seeds: a comparative study. J Sci Food Agric 91(11): 2100–2107

Bhardwaj RK, Sikka BK, Singh A et al (2011) Challenges and constraints of marketing and export of Indian spices in India. Proceedings of International Conference on Technology and Business Management, 28–30 March, 2011

Bhatia A, Khetarpaul N (2012) *'Doli Ki Roti'*–An indigenously fermented Indian bread: Cumulative effect of germination and fermentation on bioavailability of minerals. Ind J Tradit Knowl 11:109

Bhaumik S, Jyothi MD, Khar A (2000) Differential modulation of nitric oxide production by curcumin in host macrophages and NK cells. FEBS Lett 483:78–82

Bhuyan DJ, Barooah MS, Bora SS et al (2014) Biochemical and nutritional analysis of rice beer of North East India. Ind J Tradit Knowl 13:142–148
Bin-Hafeez B, Haque R, Parvez S et al (2003) Immunomodulatory effects of fenugreek (Trigonella foenum graecum L.) extract in mice. Int Immunopharmacol 3(2):257–265
Blalock JE, Smith EM (2007) Conceptual development of the immune system as a sixth sense. Brain Behav Immun 21(1):23–33
Bora SS, Keot J, Das S et al (2016) Metagenomics analysis of microbial communities associated with a traditional rice wine starter culture (Xaj-pitha) of Assam, India. 3 Biotech 6(2):153
Brahmi F, Khodir M, Mohamed C et al (2017) Chemical composition and biological activities of Mentha species. Aromat Med Plants-Back to Nat 10:47–78
Brandtzaeg PE (2002) Current understanding of gastrointestinal immunoregulation and its relation to food allergy. Ann N Y Acad Sci 964:13–45
Carrasco FR, Schmidt G, Romero AL et al (2009) Immunomodulatory activity of *Zingiber officinale* Roscoe, *Salvia officinalis* L. and *Syzygium aromaticum* L. essential oils: evidence for humor- and cell-mediated responses. J Pharm Pharmacol 61:961–967
Cebula A, Seweryn M, Rempala GA et al (2013) Thymus-derived regulatory T cells contribute to tolerance to commensal microbiota. Nature 497:258–262
Chainy GB, Manna SK, Chaturvedi MM et al (2000) Anethole blocks both early and late cellular responses transduced by tumor necrosis factor: effect on NF-κβ, AP-1, JNK, MAPKK and apoptosis. Oncogene 19(25):2943–2950
Chan K (2003) Some aspects of toxic contaminants in herbal medicines. Chemosphere 52(9): 1361–1371
Chandrashekar PM, Venkatesh YP (2009) Identification of the protein components displaying immunomodulatory activity in aged garlic extract. J Ethnopharmacol 124(3):384–390
Chandrashekara PM, Venkatesh YP (2016) Immunostimulatory properties of fructans derived from raw garlic (*Allium sativum* L.). Bioact Carbohydr Diet Fibre 8(2):65–70
Chauhan PS, Satti NK, Suri KA et al (2010) Stimulatory effects of Cuminum cyminum and flavonoid glycoside on Cyclosporine-A and restraint stress induced immune-suppression in Swiss albino mice. Int J Chem Biol 185(1):66–72
Cho KH (2012) 1, 8-cineole protected human lipoproteins from modification by oxidation and glycation and exhibited serum lipid-lowering and anti-inflammatory activity in zebrafish. BMB Rep 45(10):565–570
Chouksey D, Sharma P, Pawar RS et al (2010) Biological activities and chemical constituents of *Illicium verum* hook fruits (Chinese star anise). Der Pharmacia Sinica 1(3):1–10
Chowdhury N, Goswami G, Hazarika S et al (2019) Microbial dynamics and nutritional status of namsing: a traditional fermented fish product of Mishing community of Assam. Proc Natl Acad Sci India Sect B Biol Sci 89(3):1027–1038
Collingham L (2006) Curry: a tale of cooks and conquerors. Oxford University Press, New York, NY
Conn EE (1995) The world of phytochemicals. In: Gustine DL, Flores HE (eds) Phytochemicals and health. American Society of Plant Physiologists, Rockville, MD, pp 1–14
Coombes JL, Siddiqui KR, Arancibia-Carcamo CV et al (2007) A functionally specialized population of mucosal CD103+ DCs induces Foxp3+ regulatory T cells via a TGF-{beta}- and retinoic acid-dependent mechanism. J Exp Med 204(8):1757–1764
Das AJ, Deka SC (2012) Fermented foods and beverages of the North-East India. Int Food Res J 19: 377–392
Das M, Kundu D, Singh J et al (2017) Physiology and biochemistry of indigenous tribal liquor Haria: a state of art. Adv Biotechnol Microbiol 6(2):1–5
Deng Y, Huang X, Wu H et al (2016) Some like it hot: the emerging role of spicy food (capsaicin) in autoimmune diseases. Autoimmun Rev 15(5):451–456
Devasagayam TPA, Kamat JP, Sreejayan N et al (2001) Antioxidant action of curcumin. In: Nesaretnam K, Packer L (eds) Micronutrients and health: molecular biological mechanisms. AOCS Press, Champaign, IL, pp 42–59

Devi P, Suresh Kumar P (2012) Traditional, ethnic and fermented foods of different tribes of Manipur. Indian J Tradit Knowl 11:70–77
Divakaran M, Jayasree E, Babu KN et al (2018) Legacy of Indian spices: its production and processing. In: Indian spices. Springer, Cham, pp 13–30
Do HJ, Oh TW, Park KI et al (2019) Ethanol extract of Sesamum indicum Linn. inhibits FcεRI-mediated allergic reaction via regulation of Lyn/Syk and Fyn signaling pathways in rat basophilic leukemic RBL-2H3 mast cells. Mediat Inflamm 2019:5914396
Dokhuma J (1992) Khawtlang Dan. Hmanlai Mizo pi pute kalphung. In: Dokhuma PSJ (ed) James Dokhuma. JD Press, Mizoram, pp 177–230
Donma MM, Donma O (2020) The effects of allium sativum on immunity within the scope of COVID-19 infection. Med Hypotheses 144:109934
Dubey KG (2010) The Indian Cuisine. PHI Learning Pvt Ltd, Delhi
Duke JA (ed) (2002) CRC handbook of medicinal spices. CRC Press, Inc., Boca Raton, FL
El-Ghorab AH, El-Massry KF (2003) Free radical scavenging and antioxidant activity of volatile oils of local clove and cinnamon isolated by supercritical fluid extraction [SFE]. J Essent Oil-Bearing Plants 6:9–20
Ellis AE (2001) Innate host defense mechanisms of fish against viruses and bacteria. Dev Comp Immunol 25(8–9):827–839
Faishal LF, Utomo AW, Retnoningrum D et al (2017) Pengaruh Pemberian Ekstrak Kayu Manis (Cinnamomum burmannii) Terhadap Aktivitas Dan Kapasitas Fagositosis Studi Eksperimental Pada Tikus Wistar Yang Dipapar Staphylococcus aureus. Jurnal Kedokt Diponegoro 6(2): 772–781
FAO (1998) Fermented fruits and vegetables-A global perspective, vol 134. FAO Agricultural Services Bulletin, Rome
Frick M, Chan W, Arends CM et al (2019) Role of donor clonal hematopoiesis in allogeneic hematopoietic stem-cell transplantation. J Clin Oncol 37(5):375–385
Farsani MN, Hoseinifar SH, Rashidian G et al (2019) Dietary effects of Coriandrum sativum extract on growth performance, physiological and innate immune responses and resistance of rainbow trout (Oncorhynchus mykiss) against Yersinia ruckeri. Fish Shellfish Immunol 91:233–240
Fuhrman J, Ferreri DM (2010) Fueling the vegetarian (vegan) athlete. Curr Sports Med Rep 9(4): 233–241
Galina J, Yin G, Ardo L et al (2009) The use of immunostimulating herbs in fish. An overview of research. Fish Physiol Biochem 35(4):669–676
Gavankar R, Chemburkar M (2016) Isolation and characterization of native yeast from Mahua flowers. Int J Curr Microbiol App Sci 5(11):305–314
Geetha P, Arivazhagan R, Selvam SP et al (2015) Process standardization, characterization and shelf life studies of Chhana jalebi-a traditional Indian milk sweet. Int J Food Res 22(1):155–162
Ghosh K, Maity C, Adak A (2014) Ethnic preparation of haria, a rice-based fermented beverage, in the province of lateritic West Bengal, India. Ethnobot Res Appl 12:39–49
Gomez-Flores R, Hernández-Martínez H, Tamez-Guerra P et al (2010) Antitumor and immunomodulating potential of Coriandrum sativum. J Nat Prod 3:54–63
Granier A, Goulet O, Hoarau C et al (2003) Fermentation products: immunological effects on human and animal models. Integrated Mechan Rev 74(2):238–244
Granier A, Goulet O, Hoarau C (2013) Fermentation products: immunological effects on human and animal models. Pediatr Res 74(2):238–244
Guigoz Y, Rochat F, Perruisseau-Carrier G et al (2002) Effects of oligosaccharide on the faecal flora and non-specific immune system in elderly people. Nutr Res 22(1–2):13–25
Gupta M, Khetarpaul N, Chauhan BM (1992) Preparation nutritional value and acceptability of barley rabadi—an indigenous fermented food of India. Plant Foods Hum Nutr 42(4):351–358
Hadeel W, Zaid K, Al Tae'e MF et al (2010) Immunonological evaluation and acute toxicity study with fertility examination for the effect of aqueous extract from dried fruits of Piper nigrum L. Mice Iraqi J Sci 51(3):465–470

Haghighi M, Rohani MS (2013) The effects of powdered ginger (Zingiber officinale) on the haematological and immunological parameters of rainbow trout Oncorhynchus mykiss. J Med Plant Herb Ther Res 1:8–12

Halder S, Mehta AK, Mediratta PK et al (2011) Essential oil of clove (*Eugenia caryophyllata*) augments the humoral immune response but decreases cell mediated immunity. Phyther Res 25: 1254–1256

Hao Q, Lu Z, Dong BR et al (2011) Probiotics for preventing acute upper respiratory tract infections. Cochrane Database Syst Rev 7(9):CD006895

Hubbard GP, Wolffram S, de Vos R et al (2006) Ingestion of onion soup high in quercetin inhibits platelet aggregation and essential components of the collagen-stimulated platelet activation pathway in man: a pilot study. Br J Nutr 96(3):482–488

Ishida M, Nishi K, Kunihiro N et al (2017) Immunostimulatory effect of aqueous extract of Coriandrum sativum L. seed on macrophages. J Sci Food Agric 97:4727–4736

ISO (1995) Spices definition. In: Geneva-Based International Organization for Standardisation, vol 676. ISO, Geneva

Jain A, Bagler G (2015) Spices form the basis of food pairing in Indian cuisine. arXiv preprint arXiv:1502.03815

Jamir B, Deb CR et al (2018) Nutritional assessment and molecular identification of microorganisms from Akhuni/Axone: a soybean based fermented food of Nagaland, India. J Adv Biol 11(1)

Jeyaram K, Singh TA, Romi W et al (2009) Traditional fermented foods of Manipur. Indian J Tradit Knowl 8:115–121

Joshi SR, Khongriah W, Biswas K et al (2020) Ethnic fermented foods and beverages of Meghalaya. In: Tamang J (ed) Ethnic fermented foods and beverages of India: Science history and culture. Springer, Singapore, pp 421–433

Kanwar SS, Bhushan K (2020) Ethnic fermented foods and beverages of Himachal Pradesh. In: Ethnic fermented foods and beverages of India: Science history and culture. Springer, Singapore, pp 189–208

Kanwar SS, Gupta MK, Katoch C et al (2007) Traditional fermented foods of Lahaul and Spiti area of Himachal Pradesh. Indian J Tradit Knowl 6:42–45

Keyhanmanesh R, Boskabady MH, Khamneh S et al (2010) Effect of thymoquinone on the lung pathology and cytokine levels of ovalbumin-sensitized guinea pigs. Pharmacol Rep 62(5): 910–916

Krensky AM, Bennett WM, Vincenti F et al (2006) Immunosuppressants, tolerogens and immunostimulants. In: Brunton LL, Chabner BA, Knollmann BC (eds) Goodman & Gilman's the pharmacological basis of therapeutics, 11th edn. McGraw Hill, New York, pp 1405–1432

Krishnaswamy K (2008) Traditional Indian spices and their health significance. Asia Pac J Clin Nutr 17(S1):265–268

Kuete V (2017) Myristica fragrans: a review. In: Medicinal spices and vegetables from Africa. Academic Press, pp 497–512

Kulkarni SD, Tilak JC, Acharya R et al (2006) Evaluation of the antioxidant activity of wheatgrass (Triticum aestivum L.) as a function of growth under different conditions. Phytother Res 20(3): 218–227

Kumar V, Rao RR (2007) Some interesting indigenous beverages among the tribals of Central India. Indian J Tradit Knowl 6(1):141–143

Kumar VP, Prashanth KH, Venkatesh YP et al (2015) Structural analyses and immunomodulatory properties of fructo-oligosaccharides from onion (Allium cepa). Carbohydr Polym 117:115–122

Kumari ANILA, Pandey ANITA, Ann A et al (2016) Indigenous alcoholic beverages of South Asia. CRC Press, New York, pp 501–566

Lai PK, Roy J (2004) Antimicrobial and chemopreventive properties of herbs and spices. Curr Med Chem 11(11):1451–1460

Lake DF, Briggs AD, Akporiaye ET et al (2009) Immunopharmacology. In: Katzung BG, Masters SB, Trevor AJ (eds) Basic and clinical pharmacology, vol 55. Mc Graw Hill, New York, pp 963–986
Lake IR, Hooper L, Abdelhamid A et al (2012) Climate change and food security: health impacts in developed countries. Environ Health Perspect 120(11):1520–1526
Lalonde RL (1994) Geographical, religious, and philosophical thought: following the high road and finding common ground in environmental ethics. Doctoral dissertation, Carleton University
Laskowski M Jr, Kato I (1980) Protein inhibitors of proteinases. Annu Rev Biochem 49:593–626
Lee BJ, Kim YJ, Cho DH et al (2011) Immunomodulatory effect of water extract of cinnamon on anti-CD3-induced cytokine responses and p38, JNK, ERK1/2, and STAT4 activation. Immunopharmacol Immunotoxicol 33(4):714–722
Lonkar SP, Mahajan AP, Ranveer RC et al (2011) Development of instant mattha mix. World J Dairy Food Sci 6:125–129
Lukita-Atmadja W, Ito Y, Baker GL et al (2002) Effect of curcuminoids as anti-inflammatory agents on the hepatic microvascular response to endotoxin. Shock 17:399–403
Mady RF, El-Hadidy W, Elachy S et al (2016) Effect of Nigella sativa oil on experimental toxoplasmosis. Parasitol Res 115(1):379–390
Magnadóttir B (2006) Innate immunity of fish (overview). Fish Shellfish Immunol 20:137–151
Mahendra P, Bisht S (2012) Ferula asafoetida : Traditional uses and pharmacological activity. Pharmacogn Rev 6:141–146
Majdalawieh AF, Carr RI (2010) In vitro investigation of the potential immunomodulatory and anti-cancer activities of black pepper (Piper nigrum) and cardamom (Elettaria cardamomum). J Med Food 13(2):371–381
Majumdar RK (2020) Ethnic fermented foods and beverages of Tripura. In: Tamang J (ed) Ethnic fermented foods and beverages of India: Science history and culture. Springer, Singapore, pp 583–619
Mandal S, Mandal M (2015) Coriander (Coriandrum sativum L.) essential oil: Chemistry and biological activity. Asian Pac J Trop Biomed 5:421–428
Mangalassary S (2016) Indian cuisine—the cultural connection. In: Indigenous culture, education and globalization. Springer, Berlin, Heidelberg, pp 119–134
Mastana SS (2014) Unity in diversity: an overview of the genomic anthropology of India. Ann Hum Biol 41(4):287–299
Mendes SJ, Sousa FI, Pereira DM et al (2016) Cinnamaldehyde modulates LPS-induced systemic inflammatory response syndrome through TRPA1-dependent and independent mechanisms. Int Immunopharmacol 34:60–70
Menthe SR, Vedantam G, Patil SF et al (2020) Effect of Jatiphala (Myristica fragrans houtt.) in experimentally induced allergic asthma in male wistar Rats. Int J Herb Med 8(3):90–96
Mimche PN, Taramelli D, Vivas L (2011) The plant-based immunomodulator curcumin as a potential candidate for the development of an adjunctive therapy for cerebral malaria. Malar J 10:1–9
Mirabeau TY, Samson ES (2012) Effect of Allium cepa and Allium sativum on some immunological cells in rats. Afr J Tradit Complement Altern Med 9(3):374–379
Mohanty S, Mishra S, Pradhan R et al (2018) Optimization and storage studies of palm (Borassus flabellifer) ready-to-serve (RTS) juice. In: Second International Conference on Food Quality, Safety and Security (FOOD QUALSS), Colombo, Sri Lanka, 25–26 October, 2018
Moustafa EM, Dawood MA, Assar DH et al (2020) Modulatory effects of fenugreek seeds powder on the histopathology, oxidative status, and immune related gene expression in Nile tilapia (Oreochromis niloticus) infected with Aeromonas hydrophila. Aquaculture 515:734589
Mucida D, Park Y, Kim G et al (2007) Reciprocal TH17 and regulatory T cell differentiation mediated by retinoic acid. Science 317:256–260
Muraille E, Pajak B, Urbain J et al (1999) Carbohydrate-bearing cell surface receptors involved in innate immunity: interleukin-12 induction by mitogenic and nonmitogenic lectins. Cell Immunol 191:1–9

Nampoothiri KM, Nair NR, Soumya MP et al (2020) ethnic fermented foods and beverages of Kerala. In: Ethnic fermented foods and beverages of India: Science history and culture. Springer, Singapore, pp 261–286

Nande P, Harode S, Valle M et al (2008) Nutrient contents in selected Maharashtrian and South Indian single full meals. J Dairy Foods Home Sci 27(1):53–64

Nassan MA, Mohamed EH, Abdelhafez S et al (2015) Effect of clove and cinnamon extracts on experimental model of acute hematogenous pyelonephritis in albino rats: Immunopathological and antimicrobial study. Int J Immunopathol Pharmacol 28(1):60–68

Noroozi M, Burns J, Crozier A et al (2000) Prediction of dietary flavonol consumption from fasting plasma concentration or urinary excretion. Eur J Clin Nutr 54(2):143–149

Nybe EV, Mini Raj N, Peter KV (2007) Spices, vol 5. New India Publishing Agency, New Delhi, p 6

Opara EI, Chohan M (2014) Culinary herbs and spices: their bioactive properties, the contribution of polyphenols and the challenges in deducing their true health benefits. Int J Mol Sci 15(10): 19183–19202

Orhan IE, Mesaik MA, Jabeen A et al (2016) Immunomodulatory properties of various natural compounds and essential oils through modulation of human cellular immune response. Ind Crop Prod 81:117–122

Palika R, Dasi T, Kulkarni B et al (2020) Ethnic fermented foods and beverages of Telangana and Andhra Pradesh. In: Ethnic fermented foods and beverages of India: Science history and culture. Springer, Singapore, pp 561–582

Patel D, Shukla S, Gupta S et al (2007) Apigenin and cancer chemoprevention: progress, potential and promise (Review). Int J Oncol 30(1):233–245

Pathak N, Khandelwal S (2007) Cytoprotective and immunomodulating properties of piperine on murine splenocytes: an in vitro study. Eur J Pharmacol 576(1–3):160–170

Peter KV, Shylaja MR (2012) Introduction to herbs and spices: definitions, trade and applications. In: Peter KV (ed) Handbook of herbs and spices. Woodhead Publishing, Cambridge, UK, pp 1–24

Petrunov B, Nenkov P, Shekerdjiisky R et al (2014) The Role of Immunostimulants in Immunotherapy and Immunoprophylaxis. Biotechnol Biotechnol Equip 21(4):445–462

Prajapati JB, Sreeja V (2013) Dahi and related products. New Delhi Publishers and SASNET-Fermented Foods, New Delhi

Prasad P (2005) Crafting qualitative research: Beyond positivist traditions. ME Sharpe, Inc.

Prasanna VK, Venkatesh YP (2015) Characterization of onion lectin (Allium cepa agglutinin) as an immunomodulatory protein inducing Th1-type immune response in vitro. Int Immunopharmacol 26(2):304–313

Rahman MD, Lou Z, Zhang J et al (2017) Star anise (Illicium verum Hook. f.) as quorum sensing and biofilm formation inhibitor on foodborne bacteria: study in milk. J Food Prot 80(4):645–653

Raksamiharja R, Sy K, Novarina A et al (2012) Elettaria cardamomum Distillate Increases Cellular Immunity in Doxorubicin Treated Rats. Indones J Cancer Chemoprevent 3(3):437–443

Ramadan G, Nadia M, Abd El-Kareem HF et al (2011) Anti-metabolic syndrome and immunostimulant activities of Egyptian fenugreek seeds in diabetic/obese and immunosuppressive rat models. Br J Nutr 105(7):995–1004

Rathore MS, Shekhawat NS (2008) Incredible spices of India: from traditions to cuisine. Am Euras J Bot 1(3):85–89

Rawat K, Kumari A, Kumar S et al (2018) Traditional fermented products of India. Int J Curr Microbiol App Sci 7(4):1873–1883

Ray M, Ghosh K, Singh S et al (2016) Folk to functional: an explorative overview of rice-based fermented foods and beverages in India. J Ethn Food 3(1):5–18

Rengaraj A (2014) Screening of immunomodulatory efficiency of *Laurus nobilis*. Int J Pharma Bio Sci 5(1):978–952

Renu R, Waghray K (2016) Development of papads: a traditional savoury with purslane, Portulaca oleracea, leaves. Health Scope 5(1)

Rhee KJ, Sethupathi P, Driks A et al (2004) Role of commensal bacteria in development of gut-associated lymphoid tissues and preimmune antibody repertoire. J Immunol 172:1118–1124
Romagnani S (1999) Th1/Th2 cells. Inflamm Bowel Dis 5(4):285–294
Sachan AK, Kumar S, Kumari K et al (2018) Medicinal uses of spices used in our traditional culture: worldwide. J Med Plants Studies 6(3):116–122
Safaei-Cherehh A, Rasouli B, Alaba PA et al (2018) Effect of dietary Foeniculum vulgare Mill. extract on growth performance, blood metabolites, immunity and ileal microflora in male broilers. Agrofor Syst 94:1269–1278
Sanmugam D, Kasinathan S (2011) Indian heritage cooking. Marshall Cavendish International Asia Pvt Ltd., Singapore
Sarkar P, Lohith Kumar DH, Dhumal C et al (2015) Traditional and ayurvedic foods of Indian origin. J Ethn Food 2(3):97–109
Sathe GB, Mandal S (2016) Fermented products of India and its implication: a review. Asian J Dairy Food Res 35(1):1–9
Savan EK, Küçükbay FZ (2013) Essential oil composition of Elettaria cardamomum Maton. J Appl Biol Sci 7(3):42–45
Savitri BTC (2007) Traditional foods and beverages of Himachal Pradesh. Indian J Tradit Knowl 6: 17–24
Sekar S, Mariappan S (2007) Usage of traditional fermented products by Indian rural folks and IPR. Indian J Tradit Knowl 6(1):111–120
Sethi J, Singh J (2015) Role of medicinal plants as immunostimulants in health and disease. Ann Med Chem Res 1(2):1009–1014
Sha SP, Ghatani K, Tamang JP et al (2013) Dalbari, a traditional pulses based fermented food of West Bangal. Int J Food Sci Tech 4:6–10
Shah AS, Juvekar AR (2010) Immunostimulatory activity of aqueous extract of Murraya koenigii (Linn.) Spreng. leaves. Indian J Nat Prod Resour 1(4):450–455
Shah AS, Wakade AS, Juvekar AR et al (2008) Immunomodulatory activity of methanolic extract of Murraya koenigii (L) Spreng. Leaves Indian J Exp Biol 46(7):505–509
Shakya SR (2015) Medicinal uses of ginger (Zingiber officinale Roscoe) improves growth and enhances immunity in aquaculture. Int J Chem Stud 3(2):83–87
Shakya SR, Labh SN (2014) Medicinal uses of garlic (Allium sativum) improves fish health and acts as an immunostimulant in aquaculture. Eur J Biotechnol Biosci 2(4):44–47
Shan B, Cai YZ, Brooks JD (2011) Potential application of spice and herb extracts as natural preservatives in cheese. J Med Food 14(3):284–290
Sharangi AB, Pandit MK (2018) Supply chain and marketing of spices. In: Indian spices: the legacy, production and processing of India's treasured export. Springer International Publishing, Switzerland, pp 341–357
Sharar AH (1975) Lucknow: the last phase of an oriental culture, trans. ES Harcourt and Fakhir Hussain, Boulder, CO: Westview Press, pp. 62–63
Sharma I, Kapale R, Kango N et al (2020) Ethnic fermented foods and beverages of Madhya Pradesh. In: Ethnic fermented foods and beverages of India: science history and culture. Springer, Singapore, pp 287–303
Shinde SV (2011) Development and quality characterization of traditional sorghum (Sorghum bicolor L.) fermented'Ambil'bevarage (Doctoral dissertation, Vasantrao Naik Marathwada Krishi Vidyapeeth, Parbhani)
Shrivastava V, Bhardwaj U, Sharma V et al (2012) Antimicrobial activities of Asafoetida resin extracts (a potential Indian spice). J Pharm Res 5(10):5022–5024
Shrivastava K, Pramanik B, Sharma BJ et al (2020) Ethnic fermented foods and beverages of Arunachal Pradesh. In: Ethnic fermented foods and beverages of India: science history and culture. Springer, Singapore, pp 41–84
Shruthi RR, Venkatesh YP, Muralikrishna G et al (2016) In vitro immunomodulatory potential of macromolecular components derived from the aqueous extract of ajowan [Trachyspermum ammi (L.) Sprague]. Indian J Tradit Knowl 16(3):506–513

Shruthi RR, Venkatesh YP, Muralikrishna G et al (2017) Structural and functional characterization of a novel immunomodulatory glycoprotein isolated from ajowan (Trachyspermum ammi L.). Glycoconj J 34(4):499–514

Singh D, Singh J (2014) Shrikhand: a delicious and healthful traditional Indian fermented dairy dessert. Trends Biosci 7:153–155

Singh TA, Sarangi PK, Singh NJ et al (2018) Traditional process foods of the ethnic tribes of western hills of Manipur, India. Int J Curr Microbiol App Sci 7:1100–1110

Singh U, Singh S, Kamal SK (2020) Ethnic fermented foods and beverages of Bihar and Jharkhand. In: Ethnic fermented foods and beverages of India: science history and culture. Springer, Singapore, pp 105–120

Srinivasan K (2005) Spices as influencers of body metabolism: an overview of three decades of research. Int Food Res J 38(1):77–86

Stojanović-Radić Z, Pejčić M, Dimitrijević M et al (2019) Piperine-A major principle of black pepper: a review of its bioactivity and studies. Appl Sci 9(20):4270

Sturge CR, Benson A, Raetz M et al (2013) TLR-independent neutrophil-derived IFN-γ is important for host resistance to intracellular pathogens. Proc Natl Acad Sci U S A 110(26): 10711–10716

Sugasini D, Yalagala PC, Kavitha B (2018) Indian culinary ethnic spices uses in foods are palate of paradise. ASNH 2:22–28

Sung YY, Kim YS, Kim HK et al (2012) Illicium verum extract inhibits TNF-α-and IFN-γ-induced expression of chemokines and cytokines in human keratinocytes. J Ethnopharmacol 144(1): 182–189

Takeda K, Harada Y, Watanabe R et al (2008) CD28 stimulation triggers NF-κβ activation through the CARMA1–PKCθ–Grb2/Gads axis. Int Immunol 20(12):1507–1515

Tamang JP (2010) Himalayan fermented foods: microbiology, nutrition and ethnic values. CRC Press, Taylor & Francis Group, New York, p 295

Tamang JP (2012) Plant-based fermented foods and beverages of Asia. CRC. In: Handbook of plant-based fermented food and beverage technology. CRC Press, Boca Raton, FL

Tamang JP, Tamang N, Thapa S et al (2012) Microorganisms and nutritional value of ethnic fermented foods and alcoholic beverages of North East India. Indian J Tradit Knowl 11(1):7–25

Tembhurne SV, Feroz S, More BH et al (2014a) A review on therapeutic potential of Nigella sativa (kalonji) seeds. J Med Plant Res 8(3):167–177

Tembhurne SV, Feroz S, More BH et al (2014b) A review on therapeutic potential of Nigella sativa (kalonji) seeds. J Med Plant Res 8(3):167–177

Teramoto Y, Yoshida S et al (2002) Characteristics of a rice beer (zutho) and a yeast isolated from the fermented product in Nagaland, India. World J Microbiol Biotechnol 18(9):813–816

Thakur N, Bhalla TC (2004) Characterization of some traditional fermented foods and beverages of Himachal Pradesh. Indian J Tradit Knowl 3:325–335

Thanzami K, Lalhlenmawia H (2020) Ethnic fermented foods and beverages of Mizoram. In: Ethnic fermented foods and beverages of India: science history and culture. Springer, Singapore, pp 435–457

Tiwari S, Jadhav SK, Beliya E et al (2020) Ethnic fermented beverages and foods of Chhattisgarh. In: Ethnic fermented foods and beverages of India: science history and culture. Springer, Singapor, pp121–138

Tripathi S, Maurya AK, Kahrana M et al (2012) Immunomodulatory property of ethanolic extract of Trigonella foenum-Graeceum leaves on mice. Der Pharm Lett 4(2):708–713

Tungoe PC (2016) Isolation and characterization of bacteria from selected traditional fermented foods of Mizoram. M. Pharm Thesis, Department of Pharmacy, Regional Institute of Paramedical & Nursing Sciences (RIPANS), Aizawl, pp. 46–47

Upwar N, Patel R, Waseem N et al (2011) In vitro anthelmintic activity of Brassica nigra Linn. seeds. Int J Nat Prod Res 1(1):1–3

Uraku AJ, Nwankwo VO (2015) Phytochemical and nutritional composition analysis of murraya Koenigii Linn leaves. Int J Pharm Res 6(3):174–180

Utomo AW, Retnoningrum D, Gumay AR et al (2020) The immunomodulatory effect of Cinnamon (Cinnamomum Burmannii) bark extract on the C-Reactive Protein (CRP) Level, Leukocyte Count and Leukocyte Type Count of Wistar Rats Exposed to *Staphylococcus aureus*. Sains Med J Kedokt dan Kesehat 11(1):1–6

Van Hoffen E, Ruiter B, Faber J et al (2009) A specific mixture of short-chain galacto-oligosaccharides and long-chain fructo-oligosaccharides induces a beneficial immunoglobulin profile in infants at high risk for allergy. Allergy 64(3):484–487

Vijayabhargava K, Asad M (2012) Effect of stigmas of Crocus sativus L. (saffron) on cell mediated and humoral immunity. Nat For Prod J 1(2):151–155

Vitali LA, Beghelli D, Nya PCB et al (2016) Diverse biological effects of the essential oil from Iranian Trachyspermum ammi. Arab J Chem 9(6):775–786

Wahengbam R, Thangjam AS, Keisam S et al (2020) Ethnic fermented foods and alcoholic beverages of Manipur. In: Ethnic fermented foods and beverages of India: science history and culture. Springer, Singapore, pp 349–419

Wakil AE, Wang ZE, Ryan JC et al (1998) Interferon γ derived from CD4+ T cells is sufficient to mediate T helper cell type 1 development. The J Exp Med 188(9):1651–1656

Watzl B, Girrbach S, Roller M et al (2005) Inulin, oligofructose and immunomodulation. Br J Nutr 93(S1):S49–S55

Yadav VS, Mishra KP, Singh DP et al (2005) Immunomodulatory effects of curcumin. Immunopharmacology Immunotoxicol 27:485–497

Yashin A, Yashin Y, Xia X, Nemzer B et al (2017) Antioxidant activity of spices and their impact on human health: A review. Antioxid 6(3):70

Yu W, Yao D, Yu S et al (2018) Protective humoral and CD4+ T cellular immune responses of Staphylococcus aureus vaccine MntC in a murine peritonitis model. Sci Rep 8(1):1–13

Zhang XY, Li WG, Wu YJ et al (2005) Amelioration of doxorubicin-induced myocardial oxidative stress and immunosuppression by grape seed proanthocyanidins in tumour-bearing mice. J Pharm Pharmacol 57(8):1043–1051

Ziegler SJ, Sticher O (1989) HPLC of S-alk (en) yl-L-cysteine derivatives in garlic including quantitative determination of (+)-S-allyl-L-cysteine sulphoxide (alliin). Planta Med 55(4): 372–378

Traditional Indian Knowledge of Immunity from Plants

9

Nagendra Singh Chauhan, Manju Rawat Singh, Vikas Sharma, Nisha Yadav, Neelam S. Sangwan, and Deependra Singh

Abstract

The traditional Indian medicine like Ayurveda and Siddha and traditional Chinese medicine remain the most ancient therapies. While Ayurveda originated in North and is popular in entire India and world, Siddha system of medicine is majorly practiced in South of India. Similarly, Chinese system of medicine is also evolved and established in thousands of years. There is an important aspect of maintaining wellness from these traditional systems which includes the concept of development of inherent immunity to sustain better health throughout the life. There has been increased surge of global interest in traditional medicines' probable prophylactic/therapeutic measure for the modulation of immune response. Herbs have shown potential of being used as immunomodulators especially the drugs used in alternative system of medicine have been documented to possess immunomodulatory properties. Modulation of immune response to mitigate the diseases was the approach of Rasayana in Ayurveda is based on interrelated principles. Rasayana drugs in Ayurveda are known to increase longevity, stop aging, and offer resistance to disease by increasing the immune system. Many plants like *Terminalia arjuna*, *Glycyrrhiza glabra*, *Withania somnifera*, *Swertia chirayita*,

N. S. Chauhan
Drugs Testing Laboratory Avam Anusandhan Kendra (State Government Lab of AYUSH), Government Ayurvedic College, Raipur, Chhattisgarh, India

M. R. Singh · D. Singh (✉)
University Institute of Pharmacy, Pt. Ravishankar Shukla University, Raipur, Chhattisgarh, India

V. Sharma
Shri Rawatpura Sarkar Institute of Pharmacy, Datiya, Madhya Pradesh, India

N. Yadav · N. S. Sangwan
School of Interdisciplinary and Applied Sciences, Department of Biochemistry, Central University of Haryana, Mahendergarh, Haryana, India

N. S. Sangwan et al. (eds.), *Plants and Phytomolecules for Immunomodulation*,
https://doi.org/10.1007/978-981-16-8117-2_9

Nardostachys jatamansi, *Picrorhiza* sps., *Chlorophytum borivilianum*, *Asparagus racemosus*, etc. used in Ayurvedic system of medicine for Rasayana therapy have been investigated for their therapeutic effects on the immune system. There have been constant efforts to screen and standardize herbal drugs and traditional medicines from Ayurvedic and Himalayan plants. Traditional system of medicine still necessitates more extensive evidence-based scientific research. This chapter gives an overview of basic principles and cohesions of different traditional medicine system and key elements of these great plant-based traditions need to address to compete in global market. Further it incorporates an extensive data on plants used under these traditional systems as well the mechanism involved in the pharmacological findings with a view to appreciate immunomodulation properties of phytomolecules from these plants.

Keywords

Traditional knowledge · Ayurveda · immunomodulating · Indian medicine · Herbal · Pharmacological · Rasayana · Phytopharmaceutical

9.1 Introduction

Traditional medical system plays a very important role in the Indian subcontinent, where the population exceeds 1250 million. Development of traditional healthcare system thus involves great knowledge based on diverse cultural practices and availability of resources. Though a great deal of development is taking place and availability of modern medicine, healthcare, and facilities are being made available. Many peoples live in rural or remote areas, which are not accessible to modern medical facilities. For medical cure most of the rural communities depends on alternative system of medicines like folklore therapy or Ayush or Chinese system of medicine.

Interest in botanical, herbal, and compound formulation treatment has been increasing in Western (Routh and Bhowmik 1999).

According to ancient literature on traditional medicinal system based on plants such as the Rigveda (4500–1600 BC), Charak Samhita (1000–800 BC), and Sushruta Samhita (800–700 BC), Indians have been employing plants to cure diseases since ancient history. During the nineteenth century, advances in botanical sciences had a dramatic impact on the use of plants for new pharmaceuticals. However, it wasn't until 1956 that this was properly documented (Dhyani et al. 2010). Ayurveda plants are found to recover health quickly as well as acts as antioxidants. These plants are used in various formations like oils, clarified butters, powder, as well as very popular Chyavanprash. and Ayurvedic medicine system narrates a novel idea named as Rasayana; here plants that have rejuvenating properties are reported via stressing on encouraging the health by firming individual defense mechanism in relation to various ailments. Rasayan is named as life's most important channel (Stites et al. 1982). These plants play an important role in the immune system. The immune system plays role in protecting from foreign substance as well as other communicable vehicle. There are some forces which are known for increasing and decreasing of immune response by obstructing in the mechanism of direction named as immunomodulatory agents. Immunomodulators are chemicals that have the ability to induce, enhance, or suppress any component or phase of the immune system. Immunomodulators and immunosuppressants are two different forms of immunomodulators. Immunopharmacology, in fact, is a relatively recent field of pharmacology that deals with immunomodulators. Immunomodulators, as in the case of an excessive immune response, are both beneficial in restoring the normal immune system and extending life expectancy. Immunomodulator ingestion in conjunction with antigen is intended to stimulate the immune system, and the modulatory is referred to as an immunological adjuvant (Sharma et al. 2017). These prompt humoral or cell-mediated immunity with T-cell response. Rasayana medicines in consideration with immunomodulation perhaps are possible substitute to the traditional chemotherapy in cases when the immune system is compromised.

The word Ayurveda is a Sanskrit word that is defined as Ayus(r) meaning life and Veda meaning science or knowledge and which accurately means the science of life. It is one of the oldest traditional healthcare systems based on sound scientific principles (Ashok and Ali 2003; Jee 1927). It is an integrated system of healthcare with the idea that the human body is a matrix of seven basic tissues ("Asthi," "Meda," "Majja," "Shukra," "Rasa," "Rakta," "Mansa,") and the desecrate products of the body, such as urine, sweat, and feces which are resulting by the five basic elements water fire, air, earth, and ether and three basic types of energies or functional principles "kapha, vata, and pitta" (Tridosha). Any inconsistency or disorder in these basic principles of body generates disease. The uniform of functional principles and basic elements of the body is fundamental concept in Ayurveda therapy. Plants which are also supposed to be compose of these five elements, and three useful principles are used for curing the diseases in Ayurveda (Lad 2002). The Ayurveda is divided into eight prime disciplines known as "Ashtanga Ayurveda," which compose eight different branches as shown in Fig. 9.1.

9.2 Rasayana

Rasayana also known as geriatrics in Ayurveda system used to cure aging- and age-related disorders is well-known for attaining intelligence, improving memory, freedom from age-related disorders, preserving vigor, longevity, general healthy state of organs, and restoring youth (Jee 1927).

Rasayana therapy is a unique specialty of Ayurveda dealing with the aspect of rejuvenation.

The literal meaning of *Rasayana* is the path that *rasa* takes. Rasayana therapy enriches nutritional quality of Rasa, a defined process concerning transformation, conservation, and revitalization of energy. Rasayana *therapy drug* may have its effect on our *ojas*, *immunity*, and *resistance* (Fig. 9.2). Rasa nurtures body, boosts

Fig. 9.1 Ashtang (eight) system of classification in Ayurveda

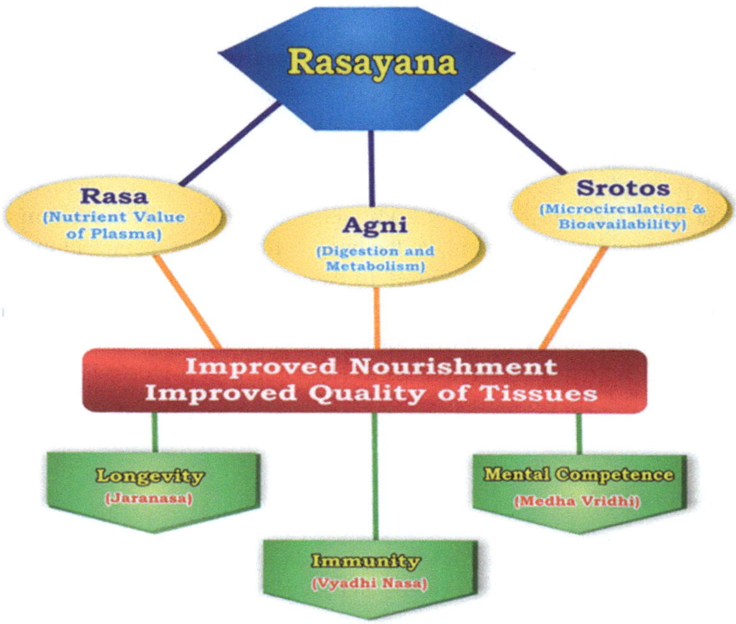

Fig. 9.2 The mode of action of Rasayana drugs

immunity, and helps to keep the body and mind in the best of health (Chaturvedi 2008; Puri 2002).

Recently lots of work done on this related issue in Western countries, due to a new branch of medical science called "psycho-neuro-endocrine-immunology," has been developed. In this branch of medical science, the functions of brain and peripheral system have been linked so as to explain the etiopathogenesis of various diseases. Based on a previous research, it has been evident that the immune system can influence the brain because sends impulse to the brain by means of secreting hormones and peptides. By this way brain endocrine system and immune system are linked together in view of fighting against diseases (Fig. 9.3). Rasayana therapeutic activity is based on the neuroendocrine immune axis theory as proved by various research works reported (Patwardhan 2005; Kokate et al. 2010; Chauhan et al. 2010).

The principal function of Rasayana therapy is to delay the aging process and to retard the degenerative process in the body. It boosts the body strength, memory, intelligence, shine of the skin, and tone of voice. It nurtures the muscles, tissues, semen, blood, and lymph that prevents degenerative disorders like Parkinson's and osteoarthritis. It ameliorates quality of body tissues and metabolic process and eliminates diseases of aging. This assists to achieve the best physical strength and keenness of sense organs. Rasayana has potential activity on reproductive organs and also nurture shukra dhatu (semen). Rasayana nurtures the complete body and ameliorates the immune system, and that's the reason for the increase in natural resistance to infections.

The Ayurvedic books explained the sequential decline of biological qualities and its treatment with the help of Rasayana, which act as anti-decline Medicine (Preventive Geriatrics) Rasayana show action as geriatrics to act on dhatu (saptadhatu) decline, tissue (epithelia, nervous, etc.) decline, organ decline (Lungs, Heart, Liver, etc.), senses decline (Hearing, Vision etc.), functional decline (Sexual, Psychomotor etc.) and expressional decline (Luster, Growth, etc.).

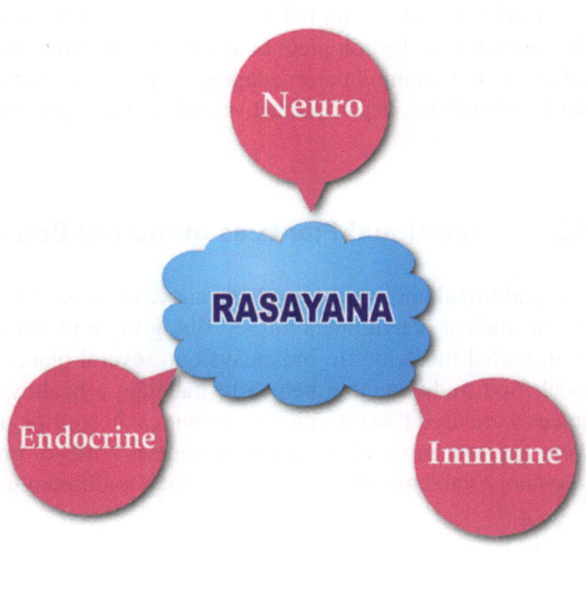

Fig. 9.3 Target sites of Rasayana

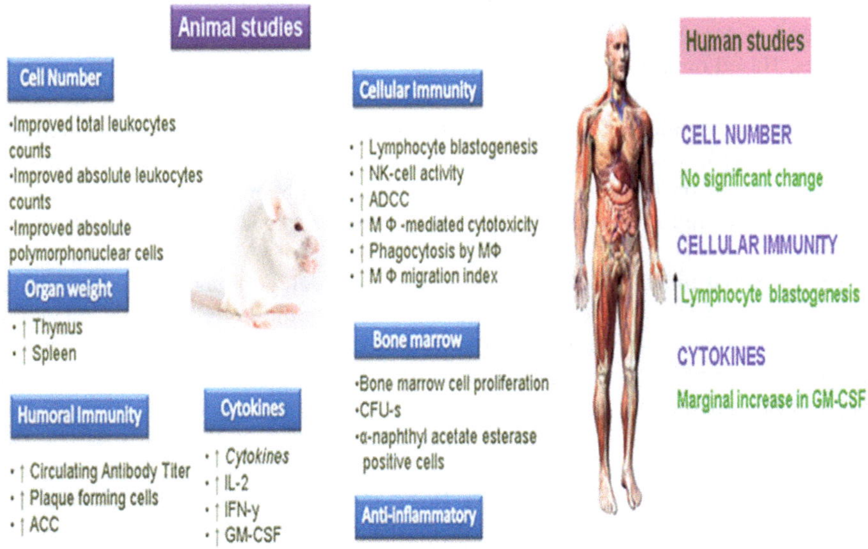

Fig. 9.4 Immunomodulatory activities of Rasayanas

Rasayana therapy delays aging and improves memory, strength, intelligence, luster, youth, sweetness of voice, and vigor. It is believed to nurture lymph, blood, flesh, semen, and adipose tissue and thus intercepts degenerative changes and illness (Vayalil et al. 2002). *Rasayana* is thought to recover metabolic processes by best possible biotransformation and create the best-quality body tissues and eliminates senility and other diseases of aging. Most of these *Rasayanas* can be used commonly as a food for maintaining balanced physical and mental health. Rasayana improves the functions of the complete body system. In modern concept the Rasayana herbs act as an immunomodulator, adaptogen, pro-host probiotic, antimutagenic, antioxidant, antidiabetic, hepatoprotective, aphrodisiac, and memory enhancer (Fig. 9.4).

9.3 Traditional Plants as Immunity Booster

The traditional knowledge on plants has existed since prehistoric times and practices by the ancient practitioners in prescribing these plants either as food preparation or as in herbal medicine. In Indian system, several plants are part of the regional and local food traditions that helped to maintain a healthy lifestyle. A combination of spices were used that provided nutraceutical in well-being. Many of such ingredients were required in small quantities in various kinds of preparation. Recent studies in laboratory validation have confirmed such traditional hypothesis and practices.

9.3.1 Allium sativum

This herb is commonly known as Lahsan. Garlic belongs to family Alliaceae; its rhizome part is used (Ghazanfaria et al. 2002). Garlic is revealed to have a Rasayana effect in Ayurveda; it is one of the elderly raised plants on the planet and shows potential curing of various ailments like antimicrobial, antithrombotic, antitumor, hypolipidemic, antiarthritic, and hypoglycemic; such outcomes have been connected to their impact on the metabolism of eicosanoids, which affect the immune system in a variety of ways. After extensive reports of research, it was validated that garlic showed varied scales of health properties including increase in immunity (Table 9.1) (Das 2021). Garlic is possibly one of the most well-known medicinal treatments and has been used by humans from long back history. It has been a fascinating plant as a therapeutic remedy from generations, and still newer evidences of its therapeutic roles are emerging (Dangar et al. 2021).

9.3.2 *Aloe vera*

It is commonly known as Kware. *Aloe vera* leaves are used mainly for health benefits. Juice of *Aloe vera* is prepared from pulpy stem which helps to enhance immunity when consumed with empty stomach (Vázquez et al. 1996); also juice help in pain relieving and reduction of inflammation. In terms of pharmacology, it acts as an immune booster and detoxifier. It's used as a supplement to antibiotics, NSAIDs (nonsteroidal anti-inflammatory drugs), and chemotherapy to prevent drug-induced gastritis and other side effects. As presence of polysaccharides in aloe vera juice stimulates macrophages, the WBC that fights viruses, it is particularly beneficial for patients with chronic immunological illnesses like polysaccharide or fibromyalgia. It was also concluded that when aloe vera gel is taken orally, it decreases indication and inflammation in individual (Zhang et al. 2006).

9.3.2.1 Aloin

Aloin is a bitter, yellow-brown-colored compound and the main anthraquinone glycoside found from the Aloe species showing anti-oxidative and anti-inflammatory activities (Lee et al. 2019). The immunostimulatory properties of aloin are proofed by various researchers (Harlev et al. 2012). Aloin acted upon HO-1 induction and decreased LPS-activated NF-κB-luciferase activity showed to specific inhibition of iNOS/NO and COX-2/PGE2 that was partly linked to inhibition of STAT-1 phosphorylation (Lee et al. 2019). Aloin may stop the inflammatory responses by blocking COX-2 mRNA and iNOS expression (Park et al. 2009). Srivastava et al. evaluated the immunostimulant of aloin; it showed a significant enhancement in the activity of enzymes – lysozyme, carboxylesterase, protease, acid phosphatase, catalase, peroxidase, and alkaline phosphatase – and alternate complement and nonenzymatic factor hemagglutinin correlated with that of the controls

Table 9.1 Plants used as immunostimulant

S. no	Botanical name	Plant part	Extract	Activity	References
1.	*Acacia catechu* L., (Fabaceae)	Heartwood	Ethanolic	Immunostimulant	Sunil et al. (2019)
2.	*Acacia nilotica* (Fabaceae)	Aerial part	Methanolic	Immunostimulant	Ahmad et al. (2012)
3.	*Acacia ferruginea* (Mimosaceae)	Aerial part	Methanolic	Immunostimulant	Sakthivel and Guruvayoorappan (2015)
4.	*Anacyclus pyrethrum DC* (Compositae)	Roots	Petroleum ether	Immunostimulant	Sharma et al. (2010)
5.	*Argyreia speciosa* Sweet (Convolvulaceae)	Roots	Ethanolic	Immunostimulant	Gokhale et al. (2003)
6.	*Aloe vera* Linn. (Aloaceae)	Leaves	Aqueous	Immunostimulant	Pugh et al. (2001)
7.	*Aerva lanata* Juss. (Amaranthaceae)		Pet ether	Immunomodulatory	Nevin and Vijayammal (2005)
8.	*Asparagus racemosus* Linn. (Liliaceae)	Roots	Aqueous	Immunostimulant	Thakur et al. (2012)
9.	*Baliospermum montanum* (Willd.) Müll. Arg. (Euphorbiaceae)	Roots	Ethanolic	Immunostimulant	Wadekar et al. (2008)
10.	Barleria lupulina Lindl.	Aerial parts	Methanol	Immunomodulating	Kumari et al. (2017)
11.	*Biophytum sensitivum* (L) DC (Oxalidaceae)	Aerial part	Alcoholic	Immunomodulatory	Guruvayoorappan and Kuttan (2007)
12.	*Boerhaavia diffusa* Linn. (Nyctaginaceae)	Roots	Aqueous and ethanolic	Antilymphoproliferative	Mungantiwar et al. (1999); Mehrotra et al. (2002)
13.	*Boswellia serrata* Roxb. (Burseraceae)	Bark	Oleo gum resin	Immunomodulating	Sharma and Dash (1996)
14.	*Buchanania lanzan* Spreng. (Anacardiaceae)	*Seed*	Ethanolic	Immunostimulant	Puri et al. (2000)

15.	*Calamus rotang (Arecceae)*	*Root*	Aqueous	Immunomodulating	Gupta and Chaphalkar (2017)
16.	*Capparis zeylanica* Linn. (Capparidaceae)	Leaves	Ethanolic and water	Immunomodulatory	Ghule et al. (2006)
17.	*Caesalpinia bonducella* FLEMING (Caesalpiniaceae)	Seed	Ethanolic	Immunomodulating	Shukla et al. (2009)
18.	*Cedrus deodara* (Roxb.) Loud. (Pinaceae)	Wood	Volatile oil	Immunomodulatory	Shinde et al. (1999)
19.	*Centella asiatica* Linn. (Apiaceae)	Whole herbs	Methanolic, aqueous	Immunomodulatory activities	Jayathirtha and Mishra (2004), Patil et al. (1998)
20.	*Carica papaya* Linn. (Caricaceae)	Seed	Hexane and aqueous extract	Lymphocyte stimulation	Mojica-Henshaw et al. (2003)
21.	Carica papaya (Caricaceae)	Leaves	Aqueous	Antitumor activity and immunomodulatory	Otsuki et al. (2010)
22.	*Chlorophytum borivilianum* Sant. and F (Liliaceae)	Roots	Ethanolic	Immunomodulatory	Thakur et al. (2006)
23.	*Cissampelos pareira* Linn. (Menispermaceae)	Roots	Methanol	Immunostimulant	Bafna and Mishra et al. (2005)
24.	*Cicer arietinum* (**Fabaceae**)	Seeds	Methanol (75%)	Immunostimulant	Jayaprakash and Das (2018)
25.	*Commiphora mukul* Engl. (Burseraceae)	Gum	Ethyl Acetate	Immunostimulant	Manjula et al. (2006)
26.	Coccinea cordifolia (Cucurbitaceae)	Root	Methanolic	Immunostimulant	Roy Deb et al. (2019)
27.	*Curculigo orchioides* Gaertn. (Amaryllidaceae)	Rhizomes	Methanol	Immunomodulatory	Lakshmi et al. (2003)
28.	Cuscuta chinensis (Convolvulaceae)	Whole plant	Methanol (70%)	Immunostimulant	Raju et al. (2015)

(continued)

Table 9.1 (continued)

S. no	Botanical name	Plant part	Extract	Activity	References
29.	*Cryptolepis buchanani* Roem. and Schult. (Asclepiadaceae)	Roots	Ethanolic	Immunostimulant	Kaul et al. (2003)
30.	*Eclipta alba* (Linn.) Hassk. (Asteraceae)	Whole herbs	Methanol	Immunomodulatory	Jayathirtha and Mishra (2004)
31.	*Euryale ferox* Salisb. (Nymphaeaceae)	Seed	Ethanolic	Immunostimulation	Puri et al. (2000)
32.	*Ficus benghalensis* Linn. (Moraceae)	Aerial Root	Methanolic	Immunostimulation	Gabhe et al. (2006)
33.	Heracleum nepalense D. Don. (Apiaceae)	Roots	Methanolic	Immunostimulation	Dash et al. (2006)
34.	Hippophae rhamnoides Linn. (Elaeagnaceae)	Leaves and fruits	Alcoholic	Cytoprotective and antioxidant	Geetha et al. (2002)
35.	*Indigofera tinctoria* (Fabaceae)	Leaves	**Aqueous**	Immunostimulant	Madakkannu and Ravichandran (2017)
36.	Ipomoea obscura L. (Convolvulaceae)		**t (10 mg/kg)**	Reduced myelosuppression caused by cyclophosphamide	Hamsa and Kuttan (2010)
37.	*Lawsonia alba* Lam. (Lytheraceae)	Leaves	Naphthoquinone	Antiviral and antitumoral	Kulkarni and Karande (1998)
38.	*Luffa cylindrica* Linn. (Cucurbitaceae)	Seeds	Methanolic	Immunostimulatory effect	Khajuria et al. (2007)
39.	*Leptadenia reticulata* Linn. (Asclpiadaceae)	Leaves	Ethanolic	Immunostimulatory effect	Pravansha et al. (2012)
40.	*Mangifera indica* Linn. (Anacardiaceae)	Stem bark	Alcoholic	Immunostimulant	Makare et al. (2001)

41.	*Mucuna pruriens* (L.) (Fabaceae)	Seed	Methanolic	Immunostimulant	Yousaf et al. (2017)
42.	*Nyctanthes arbor-tristis* Linn. (Oleaceae)	Whole plant	Ethanolic	Immunostimulant	Puri et al. (1994)
43.	*Psoralea corylifolia* Linn. (Fabaceae)	Seeds	Ethyl alcohol	Antitumor	Latha et al. (2000)
44.	*Phoenix dactylifera* Linn. (Arecaceae)	Fruit	Ethanolic	Immunostimulant	Puri et al. (2000)
45.	*Premna tomentosa* Willd. (Verbenaceae)	Leaves	Methanolic	Immunomodulatory and cytoprotective	Pandima et al. (2003, 2004)
46.	Premna integrifolia (Verbenaceae)	Root	Petroleum ether	Immunomodulatory	Azad et al. (2018)
47.	*Piper longum* Linn. (Piperaceae)	Fruits	Alcoholic	Immunomodulatory and antitumor	Sunila and Kuttan (2004)
48.	*Prunus amygdalus* Batsch. (Rosaceae)	Seed	Ethanolic	Immunostimulant	Puri et al. (2000)
49.	*Punica granatum* Linn. (Punicaceae)	Fruit	Aqueous	Immunostimulant	Gracious Ross et al. (2001)
50.	*Pistacia integerrima* J. L. Stewart ex Brandis (Anacardiaceae)	Leaf	Aqueous and methanolic	Immunomodulatory and adaptogenic	Joshi and Mishra (2008)
51.	*Scoparia dulcis* (Scrophulariaceae)	Leaves	Aqueous	Immunostimulant	Madakkannu and Ravichandran (2017)
52.	*Semecarpus anacardium* Linn. (Anacardiaceae)	Nut	Milk	Immunomodulatory; anti-inflammatory	Ramprasath et al. (2006)
53.	*Sphaeranthus indicus* Linn. (Compositae)	Flower heads	Methanolic	Immunostimulant	Bafna and Mishra (2004)
54.	*Sphaeranthus indicus* Linn. (Compositae)	Flower	Petroleum ether	Immunostimulant	Bafna and Mishra et al. (2007)

(continued)

Table 9.1 (continued)

S. no	Botanical name	Plant part	Extract	Activity	References
55.	*Selaginella involvens* (Swartz) Spring. (Selaginellaceae)	Whole plant	Aqueous	Immunostimulant and antioxidant	Gayathri et al. (2005)
56.	*Rhizophora apiculata* (Rhizophoraceae)	Whole plant	Methanolic	Immunostimulant	Prabhu and Guruvayoorappan (2012)
57.	*Rubia cordifolia* Linn. (Rubiaceae)	Root	Ethanolic	Immunomodulatory	Joharapurkar et al. (2003)
58.	*Taxus wallichiana* Zucc. (Taxaceae)	Needles	Methanolic	Immunomodulatory and cytotoxic	Chattopadhyay et al. (2006)
59.	*Tephrosia purpurea* Linn. (Leguminosae)	Aerial parts	Ethanolic	Immunostimulant	Damre et al. (2003)
60.	*Trigonella foenum graecum* Linn. (Leguminosae)	Whole plant	Aqueous	Immunostimulatory effect	Hafeez et al. (2003)
61.	*Tinospora sinensis* (Lour.) Merrill (Menispermaceae)	Stems	Water and ethanolic	Immunostimulatory effect	Manjrekar et al. (2000)
62.	*Tridax procumbens* Linn. (Compositae)	Aerial parts	Aqueous	Immunostimulatory effect	Tiwari et al. (2004)
63.	*Zingiber officinale* Roscoe (Zingiberaceae)	Rhizome	Ethanolic	Immunostimulant	Puri et al. (2000)

(Srivastava et al. 2018). Aloin is a strong anti-influenza phytochemical that reduces viral neuraminidase activity, even of the oseltamivir-resistant influenza virus. With preventing action of this virus machinery, aloin improves host immunity with increased hemagglutinin-specific T-cell response to the infection (Huang et al. 2019). Aloin ameliorates the progression of osteoarthritis via the PI3K/Akt/NF-κB signaling pathways (Zhang et al. 2020). Aloin applies its shielding effects on lipopolysaccharide-induced acute lung injury by stimulation of SIRT1, which subsequently results in the suppression of NLRP3 and NF-κB inflammasome (Lei et al. 2020). Aloin can reduce inflammation, d-gal-induced oxidative stress, and cognitive impairment, possibly via mediating the NF-κB and p38 signaling and ERK pathways (Zhong et al. 2019).

9.3.3 *Cinnamomum cassia*

This plant is commonly known as cinnamon, which belongs to family Lauraceae. This common spice is utilized in possibly all the meal preparation. It has high amount of cinnamic acid, cinnamaldehyde, and eugenol due to its chemical composition. It also enhances sensitivity to foreign particles in the body. Its essential oil and powder demonstrated antiviral properties as well as immunostimulatory properties (Dangar et al. 2021). Cinnamon contains cinnamaldehydes, polyphenols, carbohydrates, flavonoids, etc. which work as immune booster in aquatic animals. Cinnamon can be used as a cost-effective immunity supporter (Ghafoor 2020). Cinnamon tree's bark or leaves used for extracting oils which later on utilized for enhancing immunity act as immunomodulator by interacting with lymphocytes and macrophages. According to a report, cinnamon along with honey makes immune system robust (Singh et al. 2020).

9.3.4 *Curcuma longa*

It is commonly known as turmeric, which belongs to family Zingiberaceae, and its powdered form is observed as one of the finest immune boosters (Asteriou et al. 2018). A phenolic antioxidant component curcumin is derived from the rhizome of plant Curcuma longa, which provides multiple immunomodulatory effects. *Curcuma* spp. dried rhizome consists of turmerin, d-a-phellandrene, cineol, d-borneol, d-sabinene, and valeric acid along with curcuminoids which are considered as principal phenolic components having curcumin, demethoxycurcumin, cyclocurcumin, and bisdemethoxycurcumin (Mollazadeh et al. 2017). Studies done on both animal and humans revealed that antioxidants lead to change in inflammatory responses, cell-mediated immunity, as well as production of cytokines (Abu-Rizq et al. 2008). Curcumin is an effective immunomodulator. T and B cells, macrophages, neutrophils, dendritic cells, and natural killer cells may all be affected by curcumin. In various pathological circumstances, curcumin inhibits the synthesis of pro-inflammatory cytokines including IL-1, IL-2, IL-6, and IL-8 and interferon (IFN-γ), along with mast cell protease-1 (MCP-1) and NO (Mollazadeh et al. 2017).

Curcumin, a natural polyphenol phytoconstituent obtained from various curcuma species, has been shown to be anti-inflammatory activity (Zhao et al. 2019). Curcumin showed a strong anti-inflammatory property involved in the PPARγ-dependent NF-κB signaling pathway. This explores the value of curcumin in mucus secretion and asthmatic airway inflammation (Zhu et al. 2019).The curcumin-stabilized silver nanoparticles (Cur-AgNP) have pharmacological potential as immunomodulatory activities, as well as direct antiretroviral agents having suppressed the expression of pro-inflammatory mediators induced by infection with HIV-1. Action of HIV-1-infected cells with Cur-AgNP significantly decreased replication of HIV by inhibition of NF-κB nuclear translocation and the downstream expression of the pro-inflammatory cytokines IL-1β, IL-6, and TNF-α (Sharma et al. 2017).

Trivedi et al. investigated immunomodulatory effect of a novel nanocurcumin-based formulation with curcumin in NK cells' activity, phagocytosis, and LPS-induced cytokine expression. This study informed that nanocurcumin-based formulation shows expression on IL-1β, TNF-α, and MIP-1α in splenocytes was considerably inhibited, and the phagocytosis (macrophage) and NK cells' activities were increased significantly. The effect of nanocurcumin-based formulation compared to the curcumin in various inflammatory and autoimmune diseases justifies its immunomodulatory activity (Trivedi et al. 2017).

A study showed that curcumin would reduce symptoms of IPEX (immunodysregulation polyendocrinopathy enteropathy X-linked syndrome) in mouse model and would extend its survival period. Curcumin diet reduced all of the Th17/Th2/Th1cell populations and cytokine production of IFN-γ, IL-4, and IL-17A in $CD4^+$ T cells (Lee et al. 2017). Curcumin is investigated for the anticancer effect which has an immune-stimulatory effect on NK-92 by raising the surface expression of the $CD56^{dim}$ and $CD16^+$population of NK-92 (Lee et al. 2017). The curcumin nanoparticulate enhanced immunomodulatory activity which significantly stimulated primary humoral immune response by increasing antibody titer when compared to control, and it also enhanced bioavailability of curcumin for enhanced immunity (Afolayan et al. 2018). The nanocurcumin-based formulation was non-toxic and showed significant immunomodulatory effect in LPS-induced NK cells' activity, cytokine expression, and phagocytosis model (Trivedi et al. 2017). Curcumin works as a strong inhibitor of Toll-like receptor-2 (TLR2)-mediated inflammatory responses (Shuto 2013).

9.3.5 *Emblica officinalis*

This is commonly known as amla, an Indian gooseberry that belongs to family Phyllanthaceae. This Indian herb is broadly utilized in Ayurvedic medicinal system.

This esteemed herb taken place in Charak Samhita as Rasayan helps in stimulating liveliness. This herb is known for its role in providing individual robustness and retains person's body disease-free (Kumar et al. 2012) and acts as one of the best immunity boosters (Sai Ram et al. 2002). This herb contains multiple constituents including apigenin, gallic acid, ellagic acid, chebulinic acid, quercetin, chebulagic acid, corilagin, isostrictiniin, methyl gallate, and luteolin. Amla's effect on the immune system may be one of its reputations as an energy-promoting tonic. In rodents fed with amla, multiple studies have revealed considerable enhances in white blood cell counts and other indicators of improved immunity (Kumar et al. 2012).

9.3.6 *Glycyrrhiza glabra*

It is commonly known as licorice that belongs to family Fabaceae (Elabd et al. 2016). As per traditional medicinal system of China, licorice is the oldest as well as highly authorized herb for curing several diseases including tuberculosis and peptic ulcers. A triterpene glycoside obtained from the root of licorice named as glycyrrhizin has immunomodulatory properties. Studies indicated that glycyrrhizin show a prominent decrease in ovalbumin (OVA)-induced reduces OVA-specific IG, IL-4, IL-5 and IL-6 level in the blood. Along with these is the inhibition in the reduction of total IgA, IgG, IgM, IL-2, and IL-12 serum level in AR mice. According to this it was concluded that glycyrrhizin helps in harmonizing immunity in AR (allergic rhinitis) mice (Jose Cheel et al. 2010). Various studies also revealed that *Glycyrrhiza glabra* polysaccharides (GGP) play an important role in the stimulation of the immune system (Alexyuk et al. 2009). Glycyrrhizin contains one unit of glycyrrhetinic acid and two units of glucuronic acid. This glycyrrhetinic acid has corticosteroid-like structure, which has various useful pharmacological properties, and immunomodulation is one of them (Elabd et al. 2016).

Glycyrrhizin is the chief active constituent *G. glabra* also known as liquorice root and has been used in alternative medicine to alleviate gastritis, jaundice, and bronchitis. Glycyrrhizic acid stimulated both T and B lymphocytes equally (Chavali et al. 1987). Glycyrrhizic acid acts as a supporter of the late signal transduction of T lymphocytes for IL-2 production and growth response (Zhang et al. 1993). The beta-glycyrrhetinic acid a aglycone part has a steroid-like structure and is supposed to have immunomodulatory properties. Beta-glycyrrhetinic acid is a strong restriction of the classical complement pathway (Kroes et al. 1997). Glycyrrhizic acid prohibited delayed-type hypersensitivity reaction (Raphael and Kuttan 2003). Glycyrrhizin reduced H5N1-induced generation of interleukin-6, CXCL10, and CCL5 and inhibited H5N1-induced apoptosis but did not interfere with H5N1 replication. Global prohibition of immune responses may be the outcome in the loss of control of virus replication by cytotoxic immune cells as well as cytotoxic CD8(+) T lymphocytes and natural killer cells (Michaelis et al. 2010). Glycyrrhizic acid treatment caused an ameliorate expression of iNOS2 along with inhibition of COX-2 in *Leishmania donovani*-infected macrophages. Glycyrrhizic acid also reduced hepatic and splenic parasite burden and enhanced T-cell proliferation in Leishmania-infected BALB/c mice (Bhattacharjee et al. 2012).

9.3.7 *Moringa oleifera*

This plant is known by the name drumstick, which belongs to family Moringaceae. *M. oleifera* is known as miracle tree as it is full of rich assets with multiple nutrients of great biological importance. It is utilized to enhance immunity (Mahfuz and Piao 2019). A study done in Africa disclosed that powder of moringa works as immune stimulant, from back in ancient; it was broadly utilized in the condition of suppressing immune system as moringa worked as immunostimulant (Sudha et al. 2010). *M. oleifera* contains multiple proteins and peptides like isothiocyanates, glycoside, cyanides, etc. which shows productive result in reference with immunity. An experiment was carried on mice model; persistent use of drumstick enhanced WBC count along with increase of neutrophil concentration (Mahfuz and Piao 2019). Moringa also has significant inhibitory role in immune-related ailments like rheumatoid arthritis, atopic dermatitis, as well as multiple sclerosis; moringa also helps in weakening radiation-induced immune diseases (Sudha et al. 2010).

9.3.8 *Piper longum*

An Ayurvedic herbal medicine Pippali Rasayana is produced from *Piper longum*, which belongs to family Piperaceae. This plant has multiple secondary metabolites including alkaloids, flavonoids, steroids, essential oils, cardiac glycosides, anthraquinones, arbutin, and anthraglycosides. It provides a distinctive powerful smell due to the presence of alkaloid piperine in it (Das 2021). *Piper longum* contains majority of key phytoconstituents which increases macrophage movement index and phagocytosis as well as boosts host's resistance strength. Scientists used various assays which are used to confirm immunostimulatory activity of *Piper longum* fruits, including hemagglutination titer test, macrophage migration index, and phagocytic index in mice. Balb/c mice were inoculated with an alcoholic extract of *P. longum* fruit as well as piperine to signify the capability to modulate the immune system. The influence of introducing *P. longum* and its ingredient piperine on hematological parameters, bone marrow cellularity, esterase-positive cells, circulating Ab titer, and plaque-producing cells was investigated. The results of the studies showed that injecting the extracts of *Piper longum* and piperine increased the overall WBC count in Balb/c mice, indicating that they can stimulate the hematopoietic system. They also affected bone marrow cellularity and esterase-positive cells, indicating stem cell proliferation. Extracts of the plant have plate-producing cells, and these cells form antibody as well as antibody titer in circulation, which in turn give a demonstration of the ability of energizing humoral immunity (Sunila and Kuttan 2004).

9.3.9 *Zingiber officinale*

It is commonly known as adrak. Ginger belongs to family Zingiberaceae. It is taken as spice as well as for medication and work as immunity booster (Mashhadi et al. 2013). Ginger is a popular spice in Indian cuisine, as well as an old medicinal plant. It consists of spicy substances like gingerol and 6-paradol, along with this ginger rhizome which contains nonvolatile phenolic components including gingerols, paradols, shogaols, and zingerones. It is one of the highly biologically active phyto-compounds in traditional healthcare, according to Ayurvedic research (Dangar et al. 2021). Methanolic extract of ginger on doing GC-MS analysis shows the existence of carotenoids along with other antioxidant components which shows antioxidant as well as immunostimulant prospects (Puri et al. 2000). There are studies reported that in fishes, diet having ginger powder for 30 days results in increasing growth as well as immunity. It was found out that ginger powder helped in stimulating innate immunity. Active chemical constituents of ginger like gingerol, shogaol, paradol, methyl isogingerol, isoshogaol, gingerdione strengthen immunity (Abu-Rizq et al. 2008).

9.3.10 *Asparagus racemosus*

It is commonly known as Satawar, which belongs to family Asparagaceae. Charak Samhita communicated by Charak and Ashtanga Hridayam was written down by Vagbhata; the two chief works on Ayurvedic medicine list Satawar as segment of formula used to cure women's well-being ailments. In Ayurvedic Rasayana this plant is studied for its commence in immunity, having immunomodulatory role (Thakur et al. 2012). Previous research reported that shatavari shows remarkable immunostimulatory effect (Table 9.2) at humoral level along or without induced immunosuppression. Shatavari aqueous extracts have three main attributes such as immunostimulation, immunoprotection (Thakur et al. 2012), and adjuvant activity. Scientists found that steroidal saponins present in satavari known as shatavarins show significant immunostimulation activity. Aqueous extract of shatavari is also used in immunochemical industry to acquire much better as well as comfort outcome in immunostimulation (Gautam et al. 2004). Principal components of this plant are steroidal saponins (shatavarins I–IV) found in the root region. Shatavarin IV shows immunomodulation activity in opposition to T-dependent antigens in the case of animals those that are immuno-compromised (Thakur et al. 2012).

9.3.11 *Azadirachta indica*

Neem arises to have a variety of medicinal properties. For generations, the Indian subcontinent has referred the neem tree as "wonder tree." It has grown increasingly significant in today's global setting since it provides solutions to the world's most pressing problems. The US Environmental Protection Agency has permitted the use

Table 9.2 Major phytoconstituents as immunomodulators

Biological source	Chemical constituent	Mechanism of action	References
Aloe vera Mill. (Liliaceae)	Dihydrocoumarin derivatives, acemannan	Immunomodulation, anticomplement activity	Zhang et al. (2006), Lee et al. (1997)
Apium graveolens L. (Apiaceae)	Quercetin, rutin	Immunomodulatory activities	Cherng et al. (2007)
Astragalus membranaceus Fisch. (Leguminosae)	Daucosterol	Immunoregulatory activity	Lee et al. (2007)
Asparagus racemosus Willd. (Liliaceae)	Shatavarin IV and immunoside	Immunomodulating	Gautam et al. (2004)
Aegle marmelos (Linn.) Corr. (Rutaceae)	Propelargonidin	Anticomplement activity	Abeysekera et al. (1996)
Allium sativum Linn. (Amaryllidaceae)	14-kDa glycoprotein	Immunomodulatory activity	Ghazanfaria et al. (2002)
Artemisia annua Linn. (Asteraceae)	Artemesin, dihydroartemesin, and arteether	Immunomodulating	Tawfik et al. (1990)
Berberis aristata DC. (Berberidaceae)	Berberine	Enhance phagocytosis	Ehteshamfar et al. (2020)
Boswellia serrata Roxb. Ex Colebr. (Burseraceae)	BOS 2000, Boswellic acid	Immunostimulatory	Khajuria et al. (2008), Sharma et al. (1996)
Brassica oleracea Linn. (Brassicaceae)	Sulforaphane	Immunomodulation	Thejass and Kuttan (2007)
Cajanus indicus Spreng. (Leguminosae)	CI-1	Immunopotentiation	Datta et al. (1999)
Carica papaya Linn. (Caricaceae)	Myricetin, caffeic acid, trans-ferulic acid, and kaempferol	Increase in thrombocyte DTH response and phagocytic index	Anjum et al. (2017)
Coriandrum sativum L. (Umbelliferae)	Quercetin, rutin	Immunomodulatory activities	Cherng et al. (2007)
Curculigo orchioides Gaetrn. (Hypoxidaceae)	Curculigoside	Increase phagocytosis	Kubo et al. (1983)
Catharanthus roseus Linn. G. Donn (Apocynaceae)	Vincristine	Induce antibody production	Wagner (1986)
Camptotheca acuminata Decne (Nyssaceae)	Camptothecin, tenulin	Induce interferon, production of antibodies	Wagner (1986)
Curcuma longa Linn. (Zingiberaceae)	Curcumin	Inhibits cell proliferation	Abu-Rizq et al. (2008)
Cedrus deodara (Roxb.) Loud. (Pinaceae)	Wood oil	Inhibit neutrophil adhesion	Shinde et al. (1999)
Crocus sativus Linn. (Iridaceae)	Proteoglycan	Induced nitric oxide production	Escribano et al. (1999)

(continued)

Table 9.2 (continued)

Biological source	Chemical constituent	Mechanism of action	References
Datura quercifolia Kunth. (Solanaceae)	1b,5a,12a-trihydroxy-6a,7a,24a,25a-diepoxy-20S,22R with-2-enolide (Lactones)	Stimulate antibody production	Bhat et al. (2005)
Dysoxylum binectariferum Roxb. Hook. F. Ex Bedd. (Meliaceae)	(+)-cis-5,7-dihydroxy-2-methyl-8-[4, (3-hydroxy-1-methyl) piperid	Immunomodulatory and anti-inflammatory	Naik et al. (1988)
Daucus carota L. (Apiaceae)	Quercetin, rutin	Immunomodulatory activities	Cherng et al. (2007)
Desmodium gangeticum (L.) DC. (Fabaceae)	Aminoglucosyl glycerolipid, glycosphingolipid	Increase nitric oxide level	Mishra et al. (2005)
Enicostema axillare (Lam) A. Raynal (Gentianaceae)	Swertiamarin	Immunostimulating	Saravanan et al. (2014)
Foeniculum vulgare Mil. (Umbelliferae)	Quercetin, rutin	Immunomodulatory activities	Cherng et al. (2007)
Ipomoea batatas (l.) Lam. Var. Cannabina Hallier F. (Convolvulaceae)	(1→>6)-alpha-D-glucan	Immunostimulating	Zhao et al. (2005)
Luffa cylindrica (L.) M. J. Roem. (Cucurbitaceae)	Sapogenins 1 and 2	Immunostimulant	Khajuria et al. (2007)
Picrorhiza kurroa Royle Ex Benth (Scrophulariaceae)	Picroside I, II, kutkoside, vanillic acid, apocynin; iridoid glycoside	Anticomplement activity; Immunostimulant	Wagner and Prokesh (1985), Puri et al. (1992)
Plumbago zeylanica (Plumbaginaceae)	Plumbagin	Improve macrophage function	Abdul and Ramchender (1995)
Petroselinum crispum (Umbelliferae)	Quercetin, rutin	Immunomodulatory activities	Cherng et al. (2007)
Stevia rebaudiana Bertoni (Asteraceae)	Stevioside	Immunostimulant	Sehar et al. (2008)
Sterculia villosa (Malvaceae)	Lupeol	Immunomodulatory by inhibiting pro-inflammatory responses	Das et al. (2017)
Tinospora cordifolia (Willd.) Hook f. and Thoms. (Menispermaceae)	(1,4)-alpha-D-glucan, cordifolioside A and B	Activation of macrophage, immunostimulating	Nair et al. (2006), Maurya et al. (1996)
Withania somnifera (L.) Dunn. (Solanaceae)	Sitoindosides IX and X withanolide-A	Immunostimulating Immunostimulating	Ghosal et al. (1989) Malik et al. (2007)

(continued)

Table 9.2 (continued)

Biological source	Chemical constituent	Mechanism of action	References
Vigna radiata (Fabaceae)	Non-starch polysaccharides	Splenocyte proliferation and macrophage activation	Ketha and Gudipati (2018)

of neem on food crops since it is considered harmless to humans, animals, birds, beneficial insects, and earthworms. Neem bark aqueous extracts behave as immunomodulator. Neem shows positive results against various communicable and noncommunicable diseases. Experiment on BALB/C mice reveals bark of neem shows increased immune response. Aqueous extract of neem flower exhibits cell-mediated and humoral immunity as well as it also shows non-targeted immune response via phagocytic activity of macrophages (Chakraborty et al. 2008; Narnoliya et al. 2021). It aids in the maintenance of a healthy immune system and is used to treat inflammatory skin problems. Neem is used for blood purification from centuries; it not only aids in the treatment of ailments but also gives us the strength to resist them by boosting the immune system due to presence of various active biomolecules NB-II peptidoglycan, gallic acid, epicatechin, and catechin (Upadhyay et al. 1992).

9.3.12 *Hemidesmus indicus*

It is commonly known as Anantmool which means "the eternal root," as their roots are expanding to long distance. This plant belongs to family Asclepiadaceae. In vitro, *H. indicus* extract stimulates the synthesis of IgG and the activity of adenosine deaminase (ADA) in human peripheral blood cells. *H. indicus* has immunomodulatory properties linked to IgG secretion and ADA activity. The herbal extract increased lymphocyte IgG production as well as ADA activity (Chakrabortty and Choudhary 2014). Aromatic aldehydes and their derivatives, phenolics, triterpenoids, and a variety of different chemicals have been discovered in the root's phytochemistry, few from them have linked to its bioactivity (Nandy et al. 2020). In mice the effect of ethanolic extract was studied on delayed-type hypersensitivity, humoral responses to sheep red blood cells, skin allograft rejection, and reticuloendothelial system phagocytic activity, and it was discovered that both the cell-mediated and humoral components of the immune system were suppressed. The XTT assay for cell proliferation after 24 h in culture revealed that plant extract interacts with cyclosporine A in a way that reduces its immunosuppressive effect at 1000 g/mL in vitro culture lymphocytes (Manjulatha et al. 2014).

9.3.13 *Ocimum sanctum*

It is commonly known as tulsi. Basil belongs to family Lamiaceae (Maurya and Sangwan 2019). Tulsi herb is utilized in Ayurveda for its powerful meditative powers and lack of side effects. It has a strong and bitter flavor that Ayurveda claims can enter deep tissues, dry tissue secretions and balance kapha and vata. It is used to prevent sickness; increases overall health, well-being, and survival; and helps people cope with the demands of everyday life. Eating fresh tulsi leaves is thought to be immunity booster, according to folklore. This declaration was tested in animals and found to result in a considerable increase in antibody titer (Dangar et al. 2021). Basil has the ability to modify humoral immune responses by working at multiple levels in the immune system, including antibody formation, hypersensitive mediator release, and tissue response to these mediators in the target organs. The methanol extract as well as aqueous suspension of tulsi leaves can stimulate humoral response. Seed oils of this plant showed to regulate cell-mediated and humoral response in the case of studied model; these immunomodulatory properties showed perhaps because of GABA-ergic process (Logambal et al. 2000).

9.3.14 *Picrorhiza kurroa*

This plant is known as Kutki commonly which belongs to family Scrophulariaceae. This plant has ability to modify humoral immune responses by working at multiple levels in the immune system, including antibody formation, hypersensitive mediator release, and tissue responses to these mediators in the target organs. Kutki rhizome aqueous and ethanolic extracts have the capacity to activate humoral reaction. B lymphocytes and plasma cells produce antibody molecule, which are essential for humoral immune responses. Primary immunoglobulins, IgG and IgM, are elaborated in activation of complement system, opsonization, and neutralization of toxins (Wagner and Prokesh 1985). Antigen-nonspecific defense is aided by two potent anticomplementary polymeric fractions that have been identified. This finding supports the hypothesis that medicinal formulations made from this plant can have an effect on immunological processes (Puri et al. 1992).

9.3.15 *Tinospora cordifolia*

It is a widely used Ayurvedic plant of family Menispermaceae commonly known as Amrita/Gulvel/Gulancha/Guduchi. This plant is useful in treating jaundice, skin disease and infection, etc. (Nair et al. 2006). This herb is used to enhance liveliness of an individual and also works as immunostimulant. As per earlier reported work, this plant shows inhibition of classical complement pathway, increasing cell-mediated and humoral immunity and enhanced IgG antibodies. On introducing Guduchi in cholestasis-induced immunosuppressive rats, there is a notable upgradation in immunomodulatory role, along with this presence of active components like arabinogalactan, and novel alpha-D-glucan results in stimulating immune system via macrophage activation through Toll-like receptor-6 signaling and cytokine production (Maurya et al. 1996). Stem of this plant boiled in water works as best cleaner as well as immune booster (Manjrekar et al. 2000).

9.3.16 *Withania somnifera*

It is commonly known as ashwagandha, which belongs to family Solanaceae. Leaves of this plant help in boosting immunity along with several other pharmacological properties (Pathak et al. 2021; Sangwan and Sangwan 2014, Sangwan et al. 2017; Mishra et al. 2005; Misra et al. 2008). Many names are provided to this woody shrub such as winter cherry, punir, ghodakun, fatarfoda, asgund, and chirpotan. It is shown to successfully impede process of inflammation (Das 2021). This shrub has robust, fleshy roots, simple oblong opposite leaves, and inconspicuous greenish or lurid yellow flower in axillary, umbellate cymes that grow up to 2 ft. tall. The roots are quite beneficial and are utilized to treat various ailments. This therapeutic shrub's entire anatomy is valuable (Malik et al. 2007). It is one of the best-known herbs for providing youthful vigor, having the ability to behave as an adaptogen which employs as a great immunostimulator. Ashwagandha-containing components such as cuscohygrine, anahygrine, tropine, anaferine, and glycoside along with withanolide all these help in their anti-inflammatory and immune-boosting capability; this is helpful in acknowledging it as blood tonic (Pathak et al. 2021).

9.3.17 *Andrographis paniculata*

Andrographolide is the main diterpenoid present in *Andrographis paniculata* Nees (family Acanthaceae). It is abundantly found in India, Sri Lanka, and Pakistan, growing in hot and shady places. Its synonyms are kalmegh, kiriat, kalupnath, and mahatila, also known as the "king of bitters" (Mishra et al. 2007; Sharma 1983). It is well-known by traditional medicine as immune enhancer. The phytoconstituent andrographis has confirmed significant activity in fighting upper respiratory infections common cold and flu (Xu et al. 2007, Melchior et al. 2000, Kumar et al. 2004). *Andrographolide paniculata* exhibits immunostimulatory activity by increasing antibody and delayed-type hypersensitivity (Akram et al. 2014; Puri et al. 1993).

Andrographolide inhibits CHIKV by suppressing viral protein expression (by reducing viral RNA copy number, and it increases cytotoxic T lymphocytes) and upregulating host innate immunity (via protein kinase R, phosphorylated eukaryotic initiation factor 2 alpha, retinoic acid inducible gene-I, and interferon regulatory factor) thereby increasing IFN-alpha secretion and hence could be an effective therapeutic agent against CHIKV infection (Gupta et al. 2018, Wang et al. 2010, Jantan et al. 2015). Andrographolide therapy could decrease the virus loads, improve the survival rate, and diminish lung. The mode of action of andrographolide is due to its action on JAK-STAT and NF-κB signaling pathway. Its anti-influenza

Table 9.3 Chemical constituent's based classification of plants

S. no.	Chemical class	Examples
1.	Alkaloids	Berberine, chelerythrine, gelselegine, pseudocoptisine, leonurine, piperine, sinomenine, koumine, lycorine, sophocarpine, rhynchophylline, tetrandrine, matrine, colchicine, capsaicin, andrographolide, genistein
2	Essential oil	Z-ligustilide, tetramethylpyrazine
3.	Glycosides	Isorhamnetine-3-O-glycoside, eupalitin-3-O-β-galactopyranoside, aucubin, mangiferin, dendroside A and dendronobilosides A and B, proanthocyanidins, anthocyanidins, auraptene, esculetin (6,7dihydroxycoumarin), iridoid glycosides
4.	Flavonoids	Chalcone – butein, xanthohumol, dihydroxanthohumol, mallotophilippen C, D, E, licochalcone E Flavones – luteolin, apigenin, chrysin, nobiletin, baicalein, oroxylin A, wogonin Flavonols – quercetin, kaempferol, rutin Flavanols – epigallocatechin-3-gallate Isoflavones – daidzein, genistein, puerarin Phloroglucinols – myrtucommulone, arzanol, Quinones – thymoquinone, shikonin, emodin-8-O-β-D glucoside Phenolic – gallic acid, ellagic acid, chlorogenic acid, ferulic acid, p-coumaric acid, vanillic acid, coumarin, esculetin (6,7-dihydroxycoumarin), curcumin Others – apocynin Stilbenes – resveratrol, piceatanno
5.	Tannins	Chebulagic acid, corilagin, punicalagin
6.	Saponins	Asiaticoside, glycyrrhizine
7.	Terpenoids	14-deoxyandrographolide, 14-deoxy-11,12-didehydroandrographolide, ginsan, oleanolic acid, echinocystic acid, triptolide, demethylzelasteral, celastrol, asiaticoside, madecassoside, 11-keto-β-Boswellic acid, andrographolide, Boswellic acid, ursolic acid, lupeol, amyrine
8.	Sterols/ steroids	Withanolides, β-sitosterol

activity is due to combination of virus entry inhibitor (Ding et al. 2017). Andrographolide can regulate the innate and adaptive immune responses by modulating Ag-specific antibody production and macrophage phenotypic polarization. Decreased phosphorylation levels of AKT, ERK1/2, and IL-4-producing splenocytes were observed in macrophages treated with andrographolide (Wang et al. 2010). Andrographolide reduced IL-12a, IL-12b, and TNF-alpha at mRNA level and reduced the production of TNF-alpha and IL-12p70 proteins in a dose-dependent manner. Andrographolide stopped the activation of ERK1/2 MAP kinase, but not that of p38, JNK, or NF-kappa B. These results propose that andrographolide impede LPS-induced production of TNF-alpha via stopping of the ERK1/2 signaling pathway (Qin et al. 2006).

Andrographolide showed interleukin-2 (IL-2) induction and enhanced proliferation in human peripheral blood lymphocytes (HPBLs) (Kumar et al. 2004). Andrographolide is an immunoenhancing agent which can control both nonspecific immune function and antigen-specific by promoting TNF-alpha, IFN-alpha, and IFN-gamma inductions of the peripheral blood mononuclear cells, but had no effect on IL-8 (Peng et al. 2002). Clinical research reported that andrographolide reduced human immunodeficiency virus-1 (HIV-1)-induced cell cycle dysregulation and to increase CD4+ lymphocytes in HIV-1-infected patients (Calabrese et al. 2000, Mishra et al. 2007) (Table 9.3).

9.4 Conclusions

The chapter brings about the various details of plants, their major constituents and applications in Ayurvedic traditional system of medicine of India. The traditional system entails various herbs and their preparation in different manners to be used for various ailments and health conditions. Some of the herbs of Ayurveda are being experimented world over for their traditional claims and therapeutic effects. Many studies under in vitro conditions using cell lines and in vivo by using animal models revealed healing and therapeutic actions of herb-based preparations. Several studies stated above used the pure molecule isolated from the medicinal herbs and validated the functions of the preparations. Boosting of immunity has been one of the fundamental attributes of these traditional herbal preparations. Studies with major plants have attributed the role of herbs in immuno-regulation and in overall immune boosting. Such properties are key to fight with infection and in better recovery from ailment conditions. Many of such traditional herbs were adopted as specific treatment or as in part in traditional food. The specific treatment included the specific processes and procedures so as to have the bioactives in the enriched preparations. The scientific validation by several experiments in laboratories have authenticated the examples of turmeric, ginger, licorice, kalmegh, tulsi, ashwagandha, arjuna, giloy, etc. where the herbs possess immune-boosting properties.

Acknowledgments The authors would like to acknowledge DHR-ICMR project no. V.25011/286-HRD/2016-HR for financial assistance.

Declaration of Interest The authors report no conflicts of interest.

References

Abdul KM, Ramchender RP (1995) Modulatory effect of plumbagin (5-hydroxy-2-methyl-1,4-naphthoquinone) on macrophage funct. Immunopharmacology 30:231–236

Abeysekera AM, De Silva KTD, Samarasinghe S, Seneviratne PAK, Van Den Berg AJJ, Labadie RP (1996) An immunomodulatory C-glucosylated propelargonidin from the unripe fruit of Aegle marmelos. Fitoterapia 67:367–370

Abu-Rizq HA, Mansour MH, Safer AM, Afzal M (2008) Cyto-protective and immunomodulating effect of Curcuma longa in Wistar rats subjected to carbon tetrachloride-induced oxidative stress. Inflammopharmacology 16:87–95

Afolayan FID, Erinwusi B, Oyeyemi OT (2018) Immunomodulatory activity of curcumin-entrapped poly d,l-lactic-co-glycolic acid nanoparticles in mice. Integr Med Res 7(2):168–175

Ahmad S, Mika D, Guruvayoorappan C (2012) Chemoprotective and immunomodulatory effect of Acacia nilotica during cyclophosphamide induced toxicity. J Exp Ther Oncol 10(2):83–90

Akram M, Hamid A, Khaliu A, Ghaffar A, Tayyaba N, Saeed A, Alf M, Naveed A (2014) Review on medicinal uses, pharmacological, phytochemistry and immunomodulatory activity of plants. Int J Immunopathol Pharmacol 27(3):313–319

Alexyuk PG, Bogoyavlenskiy AP, Alexyuk MS, Turmagambetova AS, Zaitseva IA, Omirtaeva ES, Berezin VE (2009) Adjuvant activity of multimolecular complexes based on Glycyrrhiza glabra saponins, lipids, and influenza virus glycoproteins

Anjum V, Arora P, Ansari SH, Najmi AK, Ahmad S (2017) Antithrombocytopenic and immunomodulatory potential of metabolically characterized aqueous extract of Carica papaya leaves. Pharm Biol 55(1):2043–2056. https://doi.org/10.1080/13880209.2017.1346690. PMID: 28836477; PMCID: PMC6130488

Ashok BT, Ali R (2003) Aging research in India. Exp Gerontol 38:597–603

Asteriou E, Gkoutzourelas A, Mavropoulos A, Katsiari C, Sakkas LI, Bogdanos DP (2018) Curcumin for the management of periodontitis and early ACPA-positive rheumatoid arthritis: killing two birds with one stone. Nutrients 10(7):908

Azad R, Babu NK, Gupta AD, Reddanna P (2018 Nov) Evaluation of anti-inflammatory and immunomodulatory effects of Premna integrifolia extracts and assay-guided isolation of a COX-2/5-LOX dual inhibitor. Fitoterapia 131:189–199. https://doi.org/10.1016/j.fitote.2018.10.016

Bafna AR, Mishra SH (2004) Immunomodulatory activity of methanol extract of flower-heads of Sphaeranthus indicus Linn. Ars Pharmac 45:281–291

Bafna AR, Mishra SH (2005) Immunomodulatory activity of methanol extract of roots of *Cissampelos pareira* Linn. Ars Pharm 46(3):253–262

Bhat BA, Dhar KL, Puri SC, Qurishi MA, Khajuria A, Gupta A, Qaz GN (2005) Isolation, characterization and biological evaluation of Datura lactones as potential immunomodulators. Bioorg Med Chem 13:6672–6677

Bhattacharjee S, Bhattacharjee A, Majumder S, Majumdar SB, Majumdar S (2012) Glycyrrhizic acid suppresses Cox-2-mediated anti-inflammatory responses during Leishmania donovani infection. J Antimicrob Chemother 67(8):1905–1914. https://doi.org/10.1093/jac/dks159. Epub 2012 May 15

Calabrese C, Berman SH, Babish JG et al (2000) A phase I trial of andrographolide in HIV positive patients and normal volunteers. Phytother Res 14(5):333–338

Chakraborty K, Bose A, Pal S, Sarkar K, Goswami S, Ghosh D, Laskar S, Chattopadhyay U, Baral R (2008) Neem leaf glycoprotein restores the impaired chemotactic activity of peripheral blood mononuclear cells from head and neck squamous cell carcinoma patients by maintaining CXCR3/CXCL10 balance. Int Immunopharmacol 8:330–340

Chattopadhyay SK, Pal A, Maulik PR, Kaur T, Garga A, Khanuja SPS (2006) Taxoid from the needles of the Himalayan yew Taxus wallichiana with cytotoxic and immunomodulatory activities. Bioorg Med Chem Lett 16:2446–2449

Chaturvedi VS (2008) Fit for life through ayurveda. Sterling Publishers Pvt. Ltd., New Delhi

Chauhan NS, Saraf DK, Dixit VK (2010) Effect *of* vajikaran rasayana herbs *on* pituitary-gonadal axis. Eur J Integr Med 2(2):89–91

Chavali SR, Francis T, Campbell JB (1987) An in vitro study of immunomodulatory effects of some saponins. Int J Immunopharmacol 9(6):675–683. https://doi.org/10.1016/0192-0561(87)90038-5

Chakrabortty S, Choudhary R (2014) Hemidesmus indicus (anantmool): rare herb of Chhattisgarh. Indian J Sci Res 4(1):89–93

Cheel J, Van Antwerpen P, Tůmová L, Onofre G, Vokurková D, Zouaoui-Boudjeltia K, Vanhaeverbeek M, Nève J (2010) Free radical-scavenging antioxidant and immunostimulating effects of a licorice infusion (Glycyrrhiza glabra L.). Food Chem 122(3):508–517. https://doi.org/10.1016/j.foodchem.2010.02.060

Cherng JM, Chiang W, Chian LC (2007) Immunomodulatory activities of common vegetables and spices of Umbelliferae and its related coumarins and flavonoids. Food Chem. https://doi.org/10.1016/j.foodchem.2007.07.005

Damre AS, Gokhale AB, Phadke AS, Kulkarni KR, Saraf MN (2003) Studies on the immunomodulatory activity of flavonoidal fraction of Tephrosia purpurea. Fitoterapia 74:257–261

Dangar K, Varsani V, Vyas S (2021) To review on Indian spices and herbs has a promising activity to boosting immunity against deadly coronavirus infection. Int J Bot Stud 6(1):89–91

Das K (2021) Herbal plants as immunity modulators against COVID-19: A primary preventive measure during home quarantine. J Herb Med 100501

Das A, Jawed JJ, Das MC, Sandhu P, De UC, Dinda B, Akhter Y, Bhattacharjee S (2017) Antileishmanial and immunomodulatory activities of lupeol, a triterpene compound isolated from Sterculia villosa. Int J Antimicrob Agents 50(4):512–522. https://doi.org/10.1016/j.ijantimicag.2017.04.022. Epub 2017 Jun 29

Dash S, Nath LK, Bhise S, Kar P, Bhattacharya S (2006) Stimulation of immune function activity by the alcoholic root extract of Heracleum nepalense D. Don. Indian J Pharmacol 38:336–340

Datta S, Sinha S, Bhattacharyya P (1999) Effect of a herbal protein, CI-1, isolated from Cajanus indicus on immune response of control and stressed mice. J Ethnopharmacol 67:259–267

Dhyani A, Nautiyal BP, Nautiyal MC (2010) Importance of Astavarga plants in traditional systems of medicine in Garhwal, Indian Himalaya. Int J Biodivers Sci Ecosyst Serv Manag 6(1-2):13–19

Ding Y, Chen L, Wu W, Yang J, Yang Z (2017) Liu S Andrographolide inhibits influenza A virus-induced inflammation in a murine model through NF-κB and JAK-STAT signaling pathway. Microbes Infect 19(12):605–615

Elabd H, Wang HP, Shaheen A, Yao H, Abbass A (2016) Feeding Glycyrrhiza glabra (liquorice) and Astragalus membranaceus (AM) alters innate immune and physiological responses in yellow perch (Perca flavescens). Fish Shellfish Immunol 54:374–384. https://doi.org/10.1016/j.fsi.2016.04.024

Ehteshamfar SM, Akhbari M, Afshari JT et al (2020) Anti-inflammatory and immune-modulatory impacts of berberine on activation of autoreactive T cells in autoimmune inflammation. J Cell Mol Med 24(23):13573–13588. https://doi.org/10.1111/jcmm.16049

Escribano J, Daz-Guerra MJM, Riese HH, Ontanon J, Garca-Olmo D, Garca-Olmo DC, Rubio A, Fernandez JA (1999) In vitro activation of macrophages by a novel proteoglycan isolated from corms of Crocus sativus L. Cancer Lett 144:107–114

Gabhe SY, Tatke PA, Khan TA (2006) Evaluation of the immunomodulatory activity of the methanol extracts of Ficus benghalensis roots in rats. Indian J Pharmacol 38:271–275

Gautam M, Diwanay S, Gairola S, Shinde Y, Patki P, Patwardhan B (2004) Immunoadjuvant potential of Asparagus racemosus aqueous extract in experimental system. J Ethnopharmacol 91:251–255

Gayathri V, Asha VV, Subramoniam A (2005) Preliminary studies on the immunomodulatory and antioxidant properties of Selaginella species. Indian J Pharmacol 37:381–385

Geetha S, Sai Ram M, Singh V, Ilavazhagan G, Sawhney RC (2002) Anti-oxidant and immunomodulatory properties of sea buckthorn (Hippophae rhamnoides)—an in vitro study. J Ethnopharmacol 79:373–378

Ghafoor F (2020) Importance of herbs in aquaculture; Cinnamon a potent enhancer of growth and immunity in fish, Ctenopharyngodon idella. Iran J Aquat Anim Health 6(1):78–92

Ghazanfaria T, Hassanb ZM, Ebrahimi M (2002) Immunomodulatory activity of a protein isolated from garlic extract on delayed type hypersensitivity. Int Immunopharmacol 2:1541–1549

Ghosal S, Lal SJ, Srivastava R, Bhattacharya SK, Upadhyay SN, Jaiswal AK, Chattopadhyay U (1989) Immunomodulatory and CNS Effects of Sitoindosides IX and X, Two New Glycowithanolides from Withania somnifera. Phytother Res 3:201–206

Ghule BV, Murugananthan G, Nakhat PD, Yeole PG (2006) Immunostimulant effects of Capparis zeylanica Linn. Leaves. J Ethnopharmacol 24:311–315

Gokhale AB, Damre AS, Saraf MN (2003) Investigation into the immunomodulatory activity of Argyreia speciosa. J Ethnopharmacol 84:109–114

Gracious Ross R, Selvasubramanian S, Jayasundar S (2001) Immunomodulatory activity of Punica granatum in rabbits—a preliminary study. J Ethnopharmacol 78:85–87

Gupta A, Chaphalkar SR (2017) Assessment of immunomodulatory activity of aqueous extract of Calamus rotang. Avicenna J Phytomed 7(3):199–205. PMID: 28748166; PMCID: PMC5511971

Gupta S, Mishra KP, Dash PK, Parida M, Ganju L, Singh SB (2018) Andrographolide inhibits chikungunya virus infection by up-regulating host innate immune pathways. Asian Pac J Trop Med 11(3):214–221

Guruvayoorappan C, Kuttan G (2007) Immunomodulatory and antitumor activity of Biophytum sensitivum extract. Asian Pac J Cancer Prev 8:27–32

Hafeez BB, Haque R, Parvez S, Pandey S, Sayeed I, Raisuddin S (2003) Immunomodulatory effects of fenugreek (Trigonella foenum graecum L.) extract in mice. Int Immunopharmacol 3:257–265

Hamsa TP, Kuttan G (2010) Ipomoea obscura ameliorates cyclophosphamide-induced toxicity by modulating the immune system and levels of proinflammatory cytokine and GSH. Can J Physiol Pharmacol 88(11):1042–1053. https://doi.org/10.1139/y10-086

Harlev E, Nevo E, Lansky EP, Ofir R, Bishayee A (2012) Anticancer potential of aloes: antioxidant, antiproliferative, and immunostimulatory attributes. Planta Med 78(9):843–852

Huang CT, Hung CY, Hseih YC, Chang CS, Velu AB, He YC, Huang YL, Chen TA, Chen TC, Lin CY, Lin YC, Shih SR, Dutta A (2019 Nov) Effect of aloin on viral neuraminidase and hemagglutinin-specific T cell immunity in acute influenza. Phytomedicine 64:152904

Jantan I, Ahmad W, Bukhari SNA (2015) Plant-derived immunomodulators: an insight on their preclinical evaluation and clinical trials. Front Plant Sci 6:655

Jayaprakash B, Das A (2018) Extraction and characterization of chick PEA (Cicer arietinum) extract with immunostimulant activity in BALB/C MICE. Asian Pac J Cancer Prev 19(3): 803–810. https://doi.org/10.22034/APJCP.2018.19.3.803. PMID: 29582638; PMCID: PMC5980859

Jayathirtha MG, Mishra SH (2004) Preliminary immunomodulatory activities of methanol extracts of Eclipta alba and Centella asiatica. Phytomedicine 11:361–365

Jee BS (1927) A short history of aryan medical science, 2nd edn. Shree Bhabyat Singh Jee Electric Printing Press, Gondal

Joharapurkar AA, Deode NM, Zambad SP, Umathe SN (2003) Immunomodulatory activity of alcoholic extract of Rubia cordifolia Linn. Indian Drugs 40:179–181

Joshi UP, Mishra SH (2008) Evaluation of aqueous and methanol extracts of Pistacia integerrima galls as potential immunomodulator. Pharmacogn Mag 4:126–131

Kaul A, Bani S, Zutshi U, Suri KA, Satti NK, Suri OP (2003) Immunopotentiating properties of cryptolepis buchanani root extract. Phytother Res 17:1140–1144

Ketha K, Gudipati M (2018 Nov) Immunomodulatory activity of non starch polysaccharides isolated from green gram (Vigna radiata). Food Res Int 113:269–276. https://doi.org/10.1016/j.foodres.2018.07.010

Khajuria A, Gupta A, Garaib S, Wakhloo BP (2007) Immunomodulatory effects of two sapogenins 1 and 2 isolated from Luffa cylindrica in Balb/C mice. Bioorg Med Chem Lett 17:1608–1612

Khajuria A, Gupta A, Suden P, Singh S, Malik F, Singh J, Gupta BD, Suri KA, Srinivas VK, Ella K, Qazi GN (2008) Immunomodulatory activity of biopolymeric fraction BOS 2000 from Boswellia serrata. Phytother Res 22:340–348

Kokate CK, Purohit AP, Gokhale SB (2010) Pharmacognosy. Publisher Nirali Prakashan, New Delhi

Kroes BH, Beukelman CJ, van den Berg AJ, Wolbink GJ, van Dijk H, Labadie RP (1997) Inhibition of human complement by beta-glycyrrhetinic acid. Immunology 90(1):115–120. https://doi.org/10.1046/j.1365-2567.1997.00131.x. PMID: 9038721; PMCID: PMC1456728

Kubo M, Namba K, Nagamoto N, Nagao T, Nakanishi J, Uno H, Nishimura H (1983) A new phenolic glucoside, curculigoside from rhizomes of Curculigo orchioides. Planta Med 47:52–55
Kulkarni SR, Karande VS (1998) Study of the immunostimulant activity of Naphthoquinone extract of leaves of Lawsonia alba Linn. Indian Drugs:427–433
Kumar KS, Bhowmik D, Dutta A, Yadav AP, Paswan S, Srivastava S, Deb L (2012) Recent trends in potential traditional Indian herbs Emblica officinalis and its medicinal importance. J Pharmacogn Phytochem 1(1):18–28
Kumar RA, Sridevi K, Kumar NV, Nanduri S, Rajagopal S (2004) Anticancer and immunostimulatory compounds from Andrographis paniculata. J Ethnopharmacol 92(2–3): 291–295
Kumari R, Kumar S, Kumar A, Goel KK, Dubey RC (2017) Antibacterial, antioxidant and Immuno-modulatory properties in extracts of Barleria lupulina Lindl. BMC Complement Altern Med 17(1):484. https://doi.org/10.1186/s12906-017-1989-4. PMID: 29100518; PMCID: PMC5670697
Lad VS (2002) Fundamental Principles of Ayurveda. The Ayurvedic Press, Albuquerque, New Mexico
Lakshmi V, Pandey K, Puri A, Saxena RP, Saxena KC (2003) Immunostimulant principles from Curculigo orchioides. J Ethnopharmacol 89:181–184
Latha PG, Evans DA, Panikar KR, Jayavardhan KK (2000) Immunomodulatory and antitumour properties of Psoralea corylifolia seeds. Fitoterapia 71:223–231
Lee SY, Ryu IW, Shim CS (1997) Purification and characterization of bioactivity compound acemannan from Aloe vera. Korean J Pharmac 28:65–71
Lee JH, Lee JY, Park JH, Jung S, Kim JS, Kang SS, Kim YS, Han Y (2007) Immunoregulatory activity by daucosterol, a sitosterol glycoside, induces protective Th1 immune response against disseminated Candidiasis in mice. Vaccine 25:3834–3840
Lee G, Chung HS, Lee K, Lee H, Kim M, Bae H (2017 Sep) Curcumin attenuates the scurfy-induced immune disorder, a model of IPEX syndrome, with inhibiting Th1/Th2/Th17 responses in mice. Phytomedicine 15(33):1–6
Lee W, Yang S, Lee C, Park EK, Kim KM, Ku SK, Bae JS (2019) Aloin reduces inflammatory gene iNOS via inhibition activity and p-STAT-1 and NF-κB. Food Chem Toxicol 126:67–71. https://doi.org/10.1016/j.fct.2019.02.025. Epub 2019 Feb 13
Lei J, Shen Y, Xv G, Di Z, Li Y, Li G (2020) Aloin suppresses lipopolysaccharide-induced acute lung injury by inhibiting NLRP3/NF-κB via activation of SIRT1 in mice. Immunopharmacol Immunotoxicol 42(4):306–313
Logambal SM, Venkatalakshmi S, Dinakaran Michael RD (2000) Immunostimulatory effect of leaf extract of Ocimum sanctum Linn in Oreochromis mossambicus (Peters). Hydrobiologia 430: 113–120
Madakkannu B, Ravichandran R (2017) In vivo immunoprotective role of Indigofera tinctoria and Scoparia dulcis aqueous extracts against chronic noise stress induced immune abnormalities in Wistar albino rats. Toxicol Rep 6(4):484–493. https://doi.org/10.1016/j.toxrep.2017.09.001. PMID: 28959678; PMCID: PMC5615165
Makare N, Bodhankar S, Rangari V (2001) Immunomodulatory activity of alcoholic extract of Mangifera indica L in Mice. J Ethnopharmacol 78:133–137
Malik F, Singh J, Khajuria A, Suri KA, Satti NK, Singh S, Kaul MK, Kumar A, Bhatia A, Qazi GH (2007) A standardized root extract of Withania somnifera and its major constituent withanolide-A elicit humoral and cell-mediated immune responses by up regulation of Th1-dominant polarization in BALB/c mice. Life Sci 80:1525–1538
Manjrekar PN, Jolly CI, Narayanan S (2000) Comparative studies of the immunomodulatory activity of Tinospora cordifolia and Tinospora sinensis. Fitoterapia 71:254–257
Manjula N, Gayathri B, Vinaykumar KS, Shankernarayanan NP, Vishwakarmab RA, Balakrishnan A (2006) Inhibition of MAP kinases by crude extract and pure compound isolated from Commiphora mukul leads to down regulation of TNF-a, IL-1h and IL-2. Int Immunopharmacol 6:122–132

Manjulatha K, Saritha K, Setty OH (2014) Phytochemistry, Pharmacology and Therapeutics of Hemidesmus indicus (L.) R. Br

Mashhadi NS, Ghiasvand R, Askari G, Hariri M, Darvishi L, Mofid MR (2013) Anti-oxidative and anti-inflammatory effects of ginger in health and physical activity: review of current evidence. Int J Prev Med 4(Suppl 1):S36

Maurya S, Sangwan NS (2019) Profiling of essential oil constituents in Ocimum species. Proc Ind Natl Acad Sci India Sect B Biol Sci 90(3):577–583. https://doi.org/10.1007/s40011-019-01123-8

Maurya R, Wazir V, Kapil A, Kapil S (1996) Cordifolioside A and B two new phenylpropene disaccharides from Tinospora cordifolia possessing immunostimulating activity. Nat Prod Lett 8:7–10

Mehrotra S, Mishra KP, Maurya R, Srimal RC, Singh VK (2002) Immunomodulation by ethanolic extract of Boerhaavia diffusa roots. Int Immunopharmacol 2:987–996

Melchior J, Spasov AA, Ostrovskij OV (2000) Double blind placebo controlled pilot and phase III study of activity of standardized Andrographis paniculata Herba Nees extract fixed combination (Kan jang) in the treatment of uncomplicated upper respiratory tract infection. Phytomedicine 7: 341–350

Mishra PK, Singh N, Ahmad G, Dubeb A, Maurya R (2005) Glycolipids and other constituents from Desmodium gangeticum with antileishmanial and immunomodulatory activities. Bioorg Med Chem Lett 15:4543–4546

Misra L, Mishra P, Pandey A, Sangwan RS, Sangwan Neelam S, Tuli R (2008) Withanolides from Withania somnifera roots. Phytochem 69:1000–1005

Mishra SK, Sangwan NS, Sangwan RS (2007) Andrographis paniculata (Kalmegh): a review. Pharmacogn Rev 1(2):283–298

Michaelis M, Geiler J, Naczk P, Sithisarn P, Ogbomo H, Altenbrandt B, Leutz A, Doerr HW, Cinatl J Jr (2010) Glycyrrhizin inhibits highly pathogenic H5N1 influenza A virus-induced pro-inflammatory cytokine and chemokine expression in human macrophages. Med Microbiol Immunol 199(4):291–297. https://doi.org/10.1007/s00430-010-0155-0. Epub 2010 Apr 13. PMID: 20386921; PMCID: PMC7087222

Mojica-Henshaw M, Francisco AD, Guzman FD, Tigno XT (2003) Possible immunomodulatory actions of Carica papaya seed extract. J Clin Hemorheol Microcirc 29:219–229

Mahfuz S, Piao XS (2019) Application of Moringa (Moringa oleifera) as Natural Feed Supplement in Poultry Diets. Animals. 9(7):431. https://doi.org/10.3390/ani9070431

Mollazadeh H, Cicero AFG, Blesso CN, Pirro M, Majeed M, Sahebkar A (2017) Immune modulation by curcumin: the role of interleukin-10. Crit Rev Food Sci Nutr. 11:1–13. https://doi.org/10.1080/10408398.2017.1358139

Mungantiwar AA, Nair AM, Shinde UA, Dikshit VJ, Saraf VS, Thakur VS (1999) Studies on the immunomodulatory effects of Boerhaavia diffusa alkaloid fraction. J Ethnopharmacol 65:125–131

Naik RG, Kattige SL, Bhat SV, Alreja B, Desouza NJ, Rupp RH (1988) An inflammatory piperidyriylbenzopyranone from Dysoxylum binectariferum. Tetrahedron 44:2081–2086

Nair PK, Melnick SJ, Ramachandran R, Escalon E, Ramachandran C (2006) Mechanism of macrophage activation by (1,4)-alpha-D-glucan isolated from Tinospora cordifolia. Int Immunopharmacol 6:1815–1824

Narnoliya LK, Sangwan NS, Misra L et al (2021) Processes of high-yield isolation and flash chromatographic purification of azadiradione from neem fruits. Proc Natl Acad Sci, India Sect B Biol Sci 91(4):847–853. https://doi.org/10.1007/s40011-021-01254-x

Nandy S, Mukherjee A, Pandey DK, Ray P, Dey A (2020) Indian Sarsaparilla (Hemidesmus indicus): Recent progress in research on ethnobotany, phytochemistry and pharmacology. J Ethnopharmacol 254:112609

Nevin KG, Vijayammal PL (2005) Pharmacological and immunomodulatory effects of Aerva Lanata in Daltons lymphoma ascites-bearing mice. Pharm Biol 43:640–646

Otsuki N, Dang NH, Kumagai E, Kondo A, Iwata S, Morimoto C (2010) Aqueous extract of Carica papaya leaves exhibits anti-tumor activity and immunomodulatory effects. J Ethnopharmacol 127(3):760–767. https://doi.org/10.1016/j.jep.2009.11.024. Epub 2009 Dec 2

Pathak P, Shukla P, Kanshana JS, Jagavelu K, Sangwan NS, Dwivedi AK, Dikshit M (2021) Standardized root extract of Withania somnifera and Withanolide A exert moderate vasorelaxant effect in the rat aortic rings by enhancing nitric oxide generation. J Ethnopharmacol 278: 114296. https://doi.org/10.1016/j.jep.2021.114296

Pandima DK, Sai Ram M, Sreepriya M, Ilavazhagan G, Devaki T (2003) Immunomodulatory effects of Premna tomentosa extract against Cr (VI) induced toxicity in splenic lymphocytes-an in vitro study. Biomed Pharmacother 57:105–108

Pandima DK, Sai Ram M, Sreepriya M, Devaki T, Ilavazhagan G, Selvamurthy W (2004) Immunomodulatory effects of Premna tomentosa (L. Verbenaceae) extract in J 779 macrophage cell cultures under chromate (VI)-induced immunosuppression. J Altern Complement Med 10: 535–539

Park MY, Kwon HJ, Sung MK (2009) Evaluation of aloin and aloe-emodin as anti-inflammatory agents in aloe by using murine macrophages. Biosci Biotechnol Biochem 73(4):828–832

Patil JS, Nagavi BG, Ramesh M, Vijayakumar GS (1998) A study on the Immunostimulant activity of Centella asiatica in rats. Indian Drugs 38:711–714

Patwardhan B (2005) Botanical immunodrugs: scope and opportunities. Drug Discov Today 10: 495–502

Peng GY, Zhou F, Ding RL, Li HD, Yao K (2002) Modulation of lianbizi injection (Andrographolide) on some immune functions. China J Chin Materia Medica 27(2):147–150

Prabhu V, Guruvayoorappan C (2012) Evaluation of immunostimulant activity and chemoprotective effect of mangrove Rhizophora apiculata against cyclophosphamide induced toxicity in Balb/c mice. Immunopharmacol Immunotoxicol 34:608–615

Pravansha S, Thippeswamy BS, Veerapur VP (2012 Dec) Immunomodulatory and antioxidant effect of Leptadenia reticulata leaf extract in rodents: possible modulation of cell and humoral immune response. Immunopharmacol Immunotoxicol 34(6):1010–1019

Pugh N, Ross SA, ElSohly MA, Pasco DSJ (2001) Characterization of Aloeride, a new high-molecular-weight polysaccharide from Aloe vera with potent immunostimulatory activity. J Agric Food Chem 49:1030–1034

Puri A, Sahai R, Singh KL, Saxena RP, Tandon JS, Saxena KC (2000) Immunostimulant activity of dry fruits and plant materials used in Indian traditional medical system for mothers after child birth and invalids. J Ethnopharmacol 71:89–92

Puri A, Saxena R, Saxena RP, Saxena KC (1993) Immunostimulant agents from Andrographis paniculata. J Nat Prod 56(7):995–999

Puri A, Saxena RP, Guru PY, Kulshreshtha DK, Saxena KC, Dhawan BN (1992) Immunostimulant activity of Picroliv, the iridoid glycoside fraction of Picrorhiza kurrooa, and its protective action against Leishmania donovani infection in hamsters. Planta Med 58:528–532

Puri A, Saxena R, Saxena RP, Saxena KC, Srivastava V, Tando JS (1994) Immunostimulant activity of Nyctanthes arbor tristis. J Ethnopharmacol 42:31–37

Puri HS (2002) Rasayana: ayurvedic herbs for longevity and rejuvenation. CRC Press, New Delhi

Qin LH, Kong L, Shi GJ, Wang ZT, Ge BX (2006) Andrographolide inhibits the production of TNF-alpha and interleukin-12 in lipopolysaccharide-stimulated macrophages: role of mitogen-activated protein kinases. Biol Pharm Bull 29(2):220–224

Raju N, Sakthivel KM, Kannan N, Vinod Prabhu V, Guruvayoorappan C (2015) Cuscuta chinensis ameliorates immunosuppression and urotoxic effect of cyclophosphamide by regulating cytokines - GM-CSF and TNF-alpha. Appl Biochem Biotechnol 176(3):742–757. https://doi.org/10.1007/s12010-015-1608-0

Ramprasath V, Shanthi P, Sachdanandam P (2006) Immunomodulatory and Anti-inflammatory Effects of Semecarpus anacardium Linn. Nut Milk Extract in Experimental Inflammatory Conditions. Biol Pharm Bull 29:693–700

Raphael TJ, Kuttan G (2003) Effect of naturally occurring triterpenoids glycyrrhizic acid, ursolic acid, oleanolic acid and nomilin on the immune system. Phytomedicine 10(6–7):483–489. https://doi.org/10.1078/094471103322331421

Routh HB, Bhowmik KR (1999) Traditional Indian medicine in dermatology. Clin Dermatol 17: 41–47

Roy Deb S, Dash S, Dasgupta RK, Chakraborty J (2019) Modulatory potential on immune system of a new alkaloid isolated from Coccinea cordifolia root. Int J Biol Macromol 121:643–649. https://doi.org/10.1016/j.ijbiomac.2018.10.104

Sai Ram M, Neetu D, Yogesh B, Anju B, Dipti P, Pauline T, Sharma SK, Sarada SKS, Ilavazhagan G, Kumar D, Selvamurthy W (2002) Cyto-protective and immunomodulating properties of Amla (Emblica officinalis) on lymphocytes: an in-vitro study. J Ethnopharmacol 81:5–10

Sakthivel KM, Guruvayoorappan C (2015) Acacia ferruginea inhibits cyclophosphamide-induced immunosuppression and urotoxicity by modulating cytokines in mice. J Immunotoxicol 12(2): 154–163. https://doi.org/10.3109/1547691X.2014.914988

Saravanan S, Pandikumar P, Prakash Babu N, Hairul Islam VI, Thirugnanasambantham K, Gabriel Paulraj M, Balakrishna K, Ignacimuthu S (2014) In vivo and in vitro immunomodulatory potential of swertiamarin isolated from Enicostema axillare (Lam.) A. Raynal that acts as an anti-inflammatory agent. Inflammation 37(5):1374–1388

Sangwan NS, Sangwan RS (2014) In: Opatrný Z, Nick P (eds) Secondary metabolites of traditional medical plants – case study Ashwaganda in applied cell technology in plant cell monographs. Springer, Germany, pp 325–367. ISBN 978-3-642-41786-3

Sangwan NS, Tripathi S, Srivastava Y, Mishra B, Pandey N (2017) Phytochemical genomics of ashwagandha. In: Kaul A, Wadhwa R (eds) Science of ashwagandha: preventive and therapeutic potentials. Springer, Cham, pp 3–36

Sehar I, Kaul A, Bani S, Pal HC, Saxena AK (2008) Immune up regulatory response of a non-caloric natural sweetener, stevioside. Chem Biol Interact 28:115–121

Singh SK, Valicherla GR, Bikkasani AK, Cheruvu SH, Hossain Z, Taneja I, Ahmad H, Raju SKR, Sangwan NS, Singh SK, Dwivedi AK, Wahajuddin M, Gayen JR (2020) Elucidation of plasma protein binding, blood partitioning, permeability, CYP phenotyping and CYP inhibition studies of Withanone using validated UPLC method: An active constituent of neuroprotective herb Ashwagandha. J Ethnopharmacol 270(113819):0378–8741. https://doi.org/10.1016/j.jep.2021.113819

Shukla S, Mehta A, John J, Mehta P, Vyas SP, Shukla S (2009) Immunomodulatory activities of the ethanolic extract of Caesalpinia bonducella seeds. J Ethnopharmacol. 125(2):252–256. https://doi.org/10.1016/j.jep.2009.07.002

Sharma RK, Cwiklinski K, Aalinkeel R, Reynolds JL, Sykes DE, Quaye E, Oh J, Mahajan SD, Schwartz SA (2017 Nov) Immunomodulatory activities of curcumin-stabilized silver nanoparticles: Efficacy as an antiretroviral therapeutic. Immunol Investig 46(8):833–846

Sharma V, Thakur M, Chauhan NS, Dixit VK (2010) Immunomodulatory activity of petroleum ether extract of Anacyclus pyrethrum. Pharm Biol 48(11):1247–1254

Sharma ML, Kaul A, Khajuria A, Singh S, Singh GB (1996) Immunomodulatory activity of boswellic acids (pentacyclic triterpene acids) from Boswellia serrata. Phytother Res 10:107–112

Sharma PV (1983). Charka Samhita. Ed. Chankhambhia orientalia, Vol. II, Varanasi

Sharma RK, Dash VB (1996) Agnivesha "Charak Samhita", vol I, 1st edn. Chawkhamba Sanskrit Series Office, Varanasi

Shinde UA, Phadke AS, Nair AM, Mungantiwar AA, Dikshit VJ, Saraf MN (1999) Preliminary studies on the immunomodulatory activity of Cedrus deodara wood oil. Fitoterapia 70(4):333–339

Shuto T (2013) Regulation of expression, function, and inflammatory responses of innate immune receptor Toll-like receptor-2 (TLR2) during inflammatory responses against infection. Yasuaki zasshi 133(12):1401–1409

Srivastava A, Nigam AK, Mittal S, Mittal AK (2018) Role of aloin in the modulation of certain immune parameters in skin mucus of an Indian major carp, Labeo rohita. Fish Shellfish Immunol 73:252–261

Stites DP, Stobo JD, Fundenberg HH (1982) Basic and clinical immunology. Lange, Los Altos, CA

Sudha P, Asdaq SM, Dhamingi SS, Chandrakala GK (2010) Immunomodulatory activity of methanolic leaf extract of Moringa oleifera in animals. Indian J Physiol Pharmacol 54(2): 133–140

Sunil MA, Sunitha VS, Ashitha A, Neethu S, Midhun SJ, Radhakrishnan EK, Jyothis M (2019 Jan) Catechin rich butanol fraction extracted from Acacia catechu L. (a thirst quencher) exhibits immunostimulatory potential. J Food Drug Anal 27(1):195–207. https://doi.org/10.1016/j.jfda.2018.06.010

Sunila ES, Kuttan G (2004) Immunomodulatory and antitumor activity of Piper longum Linn. and piperine. J Ethnopharmacol 90:339–346

Tawfik AF, Bishop SJ, Ayalp A, El-feraly FS (1990) Effects of artemesin, dihydroartemesin and arteether on immune responses of normal mice. Int J Immunopharmacol 12:385–389

Thakur M, Connellan P, Deseo MA, Morris C, Praznik W, Loeppert R, Dixit VK (2012) Characterization and in vitro immunomodulatory screening of fructo-oligosaccharides of Asparagus racemosus Willd. Int J Biol Macromol 50(1):77–81. https://doi.org/10.1016/j.ijbiomac.2011.09.027

Thakur M, Bhargava S, Dixit VK (2006) Immunomodulatory activity of chlorophytum borivilianum Sant.F. Evid Based Complement Alternat Med 4:419–423

Thejass P, Kuttan G (2007) Immunomodulatory activity of Sulforaphane, a naturally occurring isothiocyanate from broccoli (Brassica oleracea). Phytomedicine 14:538–545

Tiwari U, Rastogi B, Singh P, Saraf DK, Vyas SP (2004) Immunomodulatory effects of aqueous extract of Tridax procumbens in experimental animals. J Ethnopharmacol 92:113–119

Trivedi MK, Mondal SC, Gangwar M, Jana S (2017 Dec) Immunomodulatory potential of nanocurcumin-based formulation. Inflammopharmacology 25(6):609–619

Upadhyay SN, Dhawan S, Garg S, Talwar GP (1992) Immunomodulatory effect of neem (Azadirachta indica) oil. Int J Immunopharmacol 14:1187–1193

Vayalil PK, Kuttan G, Kuttan R (2002) Rasayanas: evidence for the concept of prevention of diseases. Am J Chin Med 30:155–171

Vázquez B, Avila G, Segura D, Escalante B (1996) Antiinflammatory activity of extracts from Aloe vera gel. J Ethnopharmacol. 55(1):69–75. https://doi.org/10.1016/s0378-8741(96)01476-6. PMID: 9121170

Wadekar RR, Agrawal SV, Teweri KM, Shinde RD, Mate S, Patil K (2008) Effect of Baliospermum montanum root extract on phagocytosis by human neutrophils. Int J Green Pharm 2:46–49

Wagner H (1986) Immunostimulants from higher plants. In: Hostettmann K, Lea PJ (eds) Biologically active natural products. Blackwell Scientific, Oxford, pp 127–128

Wagner H, Prokesh A (1985) Economic and medicinal plant research, vol 1. Academic Press, London, p 113

Wang W, Wang J, Dong SF, Liu CH, Italiani P, Sun SH, Xu J, Boraschi D, Ma SP, Qu D (2010) Immunomodulatory activity of andrographolide on macrophage activation and specific antibody response. Acta Pharmacol Sin 31(2):191–201

Xu Y, Chen A, Fry S, Barrow RA, Marshall RL, Mukkur TK (2007) Modulation of immune response in mice immunised with an inactivated Salmonella vaccine and gavaged with Andrographis paniculata extract or andrographolide. Int Immunopharmacol 7(4):515–523

Yousaf F, Shahid M, Riaz M, Atta A, Fatima H (2017) Immunomodulatory potential of Anacyclus pyrethrum (L.) and Mucuna pruriens (L.) In male albino rats. J Biol Regul Homeost Agents 31(2):425–429

Zhang C, Shao Z, Hu X, Chen Z, Li B, Jiang R, Bsoul N, Chen J, Xu C, Gao W (2020) Inhibition of PI3K/Akt/NF-κB signaling by Aloin for ameliorating the progression of osteoarthritis: In vitro and in vivo studies. Int Immunopharmacol 89(Pt B):107079

Zhang XF, Wang HM, Song Y, Nie L, Wang L, Liu B, Shenb., P., and Liu, Y.. (2006) Isolation, structure elucidation, antioxidative and immunomodulatory properties of two novel dihydrocoumarins from Aloe vera. Bioorg Med Chem Lett 16:949–953

Zhang YH, Isobe K, Nagase F, Lwin T, Kato M, Hamaguchi M, Yokochi T, Nakashima I (1993) Glycyrrhizin as a promoter of the late signal transduction for interleukin-2 production by splenic lymphocytes. Immunology 79(4):528–534. PMID: 8406577; PMCID: PMC1421919

Zhao G, Kan J, Li Z, Chen Z (2005) Characterization and immunostimulatory activity of an (1--> 6)-a-D-glucan from the root of Ipomoea batatas. Int Immunopharmacol 5:1436–1445

Zhao J, Wang J, Zhou M, Li M, Li M, Tan H (2019) Curcumin attenuates murine lupus via inhibiting NLRP3 inflammasome. Int Immunopharmacol 69:213–216

Zhong J, Wang F, Wang Z, Shen C, Zheng Y, Ma F, Zhu T, Chen L, Tang Q, Zhu J (2019) Aloin attenuates cognitive impairment and inflammation induced by d-galactose via down-regulating ERK, p38 and NF-κB signaling pathway. Int Immunopharmacol 72:48–54

Zhu T, Chen Z, Chen G, Wang D, Tang S, Deng H, Wang J, Li S, Lan J, Tong J, Li H, Deng X, Zhang W, Sun J, Tu Y, Luo W, Li C (2019) Mediat Inflamm 3(2019):4927430

Select Global Immune-Boosting Plants Used in Folklore Medicine

10

Raymond Cooper and Ajay Sharma

Abstract

There are numerous foods and herbal plants that boost the immune system. They stimulate the activity of cells responsible for fighting infections. In fact, over centuries, people have relied on herbs and other plants for treating medical conditions and boosting immunity. Considering the worldwide coronavirus pandemic, natural immune boosters are being sought after in the current war against this viral infection. Most likely, immune-boosting plants help human health by tackling viruses, bacteria, and abnormal cells in the form of prevention, to support and strengthen the body's natural immune system. In this chapter, we review several terrestrial species and plants from various sources including China, India, Europe, and Africa, which have long folklore use, and we provide information on the chemistry and biological activity where available.

Keywords

Folklore medicines from Africa and China · Immune boost plants · Elderberry (*Sambucus nigra*) · *Echinacea* spp. · *Astragalus* · *Ganoderma lucidum* · India—Astavarga plants

R. Cooper (✉)
Department of Applied Biology and Chemical Technology, The Hong Kong Polytechnic University, Hung Hom, Hong Kong

A. Sharma
Department of Chemistry, Chandigarh University, Mohali, Punjab, India

N. S. Sangwan et al. (eds.), *Plants and Phytomolecules for Immunomodulation*,
https://doi.org/10.1007/978-981-16-8117-2_10

10.1 Introduction to Immune-Boosting Medicinal Plants

Human beings from primeval time are dependent on nature for managing their basic needs such as foods, medicines shelters, clothing, flavors, means of transport, fragrances, and fertilizers. Medicinal plants have been a rich source of numerous bioactive agents that are used in several traditional medicinal systems from prehistoric time to cure diverse kinds of ailments and diseases (Sharma et al. 2008: Newman and Cragg 2020; Cragg and Newman 2013 and references therein). The World Health Organization (WHO) has been revealed that most of the world's occupants (>80%) basically utilize folklore medicines for their primary medical needs. According to another survey in the USA, 60–70% patients residing in rural areas have been using phytomedicines for their primary healthcare (Krishnaraju et al. 2005; Yuan et al. 2016).

People from all around the world use the products or phytoconstituents obtained from medicinal plants on a daily basis to keep good health and to enhance their immunity against various diseases (Krishnaraju et al. 2005; Kurhekar 2014). Plants contain a multitude of small molecules (SMs), such as flavonoids, tannins, phenolic acids, phytosterol, terpenoids, saponins, etc. Several of these classes possess immune-boosting and antioxidant potential (Sen et al. 2010). The plants are first extracted and taken orally in the form of a decoction or infusion, depending on the purported biological activity. One important group of molecules found in both plant and mushrooms is the polysaccharides. These compounds affect the immune system (Nile et al. 2017; Sen et al. 2010).

Based on recent, published data (Smith et al. 2020) in 2019, the USA saw a significant increase in dietary supplements. The data showed consumers spent approximately $9.6 billion climbing even more in the first half of 2020 as the COVID-19 pandemic took hold. The demand for herbal and nutritional supplements was triggered by immune-enhancing effects (Krawiec 2020). One popular herbal is elderberry commonly used to boost immunity and help against cold and flu symptoms. A second plant is echinacea. Also, Ashwagandha (*Withania somnifera*, Solanaceae), an Ayurvedic botanical medicine, (Gupta and Rana 2007) and with traditional medical use in India enjoyed strong sales. Lastly mushroom products with immune-boosting functions saw increased sales.

Research by Sultan et al. (2014) and Wilasrusmee et al. (2002) looked at immune-boosting plants. Their work looked at nutrition and links to diet and health by examining the immune system in humans. As noted above, several plants and their phytoconstituents hold immunomodulating potential. Thus, in this chapter some plants (but by no means an exhaustive list) are presented: specifically, Elderberry (*Sambucus nigra*), Echinacea (*Echinacea purpurea*), select whole grains, reishi (*Ganoderma lucidum*), astragalus (*Astragalus membranaceus*), select African plants and Astavarga plants from India. They represent plants and mushrooms with long folklore history and used by people in various regions of the globe including Europe, China, Africa, and India (Sultan et al. 2014; Wilasrusmee et al. 2002).

10.2 Europe and North America

10.2.1 Elderberry (*Sambucus nigra*) Effectiveness and Health Benefits

Elderberry has immune-boosting properties. Antioxidants and vitamins contribute to the immune-boosting properties and are found mainly in the berries and flowers (Engels and Brinckmann 2019; Viapiana and Wesolowski 2017). Research suggests elderberry prevents and eases cold and flu symptoms. Globally, about 30 types of elderberry plants have been identified, and *Sambucus nigra* is most identified with the health benefits.

Several human clinical trials support Elderberry's immune health benefits. From a meta-analysis study on elderberry by Hawkins et al. (2019), the herb reduced upper respiratory symptoms (Fig. 10.1).

10.2.2 Echinacea

Echinacea plants are mainly represented by three species: *E. purpurea* (L.) Moench, *E. pallida* (Nutt.), and *E. angustifolia* DC. They all possess immunostimulatory activity. Each plant part contains two recognized sets of biological constituents. One set of constituents contains lipid-like small molecules, and the second contains high-molecular-weight water-soluble polysaccharides. Using flow cytometry, the polysaccharides were substantially more potent as immunostimulants (Pillai et al. 2007). When human leukocytes, ex vivo, were exposed to echinacea extracts, stimulation and immune stimulation were observed (Dobrange et al. 2019).

E. purpurea in the Asteraceae (Compositae) family is the most commonly cultivated medicinal plant (McKeown 1999) and well-reviewed for its immune stimulatory benefits (Patel et al. 2008; Grimm and Muller 1999). Some of the small-molecule secondary metabolites obtained from this species include

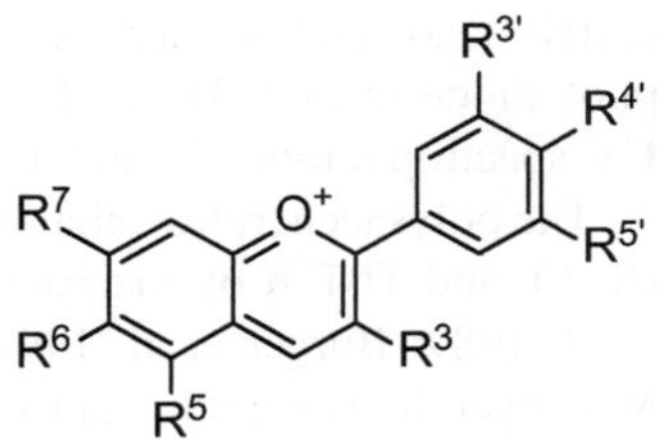

Fig. 10.1 General structure of blue pigments (anthocyanins) present in elderberries

$R^{3'} = H/OH/OCH_3$
$R^{4'} = OH$
$R^{5'} = H/OH/OCH_3$
$R^{3} = OH$
$R^{5} = OH/OCH_3$
$R^{6} = H/OH$
$R^{7} = OH/OCH_3$

alkylamides, chicoric acid, and caffeic acid derivatives. The alkylamides possess immunomodulatory properties both in vitro and in vivo (Goel et al. 2002; Gertsch et al. 2004). The large biologically active molecules are glycoproteins and polysaccharides (Coelho et al. 2020; Manayi et al. 2015), and the latter also has an imperative role in the anti-inflammatory activity of echinacea preparations (Laasonen et al. 2002). Taxonomic, pharmacological, clinical, and chemical features of few echinacea genus species including *E. pallida*, *E. purpurea*, and *E. angustifolia* were analyzed presently (Barrett 2003; Barnes et al. 2005; McKeown 1999).

E. purpurea, *E. pallida*, and *E. angustifolia* roots, stems, leaves, and flowering tops produce substantial immunostimulatory activity explained by three mechanisms: fibroblast stimulation, enhancement of respiratory activity, and phagocytosis activation (Geneva 1999). Several in vivo analyses related to anti-inflammatory and immunomodulatory potential of *E. purpurea* suggest that inborn immunity is improved by echinacea extract, whereby the immune system is strengthened against pathogenic infections. This effect may be due to the stimulation of the polymorphonuclear leukocytes (PMN), macrophages, neutrophils, and natural killer (NK) cells (Barnes et al. 2005).

The small molecules in the roots and herbs of *Echinacea* are mainly ketoalkenes, alkylamides (Fig. 10.2), and caffeic acid derivatives. Analysis of chicoric acid and alkylamides has been effectively established using high-performance liquid chromatography (HPLC) united with diverse detectors such as coulometric electrochemical, electrospray ionization mass spectrometric, and UV spectrophotometric detectors (Spelman et al. 2009; He et al. 1998). Furthermore, caffeic acid derivatives have been analyzed with the help of capillary electrophoresis (CE) or reverse-phase HPLC coupled with photodiode array (FDA) and UV spectrophotometric detector (Cech et al. 2006; Mancek and Kreft 2005; Brown et al. 2011). The composition of phenolic acids in the *E. purpurea* extract was mainly determined by micellar benzoic acid electrokinetic chromatography (MEKC) based on the use of sodium deoxycholate (SDC) (Pomponio et al. 2002; Cech et al. 2006). Reverse-phase HPLC with gradient elution has been used for simultaneous analysis of alkylamides, caffeic acid, and its derivatives from *E. purpurea* using different detectors like photodiode array (PDA), electrospray ionization mass spectrometry (EIMS), and UV spectrophotometric and detector (Gotti et al. 2002; Cech et al. 2006).

The polysaccharides enhance the production of interleukin-1 (IL-1), interleukin-6 (IL-6), and TNF-α by macrophages, both in vitro and in vivo in the body (Luettig et al. 1989; Burger et al. 1997; Wagner et al. 1988; Proksch and Wagner 1987). Moreover, in an experiment using polysaccharide application in lesser immune mice by cyclophosphamide or cyclosporin A, the mice became resistant toward *L. monocytogenes* and *C. albicans* (these pathogens generate macrophage and granulocyte-based infections, respectively) and enhanced in vitro cytotoxicity against WEHI 164 tumor cells in macrophages along with growing the number of neutrophil granulocytes and leukocytes in lesser immune mice was reported (Steinmuller et al. 1993). *E. purpurea* root extract is also known to increase the

Fig. 10.2 Key phytochemicals obtained from *E. purpurea*

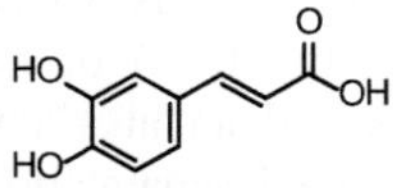

Caffeic acid

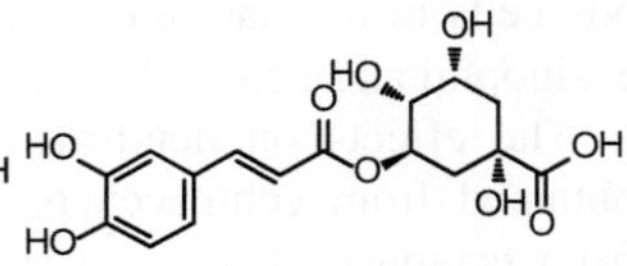

Chlorogenic acid

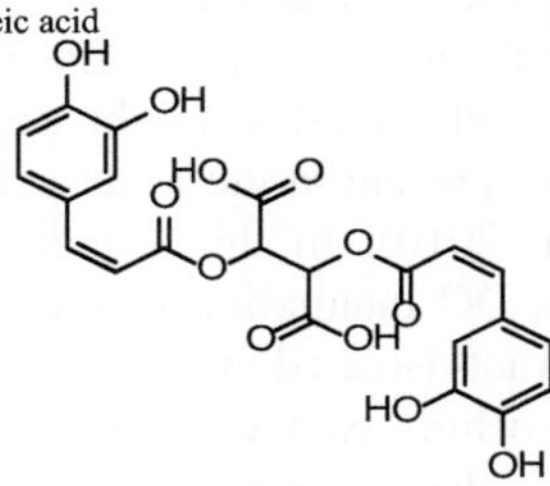

Cicoric acid

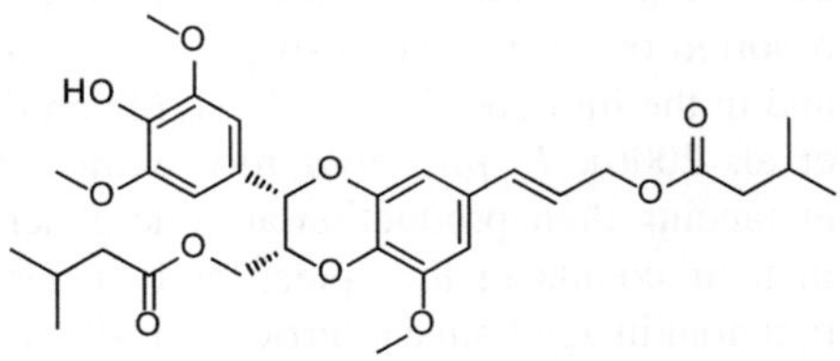

Nitidanin diisovalerianate

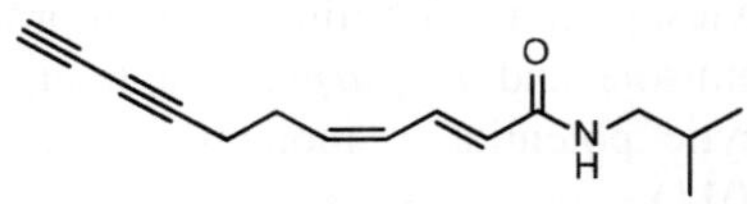

Undeca -2*E*,4*Z*-dien -8,10 -diynoic acid isobutylamide

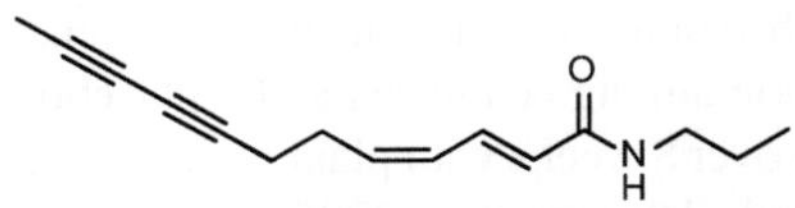

Dodeca -2*E*,4*Z*-dien -8,10 -diynoic acid isobutylamide

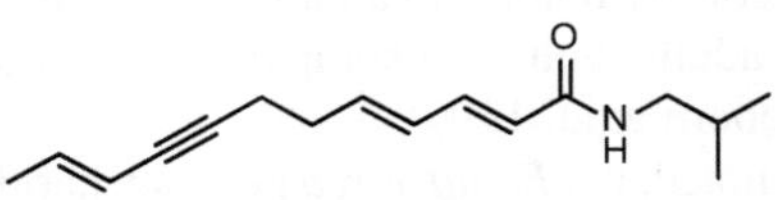

Dodeca -2*E*,4*E*,10*E*-trien -8-ynoic acid isobutylamide

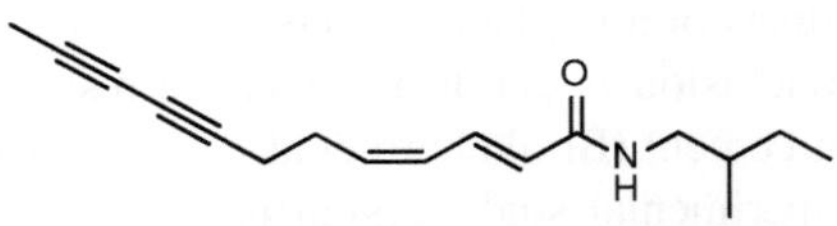

Dodeca -2*E*,4*Z*-dien -8,10 -diynoic acid 2 -methylbutylamide

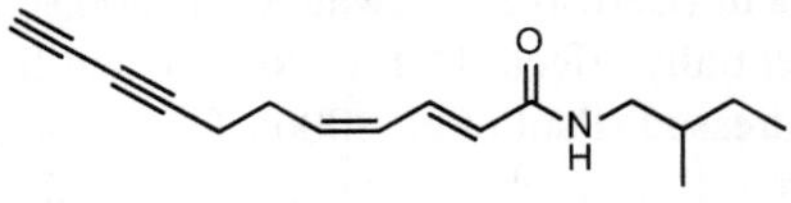

NK cells in normal, aging, and leukemic mice. It was also known to increase the eosinophil and neutrophil count in rabbits (Dussault and Miller 1993).

The effects on dendritic cells (DCs) of a polysaccharide-rich water extract obtained from echinacea root and an alkylamide-rich ethanolic extract isolated from echinacea leaf were assessed separately. The study revealed that the DCs played a significant role in inborn as well as adaptive immunity. The results indicated that alkylamides might suppress the working of DCs. Also, the polysaccharides present in the water extract might enhance the functioning of DCs (Benson et al. 2010). In this experiment, the expression of the CD83 marker (recognized as DC maturation marker) was enhanced meaningfully by both root and flower extracts isolated from echinacea plant. Conversely, leaf and stem extracts obtained from echinacea plant remarkably decreased CD83 levels.

The arabinogalactan proteins isolated from *E. purpurea* as a suspension culture did not affect the lymphocytes proliferation in mouse. There was low activity reported on nitrite and IL-6 production in macrophage cultures of alveolar mouse and in the immunoglobulin M (IgM) production in lymphocytes of mouse (Classen et al. 2006). *E. purpurea* root extract noted to increase the NK cell count by enhancing their production at bone generation site. This effect causes the upsurge in their counts in the spleen and in turn enhances their cytolytic and antitumor potential in aged mice (Currier and Miller 2000). Treatment of rats with *E. purpurea* extracts resulted in a marked rise in the white blood cells (WBC), neutrophils, monocytes, gamma globulin, and total protein. In an experiment with a combination of levamisole and *E. purpurea* extract, the synergistic effects were reported in phagocytic potential at monocyte and gamma globulin levels (Sadigh-Eteghad et al. 2011).

Results of human clinical trials remain ambiguous. In a randomized blinded trial, no substantial alteration in the prevalence and sternness of colds and respiratory infection among echinacea and a placebo group was seen. There were 108 patients dosed over 8 weeks with plant juice. A small decline in the lymphocyte numbers was reported (Schwarz et al. 2005). However, when the treatment was given in advance of the cold symptoms in another human trial, *E. purpurea* juice had a useful effect on tested adults with cold symptoms in comparison to placebo (Woelkart et al. 2008; Brinkeborn et al. 1999).

As noted, the *E. purpurea* polysaccharides enhanced the production of IL-1, IL-6, and TNF-α by macrophages (Roesler et al. 1991). However, the advantageous effects of any plant extracts for patients with cold symptoms remain arguable. This conclusion is due to the fact the results of diverse analysis are not completely in agreement (Brinkeborn et al. 1999; Woelkart et al. 2008; Schwarz et al. 2005). The experimental study design has been questioned with regard to the common cold treatment (Caruso and Gwaltney Jr 2005). In particular, questions on how the plant differentially affects TNF-α, NF-κB, and IL-6 in the different functional cells need to be addressed (Paul et al. 2006). Lastly, standardized doses of bioactive extracts are required. Polysaccharide versus alkylamide or a mixture need further study.

10.3 Whole Grains

Whole grains belong to Gramineae and contribute to providing a balanced immune system in humans. Plant species include but are not limited to members of *Triticum* and *Aegilops* species, and much attention is given to their fiber content. Improving immune health with grains may be due to a healthy human digestive system (the gut microbiome) and reduction of inflammation in the GI tract. A healthy gut can be maintained with the fiber from whole grains.

Whole grains are not only full of beneficial vitamins and fiber; they are also known to have benzoxazinoids (BX), (Fig. 10.3). In a human trial, the study participants received foods rich in BX. From subsequent blood marker tests, an increased immune system reaction was observed (Adhikari et al. 2015). The two groups of participants were given two different diets. Both groups ate diet rich in whole grains and fiber, containing different amounts of BX. After 3 weeks the high BX group responded better to suppressing pathogenic bacteria. Although preliminary this study may indicate a possible mechanism for BX, thus showing that whole grains may benefit in boosting the immune system.

Furthermore, the grains contain the fructans, an important source in our daily diet. Fructans are carbohydrates (in the form of a complex mixture of mostly branched structural sugars): mainly or exclusively of fructose. The yeast in bread baking causes considerable fructan degradation. Data suggest that cereal grain fructans are not only dietary fibers but may have prebiotic effects and impact colon health (Verspreet et al. 2015). The replacement of refined grains with whole grains in a

Benzoxazolinones | **Lactams** | **Hydroxamic acids**

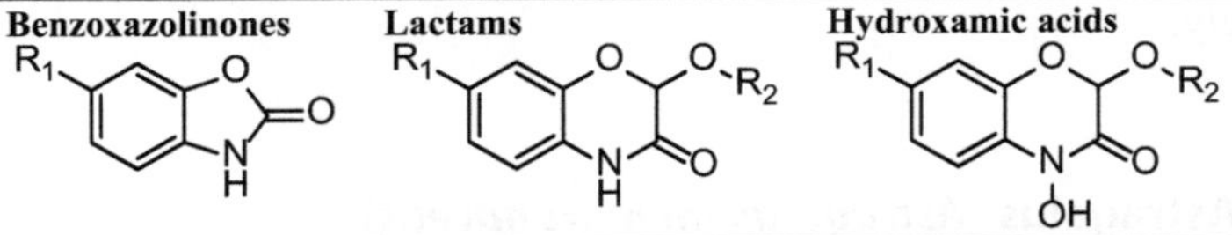

R1		R1	R2		R1	R2	
H	BOA	H	H	HBOA	H	H	DIBOA
OCH_3	MBOA	H	Glc	HBOA-Glc	H	Glc	DIBOA-Glc
		OCH_3	H	HMBOA	OCH_3	H	DIMBOA
		OCH_3	Glc	HMBOA-Glc	OCH_3	Glc	DIMBOA-Glc
		H	Glc-Hex[3]	HBOA-Glc-Hex	H	Glc-Hex[3]	DIBOA-Glc-Hex

Fig. 10.3 Most commonly found BXs in cereal. *MBOA* 6-methoxy-benzoxazolin-2-one, *BOA* benzoxazolin-2-one, *DIBOA* 2,4-dihydroxy-1,4-benzoxazin-3-one, *DIMBOA-Glc* 2-β-D-glucopyranosyloxy-4-hydroxy-7-methoxy-1,4-benzoxazin-3-one, *DIBOA-Glc-Hex* double-hexose derivative of DIBOA, *DIBOA-Glc* 2-β-D-glucopyranosyloxy-4-hydroxy-1,4-benzoxazin-3-one, *DIMBOA* 2,4-dihydroxy-7-methoxy-1,4-benzoxazin-3-one, *HBOA* 2-hydroxy-1,4-benzoxazin-3-one, *HMBOA* 2-hydroxy-7-methoxy-1,4-benzoxazin-3-one, *HBOA-Glc* 2-β-D-glucopyranosyloxy-1,4-benzoxazin-3-one, *HBOA-Glc-Hex* double-hexose derivative of HBOA, *HMBOA-Glc* 2-β-D-glucopyranosyloxy-7-methoxy-1,4-benzoxazin-3-one

6-week randomized trial showed a modest result on the gut microbiota, inflammatory and immune markers of healthy adults (Vanegas et al. 2017).

10.4 Traditional Chinese Medicine

10.4.1 Reishi

Ganoderma lucidum (W. Curt.: Fr.) Karst. usually known as "reishi" is a wood rotting fungus commonly grow on the stumps and trees (Hobbs 1995). It is found all around the world and has been extensively utilized in traditional Chinese medicine (TCM) from ancient times as herbal tonic that helps to promotes longevity. *G. lucidum* is mainly distributed in Asia and popularly known as "reishi," "Young Zhi," and "Ling Zhi," in Japan, Korea, and China, respectively.

Chinese pharmacopoeia of sixteenth century, *Pen T'sao Kang Mu*, describes Ling Zhi "as being useful for preventing forgetfulness, enhancing vital energy, producing longevity and increasing intellectual capacity" (Jones 1996). Today TCM doctors prescribe this mushroom for the cure of neurasthenia, hypercholesterolemia, dizziness, anorexia, debility from prolonged illness, fatigue, coronary heart disease, chronic hepatitis, insomnia, altitude sickness, carcinoma, hypertension, and bronchial cough (McKenna 1998; Chang and But 1986).

G. lucidum shows immune modulating activity (Wang et al. 1997; Ma et al. 2002). A hydroalcoholic extract of *G. lucidum* induces the modulation of cytokine secretion (Welker et al. 1996) including IFN-ç, IL-4, and IL-2 from normal human peripheral blood mononuclear leukocytes with IC50 values of 3.7, 1.8, and 6.2 íM, respectively.

10.4.2 Astragalus (*Astragalus membranaceus*)

Astragalus species contain antioxidants, and the plant has immune-boosting properties preventing colds and upper respiratory infections. (Zou et al. 2013). As a traditional Chinese medicine, Astragalus has been shown to lower blood pressure, treat diabetes, and offer liver protection. The astragalus roots are often used in combination with other herbs and promoted as a dietary supplement to relieve upper respiratory infections, chronic fatigue syndrome, chronic kidney disease, allergic rhinitis (hay fever), and asthma, among others.

Chemical analysis revealed that astragaloside II (Fig. 10.4) is a major active component of *A. membranaceus* (the Asian member of the Papilionaceous family) (Wan et al. 2013). Astragaloside II increases CD45 phosphatase enzyme activity, thereby activating certain white blood cells which support the body's immune functioning (Th1 cell activation). In addition, there is inhibition of pro-inflammatory cytokines, providing regulation and balance between effector cells and regulatory T cells. Several studies support the immunoregulatory effects

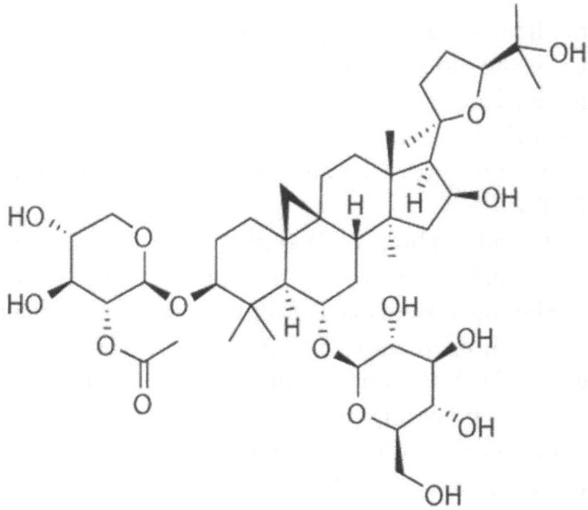

Fig. 10.4 Chemical structure of astragaloside II

of *A. membranaceus* (Astragalus 2019) and the body's natural defense mechanisms (Qi et al. 2017).

10.5 African Plants

The African continent has an extensive use of folklore medicine. One example is the use of therapeutic plants by traditional medicine practitioners to improve the immune functioning in people infected with HIV/AIDS. Most traditional medicine practitioners use herbs for enhancing immunity, whether or not the patients were on an antiretroviral treatment (Anywar et al. 2020; Pan et al. 2019).

In one study 71 therapeutic plant species from 64 genera and 37 families were identified. Trees contributed to 38% of the species used and herbs 35%. Many herbal medicines were prepared from roots (20%), bark (24%), and leaves (35%). *Zanthoxylum chalybeum* Engl. and *Psidium guajava* L. were the usually used species, followed by *Warburgia ugandensis* Sprague, *Markhamia lutea* (Benth.) K. Schum., *Bridelia micrantha* (Hochst.) Baill, *Aloe vera* (L.) Burm. f., *Acacia hockii* De Wild, *Mangifera indica* L., and *Erythrina abyssinica* DC.

10.6 India: Astavarga Plants—Importance in Traditional Systems of Medicine in Indian Himalaya

Materia Medica developed by Babylonians, Chinese, Egyptian, Greeks, Indians, Romans, and many others played an imperative role in the growth of present-day medical and healthcare systems. Ayurveda and Rig Veda are assumed to be the source of the Indian traditional medicine system. Rig Veda was written during 1600–4500 B.C., whereas Ayurveda was compiled between 2500 and 600 B.C.

Charak and Sushruta Samhita are also well-known for the description of various medicinal plants (Bhagwan and Sharma 2001; Satyavati 2003). Ayurveda (5000 B. C.) is the oldest medical system on the earth, which means "science of life." There is no denying the potential advantages of Ayurvedic medications, which many Indians and others around the globe have encountered. The analytical and treatment methods utilized in Ayurveda are inimitable and still significant today (Venkatasubramanian 2007).

In India, about 25,000 plant-based formulations are utilized in different ancient medicinal systems. The number of plant species exploited in different traditional medicinal systems is as follows: 2000 in Ayurveda, 1300 in Siddha, 1000 in Unani, 800 in homeopathy, 500 in Tibetan, 200 in current, and 4500 in pristine (Pandey et al. 2013). Presently, more than 7500 plant species are used in the Indian medicinal system, including tonics, antipyretics, antimalarials, expectorants, aphrodisiacs, anticancer, antidiabetic, hepatoprotectants, diuretics, and antirheumatics, also for the cure of various heart, liver, and central nervous system disorders (Pan et al. 2014). About 46% herbal drugs used in Unani, 80% in Ayurveda, and 33% in allopathic systems are found in the Western Himalayan region. Also, around 50% of drugs described in the British Pharmacopoeia are obtained from medicinal plants found in this region (Pan et al. 2014 and references therein).

Astavarga is a group of eight *Himalayan* medicinal plants, viz., *Crepidium acuminatum* (D. Don) Szlach (Jeevak), *Habenaria edgeworthii* Hook. f. ex Collett (Vriddhi), *Habenaria intermedia* D. Don (Riddhi), *Lilium polyphyllum* D. Don (Kshirkakoli), *Malaxis muscifera* (Lindl.) Kunt (Rishbhak), *Polygonatum cirrhifolium* (Wall.) Royle (Mahameda), *Polygonatum verticillatum* (Linn.) Allioni (Meda), and *Roscoea purpurea* Smith (Kakoli) (Dhyani et al. 2010; Balkrishna et al. 2012; Giri et al. 2017; Balkrishna et al. 2018). These plants belong to the family Liliaceae, Orchidaceae, and Zingiberaceae (Table 10.1). Astavarga plants are significant constituents of numerous Ayurvedic preparations such as Chyavanprasha (Table 10.2). Paryayaratnamala (dictionary dealing with botanical terms) was the first primeval manuscript to use the name Astavarga and made available its herbal portrayal (Joshi 1983; Dhyani et al. 2010). These plants are well-known for their health-supporting/vitality-strengthening properties. Astavarga plants are well-known for enhancing cell rejuvenation potential, rejuvenating health-promoting potential, immunity-elevating potential, and strengthening vibrant force of the body (Mathur 2003; Pandey 2005; Singh and Duggal 2009; Dhyani et al. 2010; Balkrishna et al. 2012; Balkrishna et al. 2018). Further, these plants are useful in seminal weakness, promoting body fat, diabetic conditions, abnormal thirst, healing fractures, and fever and as a cure for various illnesses described in Ayurveda such as *pitta*, *vata*, and *rakta doshas*. These plants are also known to heal the body immediately after illness and work as a source of natural antioxidants in the body (Balkrishna et al. 2018; Giri et al. 2017; Balkrishna et al. 2012; Dhyani et al. 2010; Sharma and Balkrishna 2005; Pandey 2005; Mathur 2003).

Owing to their high medicinal value, Astavarga plants are utilized in a variety of forms such as powder (*Churana*), oil (*Taila*), medicated clarified butter (*ghritam*), formulations, and decoctions in different traditional medicinal system across India,

Table 10.1 Taxonomic/botanical description of Astavarga plants (Dhyani et al. 2010; Balkrishna et al. 2012; Kant et al. 2012; Rawat et al. 2014; Giri et al. 2017; Balkrishna et al. 2018)

Botanical name	Decision	Class	Order	Family	Genus	Species	Author name
Crepidium acuminatum (D. Don) Szlach.	Tracheophyta	Liliopsida	Asparagales	Orchidaceae	*Crepidium*	*acuminatum*	(D. Don) Szlach.
Habenaria edgeworthii Hook. f. ex Collett	Tracheophyta	Liliopsida	Asparagales	Orchidaceae	*Habenaria*	*edgeworthii*	Hook. f. ex Collett
Habenaria intermedia D. Don	Tracheophyta	Liliopsida	Asparagales	Orchidaceae	*Habenaria*	*intermedia*	D. Don
Lilium polyphyllum D. Don	Tracheophyta	Liliopsida	Liliales	Liliaceae	*Lilium*	*polyphyllum*	D. Don
Malaxis muscifera (Lindl.) Kunt	Tracheophyta	Liliopsida	Asparagales	Orchidaceae	*Malaxis*	*muscifera*	(Lindl.) Kunt
Polygonatum cirrhifolium (Wall.) Royle	Tracheophyta	Liliopsida	Liliales	Liliaceae	*Polygonatum*	*cirrhifolium*	(Wall.) Royle
Polygonatum verticillatum (Linn.) Allioni	Tracheophyta	Liliopsida	Liliales	Liliaceae	*Polygonatum*	*verticillatum*	(Linn.) Allioni
Roscoea purpurea Smith	Tracheophyta	Liliopsida	Zingiberales	Zingiberaceae	*Roscoea*	*purpurea*	Smith

Table 10.2 Ayurveda formulations prepared from Astavarga plants and disorders treated (Pandey 2005; Dhyani et al. 2010)

Formulations	Uses
Amritaprasa ghritam	Aphrodisiac, cough, burning sensation, dehydration, hiccough, hemorrhagic, fever, vaginal problems, urinary diseases
Ashok ghritam	Breathing problems, indigestion, strengthening, vaginal pain
Balarishta	Increase physical strength and digestion
Balatailam	Sexual debility and physical weakness
Bhallatkam ghritam prathamam	Urinary problems, memory, weakness due to hard work, heart problems, pain, virility
Chaglangham ghritam	Cough, power, anorexia, asthma
Chyavanprasha	Antiaging, sexual satisfaction, heart, urinary, memory
Dahsmooltailam swalpam dwitiyam	Headache, pain
Ekadashsateekprasarni tailam	Infertility in older women, youthness, flesh increase
Kumarkalpdrum ghritam	Aphrodisiac, infertility in women
Kamdev ghrit	Physical strength, increase sperm count, temperature
Kumkumadi ghrita	Cough, hemorrhagic disease, urinary disorders, breathing problems
Maashtailam dwitiyam	Poor hearing, earache
Mahachandanadi thailam	Physical weakness, arthritis
Mahamasha thailam	Headache, earache
Mahapaishachika ghritham	Improve growth of young children, memory, insanity, fever
Nakulangh ghritam	Constipation, epilepsy, tremor
Narayan tailam	Regain youthfulness, fever, toothache, constipation
Piplayangh ghritam	Vomiting and diarrhea
Saarivadhleh	Power, digestion
Saptsateekprasarni tailam	Physical weakness, virility
Shatavari ghrita	Seminal weakness, virility, hemorrhagic disease, gout
Shiva ghritam	Power, urinary disorders, heart problems, fever, infertility in women
Srigopal tailam	Memory, cough, impotence
Sudarshan Churna	Fever, jaundice
Sudhakar tailam	All vaginal problems
Trisatiprasaarni tailam	Antiaging, indigestion, arthritis
Vishnu tailam	Tremor
Vrhchyamadi ghritam	Heart problems, sperm loss
Yamnayadi churnam	Impotence

China, Pakistan, Nepal, and Bhutan (Dhyani et al. 2010; Balkrishna et al. 2018). More significantly, formulations like Chyavanprasha and Sudarshan Churna prepared from Astavarga plants are accessible in Indian markets as pharmaceutical products. These products are available in nearly every primary healthcare center in rural areas across India (Dhyani et al. 2010). Across India, tribal people and people residing in remote areas also use Astavarga plants in the primary healthcare system to improve the immune system, to treat breathing and urinary problems, and to get rid of various body pains and for physical and sexual debility. Further, Astavarga plants are known to have variety of therapeutic properties such as aphrodisiac, refrigerant, hemostatic, styptic, antidiarrheal, febrifuge, antidysentric, galactagogue, mild laxative, depurative, expectorant, and emolient (Table 10.3) (Sharma and Balkrishna 2005; Dhyani et al. 2010; Giri et al. 2017; Balkrishna et al. 2018). Most Astavarga plants are either rare, threatened, vulnerable, or critically endangered, and their availability is also limited, owing to these different substitutes of these plants, which are presently employed in the traditional medicinal system (Table 10.2) (Rajashekhar et al. 2015; Balkrishna et al. 2018).

These Astavarga plants are also well-known for their pharmacological potential such as antioxidant, antimutagenic, immunomodulatory, antimalarial, antibacterial, antipyretic, anticonvulsant, antinociceptive, anti-inflammatory, and cytotoxic (Table 10.4) (Dhyani et al. 2010; Kant et al. 2012; Rawat et al. 2014; Sharma et al. 2014; Rawat et al. 2016; Giri et al. 2017; Balkrishna et al. 2018). Astavarga plants also act as a rich source of vitamins, minerals, and a variety of bioactive secondary metabolites such as alkaloids, saponins, flavonoids, tannins, phenolic acids, steroids, terpenoids, glycosides, and polysaccharides (Table 10.3) (Dhyani et al. 2010; Balkrishna et al. 2012; Kant et al. 2012; Rawat et al. 2014; Sharma et al. 2014; Rawat et al. 2016; Giri et al. 2017; Balkrishna et al. 2018). Usually, vitality-strengthening, rejuvenating health-promoting, immunity-enhancing, and antiaging attributes are mainly controlled by a variety of natural antioxidants (polyphenolics, flavonoids, isoflavones, flavones, anthocyanin, tannins, saponin, steroidal saponins, etc.) present in these plants. These natural antioxidants are known to neutralized/ scavenge variety of FRs (free radicals), ROS (reactive oxygen species), and RNS (reactive nitrogen species) (Dimitrios 2006; Giri et al. 2017). Overproduction of RNS, ROS, and FRs cause severe injury to different kinds of biomolecules, which leads to defective metabolic pathways inside the living body. These conditions have been connected to the origin of more than a hundred kinds of deteriorating diseases such as atherosclerosis, neural disorders, cancer, asthma, hypertension, acute respiratory distress syndrome, diabetes, Alzheimer's disease, cardiovascular disease, etc. (Sharma and Cannoo 2016; Sharma and Cannoo 2016a; Sharma and Cannoo 2017; Bhardwaj et al. 2019). Although, the Astavarga plants are the substantial source of health-promoting, vitality-strengthening, rejuvenating, and immunity-elevating potential, but the systematic research with respect to these properties are still poorly known. Further, these plants are also a subject of extensive biochemical examination, but inadequate literature is recognized about the phytochemical composition of these plants. With respect to immense medicinal significance of Astavarga plants, the phytochemical and bioactivity investigations have not yet been investigated appropriately, and it is undeniable to precisely validate the ancient claims.

Table 10.3 Properties, action, part used, chemical constituents, and dosages of Astavarga plants (Balkrishna et al. 2018; Giri et al. 2017; Rawat et al. 2016; Sharma et al. 2014; Rawat et al. 2014; Kant et al. 2012; Balkrishna et al. 2012; Dhyani et al. 2010)

Botanical name	Properties and action	Part used	Chemical constituents	Minerals	Dosages
Crepidium acuminatum (D. Don) Szlach.	Pseudobulbs are sweet, febrifuge, aphrodisiac, refrigerant, and tonic	Pseudobulb	Alkaloid, flavonoid, phenol, tannin glycoside and β-sitosterol. Piperitone, eugenol,1,8-cineole, citronellal, limonene, p-cymene, ceryl alcohol. Choline, O-methylbatatasin, glucose and rhamnose (Adams et al. 2017)	Calcium (Ca), cadmium (Cd), iron (Fe), chromium (Cr), copper (Cu), potassium (K), magnesium (Mg), zinc (Zn), manganese (Mn), sodium (Na), selenium (Se), phosphorus (P), lead (Pb)	As directed by the physician
Habenaria edgeworthii Hook. f. ex Collett	Cooling, emolient, blood purifier, appetizer and brain tonic Tubers are useful in excessive thirst, burning sensation, fever, asthma, cough, leprosy, skin diseases, worms, emaciation, anorexia, general debility, and gout	Tubers and leaves	Alkaloids, tannins, phenolic, flavonoids, fiber, fat, and starch Rutin, 3-hydroxy cinnamic acid, caffeic acid, chlorogenic acid, p-coumaric acid, ellagic acid, gallic acid hydroxybenzoic acid, thiamine, riboflavin	Copper (Cu), magnesium (Mg), zinc (Zn), iron (Fe), cobalt (Co), sodium (Na), calcium (Ca), potassium (K), and lithium (Li)	2–3 gm powder or as advise by the physician
Habenaria intermedia D. Don	Sweet, cooling, emolient, depurative, anthelmintic, aphrodisiac, appetizer, and brain tonic. (Habbu et al. 2012)	Tubers and leaves	Alkaloids, tannins, phenolic, flavonoids, fiber, fat and starch Rutin, caffeic acid, chlorogenic acid, 4-hydroxybenzoic acid, gallic acid, vanillic acid, 3-hydroxybenzoic acid, 3-hydroxy cinnamic acid, thiamine, riboflavin, scopoletin, and taxol	Copper (Cu), zinc (Zn), iron (Fe), magnesium (Mg), cobalt (Co), potassium (K), sodium (Na), calcium (Ca), and lithium (Li)	2–3 gm powder or as directed by the physician

Lilium polyphyllum D. Don	The bulbs of the plant are sweet, bitter, galactagogue, refrigerant, expectorant, diuretic, antipyretic, aphrodisiac and tonic. Bulbs primarily show soothing, anti-inflammatory, and astringent properties	Bulbs	Phenol, flavonoids, tannins Rutin, phloridzin, 3-hydroxybenzoic acid, caffeic acid, vanillic acid, 3-hydroxy cinnamic acid ferulic acid, chlorogenic acid, gallic acid, linalool, α-terpineol, and steroidal glyceride (Javed et al. 2012)	Calcium (Ca), potassium (K), copper (Cu), iron (Fe), sodium (Na), manganese (Mn), magnesium (Mg), phosphorus (P)	3–6 gm powder daily or as advised by the physician
Malaxis muscifera (Lindl.) Kunt	Pseudobulbs are sweet, aphrodisiac, refrigerant, hemostatic, styptic, antidiarrheal, febrifuge, antidysentric, and tonic	Pseudobulbs	Alkaloid, flavonoid, phenol, tannin, and glycoside Catechin, rutin, gallic acid, chlorogenic acid, ellagic acid, vanillic acid, 3-hydroxy cinnamic acid, ferulic acid, 3-hydroxybenzoic acid	–	As directed by the physician
Polygonatum cirrhifolium (Wall.) Royle	Cooling, galactagogue, mild laxative, depurative, febrifuge, aphrodisiac, expectorant, wound healer, and tonic	Rhizomes	Digitalis glucoside, steroidal saponin, phenol, flavonoids, tannins, steroid, terpenoid, polysaccharides, and stannin Steroidal saponins sibiricoside A and B, p-coumaric acid, 3-hydroxy cinnamic acid, caffeic acid, ellagic acid, β-sitosterol, glucose, sucrose α-L rhamnopyranosyl, β-D glucopyranoside, 6-stearic acid, and 6-nonadecenoic acid	Calcium (Ca), potassium (K), iron (Fe), copper (Cu), magnesium (Mg), phosphorus (P), sodium (Na), manganese (Mn)	As directed by the physician
Polygonatum verticillatum	Rhizomes are aphrodisiac, galactagogue, emollient,	Rhizomes	Steroidal saponins, phenol, flavonoids, tannins, lectins, and	Copper (Cu), zinc (Zn), iron (Fe), magnesium (Mg), cobalt	2–3 gm powder daily or as

(continued)

Table 10.3 (continued)

Botanical name	Properties and action	Part used	Chemical constituents	Minerals	Dosages
(Linn.) Allioni	diuretic, appetizer, and tonic Plant has burning sensation		polysaccharides. Caffeic acid, 3-hydroxy cinnamic acid, p-coumaric acid, 4-hydroxybenzoic acid, quinine 3-hydroxybenzoic acid Rhizome contains β-sitosterol, diosgenin, lysine, aspartic acid, threonine, serine, sucrose, and glucose, while leaves contain glucofractone, glucomanone, and hemicellulose (Virk et al. 2016; Saboon et al. 2016)	(Co), potassium (K), sodium (Na), calcium (Ca), lithium (Li) Antimony (Sb) and manganese (Mn)	directed by the physician
Roscoea purpurea Smith	Antirheumatic, galactagogue, expectorant, febrifuge, diuretic, hemostatic, sexual stimulant, and tonic	Rhizome	Alkaloid, flavonoids, phenolic acids, tannins, saponin, glycosides, thiamine, riboflavin, fiber and fat Catechin, kaempferol, rutin, ellagic acid, 3-hydroxybenzoic acid, caffeic acid, chlorogenic acid, gallic acid, protocatechuic acid, vanillic acid, 3-hydroxy cinnamic acid, ferulic acid, p-coumaric acid, syringic acid (Misra et al. 2015)	Copper (Cu), zinc (Zn), iron (Fe), magnesium (Mg), cobalt (Co), potassium (K), sodium (Na), calcium (Ca) and lithium (Li)	As directed by the physician

Table 10.4 Therapeutic and pharmacological uses of Astavarga plants in North-West Himalaya (Balkrishna et al. 2018; Giri et al. 2017; Rawat et al. 2016; Sharma et al. 2014; Rawat et al. 2014; Kant et al. 2012; Balkrishna et al. 2012; Dhyani et al. 2010)

Botanical name	Pharmacological potential	Therapeutic uses	Mode of use
Crepidium acuminatum (D. Don) Szlach.	Antioxidant, antimutagenic, antifungal, antibacterial, antimicrobial, anti-inflammatory (Gupta et al. 2015; Arora et al. 2017)	Debility Seminal weakness	Decoction prepared from pseudobulbs are taken Pseudobulbs powder is consumed
Habenaria edgeworthii Hook. f. ex Collett	Antioxidant, antimutagenic, immunomodulatory	Galactagogue	Tuber powder mixed with swarna bhasma (from gold calcinations) and Astavarga
Habenaria intermedia D. Don	Antioxidant, antimicrobial, antifungal, antibacterial, antimutagenic, antianxiety, immunomodulatory	Energy promoter, promote longevity, vitalizer tonic	Boiled as vegetable In morning, 1 g powder along with Astavarga plant consumed with milk
Lilium polyphyllum D. Don	Antioxidant, antimutagenic, immunomodulatory, antimalarial, antimicrobial, antipyretic, anticonvulsant, actinomorphic, antinociceptive, and cytotoxic	Promote strength and disease resistance Heat promoter, weakness Pain Impotence Virility problems Pain during menstrual cycle Kidney problems and giddiness	1 g powder mixed with *Desmodium gangeticum*, *Withania somnifera*, *Asparagus racemosus*, and *Nardostachys jatamansi* and consumed with cold milk Shade dried, cooked as vegetable along with potato Oil prepared with powder and *Asparagus racemosus* was applied daily after bath After night meal, 2 g powder along with Astavarga plants consumed with milk Powder with *Acorus calamus* and *Piper betle* consumed with milk 1 g *Smilax ovalifolia* along with 0.5 g powder boiled in water and consumed Used for preparing Maheshwari vati

(continued)

Table 10.4 (continued)

Botanical name	Pharmacological potential	Therapeutic uses	Mode of use
Malaxis muscifera (Lindl.) Kunt	Antioxidant, antimutagenic	Tonic	Dried powder consumed with boiled milk Powder of both *Polygonatum* species, *Desmodium gangeticum*, and *M. muscifera* consumed with milk
Polygonatum cirrhifolium (Wall.) Royle	Antioxidant, antimutagenic, antimalarial, antimicrobial, antipyretic, anticonvulsant, actinomorphic, antinociceptive, and cytotoxic	Suboleative, aphrodisiac, blood Purifier, cuts and wounds, energy tonic	Powder of both *Polygonatum* species, *Desmodium gangeticum* and *M. muscifera* consumed with milk Tuber infusion also taken with milk Leaves cooked with onion and potato used as energy tonic
Polygonatum verticillatum (Linn.) Allioni	Antioxidant, antimutagenic, antinociceptive, antispasmodic, antidiarrheal, antipyretic, tracheorelaxant, anti-inflammatory, antimicrobial (Virk et al. 2016; Saboon et al. 2016; Khan et al. 2010)	Tonic, promote body heat, increase sexual potency, gastric trouble and wound and cut healing (Khan et al. 2010; Saboon et al. 2016)	1 g powder consumed with milk Dried powder baked with ghee taken with milk Tuber powder also taken with water Tuber paste applied on wounds and cuts
Roscoea purpurea Smith	Antioxidant, antimutagenic, antidiabetic, antioxidant, antimicrobial, antifungal, antibacterial, immunomodulatory (Misra et al. 2015)	Debility, tonic in impotency, leucorrhea, diarrhea, malaria, dysentery	Root powder mixed with sugar and consumed with milk as tonic

10.6.1 Taxonomy and Habitat

Out of eight Astavarga plants, four [*Crepidium acuminatum* (D. Don) Szlach (Jeevak), *Habenaria edgeworthii* Hook. f. ex Collett (Vriddhi), *Habenaria intermedia* D. Don (Riddhi), *Malaxis muscifera* (Lindl.) Kunt (Rishbhak)] belong to family Orchidaceae, three [*Lilium polyphyllum* D. Don (Kshirkakoli), *Polygonatum cirrhifolium* (Wall.) Royle (Mahameda), *Polygonatum verticillatum*

(Linn.) Allioni (Meda)] to family Liliaceae, and one [*Roscoea purpurea* Smith (Kakoli)] to family Zingiberaceae. All the Astavarga plants have their usual habitats in the Himalayan region specifically in the North-West Himalaya (India, China, Nepal, Afghanistan, and Pakistan) between elevation of 800 and 4500 m. Their natural habitats are precise to their ecological necessities, and therefore these plants exist only in small patches across the all the Himalayas. In India, these plants particularly found in the North-West Himalayan states like Leh Ladakh, Himachal Pradesh, Uttarakhand, Jammu, and Kashmir between altitude of 800 and 4500 m (Tables 10.1 and 10.5) plants (Balkrishna et al. 2018; Giri et al. 2017; Rawat et al. 2014; Balkrishna et al. 2012; Kant et al. 2012; Dhyani et al. 2010;).

10.6.2 Phytochemistry

The Astavarga species are mainly rich in alkaloid, flavonoid, polyphenolics, tannin, glycoside, terpenoids, steroids, steroidal glyceride, steroidal saponin, fiber, fat, starch, and minerals (Table 10.3). β-Sitosterol, piperitone, O-methylbatatasin, eugenol,1,8-cineole, citronellal, limonene, and p-cymene were the major SMs found in *Crepidium acuminatum* (Adams et al. 2017). Rutin, caffeic acid, 3-hydroxy cinnamic acid, ellagic acid, chlorogenic acid, p-coumaric acid, gallic acid, and hydroxybenzoic acid were the key SMs found in *Habenaria edgeworthii* (Giri et al. 2017; Balkrishna et al. 2018). Rutin, catechin, caffeic acid, 3-hydroxy cinnamic acid, chlorogenic acid, gallic acid, 3-hydroxybenzoic acid, vanillic acid, 4-hydroxybenzoic acid, and scopoletin were the principal SMs present in *Habenaria intermedia* (Giri et al. 2017; Balkrishna et al. 2018). Rutin, phloridzin, caffeic acid, 3-hydroxy cinnamic acid, 3-hydroxybenzoic acid, vanillic acid, chlorogenic acid, gallic acid, ferulic acid, linalool, α-terpineol, and steroidal glyceride were the main SMs present in *Lilium polyphyllum* (Javed et al. 2012). Catechin, rutin, chlorogenic acid, gallic acid, 3-hydroxy cinnamic acid, vanillic acid, ellagic acid, ferulic acid, and 3-hydroxybenzoic acid were the major SMs present in *Malaxis muscifera*. Steroidal saponins sibiricoside A and B, caffeic acid, ellagic acid, 3-hydroxy cinnamic acid, p-coumaric acid, β-sitosterol, stearic acid, and 6-nonadecenoic acid were the chief SMs occur in *Polygonatum cirrhifolium* (Giri et al. 2017; Balkrishna et al. 2018). Caffeic acid, 3-hydroxy cinnamic acid, 4-hydroxybenzoic acid, p-coumaric acid, 3-hydroxybenzoic acid, quinine, diosgenin, santonin, and β-sitosterol were the principal SMs found in *Polygonatum verticillatum* (Giri et al. 2017; Balkrishna et al. 2018), while catechin, kaempferol, rutin, caffeic acid, 3-hydroxybenzoic acid, chlorogenic acid, gallic acid, protocatechuic acid, 3-hydroxy cinnamic acid, ellagic acid, p-coumaric acid, syringic acid, vanillic acid, and ferulic acid were the topmost SMs found in *Roscoea purpurea* (Balkrishna et al. 2018; Misra et al. 2015; Giri et al. 2017). Structures of all the key SMs found in Astavarga species are shown in Fig. 10.5.

Table 10.5 General description of Astavarga plants (Balkrishna et al. 2018; Giri et al. 2017; Rawat et al. 2014; Balkrishna et al. 2012; Kant et al. 2012; Dhyani et al. 2010)

Botanical name	Current status	Sanskrit/Hindi name	English/common name	Main characteristic features for identification	Habit	Habitat	Distribution range	Flowering	Fruiting
Crepidium acuminatum (D. Don) Szlach.	Rare	Jeevak	The gradually tapering malaxis	Raceme is pencillate and tip of lip edge is straigh*t*—linear	Pseudo bulbous, terrestrial, perennial, tender herb 5–25 cm height	China, Cambodia, and Southeast Asia In India, Himachal Pradesh, Assam, Uttarakhand, Nagaland, Mizoram, Arunachal Pradesh, Tripura Manipur, especially in subtropical and temperate Himalayas. Also found in Travancore, Andaman Islands, Madhya Pradesh and Anaimalai Hills	1200–2100 m	July–August	September–October

Habenaria edgeworthii Hook. f. ex Collett	Rare	Vriddhi, Himadrija, Dakshinavarta	Edgeworth's habenaria	Petals and ligules are yellow in color. Spur is truncated above	Terrestrial tuberous orchid 30–60 cm in height	Nepal, China, and Pakistan In India, Himachal Pradesh, and Uttarakhand especially in North-West Himalaya	800–2500 m	July–August	September–October
Habenaria intermedia D. Don	Critically endangered	Riddhi, Sidhi, Rathanga, Rishisrista, Yuga, Pranda, Sukh, Talgranthisamakand, Mangal, Vrisya, Vasu, Laksami, Vamavartal,	Intermediate habenaria, the in-between habenaria, white wild orchid, rain deer orchid	Flowers are long in size, white in color, lobes of petals fimbriated	Perennial terrestrial herb.25–50 cm in height	Pakistan, China, Nepal, and Bhutan up to an elevation of 2000–3300 m asl In India, Jammu and Kashmir, Uttarakhand, Sikkim, and Himachal Pradesh particularly in temperate Himalayas up to an altitude of 1500–2400 m	800–3300 m	July–August	September–October
Lilium polyphyllum D. Don	Critically endangered	Kshirakakoli, Kshir madhura, Ksirani, Kshirshukla, Payaswani, Sukla,	White Himalayan lily, many-leaved lily,	Flowers are long linear, dotted, purple colored	Perennial herb, 60–120 cm in height	Nepal, Pakistan, Afghanistan, Tibet, and	1500–4000 m	Mid-June to mid-July	July–September

(continued)

Table 10.5 (continued)

Botanical name	Current status	Sanskrit/Hindi name	English/ common name	Main characteristic features for identification	Habit	Habitat	Distribution range	Flowering	Fruiting
		Sundari, Salamkantha, Veera, Vayasya, Vayastha	and white lily	within, and white in color from outside		West China In India, Jammu and Kashmir, Uttarakhand, and Himachal Pradesh particularly in Western temperate Himalayas			
Malaxis muscifera (Lindl.) Kunt	Vulnerable	Rishbhak, Bandhur, Durdhar, Dheer, Gopati, Indraksa, Kakud, Matrik, Lashunkand, Nissar, Suksampatrak, Vrisha, Vrishabha, Vishani	Snake mouth orchid and Adder mouth orchid	Pseudobulb resembles with the bull's horn, leaves two, three lipped or centrally bifurcated	Terrestrial perennial herb 15–45 cm in height	Afghanistan, Pakistan, China, Bhutan, and Nepal In India, Uttarakhand, Himachal Pradesh, Jammu and Kashmir, and Sikkim especially in temperate Himalayas	1800–4100 m	July–August	September–October
	Rare	Mahameda, Basuchidra,	King's Solomon's	Leaves are three to six in		China, Bhutan,	2000–4000 m	July–August	September–October

Polygonatum cirrhifolium (Wall.) Royle		Nakhchechi, Dharaa, Manichhidra, Subrakand, Shalyaparni	seal, coiling leaf Solomon's seal, tendril leaf Solomon's seal, coiling leaf *Polygonatum*	number, arranged in whorls, tendril like tip. Ascending, occurs with the support of nearby shrubs	Perennial herb 30–120 cm in height	Nepal, and Pakistan In India, Himachal Pradesh, Uttarakhand, Sikkim, and Manipur especially in temperate Himalayas			
Polygonatum verticillatum (Linn.) Allioni	Threatened	Meda, Mitha Dudhia, Basuchidra, Devamani, Pandura, Salam Mishri, Shakakul, seal, Saat Asher	Whorled leaf Solomon's seal and whorled Solomon's seal	Leaves are four to eight in number, arranged in whorls, acute, pointed tips	Perennial herb. 0.3–1.2 m in height	Pakistan, Afghanistan, China, Tibet, North and Central Asia, Turkey and Europe In India, Jammu and Kashmir, Himachal Pradesh, Uttarakhand, and Sikkim especially in Temperate Himalayas	1600–4500 m	July–August	September–October
Roscoea purpurea Smith	Commonly available	Kakoli	Roscoe's lily, Roscoe's purple lily, purple	Stem is slightly purple in color, anthers are bifurcated	Perennial rhizomatous herb, 15–30 cm in height	Bhutan, Pakistan, China, and Tibet In India,	1500–3300 m	June–July	August–September

(continued)

Table 10.5 (continued)

Botanical name	Current status	Sanskrit/Hindi name	English/ common name	Main characteristic features for identification	Habit	Habitat	Distribution range	Flowering	Fruiting
			Roscoea, cinnamon stick, hardy ginger	at the tip, and flowers are purple		Uttarakhand, Sikkim, and Assam Mainly found in Eastern and Central Himalaya			

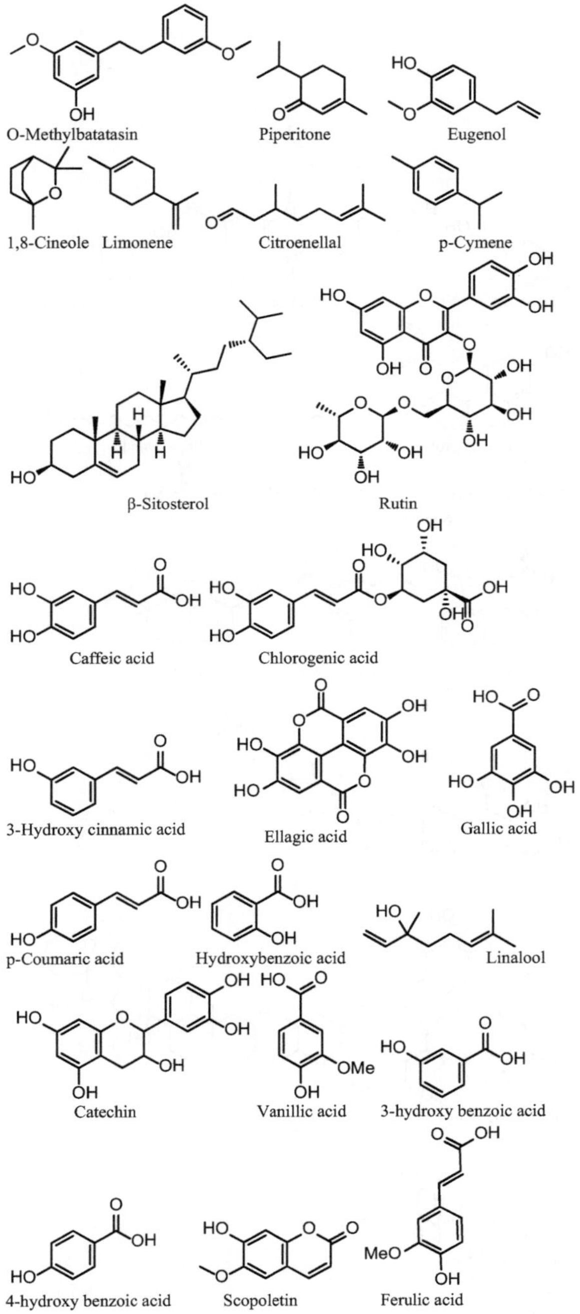

Fig. 10.5 Structures of key SMs found in Astavarga species

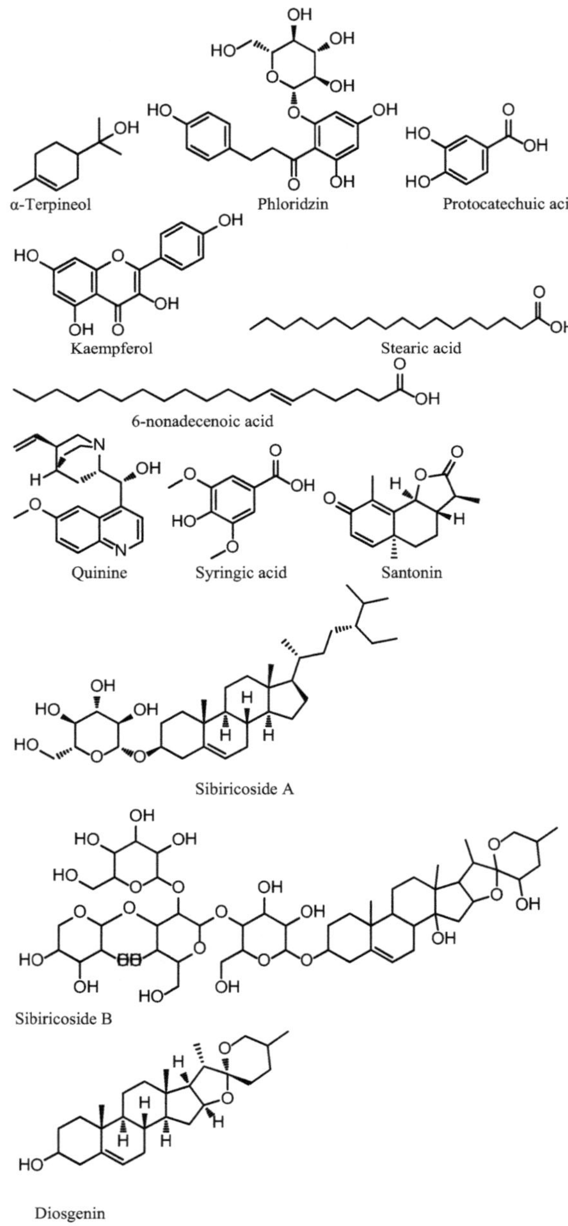

Fig. 10.5 (continued)

10.6.3 Biological Potential of Astavarga Plants: Immunomodulating and Antioxidant Potential

The therapeutic and pharmacological uses of Astavarga are presented in Table 10.4. In Astavarga plants there are immunomodulators compounds/agents that have the potential to modulate or normalize pathophysiological processes. These compounds/agents have biphasic effects with a resulting improvement of immune reaction, i.e., immunostimulating agents, which simply indicates the stimulation of nonspecific immune system such as certain T lymphocytes, macrophages, granulocytes, and different effector substances. In the case of immunosuppression, this simply indicates to the reduction in resistance against infections and stress, which may happen due to chemotherapeutic or environmental factors (Sahu et al. 2010; Sahu et al. 2013; Sahu et al. 2016). The mechanisms are explored below.

The immunomodulatory potential of *Habenaria Intermedia* ethanolic extract was evaluated by Sahu et al. (2013) using hematological parameters, delayed-type hypersensitivity response (DTH), and carbon clearance assay (phagocytic index). Swiss albino mice of both genders were used during the study. The results revealed that the mice treated with doses of 600 mg/kg (Group III) and 300 mg/kg (Group II) resulted in enhanced response in foot pad edema that was statistically significant ($p < 0.05$) as compared to control (Group I). In case of hematological parameters, an ethanolic extract also resulted in statistically significant ($p < 0.05$) increase in WBC and RBC count when compared to control and cyclophosphamide (20 mg/kg)-treated mice. Alike results were also noted in the case of phagocytic activity of ethanolic extract of *Habenaria intermedia* in tested mice. This substantial increase in immunomodulatory potential of *Habenaria intermedia* ethanolic extracts could be attributed to presence of phenolic, tannins, alkaloids, flavonoids, and steroids, compounds (Sahu et al. 2013).

The adaptogenic potential of *Habenaria intermedia* tubers ethanol and ethyl acetate fractions was evaluated by Habbu et al. (2012) using chronic stress, swimming-induced stress, and immobilization-induced acute stress in rats. The results revealed that the pre-treatment with ethyl acetate and ethanol (200 mg/kg) extracts significantly reduced the higher levels of total cholesterol, glucose, and triacylglyceride in chronic and acute stress. Further, pre-treatment with ethyl acetate and ethanol (200 mg/kg) extracts also significantly reduced the increased adrenal weight in acute stress. Furthermore, the survival time of swimming mice was improved and suggestive of a dose-dependent manner in the case of pre-treatment with ethyl acetate and ethanol (200 mg/kg) extracts in contrast to normal (nondrug-treated) mice. The anti-stress potential of *H. intermedia* could be ascribed to the existence of gallic acid and scopoletin or their synergistic properties that might be facilitated by an antioxidant mechanism.

The immunomodulatory potential of *Lilium polyphyllum* and *Roscea procera* ethanolic extract was evaluated by Sahu et al. (2016) on mice for a delayed-type hypersensitive reaction, hematological parameters, and lymphoid organ weighing. Among the different extracts tested, ethanol extract of *R. procera* (at dose of 200 mg per kg of body weight) was the most effective and produced a characteristic increase in delayed-type hypersensitive (DTH) reactivity (1.16 ± 0.03 mm, $P < 0.01$) in mice in comparison to control (0.53 ± 0.15 mm). The upsurge in DTH reaction in tested animal as result of T cell-based antigen showed the stimulatory consequence of ethanolic extract of *R. procera*. In the case of *L. polyphyllum*, hydroalcoholic extracts (at a dose of 200 mg per kg of body weight) showed the best results (0.795 ± 0.0 mm). In the hematological parameters, the ethanol extract of *R. procera* (at a dose of 200 mg per kg of body weight) exhibited characteristic enhancement in lymphocyte, platelet, and WBC count, whereas there is a decrease in mean corpuscular hemoglobin in mice in comparison to the control. Whereas treatment of the tested animal with ethyl acetate extract of *L. polyphyllum* increased the lymphocyte and WBC count, it caused significant decrease in mean corpuscular hemoglobin, platelet count, and mean corpuscular volume. In lymphoid organ weights tests, none of the *R. procera* and *L. polyphyllum* extracts exhibited mortality or toxicity in tested animals. No characteristic body weight gain alterations were noted in different groups of tested animals. The *R. procera* ethanol extract and *L. polyphyllum* ethyl acetate extract displayed characteristic variation of immune reactivity in the three tested animal models (Sahu et al. 2016).

The immunomodulatory potential of *R. procera* rhizomes ethanolic extract was evaluated by Sahu et al. (2010) using carbon clearance and delayed-type hypersensitivity (DTH) method in comparison to cyclophosphamide (standard drug) in mice. In the ethanolic extract group, the foot pat thickness ($P < 0.05$) significantly boosted the production of circulating antibody titer in response to sheep RBC, nonspecific immunity, and phagocytic functions of mononuclear macrophages. These investigations were also supported by hematological and serological tests data, and the *Roscoea procera* rhizome's ethanolic extract possesses immunostimulant potential. The characteristic enhancement in immunostimulatory potential of *Roscoea procera* rhizome's ethanolic extract could be ascribed to the presence of flavonoids, tannins, alkaloids, phenolic, glycosides, and saponin (Sahu et al. 2010).

Among the diverse extracts of (petroleum ether, chloroform, ethanol, and dichloromethane) rhizome of *P. verticillatum*, the ethanol extract exhibited the highest total phenolic content (0.126 ± 0.05 mg GAE (gallic acid equivalent)/g of dry weight), total flavonoid content (0.094 ± 0.004 mg RE (rutin equivalent/g of dry weight), and total tannin content (29.32 ± 1.25 mg CE (catechin equivalent)/g dw). The ethanol extract also showed highest antioxidant potential (DPPH, IC50 42.07 ± 1.12 μg/mL; ABTS, IC50 62.46 ± 0.51 μg/mL) among all the extracts, while the rest of the extracts exhibited much lower ABTS and DPPH radical scavenging potential than the reference compound (ascorbic acid) (ABTS, IC50 49.83 ± 0.63 μg/mL; DPPH, IC50 33.48 ± 0.58 μg/mL). In the case of anti-inflammatory potential, the tested extracts showed a dose-based retardation of protein denaturation. An ethanol extract resulted in the highest anti-inflammatory potential over other tested extracts at all the doses excluding the maximum dose (100 μg/mL), whereas petroleum ether extract produced the uppermost anti-inflammatory potential. The order of protein denaturation inhibition of different extracts at dose of 100 μg/mL was petroleum ether (78.1%), ethanol (67.0%), dichloromethane (52.78%), aqueous (41.37%), and chloroform (41.36%) extract, while the standard drug diclofenac sodium showed the protein denaturation inhibition of 79.1% at 100 μg/mL. The anticancer potential of different extracts of *P. verticillatum* rhizome was also evaluated at different doses (25–400 μg/mL) against the breast cancer cell line (MCF-7) using the MTT test. Chloroform, aqueous extracts, and dichloromethane showed notable anticancer potential in a dose-based manner. The order of cytotoxicity potential of different extracts was chloroform > aqueous > dichloromethane > pet ether > ethanol extract, respectively. Further, GC/MS analysis of different tested extracts revealed the existence of some constituents that are recognized to have anti-inflammatory, anticancer, and antioxidant potential. DPPH radical scavenging potential of *P. verticillatum* aerial parts was also evaluated by Khan et al. (2012). The result revealed that the crude extract (IC50, 122 μg/mL) had the highest antioxidant potential followed by ethyl acetate (IC50, 137 μg/mL), and n-butanol (IC50, 167 μg/mL) fractions. In case of rhizome extracts, chloroform (IC50, 90 μg/mL) extract showed highest DPPH radical scavenging potential followed by ethyl acetate (IC50, 93 μg/mL) and then n-butanol (IC50, 95 μg/mL) extract. Apart from these, *P. verticillatum* also is known to have antipyretic, anticonvulsant, antinociceptive, bronchodilator, anti-inflammatory, and antimalarial potential (Khan et al. 2011, 2012, 2013, 2013a).

The antioxidant and antimutagenic potentials of a methanol extract obtained from different Astavarga species were evaluated by Giri et al. (2017). The results of antioxidant potential measured by the ABTS method varied from 3.39 to 4.42 mM ascorbic acid equivalent (AAE)/100 g of dry weight. The supreme activity was reported in the case of the methanol extract of *Polygonatum cirrhifolium* (4.42 mM AAE/100 g of dry weight) followed by the methanol extract of *Malaxis muscifera* (4.18 mM AAE/100 g of dry weight), while the lowest activity was reported for the *Roscoea procera* extract (3.39 mM AAE/100 g of dry weight). The results of the DPPH assay ranged from 3.87 to 1.03 mM AAE/100 g of dry weight. *Polygonatum cirrhifolium* methanol extract showed maximum antioxidant activity (3.87 mM AAE/100 g of dry weight), while least activity was observed in the *Habenaria intermedia* extract (1.03 mM AAE/100 g of dry weight). In the FRAP assay *Lilium polyphyllum* extract (1.99 mM AAE/100 g of dry weight) presented highest antioxidant potential, while minimum antioxidant potential was observed in *Malaxis muscifera* extract (0.96 mM AAE/100 g of dry weight).

The antimutagenic potential of Astavarga species at different concentrations (4, 2.66 and 1.33 μg/mL) resulted in retrieval of damaged DNA from 35.58% (*Habenaria edgeworthii* extract, 1.33 μg/mL) to 82.88% (*Roscoea procera* extract, 4 μg/mL). The order of DNA recovery at 4 μg/mL concentration in the case of different Astavarga species extracts was *Roscoea procera* extract (82.88%) > *Lilium polyphyllum* extract (81.45%) > *Polygonatum verticillatum* extract (77.16%) > *Malaxis muscifera* extract (72.95%) > *Polygonatum cirrhifolium* extract (70.59%) *Habenaria edgeworthii* extract (54.24%) > *Habenaria intermedia* extract (44.01%) > *Malaxis acuminata* extract (41.70%). The antioxidant and antimutagenic potential of Astavarga species may be attributed to the phenolic constituents such as vanillic acid, p-hydroxybenzoic acid, chlorogenic acid, caffeic

Table 10.6 Substitutes for Astavarga plants in commerce and Ayurveda (Balkrishna et al. 2012; Balkrishna et al. 2018)

Botanical name	Substitute
Crepidium acuminatum (D. Don) Szlach.	*Pueraria tuberosa* (Wild.) DC (Vidarikand), *Malaxis cylindrostachya* (Lindl.), *Centaurea behen* Linn. (Safed Behman), *Tinospora cordifolia* (Willd.) (Guduchi), and *Malaxis mackinnoni* (Duthie) Ames
Habenaria edgeworthii Hook. f. ex Collett	*Sida acuta Burm* f. (Mahabala), *Dactylorhiza hatagirea* (D. Don) Soo (Salam panja), *Tacca integrifolia* Ker Gawl. (Varahikand), and *Habenaria griffithii* Hook.f.
Habenaria intermedia D. Don	*Sida cordifolia* L. (Varahikand Bala) and *Asparagus filicinus* Buch-Ham. ex D. Don (Chiriya musali)
Lilium polyphyllum D. Don	*Withania somnifera* (L.) Dunal (Ashwagandha), *Chlorophytum arundinaceum Baker* (Safed musali), *Fritillaria oxypetala* D. Don., and *Fritillaria roylei* Hook
Malaxis muscifera (Lindl.) Kunt	*Pueraria tuberose* (Wild.) DC (Vidarikand) and *Centaurium roxburghii* (D. Don) Druce (Lal behmen)
Polygonatum cirrhifolium (Wall.) Royle	*Paederia foetida* L. (Prasarini), *Sida veronicifolia* Lam. (Nagbala), *Asparagus racemosus* Willd. (Shatavari), and *Polygonatum multiflorum* (L.) All. (Shakakul mishri)
Polygonatum verticillatum (Linn.) Allioni	*Eulophia campestris* Wall. (Salam Mishri) and *Asparagus racemosus* Willd. (Shatavari)
Roscoea purpurea Smith	*Withania somnifera* (L.) Dunal (Ashwagandha) and *Curculigo orchioides* Gaertn (Kali musali)

acid, cinnamic acid, gallic acid, etc. (Giri et al. 2017). These phenolics are usually well-known for their antioxidant potential, DNA protection, and antiaging potential (Sevgi et al. 2015; Sharma et al. 2020a, b). Further, the existence of optimal quantity of these phytoconstituents in Astavarga species may verify their part in antiaging and vitality-strengthening activities (Giri et al. 2017) (Table 10.6).

10.7 Conclusion

In conclusion, several terrestrial species with immune-boosting activity and used in various countries globally are reviewed. Even as several of these plants are highly regarded and used in many commercial preparations; the medical evidence of their substantiation still requires further studies. However, even with these limitations, these plants are valued by indigenous population groups and may offer prevention over cure.

References

Adams SJ, Kumar TS, Muthuraman G, Majeed A, Majeed M (2017) Mineral elements and their ICP-MS validation in Crepidium acuminatum (D.Don) Szlach.-An Ashtavarga plant. Asian J Biochem Pharmac Res 4(7):38–46

Adhikari KB, Tanwar F, Gregersen PL, Steffensen SK, Jensen BM, Paulson LK, Nielsen CH, Hoyer S, Borre M, Fomsgaard IS (2015) Benzoxazinoids: cereal phytochemicals with putative therapeutic and health-protecting properties. Mol Nutrit Food Res 59(7):1324–1338

Anywar G, Kakudidi E, Mukonzo BR, Schubert J, Oryem-Origa A, H. (2020) Indigenous traditional knowledge of medicinal plants used by herbalists in treating opportunistic infections among people living with HIV/AIDS in Uganda. J Ethnopharmacol 246:112205. https://doi.org/10.1016/j.jep.2019.112205

Arora M, Kaur G, Kahlon PS, Mahajan A, Sembi JK (2017) Pharmacognostic evaluation & antimicrobial activity of endangered ethnomedicinal plant Crepidium acuminatum (D. Don) Szlach. Pharmacogn J 9(6s):s56–s63. https://doi.org/10.5530/pj.2017.6s.158

Astragalus (2019). Natural medicines website. naturalmedicines.therapeuticresearch.com. Accessed 14 Oct 2019

Balkrishna A, Srivastava A, Mishra RK, Patel SP, Vashistha RK, Singh A, Jadon V, Saxena P (2012) Astavarga plants-threatened medicinal herbs of the North-West Himalaya. Int J Med Arom Plants 2:661–676

Balkrishna A, Mishra RK, Sharma N, Sharma VK, Misra L (2018) Phytochemical, botanical and biological paradigm of astavarga plants-the ayurvedic rejuvenators. J Nat Ayurvedic Med 2(6): 000145

Barnes J, Anderson LA, Gibbons S, Phillipson JD (2005) *Echinacea* species (*Echinacea angustifolia* (DC.). Hell *Echinacea pallida* (Nutt.) Nutt *Echinacea purpurea* (L.) Moench): a review of their chemistry, pharmacology and clinical properties. J Pharm Pharmacol 57:929–954

Barrett B (2003) Medicinal properties of *Echinacea*: a critical review. Phytomedicine 10:66–86

Benson JM, Pokorny AJ, Rhule A, Wenner CA, Kandhi V, Cech NB et al (2010) *Echinacea purpurea* extracts modulate murine dendritic cell fate and function. Food Chem Toxicol 48: 1170–1177

Bhagwan D, Sharma BK (2001) Charak Samita, 7th edn. Chaukhamba Sanskrit Series Office, Varanasi

Bhardwaj P, Thakur MS, Kapoor S, Bhardwaj AK, Sharma A, Saxena S, Chaurasia OP, Kumar R (2019) Phytochemical screening and antioxidant activity study of methanol extract of stems and roots of *Codonopsis clematidea* from trans-himalayan region. Pharmacogn J 11(3):536–546

Brinkeborn RM, Shah DV, Degenring FH (1999) Echinaforce and other *Echinacea* fresh plant preparations in the treatment of the common cold. A randomized, placebo controlled, double-blind clinical trial. Phytomedicine 6:1–6

Brown PN, Chan M, Paley L, Betz JM (2011) Determination of major phenolic compounds in *Echinacea* spp. raw materials and finished products by high-performance liquid chromatography with ultraviolet detection: Single-laboratory validation matrix extension. J AOAC Int 94: 1400–1410

Burger RA, Torres AR, Warren RP, Caldwell VD, Hughes BG (1997) *Echinacea* -induced cytokine production by human macrophages. Int J Immunopharmacol 19:371–379

Caruso TJ, Gwaltney JM Jr (2005) Treatment of the common cold with *Echinacea*: a structured review. Clin Infect Dis 40:807–810

Cech NB, Eleazer MS, Shoffner LT, Crosswhite MR, Davis AC, Mortenson AM (2006) High performance liquid chromatography/electrospray ionization mass spectrometry for simultaneous analysis of alkamides and caffeic acid derivatives from *Echinacea purpurea* extracts. J Chromatogr A 1103:219–228

Chang HM, But RP (1986) Pharmacology and applications of Chinese Materia Medica, 1. World Scientific, Singapore, pp 642–653

Classen B, Thude S, Blaschek W, Wack M, Bodinet C (2006) Immunomodulatory effects of arabinogalactan-proteins from Baptisia and *Echinacea*. Phytomedicine 13:688–694

Coelho J, Barros L, Dias MI, Finimundy TC, Amaral JS, Alves MJ, Calhelha RC, Santos PF, Ferreira ICFR (2020) *Echinacea purpurea* (L.) Moench: chemical characterization and bioactivity of its extracts and fractions. Pharmaceuticals (Basel) 13(6):125

Cragg GM, Newman DJ (2013) Natural products: a continuing source of novel drug leads. Biochim Biophys Acta 1830:3670–3695

Currier NL, Miller SC (2000) Natural killer cells from aging mice treated with extracts from *Echinacea purpurea* are quantitatively and functionally rejuvenated. Exp Gerontol 35:627–639

Dhyani A, Nautiyal BP, Nautiyal MC (2010) Importance of Astavarga plants in traditional systems of medicine in Garhwal, Indian Himalaya. Int J Biodiver Sci Ecosyst Serv Manag 6(1–2):13–19

Dimitrios B (2006) Sources of natural phenolic antioxidants. Trends Food Sci Technol 17:505–512

Dobrange E, Peshev D, Loedolff B, Van den Ende W (2019) Fructans as immunomodulatory and antiviral agents: the case of*Echinacea*. Biomolecules (9, 10):615. https://doi.org/10.3390/biom9100615

Dussault I, Miller SC (1993) Stimulation of natural killer cell numbers but not function in leukemic infant mice: a system primed in infancy allows survival in adulthood. Nat Immun 12:66–78

Engels G, Brinckmann J (2019) European elder,*Sambucus nigra*, L. HerbalGram, American Botanical Council. October volume

Geneva (1999) WHO monographs on selected medicinal plants. World Health Organization, pp 136–145

Gertsch J, Schoop R, Kuenzle U, Suter A (2004) *Echinacea* alkylamides modulate TNF-alpha gene expression via cannabinoid receptor CB2 and multiple signal transduction pathways. FEBS Lett 577:563–569

Giri L, Belwal T, Bahukhandi A, Suyal R, Bhatt ID, Rawal RS, Nandi SK (2017) Oxidative DNA damage protective activity and antioxidant potential of Ashtvarga species growing in the Indian Himalayan Region. Ind Crop Prod 102:73–179

Goel V, Chang C, Slama JV, Barton R, Bauer R, Gahler R et al (2002) Alkylamides of *Echinacea purpurea* stimulate alveolar macrophage function in normal rats. Int Immunopharmacol 2:381–387

Gotti R, Pomponio R, Bertucci C, Cavrini V (2002) Simultaneous analysis of the lipophilic and hydrophilic markers of *Echinacea* plant extracts by capillary electrophoresis. J Sep Sci 25: 1079–1086

Grimm W, Muller HH (1999) A randomized controlled trial of the effect of fluid extract of *Echinacea purpurea* on the incidence and severity of colds and respiratory infections. Am J Med 106:138–143

Gupta GL, Rana AC (2007) Withania somnifera (Ashwagandha): a review. Pharmacogn Rev 1: 129–136

Gupta A, Mishra RK, Bhati MK (2015) Traditional Medicinal Uses, Phytochemical Profile and Pharmacological Activities of Crepidium acuminatum (D. Don) Szlach. Indian J Ancient Med Yoga 8(4):179–183

Habbu PV, Smita DM, Mahadevan KM, Shastry RA, Biradar SM (2012) Protective effect of *Habenaria intermedia* tubers against acute and chronic physical and psychological stress paradigms in rats. Rev Bras 22(3):568–579

Hawkins J, Baker C, Cherry L, Dunne E (2019) Black elderberry (*Sambucus nigra*) supplementation effectively treats upper respiratory symptoms: a meta-analysis of randomized, controlled clinical trials. Complement Ther Med 42:361–365

He X, Lin L, Bernart MW, Lian L (1998) Analysis of alkamides in roots and achenes of *Echinacea purpurea* by liquid mass spectrometry. J Chromatogr A 815:205–211

Hobbs C (1995) Medicinal mushroom: an exploration of tradition, healing and culture, 2nd edn. Botanica Press, Santa Cruz, CA

Javed NK, Ansari SH, Mohammad A, Nazish (2012) Phyto- constituents from the bulb of Lilium polyphyllum D Don. Int Res J Pharm 3:146–148

Jones K (1996) Reishi: ancient herb for modern times, 2nd edn. Sylvan Press, Seattle, WA

Joshi VK (1983) Evolution of the concept of Astavarga. Indian J Hist Sci 18(1):9–14

Kant R, Verma J, Thakur K (2012) Distribution Pattern, Survival Threats and Conservation of 'Astavarga' Orchids in Himachal Pradesh, Northwest Himalaya. Plant Archiv 12(1):165–168

Khan H, Saeed M, Khan MA, Dar A, Khan I (2010) The antinociceptive activity of Polygonatum verticillatum rhizomes in pain models. J Ethnopharmacol 127(2):521–527

Khan H, Saeed M, Gilani AUH, Khan MA, Khan I, Ashraf N (2011) Antinociceptive activity of aerial parts of *Polygonatum verticillatum*: attenuation of both peripheral and central pain mediators. Phytother Res 25(7):1024–1030

Khan H, Saeed M, Khan MA et al (2012) Antimalarial and free radical scavenging activities of rhizomes of *Polygonatum verticillatum* supported by isolated metabolites. Med Chem Res 21: 1278–1282

Khan H, Saeed M, Gilani AH, Mehmood MH, Rehman NU, Muhammad N, Abbas M, Haq IU (2013) Bronchodilator activity of aerial parts of *Polygonatum verticillatum* augmented by anti-inflammatory activity: attenuation of Ca2+ channels and lipoxygenase. Phytother Res 27(9): 1288–1292

Khan H, Saeed M, Gilani AH, Muhammad N, Haq IU, Ashraf N, Rehman NU, Haleemi A (2013a) Antipyretic and anticonvulsant activity of *Polygonatum verticillatum*: comparison of rhizomes and aerial parts. Phytother Res 27(3):468–471

Krawiec S (2020) Immune health dietary supplement sales and ingredient innovation are skyrocketing. Nutritional Outlook. February 26. www.nutritionaloutlook.com/view/immune-health-dietary-supplement-sales-and-ingredient-innovation-are-skyrocketing

Krishnaraju AV, Rao TVN, Sundararaju D, Vanisree M, Tsay HS, Subbaraju GV (2005) Assessment of bioactivity of Indian medicinal plants using brine shrimp (Artemia salina) lethality assay. Int J Appl Sci Eng 2:125–134

Kurhekar JV (2014) Conservation of biodiversity with reference to indigenous herbal therapeutic agents. J Appl Environ Microbiol 2(2):42–45

Laasonen M, Wennberg T, Harmia-Pulkkinen T, Vuorela H (2002) Simultaneous analysis of alkamides and caffeic acid derivatives for the identification of *Echinacea purpurea Echinacea angustifolia Echinacea pallida* and *Parthenium integrifolium* roots. Planta Med 68:572–574

Luettig B, Steinmüller C, Gifford GE, Wagner H, Lohmann-Matthes ML (1989) Macrophage activation by the polysaccharide arabinogalactan isolated from plant cell cultures of *Echinacea purpurea*. J Natl Cancer Inst 81:669–675

Ma J, Ye Q, Hua Y, Zhang D, Cooper R, Chang MN, Chang JY, Sun HH (2002) New Lanostanoids from the Mushroom *Ganoderma lucidum*. J Nat Prod 65:72–75

Manayi A, Vazirian M, Saeidnia S (2015) *Echinacea purpurea*: Pharmacology, phytochemistry and analysis methods. Pharmacogn Rev 9(17):63–72

Mancek B, Kreft S (2005) Determination of cichoric acid content in dried press juice of purple coneflower (*Echinacea purpurea*) with capillary electrophoresis. Talanta 66:1094–1097

Mathur DR (2003) Yogtarangini. Chaukhamba Amarabharati Prakashan, Varanasi

McKenna DJ (1998) Natural dietary supplements: a desktop reference. Institute For Natural Products Research, St. Croix, U.S. Virgin Islands

McKeown KA (1999) A review of the taxonomy of the genus *Echinacea*. In: Janick J (ed) Perspectives on new crops and new uses. ASHS Press, Alexandria, VA, pp 482–498

Misra A, Srivastava S, Verma S, Rawat AKS (2015) Nutritional evaluation, antioxidant studies and quantification of poly phenolics, in Roscoea purpurea tubers. BMC Res Notes 8(1):1–7

Newman DJ, Cragg GM (2020) Natural products as sources of new drugs over the nearly four decades from 01/1981 to 09/2019. J Nat Prod 83(3):770–803

Nile SH, Nile AS, Keum YS (2017) Total phenolics, antioxidant, antitumor, and enzyme inhibitory activity of Indian medicinal and aromatic plants extracted with different extraction methods. Biotech 7:76–85

Pan SY, Litscher G, Gao S, Zhou SF, Yu ZL, Chen HQ, Zhang SF, Tang MK, Sun JN, Ko KM (2014) Historical perspective of traditional indigenous medical practices: the current renaissance and conservation of herbal resources. Evid Based Complement Alternat Med 2014:1–20

Pan P, Huang YW, Oshima K, Yearsley M, Zhang J, Arnold M, Yu J, Wang LS (2019) The immunomodulatory potential of natural compounds in tumor-bearing mice and humans. Crit Rev Food Sci Nutr 59(6):992–1007

Pandey D (2005) Sarangadhara Samhita. Chaukhamba Amarabharati Prakashan, Varanasi

Pandey MM, Rastogi S, Rawat AKS (2013) Indian traditional Ayurvedic system of medicine and nutritional supplementation. Evid Based Complement Altern Med 2013:1–12. https://doi.org/10.1155/2013/376327

Patel T, Crouch A, Dowless K, Freier D (2008) Acute effects of oral administration of a glycerol extract of *Echinacea purpurea* on peritoneal exudate cells in female swiss mice. Brain Behav Immun 22:39. https://doi.org/10.1016/j.bbi.2008.04.124

Paul AT, Gohil VM, Bhutani KK (2006) Modulating TNF-alpha signaling with natural products. Drug Discov Today 11:725–732

Pillai S, Pillai C, Mitscher LA, Cooper R (2007) Use of quantitative flow cytometry to measure ex vivo immunostimulant activity of echinacea: the case for polysaccharides. J Altern Complement Med 13(6):625–634

Pomponio R, Gotti R, Hudaib M, Cavrini V (2002) Analysis of phenolic acids by micellar electrokinetic chromatography: Application to *Echinacea purpurea* plant extracts. J Chromatogr A 945:239–247

Proksch A, Wagner H (1987) Structural analysis of a 4-O-methyl-glucuronoarabinoxylan with immuno-stimulating activity from *Echinacea purpurea*. Phytochemistry 26:1989–1993

Qi Y et al (2017) Anti-inflammatory and immunostimulatory activities of astragalosides. Am J Chin Med 45(6):1157–1167

Rajashekhar I, Rathod H, Desai H (2015) A short review on astavarga plants-losing their existence. Int J Ayur Pharma Res 3(7):32–38

Rawat S, Andola H, Giri L, Dhyani P, Jugran A, Bhatt ID, Rawal RS (2014) Assessment of nutritional and antioxidant potential of selected vitality strengthening Himalayan medicinal plants. Int J Food Prop 17(3):703–712

Rawat S, Jugran AK, Bahukhandi A, Bahuguna A, Bhatt ID, Rawal RS, Dhar U (2016) Anti-oxidant and anti-microbial properties of some ethno-therapeutically important medicinal plants of Indian Himalayan Region. *3*. Biotech 6:154. https://doi.org/10.1007/s13205-016-0470-2

Roesler J, Steinmüller C, Kiderlen A, Emmendörffer A, Wagner H, Lohmann-Matthes ML (1991) Application of purified polysaccharides from cell cultures of the plant *Echinacea purpurea* to mice mediates protection against systemic infections with Listeria monocytogenes and Candida albicans. Int J Immunopharmacol 13:27–37

Saboon YB, Arshad M, Sabir S, Amjad MS, Ahmed E, Chaudhari SK (2016) Pharmacology and biochemistry of *Polygonatum verticillatum*: a review. J Coastal Life Med 4(5):406–415

Sadigh-Eteghad S, Khayat-Nuri H, Abadi N, Ghavami S, Golabi M, Shanebandi D (2011) Synergetic effects of oral administration of levamisole and *Echinacea purpurea* on immune response in Wistar rat. Res Vet Sci 91:82–85

Sahu MS, Mali PY, Waikar SB, Rangari VD (2010) Evaluation of immunomodulatory potential of ethanolic extract of *Roscoea procera* rhizomes in mice. J Pharm Bioall Sci 2(4):346–349

Sahu MS, Sahu RA, Verma A (2013) Immunomodulatory activity of alcoholic extract of *Habenaria intermedia* in mice. Int J Pharm Pharm Sci 5:406–409

Sahu RA, Itankar PR, Mishra RR, Maliye AN (2016) Pharmacological evaluation of Roscea procera (Kakoli) and Lilium polyphyllum (Kshirkakoli) extracts for immunomodulatory activity. Int J Pharmacogn Phytochem Res 8(11):1788–1794

Satyavati GV (2003) The legacy of charaka. Curr Sci 85:1087–1090

Schwarz E, Parlesak A, Henneicke-von Zepelin HH, Bode JC, Bode C (2005) Effect of oral administration of freshly pressed juice of *Echinacea purpurea* on the number of various subpopulations of B- and T-lymphocytes in healthy volunteers: results of a double-blind, placebo-controlled cross-over study. Phytomedicine 12:625–631

Sen S, Chakraborty R, Sridhar C, Reddy YSR, De B (2010) Free radicals, antioxidants, diseases and phytomedicines: current status and future prospects. Int J Pharm Sci Rev Res 3:91–100

Sevgi K, Tepe B, Sarikurkcu C (2015) Antioxidant and DNA damage protection potential of selected phenolic acids. Food Chem Toxicol 77:15–21

Sharma BD, Balkrishna AV (2005) Vitality strengthening Astavarga plants (Jeevaniya & Vayasthapan paudhe). Divya Publishers, Divya yog mandir, Uttaranchal

Sharma A, Cannoo DS (2016) Effect of extraction solvents/techniques on polyphenolic contents and antioxidant potential of the aerial parts of *Nepeta leucophylla* and the analysis of their phytoconstituents using RP-HPLC-DAD and GC-MS. RSC Adv 6:78151–78160

Sharma A, Cannoo DS (2016a) Comparative evaluation of extraction solvents/techniques for antioxidant potential and phytochemical composition from roots of Nepeta leucophylla and quantification of polyphenolic constituents by RP-HPLC-DAD. Food Measure 10:658–669

Sharma A, Cannoo DS (2017) A comparative study of effects of extraction solvents/techniques on percentage yield, polyphenolic composition and antioxidant potential of various extracts obtained from stems of *Nepeta leucophylla*: RP-HPLC-DAD assessment of its polyphenolic constituents. J Food Biochem 41:12337–12348

Sharma A, Shanker C, Tyagi L, Singh M, Rao CV (2008) Herbal medicine for market potential in India: an overview. Acad J Plant Sci 1:26–36

Sharma BD, Singh L, Kaur MJ (2014) Nutritional composition of rare himalayan herbs constituting the world's first health food. Int J Agric Food Sci Technol 5:75–80

Sharma A, Bhardwaj G, Gaba J, Cannoo DS (2020a) Natural antioxidants: assays and extraction methods/solvents used for their isolation. In: Antioxidants in fruits: properties and health benefits. Springer, Singapore, pp 1–33

Sharma A, Cooper R, Bhardwaj G, Cannoo DS (2020b) The genus Nepeta: traditional uses, phytochemicals and pharmacological properties. J Ethnopharmacol 268:113679

Singh A, Duggal S (2009) Medicinal orchids-an overview. Ethnobotanical Leaflets Issue #3, Article 3. https://opensiuc.lib.siu.edu/ebl/vol2009/iss3/3

Smith T, May G, Eckl V, Reynolds CM (2020) US sales of herbal supplements. Herbal Gram ABC 127:54–56

Spelman K, Wetschler MH, Cech NB (2009) Comparison of alkylamide yield in ethanolic extracts prepared from fresh versus dry *Echinacea purpurea* utilizing HPLC-ESI-MS. J Pharm Biomed Anal 49:1141–1149

Steinmuller C, Roesler J, Gröttrup E, Franke G, Wagner H, Lohmann-Matthes ML (1993) Polysaccharides isolated from plant cell cultures of Echinacea purpurea enhance the resistance of immunosuppressed mice against systemic infections with Candida albicans and Listeria monocytogenes. Int J Immuno Pharmcol 15:605–614

Sultan MT, Butt MS, Qayyum MMN, Suleria HAR (2014) Immunity: plants as effective mediators. Crit Rev Food Sci Nutr 54(10):1298–1308

Vanegas SM, Meydani M, Barnett JB, Goldin B, Kane A, Rasmussen H, Brown C, Vangay P, Knights D, Jonnalagadda S (2017) Substituting whole grains for refined grains in a 6-wk randomized trial has a modest effect on gut microbiota and immune and inflammatory markers of healthy adults. Am J Clin Nutr 105(3):635–650

Venkatasubramanian P (2007) Drug discovery in ayurveda - different ways of knowing. Pharmacogn Mag 3:64

Verspreet J et al (2015) Cereal grain fructans: structure, variability and potential health Effects. Trends Food Sci Technol 43:32–42

Viapiana A, Wesolowski M (2017) The phenolic contents and antioxidant activities of infusions of Sambucus nigra L. Plant Foods Hum Nutr 72(1):82–87

Virk JK, Kumar S, Singh R, Tripathi AC, Saraf SK, Gupta V, Bansal P (2016) Isolation and characterization of quinine from Polygonatum verticillatum: a new marker approach to identify substitution and adulteration. J Adv Pharm Technol Res 7(4):153–158

Wagner H, Stuppner H, Schäfer W, Zenk M (1988) Immunologically active polysaccharides of *Echinacea purpurea* cell cultures. Phytochemistry 27:119–126

Wan CP et al (2013) Astragaloside II triggers T cell activation through regulation of CD45 protein tyrosine phosphatase activity. Acta Pharmacol Sin 34(4):522–530

Wang, S. Y.; Hsu, M. L.; Hsu, H. C.; Tzeng, C. H.; Lee, S. S.; Shiao, M. S.; Ho, C. K. 1997. Int J Cancer 70: 699–705

Welker P, Lippert U, Nurnberg W, Kruger-Krasagakes S, Moller A, Czarnetzki B (1996) Glucocorticoid-lnduced modulation of cytokine secretion from normal and leukemic human myelomonocytic cells. Int Arch Allergy Immunol 109:110–115

Wilasrusmee C, Siddiqui J, Bruch D, Wilasrusmee S, Kittur S, Kittur DS (2002) In vitro immunomodulatory effects of herbal products. Am Surg 68(10):860–864

Woelkart K, Linde K, Bauer R (2008) *Echinacea* for preventing and treating the common cold. Planta Med 74:633–637

Yuan H, Ma Q, Ye L, Piao G (2016) The traditional medicine and modern medicine from natural products. Molecules 21:559–576

Zou C, Su G, Wu Y et al (2013) *Astragalus* in the prevention of upper respiratory tract infection in children with nephrotic syndrome: evidence-based clinical practice. Evid Based Complement Altern Med 2013:352130

Immunomodulatory Role of Terpenoids and Phytosteroids

11

Paula Mendonça Leite, Juliana Mendes Amorim, and Rachel Oliveira Castilho

Abstract

The immune system in vertebrates comprises a complex and sophisticated defense system and is traditionally divided in two types: innate immunity and adaptive immunity. It involves a series of factors for its functioning, such as physical, chemical, physical-chemical barriers, phagocytic system, cytotoxicity mediated by cytotoxic cells (NK cells), circulating molecules (antibodies and complement), and soluble mediators produced by cell activation (cytokines), in addition to antigen-specific T and B lymphocytes. The function of the immune system is to maintain body homeostasis against endogenous and exogenous factors, which results in an immunosuppressive or stimulatory activity. Thus, any influence on the immune system response is defined as immunomodulation; consequently, immunomodulators are agents that have the capacity to modulate or regulate pathophysiological activities. Therefore, possible mechanisms of herbal medicines in immune systems can be generally categorized into two major groups: to suppress immune responses and to enhance them, which can be applied as a treatment for autoimmune diseases or in immunodeficiency cases and infectious diseases, respectively. Many cytotoxic drugs and antibiotics have been developed from natural products derived from plants and microorganisms, but their use as immunomodulators for treating immune diseases is still limited, and new researches and approaches are necessary. In this chapter, terpenoids and steroids derived from medicinal plants will be discussed, focusing on their immunomodulatory bioactivities. Analyses of scientific literature were conducted to assess this type of activity. Among the natural products, terpenoids are one of

P. M. Leite · J. M. Amorim · R. O. Castilho (✉)
Laboratório de Farmacognosia, Programa de Pós-graduação em Ciências Farmacêuticas, Departamento de Produtos Farmacêuticos, Faculdade de Farmácia, Universidade Federal de Minas Gerais. Faculdade de Farmácia, Belo Horizonte, Minas Gerais, Brazil
e-mail: roc2006@farmacia.ufmg.br

N. S. Sangwan et al. (eds.), *Plants and Phytomolecules for Immunomodulation*,
https://doi.org/10.1007/978-981-16-8117-2_11

the major classes of substances present in vegetables, fruits, and other dietary food or medicinal plants. Terpenoids comprises a large family of substances derived from isoprene units that contain carbon skeletons represented by $(C5)_n$. They have diverse structures classified as hemiterpenes (C5), monoterpenes (C10), sesquiterpenes (C15), diterpenes (C20), sesterpenes (C25), triterpenes (C30), and tetraterpenes (C40). Plant steroids, also known as phytosteroids, are modified triterpenoids containing the tetracyclic ring system of lanosterol. The terpenoids and phytosteroids were classified according to their carbon skeleton types and immune modulation activities. The data for each type of terpene was summarized, and the relevant terpene-producing plants were described accordingly and their effects discussed to better consolidate isolated information available in the literature. Focus on saponins, steroid heterosides, or triterpenes was also given, highlighting their potential for the development of immunomodulatory or inflammation-regulatory therapeutics agents.

Keywords

Terpene immunomodulators · Phytosteroid immunomodulators · Saponin immunomodulators · Plant immunomodulators · Natural product immunomodulators

11.1 Introduction

Immunity is a set of actions that the organism conducts in an orchestrated way between molecules, cells, and tissues, to guarantee hemostasis. It is activated in response to the presence of pathogens, so, in this sense, the immune system has evolved to differentiate what is inherent from what is strange to the organism. The challenge involved in this defense system is to destroy invaders or their own unregulated components, such as tumors, keeping normal cells intact. The immune system in vertebrates comprises a complex and sophisticated defense system which encompasses sensory and effector mechanisms mediated by receptors. It is traditionally divided into two types, innate immunity and adaptive immunity (Abbas 2019; Bruton et al. 2010).

The body's first defense against infection by microorganisms, for example, is called innate immunity and consisting of cellular and biochemical defense mechanisms that react to the products of these injured microorganisms and cells. Some important characteristics of this immunity are anticipation and speed, in addition to the lack of memory generation and specificity, in addition to not attacking the organism itself and little diversity. The main agents of innate immunity are physical and chemical barriers; phagocytic cells such as neutrophils, monocytes, and macrophages; natural killer cells (NK); dendritic cells; and blood proteins, such as the complement system and mediators of inflammation (Abbas 2019; Netea et al. 2019; Yatim and Lakkis 2015).

The acquired immunity, also called adaptive or specific, encompasses the immune response stimulated by exposure to antigens. Therefore, the mechanisms of this immunity do not provide anticipated defenses, since they are activated only in the presence of agents, and will consequently have a lower speed of formation. However, they can be very specific and diversified, capable of differentiating microorganisms and substances, of generating memory, which makes it possible for secondary responses to happen quickly, all while not harming the organism itself. The main agents of acquired immunity are the various classes of lymphocytes (B lymphocyte, helper T lymphocyte, cytotoxic T lymphocyte, regulatory T lymphocyte, and memory T lymphocyte), antibodies, and antigen-presenting macrophages (Abbas 2019; Netea et al. 2019; Nicholson 2016).

Innate immunity is the first to be activated during an infection, and it takes from minutes to a few hours to have a complete operation, which is essential for the first stage of a new infection. Although efficient, innate immunity may not be sufficient due to aggressiveness and/or the number of pathogens. In these cases, adaptive immunity is activated, enabling specific mechanisms for the recognition and elimination of the pathogen. Some exogenous and endogenous factors, however, can affect the immune system and end up resulting in immunosuppressive or stimulatory function. A few examples are climate change, quality of sleep, food, physical activity, and stress, among others (Abbas 2019; Netea et al. 2019).

In this sense, substances that influence the immune system can be classified into immunomodulators, that act by stimulating, suppressing, or modulating the innate and/or adaptive immune system; immunoadjuvants, considered to be specific immune stimulants, that are used to improve the effectiveness of vaccines; immunostimulants, that increase the body's resistance against infections through of the innate and adaptive immunity; and immunosuppressants, that are applied in the prevention of transplant rejection and in the treatment of autoimmune diseases. Several natural products are known to act on the immune system, helping to maintain hemostasis. In this context, terpenes are substances isolated mainly from plants and medicinal plants that can be applied in this area and will be studied further (Ortuño-Sahagún et al. 2017; Shantilal et al. 2018).

11.2 Terpenes

Terpenes, also known as terpenoids or isoprenoids, are naturally occurring chemicals produced by plants, microorganisms, and some insects. Usually, plants biosynthesize a diverse types of terpenoids, some of them are common to many species, and others are characteristic of certain taxa and probably involved in ecological adaptations (Kiyama 2017; Tholl 2006). As example, phytohormones gibberellins are important in interaction of the plants with the environment and for growth and development. Terpenes play a role in the defense against microorganisms and insects, in protecting against oxidative stress, thermal tolerance control, and attraction of both pollinators and herbivore predators. Besides, terpenes extracted from plants have been used as fragrances and flavors, insecticides,

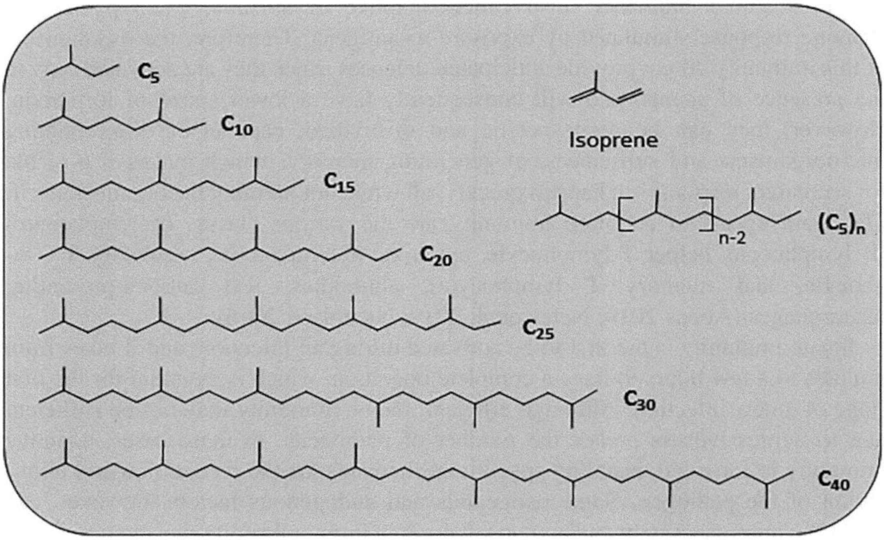

Fig. 11.1 Isoprene unit and the relationship to terpene backbones. The core structures of terpenes, hemi- (C5), mono- (C10), sesqui- (C15), di- (C20), sester- (C25), tri- (C30), tetra- (C40), and polyterpenes ((C5)n; $n > 8$)

antimicrobial, and herbal remedies used directly in the pharmaceutical industry or for the semi-synthesis or synthesis of important products, such as synthetic hormones (Pichersky and Gershenzon 2002; Loreto et al. 2004; Sharkey and Yeh 2001; Dhandapani et al. 2020). In addition, they are a class of hydrocarbons, recognized as novel sources of renewable chemical source that could be used as value-added substitutes to petrochemical equivalents (Tsolakis et al. 2019).

Terpenoids comprise a large family of secondary metabolites derived from isoprene units, with more than 80,000 compounds already known to date, and structures containing isoprene units and $(C5)_n$ carbon skeletons (Christianson 2017).

According to the number of isoprene units, the terpenes can be classified as hemiterpenes (C5), monoterpenes (C10), sesquiterpenes (C15), diterpenes (C20), sesterpenes (C25), triterpenes (C30), tetraterpenes (C40), and polyterpenes $(C5)_n$, $n > 8$). Plant steroids, also known as phytosteroids, are modified triterpenoids containing the tetracyclic ring system of lanosterol with losses of some alkyl groups, while saponins could be steroid or triterpene heterosides (Dewick 2009a) (Fig. 11.1). This class of compound has been described as anti-inflammatory, analgesic, antioxidant, neuroprotective and cardioprotective, antiviral, estrogenic, antimicrobial, insecticidal, and antitumoral, as well as being immunomodulators. Although their effects have not always been clearly explained, several of them are implicated in the resolution of immune diseases.

The two $(C5)_n$ fundamental building blocks that biosynthesize terpenes are isopentenyl diphosphate (IPP) and its isomer dimethylallyl diphosphate (DMAPP), both also referred to as isoprene activated (Christianson 2017). Isoprene is a natural product but is not direct involved in the formation of terpenes. These C5 compounds are generated by both the mevalonate (MVA) and 2-C-methyl-D-erythritol-4-phosphate (MEP) pathways. The first, occurs among the cytoplasm, endoplasmic reticulum, and peroxisomes and the second in the plastids (Vranová et al. 2013; Zhou and Pichersky 2020). IPP and DMAPP (C5) are activated hemiterpene present in the pathways that generate more complex terpenes through simple biosynthetic roots that couple activated isoprene (IPP and DMAPP) head-to-tail. These reactions yield linear structures, which can undergo cyclization reactions that, together with methyl migration, rearrangements, and proton and hydride transfers, originate the great structural complexity of this class of compounds. The corresponding substrates of each addition of C5 units are called geranyl diphosphate (GPP-C10), farnesyl diphosphate (FPP-C15), and geranylgeranyl diphosphate (GGPP-C20) that will give rise to monoterpenes, sesquiterpenes, and diterpenes, respectively. Squalene (C30) and phytoene (C40), both are formed by the tail-to-tail of the molecules of farnesyl diphosphate and geranylgeranyl diphosphate to form triterpenes and carotenoids, respectively (Fig. 11.2).

The enzymes called terpenoid synthases or terpenoid cyclases (TPS) catalyze terpene biosynthesis reactions. Recent discoveries have revealed that there are a great number of substrates that can be used by enzymes of TPS family for terpenes biosynthesis, non-TPS enzymes can also be used in the synthesis, and novel transport mechanisms can also be present on cell organelles (Dhandapani et al. 2020; Christianson 2017; Dewick 2009a; Vranová et al. 2013; Zhou and Pichersky 2020; Raz et al. 2020). They can occur free or linked to sugar as saponins and cardiotonic glycosides. In addition, there are many derivatives as iridoids, limonoids, quassinoids, vitamins, and others.

11.3 Monoterpenes

Monoterpenes are formed by two units of five carbons (isoprene) joined following the head-to-tail pattern. They are one of the main classes of substances present in essential oils and are classified according to their chemical structure into acyclic, monocyclic, and bicyclic. Reports from the literature relate monoterpenes to anticancer, antimicrobial, antioxidant, antiviral, analgesic, and anti-inflammatory activities, in addition to several connections with the immune system (Dewick 2009a; Quintans et al. 2019). The main target of action of monoterpenes in immunity are dendritic cells and the studies found in the literature mainly addressed the application of these substances in the treatment of asthma.

Ascaridole **1** is one of the substances responsible for allergic contact dermatitis to tea tree oil. An experiment with skin cell culture and this oil demonstrated the ability to activate dendritic cells, which are part of the innate immune system and have function in phagocytize pathogens and process the antigenic material to present to the adaptive immune system (Sahli et al. 2019). In a mouse model, 1,8-cineole **2** demonstrated capacity to be used as an adjuvant for the influenza vaccine by

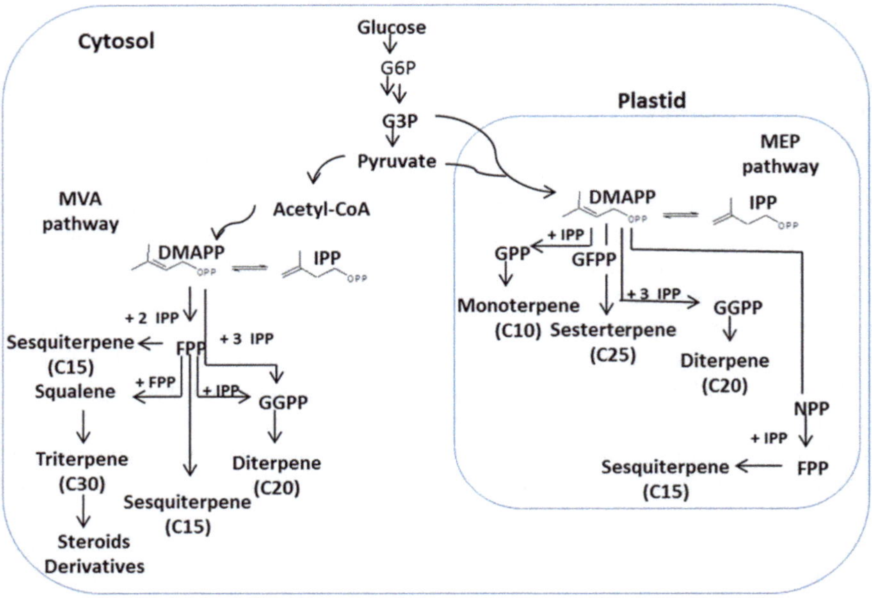

Fig. 11.2 Pathway map for the production of terpenoids in cell compartments. The mevalonate (MVA) pathway takes place in the cytoplasm, endoplasmic reticulum, and peroxisome, while 2-C-methyl-D-erythritol-4-phosphate (MEP) pathway in the plastids. Isopentenyl diphosphate (IPP) and the isomer dimethylallyl diphosphate (DMAPP) (C5) activate isoprene present in the pathways that generate more complex terpenes through simple coupling reactions head-to-tail of these C5 precursors. The corresponding substrates of each addition of C5 units are called geranyl diphosphate (GPP-C10), farnesyl diphosphate (FPP-C15), and geranylgeranyl diphosphate (GGPP-C20) that will give rise to monoterpenes, sesquiterpenes, and diterepenes, respectively. Squalene (C30) and phytoene (C40) both are formed by the tail-to-tail of the molecules of farnesyl diphosphate and geranylgeranyl diphosphate to form triterpenes and carotenoids, respectively

increasing specific IgG antibodies in the serum. This can be explained because this monoterpene also activates mature dendritic cells to express high levels of co-stimulatory molecules, CD40, CD80, and CD86, promoting signals for T-cell activation (Li et al. 2017a). Another study with lung macrophages demonstrated anti-inflammatory activity of 1,8-cineole by selective modulation via downregulation of the pattern recognition receptors (PRR) routes and common downstream signaling cascade partners (nuclear factor-κB, NF-κB and mitogen-activated protein kinases, MAPKs) (Yadav and Chandra 2017).

Carvacrol **3** and thymol **4**, in turn, demonstrated an opposite effect on dendritic cells. These two monoterpenes, found in the species *Thymus vulgaris* L., *Thymus daenensis* Celak, and *Zataria multiflora* Boiss (Labiatae family), were tested for immunomodulatory activity in splenic dendritic cells. These monoterpenes can inhibit the maturation of these cells, in addition to inhibiting the T-cell response and secretion of interleukin-4 (IL-4) and interferon-γ (IFN-γ) cytokines (Amirghofran et al. 2016). Both molecules were also tested in mice immunized with ovalbumin and were able to inhibit antigen-specific immune response in vivo, reducing cytokines and key transcription factors related to T helper 1, 2, and 17 (Th1, Th2, and Th17) responses (Gholijani and Amirghofran 2016). Another experiment using this same ovalbumin model demonstrated that myrcene **5** is also able to enhance Th2 and Th1 immune response, in addition of rapidly increasing IgG levels (Uyeda et al. 2016).

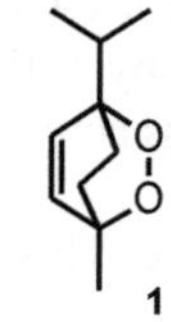
1
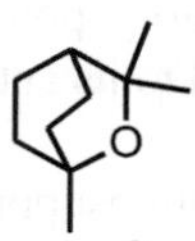
2
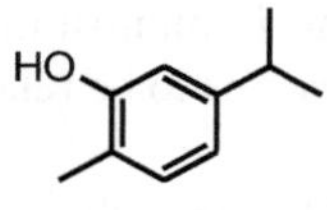
3
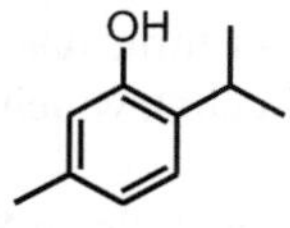
4

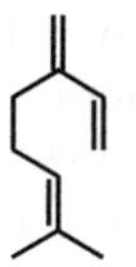
5
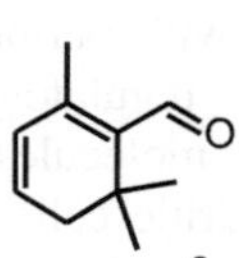
6
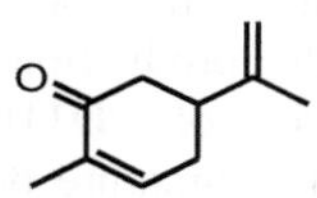
7
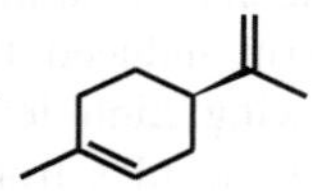
8

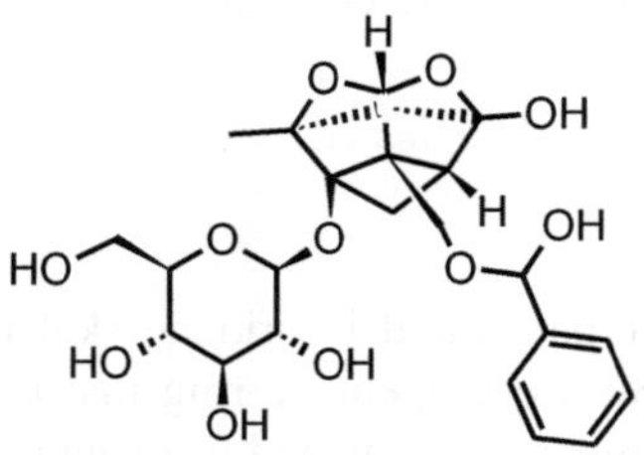
9

Similarly, carvacrol such as suppression of 3 alone protected macrophages against inflammation induced by *lipopolysaccharides* (LPS) in an in vitro experiment due to the reduction of extracellular signal-regulated kinase 1/2 (***ERK1/2***) phosphorylation, NF-kB p65 translocation to nucleus and transcriptional transactivation, phagocytic activation, and interleukin-1β (IL-1β), and tumor necrosis factor-alpha (TNF-α) release (Somensi et al. 2019). In an asthma model in rats, carvacrol **3** also demonstrated its immunomodulatory activity by reducing inflammatory reactions in several ways eosinophils and reduction of serum IgE levels, largely related to allergic responses. A reduction in IL-4, IL-5, and IL-13 was also observed that can be explained by the fact that carvacrol **3** has an excitatory effect on Th1 and inhibitory effect on Th2. Carvacrol **3** also reduced TNF-α and IFN-γ by suppressing the T-cell response and limiting NF-kB expression, leading to iNOS suppression and relieving oxidative stress (Ezz-Eldin et al. 2020).

Another monoterpene that seems to act in the Th1/Th2 balance is the safranal **6**, a constituent from *Crocus sativus* L. In a study with peripheral blood mononuclear cells, it has increased IFN-γ/IL-4 ratio, indicating an increase in the Th1/Th2 balance. This suggests potential application in Th1/Th2 imbalance disorders, such as asthma (Feyzi et al. 2016).

Thymol **4** itself also demonstrated a less specific action on the immune system by showing protective activity in the oxidative stress and asthma-induced inflammation in an ovalbumin model, since it increased the antioxidant capacity and reduced lipid peroxidation and oxidative damage (Mohammadi et al. 2018). Another monoterpene that demonstrated immunomodulatory activity was carvone **7**. Its inhalation induced an increase in IL-1β, IL-6, and TNF-α in BALB/c mice (Lasarte-Cia et al. 2018). R-limonene **8**, in turn, has an immunostimulatory potential by activating natural killer cells (NK cells), which seems to be related to its antistress activity (Terao et al. 2019).

In contrast, paeoniflorin **9** is a potential immunosuppressive. This substance is a terpene glucoside extracted from the paeony root (*Paeonia lactiflora* Pall.) and has been approved by the State of Food and Drug Administration of China for the treatment of rheumatoid arthritis. In a study with peripheral blood mononuclear cells, paeoniflorin induced the production of regulatory dendritic cells, with a phenotype showing high levels of CD11b/c molecules and low expression of co-stimulatory molecules from immature dendritic cells. Besides, it also induced the expression of an enzyme that catabolizes tryptophan and leads to inhibition of T lymphocyte proliferation, IDO (Kasahara et al. 2017).

11.4 Iridoids

Iridoids are also 10-carbon terpenes and its iridane skeleton has a monoterpenic origin. However, they do have a cyclopentane ring that is normally fused into an oxygen heterocycle of six-membered, originated from geraniol by a different type of folding. Another characteristic of this class is that they are often found in the form of glycosides (Dewick 2009a). Studies show that iridoids have neuroprotective, anti-inflammatory, immunomodulatory, hepatoprotective, and cardioprotective effects (Tundis et al. 2008; Dewick 2009b). Among the main mechanisms involved in the immunomodulation of iridoids are the action on T lymphotics and the release of cytokines.

The *Verbascum* species are rich in iridoids and commonly used in the treatment of respiratory and inflammatory conditions. *Verbascum nobile* Velen. extract as well as isolated iridoids were tested for their effects on lectin-induced T-cell activation and proliferation and demonstrated potential to modulate T-cell-related pathologies. This is due to the inhibition of lectin-induced cell growth of Jurkat T-cell line and the suppression of the early activation marker CD69 expression, the intracellular level of IFN-γ, and the generation of CD3 + IFN-γ + effectors (Dimitrova et al. 2018).

Geniposide **10** is the main substance of the fruit of *Gardenia jasminoides* Ellis. Studies have shown that it controls cytokine activity and decreases inflammation. In *adjuvant arthritis rats* geniposide demonstrated immunomodulatory and anti-inflammatory activity by the induction of immune tolerance in MLN lymphocytes and Th1; increased Th2 activity; increased production of anti-inflammatory cytokines IL-4 and TGF-β1; reduction of the pro-inflammatory cytokine IL-2 (Zhang et al. 2017). Another study, also with geniposide as an adjuvant in arthritis, also showed hyperproliferation of fibroblast-like synoviocytes (FLS), downregulation of the IFN-γ and IL-17 production, in addition to MAPK signaling pathway blockade. Considering this, the use of MAPK as a therapeutic arthritis model in rats, the geniposide **10** tested as a target may be a new strategy in the management of inflammatory/autoimmune diseases (Li et al. 2018a).

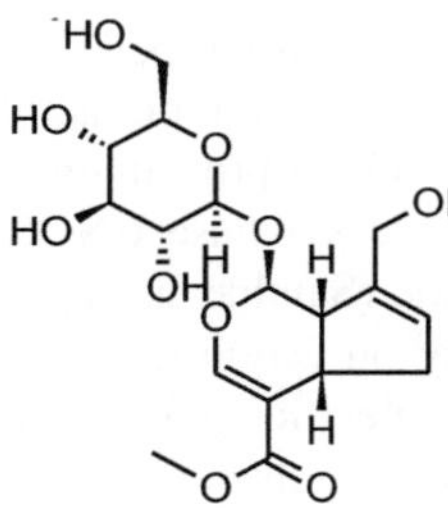

10

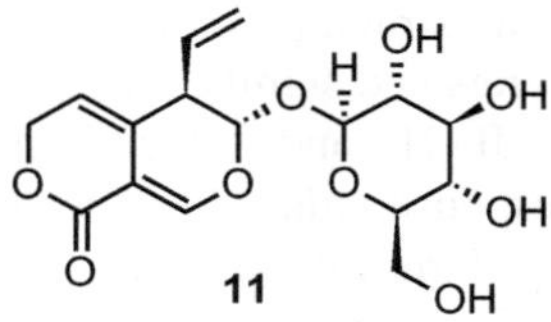

11

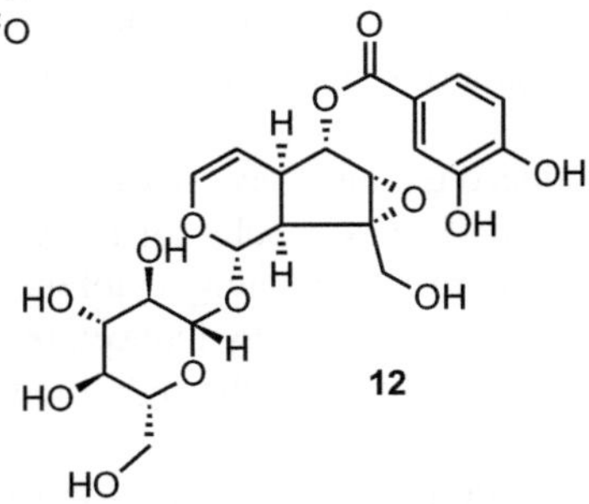

12

Gentiopicroside **11** is a natural secoiridoid glycoside from gentian species used as medicinal herbs in China and Europe. It has been tested in the treatment of bleomycin-induced pulmonary fibrosis in mice and has demonstrated anti-inflammatory and anti-fibrotic potential by significantly decreasing levels of TNF-α and IL-1β in BALF and Hyp, TGF-β1, and CTGF in lung tissues, respectively (Chen et al. 2018).

Verproside **12** is a catalpol derivative iridoid glycoside common in the genus *Pseudolysimachion*. It is the main substance of *Pseudolysimachion rotundum* var. *subintegrum* (Nakai) D.Y. Hong extract and is a potent anti-asthmatic/COPD drug candidate in vivo. Although the mechanism of action is still unknown, a study showed that verproside **12** reduced the expression of TNF-α by inhibiting both nuclear factor kappa B (NF-κB) transcriptional activity and the phosphorylation of its upstream effectors such as IκB kinase (IKK) β, IκBα, and TGF-β-activated kinase

1 (TAK1) in NCI-H292 cells. Molecular docking studies have also shown that verproside binds between TRADD and TRAF2 subunits (Lee et al. 2016).

11.5 Sesquiterpenes

Sesquiterpenes are made up of 15 carbon units, meaning they are formed by 3 isoprene units. This class of terpenes also makes up the essential oil while presenting promising anti-inflammatory, antiparasitic, and anticarcinogenic activities, but they can be toxic and cause other adverse effects (Dewick 2009a; Bartikova et al. 2014). Some sesquiterpenes, mainly the sesquiterpene lactones group such as artemisinin **13**, have been associated to immune system effects. Most of them also demonstrated immunomodulatory activity involving T lymphocytes.

The first example is farnesol **14**, an acyclic sesquiterpene alcohol with the simplest chemical structure in this class. It was tested on acrylamide-induced neurotoxicity in rats and significantly reduced levels of pro-inflammatory cytokines such as TNF-α, IL-1β, and inducible nitric oxide synthase (iNOS), reducing the inflammatory response induced by the acrylamide (Santhanasabapathy et al. 2015). In another study, farnesol demonstrated activity on acquired immunity in allergic asthmatic mice. Its supplementation increased specific IgG2a/IgE antibodies and reduced nonspecific IgE, IgA, IgM, and IgG. This results demonstrated that farnesol might regulate Th2 immune responses toward Th1 immune balance (Ku and Lin 2016).

Another study of induced toxicity was carried out with nerolidol **15**. Supplementation with this sesquiterpene was capable of decreasing activated astrocytes and microglia and the levels of pro-inflammatory cytokines IL-1 [beta], IL-6, and TNF-α in mice (Javed et al. 2016). In the toxicity induced by cyclophosphamide, however, nerolidol reduced neuroinflammation through modulation of Nrf2 and NF-κB (Iqubal et al. 2019).

A study of several sesquiterpenes isolated from Chinese agarwood (*Aquilaria malaccensis* Benth.) showed that several of them had effects on the immune system. Among them, (5S, 7S, 9S, 10S)-(+)-9-hydroxyseline-3, 11-dien-12-al (HHX-5) was the most active, suppressing the innate and adaptive immunity by inhibiting signal transducer and activator of transcription 1, 4, 5, and 6 (STAT1, STAT4, STAT5, and STAT6) signaling pathways. However, the mechanism involved in inhibiting the differentiation of naive CD4+ T cells into Th17 cells is still being studied (Zhu et al. 2016).

An important example of sesquiterpenes that act on the immune system is the zerumbone **16**. This substance is a cyclic 11-membered sesquiterpenoid and has been widely studied for its effect on the immune system. In fact, its immunomodulatory activity appears to be related to the presence of α,β-unsaturated carbonyl-based moiety in its structure. A literature review described the signaling events modulated by zerumbone **16** in cellular models as suppression of NO, PGE2, iNOS, COX-2, VEGF, AP-1, NF-κB, eotaxin, IL-1-beta, IL-2, IL-4, IL-5, IL-6, IL-10,

IL-12, and IL-13 production and increased production of IFN-γ and HO-1 (Haque et al. 2017).

The starting example of sesquiterpene lactone, artemisinin **13**, is one of the best studied substances. It is used as an antimalarial, and several derivatives have been synthesized. In addition to antimalarial activity, they have also shown promising activity as immunoregulators, since artemisinin **13** and its derivatives regulate innate and adaptive immunity in several ways, such as inhibition of TNF-α production through the nuclear suppression of NF-κB translocation; suppression of pro-inflammatory cytokine production and induction of the anti-inflammatory cytokine production, such as IL-10; suppression of T-cell activation; and suppression of the autoantibody production in autoimmune disease models such as arthritis and lupus (Hou and Huang 2016; Mu and Wang 2018).

Another work emphasized the potential of artemisinin **13** in the immunomodulation of T lymphocytes, including activation, proliferation, differentiation, apoptosis, and subset function. To discover the mechanisms involved, studies have been done, and its use seems very promising for the treatment of malaria, autoimmune disease, hypersensitivity reaction, cancer schistosomiasis, and AIDS (Yan et al. 2019a). An in vivo study with 4 T1 mouse with mammary carcinoma cells also demonstrated the effect of artemisinin **13** on T lymphocytes. In it, artemisinin demonstrated antitumor immune response by increasing cytotoxic T lymphocytes and T helper lymphocyte 1 and IFN-γ and TNF- α expression levels in the tumor (Cao et al. 2019). One study also demonstrated that artemisinin enhances human NK cell cytotoxicity and degranulation and can present antitumor potential (Houh et al. 2017).

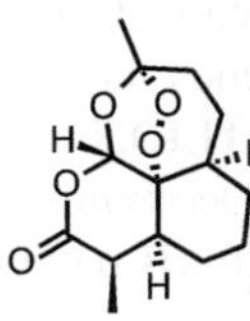
13

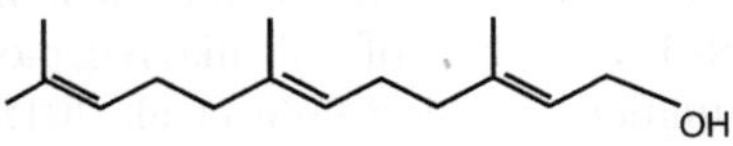
14

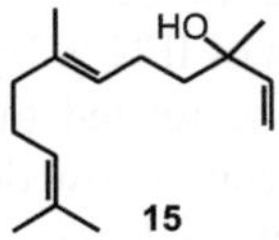
15

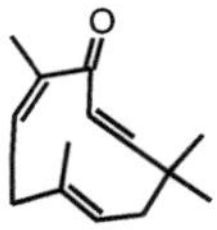
16

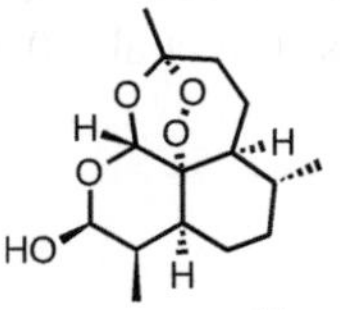
17

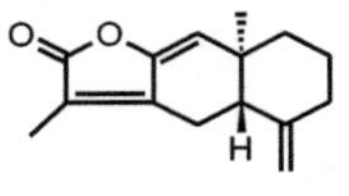
18

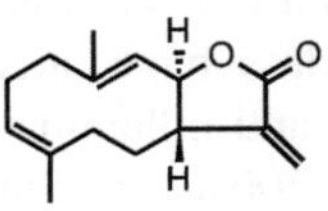
19

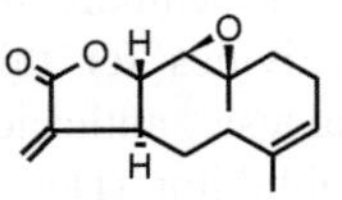
20

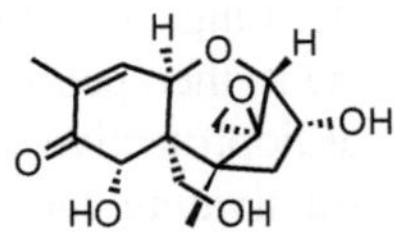
21

Dihydroartemisinin **17** also demonstrated immunomodulatory potential through regulation of the level Th/Treg. The CD4 + T-cell apoptosis is induced by heme oxygenase-1 in mouse models of inflammatory bowel disease; promoting B-cell suppression and CD8 + T lymphocyte responses in mice infected with two protozoan parasites (Yan et al. 2019b; Zhang et al. 2020a).

Atractylenolide I **18**, another sesquiterpene lactone, was able to reverse the immunosuppression caused by epithelial ovarian cancer by acting through NF-κB signaling (Liu et al. 2016a). Costunolide **19**, also a sesquiterpene lactone, is able to suppress innate immunity by preventing the production of nitric oxide and IL-1β which leads to the inhibition of macrophages pro-inflammatory effects and RAW 264.7 cells. In addition to this effect on innate immunity, costunolide **19** also has an effect on immunity acquired through the suppression of T helper lymphocytes differentiation and production of inflammatory cytokines, preventing T-cell-mediated responses in vivo, which may be useful in the treatment of CD4+ T-cell-mediated immune disorders (Park et al. 2016).

Finally, parthenolide **20**, another sesquiterpene lactone, is known for its anti-inflammatory and anticancer activities. Recently, its in vitro immunomodulatory activity has also been demonstrated, and it seems to involve the regulation of the Th1 and Th17 cells' activity, since the parthenolide **20** decreased the production of IL-12, TNF-α, and IFN-γ (Th1 response) and also reduced IL-6 and IL-17 production (Th17 response) (de Carvalho et al. 2017).

Another group of interesting sesquiterpenes related to the immune system are trichothecenes, a kind of mycotoxins produced by different species of fungi. They have both immunosuppressive and immunostimulatory effects, which will be determined according to the dose, exposure period, and the time at which immune function is assessed (Wu et al. 2017). The mechanisms involved are still being studied. In the case of immunosuppression, some trichothecenes have reduced the expression of IFN-γ, IFN-β, and others of Toll-like receptors (TLR), but it generally seems to be related to immune evasion (Wu et al. 2018). Deoxynivalenol **21**, a trichothecene, at low doses showed activation of the TLR4/NF-κB pathway and, at high doses, induced mitophagy-mediated mitochondrial dysfunction in weaned piglets and porcine alveolar macrophages (Liu et al. 2020a).

11.6 Diterpenes

The diterpenes are made up of four isoprene units, thus counting 20 carbons. They can be found in plants, fungi, bacteria, and animals in both terrestrial and marine environments. Important therapeutic activities for this class have already been described, such as cardioprotective, analgesic, anticancer, anti-inflammatory, anti-oxidant, and platelet-activating factor inhibitor (Dewick 2009a; de Sousa et al. 2018). Diterpenes are structurally diverse and that is why they have already been described as acting on very different targets from the immune system.

Phytol (PYT) **22** is a diterpene member of the long-chain unsaturated acyclic alcohols, one of the diterpenes with the simplest chemical structure. There is evidence that it has anti-inflammatory and immune-modulating activity. Some mechanisms involved include induction of apoptosis and protective autophagy; decrease in IFN-γ, IL-4, IL-10, and IL-17A production; and reduction in TNF-α and IL-1β levels. In general, these results indicate a benefit in relation to T-cell-mediated autoimmune diseases (Islam et al. 2018).

Rosmarinus officinalis L. and *Salvia officinalis* L. are rich in phenolic diterpenes such as carnosol **23** and derivatives. In general, they demonstrate potential anti-inflammatory activity. These types of phenolic diterpenes reduced gene expression levels of iNOS, cytokines/interleukins (IL-1α, IL-6) and chemokines and NO and PGE2 production in LPS-stimulated macrophages. These results showed improvement in acute and chronic inflammatory processes (Schwager et al. 2016). Another study conducted only with carnosol **23** showed that it inhibited the activation of transcription 3 phosphorylation, Th17 cell differentiation, and signal transducer and also blocked transcription factor NF-κB nuclear translocation (Li et al. 2018b).

Carnosic acid **24**, one of these phenolic diterpenes, with known cytoprotective activity, caused pro-inflammatory cytokine downregulation by deactivating MAPK and inhibiting NF-κB phosphorylation, in a model of collagen-induced arthritis in mice (Xia et al. 2017).

Tanshinone IIA **25**, an abietane-type diterpenoids, found in *Salvia miltiorrhiza* Bunge showed in an experiment with autoimmune encephalomyelitis rat model, reduction of macrophages/microglia in the spinal cord, CD4 + T cells, and CD8 + T cells, and it was able to downregulate IL-17 and IL-23 levels (Yan et al. 2016a). In another study, with the mouse peritonitis model, tanshinone IIA **25** was able to completely reverse the changes caused by sepsis achieved by cecal ligation and puncture, by apoptosis and decreased percentages of splenic CD4 + and CD8 + T cells, formation of regulatory T cells, reduction levels of interferon-γ and IL-2 in the spleen, and increased IL-4 and IL-10. These results originated a more balanced Th1/Th2 response, demonstrating the immune modulatory effect (Gao et al. 2019). Finally, tanshinone IIA activated protein 1 (AP-1) and transcription 1 (STAT1) and provided inhibition of transcription factors such as NF-κB in a dextran sulfate sodium-induced colitis in mice model (Liu et al. 2016b).

Andrographolide **26**, a labdane diterpenoid, which is the main compound of *Andrographis paniculata* (Burm. F.) Nees, showed anti-inflammatory activity; however, there is more to be discovered regarding the mechanism involved. Andrographolide has been described as an NF-κB activation inhibitor, inducer of iNOS expression, and inductor of prostaglandin E2 production by lipopolysaccharide. A study also demonstrated this diterpene downregulation of IRF3, including IFN-β and IP-10 by suppression of TRIF-dependent TLR signaling pathways (Kim et al. 2018). Another study also shows that in addition to these targets involved in inflammation, andrographolide negatively regulates glial cell-derived neurodegeneration and may be a candidate in the treatment of neurodegenerative diseases (Das et al. 2017).

Ent-kaurane **27**, in turn, represents another group of diterpenes, such as glaucocalyxin A, that showed anti-fibrotic activity in mice pulmonary fibrosis. In that study, glaucocalyxin A **28** alleviated the infiltration of macrophages and neutrophils in the lungs, inhibited the activation of NF-κB, and consequently decreased levels of cytokines as TNF-α, IL-1β, IL-6, and MCP-1 in the lungs of mice receiving bleomycin (Yang et al. 2017a). Oridonin **29** is another ent-kaurane-type diterpene with activity in the immune system. A research group found that this diterpene causes phagocytosis by macrophage-like U937 cells by autophagy. Later, the same research group determined that oridonin activates TLR4-autophagy pathway, which then leads to phagocytosis of apoptotic cells. This discovery can be a new strategy for immune disease and cancer therapy (Zang et al. 2019).

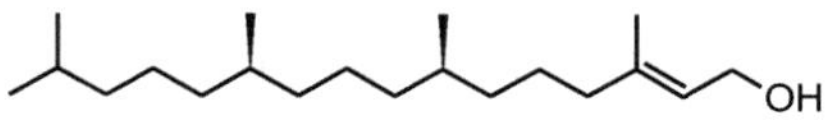

22

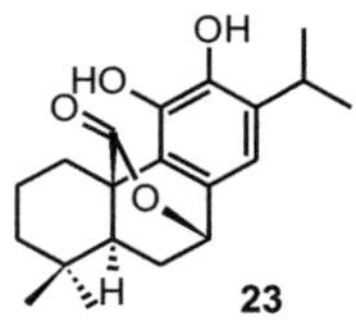

23

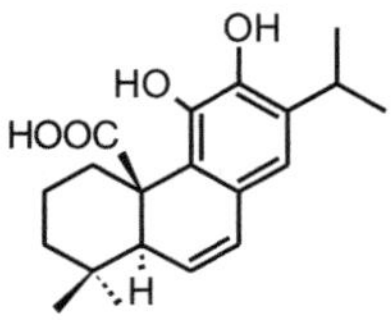

24

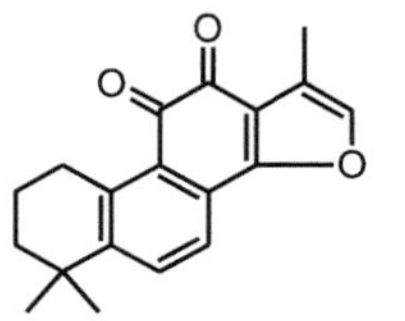

25

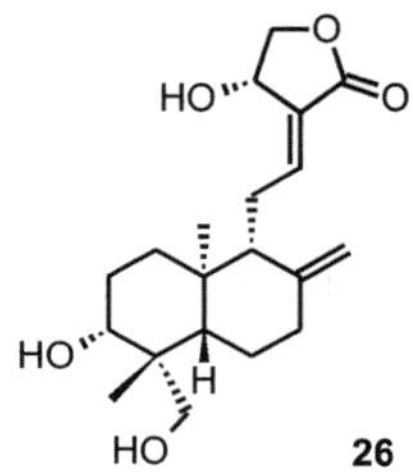

26

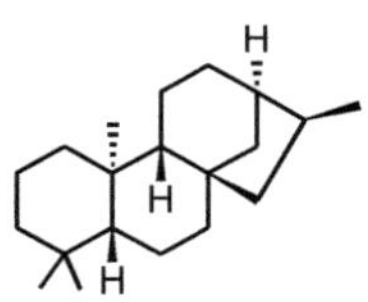

27

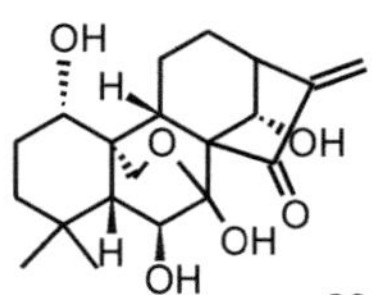

28

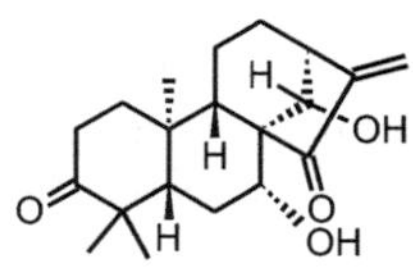

29

Paclitaxel **30**, a diterpene with a taxadiene skeleton, is an antitumor agent that has been widely studied for its effect on the immune system in the regulation of various cells, such as regulatory T cells, dendritic cells, natural killer cells, and macrophages (Zhu and Chen 2019). In preclinical studies, paclitaxel improved innate immune responses, including IL-1β release, NLRP3 inflammasome, and pyroptosis activation, reducing bacterial load and improving survival of animal (Zeng et al. 2019).

Another well-studied substance is triptolide **31**, from *Tripterygium wilfordii* Hook. F., a plant from Chinese traditional medicine herb known as thunder god vine. Its crude ethanol extract has already demonstrated great therapeutic potential, and one of its main components is triptolide, an abietane diterpenoid epoxide. It has already been extensively studied, and its effect on the immune system shows that triptolide stimulates cyto-protection by inhibiting pro-inflammatory cytokines and chemokines and thus suppressing inflammation. However, due to the narrow therapeutic window, imprecise mechanisms of action, poor water solubility, multi-organ toxicity, and its clinical development were embargoed and with some derivatives been produced for research in this same field (Hou et al. 2019; Xu et al. 2016; Pan et al. 2019; Yan et al. 2016b; Cheng et al. 2018; Guo et al. 2016a; Noel et al. 2019; Huang et al. 2019; Yuan et al. 2019; Guo et al. 2016b).

Finally, several diterpenes of marine origin with striking activity have been discovered in recent years. One example with promising activity in the immune system is cembranoids. A study with the species *Sinularia scabra* Tixier-Durivault led to the isolation of several diterpenes of this type, and they were studied for their activity on T and B lymphocytes. The substances xiguscabrate B **32**, xiguscabrols A **33** and B, 8-epi-xiguscabrol B **34**, sinulariol C **35**, and 14-deoxycrassin **36** demonstrated specific immunosuppressive potential in T cells. The substances (2R, 11S, 12S)-isosarcophytoxide and sinulariolide selectively inhibited B cells. In addition, a structure-activity study showed that the 11,12-oxirane ring and the oxidation on the C-19 position could be important on the bioactivity of cembranoids in T and B cells (Yang et al. 2019). An additional study carried out with sinulariolide evaluated dendritic cells stimulated by LPS and detected that their functional maturation was reduced by this concentration-dependent manner. Beside, the expression of co-stimulatory molecules CD40, CD80 and CD86; the release of TNF-α, IL-6, IL-12 and NO; T-cell proliferation; and the induction of factor-κB pathways (Chung et al. 2017).

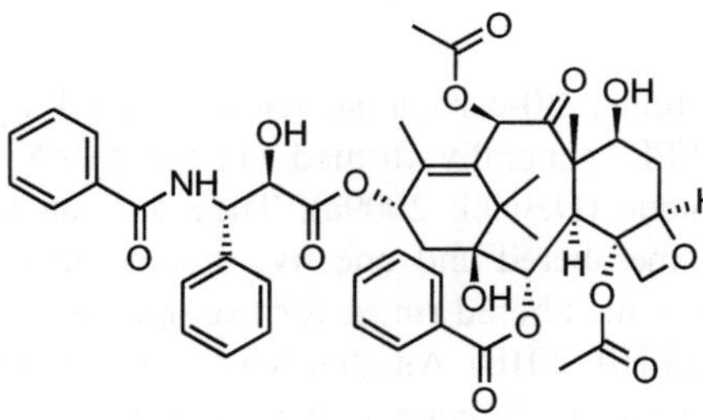

30

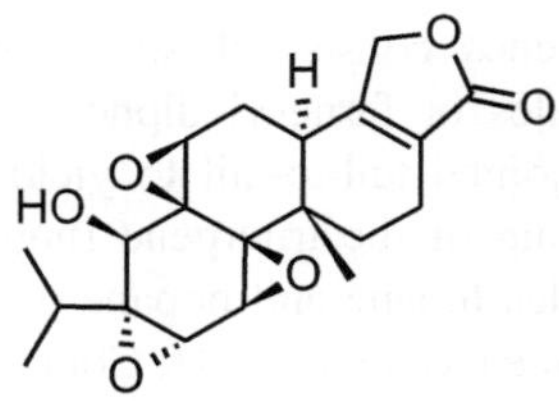

31

One more marine diterpene with an immunomodulator potential is dihydrogracilin A **37**. In in vitro and in vivo studies, dihydrogracilin A was able to downregulate the phosphorylation of NF-κB, signal transducer and activator of transcription (STAT), and extracellular signal-regulated kinase (ERK), accompanied by downregulation of IL-6 production and slight decrease of IL-10 (Ciaglia et al. 2017).

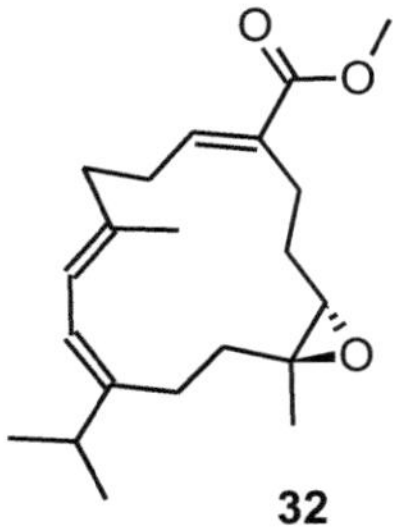

32

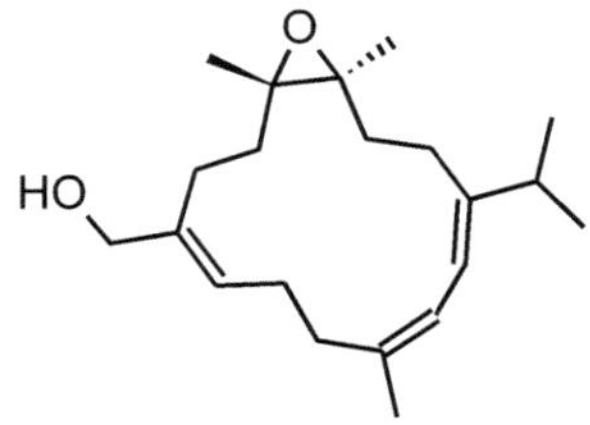

33

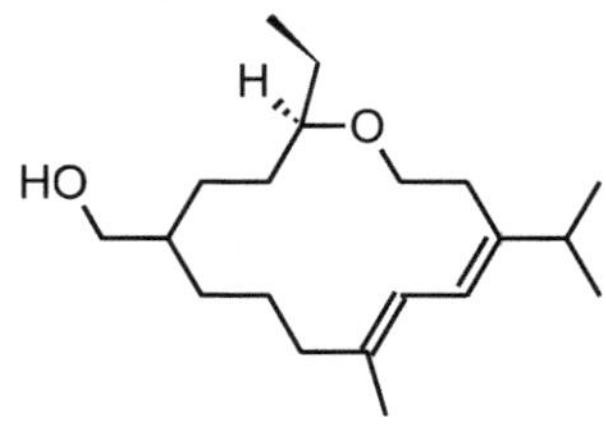

34

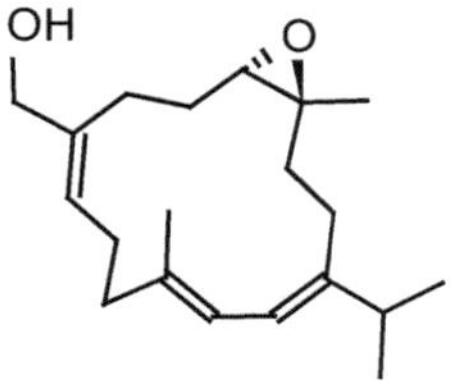

35

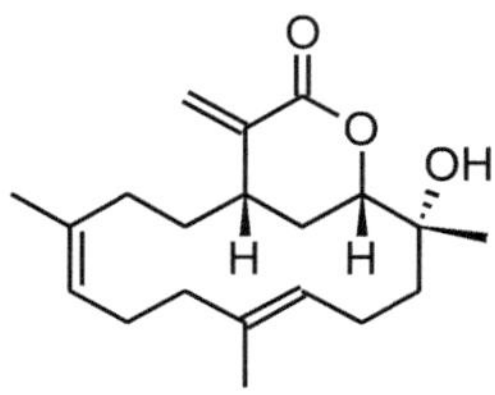

36

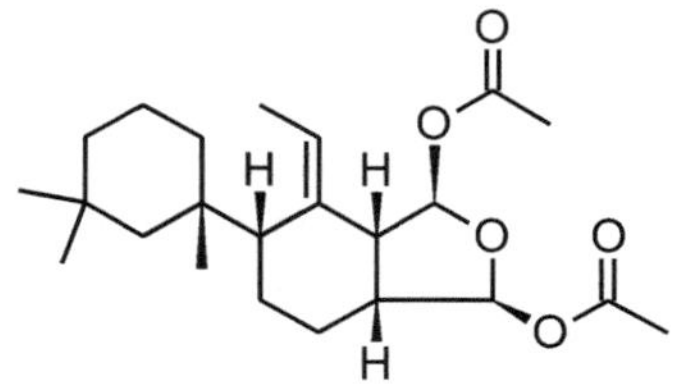

37

11.7 Triterpenes

Triterpenes consist of six isoprene units, 30-carbon skeleton, formed by two molecules of farnesyl diphosphate (FPP, generally formed via the MVA route) units, joined tail-to-tail to yield squalene (Dewick 2009a). They are categorized according of the triterpene rings: six-membered and one five-membered ring, as examples lupane and hopane or five six membered rings, for example, ursane and lanostane (Xu et al. 2004; Connolly and Hill 2010). Another feature is that they can exist in free form or in conjunction with sugars. In general, they are usually found in leaves, stem bark, and fruit and possess antimicrobial, insecticidal, and antifeedant activities against herbivores (González-Coloma et al. 2011), in addition to immunomodulatory, antitumor, anti-oxidant, and anti-inflammatory properties (Harun et al. 2020).

Lanosterol **38**, in animals and fungi, is the first sterol intermediate in the cholesterol biosynthetic pathway (Dewick 2009a). It achieved functions in both cholesterol metabolism and innate immunity by modulation of TLR4 innate immune responses in macrophages. Macrophages, adapting their gene expression profile, phagocytize bacteria and secrete both pro-inflammatory and antimicrobial mediators. The effect in mouse macrophages in vitro is to decrease IFNβ-mediated signal transducer and activate the transcription (STAT)1-STAT2. On the other hand, lanosterol accumulation increases membrane fluidity, allowed phagocytosis, increasing the capability to kill bacteria (Araldi et al. 2017).

Pentacyclic triterpenes are formed in a different sequence than lanosterol **38** (Dewick 2009a). They have several pharmacological activities, such as antimicrobial, antitumor, antioxidant, and anti-inflammatory properties. There is a hypothesis that these activities are due to modulation of the immune system (Harun et al. 2020). Betulin **39** isolated from *Hedyotis hedyotidea* (DC.) Merr. Betulin, suppressed the proliferation of T cell, in vitro, of mice splenocytes stimulated by concanavalin A (autoimmune hepatitis). In vivo, betulin significantly decreased the levels of IFN-γ, TNF-α, and IL-6 and improved liver injury. In addition, there are inhibition of activation of NKT and conventional T cells and decreased production of pro-inflammatory cytokines (Zhou et al. 2017).

Another pentacyclic triterpene, ursolic acid **40**, showed both immunomodulation and direct remyelination in the treatment of multiple sclerosis. Oral treatment with these compound significantly decreased disease severity, demyelination, and central nervous system inflammation in autoimmune encephalomyelitis (EAE) in animal multiple sclerosis model. Ursolic acid **40**-induced promyelinating neurotrophic factor CNTF in astrocytes by peroxisome proliferator-activated receptor γ(PPARγ)/CREB signaling, as well as by upregulation of myelin related gene expression during oligodendrocyte maturation via PPARγ activation (Zhang et al. 2020b). In a model of multiple sclerosis, autoimmune encephalomyelitis, dietary supplementation with an ursolic acid derivative (23-hydroxy ursolic acid), also accelerates the recovery with significantly decrease of the frequency of T cells producing IL-17 or IFN-γ (Chase et al. 2020). In one study, the protective effect of ursolic acid (10 mg/kg and 20 mg/kg day) against the LPS-induced cognitive deficits in mice was observed. The tests used to evaluate cognitive deficits were open field, step-through passive avoidance and Morris water maze tasks. The inhibition of IjBa phosphorylation and degradation, NF-jB p65 nuclear translocation, and p38 activation in mouse brain, without activation of TLR4, MyD88, ERK, JNK, and Akt were attributed mechanism of action. These results showed that ursolic acid **40** may be useful for mitigating inflammation-associated brain disorders (Wang et al. 2011).

Celastrol (tripterine) **41** is a pentacyclic triterpenoid of the quinone methides type, isolated from the root extracts of *Tripterygium wilfordii* Hokki. F. (thunder god vine), a traditional Chinese herb grown in the east and south of China, Japan, and Korea. This species exhibited multiple pharmacological activities, such as anti-inflammatory, immune modulation, and antitumor properties, including rheumatoid arthritis (Wang et al. 2018). IL-1β and TNF-α are inhibited by Celastrol **41**, and it is used in rheumatoid arthritis. Celastrol **41** significantly decreases the number of macrophages and ceased joint destruction without side effects, being validated as a *p* compound for the arthritis treatment (Cascão et al. 2015). Another research demonstrated that celastrol (1–15 mg/kg/day) improves autoimmune disorders in Trex1-deficient mice, by downregulation of the interferon response in vitro and in vivo. Thus celastrol improves the myocarditis that has high interferon response and autoantibody production (Liu et al. 2020b).

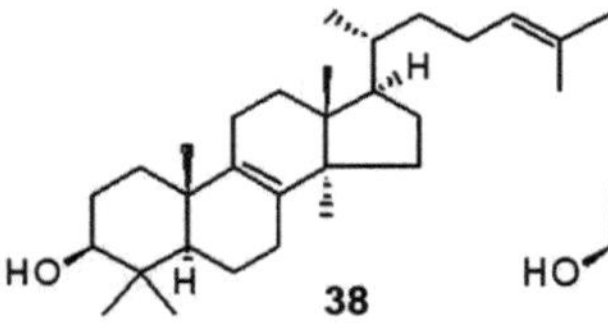
38

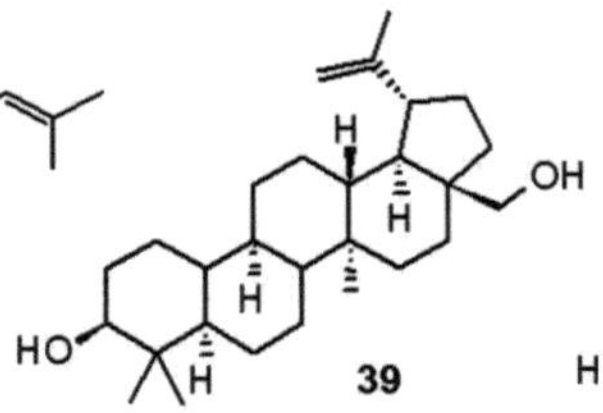
39

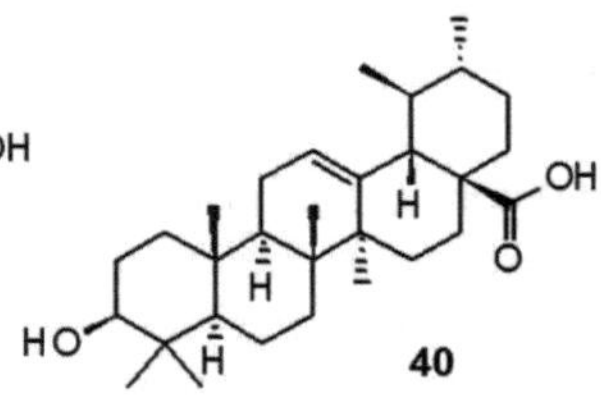
40

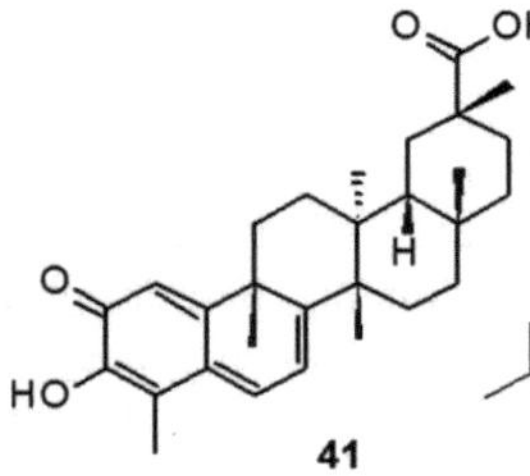
41

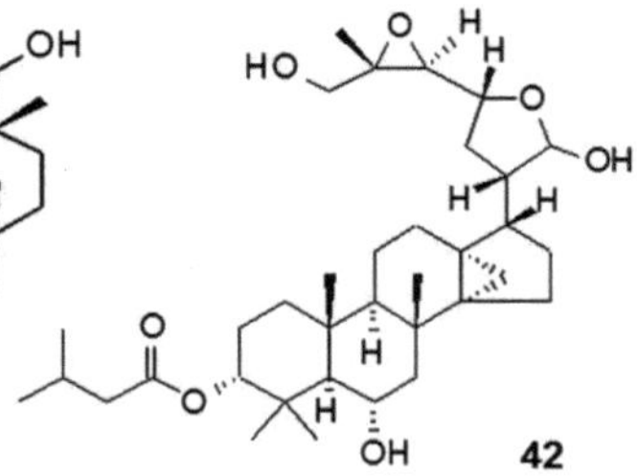
42

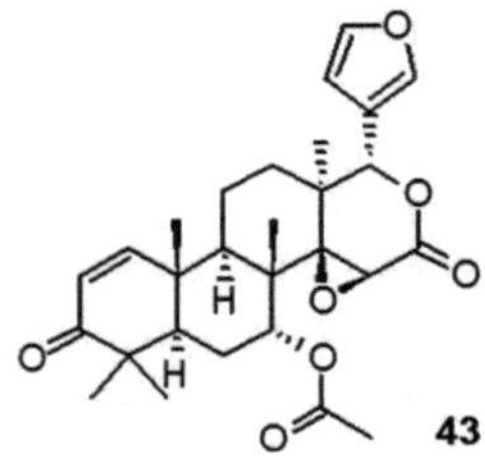
43

A glabretal type triterpenoid named dictabretol A **42** to D was isolated *from Dictamnus dasycarpus* Turcz. (Rutaceae). This specie has been used as a traditional herbal medicine and has immunosuppressive activity. As a result it was found that dictabretol A blocked the cell cycle transition from the G1 to the S phase and consequently inhibited lymphocyte proliferation. Additional the treatment with dictabretol A improved the severity of collagen-induced arthritis (Choi et al. 2016).

The limonoids, also known as tetranortriterpenoids, are derivative from the triterpenes and are usually investigated due to the bitterness, mainly in *Citrus* fruit. The limonoid name is derived from limonin, the first bitter tetranortriterpenoid isolated (Tundis et al. 2014). Gedunin **43** is a limonoid isolated from *Azadirachta indica* (Neem) and *Carapa guianensis* (andiroba) and showed protective effect on TLR-mediated inflammation by modulation cytokine production and inflammasome activation. Gedunin inhibited LPS-induced caspase-1 activation, production of IL-1β, decreased TNF-α, IL-6, NO, NLRP3 and IL-1β levels, and enhanced IL-10 and HO-1 levels by macrophages stimulated with the TLR2 and TLR3 agonists, in vivo and in vitro (Borges et al. 2017).

11.8 Phytosteroids

Steroids or phytosteroids are modified triterpenoids containing the tetracyclic ring of lanosterol. A wide range of biologically important activities are brought by modification, mainly, in the side chain and by the presence of substituents, e.g., sterols, steroidal saponins, cardioactive glycosides, bile acids, corticosteroids, and mammalian sex hormones. Many natural, synthetic, and semisynthetic steroids are normally used as therapies in medicine. The cholesterol is the main sterol found in mammals, which acts as a precursor for other steroid structures such as sex hormones and corticosteroids. In plants, fungi, and algae, the sterols have extra one-carbon or two-carbon substituents on the side chain, attached at C-24. The campesterol **44** and β-sitosterol **45,** widespread plant sterols, are respectively 24-methyl and 24-ethyl analogues of cholesterol **46**. Stigmasterol **47** contains additional unsaturation in the side chain (Dewick 2009a). Sitosterol (50%) and stigmasterol **45** (20%) are produced commercially from soy beans (*Glycine max* (L.) Merr.) as raw materials for the semi-synthesis of medicinal steroids. This class of compounds has been described as presenting estrogenic, anticancer, anti-inflammatory, anti-allergic, and immunomodulatory activities. The immunosuppressive property of this class, acting on different immune system targets, has already been described.

Stigmasterol **47** has shown beneficial health effects, including lipid lowering, anticancer, anti-inflammatory, immunomodulatory, and anti-allergic effects. Pretreatment with stigmasterol reduced fever induced by LPS and reversed the neutrophil proliferation both in blood and peritoneal fluid, controlling lung and liver damage and protecting mice from mortality (Antwi et al. 2017).

β-Sitosterol **45**, isolated from chloroform extract of *Nitraria retusa* (Forssk.) Asch. leaves, presented antitumoral activity through immunomodulation. β-Sitosterol remarkably protected the lung parenchyma, increased splenocyte proliferation, and enhanced lysosomal activity of host macrophages and present antioxidant cellular activity (Boubaker et al. 2018).

In another study, the effect of dietary β-sitosterol **45** addition at different levels could regulate serum lipid levels, promote immune function, and improve intestinal oxidative status and morphology in chickens. β-Sitosterol linearly decreased concentrations of total serum cholesterol, tumor necrosis factor-α (TNF-α), interleukin-1β (IL-1β), and mRNA relative expression of TLR4 and MyD88 (Cheng et al. 2020).

Regular consumption of plant stanol **48** esters is shown to reduce cholesterol levels, but it is important in the immunomodulation. Recently the introduction of plant sterol and stanol esters as food additives was performed due to lowering cholesterol. Stanols are obtained by hydrogenation of plant sterols, mainly from sitosterol, campesterol **46**, and stigmasterol **44** and then esterification with fatty acids (Dewick 2009a). In a clinical study, stanol ester **48** consumption improved the immune function in asthma patients in vivo. Asthma patients (58) participated in a randomized, double-blind, placebo-controlled interventional study for 4 weeks with daily plant stanol consumption. The group of asthma patients that consumed plant stanol ester **48** showed reduction in plasma total immunoglobulin E, interleukin (IL)-1β, and tumor necrosis factor-α levels. The concentrations of plant stanol was increased in serum, and it was correlated with the decrease in IL-13 concentrations and the Th1 (Brüll et al. 2016).

One example of immunosuppressive activity is stephanthraniline A **49**, a C21 steroid isolated from roots of *Stephanotis mucronata* (Blanco) Merr., that is used in Chinese folk medicine for the treatment of rheumatoid arthritis. Stephanthraniline A inhibited T-cell activation and proliferation through proximal T-cell receptor (TCR) signaling- and Ca(2+) signaling-independent way. Thus, stephanthraniline A **49** has potential application in the treatment of CD4(+) T cell-mediated inflammatory and autoimmune diseases (Chen et al. 2016).

Withasteroids isolated from flowers of *Datura metel* L. showed efficiency to inhibit immune responses in vitro and in vivo experiments. Withasteroid B2 **50** showed inhibitory effect on immune responses by suppress the differentiation of CD4(+) T cells and cytokines and could be responsible by the therapeutic effects observed in psoriasis patients (Su et al. 2017).

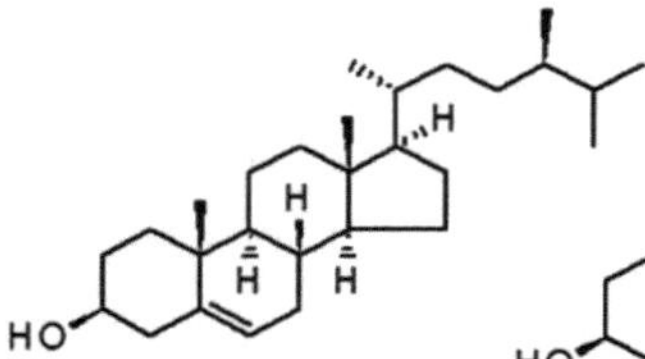

44

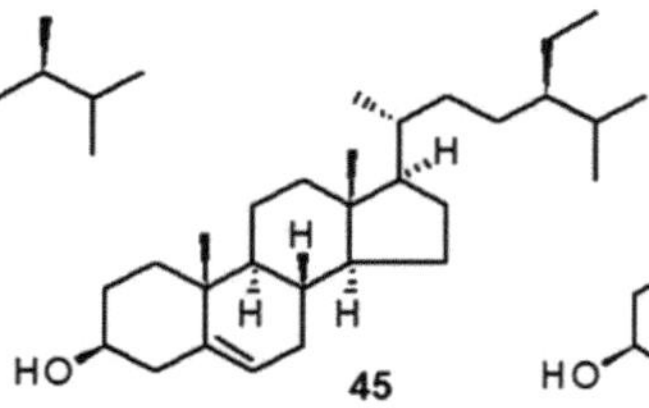

45

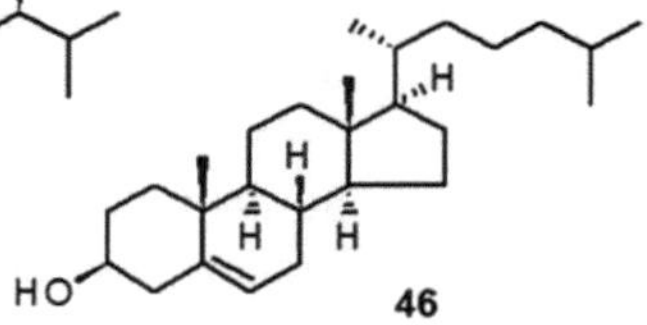

46

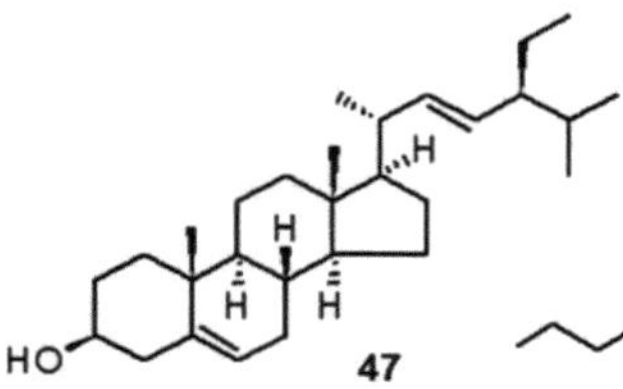

47

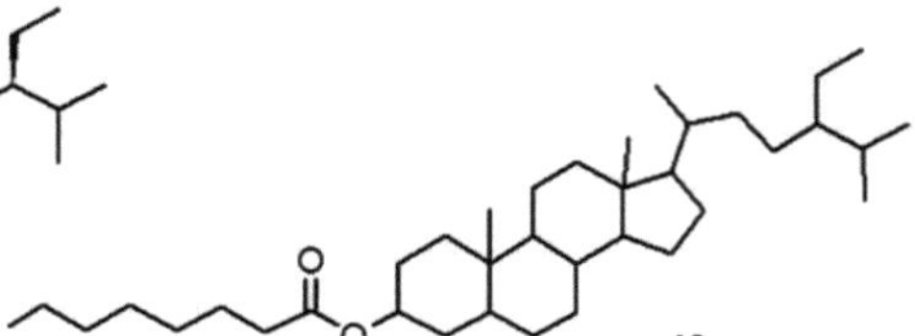

48

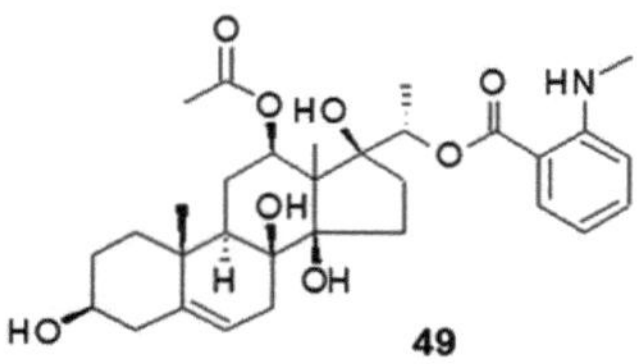

49

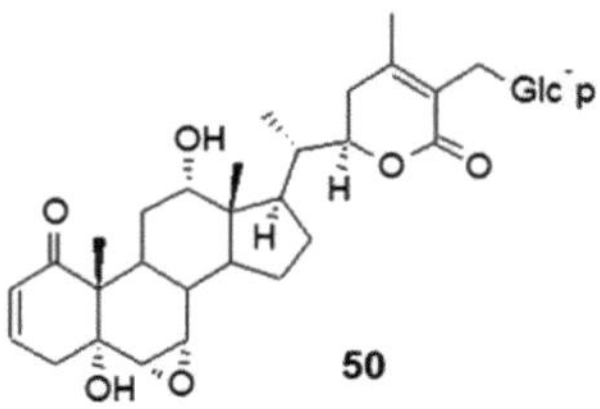

50

Vitamin D, one metabolite of steroids, is also known to have physiological functions, including immunosuppression and differentiation of both normal and malignant cells and hormone secretion. Vitamin D_3 (cholecalciferol) is a sterol metabolite photochemically formed on the skin in animals from 7-dehydrocholesterol by the sun's irradiation (Fig. 11.3). These photochemical reaction allows ring opening to generate precholecalciferol (Dewick 2009a), that is hydroxylated into calcidiol and them into calcitriol. Cholecalciferol and calcitriol have also been found in several plant species.

Vitamin D_3 activated T lymphocytes, macrophages, and dendritic cells, leading to antiproliferative and immunosuppressive effects on the immune system. It also inhibits the secretion of IL-12, an essential cytokine for the differentiation of CD4+ T lymphocytes in Th1 cells, and consequent decrease of IL-2, interferon, and TNF-α levels, contributing to the activation of Th2 response, and the production of cytokines such as IL-4 (Sadeghi et al. 2006). Vitamin D also can reduce the risk of infections through several mechanisms, such as the induction of cathelicidins and defensins that reduce viral replication and concentration of pro-inflammatory

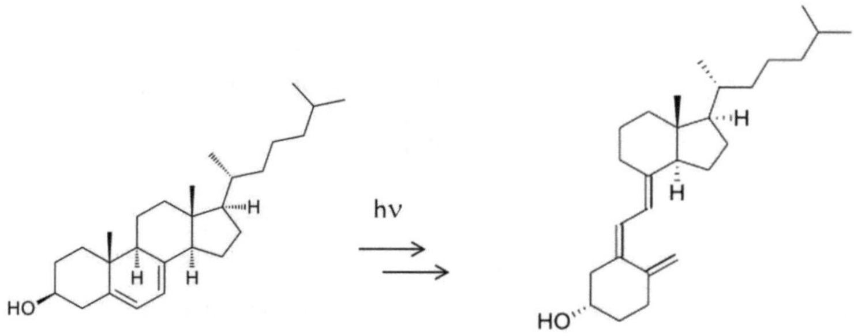

7-dehydrocholesterol

Colecalciferol
(Vitamin D_3)

Fig. 11.3 Cholecalciferol (vitamin D_3) biotransformation by UV radiation

cytokines, with a potential therapeutic use in lung injury in cases of COVID-19 (Grant et al. 2020). Additionally, hypovitaminosis D can be related to disorders with potential impact on COVID-19, such as arterial hypertension, hepatic steatosis, and hyperuricemia (Nogueira-de-Almeida et al. 2020).

11.9 Tetraterpenes

The tetraterpenes are represented by carotenoids. These compounds are found in plant tissues, in fungi, and in bacteria and play an important role in the photosynthesis. Formation of this kind of skeleton involves tail-to-tail coupling of the two molecules of geranylgeranyl diphosphate (C20). After, the conjugation of double bonds is extended by a sequence of desaturation reactions, eventually generating lycopene (Fig. 11.4), that presents an all trans configuration, common in most carotenoids (Dewick 2009a). Tomato (*Solanum lycopersicum* L.) and processed tomato are the main source of carotenoids, more specifically lycopene. The orange color of carrots (*Daucus carota* L.) is due to β-carotene, a compound widespread in higher plants. Recent studies suggest that this class of substances represents important antioxidant molecules in plants and in humans. They protect plants and algae of photooxidative damage and mitigation of toxic oxygen species (ROS), and in humans they minimize cell damage and protect against some types of cancer.

A dietary supplementation with astaxanthin reduced immunopathology and improves longevity through immunosuppression in an insect model (mealworm beetle, *Tenebrio molitor*). This carotenoid is isolated from a green alga *Haematococcus pluvialis* and has a clinically proven safety profile in addition to immunomodulatory properties, anti-inflammatory, and antioxidant (Jiang et al. 2016; Talukdar et al. 2020).

Experimental data brought promising results about immune-stimulating and antioxidant properties of astaxanthin. All the immune parameters were influenced

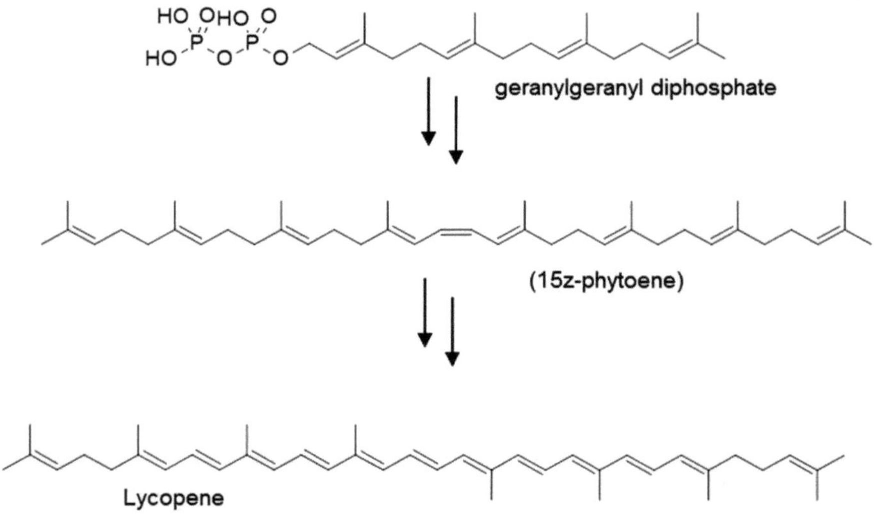

Fig. 11.4 Schematic route of lycopene formation

by the dietary supplementation with astaxanthin; concentration of carotenoids was increased in the hemolymph of *T. molitor* larvae after 3 weeks (Dhinaut et al. 2017). Another study showed that astaxanthin (70–300 nM), in primary cultured lymphocytes, modulates immune responses in vitro and exerts immunomodulatory effects, ex vivo, by increasing INF-γ and IL-2 production without inducing cytotoxicity (Lin et al. 2015). Additionally, there is evidence that astaxanthin could support COVID-19 treatment, once the excessive inflammatory immune response to SARS-CoV-2 infection is associated to severe pneumonitis and acute lung injury (ALI) or acute respiratory distress syndrome (ARDS). Preclinical studies support astaxanthin preventive actions against ALI/ARDS due to regulation of the expression of IL-1β, IL-6, IL-8, and TNF-α and the signaling pathways (Talukdar et al. 2020).

The carotenoid (4,4′-diaponeurosporene) **51** produced by *Bacillus subtilis* (B.s-Dia) was tested on intestinal mucosal immunity in piglets. Oral administration of B. s-Dia improved the development of Peyer's patches and increased villus height, colon crypt depth, and the number of intraepithelial lymphocytes in piglets and improved the immune system. This fact is important because piglets have an incompletely development of mucosal immune system, and, consequently, they are sensitive to intestinal infections; thus B.s-Dia can be considered as an immunopotentiator candidate (Jing et al. 2019).

The vitamins of the group A are important metabolites of carotenoids and are considered provitamins. Retinol **52** and dehydroretinol (vitamin A_1 and A_2) **53**, in mammals, are derived of tetraterpenoid oxidative metabolism, obtained from the diet. In the intestine, cleavage occurs in the mucosal cells, resulting in two molecules of retinal, that will be subsequently reduced to alcohol retinol (Dewick 2009a). Vitamins A_1 and A_2 are vitamins found in animal products, such as eggs, dairy, and animal liver and kidneys. They are now known as regulator of vision, growth, cell differentiation, and embryonic development. Retinoic acid **54** is the most active vitamin A metabolite, and it has been found in the cellular differentiation and proliferation and in regulation of gene expression (Bono et al. 2016). Synthetic retinoic acid (tretinoin) and isotretinoin are used as topical or oral acne vulgaris treatment. (Yin et al. 2017).

It was suggested that at pharmacological doses of retinoic acid affect immune functions after a depletion period of β-carotene and lycopene diet. In the gut-associated lymphoid tissue, the production of retinoic acid induce local inflammation, so it supplements with β-carotene and lycopene supplementation, might improve immune functions without increasing risk for cancers (Toti et al. 2018). One study shows that vitamin A dietary status and the consequent epithelial cell expression of the transcription factor retinoic acid receptor β (RARβ) are essential to regulate adaptive immunity (Gattu et al. 2019).

Retinoic acid **54** is important in the regulation of intracellular antiviral innate immunity in hepatocytes in persistent infection by the hepatitis C virus. The mechanism involves the alcohol dehydrogenase (ADH) and aldehyde dehydrogenase (ALDH) pathway, being antiviral in regulating hepatocytes of the expression of interferon (Cho et al. 2016).

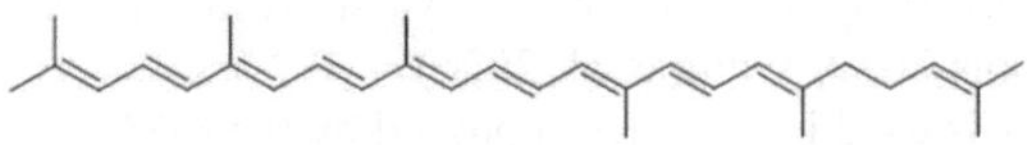

51

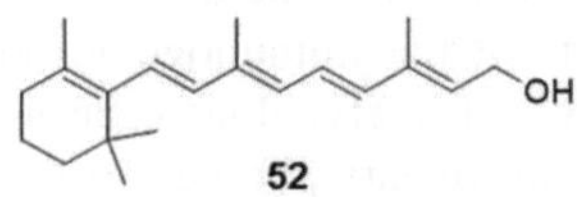

52

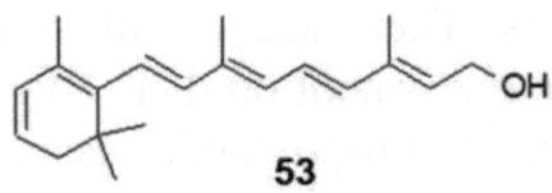

53

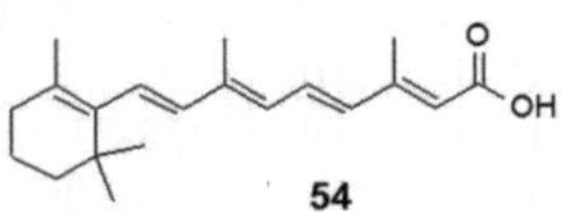

54

11.10 Saponins

Saponins are glycosides of polycyclic steroids or triterpenes characterized by their surfactant and detergent properties, as they are able to form abundant and persistent foam in water. The physical-chemical characteristics of these metabolites, endowed with a nonpolar portion linked to polar chains of sugar, are related to the various biological activities reported for the class, especially their action on membranes, their hemolytic activity, and their ability to complex with steroids. These properties, in turn, determine their therapeutic applications, such as their use as adjuvants in vaccines due to the immunomodulatory activity presented by them (Fleck et al. 2019).

Adjuvants of vaccines are components that have the function to enhance and/or direct the patient's immune response to a certain antigen. Isolated saponins of the genus *Quillaja* are the main example of substances in this class that are used for this purpose (Magedans et al. 2019; Lacaille-Dubois 2019). The genus is composed of the species *Quillaja saponaria* Molina and *Quillaja brasiliensis* Mart., both known as "soap trees," precisely because of the significant presence of saponins in their composition. Previous studies analyzing extracts, fractions, and isolated saponins from these two species demonstrated a potent immunogenic activity that seems to be mediated by stimulating cellular and humoral immune responses, related to vaccines with a wider spectrum of application. The fraction called Quil A, for example, is a mixture of 20 saponins obtained from *Q. saponaria* and has already been evaluated as an adjunct to several veterinary vaccines and human vaccines undergoing pre-clinical tests. However, due to its heterogeneous composition and toxicity profile, they have not been used in human vaccines so far (Fleck et al. 2019; Magedans et al. 2019).

Several efforts have been made to fractionate and purify Quil A, resulting in the isolation of the saponin QS-21, which is a mixture of the isomeric saponins QS-21 apiose and QS-21 xylose **55.** The difference between these two saponins is the terminal sugar of the linear tetrasaccharide linked at C-28 through an ester bond, which can be an apiose or xylose molecule. The proportion of saponins in the mixture is 2:1 of QS-21 apiose and QS-21 xylose, respectively (Magedans et al. 2019; Lacaille-Dubois 2019; Wang et al. 2020).

Currently, QS-21 is used in vaccines approved by the FDA and EMA for human use, in addition to its use in veterinary vaccines and formulations still in preclinical and clinical tests. There are several reports of the immunogenic potential of the adjuvants, with assessment on their efficacy and safety, alone or as a component of adjuvant systems, in prophylactic vaccines for malaria, herpes zoster, tuberculosis, and HIV and also in therapeutic vaccines for cancer treatment (Magedans et al. 2019; Lacaille-Dubois 2019; Wang et al. 2020).

Although the immunogenic potential of QS-21 has already been extensively reported, the mechanisms of action related to this activity are still not well characterized. In this sense, the evaluation of the immunogenic activity of several synthetic derivatives of QS-21 contributed to the study of its mechanism of action and structure-activity relationship. The search for synthetic derivatives also aims to develop safer and more effective compounds that can be easier to obtain, since the isolation of QS-21 is arduous and generates economic and ecological consequences. Additionally, due to the toxicity profile of saponin mixtures, mainly associated with hemolytic activity, its application is dose-limited when in concentrations above those are recommended, which vary between 50 and 100 μg (Magedans et al. 2019; Wang et al. 2020; Ghirardello et al. 2020).

In any case, some progress has already been made regarding the comprehension of the QS-21 mechanism of action, and it is known that it is probably related to its immunomodulatory effect. QS-21 seems to stimulate both humoral and cellular immunity, which enables its action against a wide variety of antigens. Its broad spectrum is an advantage over other commonly used adjuvants, such as aluminum salts, for example. It is known that QS-21 is capable of inducing Th1-type immunity, which is extremely important in controlling most intracellular pathogens and is not promoted by aluminum salts (Lacaille-Dubois 2019).

According to Marciani et al. (Marciani 2018), one of the mechanisms of action of QS-21 involves its effect in dendritic cells (DC) and Th1 cells. Apparently, exogenous antigens are phagocytosed by DCs along with saponin cholesterol-dependent by endocytosis. The saponin mixture has a high affinity for the cholesterol molecules present in the endosomal membrane, resulting in membrane destabilization and pore formation, which favors the antigen escape into the cytosol and its further processing into peptides. These are then carried to classic class 1 histocompatibility molecules (MHC-i) and presented on the DC surface for CD8 T cells, which form cytotoxic T lymphocytes. It is also noteworthy that the aldehyde group present in the triterpene portion of QS-21 appears to bind with the amino group of the inactivated T-cell surface receptor through the formation of an imine, a process that generates a signal to stimulate the T cell. This signal contributes to T-cell activation and cytokine secretion, which is important for eliminating infected cells (Lacaille-Dubois 2019; Marciani 2018).

In studies on DCs and macrophages, another mechanism suggests that QS-21 activates multiprotein complexes responsible for triggering the cascade of caspase-1, called NLRP3 inflammasome. As a result, the release of active pro-inflammatory cytokines, especially IL-1 and IL-18, promotes the maturation of Th17 lymphocytes and is capable of inducing Th1 cell response. However, in a study in NLRP3-deficient mice that received a vaccine containing the adjuvant, an increase in the responses of specific T cells was observed, both Th1 and Th2 cells, along an increase in IgG1 and IgG2c, suggesting that the role of QS-21 still needs to be further characterized (Marty-Roix et al. 2016).

In addition to the application of QS-21, the development of adjuvant systems (AS) and the incorporation of saponins in liposomes have shown to be good alternatives to reduce its toxicity and hemolytic effect, while also increasing the immunogenic potential. One of the greatest examples is the development of the AS01 system containing QS-21 and the immunostimulant 3-O-deacyl-4-monophosphoryl lipid (MPL). AS has been able to elicit a more robust immune response while at the same time generating fewer adverse effects. Reports from the literature suggest that the effect of AS01 is due to the synergistic effects of MPL and QS-21. The system is believed to induce a rapid and transient innate immune response at the injection site, resulting in the activation of a wide range of antigen-presenting cells (APC), leading to a modulation and an increase of the protective immune response (Didierlaurent et al. 2017; Netea et al. 2019; Ortuño-Sahagún et al. 2017; Kiyama 2017; Tholl 2006).

AS01 is being tested as an adjunct to several vaccine candidates in clinical trials. The most advanced studies include the malaria vaccines and the prevention of herpes zoster in adults over 50 years old. The evidence available regarding these vaccines suggests that AS01 has a clinically acceptable safety profile and a positive risk-benefit ratio. Thus, the prospects for applying AS01 as an adjuvant in vaccines against diseases caused by complex pathogens and those targeting populations with challenging immune states are promising (Sarikahya et al. 2018; Netea et al. 2019; Ortuño-Sahagún et al. 2017; Kiyama 2017; Tholl 2006).

Another strategy adopted to minimize the toxicity of adjuvants containing saponins is to combine them with cholesterol and phospholipids, forming a spherical structure of 40 nm called immune-stimulating complexes (ISCOM). These complexes enhance the immunogenic activity of saponins, and, therefore, a smaller amount of metabolites is necessary to obtain the expected response, decreasing system toxicity. In addition, the fact that the saponins are packed within the structure also seems to reduce the adverse events reported for these substances. Hydrophobic or amphipathic antigens can be incorporated into the complex, increasing the cellular and humoral immune response. Studies with ISCOMs based on Quil A and QS-21 using a wide range of antigens showed satisfactory immunogenic activities and safety (Magedans et al. 2019; Garcia and Lema 2016). A meta-analysis assessing the safety and tolerability of the QS-21 and other systems reinforces the potential of *Quillaja* saponins, especially QS-21, both isolated and incorporated in adjuvant systems, for the development of new vaccines (Bigaeva et al. 2016).

Other saponins obtained from the genus *Quilaia*, such as those isolated from the species *Q. brasiliensis*, have also been considered as a promising alternative for the development of vaccines. Preclinical studies demonstrated that fractions of these species that are rich in saponins showed similar immunogenic effect, but with a better safety profile if compared to the Quil-A fraction (Magedans et al. 2019; Cibulski et al. 2018).

In addition to the species of the genus *Quilaia*, other medicinal plants are known for the distinctive presence of saponins with immunomodulatory potential. Examples include saponins isolated from species of the genus *Astragalus*, which have already demonstrated immunogenic potential with very low hemolytic activity, being a possible alternative to *Quillaja* saponins as adjuvants in vaccines. Zhang and colleagues (Zhang et al. 2018) evaluated the application of saponins isolated from *Astragalus* in a formulation containing cholesterol and liposomes, in addition to the antigen, aiming the future development of a cancer vaccine. The main goal of this type of vaccines is to induce cytotoxic T lymphocytes against tumors and trigger the cellular immune response. As a result, the authors observed that the complex promoted a robust antibody response and an increase in interferon-γ levels in mice, which presupposes antitumor activity and the prospect that *Astragalus* saponins have the potential to be used in vaccines for tumor treatment (Zhang et al. 2018).

The use of saponins as adjuvants in vaccines is largely reported and described in the literature, but the immunomodulatory activity of this class of metabolites also has the potential to be used as prevention and treatment of many diseases. There is evidence, for example, that the saponin Astragaloside IV (AST IV) **56**, also isolated from species of the genus *Astragalus*, can be used in the treatment of immunological disorders, what has already been demonstrated in an in vivo study where the substance increased immunological response. Regarding the mechanism of action, some studies have investigated the effect of these saponins in specific pathways. Li and collaborators (Li et al. 2016) evaluated the effect of AST IV on the immune response of regulatory T cells inhibited by HMGB1 in mice. This protein plays an essential role in triggering the characteristic events of sepsis in which immunity is compromised resulting in immunosuppression. The results indicated that AST IV modulated the host's immune system as it controlled the inflammation promoted by HMGB1, suggesting it may be a future alternative for the treatment of sepsis (Li et al. 2016). A second study demonstrated that AST IV improves the immune function of RAW264.7 cells by activating the NF-κB/MAPK signaling pathway (Li et al. 2017b).

Other species rich in saponins with immunomodulatory potential are those popularly known as ginseng. The popular name is attributed to species that belong to the genus *Panax*, historically and widely used in traditional Chinese medicine. Among these, the best known are *Panax ginseng* C.A. Mey, *Panax notoginseng* J. Wen, and *Panax quinquefolius* L., due to the significant presence of specific triterpenes saponins exclusive to the genus *Panax*, called ginsenosides (Shi et al. 2019).

More than 100 different saponins have already been identified in ginseng species, and many of them have demonstrated high immunomodulatory activity, which is often associated with the potential application of these metabolites in cancer therapies. Studies with the ginsenoside Rg1 **57**, for example, have shown this saponin increases humoral and cellular immune responses inducing the release of several cytokines, such as IL-6, tumor necrosis factor-alpha (TNF-α), and IL-1β and chemokines such as IL-8 and IP-10. In addition, ginsenoside Rg1 **57** also stimulates the expression of surface molecules CD83, CD80, and leukocyte antigen D (HLA-DR) (Shi et al. 2019; Huang et al. 2017). Huang et al. (Huang et al. 2017) believe that these mechanisms explain the higher survival rate of mice immunized with a formulation containing Rg1 in an experimental lymphoma induction test with E.G7-OVA cells, suggesting a potential application of Rg1 **57** as an adjunct to prophylactic vaccines for this type of tumors.

In contrast, ginseng saponins such as RD **58** and CK **59** have shown positive regulation of T-cell differentiation in in vitro assays and an increase in the production of immunosuppressive cytokines TGF-β and IL-35. These findings suggest a potential application of these saponins as immunosuppressive agents that can be useful in transplants and autoimmune diseases (Shi et al. 2019).

YI and colleagues (Yi 2019) investigated the role of ginseng saponins in the innate immune response due to their inhibitory effect on inflammasomes. Some types of ginsenosides inhibited inflammatory responses by suppressing the activation of several inflammasomes, including NLRP3 and NLRP1. Additionally, unlike other findings, the ginsenosides evaluated inhibited caspase-1 and decreased the expression of IL-1β and IL-18. According to the authors, these results may support future development of drugs to prevent and treat inflammatory diseases that have a mechanism of action involving the inhibition of inflammasomes. Considering this, there is promising and varied evidence about the immunomodulatory properties of ginseng saponins, but more studies are needed to better understand the mechanism of action and effective clinical application of these substances (Shi et al. 2019; Yi 2019).

Lastly, it is important to highlight that some saponins have recently been cited in the literature as promising substances for the development of drugs to prevent and treat COVID-19. In a literature review on natural products with the potential to prevent infections by the new coronavirus, glycyrrhizic acid **60** was identified as an alternative that deserves to be investigated (Chen and Du 2020). This saponin has already been evaluated in several preclinical models, demonstrating anti-inflammatory and immunomodulatory activities. Chen and colleagues (Chen et al. 2020) evaluated the effect of a formulation containing glycyrrhizic acid saponin, in addition to vitamin C and curcumin, which regulated the immune response against SARS-CoV-2 infections, inhibiting the excessive inflammatory response and preventing the beginning of the cytokine storm, characteristics of the disease (Chen et al. 2020).

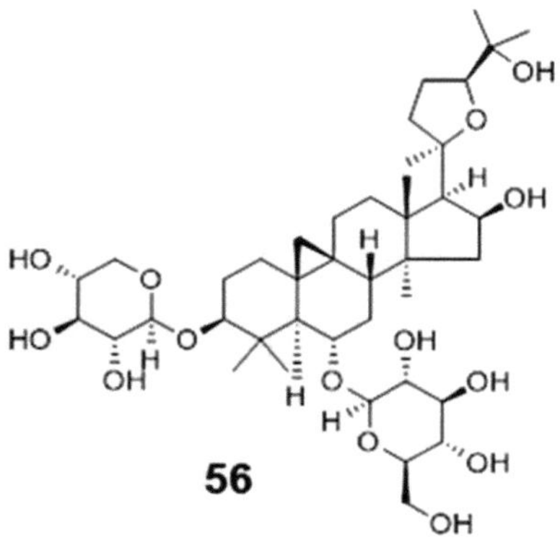

56

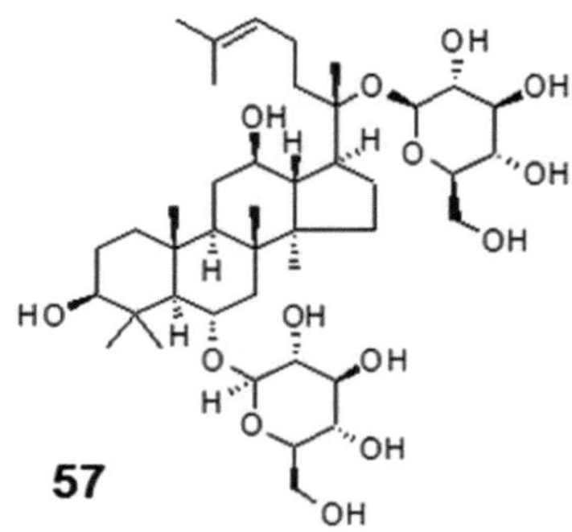

57

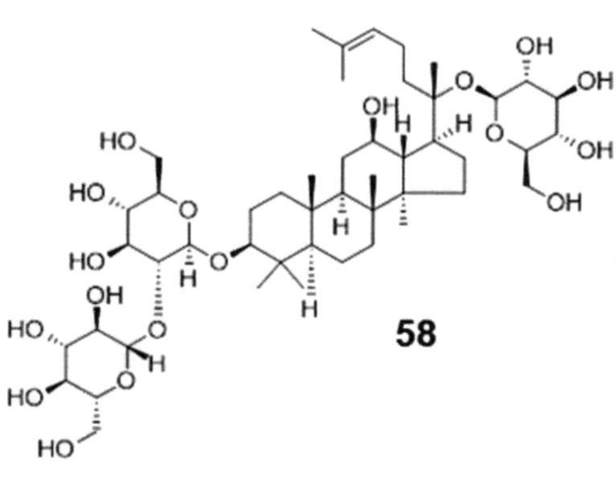

58

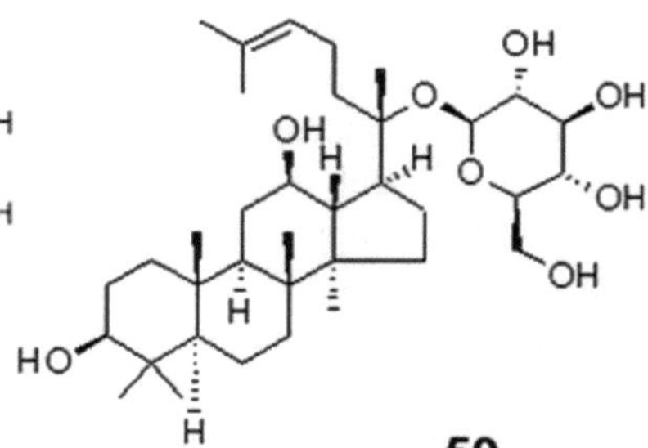

59

The so-called saikosaponins isolated from the species *Bupleurum falcatum* L. were also mentioned as promising alternatives for the treatment of COVID-19. According to Bahbah and collaborators (Bahbah et al. 2020), these saponins exhibit immunomodulatory activities that could be interesting in cases of infections by the new coronavirus, in addition of its effectiveness against various types of viruses.

Saikosaponin A **61,** for example, inhibits the production of several inflammatory mediators, including reactive oxygen species, TNF-α, prostaglandins, nitric oxide, and IL-6, IL-8, and IL-1, responsible for the cytokine storm observed in critically ill patients diagnosed with COVID-19 (Bahbah et al. 2020).

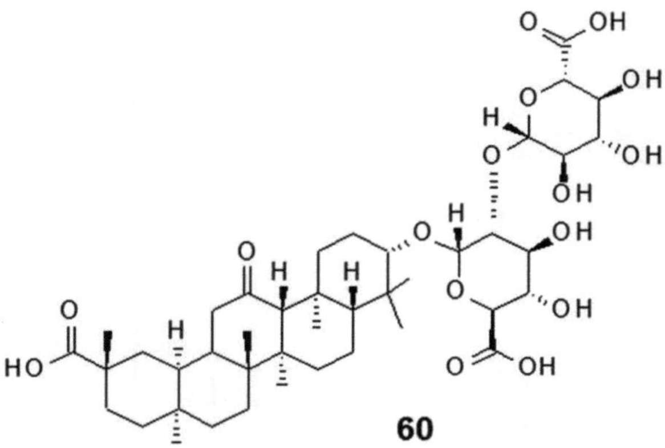

60

Another example would be saikosaponin D **62**, which seems to promote an antiproliferative effect in activated T lymphocytes, due to the suppression of nucleotide-binding oligomerization domain containing 2 (NOD2) and NF-κB, NF-AT, AP-1, and MAPK signaling that are pathways associated with greater disease severity (Bahbah et al. 2020; Yuan et al. 2017; Yang et al. 2017b). In line with these findings, a systematic review evaluated the therapeutic application of extracts and saikosaponins isolated from *Radix bupleuri* (*Bupleurum chinense* D.C.) and concluded that the species extracts and the isolated substances saikosaponin A, saikosaponin D, saikosaponin C **63**, and saikosaponin B2 **64** exhibit anti-inflammatory, antitumor, antiviral, anti-allergic, immune, and neuroregulatory activities, which involve, among others, the NF-κB and MAPK pathways (Yuan et al. 2017).

In conclusion, a significant number of saponins have already been investigated and present interesting immunomodulatory activities that justify their current or future therapeutic application. Although these activities mainly implicated in the use of saponins as adjuvants in vaccines, their use for the treatment of inflammatory and autoimmune diseases is also promising, but the mechanisms of action still need to be better characterized.

11.11 Cardiotonic Heterosides

Cardiotonic heterosides are natural steroids combined in this class due to their potent and specific effects on the heart muscle. Despite this, other activities have been reported for these compounds including their effect in the immune system, although the information available is not as extensive as for other terpenoids. As seen for some saponins, the immunomodulatory activity of cardiotonic heterosides is often investigated in the development of cancer therapies (Reddy et al. 2020).

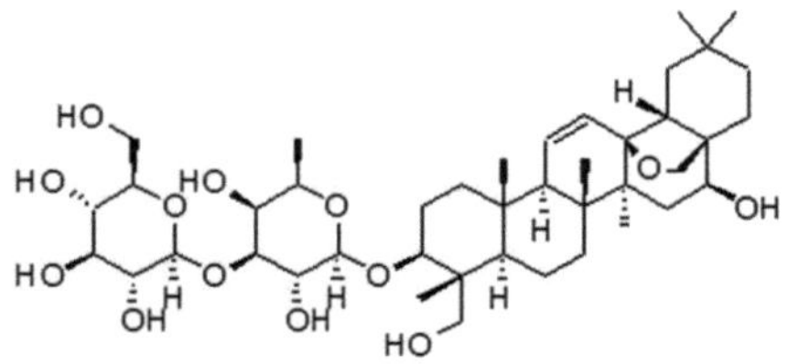

61

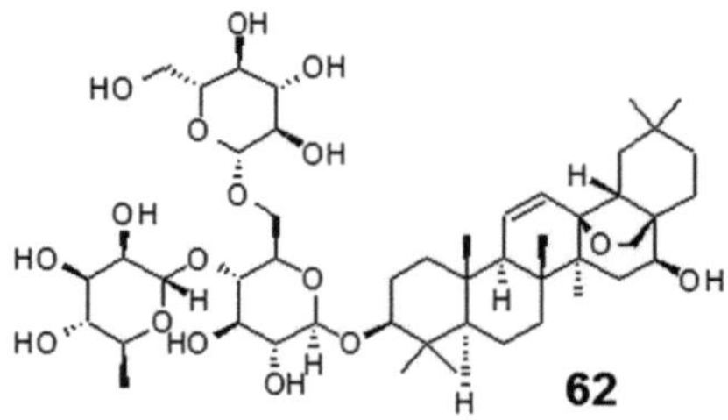

62

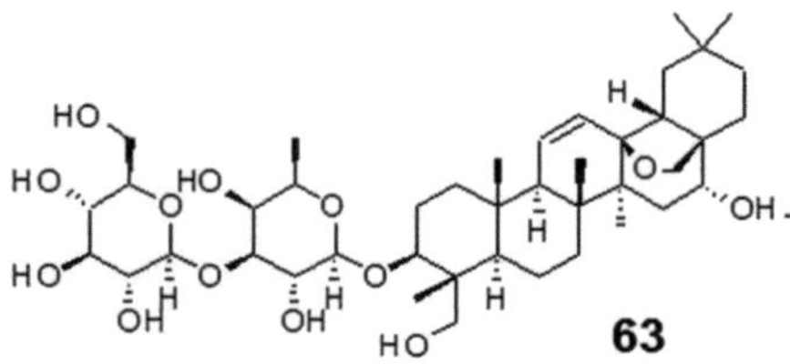

63

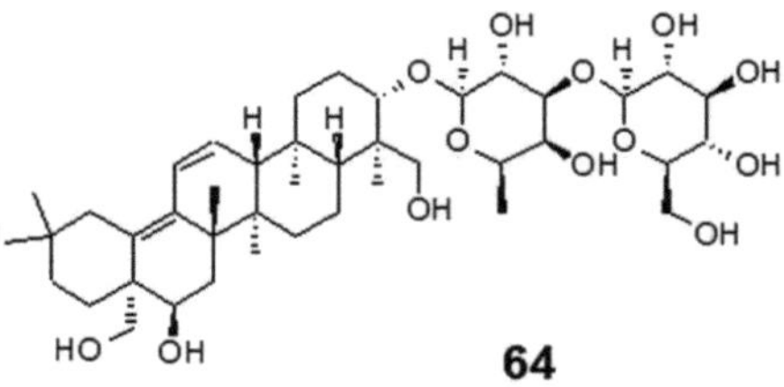

64

A literature review about the pharmacological activities of this class of compounds brought together studies elucidating several mechanisms of action of cardiotonic heterosides that are responsible for stimulating the immune response to various diseases, including the activation of immunogenic cell death (ICD) of various cancer cells and the stimulation of CD4 + or CD8 + T lymphocytes (γ) (Reddy et al. 2020).

In this sense, an example is the action of digoxin **65**, one of the most well-known cardiotonic heterosides, in inhibiting HIV-1 infection in a highly selective way. The substance inhibited the expression of genes involved in T-cell activation and cell metabolism. Two main networks of genes negatively regulated by digoxin have been identified, namely, CD40L and CD38 (Zhyvoloup et al. 2017).

A second study investigated the effect of the cardiotonic steroid ouabain **66** in a leukemia model in vivo. The results indicated positive effects in the increased proliferation of B and T cells and decreased levels of CD3, CD11b, and Mac-3 cells compared to the positive control, induction of phagocytosis by cell macrophages, mononuclear cells of peripheral blood, and peritoneal cavity, along with an increase in the activities of natural killer cells (Shih et al. 2019).

Therefore, based on reports about the activity of these compounds, while there are fewer of them when compared to other classes of terpenes, it is possible to suggest that cardiotonic heterosides may be interesting alternatives for therapeutic agents acting on the immune system.

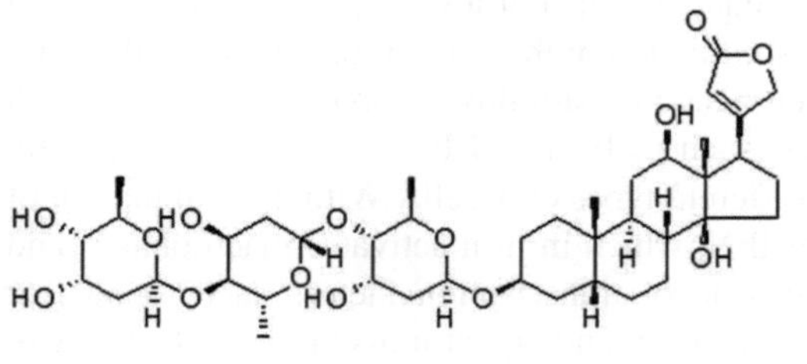

65

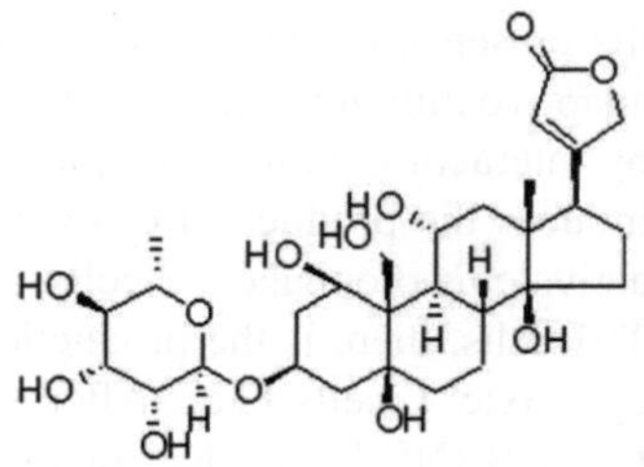

66

11.12 Main Knowledge of the Chapter

Terpenes, which comprise a broad chemical group of active principles, are natural products derived from isoprene units (C5)n. They are a chemical group classified as hemiterpenes (C5), monoterpenes (C10), sesquiterpenes (C15), diterpenes (C20), sesterpenes (C25), triterpenes (C30), tetraterpenes (C40), and polyterpenes $(C5)_n$, $n > 8$). All terpenes can be found either in their free form or linked to sugar, such as saponins and cardiotonic glycosides. In addition, there are many derivatives such as iridoids, limonoids, quassinoids, vitamins, etc. They are active substances presented in many medicinal plants used in traditional medicine against diseases in which the immune system is involved. This class of compound has been described as anti-inflammatory, analgesic, antioxidant, neuroprotective, cardioprotective, antiviral, estrogenic, antimicrobial, insecticidal, and antitumoral, as well immunomodulator. Despite their effects have not always been clearly described, many of them have been related to the resolution of immune diseases.

The immune system in vertebrates comprises a complex and sophisticated defense system, including sensory and effector mechanisms mediated by receptors. It is traditionally divided into two types, innate immunity and adaptive immunity. Innate immunity provides the first line of defense, consisting of cellular and biochemical defense mechanisms that react to the products of injured microorganisms and cells.

Macrophages and antigen-presenting cells (APC) are involved in this immune response, being responsible for phagocytosis, secretion of cytokines, antigen processing and presentation, and cytotoxicity. There are other cells, such as accessory dendritic cells, that have different functions, including the activation of native T cells and memory B cells, and regulation of innate immunity effectors such as natural killer (NK) cells, production of interferon (IFN), TNF, and granulocyte macrophage colony-stimulating factor (GM-CSF). Lastly, the complement system is important effector of humoral immunity, involved in the defense.

The acquired immunity, also called adaptive or specific, encompasses the immune response stimulated by exposure to antigens. Therefore, the mechanisms of this immunity do not provide anticipated defenses, since they are activated only in the presence of agents, and will consequently have a lower speed of formation. In adaptive immunity, there is initially stimulation of the antigen-presenting cell (APC) by interacting with the major antigen-histocompatibility complex (MHC) II, which induces the production of cytokines, such as IL-4 and IL-12, which consequently activate macrophages, B cells, and different types of T cells. With the stimulation of Th1 cells, there is the production of IFN, which in turn activates macrophages and cytotoxic T cells (Tc). After stimulation of transcription factors such as nuclear factor-B (NF-B), nuclear factor AT (NF-AT), and signal transducers and transcription activators (STATs), there is the production of peptide mediators and inducible enzymes.

Consequently, the immune system maintains normal homeostasis within the body, which results in either immunosuppressive or stimulatory function. Therefore, immune modulation represents any influence on the immune system response, and immunomodulators are agents that have the capacity to modulate or regulate pathophysiological activities. Hence, the mechanisms of herbal medicines natural terpenes in immune systems can be generally categorized into two major groups, with one of them enhancing the immune responses, being applied in immunodeficiency cases and infectious diseases, and the other suppressing them as a treatment for auto immune diseases.

The main target of action of monoterpenes in immunity are dendritic cells, and the studies found in the literature mainly addressed the application of these substances in the treatment of asthma. Among the main mechanisms involved in the immunomodulation of iridoids are the action on T lymphotics and the release of cytokines. Some sesquiterpenes, mainly the sesquiterpene lactones group such as artemisinin, have been associated with immune system effects. Most of them also demonstrated immunomodulatory activity involving T lymphocytes. Diterpenes and triterpenes are structurally diverse, and that is why they have already been described as acting on very different targets from the immune system. Steroids have already been associated with immunosuppressive action on different targets from the immune system, while vitamin D_3, one metabolite of steroids, is also known to have various physiological functions, including hormone secretion, immunomodulatory, and differentiation of both normal and malignant cells. Tetraterpene presents an immune depressive effect. Saponins are characterized by their surfactant and detergent properties, related to the various biological activities reported for the class, especially their action on membranes, their hemolytic activity, and their ability to complex with steroids. These properties, in turn, determine their therapeutic applications, such as their use as adjuvants in vaccines due to the immunomodulatory activity presented by them. Several examples that describe these activities are given for all types of terpenes.

References

Abbas AK (2019) Imunologia celulare molecular, 9th edn. GEN Guanabara Koogan, São Paulo, p 576

Amirghofran Z, Ahmadi H, Karimi MH, Kalantar F, Gholijani N, Malek-Hosseini Z (2016) In vitro inhibitory effects of thymol and carvacrol on dendritic cell activation and function. Pharm Biol 54(7):1125–1132

Antwi AO, Obiri DD, Osafo N, Forkuo AD, Essel LB (2017) Stigmasterol inhibits lipopolysaccharide-induced innate immune responses in murine models. Int Immunopharmacol 53:105–113

Araldi E, Fernández-Fuertes M, Canfrán-Duque A, Tang W, Cline GW, Madrigal-Matute J et al (2017) Lanosterol modulates TLR4-mediated innate immune responses in macrophages. Cell Rep 19(13):2743–2755

Bahbah EI, Negida A, Nabet MS (2020) Purposing Saikosaponins for the treatment of COVID-19. Med Hypotheses 140:109782

Bartikova H, Hanusova V, Skalova L, Ambroz M, Bousova I (2014) Antioxidant, pro-oxidant and other biological activities of sesquiterpenes. Curr Top Med Chem 14(22):2478–2494

Bigaeva E, van Doorn E, Liu H, Hak E (2016) Meta-analysis on randomized controlled trials of vaccines with QS-21 or ISCOMATRIX adjuvant: safety and tolerability. PLoS One 11(5): e0154757

Bono MR, Tejon G, Flores-Santibañez F, Fernandez D, Rosemblatt M, Sauma D (2016) Retinoic acid as a modulator of T cell immunity. Nutrients 8(6):349

Borges PV, Moret KH, Raghavendra NM, Maramaldo Costa TE, Monteiro AP, Carneiro AB et al (2017) Protective effect of gedunin on TLR-mediated inflammation by modulation of inflammasome activation and cytokine production: evidence of a multitarget compound. Pharmacol Res 115:65–77

Boubaker J, Ben Toumia I, Sassi A, Bzouich-Mokded I, Ghoul Mazgar S, Sioud F et al (2018) Antitumoral potency by immunomodulation of chloroform extract from leaves of nitraria retusa, tunisian medicinal plant, via its major compounds β-sitosterol and palmitic acid in BALB/c mice bearing induced tumor. Nutr Cancer 70(4):650–662

Brüll F, De Smet E, Mensink RP, Vreugdenhil A, Kerksiek A, Lütjohann D et al (2016) Dietary plant stanol ester consumption improves immune function in asthma patients: results of a randomized, double-blind clinical trial. Am J Clin Nutr 103(2):444–453

Bruton LL, Lazo JS, Parker KL (2010) Goodman e Gilman: as bases farmacológicas da terapêutica. AMGH, Porto Alegre

Cao Y, Feng Y-H, Gao L-W, Li X-Y, Jin Q-X, Wang Y-Y et al (2019) Artemisinin enhances the anti-tumor immune response in 4T1 breast cancer cells in vitro and in vivo. Int Immunopharmacol 70:110–116

Cascão R, Vidal B, Lopes IP, Paisana E, Rino J, Moita LF et al (2015) Decrease of CD68 synovial macrophages in celastrol treated arthritic rats. PLoS One 10(12):e0142448

Chase C, Abdul-Baki N, Negron A, Joern R, Forsthuber T, Asmis R (2020) Dietary supplementation with 23-hydroxy ursolic acid accelerates the recovery from acute experimental autoimmune encephalomyelitis (EAE) in a murine model of multiple sclerosis. Free Radic Biol Med 159: S113. [cited 2020 Nov 30]; https://linkinghub.elsevier.com/retrieve/pii/S0891584920315598

Chen H, Du Q (2020) Potential natural compounds for preventing 2019-nCoV infection. Preprints

Chen F-Y, Zhou L-F, Li X-Y, Zhao J-W, Xu S-F, Huang W-H et al (2016) Stephanthraniline A suppressed CD4(+) T cell-mediated immunological hepatitis through impairing PKCθ function. Eur J Pharmacol 789:370–384

Chen C, Wang Y-Y, Wang Y-X, Cheng M-Q, Yin J-B, Zhang X et al (2018) Gentiopicroside ameliorates bleomycin-induced pulmonary fibrosis in mice via inhibiting inflammatory and fibrotic process. Biochem Biophys Res Commun 495(4):2396–2403

Chen L, Hu C, Hood M, Zhang X, Zhang L, Kan J et al (2020) A novel combination of vitamin c, curcumin and glycyrrhizic acid potentially regulates immune and inflammatory response associated with coronavirus infections: a perspective from system biology analysis. Nutrients 12(4):1193

Cheng Y, Chen G, Wang L, Kong J, Pan J, Xi Y et al (2018) Triptolide-induced mitochondrial damage dysregulates fatty acid metabolism in mouse sertoli cells. Toxicol Lett 292:136–150

Cheng Y, Chen Y, Li J, Qu H, Zhao Y, Wen C et al (2020) Dietary β-sitosterol regulates serum lipid level and improves immune function, antioxidant status, and intestinal morphology in broilers. Poult Sci 99(3):1400–1408

Cho NE, Bang B-R, Gurung P, Li M, Clemens DL, Underhill TM et al (2016) Retinoid regulation of antiviral innate immunity in hepatocytes. Hepatology 63(6):1783–1795

Choi S-P, Choi C-Y, Park K, Kim N, Moon H-S, Lee D et al (2016) Glabretal-type triterpenoid from the root bark of Dictamnus dasycarpus ameliorates collagen-induced arthritis by inhibiting Erk-dependent lymphocyte proliferation. J Ethnopharmacol 178:13–16

Christianson DW (2017) Structural and chemical biology of terpenoid cyclases. Chem Rev Am Chem Soc 117:11570–11648

Chung T-W, Li Y-R, Huang WY, Su J-H, Chan H-L, Lin S-H et al (2017) Sinulariolide suppresses LPS-induced phenotypic and functional maturation of dendritic cells. Mol Med Rep 16(5): 6992–7000

Ciaglia E, Malfitano AM, Laezza C, Fontana A, Nuzzo G, Cutignano A et al (2017) Immunomodulatory and anti-inflammatory effects of dihydrogracilin A, a terpene derived from the marine sponge Dendrilla membranosa. Int J Mol Sci 18(8):1643

Cibulski S, Rivera-Patron M, Suárez N, Pirez M, Rossi S, Yendo AC et al (2018) Leaf saponins of Quillaja brasiliensis enhance long-term specific immune responses and promote dose-sparing effect in BVDV experimental vaccines. Vaccine 36(1):55–65

Connolly JD, Hill RA (2010) Triterpenoids. Nat Prod Rep 27:79–132

Das S, Mishra KP, Ganju L, Singh SB (2017) Andrographolide - A promising therapeutic agent, negatively regulates glial cell derived neurodegeneration of prefrontal cortex, hippocampus and working memory impairment. J Neuroimmunol 313:161–175

de Carvalho LSA, Fontes LBA, Gazolla MC, Dias DDS, Juliano MA, Macedo GC et al (2017) Parthenolide modulates immune response in cells from C57BL/6 mice induced with experimental autoimmune encephalomyelitis. Planta Med 83(8):693–700

de Sousa IP, Sousa Teixeira MV, Jacometti Cardoso Furtado NA (2018) An overview of biotransformation and toxicity of diterpenes. Molecules 23(6):1387

Dewick PM (2009a) Medicinal natural products: a biosynthetic approach: third edition. medicinal natural products: a biosynthetic approach, 3rd edn. John Wiley and Sons, Chichester, pp 1–539

Dewick P (2009b) Medicinal natural product - a biosynthetic approach, 3rd edn. Department of Ecology - Swedish University of Agricultural Science, Switzerland

Dhandapani S, Tjhang JG, Jang IC (2020 Sep) Production of multiple terpenes of different chain lengths by subcellular targeting of multi-substrate terpene synthase in plants. Metab Eng 61: 397–405

Dhinaut J, Balourdet A, Teixeira M, Chogne M, Moret Y (2017) A dietary carotenoid reduces immunopathology and enhances longevity through an immune depressive effect in an insect model. Sci Rep 7(1):12429

Didierlaurent AM, Laupèze B, Di Pasquale A, Hergli N, Collignon C, Garçon N (2017) Adjuvant system AS01: helping to overcome the challenges of modern vaccines. Expert Rev Vaccines 16(1):55–63

Dimitrova P, Alipieva K, Grozdanova T, Simova S, Bankova V, Georgiev MI et al (2018) New iridoids from Verbascum nobile and their effect on lectin-induced T cell activation and proliferation. Food Chem Toxicol 111:605–615

Ezz-Eldin YM, Aboseif AA, Khalaf MM (2020 Feb) Potential anti-inflammatory and immunomodulatory effects of carvacrol against ovalbumin-induced asthma in rats. Life Sci 242:117222

Feyzi R, Boskabady MH, Seyedhosseini Tamijani SM, Rafatpanah H, Rezaei SA (2016 Dec) The effect of Safranal on Th1/Th2 cytokine balance. Iran J Immunol 13(4):263–273

Fleck JD, Betti AH, da Silva FP, Troian EA, Olivaro C, Ferreira F et al (2019) Saponins from Quillaja saponaria and Quillaja brasiliensis: Particular Chemical Characteristics and Biological Activities. Molecules 24(1)

Gao M, Ou H, Jiang Y, Wang K, Peng Y, Zhang H et al (2019) Tanshinone IIA attenuates sepsis-induced immunosuppression and improves survival rate in a mice peritonitis model. Biomed Pharmacother 112:108609

Garcia A, Lema D (2016) An updated review of ISCOMSTM and ISCOMATRIXTM vaccines. Curr Pharm Des 22(41):6294–6299

Gattu S, Bang Y-J, Pendse M, Dende C, Chara AL, Harris TA et al (2019) Epithelial retinoic acid receptor β regulates serum amyloid A expression and vitamin A-dependent intestinal immunity. Proc Natl Acad Sci U S A 116(22):10911–10916

Ghirardello M, Ruiz-de-Angulo A, Sacristan N, Barriales D, Jiménez-Barbero J, Poveda A et al (2020) Exploiting structure-activity relationships of QS-21 in the design and synthesis of streamlined saponin vaccine adjuvants. Chem Commun (Camb) 56(5):719–722

Gholijani N, Amirghofran Z (2016) Effects of thymol and carvacrol on T-helper cell subset cytokines and their main transcription factors in ovalbumin-immunized mice. J Immunotoxicol 13(5):729–737

González-Coloma A, López-Balboa C, Santana O, Reina M, Fraga BM (2011) Triterpene-based plant defenses. Phytochem Rev 10(2):245–260

Grant WB, Lahore H, McDonnell SL, Baggerly CA, French CB, Aliano JL et al (2020) Evidence that vitamin D supplementation could reduce risk of influenza and COVID-19 infections and deaths. Nutrients 12(4):988. [cited 2020 Dec 1] https://www.mdpi.com/2072-6643/12/4/988

Guo H, Pan C, Chang B, Wu X, Guo J, Zhou Y et al (2016a) Triptolide improves diabetic nephropathy by regulating Th cell balance and macrophage infiltration in rat models of diabetic nephropathy. Exp Clin Endocrinol Diabetes 124(6):389–398

Guo X, Xue M, Li C-J, Yang W, Wang S-S, Ma Z-J et al (2016b) Protective effects of triptolide on TLR4 mediated autoimmune and inflammatory response induced myocardial fibrosis in diabetic cardiomyopathy. J Ethnopharmacol 193:333–344

Haque MA, Jantan I, Arshad L, Bukhari SNA (2017) Exploring the immunomodulatory and anticancer properties of zerumbone. Food Funct 8(10):3410–3431

Harun NH, Septama AW, Ahmad WANW, Suppian R (2020) Immunomodulatory effects and structure-activity relationship of botanical pentacyclic triterpenes: A review. Chinese Herb Med 12(2):118–124

Hou L, Huang H (2016) Immune suppressive properties of artemisinin family drugs. Pharmacol Ther 166:123–127

Hou W, Liu B, Xu H (2019) Triptolide: Medicinal chemistry, chemical biology and clinical progress. Eur J Med Chem 176:378–392

Houh YK, Kim KE, Park S, Hur DY, Kim S, Kim D et al (2017) The effects of artemisinin on the cytolytic activity of natural killer (NK). Cells Int J Mol Sci 18(7):1600

Huang Y, Zou Y, Lin L, Zheng R (2017) Ginsenoside Rg1 activates dendritic cells and acts as a vaccine adjuvant inducing protective cellular responses against lymphomas. DNA Cell Biol 36(12):1168–1177

Huang Y, Zhu N, Chen T, Chen W, Kong J, Zheng W et al (2019) Triptolide Suppressed the Microglia Activation to Improve Spinal Cord Injury Through miR-96/IKKβ/NF-κB Pathway. Spine (Phila Pa 1976) 44(12):E707–E714

Iqubal A, Sharma S, Najmi AK, Syed MA, Ali J, Alam MM et al (2019) Nerolidol ameliorates cyclophosphamide-induced oxidative stress, neuroinflammation and cognitive dysfunction: plausible role of Nrf2 and NF- κB. Life Sci 236:116867

Islam MT, Ali ES, Uddin SJ, Shaw S, Islam MA, Ahmed MI et al (2018) Phytol: a review of biomedical activities. Food Chem Toxicol 121:82–94

Javed H, Azimullah S, Abul Khair SB, Ojha S, Haque ME (2016) Neuroprotective effect of nerolidol against neuroinflammation and oxidative stress induced by rotenone. BMC Neurosci 17(1):58

Jiang H, Promchan K, Lin B-R, Lockett S, Chen D, Marshall H et al (2016) LZTFL1 upregulated by all-trans retinoic acid during CD4+ T cell activation enhances IL-5 production. J Immunol 196(3):1081–1090

Jing Y, Liu H, Xu W, Yang Q (2019) 4,4′-diaponeurosporene-producing Bacillus subtilis promotes the development of the mucosal immune system of the piglet gut. Anat Rec (Hoboken) 302(10): 1800–1807

Kasahara H, Kondo T, Nakatsukasa H, Chikuma S, Ito M, Ando M et al (2017) Generation of allo-antigen-specific induced Treg stabilized by vitamin C treatment and its application for prevention of acute graft versus host disease model. Int Immunol 29(10):457–469

Kim A-Y, Shim H-J, Shin H-M, Lee YJ, Nam H, Kim SY et al (2018) Andrographolide suppresses TRIF-dependent signaling of toll-like receptors by targeting TBK1. Int Immunopharmacol 57: 172–180

Kiyama R (2017) Estrogenic terpenes and terpenoids: pathways, functions and applications. Eur J Pharmacol Elsevier BV 815:405–415

Ku C-M, Lin J-Y (2016) Farnesol, a sesquiterpene alcohol in essential oils, ameliorates serum allergic antibody titres and lipid profiles in ovalbumin-challenged mice. Allergol Immunopathol (Madr) 44(2):149–159

Lacaille-Dubois M-A (2019) Updated insights into the mechanism of action and clinical profile of the immunoadjuvant QS-21: a review. Phytomedicine 60:152905

Lasarte-Cia A, Lozano T, Pérez-González M, Gorraiz M, Iribarren K, Hervás-Stubbs S et al (2018) Immunomodulatory Properties of Carvone Inhalation and Its Effects on Contextual Fear Memory in Mice. Front Immunol 9:68

Lee SU, Sung MH, Ryu HW, Lee J, Kim H-S, In HJ et al (2016) Verproside inhibits TNF–αinduced MUC5AC expression through suppression of the TNF-α/NF-κB pathway in human airway epithelial cells. Cytokine 77:168–175

Li J, Huang L, Wang S, Yao Y, Zhang Z (2016) Astragaloside IV attenuates inflammatory reaction via activating immune function of regulatory T-cells inhibited by HMGB1 in mice. Pharm Biol 54(12):3217–3225

Li Y, Xu Y-L, Lai Y-N, Liao S-H, Liu N, Xu P-P (2017a) Intranasal co-administration of 1,8-cineole with influenza vaccine provide cross-protection against influenza virus infection. Phytomedicine 34:127–135

Li Y, Meng T, Hao N, Tao H, Zou S, Li M et al (2017b) Immune regulation mechanism of Astragaloside IV on RAW264.7 cells through activating the NF-κB/MAPK signaling pathway. Int Immunopharmacol 49:38–49

Li F, Dai M, Wu H, Deng R, Fu J, Zhang Z et al (2018a) Immunosuppressive effect of geniposide on mitogen-activated protein kinase signalling pathway and their cross-talk in fibroblast-like synoviocytes of adjuvant arthritis rats. Molecules 23(1):91

Li X, Zhao L, Han J-J, Zhang F, Liu S, Zhu L et al (2018b) Carnosol modulates Th17 cell differentiation and microglial switch in experimental autoimmune encephalomyelitis. Front Immunol 9:1807

Lin K-H, Lin K-C, Lu W-J, Thomas P-A, Jayakumar T, Sheu J-R (2015) Astaxanthin, a carotenoid, stimulates immune responses by enhancing IFN-γ and IL-2 secretion in primary cultured lymphocytes in vitro and ex vivo. Int J Mol Sci 17(1):44

Liu H, Zhang G, Huang J, Ma S, Mi K, Cheng J et al (2016a) Atractylenolide I modulates ovarian cancer cell-mediated immunosuppression by blocking MD-2/TLR4 complex-mediated MyD88/NF-κB signaling in vitro. J Transl Med 14(1):104

Liu X, He H, Huang T, Lei Z, Liu F, An G et al (2016b) Tanshinone IIA protects against dextran sulfate sodium- (DSS-) induced colitis in mice by modulation of neutrophil infiltration and activation. Oxidative Med Cell Longev 2016:7916763

Liu D, Wang Q, He W, Chen X, Wei Z, Huang K (2020a) Two-way immune effects of deoxynivalenol in weaned piglets and porcine alveolar macrophages: Due mainly to its exposure dosage. Chemosphere 249:126464

Liu Y, Xiao N, Du H, Kou M, Lin L, Huang M et al (2020b) Celastrol ameliorates autoimmune disorders in Trex1-deficient mice. Biochem Pharmacol 178:114090

Loreto F, Pinelli P, Manes F, Kollist H (2004) Impact of ozone on monoterpene emissions and evidence for an isoprene-like antioxidant action of monoterpenes emitted by Quercus ilex leaves. Tree Physiol 24(4):361–367

Magedans YV, Yendo AC, de Costa F, Gosmann G, Fett-Neto AG (2019) Foamy matters: an update on Quillaja saponins and their use as immunoadjuvants. Future Med Chem 11(12): 1485–1499

Marciani DJ (2018) Elucidating the mechanisms of action of saponin-derived adjuvants. Trends Pharmacol Sci 39(6):573–585

Marty-Roix R, Vladimer GI, Pouliot K, Weng D, Buglione-Corbett R, West K et al (2016) Identification of QS-21 as an inflammasome-activating molecular component of saponin adjuvants. J Biol Chem 291(3):1123–1136

Mohammadi A, Mahjoub S, Ghafarzadegan K, Nouri HR (2018) Immunomodulatory effects of Thymol through modulation of redox status and trace element content in experimental model of asthma. Biomed Pharmacother 105:856–861

Mu X, Wang C (2018) Artemisinins-a Promising New Treatment for Systemic Lupus Erythematosus: a Descriptive Review. Curr Rheumatol Rep 20(9):55

Netea MG, Schlitzer A, Placek K, Joosten LAB, Schultze JL (2019) Innate and adaptive immune memory: an evolutionary continuum in the host's response to pathogens. Cell Host Microbe 25(1):13–26

Nicholson LB (2016) The immune system. Essays Biochem 60(3):275–301

Noel P, Von Hoff DD, Saluja AK, Velagapudi M, Borazanci E, Han H (2019) Triptolide and its derivatives as cancer therapies. Trends Pharmacol Sci 40(5):327–341

Nogueira-de-Almeida CA, Del Ciampo LA, Ferraz IS, Del Ciampo IRL, Contini AA, da Ued FV (2020) COVID-19 and obesity in childhood and adolescence: a clinical review. J Pediatr 96: 546–558

Ortuño-Sahagún D, Zänker K, Rawat AKS, Kaveri SV, Hegde P (2017) Natural immunomodulators. J Immunol Res 2017:7529408

Pan J, Shen F, Tian K, Wang M, Xi Y, Li J et al (2019) Triptolide induces oxidative damage in NRK-52E cells through facilitating Nrf2 degradation by ubiquitination via the GSK-3β/Fyn pathway. Toxicol In Vitro 58:187–194

Park E, Song JH, Kim MS, Park S-H, Kim TS (2016) Costunolide, a sesquiterpene lactone, inhibits the differentiation of pro-inflammatory CD4(+) T cells through the modulation of mitogen-activated protein kinases. Int Immunopharmacol 40:508–516

Pichersky E, Gershenzon J (2002) The formation and function of plant volatiles: perfumes for pollinator attraction and defense. Curr Opin Plant Biol 5(3):237–243. https://doi.org/10.1016/s1369-5266(02)00251-0

Quintans JSS, Shanmugam S, Heimfarth L, Araújo AAS, Almeida JRG et al (2019) Monoterpenes modulating cytokines - a review. Food Chem Toxicol 123:233–257

Raz K, Levi S, Gupta PK, Major DT (2020) Enzymatic control of product distribution in terpene synthases: insights from multiscale simulations. In: Current opinion in biotechnology, vol 65. Elsevier Ltd, pp 248–258

Reddy D, Kumavath R, Barh D, Azevedo V, Ghosh P (2020) Anticancer and antiviral properties of cardiac glycosides: a review to explore the mechanism of actions. Molecules 25(16)

Sadeghi K, Wessner B, Laggner U, Ploder M, Tamandl D, Friedl J et al (2006) Vitamin D3 down-regulates monocyte TLR expression and triggers hyporesponsiveness to pathogen-associated molecular patterns. Eur J Immunol 36(2):361–370

Sahli F, Sousa MSE, Vileno B, Lichter J, Lepoittevin J-P, Blömeke B et al (2019) Understanding the skin sensitization capacity of ascaridole: a combined study of chemical reactivity and activation of the innate immune system (dendritic cells) in the epidermal environment. Arch Toxicol 93(5):1337–1347

Santhanasabapathy R, Vasudevan S, Anupriya K, Pabitha R, Sudhandiran G (2015) Farnesol quells oxidative stress, reactive gliosis and inflammation during acrylamide-induced neurotoxicity: Behavioral and biochemical evidence. Neuroscience 308:212–227

Sarikahya NB, Nalbantsoy A, Top H, Gokturk RS, Sumbul H, Kirmizigul S (2018) Immunomodulatory, hemolytic and cytotoxic activity potentials of triterpenoid saponins from eight Cephalaria species. Phytomedicine 38:135–144

Schwager J, Richard N, Fowler A, Seifert N, Raederstorff D (2016) Carnosol and related substances modulate chemokine and cytokine production in macrophages and chondrocytes. Molecules 21(4):465

Shantilal S, Vaghela JS, Sisodia SS (2018) Review on immunomodulation and immunomodulatory activity of some medicinal plant. Eur J Biomed 5(8):163–174

Sharkey TD, Yeh S (2001) Isoprene emission from plants. Annu Rev Plant Biol 52:407–436

Shi Z-Y, Zeng J-Z, Wong AST (2019) Chemical structures and pharmacological profiles of ginseng saponins. Molecules 24(13):2443

Shih Y-L, Shang H-S, Chen Y-L, Hsueh S-C, Chou H-M, Lu H-F et al (2019) Ouabain promotes immune responses in WEHI-3 cells to generate leukemia mice through enhancing phagocytosis and natural killer cell activities in vivo. Environ Toxicol 34(5):659–665

Somensi N, Rabelo TK, Guimarães AG, Quintans-Junior LJ, de Souza Araújo AA, Moreira JCF et al (2019) Carvacrol suppresses LPS-induced pro-inflammatory activation in RAW 264.7 macrophages through ERK1/2 and NF-kB pathway. Int Immunopharmacol 75:105743

Su Y, Wang Q, Yang B, Wu L, Cheng G, Kuang H (2017) Withasteroid B from D. metel L. regulates immune responses by modulating the JAK/STAT pathway and the IL-17(+) RORγt(+) /IL-10(+) FoxP3(+) ratio. Clin Exp Immunol 190(1):40–53

Talukdar J, Bhadra B, Dattaroy T, Nagle V, Dasgupta S (2020) Potential of natural astaxanthin in alleviating the risk of cytokine storm in COVID-19. Biomed Pharmacother 132:110886

Terao R, Murata A, Sugamoto K, Watanabe T, Nagahama K, Nakahara K et al (2019) Immunostimulatory effect of kumquat (Fortunella crassifolia) and its constituents, β-cryptoxanthin and R-limonene. Food Funct 10(1):38–48

Tholl D (2006) Terpene synthases and the regulation, diversity and biological roles of terpene metabolism. Curr Opin Plant Biol 9:297–304

Toti E, Chen C-YO, Palmery M, Villaño Valencia D, Peluso I (2018) Non-provitamin A and provitamin A carotenoids as immunomodulators: recommended dietary allowance, therapeutic index, or personalized nutrition? Oxidative Med Cell Longev 2018:4637861

Tsolakis N, Bam W, Srai JS, Kumar M (2019) Renewable chemical feedstock supply network design: the case of terpenes. J Clean Prod 222:802–822

Tundis R, Loizzo MR, Menichini F, Statti GA, Menichini F (2008) Biological and pharmacological activities of iridoids: recent developments. Mini-Rev Med Chem 8:399–420

Tundis R, Loizzo MR, Menichini F (2014) An overview on chemical aspects and potential health benefits of limonoids and their derivatives. Crit Rev Food Sci Nutr 54:225–250. [cited 2020 Nov 29]. https://pubmed.ncbi.nlm.nih.gov/24188270/

Uyeda S, Sharmin T, Satho T, Irie K, Watanabe M, Hosokawa M et al (2016) Enhancement and regulation effect of myrcene on antibody response in immunization with ovalbumin and Ag85B in mice. Asian Pacific J allergy Immunol 34(4):314–323

Vranová E, Coman D, Gruissem W (2013) Network analysis of the MVA and MEP pathways for isoprenoid synthesis. Annu Rev Plant Biol 64:665–700

Wang YJ, Lu J, Wu D, Zheng ZH, Zheng YL, Wang XH, Ruan J, Sun X, Shan Q, Zhang ZF et al (2011) Ursolic acid attenuates lipopolysaccharide-induced cognitive deficits in mouse brain through suppressing p38/NF-κB mediated inflammatory pathways. Neurobiol Learn Mem 96(2):156–165

Wang J, Chen N, Fang L, Feng Z, Li G, Mucelli A et al (2018. [cited 2020 Dec 1]) A systematic review about the efficacy and safety of tripterygium wilfordii Hook.f. Preparations used for the management of rheumatoid arthritis. Evid Based Complement Altern Med. https://doi.org/10.1155/2018/1567463

Wang P, Ding X, Kim H, Michalek SM, Zhang P (2020) Structural effect on adjuvanticity of saponins. J Med Chem 63(6):3290–3297

Wu Q, Wang X, Nepovimova E, Miron A, Liu Q, Wang Y et al (2017) Trichothecenes: immunomodulatory effects, mechanisms, and anti-cancer potential. Arch Toxicol 91(12):3737–3785

Wu Q, Wu W, Franca TCC, Jacevic V, Wang X, Kuca K (2018) Immune evasion, a potential mechanism of trichothecenes: new insights into negative immune regulations. Int J Mol Sci 19(11):3307

Xia G, Wang X, Sun H, Qin Y, Fu M (2017) Carnosic acid (CA) attenuates collagen-induced arthritis in db/db mice via inflammation suppression by regulating ROS-dependent p38 pathway. Free Radic Biol Med 108:418–432

Xu R, Fazio GC, Matsuda SPT (2004) On the origins of triterpenoid skeletal diversity. Phytochemistry 65:261–291

Xu H, Zhao H, Lu C, Qiu Q, Wang G, Huang J et al (2016) Triptolide inhibits osteoclast differentiation and bone resorption in vitro via enhancing the production of IL-10 and TGF-β1 by regulatory T cells. Mediat Inflamm 2016:8048170

Yadav N, Chandra H (2017) Suppression of inflammatory and infection responses in lung macrophages by eucalyptus oil and its constituent 1,8-cineole: Role of pattern recognition receptors TREM-1 and NLRP3, the MAP kinase regulator MKP-1, and NFκB. PLoS One 12(11):e0188232

Yan J, Yang X, Han D, Feng J (2016a) Tanshinone IIA attenuates experimental autoimmune encephalomyelitis in rats. Mol Med Rep 14(2):1601–1609

Yan W, Peng Y, Liu C, Ye M, Chen X, Peng X et al (2016b) Efficacy of triptolide on the apoptosis of tonsillar mononuclear cells from patients with IgA nephropathy. Ren Fail 38(1):109–116

Yan S-C, Li Y-J, Wang Y-J, Cai W-Y, Weng X-G, Li Q et al (2019a) Research progress of effect of artemisinin family drugs on T lymphocytes immunomodulation. Zhongguo Zhong Yao Za Zhi 44(22):4992–4999

Yan SC, Wang YJ, Li YJ, Cai WY, Weng XG, Li Q et al (2019b) Dihydroartemisinin regulates the Th/Treg balance by inducing activated CD4+ T cell apoptosis via heme oxygenase-1 induction in mouse models of inflammatory bowel disease. Molecules 24(13):2475

Yang F, Cao Y, Zhang J, You T, Zhu L (2017a) Glaucocalyxin A improves survival in bleomycin-induced pulmonary fibrosis in mice. Biochem Biophys Res Commun 482(1):147–153

Yang H, Chen X, Jiang C, He K, Hu Y (2017b) Antiviral and immunoregulatory role against PCV2 in vivo of Chinese herbal medicinal ingredients. J Vet Res 61(4):405–410

Yang M, Li H, Zhang Q, Wu Q-H, Li G, Chen K-X et al (2019) Highly diverse cembranoids from the South China Sea soft coral Sinularia scabra as a new class of potential immunosuppressive agents. Bioorg Med Chem 27(15):3469–3476

Yatim KM, Lakkis FG (2015) A brief journey through the immune system. Clin J Am Soc Nephrol 10(7):1274–1281

Yi Y-S (2019) Roles of ginsenosides in inflammasome activation. J Ginseng Res 43(2):172–178

Yin W, Song Y, Liu Q, Wu Y, He R (2017) Topical treatment of all-trans retinoic acid inhibits murine melanoma partly by promoting CD8(+) T-cell immunity. Immunology 152(2):287–297

Yuan B, Yang R, Ma Y, Zhou S, Zhang X, Liu Y (2017) A systematic review of the active saikosaponins and extracts isolated from Radix Bupleuri and their applications. Pharm Biol 55(1):620–635

Yuan Z, Zhang H, Hasnat M, Ding J, Chen X, Liang P et al (2019) A new perspective of triptolide-associated hepatotoxicity: liver hypersensitivity upon LPS stimulation. Toxicology 414:45–56

Zang L, Wang J, Ren Y, Liu W, Yu Y, Zhao S et al (2019) Activated toll-like receptor 4 is involved in oridonin-induced phagocytosis via promotion of migration and autophagy-lysosome pathway in RAW264.7 macrophages. Int Immunopharmacol 66:99–108

Zeng Q-Z, Yang F, Li C-G, Xu L-H, He X-H, Mai F-Y et al (2019) Paclitaxel enhances the innate immunity by promoting NLRP3 inflammasome activation in macrophages. Front Immunol 10:72

Zhang Z-R, Wu H, Wang R, Li S-P, Dai L, Wang W-Y (2017) Immune tolerance effect in mesenteric lymph node lymphocytes of geniposide on adjuvant arthritis rats. Phytother Res 31(8):1249–1256

Zhang X-P, Li Y-D, Luo L-L, Liu Y-Q, Li Y, Guo C et al (2018) Astragalus saponins and liposome constitute an efficacious adjuvant formulation for cancer vaccines. Cancer Biother Radiopharm 33(1):25–31

Zhang T, Zhang Y, Jiang N, Zhao X, Sang X, Yang N et al (2020a) Dihydroartemisinin regulates the immune system by promotion of CD8(+) T lymphocytes and suppression of B cell responses. Sci China Life Sci 63(5):737–749

Zhang Y, Li X, Ciric B, Curtis MT, Chen W-J, Rostami A et al (2020b) A dual effect of ursolic acid to the treatment of multiple sclerosis through both immunomodulation and direct remyelination. Proc Natl Acad Sci U S A 117(16):9082–9093

Zhou F, Pichersky E (2020) More is better: the diversity of terpene metabolism in plants. In: Current opinion in plant biology, vol 55. Elsevier Ltd, pp 1–10

Zhou Y-Q, Weng X-F, Dou R, Tan X-S, Zhang T-T, Fang J-B et al (2017) Betulin from Hedyotis hedyotidea ameliorates concanavalin A-induced and T cell-mediated autoimmune hepatitis in mice. Acta Pharmacol Sin 38(2):201–210

Zhu L, Chen L (2019) Progress in research on paclitaxel and tumor immunotherapy. Cell Mol Biol Lett 24:40

Zhu Z, Zhao Y, Huo H, Gao X, Zheng J, Li J et al (2016) HHX-5, a derivative of sesquiterpene from Chinese agarwood, suppresses innate and adaptive immunity via inhibiting STAT signaling pathways. Eur J Pharmacol 791:412–423

Zhyvoloup A, Melamed A, Anderson I, Planas D, Lee C-H, Kriston-Vizi J et al (2017) Digoxin reveals a functional connection between HIV-1 integration preference and T-cell activation. PLoS Pathog 13(7):e1006460

Rhizomatous Plants: Curcuma longa and Zingiber officinale in Affording Immunity

12

Noha Fawzy Abdelkader and Passant Elwy Moustafa

Abstract

Rhizomatous plants have been sources of remedy and were used widely as dietary spices and flavor. Their effects as immunomodulators are also documented. Several studies have confirmed that both *Curcuma longa* (turmeric) and *Zingiber officinale* (ginger) have a vast array of medicinal and immunomodulatory properties. Curcumin, which is one of the main curcuminoids of *Curcuma longa*, possesses various pharmacological properties, including immunomodulatory activities. Similarly, *Zingiber officinale* has a history of medicinal use for over 2500 years as one of the most versatile medicinal plants, which is traced to its bioactive compounds such as gingerol, paradol, shogaols, etc. The extracts and/or bioactive compounds from those plants are promising drug candidates against various diseases such as diabetes mellitus, bacterial infection, Alzheimer's disease, rheumatoid arthritis, and cancer, partly via their immunomodulatory properties. Several studies have identified some of the bioactive compounds in *Curcuma longa* and *Zingiber officinale* rhizomes as potential inhibitors of the novel coronaviruses responsible for the COVID-19 pandemic. Such bioactive compounds are eligible for further investigation of the potential to treat COVID-19 patients effectively. This chapter describes the regulation of immune responses by rhizomatous *Curcuma longa* and *Zingiber officinale* for the treatment of diseases of diverse origin.

N. F. Abdelkader (✉)
Department of Pharmacology and Toxicology, Faculty of Pharmacy, Cairo University, Cairo, Egypt
e-mail: noha.fawzy@pharma.cu.edu.eg

P. E. Moustafa
Department of Pharmacology, Medical Research Division, National Research Centre, Giza, Egypt
e-mail: passantelwy@aucegypt.edu

N. S. Sangwan et al. (eds.), *Plants and Phytomolecules for Immunomodulation*,
https://doi.org/10.1007/978-981-16-8117-2_12

Keywords

Rhizomatous plants · Immunomodulator · *Curcuma longa* · Curcumin · Coronavirus · *Zingiber officinale* · Gingerol

Abbreviations

ACE-2	Angiotensin-converting enzyme 2
AIF	Apoptosis-inducing factor
AKT	Protein kinase B
AP-1	Activator protein-1
Bax	Bcl-2-associated X
Bcl-2	Inhibits B cell lymphoma-2
Bcl-xL	B cell lymphoma-extra-large
bFGF	Basic fibroblast growth factor
CDK	Cyclin-dependent kinase
cIAP1	Cellular inhibitor of apoptosis protein 1
COVID-19	Coronavirus disease 2019
COX	Cyclooxygenase
CXCL12	CXC chemokine ligand 12
CXCR4	CXC chemokine receptor 4
DSS	Dextran sulfate sodium
EGFR	Epidermal growth factor receptor
Egr-1	Early growth response protein-1
eIF2α	α-Subunit of eukaryotic initiation factor-2
ERK	Extracellular signal-regulated kinase
ERS	Endoplasmic reticulum stress
FPTase	Farnesyl protein transferase
GDF-15	Growth differentiation factor-15
GST	Glutathione-*S*-transferase
HO-1	Heme oxygenase-1
IAP	Inhibitor of apoptosis protein
IL	Interleukin
iNOS	Inducible nitric oxide synthase
JAK2	Janus kinase 2
JNK	c-Jun N-terminal kinase
LOX	Lipoxygenase
LPS	Lipopolysaccharide
MAPK	Mitogen-activated protein kinase
Mcl-1	Myeloid cell leukemia-1
MCP-1	Monocyte chemoattractant protein-1
MIP-1α	Macrophage inflammatory protein-1α
MMP	Matrix metalloproteinase
M-pro	Main protease

MRP1	Multidrug resistance-associated protein 1
mTOR	Mammalian target of rapamycin
NAG-1	Nonsteroidal anti-inflammatory drug-activated gene-1
NFAT	Nuclear factor of activated T cells
NF-κB	Nuclear factor kappa-B
NO	Nitric oxide
PCNA	Proliferating cell nuclear antigen
PI3K	Phosphoinositide 3-kinase
PLpro	Papain-like protease
PPAR-γ	Peroxisomal proliferator-activated receptor gamma
STAT	Signal transducer and activator of transcription
Th	T-helper
TNF-α	Tumor necrosis factor-α
TPA	12-*O*-tetradecanoylphorbol-13-acetate
TRAIL	TNF-related apoptosis-inducing ligand
TRPV1	Transient receptor potential vanilloid type-1
VEGF	Vascular endothelial growth factor

12.1 Introduction

Curcuma longa (turmeric) and *Zingiber officinale* (ginger), belonging to the family Zingiberaceae, are among the most widely consumed plants worldwide as spices, flavoring agents, and herbal remedies due to their numerous medicinal and nutritional values (Grzanna et al. 2005; Boroumand et al. 2018). Turmeric is an assortment of compounds related to curcumin identified as curcuminoids, comprising curcumin, demethoxycurcumin, bisdemethoxycurcumin and cyclocurcumin (Fig. 12.1) (Singh and Khar 2008). Curcumin is a versatile and pharmacologically nontoxic natural agent (Mullaicharam and Maheswaran 2012). It is considered the dominant active compound of turmeric and is responsible for its yellow color (Ruby et al. 1995). It has been used for its therapeutic activities for ages; however, the first documented case for its use took place in 1937 when it was used for the treatment of biliary disease. Afterward, its biological potential has been investigated in various disorders including but not limited to cancer treatment (Miriyala et al. 2007; Mansouri et al. 2015). Curcumin improved the effectiveness of other chemotherapies and mitigated their poisonous adverse effects, which are counted as the main problem of chemotherapeutic agents. Further, curcumin is generally known for its antioxidant and anti-inflammatory activities (Gonzales and Orlando 2008).

Traditionally, ginger has been used to treat urinary infections, gastrointestinal disorders, headache, rheumatism, asthma, cold, and cough (Grzanna et al. 2005; Shukla and Singh 2007). In the past few decades, the medicinal properties of ginger have been thoroughly investigated, and a variety of bioactive compounds have been isolated from different parts of the plant. Ginger contains α- and β-zingiberenes,

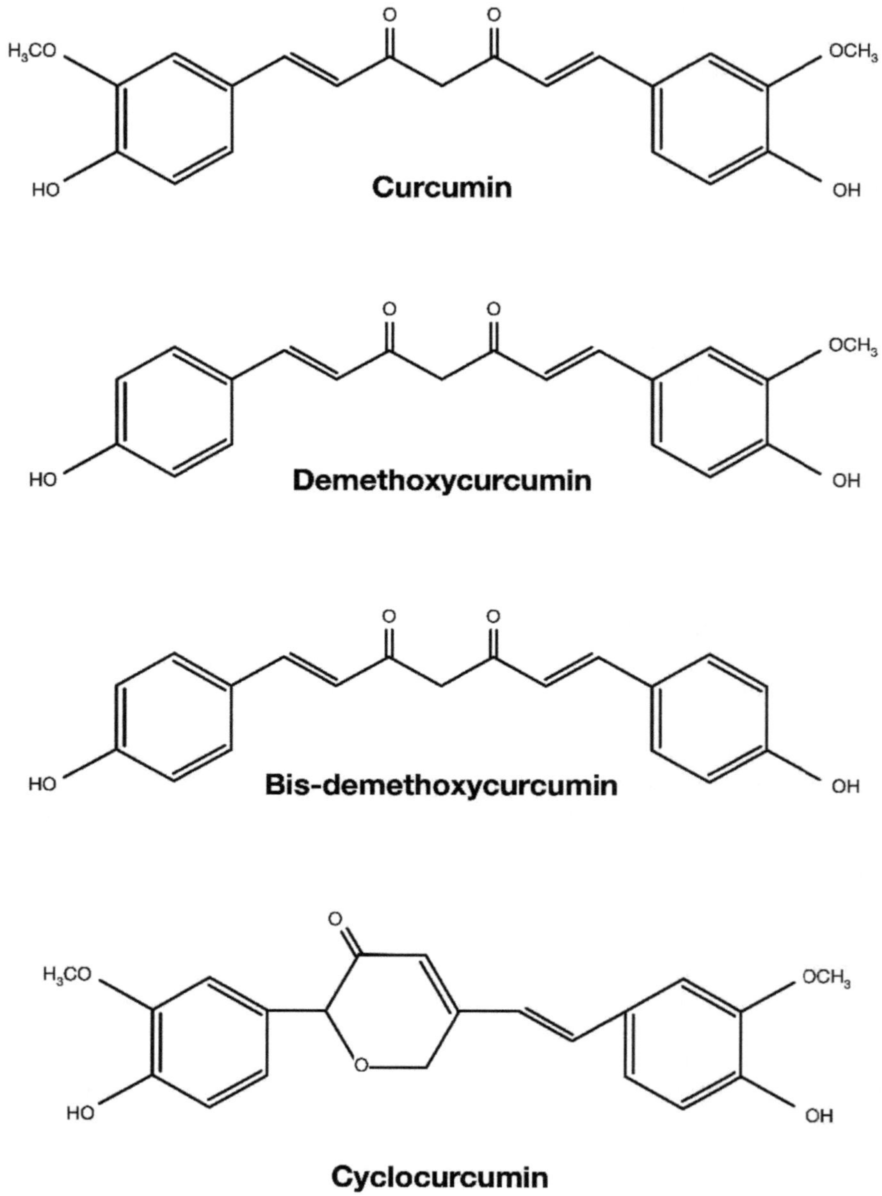

Fig. 12.1 Chemical structure of major immunomodulatory constituents of turmeric (*Curcuma longa*)

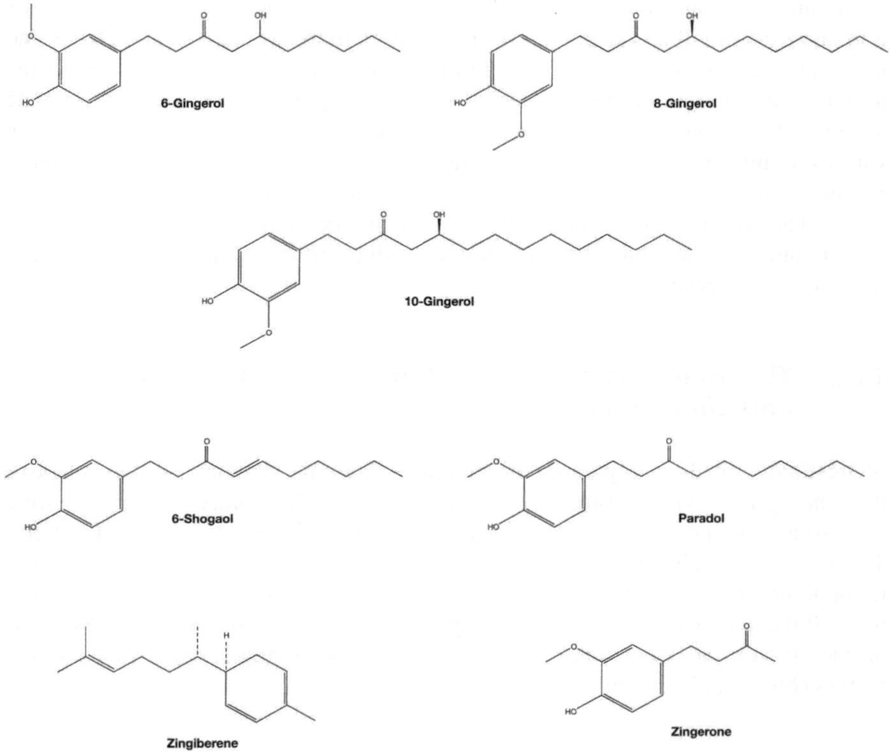

Fig. 12.2 Chemical structure of major immunomodulatory constituents of ginger (*Zingiber officinale*)

zingiberol, zingerone, gingerol, shogaol, paradol, and α-curcumene as major chemical constituents (Fig. 12.2) (Subhrajyoti and Shalini 2020). Gingerols (6-, 8-, and 10-gingerol) and shogaols (6-, 8-, and 10-shogaol) are the most potent bioactive compounds which exhibit a variety of pharmacological benefits (Syafitri et al. 2018). Interestingly, various studies have reported the antimicrobial, antioxidant, anti-inflammatory, analgesic, antidiabetic, nephroprotective, hepatoprotective, neuroprotective, anticancer, and immunomodulatory activities of ginger (Syafitri et al. 2018; Bhaskar et al. 2020).

Approximately 25% of approved drugs are derived from plants, and most pharmaceuticals are of natural origin or synthesized from natural products (Tonin et al. 2019). Medicinal plants exhibited fewer side effects and improved patient tolerance than synthetic drugs (Ali Ab et al. 2020). Usage of natural immunomodulators in the prevention and management of various diseases has increased considerably in the last few decades (Nair et al. 2018). Indeed, immunomodulatory plants have a crucial impact on the treatment of inflammation, infection, cancer, and immunodeficiency disorders (Nair et al. 2018). Numerous medicinal plants and their bioactive compounds have been discovered to support the host's immune system either via immunostimulation or immunosuppression (Thatte and Dahanukar 1986; Nair et al. 2018). Immunostimulators are used to replenish the immune response deficiency, whereas immunosuppressors hamper the immune response to restore normalcy (Nair et al. 2018). Immunomodulators can target various immune mediators and transcription factors as well as regulate many cellular processes such as apoptosis, protein processing, and antigen presentation (Nair et al. 2018). The current chapter describes the immunomodulatory activities of turmeric, ginger, and their main bioactive compounds with a particular attention on coronavirus disease and cancer.

12.2 The Immunomodulatory Effect of *Curcuma longa* and *Zingiber officinale*

Interestingly, medicinal plants can modulate both the innate and the adaptive immune systems via altering their humoral and cellular compartments including numerous immune cells and specific molecules, respectively (Samec et al. 2020). Turmeric and ginger are among the most extensively studied natural immunomodulators. They affect several components of the immune system including cellular components such as macrophages, dendritic cells, and B and T cells besides molecular components such as inflammatory cytokines and transcription factors (Momtazi-Borojeni et al. 2018).

12.2.1 Immunomodulatory Effect of Turmeric

Curcumin has been shown to regulate the immune function mainly through immunosuppressive mechanisms. It blocked the CD28 co-stimulation required for T cell activation through tyrosine kinase inhibition (Han et al. 1999). Notably, calcium is an essential element in various T cell responses. Curcumin has been shown to possess a downregulatory effect on the activity of calcium ATPase and calcium transport via impeding the formation of phosphoenzyme with ATP and Pi. Also, curcumin attached to an additional ATPase position that resulted in conformational alteration and averted ATP from binding (Cohly et al. 2003). In unprovoked T cells, the phosphorylated nuclear factor of activated T cells (NFAT) is located inside the cytosol. Calcium/calmodulin complex is formed due to receptor stimulation and an increase of cytosolic calcium induced by inositol 1,4,5 triphosphate. This complex binds to calcineurin, protein-serine phosphatase, and results in its activation. The activation form of calcineurin afterward dephosphorylates NFAT that consequently moves into the nucleus and leads to the expression of genes that are needed for the activation of T cells (Hogan 2003).

Conversely, Sikora et al. (1997) demonstrated that curcumin protected rat thymocytes and Jurkat T cells against ultraviolet- and dexamethasone-induced apoptosis, respectively. Curcumin acted mainly through the downregulation of tumor necrosis factor-α (TNF-α), interleukin (IL)-8, IL-β, macrophage inflammatory protein-1α (MIP-1α), and monocyte chemoattractant protein-1 (MCP-1) production (Abe et al. 1999). Moreover, curcumin prevented the activation of nuclear factor kappa-B (NF-κB) and eventually the production of cytokines in macrophages (Han et al. 1999). Notably, TNF-α is counted as one of the most pleiotropic cytokines that has a vital role in growth stimulation and inhibition. Treatment with curcumin reduced lipopolysaccharide (LPS)-induced production of TNF-α and IL-1 in human monocyte/macrophage cell lines and endothelial, dendritic, and bone marrow cells (López-Lázaro 2008). Besides, curcumin altered interleukin expression and activity via affecting leukocytes in different ways. It decreased LPS-induced high fever in humans by reducing the expression of serum TNF-α, IL-β, and IL-6 (Lee et al. 2003). Curcumin, likewise, inhibited dimerization of Toll-like receptor 4 induced by LPS (Youn et al. 2006). Further, it has been reported that curcumin inhibited phorbol-12-myristate-13-acetate-, TNF-α-, and hydrogen peroxide-induced NF-κB activation through blocking the activation of the inhibitor of nuclear factor kappa-B kinase subunit beta (Rossi et al. 2000).

In addition, following curcumin treatment, both T-helper (Th) 1 cell- and normal cell-medicated immune responses were reported to be very mild. Therefore, curcumin could be considered as a promising treatment for Th 1 cell-mediated autoimmune disorders (Zheng et al. 2013). Further, it has been reported that curcumin had a cytocidal effect on natural killer cells/T cell lymphoma cell lines, which have been unaffected by other remedies. Since NF-κB was activated in the aforementioned cell lines, thus curcumin could induce cell apoptosis through the suppression of NF-κB activation. It acted primarily through lessening the proliferation of immature B-lymphoma cells BKS-2 but not of normal cells (Churchill et al. 2000). Curcumin also inhibited early growth response protein-1 (Egr-1), a regulator gene that codes for transcription factors c-myc, B cell lymphoma-extra-large (Bcl-xL), NF-κB, and tumor suppressor gene p53 (Han et al. 1999). Further, Magalska et al. (2006) shed light on the fact that curcumin affected proliferating T cells' viability more rigorously than T cells. Besides, Deters et al. (2008) revealed that curcumin markedly inhibited peripheral blood mononuclear cells' prefiltration provoked by monoclonal antibody OKT3.

On the other hand, it was found that curcumin has immunostimulant properties. Curcumin treatment has been reported to augment natural killer cell activity markedly. In a research study on tumorous animals with metastatic cancer, a marked enhancement in both antibody-dependent cellular toxicity and natural killer cell activity was found after treatment with curcumin (Yadav et al. 2005). Besides, long-term therapy with curcumin indicated the upregulation of Th 1 cytokines and the production of nitric oxide (NO) in natural killer cells (Bhaumik et al. 2000). Curcumin moderated the activation and proliferation of T cells (Mehrotra et al. 2013). It has been shown to increase apoptosis-promoting activity of TNF-related apoptosis-inducing ligand (TRAIL) through inducing DNA destruction in human prostate metastatic cell lines (Deeb et al. 2003). Another study showed that curcumin could stimulate apoptosis via caspase-3 activation in both normal quiescent and proliferating human lymphocytes but without DNA fragmentation (Magalska et al. 2006).

12.2.2 Immunomodulatory Effect of Ginger

Ginger has a potential impact as an immunosuppressor in diseases such as chronic inflammation and autoimmune disorders (Zhou et al. 2006). Ginger extract prevented LPS-induced macrophage activation and antigen-presenting function and indirectly suppressed T cell proliferation (Tripathi et al. 2008). Also, it was reported that 6-gingerol interfered with macrophage function, explicitly inhibiting production of pro-inflammatory cytokines (Tripathi et al. 2007). Eun et al. (2009) showed that 6-shogaol, 1-dehydro-10-gingerdione, and 10-gingerdione prevented LPS-induced NO production in vitro. Among them, 1-dehydro-10-gingerdione exhibited an additional stimulatory effect on phagocytosis. Treatment with ginger displayed immunosuppressive effect in the in vitro lymphocyte proliferation assays (Wilasrusmee et al. 2002a) mediated by reduced production of IL-2 and IL-10 (Wilasrusmee et al. 2002b). Moreover, Zhou et al. (2006) reported that cell-mediated immune response and non-specific T cell proliferation were suppressed by ginger volatile oil in vitro and in vivo in mice. Ginger volatile oil suppressed T cell proliferation and the amount of T helper cells while increased suppressor T cells and prevented IL-1α production by mice peritoneal macrophages. Treatment of 2,4-dinitro-1-fluorobenzene-sensitized mice with the volatile oil of ginger weakened the delayed-type hypersensitivity reactions dose-dependently (Zhou et al. 2006). Administration of ginger aqueous extract before airway challenge of ovalbumin-sensitized mice resolved the airway inflammation via suppressing Th 2-mediated immune responses (Ahui et al. 2008). Moreover, Han et al. (2019) revealed that 6-gingerol mitigated the severity of experimental autoimmune encephalomyelitis model of multiple sclerosis via activation of immunomodulatory dendritic cells that prevent the differentiation of Th 17 cells.

Alternatively, numerous experimental studies revealed the immunostimulatory properties of ginger. An in vitro study demonstrated that aqueous ginger extract significantly increased splenocyte proliferation and capacity for cytokine production by activated macrophages in mice (Ryu and Kim 2004). Experimental studies on fish fed ginger supplemented diet displayed a significant increase in number as well as activity of macrophages, neutrophils, and lymphocytes (Dügenci et al. 2003). Ginger fed to rainbow trout improved their non-specific immune response and eventually controlled their experimental infection with *Aeromonas hydrophila* (Haghighi and Rohani 2013). Ginger extract enhanced the growth, immune response, and survival of tilapia Oreochromis mossambicus (Immanuel et al. 2009). It has been reported that gingerols alleviated the liver damage in mouse with heatstroke endotoxemia via enhancing the phagocytic ability of celiac macrophage through increasing its metabolic energy level (Nie et al. 2006). Ginger indirectly inhibited growth of the influenza virus via macrophage activation and consequent TNF-α production (Imanishi et al. 2006). Moreover, Carrasco et al. (2009) reported that ginger essential oil enhanced T and B cells, eventually reinstating the humoral immune response, in cyclophosphamide-immunosuppressed mice. Bhaskar et al. (2020) showed that 6-gingerol induced Th 1/Th 17 cell responses in spleens of mycobacterium tuberculosis-infected mice.

12.3 *Curcuma longa* and *Zingiber officinale* as Potential Treatments for Coronavirus Disease

In early 2020, coronavirus disease 2019 (COVID-19), a highly contagious viral infection, has emerged in Wuhan, China, and has spread quickly all over the world (Rajagopal et al. 2020). According to the WHO, as of the first of November 2020, there have been 46,412,115 confirmed cases of COVID-19, including 1,198,717 deaths worldwide, and the numbers are exponentially increasing (WHO 2020). A vaccine against COVID-19 is not yet available; hence, many approved drugs and natural products are repurposed to treat COVID-19 infection (Goswami et al. 2020). COVID-19 encodes two proteases, the main protease (M-pro), called 3-C-like protease, and the papain-like protease (PLpro), which are necessary for its survival in hosts (Li and De Clercq 2020). The M-pro and Plpro are potential drug targets against COVID-19 (Kandeel and Al-Nazawi 2020). Thus, virtual screening of many bioactive compounds from medicinal plants, including turmeric and ginger, was performed. The molecular docking studies of Khaerunnisa et al. (2020) and Suravajhala et al. (2020) revealed that the chemical constituents from turmeric like demethoxycurcumin and curcumin were among the most recommended bioactive compounds as potential inhibitors for COVID-19 M-pro. Similarly, the docking studies of Rajagopal et al. (2020) revealed that some of the chemical constituents from ginger such as 8-gingerol and 10-gingerol were significantly active against COVID-19, with elevated Glide score when compared to the currently used drug hydroxychloroquine. These compounds have a good affinity to M-pro due to their high lipophilicity and hydrogen bonding. In addition, 6-gingerol, 8-gingerol, and 10-gingerol are potent inhibitors of PLpro, having high in silico binding affinity with high ligand efficiency. Though demethoxycurcumin and curcumin have high binding affinity to PLpro than chloroquine and hydroxychloroquine, they have lower ligand efficiency than the docked compounds from ginger (Goswami et al. 2020).

Another potential drug target against COVID-19 disease is the angiotensin-converting enzyme 2 (ACE-2) receptor (Yan et al. 2020). The latest research studies showed that COVID-19 virus attacks individual host cells through targeting ACE-2 membrane receptor. The attachment of viral S protein to the ACE-2 receptor that exists in the mucus membrane results in membrane and viral fusion and later replication of the virus in the host (Jia et al. 2005; Jean et al. 2020). ACE-2 expression was noticed in nasal epithelial cells, alveolar epithelial type II cells, and luminal surface of intestinal epithelium. Therefore, the nasopharynx, lungs, and intestine enable the entry of virus and function as the possible access to viral invasion (Jia et al. 2005).

Interestingly, a research study using in silico method comprising docking and stimulation showed the dual binding affinity of polyphenolic compounds where curcumin bound to the viral S protein and ACE-2 receptor. The attachment of curcumin to the ACE-2 receptor-binding domain position for viral S protein as well as the viral-binding sites for ACE-2 receptor indicated that curcumin could antagonize the entry of COVID-19 viral protein and hence act as a possible inhibitory agent (Das et al. 2021). Besides, topical application of the emulsion form of curcumin could successfully avert COVID-19 infection in individuals because of the distribution of the viral entry site of ACE-2 receptor at the nasal cells, eye, and mucosal surface of respiratory tract (Jia et al. 2005). Moreover, dietary complements of curcumin with zinc and vitamin C exhibited encouraging results in boosting immunity and provided a natural protective guard against COVID-19 infections (Chen et al. 2020).

Further, Suravajhala et al. (2020) recommended curcumin as a therapeutic agent for anti-coronavirus treatment improvement. The authors revealed that curcumin had a high binding affinity regarding nucleocapsid and COVID-19 nsp-10 proteins with possible anti-viral effect. The binding affinity of curcumin targeted a diversity of COVID-19 proteins, namely, spike glycoproteins (PDB ID: 6VYB), membrane glycoprotein (PDB ID: 6 M17), nucleocapsid phosphoprotein (PDB ID: 6VYO), nsp10 (PDB ID: 6W4H), and RNA-dependent RNA polymerase (PDB ID: 6 M71) (Suravajhala et al. 2020). In addition, Zahedipour et al. (2020) indicated the possible efficacy of curcumin in COVID-19 therapy through impeding the virus entry into the cell, hindering the virus encapsulation and its protease, and moderating several cellular signaling mechanisms. Besides, the study of Ting et al. (2018) showed that curcumin could reduce porcine epidemic diarrhea virus at the replication phase. The inhibition of both virus titers and plaque numbers when exposed to curcumin supports the potential role of curcumin in inhibiting the replication of COVID-19 virus (Ting et al. 2018). Noteworthy, interferons have a fundamental role in the protection against infection with COVID-19. Such viruses impeded interferon stimulation in humans and antagonized signal transducer and activator of transcription (STAT)-1, which is considered a main protein in the interferon-mediated immune reaction (Kindler et al. 2016). Ting et al. (2018) revealed that the treatment with cationic carbon dots based on curcumin could stimulate the generation of interferon-stimulating genes and IL-8 and IL-6 cytokines of Vero cells through prompting innate immunity.

Likewise, Ferreira et al. (2015) elucidated the efficacy of curcumin in treating COVID-19-associated pulmonary inflammation, edema, and fibrosis. Curcumin acts mainly through blocking the necessary signals that modify the expression of many pro-inflammatory cytokines such as mitogen-activated protein kinase (MAPK) and NF-κB. In addition, Senathilake et al. (2020) mentioned that curcumin possessed anti-fibrotic and anti-inflammatory effects via inhibiting the expression of cytokines and chemokines associated with lung infections like interferon-γ and MCP-1. Curcumin reduced collagen deposition in cyclophosphamide-, bleomycin-, and irradiation-induced pulmonary fibrosis in rats (Chapman 2004). Lately, experimental evidence showed that prophylactic administration of curcumin reduced inflammation leading to diminished fluid influx in the lungs of rats under hypoxia. Similar effect could take place via modulating of NF-κB pathway, reducing the pro-inflammatory cytokines and cell adhesion molecules and stabilizing hypoxia-inducible factor 1-alpha resulting in a reduction of angiogenic factors followed by a reduction in pulmonary edema resulting from COVID-19 (Huang et al. 2020).

Further, Xu et al. (2020) stated that curcumin could be potentially effective for treating COVID-19 accompanying renal inflammation. Curcumin potentially inhibited renal fibrosis through suppression of MCP-1, NF-kB, TNF-α, IL-β, and caveolin-1 levels. Likewise, curcumin increased the expression of anti-inflammatory mediators including neural precursor cell-expressed developmentally downregulated protein 4, heme oxygenase-1 (HO-1), and mannose-6-phosphate receptor binding protein 1. In addition, curcumin targeted MAPK/extracellular signal-regulated kinase (ERK) and peroxisomal proliferator-activated receptor gamma (PPAR-γ) pathways in various experimental models of renal disorders (Xu et al. 2007).

12.4 Immunomodulatory Effect of *Curcuma longa* and *Zingiber officinale* Against Cancer

Experimental evidence accumulated over the past years signified the potential role of medicinal plants and their bioactive compounds in cancer prevention and treatment (Kapinova et al. 2019). Recently, their use has received much attention due to their wider safety margin and capacity to complement conventional anticancer drugs. Notably, the regulation of immune response by medicinal plants represents a promising approach in cancer therapy via suppression, amplification, or stimulation of cells and molecules of the innate and adaptive immune systems (Samec et al. 2020). They impact the tumor microenvironment via infiltrating innate immune cells, such as macrophages, dendritic, natural killer, and myeloid-derived suppressor cells as well as the adaptive immune cells, T cells, and B cells (Tap et al. 2015).

12.4.1 Anticancer Effect of Turmeric

Curcumin is a phytochemical that has powerful in vitro antineoplastic effects against various cancers. It acts through different biological activities. Curcumin inhibited NF-κB and suppressed angiogenesis. Besides, it possesses apoptosis-including activity, resulting in malignant cell death mainly via the mitochondrial death pathway. Further, curcumin upregulated proapoptotic protein Bcl-2-associated X (Bax) and inhibited B cell lymphoma-2 (Bcl-2) in breast cancer cells (Pongrakhananon and Rojanasakul 2011), leading to hammering of mitochondrial function, production of cytochrome c, and triggering of caspases-9 and 3. Curcumin improves the anticancer effects of many common chemotherapeutic agents including paclitaxel, cisplatin, and doxorubicin (Chan et al. 2003; Mullaicharam and Maheswaran 2012). Further, curcumin displayed a remedial effect against many different cancers such as gastro-intestinal, leukemia, lung, breast, ovarian, and neurological cancers (Kita et al.

2008). The present status of anticancer effects of curcumin against different forms of cancers is thoroughly elucidated below beneath various headings.

12.4.1.1 Blood Cancer

Curcumin prevented the growth of human myeloid ML-1a cells by reducing TNF-α-induced DNA binding of NF-κB via downregulation of the inhibitor of κB phosphorylation (Goel et al. 2008). In addition, curcumin displayed anticancer effect in B cell chronic lymphocytic leukemia through inhibition of NF-κB, protein kinase B (Akt), STAT-3, and X-linked inhibitor of apoptosis protein. Curcumin induced apoptosis via triggering the c-Jun N-terminal kinase (JNK)/ERK/AP1 signaling pathway in human acute monocytic leukemia cells THP-1 (Yang et al. 2012). Also, it has been described that curcumin inhibited carcinogenesis through downregulation of both IL-1α and IL-1β in lymphoma-bearing mice (Das and Vinayak 2014). Curcumin repressed the gene expression of Egr-1 and, in turn, inhibited the growth of immature B cell lymphoma (Han et al. 1999).

12.4.1.2 Bone Cancer

Dietary curcumin moderated tumor marker indicators of fibrosarcoma in an in vivo study on rats. Moreover, treatment with curcumin improved cancer cell killing and inhibited radiotherapy resistance in mice having fibrosarcoma through the inhibition of radiation-induced ERK and NF-κB expression (Kumar Mitra and Krishna 2004).

12.4.1.3 Brain Cancer

Malignant gliomas are severe brain tumors that are unaffected by chemotherapeutic drugs and radiation. Curcumin displayed a potent therapeutic efficacy against malignant glioblastoma cells. Curcumin downgraded the cell survival in p-53- and caspase-independent way, a consequence associated with the decrease of NF-κB and activator protein-1 (AP-1) signaling pathways through inhibition of the JNK and AKT activation in glioma cell lines (Dhandapani et al. 2007).

12.4.1.4 Esophageal Cancer

Curcumin suppressed the growth of human esophageal microvacuolar endothelial cells via preventing cytokine-induced activation of inducible nitric oxide synthase (iNOS), NK, vascular cell adhesion molecule, and NF-κB (Rafiee et al. 2003). Wax et al. (2005) indicated the efficiency of curcumin as a chemopreventive agent when examined through quantifying the modulation in the occurrence of malignant changes in rats' esophagus. Moreover, Goel et al. (2008) revealed that nutritional curcumin administrated during the initiation and post-initiation phases repressed esophageal cancer in rats.

12.4.1.5 Lung Cancer

Curcumin demonstrated anticancer activities in many cancerous lung cells through many molecular targets. It acted mainly through the inhibition of farnesyl protein transferase (FPTase) in A549 human lung cancer cells. Likewise, it downregulated AP-1 transcription as well as mediastinal lymph node metastasis in Lewis lung cancer cells (Lucile White et al. 1998). Curcumin downregulated NF-κB and acted on Janus kinase 2 (JAK2)/STAT-3 pathway through inhibiting JAK2 in A549 cells (Wu et al. 2015). In addition, curcumin induced apoptosis and cell proliferation of human lung cancer cells through stimulation of microRNA-192-5p and downregulation of phosphoinositide 3-kinase (PI3K)/Akt signaling pathway (Boroumand et al. 2018). Curcumin suppressed the expression of cyclooxygenase (COX)-2, epidermal growth factor receptor (EGFR), and extracellular signal-regulated kinase (ERK) resulting in high apoptosis of lung adenocarcinoma cells. Besides, curcumin suppressed cellular progression and stimulated G0/G1 cell cycle arrest via metastasis-associated protein 1 (MTA1)-mediated inactivation of Wnt/catenin signaling (Jin et al. 2015).

12.4.1.6 Breast Cancer

Curcumin possessed anticarcinogenic activity in various breast cancer cell lines via reducing aryl hydrocarbon receptor and cytochrome P450 1A1, the tyrosine kinase activity of p185neu, the expression of Ki-67, proliferating cell nuclear antigen (PCNA), p53 mRNAs, and COX-1 and COX-2 enzymes (Aggarwal et al. 2007). Besides, curcumin activated the expression of p53-dependent Bax protein and inhibited vascular endothelial growth factor (VEGF) and basic fibroblast growth factor (bFGF) (Shao et al. 2002; Schindler and Mentlein 2006). Likewise, curcumin downregulated telomerase activity via human telomerase reverse transcriptase, inhibited matrix metalloproteinase (MMP)-2 expression, stimulated the tissue inhibitor of metalloproteinase-1, and blocked the activation of AP-1 and NF-κB (Bobrovnikova-Marjon et al. 2004). Further, it has been shown that curcumin reduced the lipoxygenase (LOX) pathway, stimulated the destruction of cyclin E expression via a ubiquitin-dependent pathway, inhibited the insulin-like growth factor-1, and stimulated the cyclin-dependent kinase (CDK) inhibitors p21 and p27 in breast cancer cell lines (Hammamieh et al. 2007). In MDA-MB-231 breast cancer cells, curcumin also inhibited Akt protein in a time- and dose-dependent way, reduced ubiquitin-proteasome pathway, and prompted autophagy (Guan et al. 2016). Further, it has been shown that the autophagy and apoptotic abilities of curcumin were associated with blocking PI3K/Akt signaling pathway (Akkoç et al. 2015). Lately, Norouzi et al. (2018) reported that curcumin regulated the expression of oncogenic and tumor oppressive microRNAs in breast cancer cells.

12.4.1.7 Stomach Cancer

Curcumin triggered apoptosis in gastric adenocarcinoma cell line SGC-7901 by releasing cytochrome c into cytosol and causing indulgence of mitochondrial membrane potential resulting in cell death. Moreover, curcumin downregulated Bcl-2 and upregulated Bax, triggering the cleavage of caspase-3 (Xue et al. 2014).

12.4.1.8 Liver Cancer

Curcumin markedly reduced gamma-glutamyl transpeptidase-positive *foci* that is considered the sign for hepatocellular carcinoma in rats. The anticarcinogenic effect of curcumin was attributed to the simulation of glutathione-linked detoxification enzymes in rat's liver. It also acted via averting the stimulation of hepatic hyperplastic nodules, hypoproteinemia, and hepatic lipid peroxidation in different hepatic cancer models induced in rats (Dhandapani et al. 2007).

12.4.1.9 Pancreatic Cancer

Several research studies showed that curcumin possessed an anticarcinogenic effect against many pancreatic cell lines. Curcumin suppressed FPTase in human pancreatic cancer cells MIA PACa-2 (Kawamori et al. 1999). Moreover, curcumin has been found to reduce NF-κB that were overexpressed in human pancreatic cancer tissues and cell lines (Khanbolooki et al. 2006). Curcumin affected pancreatic cells' vitality through reducing COX-2, VEGF, STAT-3, and IL-8 (Khanbolooki et al. 2006). In addition, curcumin upregulated the expression of forkhead box protein O1 in pancreatic cells through acting on PI3K/Akt signaling, resulting in apoptosis and cell cycle arrest (Zhao et al. 2015).

12.4.1.10 Kidney Cancer

Kössler et al. (2012) revealed that long-term contact of a human kidney cell line to curcumin induced apoptosis and activated the production of a subpopulation of cells with augmented volume. Further, curcumin stimulated apoptotic events such as chromatic condensation, DNA fragmentation, as well as cell shrinkage and downregulated FPTase (Jiang et al. 1996). Curcumin also acted as a COX-1 and -2 downregulator and inhibitor of DNA damage, microsomal lipid peroxidation, Bcl-2, apoptosis protein (IAP), and Bcl-xl. Moreover, curcumin prodrugs hindered the growth of human renal tubular epithelial cells HKC (Ramsewak et al. 2000; Iqbal et al. 2003; Jung et al. 2005).

12.4.1.11 Bladder and Intestinal Cancer

Curcumin inhibited bladder cancer propagation either via inhibition of cyclin A and stimulation of p21 or downregulation of NF-κB (Anand et al. 2008). Investigation of intestinal tissue from animals treated with curcumin showed that cancer prevention was accompanied by boosted enterocyte proliferation and apoptosis (Anand et al. 2008).

12.4.1.12 Colorectal Cancer

Curcumin administration increased colonic mucosal $CD4^+$ T and B cells and prevented adenoma formation of C57BL/6 J-Min/$^+$ mouse bearing a germline Apc mutation, a standard model for familial adenomatous polyposis and sporadic colorectal cancer (Churchill et al. 2000). In addition, nanoparticle curcumin suppressed the development of dextran sulfate sodium (DSS)-induced colitis in BALB/c mice via preventing NF-κB activation alongside inducing $CD4^+$ $Foxp3^+$ regulatory T cells and $CD103^+$ $CD8\alpha^-$ regulatory dendritic cells in intestinal mucosa (Ohno et al. 2017). It has been reported that curcumin exhibited chemopreventive activities in IL-10-deficient mouse model for colitis-associated cancer through anti-inflammatory, anti-oxidative, and anti-proliferative properties. Curcumin reduced the intestinal tumor burden alongside maintaining microbiota diversity (Mcfadden et al. 2015). Moreover, curcumin enhanced chemosensitization to 5-fluorouracil-based chemotherapy by targeting cancer stem cell subpopulation (Shakibaei et al. 2014).

12.4.1.13 Ovarian Cancer

Treatment with curcumin either alone or combined with docetaxel inhibited proliferation and micro-vessel density and stimulated tumor cell apoptosis. In mice with multidrug-resistant ovarian cancer, treatment with curcumin alone and combined with docetaxel led to a marked reduction in cancer growth (Lin et al. 2007).

12.4.1.14 Prostate Cancer

Curcumin displayed anticancer activity against several prostate cancer cells LNCaP, DU145, C4-2B, and PC3. In addition, it prompted programmed cell death in androgen-dependent and androgen-independent prostate cancer cells. Further, curcumin downregulated capillary tube development and cell migration as well as affected markedly the actin cytoskeletons in prostate cancer cells (Shenouda et al. 2004; Shankar et al. 2007).

12.4.2 Anticancer Effect of Ginger

Excessive experimental research has indicated the potential anticancer activities of ginger and its bioactive compounds against several cancer cells (Table 12.1).

Table 12.1 Immunomodulatory effect of ginger and its bioactive compounds against cancer in experimental research

Extract/compound	Cancer type	Cell line/animal model	Mechanism of immunomodulation	References
Ginger extract	Skin	TPA-induced skin tumorigenesis in mice	Ø activities of epidermal ornithine decarboxylase, COX, and LOX	Katiyar et al. (1996)
	Esophageal	ESO26 human esophageal cancer cell line	Ø proliferation and growth, apoptosis via ↓ Bcl-2, ↑ Bax, p21, and caspase-3	Abbasi et al. (2020)
	Lung	A549 human non-small cell lung cancer cells	Paraptosis via ↑ ERS, mitochondrial dysfunction, AIF translocation, DNA damage, telomere shortening, cellular senescence	Kaewtunjai et al. (2018), Nedungadi et al. (2019)
	Breast	MDA-MB-231 human breast cancer cells	Paraptosis	Nedungadi et al. (2019)
	Stomach	*N*-nitroso-*N*-methyl urea-induced gastric cancer in rats	Anti-inflammatory and antioxidant activities	Mansingh et al. (2020)

Gingerol, paradol, zingiberene, zerumbone, and shogaol inhibited cancer growth by modulating various cell signaling pathways. They modulated tumor suppressor genes, transcription factors, growth factors, angiogenesis, and apoptosis (Almatroudi et al. 2019). Thus, ginger and its bioactive compounds could possibly prevent or control various types of cancer such as colorectal, stomach, ovarian, liver, breast, and prostate cancers (Ishiguro et al. 2007; Lee et al. 2008a; Sung et al. 2008; Kim et al. 2009; Brown et al. 2009; Hung et al. 2009).

12.4.2.1 Skin Cancer

Topical application of an ethanolic ginger extract protected against 12-*O*-tetradecanoylphorbol-13-acetate (TPA)-induced mouse skin tumorigenesis model via inhibiting epidermal ornithine decarboxylase, COX, and LOX activities (Katiyar et al. 1996). Similarly, topical application of 6-gingerol inhibited TPA-induced COX-2 expression in mouse skin through impeding p38 MAPK/NF-κB signaling pathway (Kim et al. 2005). In addition, 6-gingerol prevented the growth of human epidermoid carcinoma cells A431 via inducing mitochondrial apoptosis (Nigam et al. 2009).

12.4.2.2 Blood Cancer

The substance 6-shogaol displayed a cytotoxic effect on human promyelocytic leukemia cells HL-60 and human chronic granulocytic leukemia cells K562 (Peng et al. 2012). The in vitro and in vivo study of Liu et al. (2013) revealed that 6-shogaol provoked cellular apoptosis in transformed and primary human leukemia cells and in leukemia xenografts. Apoptosis in acute T cell leukemia Jurkat, histiocytic lymphoma U937, and acute promyelocytic leukemia cells HL-60 were also induced by 6-Shogaol. Also, it significantly repressed tumor growth and induced apoptosis in U937 xenograft mouse model. Apoptosis induced by 6-shogaol is mediated partly via the phosphorylation of α-subunit of eukaryotic initiation factor-2 (eIF2α), an inducer of programmed cell death, and the cleavage eIF2α provoked by caspase-3 activation (Liu et al. 2013).

12.4.2.3 Brain Cancer

Meningiomas are the second most common brain tumors found in adults. The in vitro study of Das et al. (2015) indicated that 6-gingerol and zerumbone, ginger bioactive compounds, induced cell death in IOMM-Lee and CH157MN human meningioma cells. They triggered apoptosis as revealed by downregulation of tetraspanin-12 protein and the survival proteins Bcl-XL and myeloid cell leukemia-1 (Mcl-1), alongside overexpression of the apoptotic factor Bax and activation of caspase-3 (Das et al. 2015). Moreover, they inhibited the Wnt/β-catenin signaling pathway, which has been linked with apoptosis in various cancer cells (Novak and Dedhar 1999; Das et al. 2015).

12.4.2.4 Esophageal and Lung Cancer

Recently, an in vitro study displayed that ginger extract induced apoptosis in the esophageal cancer cell line ESO26 leading to inhibition of their proliferation and growth. Ginger increased the cleavage of caspase-3 and decreased Bcl-2 while increased Bax and p21 gene expression in tumor cells (Abbasi et al. 2020).

Nedungadi et al. (2019) reported that ginger extract exhibited anticancer effect against A549 non-small cell lung cancer cells. Ginger extract induced caspase-independent paraptosis mediated via endoplasmic reticulum stress (ERS), mitochondrial dysfunction, apoptosis-inducing factor (AIF) translocation, and ultimately DNA damage (Nedungadi et al. 2019). Besides, it was reported that ginger extract promoted the tumor-suppressor mechanism of telomere shortening and cellular senescence in A549 cells (Kaewtunjai et al. 2018). Oral administration of 6-gingerol antagonized the cancer-promoting effect of capsaicin in the urethane-induced lung cancer in mice by elevating transient receptor potential vanilloid type-1 (TRPV1) level resulting in diminished NF-κB, EGFR, and cyclin D1 levels (Geng et al. 2016). Dietary administration of zerumbone dose-dependently repressed lung adenomas in mice through suppression of proliferation and expression of NF-κB and HO-1 along with apoptosis induction (Kim et al. 2009). Moreover, 6-shogaol suppressed A549 cell proliferation via inhibiting the AKT/mammalian target of rapamycin (mTOR) survival signaling and ultimately inducing autophagic cell death (Hung et al. 2009). Likewise, 6-shogaol suppressed the growth of H-1299 human lung cancer cells, mediated by inhibiting the release of arachidonic acid and NO synthesis (Sang et al. 2009).

12.4.2.5 Breast Cancer

Nedungadi et al. (2019) displayed that whole ginger extract induced paraptosis in human breast cancer cells MDA-MB-231, indicating its potential for breast cancer prevention (Nedungadi et al. 2019). Its active constituent 6-gingerol prevented cell adhesion, invasion, and motility in MDA-MB-231 cells in a dose-dependent manner while decreasing MMP-2 and MMP-9 expression and activity (Lee et al. 2008a). It has been shown that MMP-2 and MMP-9 were significantly implicated in cancer cell invasion due to their role in the degradation of type IV collagen (Ling et al. 2010). Similarly, 10-gingerol exhibited antitumor activities against MDA-MB-231 cells via inhibition of Akt and p38MAPK activity and downregulating the expression of EGFR. Cancer cell proliferation was prevented by 10-gingerol by downregulating cell cycle regulatory proteins such as cyclin-dependent kinases and cyclins, resulting in G1 cell arrest. Furthermore, it hindered cell invasion in response to mitogenic stimulation (Joo et al. 2016). Besides, it has been revealed that 10-gingerol prompted intrinsic apoptotic cell death in metastatic triple negative breast cancer cell lines in a dose-dependent manner (Martin et al. 2017). Noteworthy, the chemokine of CXC chemokine ligand 12 (CXCL12) and its receptor CXC chemokine receptor 4 (CXCR4) are closely linked with migration and metastasis of cancer cells. Zerumbone dose and time dependently downregulated the expression of CXCR4 and ultimately inhibited CXCL12-induced invasion of HER2-breast cancer cells (Sung et al. 2008). Recently, Bawadood et al. (2020) showed that 6-shogaol prevented the growth of breast adenocarcinoma cells MCF-7 and breast ductal carcinoma cells T47D by S-phase and G2/M-phase cell cycle arrest and apoptosis instigation via interfering with notch signaling genes Hes1 and Cyclin D1.

Moreover, 6-shogaol prevented the growth of MCF-7 breast cancer cells by activating PPAR-γ signaling pathway (Tan et al. 2013). Further, Ling et al. (2010) revealed that 6-shogaol displayed a dose-dependent anti-invasive effect on the breast cancer cells MDA-MB-231 partly via preventing NF-κB activation and reducing MMP-9 expression.

12.4.2.6 Stomach Cancer

The study of Mansingh et al. (2020) displayed the anticancer potential of aqueous ginger extract against *N*-nitroso-*N*-methylurea-induced gastric cancer in rats mediated through its anti-inflammatory and antioxidant activities. The study of Ishiguro et al. (2007) showed that 6-gingerol and 6-shogaol reduced gastric cancer cells' viability via several mechanisms. The 6-gingerol enhanced TRAIL-induced viability reduction through increasing TRAIL-induced caspase-3/7 activation. In addition, 6-gingerol suppressed the expression of cellular inhibitor of apoptosis protein 1 (cIAP1), which has an inhibitory effect on TRAIL-induced NF-κB activation and consequently on TRAIL-induced caspase-3/7 activation. Instead, 6-shogaol decreased gastric cancer cells' viability via damaging microtubules and inducing mitotic arrest (Ishiguro et al. 2007). Moreover, 6-gingerol augmented the radiosensitivity of gastric cancer cell line HGC-27 via cell cycle arrest at G2/M phase and apoptosis induction in a dose-dependent manner. It downregulated cyclin A2, cyclin B1, cyclin D1, and CDC2 expression, upregulated the mRNA expression of p27, and activated caspase-9, caspase-3, and cytochrome c (Luo et al. 2018). Moreover, 6-gingerol prevented the proliferation of AGS human gastric cancer cells through mitochondrial-dependent caspase activation. The 6-gingerol increased reactive oxygen species (ROS) production that disrupted mitochondrial membrane potential. Consequently, deregulation of Bax/Bcl-2 ratio occurred leading to upregulation of cytochrome c and ultimately triggering the caspase-dependent apoptosis (Mansingh et al. 2018).

12.4.2.7 Liver Cancer

Oral administration of ginger extract protected against ethionine-induced liver cancer in rats via blocking the activation of NF-κB (Habib et al. 2008). The proliferation and invasion of a rat ascites hepatoma AH109A cells were dose-dependently repressed by 6-gingerol. Yagihashi et al. (2008) suggested that 6-gingerol capacity to arrest cell cycle and induce apoptosis accounts for its anti-proliferative action, while 6-gingerol antioxidant capacity accounts for anti-invasive action on hepatoma cells. The 10-gingerol suppressed the proliferation of hepatocellular carcinoma HepG2 cells via Src/STAT-3 signaling pathway in a dose-dependent manner (Chen et al. 2018). It has also been observed that exposure to 6-shogaol exhibited a higher cytotoxic effect than 10-gingerol on BEL-7404 human hepatoma cells (Peng et al. 2012).

12.4.2.8 Pancreatic Cancer

Previous studies displayed that ginger extract killed PANC-1 pancreatic cancer cells after altering the antioxidant/oxidative stress pathway and inducing ROS-mediated autotic cell death (Akimoto et al. 2015; Homa and Wentz-Hunter 2018). In addition, zerumbone inhibited PANC-1 cells in a time- and dose-dependent manner. It induced apoptosis of PANC-1 cells through p53 signal pathway. Pancreatic cancer cells treated with zerumbone displayed reduced proliferation alongside augmented apoptosis that was manifested by generation of apoptotic bodies and condensed nuclei, activation of ROS and caspase-3, and upregulation of miR-34 and p21 expression (Zhang et al. 2012).

12.4.2.9 Bladder and Colorectal Cancer

Dietary intake of ginger extract protected against *N*-butyl-*N*-(4-hydroxybutyl)-nitrosamine-induced urothelial carcinogenesis in rats via reducing the development of proliferative lesions like hyperplasia and neoplasia (Ihlaseh et al. 2006).

The in vitro study of Malmir et al. (2020) reported that ginger extract repressed the growth of colorectal cancer cells HCT-116. Similarly, zingerone, a ginger constituent, dose-dependently induced oxidative stress-mediated apoptosis and suppressed proliferation in HCT-116 cells (Su et al. 2019). The combined therapy of 5-fluorouracil with some natural compounds increased its anticancer activity and possibly reduced its dose-related toxicity and resistance, both experimentally and clinically (Shakibaei et al. 2013; Hamaya et al. 2015). In context, combined treatment of 5-fluorouracil with ginger extracts enhanced its chemotherapeutic effect against the HCT-116 colon adenocarcinoma cells (Hakim et al. 2014). The in vitro study of Brown et al. (2009) revealed that ginger extract and 6-gingerol exhibited anticancer activities via directly suppressing the proliferation of rat colonic adenocarcinoma cells and preventing blood supply to tumor cells via angiogenesis. In support, Lee et al. (2008b) indicated that 6-gingerol induced apoptosis in HCT-116 cells by upregulating nonsteroidal anti-inflammatory drug-activated gene-1 (NAG-1) besides arresting cell cycle at G1 phase mediated by cyclin D1 suppression. Lately, Farombi et al. (2020) additionally reported that 6-gingerol mitigated benzo[a]pyrene and DSS-induced colorectal cancer in mice via anti-proliferative, anti-inflammatory, and apoptotic mechanisms. The 6-gingerol inhibited angiogenesis by reducing VEGF, angiopoietin-1, bFGF, and growth differentiation factor-15 (GDF-15) levels. Also, it raised the expression of β-catenin, adenomatous polyposis coli, and p53 but reduced the expression of cyclin D1, TNF-α, IL-1β, COX-2, and iNOS (Farombi et al. 2020). The 8-gingerol inhibited the proliferation and migration of colorectal cancer because of apoptosis and cell cycle. Cancer cells' proliferation and migration were inhibited by 8-gingerol through impeding EGFR/STAT/ERK signaling pathway (Hu et al. 2020). Moreover, ginger prevented colon carcinogenesis induced by 1,2-dimethylhydrazine in male Wistar rats via activating both enzymatic and non-enzymatic antioxidants (Manju and Nalini 2005). Zerumbone feeding prevented colonic adenocarcinomas in DSS-treated mice via several mechanisms as evidenced by induction of apoptosis as well as suppression of proliferation and inflammation (Kim et al. 2009).

12.4.2.10 Ovarian Cancer

The in vitro study of Pashaei-Asl et al. (2017) showed that the anticancer activity of ginger extract against the ovarian cancer cell line SKOV-3 was mediated via p53-induced apoptosis. Similarly, ginger suppressed the growth of various ovarian cancer cell lines by inhibiting NF-κB and subsequent secretion of the angiogenic factors IL-8 and VEGF (Rhode et al. 2007). In addition, 6-shogaol was the most potent component of ginger in suppressing the growth and viability of ovarian cancer cells A2780 (Rhode et al. 2007). In support, Liang et al. (2019) displayed that 6-shogaol suppressed cell growth and triggered apoptosis in A2780 ovarian cancer cell lines through in a time-dependent manner. The 6-shogaol hindered the translocation of STAT-3 and consequently suppressed PCNA, cyclin D1, and Bcl-2 expression while increased Bax, caspase-9, and caspase-3 expression in ovarian cells (Liang et al. 2019). The 10-gingerol suppressed time- and dose-dependently the growth of ovarian cancer cell lines HEY, OVCAR3, and SKOV-3 by inducing G2 phase cell cycle arrest. In addition, 10-gingerol showed decreased cyclin A, cyclin B1, and cyclin D3 expression in ovarian cancer cells (Rasmussen et al. 2019).

12.4.2.11 Cervical Cancer

Previous in vitro experiments revealed that 6-gingerol and 10-gingerol caused G0/G1 cell cycle arrest in cervical cancer HeLa cells and ultimately cell death. They downregulated cyclin A, cyclin D1, and cyclin E1 expression while slightly decreased Cdk1, p21, and p27. Moreover, death receptor- and caspase-dependent apoptoses were activated as revealed by increased Bax/Bcl-2 ratio, cytochrome c release, and cleavage of caspase-3, caspase-8, and caspase-9 (Zhang et al. 2017a, b). Furthermore, 10-gingerol inhibited PI3K/AKT and activated AMPK to induce mTOR-mediated cell apoptosis in HeLa cells (Zhang et al. 2017a, b). The study by Liu et al. (2012) revealed that 6-shogaol induced G2/M cell cycle arrest in the HeLa cells by ERS- and mitochondrial-dependent apoptosis.

12.4.2.12 Prostate Cancer

The anticancer activity of whole ginger extract against prostate cancer was demonstrated in both in vitro and in vivo studies. The in vitro experiment revealed that the extract caused inhibition of growth, arrest of cell cycle, and caspase-dependent apoptosis in human prostate cancer cells. In the in vivo study, oral administration of ginger extract inhibited the proliferation and progression of human PC-3 xenografts implanted in mice (Karna et al. 2012). Liu et al. (2017) reported that 6-gingerol, 10-gingerol, 6-shogaol, and 10-shogaol impeded the proliferation of docetaxel-resistant human prostate cancer cells PC3R and downregulated the expression of drug resistance factors, glutathione-*S*-transferase (GST) and multidrug resistance-associated protein 1 (MRP1). Recently, an in silico study by Mohamad (2019) revealed that 6-shogaol is a potential treatment for prostate cancer as evidenced by the high values of binding energy to the overexpressed androgen receptor NR3C4.

12.5 Conclusion

Over the last few decades, there is a growing experimental evidence that turmeric, ginger, and their main bioactive compounds could play a vital role in the treatment of immune imbalances in diseases of diverse nature. Noteworthy, the available recent structure-based molecular docking studies also indicated that the bioactive compound inhibitors from turmeric and ginger are potential drug candidates against COVID-19, and the use of their rhizome extracts is highly recommended. Hence, future in vitro and in vivo experiments are highly needed to elucidate their efficacy against the pandemic infection of COVID-19. As immunomodulatory agents, turmeric, ginger, and their bioactive compounds also demonstrated high potential to control a number of cancer types such as blood, breast, lung, stomach, liver, pancreatic, colorectal, ovarian, cervical, and prostate cancers. There is still, however, a lack of clinical evidence validating the outcomes of preclinical mechanistic studies. Thus, human clinical trials are required to validate their benefits for cancer patients as potent immunomodulatory agents.

References

Abbasi A, Azizi A, Nachvak S et al (2020) Apoptotic effects of ginger extract (Zingiber officinale) on esophageal cancer cells ESO26: an in vitro study. J Reports Pharm Sci 9:183–188. https://doi.org/10.4103/jrptps.JRPTPS_98_19

Abe Y, Hashimoto S, Horie T (1999) Curcumin inhibition of inflammatory cytokine production by human peripheral blood monocytes and alveolar macrophages. Pharmacol Res 39:41–47. https://doi.org/10.1006/phrs.1998.0404

Aggarwal BB, Bhatt ID, Ichikawa H et al (2007) Curcumin — biological and medicinal properties. In: Ravindran P, Nirmal Babu K, Sivaraman K (eds) Turmeric. CRC Press, Boca Raton, FL, pp 317–388

Ahui MLB, Champy P, Ramadan A et al (2008) Ginger prevents Th2-mediated immune responses in a mouse model of airway inflammation. Int Immunopharmacol 8:1626–1632. https://doi.org/10.1016/j.intimp.2008.07.009

Akimoto M, Iizuka M, Kanematsu R et al (2015) Anticancer effect of ginger extract against pancreatic cancer cells mainly through reactive oxygen species-mediated autotic cell death. PLoS One 10:e0126605. https://doi.org/10.1371/journal.pone.0126605

Akkoç Y, Berrak Ö, Arisan ED et al (2015) Inhibition of PI3K signaling triggered apoptotic potential of curcumin which is hindered by Bcl-2 through activation of autophagy in MCF-7 cells. Biomed Pharmacother 71:161–171. https://doi.org/10.1016/j.biopha.2015.02.029

Ali Ab HA, Mohamed SH, Algheshairy RM et al (2020) Immunomodulatory impact of herbs and probiotics in type 2 diabetic rat model. Syst Rev Pharm 11:278–289

Almatroudi A, Alsahli MA, Alrumaihi F et al (2019) Ginger: a novel strategy to battle cancer through modulating cell signalling pathways: a review. Curr Pharm Biotechnol 20:5–16. https://doi.org/10.2174/1389201020666190119142331

Anand P, Sundaram C, Jhurani S et al (2008) Curcumin and cancer: an "old-age" disease with an "age-old" solution. Cancer Lett 267:133–164. https://doi.org/10.1016/j.canlet.2008.03.025

Bawadood AS, Al-Abbasi FA, Anwar F et al (2020) 6-Shogaol suppresses the growth of breast cancer cells by inducing apoptosis and suppressing autophagy via targeting notch signaling pathway. Biomed Pharmacother 128:110302. https://doi.org/10.1016/j.biopha.2020.110302

Bhaskar A, Kumari A, Singh M et al (2020) [6]-Gingerol exhibits potent anti-mycobacterial and immunomodulatory activity against tuberculosis. Int Immunopharmacol 87:106809. https://doi.org/10.1016/j.intimp.2020.106809

Bhaumik S, Jyothi MD, Khar A (2000) Differential modulation of nitric oxide production by curcumin in host macrophages and NK cells. FEBS Lett 483:78–82. https://doi.org/10.1016/S0014-5793(00)02089-5

Bobrovnikova-Marjon EV, Marjon PL, Barbash O et al (2004) Expression of angiogenic factors vascular endothelial growth factor and interleukin-8/CXCL8 is highly responsive to ambient glutamine availability: role of nuclear factor-κB and activating protein-1. Cancer Res 64:4858–4869. https://doi.org/10.1158/0008-5472.CAN-04-0682

Boroumand N, Samarghandian S, Hashemy SI (2018) Immunomodulatory, anti-inflammatory, and antioxidant effects of curcumin. J Herbmed Pharmacol 7:211–219. https://doi.org/10.15171/jhp.2018.33

Brown AC, Shah C, Liu J et al (2009) Ginger's (Zingiber officinale Roscoe) inhibition of rat colonic adenocarcinoma cells proliferation and angiogenesis in vitro. Phyther Res 23:640–645. https://doi.org/10.1002/ptr.2677

Carrasco FR, Schmidt G, Romero AL et al (2009) Immunomodulatory activity of Zingiber officinale Roscoe, Salvia officinalis L. and Syzygium aromaticum L. essential oils: evidence for humor- and cell-mediated responses. J Pharm Pharmacol 61:961–967. https://doi.org/10.1211/jpp.61.07.0017

Chan MM, Fong D, Soprano KJ et al (2003) Inhibition of growth and sensitization to cisplatin-mediated killing of ovarian cancer cells by polyphenolic chemopreventive agents. J Cell Physiol 194:63–70. https://doi.org/10.1002/jcp.10186

Chapman HA (2004) Disorders of lung matrix remodeling. J Clin Invest 113:148–157. https://doi.org/10.1172/JCI20729

Chen J, Wu Y, Li S et al (2018) 10-gingerol inhibits proliferation of hepatocellular carcinoma HepG2 cells via Src/STAT3 signaling pathway. J South Med Univ 38:1002–1007. https://doi.org/10.3969/j.issn.1673-4254.2018.08.17

Chen L, Hu C, Hood M et al (2020) A novel combination of vitamin C, curcumin and glycyrrhizic acid potentially regulates immune and inflammatory response associated with coronavirus infections: a perspective from system biology analysis. Nutrients 12:1193. https://doi.org/10.3390/nu12041193

Churchill M, Chadburn A, Bilinski RT, Bertagnolli MM (2000) Inhibition of intestinal tumors by curcumin is associated with changes in the intestinal immune cell profile. J Surg Res 89:169–175. https://doi.org/10.1006/jsre.2000.5826

Cohly H, Rao M-R, Kanji V et al (2003) Effect of turmeric, turmerin and curcumin on Ca2+, Na/K+ atpases in concanavalin A-stimulated human blood mononuclear cells. Int J Mol Sci 4:34–44. https://doi.org/10.3390/i4020034

Das L, Vinayak M (2014) Curcumin attenuates carcinogenesis by down regulating proinflammatory cytokine interleukin-1 (IL-1α and IL-1β) via modulation of AP-1 and NF-IL6 in lymphoma bearing mice. Int Immunopharmacol 20:141–147. https://doi.org/10.1016/j.intimp.2014.02.024

Das A, Miller R, Lee P et al (2015) A novel component from citrus, ginger, and mushroom family exhibits antitumor activity on human meningioma cells through suppressing the Wnt/β-catenin signaling pathway. Tumor Biol 36:7027–7034. https://doi.org/10.1007/s13277-015-3388-0

Das S, Sarmah S, Lyndem S, Singha Roy A (2021) An investigation into the identification of potential inhibitors of SARS-CoV-2 main protease using molecular docking study. J Biomol Struct Dyn 39(9):3347–3357. https://doi.org/10.1080/07391102.2020.1763201

Deeb D, Xu YX, Jiang H et al (2003) Curcumin (diferuloyl-methane) enhances tumor necrosis factor-related apoptosis-inducing ligand-induced apoptosis in LNCaP prostate cancer cells. Mol Cancer Ther 2:95–103

Deters M, Knochenwefel H, Lindhorst D et al (2008) Different curcuminoids inhibit T-lymphocyte proliferation independently of their radical scavenging activities. Pharm Res 25:1822–1827. https://doi.org/10.1007/s11095-008-9579-2

Dhandapani KM, Mahesh VB, Brann DW (2007) Curcumin suppresses growth and chemoresistance of human glioblastoma cells via AP-1 and NFκB transcription factors. J Neurochem 102:522–538. https://doi.org/10.1111/j.1471-4159.2007.04633.x

Dügenci SK, Arda N, Candan A (2003) Some medicinal plants as immunostimulant for fish. J Ethnopharmacol 88:99–106. https://doi.org/10.1016/S0378-8741(03)00182-X

Eun MK, Hye JK, Kim S et al (2009) Modulation of macrophage functions by compounds isolated from Zingiber officinale. Planta Med 75:148–151. https://doi.org/10.1055/s-0028-1088347

Farombi EO, Ajayi BO, Adedara IA (2020) 6-Gingerol delays tumorigenesis in benzo[a]pyrene and dextran sulphate sodium-induced colorectal cancer in mice. Food Chem Toxicol 142:111483. https://doi.org/10.1016/j.fct.2020.111483

Ferreira VH, Nazli A, Dizzell SE et al (2015) The anti-inflammatory activity of curcumin protects the genital mucosal epithelial barrier from disruption and blocks replication of HIV-1 and HSV-2. PLoS One 10:e0124903. https://doi.org/10.1371/journal.pone.0124903

Geng S, Zheng Y, Meng M et al (2016) Gingerol reverses the cancer-promoting effect of capsaicin by increased TRPV1 level in a urethane-induced lung carcinogenic model. J Agric Food Chem 64:6203–6211. https://doi.org/10.1021/acs.jafc.6b02480

Goel A, Kunnumakkara AB, Aggarwal BB (2008) Curcumin as "Curecumin": from kitchen to clinic. Biochem Pharmacol 75:787–809. https://doi.org/10.1016/j.bcp.2007.08.016

Gonzales AM, Orlando RA (2008) Curcumin and resveratrol inhibit nuclear factor-kappaB-mediated cytokine expression in adipocytes. Nutr Metab (Lond) 5:17. https://doi.org/10.1186/1743-7075-5-17

Goswami D, Kumar M, Ghosh S, Das A (2020) Natural product compounds in alpinia officinarum and ginger are potent SARS-CoV-2 papain-like protease inhibitors. ChemRxiv. https://doi.org/10.26434/chemrxiv.12071997

Grzanna R, Lindmark L, Frondoza CG (2005) Ginger - an herbal medicinal product with broad anti-inflammatory actions. J Med Food 8:125–132

Guan F, Ding Y, Zhang Y et al (2016) Curcumin suppresses proliferation and migration of MDA-MB-231 breast cancer cells through autophagy-dependent Akt degradation. PLoS One 11:e0146553. https://doi.org/10.1371/journal.pone.0146553

Habib SHM, Makpol S, Hamid NAA et al (2008) Ginger extract (Zingiber officinale) has anti-cancer and anti-inflammatory effects on ethionine-induced hepatoma rats. Clinics 63:807–813. https://doi.org/10.1590/S1807-59322008000600017

Haghighi M, Rohani MS (2013) The effects of powdered ginger (Zingiber officinale) on the haematological and immunological parameters of rainbow trout Oncorhynchus mykiss. J Med Plant Herb Ther Res 1:8–12

Hakim L, Alias E, Makpol S et al (2014) Gelam honey and ginger potentiate the anti cancer effect of 5-FU against HCT 116 colorectal cancer cells. Asian Pacific J Cancer Prev 15:4651–4657. https://doi.org/10.7314/APJCP.2014.15.11.4651

Hamaya Y, Guarinos C, Tseng-Rogenski SS et al (2015) Efficacy of adjuvant 5-fluorouracil therapy for patients with emast-positive stage ii/iii colorectal cancer. PLoS One 10:e0127591. https://doi.org/10.1371/journal.pone.0127591

Hammamieh R, Sumaida D, Zhang XY et al (2007) Control of the growth of human breast cancer cells in culture by manipulation of arachidonate metabolism. BMC Cancer 7:138. https://doi.org/10.1186/1471-2407-7-138

Han SS, Chung ST, Robertson DA et al (1999) Curcumin causes the growth arrest and apoptosis of B cell lymphoma by downregulation of egr-1, C-myc, Bcl-X(l), NF-κB, and p53. Clin Immunol 93:152–161. https://doi.org/10.1006/clim.1999.4769

Han J, Li X, Ye Z et al (2019) Treatment with 6-gingerol regulates dendritic cell activity and ameliorates the severity of experimental autoimmune encephalomyelitis. Mol Nutr Food Res 63:1801356. https://doi.org/10.1002/mnfr.201801356

Hogan PG (2003) Transcriptional regulation by calcium, calcineurin, and NFAT. Genes Dev 17:2205–2232. https://doi.org/10.1101/gad.1102703

Homa S, Wentz-Hunter K (2018) Differential gene expression in pancreatic cancer when treated with ginger extract. Free Radic Biol Med 128:S67. https://doi.org/10.1016/j.freeradbiomed.2018.10.139

Hu SM, Yao XH, Hao YH et al (2020) 8-Gingerol regulates colorectal cancer cell proliferation and migration through the EGFR/STAT/ERK pathway. Int J Oncol 56:390–397. https://doi.org/10.3892/ijo.2019.4934

Huang C, Wang Y, Li X et al (2020) Clinical features of patients infected with 2019 novel coronavirus in Wuhan, China. Lancet 395:497–506. https://doi.org/10.1016/S0140-6736(20)30183-5

Hung JYU, Hsu YAL, Te LC et al (2009) 6-shogaol, an active constituent of dietary ginger, induces autophagy by inhibiting the AKT/mTOR pathway in human non-small cell lung cancer A549 cells. J Agric Food Chem 57:9809–9816. https://doi.org/10.1021/jf902315e

Ihlaseh SM, de Oliveira MLC, Teràn E et al (2006) Chemopreventive property of dietary ginger in rat urinary bladder chemical carcinogenesis. World J Urol 24:591–596. https://doi.org/10.1007/s00345-006-0108-9

Imanishi N, Andoh T, Mantani N et al (2006) Macrophage-mediated inhibitory effect of Zingiber officinale Rosc, a traditional oriental herbal medicine, on the growth of influenza A/Aichi/2/68 virus. Am J Chin Med 34:157–169. https://doi.org/10.1142/S0192415X06003722

Immanuel G, Uma RP, Iyapparaj P et al (2009) Dietary medicinal plant extracts improve growth, immune activity and survival of tilapia oreochromis mossambicus. J Fish Biol 74:1462–1475. https://doi.org/10.1111/j.1095-8649.2009.02212.x

Iqbal M, Okazaki Y, Okada S (2003) In vitro curcumin modulates ferric nitrilotriacetate (Fe-NTA) and hydrogen peroxide (H2O2)-induced peroxidation of microsomal membrane lipids and DNA damage. Teratog Carcinog Mutagen 23:151–160. https://doi.org/10.1002/tcm.10070

Ishiguro K, Ando T, Maeda O et al (2007) Ginger ingredients reduce viability of gastric cancer cells via distinct mechanisms. Biochem Biophys Res Commun 362:218–223. https://doi.org/10.1016/j.bbrc.2007.08.012

Jean S-S, Lee P-I, Hsueh P-R (2020) Treatment options for COVID-19: the reality and challenges. J Microbiol Immunol Infect 53:436–443. https://doi.org/10.1016/j.jmii.2020.03.034

Jia HP, Look DC, Shi L et al (2005) ACE2 receptor expression and severe acute respiratory syndrome coronavirus infection depend on differentiation of human airway epithelia. J Virol 79:14614–14621. https://doi.org/10.1128/jvi.79.23.14614-14621.2005

Jiang MC, Yang-Yen HF, Yen JJY, Lin JK (1996) Curcumin induces apoptosis in immortalized NIH 3T3 and malignant cancer cell lines. Nutr Cancer 26:111–120. https://doi.org/10.1080/01635589609514468

Jin H, Qiao F, Wang Y et al (2015) Curcumin inhibits cell proliferation and induces apoptosis of human non-small cell lung cancer cells through the upregulation of miR-192-5p and suppression of PI3K/Akt signaling pathway. Oncol Rep 34:2782–2789. https://doi.org/10.3892/or.2015.4258

Joo JH, Hong SS, Cho YR, Seo DW (2016) 10-Gingerol inhibits proliferation and invasion of MDA-MB-231 breast cancer cells through suppression of Akt and p38MAPK activity. Oncol Rep 35:779–784. https://doi.org/10.3892/or.2015.4405

Jung EM, Lee TJ, Park JW et al (2005) Curcumin sensitizes tumor necrosis factor-related apoptosis-inducing ligand (TRAIL)-induced apoptosis through reactive oxygen species-mediated upregulation of death receptor 5 (DR5). Carcinogenesis 26:1905–1913. https://doi.org/10.1093/carcin/bgi167

Kaewtunjai N, Wongpoomchai R, Imsumran A et al (2018) Ginger extract promotes telomere shortening and cellular senescence in A549 lung cancer cells. ACS Omega 3:18572–18581. https://doi.org/10.1021/acsomega.8b02853

Kandeel M, Al-Nazawi M (2020) Virtual screening and repurposing of FDA approved drugs against COVID-19 main protease. Life Sci 251:117627. https://doi.org/10.1016/j.lfs.2020.117627

Kapinova A, Kubatka P, Liskova A et al (2019) Controlling metastatic cancer: the role of phytochemicals in cell signaling. J Cancer Res Clin Oncol 145:1087–1109. https://doi.org/10.1007/s00432-019-02892-5

Karna P, Chagani S, Gundala SR et al (2012) Benefits of whole ginger extract in prostate cancer. Br J Nutr 107:473–484. https://doi.org/10.1017/S0007114511003308

Katiyar SK, Agarwal R, Mukhtar H (1996) Inhibition of tumor promotion in SENCAR mouse skin by ethanol extract of Zingiber officinale rhizome. Cancer Res 56:1023–1030

Kawamori T, Lubet R, Steele VE et al (1999) Chemopreventive effect of curcumin, a naturally occurring anti- inflammatory agent, during the promotion/progression stages of colon cancer. Cancer Res 59:597–601

Khaerunnisa S, Kurniawan H, Awaluddin R, Suhartati S (2020) Potential inhibitor of COVID-19 main protease (M pro) from several medicinal plant compounds by molecular docking study. Preprints 2020:2020030226. https://doi.org/10.20944/preprints202003.0226.v1

Khanbolooki S, Nawrocki ST, Arumugam T et al (2006) Nuclear factor-κB maintains TRAIL resistance in human pancreatic cancer cells. Mol Cancer Ther 5:2251–2260. https://doi.org/10.1158/1535-7163.MCT-06-0075

Kim SO, Kundu JK, Shin YK et al (2005) [6]-Gingerol inhibits COX-2 expression by blocking the activation of p38 MAP kinase and NF-κB in phorbol ester-stimulated mouse skin. Oncogene 24:2558–2567. https://doi.org/10.1038/sj.onc.1208446

Kim M, Miyamoto S, Yasui Y et al (2009) Zerumbone, a tropical ginger sesquiterpene, inhibits colon and lung carcinogenesis in mice. Int J Cancer 124:264–271. https://doi.org/10.1002/ijc.23923

Kindler E, Thiel V, Weber F (2016) Interaction of SARS and MERS coronaviruses with the antiviral interferon response. In: Ziebuhr J (ed) Advances in virus research. Academic Press, pp 219–243

Kita T, Imai S, Sawada H et al (2008) The biosynthetic pathway of curcuminoid in turmeric (Curcuma longa) as revealed by 13C-labeled precursors. Biosci Biotechnol Biochem 72:1789–1798. https://doi.org/10.1271/bbb.80075

Kössler S, Nofziger C, Jakab M et al (2012) Curcumin affects cell survival and cell volume regulation in human renal and intestinal cells. Toxicology 292:123–135. https://doi.org/10.1016/j.tox.2011.12.002

Kumar Mitra A, Krishna M (2004) In vivo modulation of signaling factors involved in cell survival. J Radiat Res 45:491–495. https://doi.org/10.1269/jrr.45.491

Lee JJ, Huang WT, Shao DZ et al (2003) Blocking NF-κB activation may be an effective strategy in the fever therapy. Jpn J Physiol 53:367–375. https://doi.org/10.2170/jjphysiol.53.367

Lee HS, Seo EY, Kang NE, Kim WK (2008a) [6]-Gingerol inhibits metastasis of MDA-MB-231 human breast cancer cells. J Nutr Biochem 19:313–319. https://doi.org/10.1016/j.jnutbio.2007.05.008

Lee SH, Cekanova M, Seung JB (2008b) Multiple mechanisms are involved in 6-gingerol-induced cell growth arrest and apoptosis in human colorectal cancer cells. Mol Carcinog 47:197–208. https://doi.org/10.1002/mc.20374

Li G, De Clercq E (2020) Therapeutic options for the 2019 novel coronavirus (2019-nCoV). Nat Rev Drug Discov 19:149–150. https://doi.org/10.1038/d41573-020-00016-0

Liang T, He Y, Chang Y, Liu X (2019) 6-shogaol a active component from ginger inhibits cell proliferation and induces apoptosis through inhibition of STAT-3 translocation in ovarian cancer cell lines (A2780). Biotechnol Bioprocess Eng 24:560–567. https://doi.org/10.1007/s12257-018-0502-3

Lin YG, Kunnumakkara AB, Nair A et al (2007) Curcumin inhibits tumor growth and angiogenesis in ovarian carcinoma by targeting the nuclear factor-κB pathway. Clin Cancer Res 13:3423–3430. https://doi.org/10.1158/1078-0432.CCR-06-3072

Ling H, Yang H, Tan SH et al (2010) 6-Shogaol, an active constituent of ginger, inhibits breast cancer cell invasion by reducing matrix metalloproteinase-9 expression via blockade of nuclear factor-κB activation. Br J Pharmacol 161:1763–1777. https://doi.org/10.1111/j.1476-5381.2010.00991.x

Liu Q, Peng Y-B, Qi L-W et al (2012) The cytotoxicity mechanism of 6-shogaol-treated hela human cervical cancer cells revealed by label-free shotgun proteomics and bioinformatics analysis. Evid Based Complement Altern Med 2012:1–12. https://doi.org/10.1155/2012/278652

Liu Q, Peng YB, Zhou P et al (2013) 6-Shogaol induces apoptosis in human leukemia cells through a process involving caspase-mediated cleavage of eIF2α. Mol Cancer 12:135. https://doi.org/10.1186/1476-4598-12-135

Liu C-M, Kao C-L, Tseng Y-T et al (2017) Ginger phytochemicals inhibit cell growth and modulate drug resistance factors in docetaxel resistant prostate cancer cell. Molecules 22:1477. https://doi.org/10.3390/molecules22091477

López-Lázaro M (2008) Anticancer and carcinogenic properties of curcumin: Considerations for its clinical development as a cancer chemopreventive and chemotherapeutic agent. Mol Nutr Food Res 52:S103–S127. https://doi.org/10.1002/mnfr.200700238

Lucile White E, Ross LJ, Schmid SM et al (1998) Screening of potential cancer-preventing chemicals for inhibition of induction of ornithine decarboxylase in epithelial cells from rat trachea. Oncol Rep 5:717–722. https://doi.org/10.3892/or.5.3.717

Luo Y, Chen X, Luo L et al (2018) [6]-Gingerol enhances the radiosensitivity of gastric cancer via G2/M phase arrest and apoptosis induction. Oncol Rep 39:2252–2260. https://doi.org/10.3892/or.2018.6292

Magalska A, Brzezinska A, Bielak-Zmijewska A et al (2006) Curcumin induces cell death without oligonucleosomal DNA fragmentation in quiescent and proliferating human CD8+ cells. Acta Biochim Pol 53:531–538. https://doi.org/10.18388/abp.2006_3324

Malmir S, Ebrahimi A, Mahjoubi F (2020) Effect of ginger extracts on colorectal cancer HCT-116 cell line in the expression of MMP-2 and KRAS. Gene Rep 21:100824. https://doi.org/10.1016/j.genrep.2020.100824

Manju V, Nalini N (2005) Chemopreventive efficacy of ginger, a naturally occurring anticarcinogen during the initiation, post-initiation stages of 1,2 dimethylhydrazine-induced colon cancer. Clin Chim Acta 358:60–67. https://doi.org/10.1016/j.cccn.2005.02.018

Mansingh DP, OJ S, Sali VK, Vasanthi HR (2018) [6]-Gingerol-induced cell cycle arrest, reactive oxygen species generation, and disruption of mitochondrial membrane potential are associated with apoptosis in human gastric cancer (AGS) cells. J Biochem Mol Toxicol 32:e22206. https://doi.org/10.1002/jbt.22206

Mansingh DP, Pradhan S, Biswas D et al (2020) Palliative role of aqueous ginger extract on N-nitroso-N-methylurea-induced gastric cancer. Nutr Cancer 72:157–169. https://doi.org/10.1080/01635581.2019.1619784

Mansouri MT, Hemmati AA, Naghizadeh B et al (2015) A study of the mechanisms underlying the anti-inflammatory effect of ellagic acid in carrageenan-induced paw edema in rats. Indian J Pharmacol 47:292–298. https://doi.org/10.4103/0253-7613.157127

Martin ACBM, Fuzer AM, Becceneri AB et al (2017) [10]-gingerol induces apoptosis and inhibits metastatic dissemination of triple negative breast cancer in vivo. Oncotarget 8:72260–72271. https://doi.org/10.18632/oncotarget.20139

Mcfadden RMT, Larmonier CB, Shehab KW et al (2015) The role of curcumin in modulating colonic microbiota during colitis and colon cancer prevention. Inflamm Bowel Dis 21:2483–2494. https://doi.org/10.1097/MIB.0000000000000522

Mehrotra S, Agnihotri G, Singh S, Jamal F (2013) Immunomodulatory potential of Curcuma longa: a review. South Asian J Exp Biol 3:299–307

Miriyala S, Panchatcharam M, Rengarajulu P (2007) Cardioprotective effect of curcumin. In: Aggarwal BB, Surh Y, Shishodia S (eds) The molecular targets and therapeutic uses of curcumin in health and disease. Springer, Boston, MA, pp 359–377

Mohamad EA (2019) In silico study of ginger extract and capsaicin effects on prostate cancer. Roman J Biophys 29:81–87

Momtazi-Borojeni AA, Haftcheshmeh SM, Esmaeili S-A et al (2018) Curcumin: a natural modulator of immune cells in systemic lupus erythematosus. Autoimmun Rev 17:125–135. https://doi.org/10.1016/j.autrev.2017.11.016

Mullaicharam A, Maheswaran A (2012) Pharmacological effects of curcumin. Int J Nutr Pharmacol Neurol Dis 2:92–99. https://doi.org/10.4103/2231-0738.95930

Nair A, Chattopadhyay D, Saha B (2018) Plant-derived immunomodulators. In: Khan M, Ahmad I, Chattopadhyay D (eds) New look to phytomedicine: advancements in herbal products as novel drug leads. Academic Press, London, pp 435–499

Nedungadi D, Binoy A, Vinod V et al (2019) Ginger extract activates caspase independent paraptosis in cancer cells via ER stress, mitochondrial dysfunction, AIF translocation and DNA damage. Nutr Cancer (73):1, 147–159. https://doi.org/10.1080/01635581.2019.1685113

Nie H, Meng L, Zhang H (2006) Effect of gingerol on endotoxemia mouse model induced by heatstroke. Chinese J Integr Tradit West Med 26:529–532

Nigam N, Bhui K, Prasad S et al (2009) [6]-Gingerol induces reactive oxygen species regulated mitochondrial cell death pathway in human epidermoid carcinoma A431 cells. Chem Biol Interact 181:77–84. https://doi.org/10.1016/j.cbi.2009.05.012

Norouzi S, Majeed M, Pirro M et al (2018) Curcumin as an adjunct therapy and microRNA modulator in breast cancer. Curr Pharm Des 24:171–177. https://doi.org/10.2174/1381612824666171129203506

Novak A, Dedhar S (1999) Signaling through β-catenin and Lef/Tcf. Cell Mol Life Sci 56:523–537

Ohno M, Nishida A, Sugitani Y et al (2017) Nanoparticle curcumin ameliorates experimental colitis via modulation of gut microbiota and induction of regulatory T cells. PLoS One 12:e0185999. https://doi.org/10.1371/journal.pone.0185999

Pashaei-Asl R, Pashaei-Asl F, Gharabaghi PM et al (2017) The inhibitory effect of ginger extract on Ovarian cancer cell line; Application of systems biology. Adv Pharm Bull 7:241–249. https://doi.org/10.15171/apb.2017.029

Peng F, Tao Q, Wu X et al (2012) Cytotoxic, cytoprotective and antioxidant effects of isolated phenolic compounds from fresh ginger. Fitoterapia 83:568–585. https://doi.org/10.1016/j.fitote.2011.12.028

Pongrakhananon V, Rojanasakul Y (2011) Anticancer properties of curcumin. In: Gali-Muhtasib H (ed) Advances in cancer therapy. IntechOpen. https://www.intechopen.com/chapters/23926

Rafiee P, Ogawa H, Heidemann J et al (2003) Isolation and characterization of human esophageal microvascular endothelial cells: mechanisms of inflammatory activation. Am J Physiol Liver Physiol 285:G1277–G1292. https://doi.org/10.1152/ajpgi.00484.2002

Rajagopal K, Byran G, Jupudi S, Vadivelan R (2020) Activity of phytochemical constituents of black pepper, ginger, and garlic against coronavirus (COVID-19): an in silico approach. Int J Heal Allied Sci 9:43–50

Ramsewak RS, DeWitt DL, Nair MG (2000) Cytotoxicity, antioxidant and anti-inflammatory activities of curcumins I-III from Curcuma longa. Phytomedicine 7:303–308. https://doi.org/10.1016/S0944-7113(00)80048-3

Rasmussen A, Murphy K, Hoskin DW (2019) 10-gingerol inhibits ovarian cancer cell growth by inducing G2 arrest. Adv Pharm Bull 9:685–689. https://doi.org/10.15171/apb.2019.080

Rhode J, Fogoros S, Zick S et al (2007) Ginger inhibits cell growth and modulates angiogenic factors in ovarian cancer cells. BMC Complement Altern Med 7:44. https://doi.org/10.1186/1472-6882-7-44

Rossi A, Kapahi P, Natoli G et al (2000) Anti-inflammatory cyclopentenone prostaglandins are direct inhibitors of IκB kinase. Nature 403:103–108. https://doi.org/10.1038/47520

Ruby AJ, Kuttan G, Dinesh Babu K et al (1995) Anti-tumour and antioxidant activity of natural curcuminoids. Cancer Lett 94:79–83. https://doi.org/10.1016/0304-3835(95)03827-J

Ryu H, Kim H (2004) Effect of zingiber officinale roscoe extracts on mice immune cell activation. Korean J Nutr 37:23–30

Samec M, Liskova A, Koklesova L et al (2020) The role of plant-derived natural substances as immunomodulatory agents in carcinogenesis. J Cancer Res Clin Oncol:1–18. https://doi.org/10.1007/s00432-020-03424-2

Sang S, Hong J, Wu H et al (2009) Increased growth inhibitory effects on human cancer cells and anti-inflammatory potency of shogaols from Zingiber officinale relative to gingerols. J Agric Food Chem 57:10645–10650. https://doi.org/10.1021/jf9027443

Schindler R, Mentlein R (2006) Flavonoids and vitamin E reduce the release of the angiogenic peptide vascular endothelial growth factor from human tumor cells. J Nutr 136:1477–1482. https://doi.org/10.1093/jn/136.6.1477

Senathilake K, Samarakoon S, Tennekoon K (2020) Virtual screening of inhibitors against spike glycoprotein of 2019 novel corona virus: a drug repurposing approach. Preprints 2020030042. DOI: https://doi.org/10.20944/PREPRINTS202003.0042.V1

Shakibaei M, Mobasheri A, Lueders C et al (2013) Curcumin enhances the effect of chemotherapy against colorectal cancer cells by inhibition of NF-κB and Src protein kinase signaling pathways. PLoS One 8. https://doi.org/10.1371/journal.pone.0057218

Shakibaei M, Buhrmann C, Kraehe P et al (2014) Curcumin Chemosensitizes 5-fluorouracil resistant MMR-deficient human colon cancer cells in high density cultures. PLoS One 9: e85397. https://doi.org/10.1371/journal.pone.0085397

Shankar S, Chen Q, Sarva K et al (2007) Curcumin enhances the apoptosis-inducing potential of TRAIL in prostate cancer cells: molecular mechanisms of apoptosis, migration and angiogenesis. J Mol Signal 2:10. https://doi.org/10.1186/1750-2187-2-10

Shao ZM, Shen ZZ, Liu CH et al (2002) Curcumin exerts multiple suppressive effects on human breast carcinoma cells. Int J Cancer 98:234–240. https://doi.org/10.1002/ijc.10183

Shenouda NS, Zhou C, Browning JD et al (2004) Phytoestrogens in common herbs regulate prostate cancer cell growth in vitro. Nutr Cancer 49:200–208. https://doi.org/10.1207/s15327914nc4902_12

Shukla Y, Singh M (2007) Cancer preventive properties of ginger: a brief review. Food Chem Toxicol 45:683–690. https://doi.org/10.1016/j.fct.2006.11.002

Sikora E, Bielak-Zmijewska A, Piwocka K et al (1997) Inhibition of proliferation and apoptosis of human and rat T lymphocytes by curcumin, a curry pigment. Biochem Pharmacol 54:899–907. https://doi.org/10.1016/S0006-2952(97)00251-7

Singh S, Khar A (2008) Biological effects of curcumin and its role in cancer chemoprevention and therapy. Anti Cancer Agents Med Chem 6:259–270. https://doi.org/10.2174/187152006776930918

Su P, Veeraraghavan VP, Krishna Mohan S, Lu W (2019) A ginger derivative, zingerone—a phenolic compound—induces ROS-mediated apoptosis in colon cancer cells (HCT-116). J Biochem Mol Toxicol 33:e22403. https://doi.org/10.1002/jbt.22403

Subhrajyoti C, Shalini (2020) Immunomodulatory herbs of Ayurveda and COVID-19: a review article. J Ayurveda Integr Med Sci 5:203–208

Sung B, Jhurani S, Kwang SA et al (2008) Zerumbone down-regulates chemokine receptor CXCR4 expression leading to inhibition of CXCL12-induced invasion of breast and pancreatic tumor cells. Cancer Res 68:8938–8944. https://doi.org/10.1158/0008-5472.CAN-08-2155

Suravajhala R, Parashar A, Malik B et al (2020) Comparative docking studies on curcumin with COVID-19 proteins. Preprints 2020050439. https://doi.org/10.20944/preprints202005.0439.v2

Syafitri DM, Levita J, Mutakin M, Diantini A (2018) A review: is ginger (Zingiber officinale var. Roscoe) potential for future phytomedicine? Indones J Appl Sci 8(1). https://doi.org/10.24198/ijas.v8i1.16466

Tan BS, Kang O, Mai CW et al (2013) 6-Shogaol inhibits breast and colon cancer cell proliferation through activation of peroxisomal proliferator activated receptor γ (PPARγ). Cancer Lett 336: 127–139. https://doi.org/10.1016/j.canlet.2013.04.014

Tap J, Furet JP, Bensaada M et al (2015) Gut microbiota richness promotes its stability upon increased dietary fibre intake in healthy adults. Environ Microbiol 17:4954–4964. https://doi.org/10.1111/1462-2920.13006

Thatte UM, Dahanukar SA (1986) Ayurveda and contemporary scientific thought. Trends Pharmacol Sci 7:247–251. https://doi.org/10.1016/0165-6147(86)90336-6
Ting D, Dong N, Fang L et al (2018) Multisite inhibitors for enteric coronavirus: antiviral cationic carbon dots based on curcumin. ACS Appl Nano Mater 1:5451–5459. https://doi.org/10.1021/acsanm.8b00779
Tonin LTD, Camargo JNA, Bertan AS et al (2019) Antitumoral activity, antioxidant capacity and bioactive compounds of ginger (Zingiber officinale). Acta Sci Technol 42:e45724. https://doi.org/10.4025/actascitechnol.v42i1.45724
Tripathi S, Maier KG, Bruch D, Kittur DS (2007) Effect of 6-gingerol on pro-inflammatory cytokine production and costimulatory molecule expression in murine peritoneal macrophages. J Surg Res 138:209–213. https://doi.org/10.1016/j.jss.2006.07.051
Tripathi S, Bruch D, Kittur DS (2008) Ginger extract inhibits LPS induced macrophage activation and function. BMC Complement Altern Med 8:1. https://doi.org/10.1186/1472-6882-8-1
Wax A, Pyhtila JW, Graf RN et al (2005) Prospective grading of neoplastic change in rat esophagus epithelium using angle-resolved low-coherence interferometry. J Biomed Opt 10:051604. https://doi.org/10.1117/1.2102767
WHO (2020) WHO Coronavirus Disease (COVID-19) Dashboard | WHO Coronavirus Disease (COVID-19) Dashboard. https://covid19.who.int
Wilasrusmee C, Kittur S, Siddiqui J et al (2002a) In vitro immunomodulatory effects of ten commonly used herbs on murine lymphocytes. J Altern Complement Med 8:467–475. https://doi.org/10.1089/107555302760253667
Wilasrusmee C, Siddiqui J, Bruch D et al (2002b) In vitro immunomodulatory effects of herbal products. Am Surg 68:860–864
Wu L, Guo L, Liang Y et al (2015) Curcumin suppresses stem-like traits of lung cancer cells via inhibiting the JAK2/STAT3 signaling pathway. Oncol Rep 34:3311–3317. https://doi.org/10.3892/or.2015.4279
Xu M, Deng B, Chow YL et al (2007) Effects of curcumin in treatment of experimental pulmonary fibrosis: a comparison with hydrocortisone. J Ethnopharmacol 112:292–299. https://doi.org/10.1016/j.jep.2007.03.011
Xu H, Zhong L, Deng J et al (2020) High expression of ACE2 receptor of 2019-nCoV on the epithelial cells of oral mucosa. Int J Oral Sci 12:1–5. https://doi.org/10.1038/s41368-020-0074-x
Xue X, Yu JL, Sun DQ et al (2014) Curcumin induces apoptosis in SGC-7901 gastric adenocarcinoma cells via regulation of mitochondrial signaling pathways. Asian Pacific J Cancer Prev 15:3987–3992. https://doi.org/10.7314/APJCP.2014.15.9.3987
Yadav VS, Mishra KP, Singh DP et al (2005) Immunomodulatory effects of curcumin. Immunopharmacol Immunotoxicol 27:485–497. https://doi.org/10.1080/08923970500242244
Yagihashi S, Miura Y, Yagasaki K (2008) Inhibitory effect of gingerol on the proliferation and invasion of hepatoma cells in culture. Cytotechnology 57:129–136. https://doi.org/10.1007/s10616-008-9121-8
Yan R, Zhang Y, Li Y et al (2020) Structural basis for the recognition of SARS-CoV-2 by full-length human ACE2. Science 367:1444–1448. https://doi.org/10.1126/science.abb2762
Yang CW, Chang CL, Lee HC et al (2012) Curcumin induces the apoptosis of human monocytic leukemia THP-1 cells via the activation of JNK/ERK Pathways. BMC Complement Altern Med 12:22–22. https://doi.org/10.1186/1472-6882-12-22
Youn HS, Saitoh SI, Miyake K, Hwang DH (2006) Inhibition of homodimerization of Toll-like receptor 4 by curcumin. Biochem Pharmacol 72:62–69. https://doi.org/10.1016/j.bcp.2006.03.022
Zahedipour F, Hosseini SA, Sathyapalan T et al (2020) Potential effects of curcumin in the treatment of COVID-19 infection. Phyther Res 34(11):2911–2920. https://doi.org/10.1002/ptr.6738

Zhang S, Liu Q, Liu Y et al (2012) Zerumbone, a Southeast Asian ginger sesquiterpene, induced apoptosis of pancreatic carcinoma cells through p53 signaling pathway. Evid Based Complement Altern Med 2012. https://doi.org/10.1155/2012/936030

Zhang F, Thakur K, Hu F et al (2017a) 10-gingerol, a phytochemical derivative from "tongling white ginger", inhibits cervical cancer: insights into the molecular mechanism and inhibitory targets. J Agric Food Chem 65:2089–2099. https://doi.org/10.1021/acs.jafc.7b00095

Zhang F, Zhang JG, Qu J et al (2017b) Assessment of anti-cancerous potential of 6-gingerol (Tongling White Ginger) and its synergy with drugs on human cervical adenocarcinoma cells. Food Chem Toxicol 109:910–922. https://doi.org/10.1016/j.fct.2017.02.038

Zhao Z, Li C, Xi H et al (2015) Curcumin induces apoptosis in pancreatic cancer cells through the induction of forkhead box O1 and inhibition of the PI3K/Akt pathway. Mol Med Rep 12:5415–5422. https://doi.org/10.3892/mmr.2015.4060

Zheng M, Zhang Q, Joe Y et al (2013) Curcumin induces apoptotic cell death of activated human CD4 + T cells via increasing endoplasmic reticulum stress and mitochondrial dysfunction. Int Immunopharmacol 15:517–523. https://doi.org/10.1016/j.intimp.2013.02.002

Zhou HL, Deng YM, Xie QM (2006) The modulatory effects of the volatile oil of ginger on the cellular immune response in vitro and in vivo in mice. J Ethnopharmacol 105:301–305. https://doi.org/10.1016/j.jep.2005.10.022

Fruits as Boosters of the Immune System

13

Siddhartha Kumar Mishra, Pir Mohammad Ishfaq, Swati Tripathi, and Neelima Gupta

Abstract

Immunity is the major mechanism of host defence system against infectious and chronic diseases. The recent global concern of recent viral infection of SARS-CoV-19 has raised the demand of functional foods, nutraceuticals and fruits that can boost immunity. This will help in managing the overall physiological health and prevention of infectious and chronic diseases. Medicinal plants and fruits can help in boosting immunity through modulation of immune system and changing the types of immune response such as involvement of the induction, expression or amplification of the genes and proteins in inflammation and antioxidant system. The traditional medicine systems have used a wide variety of plants and fruits as supplement for immunomodulation including those for stimulation of immune system as well as immune compromisation deemed per se. Some of the plants that have been listed for immune boosting abilities included *Curcuma longa*, *Withania somnifera*, *Phyllanthus emblica*, *Azadirachta indica*, *Panax ginseng*, *Rhododendron spiciferum*, *Caesalpinia bonducella*, *Tinospora cordifolia*, *Capparis zeylanica*, *Asparagus racemosus*, *Nelumbo nucifera*, *Arnica montana*,

S. K. Mishra (✉)
Department of Life Sciences, Chhatrapati Shahu Ji Maharaj University, Kanpur, U.P, India

Cancer Biology Laboratory, Department of Zoology, School of Biological Sciences, Dr. Harisingh Gour Central University, Sagar, M.P, India

P. M. Ishfaq
Cancer Biology Laboratory, Department of Zoology, School of Biological Sciences, Dr. Harisingh Gour Central University, Sagar, M.P, India

S. Tripathi
Amity Institute of Microbial Technology, Amity University, Noida, U.P, India

N. Gupta
Department of Life Sciences, Chhatrapati Shahu Ji Maharaj University, Kanpur, U.P, India

N. S. Sangwan et al. (eds.), *Plants and Phytomolecules for Immunomodulation*,
https://doi.org/10.1007/978-981-16-8117-2_13

Calendula officinalis, *Echinacea purpurea* and *Euphorbia tirucalli*. Reports indicate that a wide variety of phytochemicals like polysaccharide, alkaloids, flavonoids, terpenoids, lactones and glycoside have shown immunomodulatory properties under different pathophysiological conditions. Amongst the diverse chemical profile of plant extracts, polysaccharides are the water-soluble molecules that could activate immune responses when interacting directly with immune cells, while hydrophobic compounds like flavonoids such as quercetin and luteolin and terpenoids such as sesquiterpene lactones and curcumin showed potent immunomodulatory effects. Growing evidences suggest that phytochemicals from functional foods and fruits may be useful in maintaining the cytokine and chemokine balance, regulating oxidative status of cells, and targeting the specific cellular receptors as therapeutic targets. This chapter comprehensively enlists the plant resources with immune boosting abilities and explore their phytochemical characterization and molecular mechanism behind their protective effects.

Keywords

Functional food · Fruits · Nutraceuticals · Antioxidant · Anti-inflammation · Immune-stimulating

13.1 Introduction

People with a lower immunity especially the children and overage people get affected due to different diseases. The immune system is built in an association with live beneficial bacteria in the gut that helps in protecting the human body from several types of diseases, whereas a weaker or damaged immune system becomes a cause of very sensitive infections or other diseases of cardiovascular systems, diabetes, cancer, as well as COVID-19 (Iddir et al. 2020). Plant-based foods have shown potentials to increase and help in the growth of beneficial intestinal bacteria and the growing conditions of the overall gut microbiome that makes up to 85% of the human body immune system. Conversely an excess of animal foods in diet can deplete the level of good bacteria from the body and can promote inflammation and diseases of heart and diabetes, hepatitis B, chronic kidney disease, and cancer (Iddir et al. 2020; Pickard et al. 2017). Drinking water help cells to retain the oxidative state, and sustained cellular oxygen can help to protect the body from infectious agents. The Centers for Disease Control and Prevention reports suggest that hydration often plays a major role in monitoring the body temperature (Schols et al. 2009). However, a fever caused by some infection or disorder in the body can be lowered by drinking plenty of water. Drinking enough water is essential as it keeps the risk of disease to its minimal. Staying hydrated also enables to transmit nutrients to all parts of the body and helps to maintain all body functions and organs working potentially to decrease body infection. Drinking too much of water keeps the body hydrated and therefore can remove more mucus from the body. Taking all the preventive measures and keeping the body healthy and safe is crucial. Although drinking enough water does not guarantee that one would not get affected by diseases or illness, it can be effective in reducing the danger to a maximum extent and can therefore help recover from the illness.

Undernutrition insufficient because of the intake of energy and nutrients impairs the immune system, suppressing immune responses that are required for the host protection (Calder and Kew 2002). The common abnormalities are found in the cell-mediated immunity, phagocyte function, complement system, cytokine production and antibody affinity (Chandra 1997). Different physiological situations such as ageing and performance of vigorous physical exercise are linked with deterioration of some immune parameters functioning. Proper nutrition can therefore influence the extent of immune alteration in such scenarios. Certain pathological situations exist where nutrition furnishes a major role as a determinant of some fundamental immunological impairments (Di Renzo et al. 2019). The immune system plays a vital role in protecting the host from multiple infectious agents that do exist in the surrounding environment and from various other harmful agents. The appropriate effectiveness of the immune system is predominantly determined by the intake of nutrition (De Visser et al. 2006). The imbalance in the nutrient intake is linked with high risk of acquiring infections.

The immune system is evidently underprivileged of the components required to regulate an effective immune response, because of adequate nutrition. Researchers have observed that, in laboratory animals deficient of one dietary element or with a single nutrient deficit, there is a crucial role of a number of minerals, vitamins and trace elements in upkeep of immune competence (Wintergerst et al. 2007). These supplements are comprised of folic acid; riboflavin; vitamins B_6, A, E and C; beta-carotene; iron; and selenium. Antioxidant nutrients also play a vital role in wielding the redox equilibrium in the immune cells and in protecting these immune cells from oxidative stress and maintaining their cellular roles. Reinstating the level of nutrients deficient in the diet plan can restore the normal immune function and provide enough resistance to fighting infection. Moreover, consuming some nutrients in excessive amounts can lead to the malfunction of the immune function (Gombart et al. 2020).

Lipids, being the essential components in the diet, are substances that lead to a fundamental effect in maintaining the functional aspect of the immune system (Lobo et al. 2010). The fatty acid composition in the diet consequently alters the fatty acid composition of lymphocytes and other immune potentiating cells. Thus, there is a need of an immunomodulatory role for these dietary lipids that may be employed in providing resistance to some diseases involving inflammation processes like auto-immune diseases (Aslani and Ghobadi 2016). As the importance of diet on immune health is becoming increasingly well recognized, a considerable interest has grown to develop food products as aid in the immune system recovery from the potential impact of different ailments. A number of studies have reported that the functional foods may prove to be beneficial for the immune system, as these may led to recover in the impelled immune complications (Arreola et al. 2015). Medicinal plants are being used as immunomodulatory effect to provide alternative potential to conventional chemotherapy for several diseases especially in the conditions of the host defence mechanism, whereas supplementation of the plant product like flavonoids, peptides, lectins, polysaccharides and tannins could restore the immune response (Mehrotra et al. 2013; Seyed 2019).

Fruits being an essential component of a healthy diet can reduce the risk of cardiovascular disease, cancer and other diet-related diseases when taken in regular diet, as well as fruits can supplement the micronutrient in the case of deficiencies (Everitt et al. 2006; Sinha et al. 2003). Much of the potential of these fruits for disease prevention is thought to be provided by phytochemicals, among which the preventive activity of antioxidants is majorly documented. Several studies have shown variable and often contradictory results about the impact of the isolated phytochemicals on health; thus their consumption as supplements has been advised to be carried out with care because at times the dose may exceed the recommended nutritional intake (Balch 2006). But still there is a general trend that whole fruit intake is more important in providing health benefits than that of any particular constituent, because of the additive and synergistic effects. Nutrition is among the most important determinants of health, and developing evidence indicates that diet rich in fruits and vegetables may control a wide range of diseases (Boyer and Liu 2004). Many of the phytochemicals possess the potential to interfere with cellular functions, aiding in the activation of transcriptional factors that regulate the expression of the genes, and thereby change cellular metabolism in a number of ways (Chen and Liu 2018). Evidence from the earlier studies demonstrates that majority of the phytochemicals exhibit antioxidant and anti-inflammatory activities that are likely important to prevent different diseases (Li et al. 2016).

13.2 Potential of Functional Foods to Help the Immune System

There are a number of systematically approved functional foods currently available that are reported to have immune system potentiating properties, viz. functional yogurt that contain probiotics as well as micronutrient supplements (Kaur et al. 2021; Sharma et al. 2020). Many studies have been carried out to identify plant compounds that project the specific cellular events and enhance the immune actions when exposed to different allergens. Different food ingredients having immune boosting properties like probiotics, herbs, micronutrients and carotenoids have been studied and revealed to possess multiple benefits to the immune health (Dhama et al. 2016; Pratap et al. 2020). Probiotics are believed to stimulate cytokine production and to alter the gastrointestinal function, thereby boosting the immunity (Ashaolu 2020). The consumption of yoghurt containing the probiotic bacteria can enhance the phagocytic potential of the granulocytes which can support a stronger natural immunity (Isolauri et al. 2001; Matsuzaki and Chin 2000). Micronutrients are essential for boosting the immune system, and the immune boosting properties of many vitamins have been studied by previous researchers (Alpert 2017). Studies have indicated that supplementation of relatively high doses of vitamin C can reduce the symptoms associated with the common cold and duration of symptoms is reduced (Hemilä and Chalker 2013; Ran et al. 2018). Flavonoids are the biologically active compounds mostly found in the fruits, vegetables and nuts like proanthocyanidins (Harnly et al. 2006). Different studies have revealed the potent immunomodulatory actions of flavonoids that included anti-inflammation and anti-oxidant nature (Xiao 2017; Yahfoufi et al. 2018). Fruits diet of humans constitute to the major portion of the carotenoids which largely (approximately 90%) contain α-carotene, β-carotene, lycopene and cryptoxanthin (Khoo et al. 2011). Carotenoids are reported as antioxidant with abilities to regulate apoptosis-related gene function and have advanced function on the mechanisms regulating immune function (Chew and Park 2004).

13.3 Fruits as Immune Boosters

The daily intake of the fruits is required to maintain good health and to provide multiple immune benefits. Flavonoids and carotenoids are the natural food compounds from plants that function as antioxidant and anti-inflammatory agents in the host and beneficial effects on the immune system (Neri-Numa et al. 2020). Flavonoids and carotenoids are the most beneficial dietary antioxidants believed to strengthen the immune system. The levels of these antioxidants are high in fruits and vegetables and therefore referred to as natural immune boosting foods (Hughes 2000). Berry fruits like blueberries, blackberries, blackcurrants and cranberries are important sources of flavonoids having immune boosting properties, viz. anti-inflammation, antioxidant and antiproliferative properties (Nile and Park 2014; Szajdek and Borowska 2008). Antioxidant-rich fruit-based major sources are blueberries, cranberries, carrots, cocoa, cherries, pomegranate, broccoli, mango, spinach and sweet potatoes, and majority of them are also associated with anti-inflammatory properties (Zhang et al. 2019).

13.3.1 Fruits as a Source of Natural Antioxidants

Fruits are abundant in higher levels of biologically active components, which produce multiple health profits as well a rich source of nutrition. Fruits constitute major source of the dietary antioxidants that tend to increase the plasma antioxidant strength that results in the inhibition of atherosclerosis like diseases (Bankson et al. 1993; Santos-Buelga and Scalbert 2000). Earlier studies suggest the consumption of fruits is linked to lower prevalence and lower mortality rates that are caused by the cancer (Wang et al. 2014). In addition to fruits, the antitumorigenic properties of vegetables have also been evaluated through experiments using the cells and animals as models (van't Veer et al. 2000). A significant negative correlation between the total intake of fruits and vegetables, cardio- and cerebrovascular diseases and the mortality was reported (Verlangieri et al. 1985). Fibre, flavonoids, polyphenols, epigallocatechin, soya protein, isoflavanones, vitamin A, vitamin E, vitamin B, vitamin C, tocopherols, selenium, catechin, glutathione, sulphides, curcumin, sesaminol and indoles are the meticulously studied dietary components in fruits and vegetables for their antioxidant functions (Gregory-Mercado 2004; Konczak and Roulle 2011). Many of these compounds may act as an independent system or in multiple combinations as an anticancer agent by a number of mechanisms (Rabeta et al. 2013; Wargovich 2000).

13.3.2 Fruits Rich in Polyphenols

Berries are low-energy delicious foods rich in fibre, antioxidant vitamins and various other phenolic compounds. Studies have revealed that berries have similar or higher levels of phenolic acids and flavonoids than other commonly fruits used in diet (Mattila et al. 2006; Nile and Park 2014). Similarly cranberry juice has been found potent in treatment of the urinary tract infections with antioxidant effects (Vinson et al. 2008). Proanthocyanidins that are present in the blue berries are crucial for preventing urinary tract infections that are caused by *E. coli*. The phenolic acids which are present in berries are the hydroxylated derivatives of cinnamic acid and benzoic acid (Hisano et al. 2012; Jepson and Craig 2007). Studies revealed that there is a strong correlation between antioxidant activity and the total phenolics. Strawberry is reported to contain higher antioxidant potential and thus considered as a potential immune boosting fruit (Hannum 2004). Prunes and prune juices are an also efficient source of the dietary antioxidants (Gallaher and Gallaher 2008). Pomegranate juice highly rich in polyphenols has recently attained more focus because of its attribution of some of the important biological properties including antitumoral and antioxidative potential (Kaur and Kapoor 2001). Grape fruit and citrus seeds and peels have been found to have higher antioxidant activity especially the orange peel containing a number of flavones, viz. nobiletin and sinensetin (Bocco et al. 1998). Hydroxycinnamic acids, viz. p-coumaric acid, caffeic acid and ferulic acid, are the main potential phenolic components in plant tissues (Ferreira et al. 2019). Phenolic acids and flavonoids derived from apple were shown to possess strong antioxidant activity in liver and colon cancer cells (Ganesan et al. 2020; Govind 2011).

13.3.3 Immunomodulatory Potential of Plant-Based Products

The Indian traditional medicine system containing numerous medicinal plants has attracted the specific attention of scientists worldwide for their immunomodulatory, antioxidant, anti-inflammatory, hepatoprotective, antiasthmatic, antifungal, hypolipidaemic, diuretic and other medicinal activities (Hussain et al. 2017; Pengelly 2020). A healthy organism maintains the homeotic balance of the immune system by balancing between various exogenous and endogenous factors in the body which results in immunomodulation (Galluzzi et al. 2012; Jantan et al. 2015). The biomolecules of synthetic or biological origin are capable of modulating (either suppression or stimulation) any component of the adaptive or innate immune system which are known as immunomodulators, immunorestoratives, immunoaugmentors or biological response modifiers. The term immunomodulation represents strengthening or suppression of the indicators of cellular and humoral immune system and nonspecific cellular defence factors. The advantage of immunomodulation lies with a pharmacological agent acting under multiple doses and time (Brindha 2016). Food-based immunostimulants can increase the resistance against autoimmunity, cancer, allergy and infections. Since chemical-based supplements may create adverse effects, natural immunomodulators are needed to replace them in therapeutic industry (Jantan et al. 2015). The prevention and treatment of disease using plant products have been reported in human culture since centuries (Cowan 1999; Hosseinzadeh et al. 2015). Repertoire of a large number of food-based plants and fruit-based compounds have been used as drugs against a variety of diseases. Vincristine, vinblastine and galantamine from the Caucasian snowdrop (*Galanthus caucasicus*) are a few leading examples of plant-based medicines (Fürst and Zündorf 2014; Tan and Norhaizan 2018). The clinical potential of a number of plant-derived anti-inflammatory compounds such as curcumin, resveratrol, colchicine, catechins, quercetin and capsaicin are well elaborated (Fürst and Zündorf 2014). Plant-derived compounds like andrographolide and genistein have exhibited potent favourable impacts on cellular and humoral immune functions (Behl et al. 2020). Several studies have investigated the identification of the plant-based immunomodulatory compounds with their enhanced pharmacological potential and reduced toxicity/adverse effects. A large number of compounds including flavonoids, tocopherol, tannic acid, curcumin, carotenoids, polyphenols, etc. have been elaborated for their potent immunomodulatory properties (Madhuri and Pandey 2009; Škrovánková et al. 2012). Currently trending researches are focusing on herbs as immune potentiating agents because of their longer history and being widely active as a vital part of the treatment or prevention of several disease outbreaks (Khanna et al. 2020; Yin et al. 2006). The list specifically includes the names of neem, tulsi, ashwagandha, garlic and turmeric as food- and fruit-based valuable herbs that efficiently boost the immunity and therefore should be recommended to include in the regular diet (Cohen 2014; Tiwari et al. 2018). These herbal components not only enhance the immunity but also improve the gut function and metabolism as well. Spices and herbs having phenolics have shown potent antioxidatives and some antimicrobial constituents (Ceylan and Fung 2004). Many leafy spices particularly the ones belonging to the family Labiatae, viz. rosemary, sage, oregano and thyme, have shown to possess strong antioxidant potential (Shan et al. 2005; Venkateshappa and Sreenath 2013).

13.3.4 Edible Fruits with Immune Boosting Potentials

A proper diet may improve the immune system on multiple fronts, and this essentially includes a variety of foods and fruits. Amongst them reported evidences suggest that fruits with vitamins and minerals can improve the immune system as well as may help in maintaining the healthy microbiome (Fig. 13.1). Fruits of *Lycium barbarum* commonly called goji berries are typically being consumed as food supplements and for general medicinal properties that include stimulation of liver

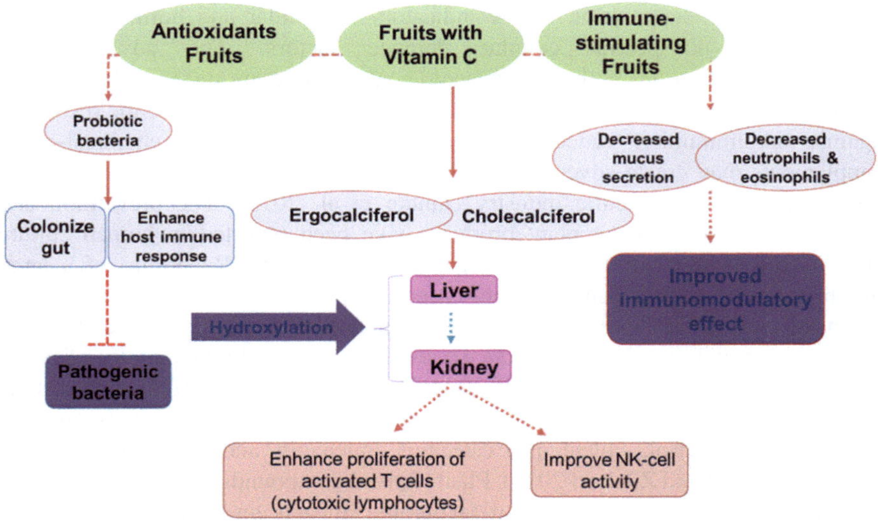

Fig. 13.1 General effect of fruits with antioxidant and immune-stimulating effects

and kidney functions. Goji berries have potent bioactive molecules, namely, carotenoids, polysaccharides, flavonoids, betaine, amino acids, cerebroside, minerals and other vitamins (Ma et al. 2019). *Lycium barbarum* polysaccharide was shown to possess several biological functions that include antioxidant and immuno-modulatory properties (Zhu et al. 2020). It has been evaluated that *Lycium barbarum* polysaccharide can act as an adjuvant, hence, could improve the immune responses against vaccine and increase humoral immunity (Su et al. 2014). Dragon fruit (*Hylocereus undatus*) has been reported to contain multiple bioactive compounds that are beneficial for health especially flavonoids, polyphenols and vitamin C with relatively high antioxidant activity (Xu et al. 2016). *Hylocereus undatus* containing oligosaccharides and other prebiotic properties has been well reported (Wichienchot et al. 2010). Dragon fruit oligosaccharide has been reported to act as a potentially novel source of prebiotic ingredients with probiotic stimulation and immune boosting properties by significantly increasing plasma immunoglobulin A and G concentrations (Pansai et al. 2020). Avocado fruit is also reported to have an anti-infective and immunomodulatory therapeutic value by virtue of their antimicrobial peptides. Avocado fruit thus may be a beneficial source of bioactive compounds with potent immune potentiating activity and with minimal side effects (Bhuyan et al. 2019). Antimicrobial peptides are the natural bioactive compounds essential for the innate immune system with beneficial values as antimicrobials, which may be used in combination with other drugs to enhance its immunomodulatory potentials (Pasupuleti et al. 2012). A general mechanism of the immune-stimulating effects of fruits is summarized in Fig. 13.1.

Earlier studies reveal that different fruits, viz. *guava, kiwi, orange, mango, pineapple, papaya, pomegranate and apple, possess antioxidant activities and are reported to have strong immune boosting potential.* Kiwi is a fruit that is rich in vitamin C, vitamin K, vitamin E, folate, magnesium and potassium nutrients, therefore possessing multiple health benefits (Drummond 2013). Kiwis have higher level of antioxidants and a good fibre content and are also considered as a potential immunity booster. Kiwis also possess various other health benefits, viz. used as a medication in treatment of hair fall, asthma, eye sight and depression (Richardson et al. 2018). Apple (*Malus domestica*) is regarded as one of the most popular fruits worldwide which has been found that its consumption can reduce the risk of various diseases as studied epidemiologically (Hyson 2011). It has been reported that apple skin contains higher levels of triterpenes which are the potential reason behind its health benefits. Investigations indicated that apple triterpenes have been linked with its multiple pharmacological properties, including anti-inflammatory and the immuno-modulatory properties (Szakiel et al. 2012). Recent studies have evaluated the potential of many triterpenes in apple as a therapeutic anticancer agent in clinical trials (Andre et al. 2013). Yet a limited number of fruits have shown direct immune boosting potentials which suggest that including them in a proper diet will be beneficial in a long term for guarding off infections and diseases. The mechanism of health-promoting effects of fruits and food components is summarized in Fig. 13.2.

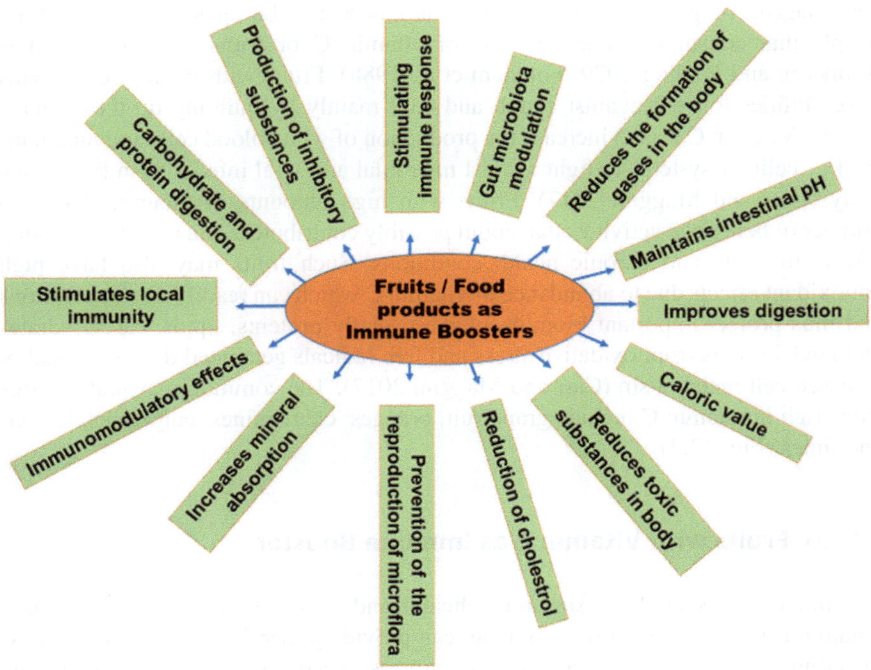

Fig. 13.2 Effects of food components and fruits with mechanisms of health-promoting benefits

13.3.5 Fruits with Vitamin C as Immune Booster

Vitamin C is a key source natural supplement to improve immunity. Several fruits with citrus properties like oranges, kiwi, papaya and guava are rich in vitamin C and thus should be recommended for inclusion in the daily diet plan. Moreover, some vegetables like beetroots, cauliflower and spinach are also known to contain significant amount of vitamin C and are good source for boosting the immunity. Green vegetables like broccoli and mushrooms are amongst the few immunities boosting foods that can be included in the daily diet plan to enhance the immunity (Baidya and Sethy 2020; Hedges and Lister 2007). These vitamin C-rich fruits and foods help in building and strengthening the immune system to a greater extent and in a natural way. Study showed that in vivo supplementation of vitamin C provided an antioxidant effect as a radical scavenger of peroxyl and oxygen. Vitamin C is easily oxidized to form a free radical, semidehydroascorbic acid being relatively stable (Bendich et al. 1986). The antioxidant activity of ascorbic acid is mediated by the loss of electrons which makes it biologically very efficient. Being an electron donor ascorbic acid serves as a reducing agent for a number of redox species and protects the water-soluble compounds in the cells and tissues from oxidation. It also reduces the tocopherol radicals back to their active form present at the cellular membranes (Ahmed et al. 2011). Studies revealed that deficiency of vitamin C leads to atherogenesis in animal models (Frikke-Schmidt and Lykkesfeldt 2009). The incidence of oesophageal, lung and pancreatic cancer is reported to be on lower side in those people that consume ample amounts of vitamin C or fruits rich in vitamin C (Cameron and Pauling 1979; Fontham et al. 1988). Fruits with vitamin C contents have abilities to fight against cough and cold mainly by building up the immune system. Vitamin C acts to increase the production of white blood cells and maintains the key cellular systems to fight against microbial and viral infections in the human body (Carr and Maggini 2017). Fruits with high amounts of vitamin C have a number of beneficial activities that could possibly contribute to the immune boosting effects in acute and chronic health conditions. Such fruits may also have high antioxidant effects due to abundance of vitamin C which can readily donate electrons and thus protect important biomolecules especially proteins, lipids, carbohydrates and nucleic acids from oxidative stress and free radicals generated during normal or diseased cell metabolism (Carr and Maggini 2017). The commonly available citrus fruits rich in vitamin C include grapefruit, oranges, clementines, tangerines, lemons and limes (Fig. 13.3).

13.3.6 Fruits with Vitamin E as Immune Booster

Vitamin E is essential for sustaining health and a booster for strong immunity. Vitamin E being a powerful antioxidant can provide protection from various illness including infections from bacteria and viruses. Seeds of almonds, peanuts and sunflower contain a notable level of vitamin E and therefore should be recommended to be consumed in daily diet plan to maintain the level of vitamin E in the body.

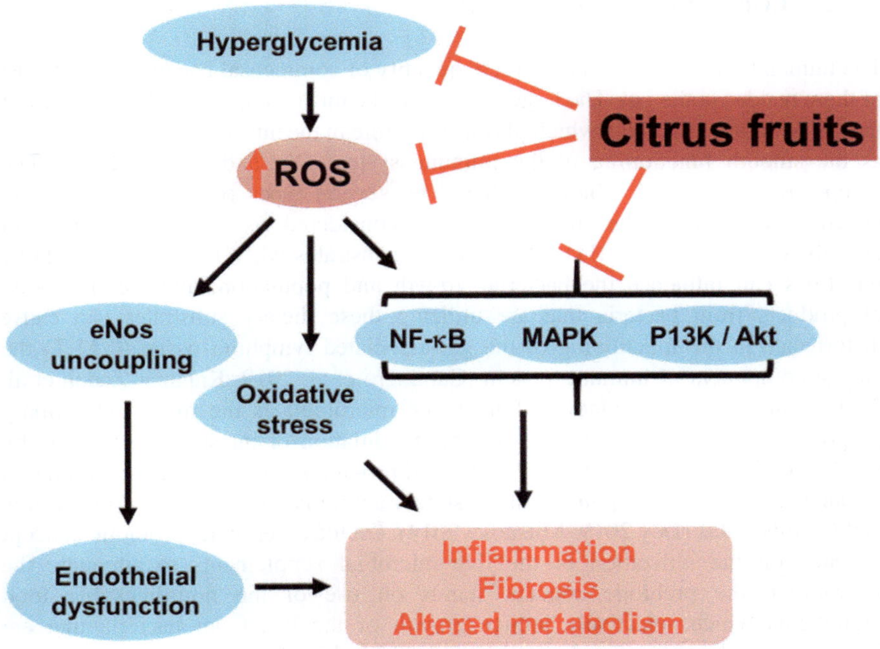

Fig. 13.3 Effect of citrus fruits with vitamin C on several cellular and molecular factors associated with immune functions

Vitamin E acts by mechanism as chain-breaking antioxidant that prohibits the spread of lipid peroxidation and acts as a radical peroxyl scavenger that protects polyunsaturated fats in plasma membranes and lipoproteins (Frei 1998; Lewis et al. 2019). Vitamin E acts to preserve the immune responses, and therefore its deficiency leads to lowered immunity and lowered humoral and cell-mediated immune response (Meydani et al. 2005; Wintergerst et al. 2007). Vitamin E supplementation was shown to alleviate physical activity-induced increase in immune alterations like oxidative stress and inflammatory cytokines (Singh et al. 2005). Vitamin E is considered to be the most abundant lipid-soluble antioxidant that protects the lipid portions of cell mainly the cellular membranes (Sies et al. 1992; Wang and Quinn 2000). Tocopherols can scavenge the free radicals by reacting with the lipid peroxyl radicals to produce the tocopheroxyl radical (Yamauchi 2007). Studies revealed that vitamin E supplement might reduce the risk of prostate cancer; also the epidemiological studies suggest a protective role of vitamin E against colon cancer (Lippman et al. 2009). γ-Tocopherol is reported to be more accurate and efficient in quenching of the dangerous radicals that are derived from peroxynitrite, that is, a product of inflammation (Jiang and Ames 2003).

13.4 Colonic Functional Foods

The human body is a host to a multiple quantity of commensal bacteria, and utmost of them resides in the gut. The large intestine is the most compactly inhabited area of the gut with the microbiota which play a major role in the nutrition and health as well as the smooth functioning of the immune system (Xu and Knight 2015). The composition of the microbiota is altered by several types of environmental and genetic factors with the dietary residues considered as the most significant contributors (Kau et al. 2011). These dietary substrates when reaching to the large intestines can influence the bacterial growth and population, and the metabolic by-products from bacteria that are utilizing these dietary substrates can cause disturbance in the functioning of the gut-associated lymphoid tissue (GALT) the largest component of immune system (Caminero et al. 2019; Kalantar-Zadeh et al. 2019). The dietary modulation of intestinal microbiota is the main and primary purpose of various functional foods. This modulation of intestinal microbiota by the dietary factors is the basis for the pre- and pro-symbiotic concepts depending on enhancing the useful components of intestinal microbiota especially *Bifidobacterium* and *Lactobacillus* (Dey 2019; Mitsuoka 2014). On the other hand, probiotic concept depends on the utilization of the live microbial supplements to attenuate the microbiota; the prebiotic concept stands on use of the non-digestible food ingredients which selectively induce growth of the beneficial bacteria that are indigenous to the colon (Fooks et al. 1999) (Table 13.1).

13.5 Conclusions and Future Perspectives

The plant-based foods and fruits vital in maintaining the health serve as key player by promoting the growth of the population of beneficial bacteria in the body and therefore help or boost the immunity to a large extent. Various vitamins are proved to provide important benefits for enhancing the immunity. Fruits like oranges, kiwi, guava and papaya and vegetables like broccoli, beetroots, spinach, mushrooms and cauliflower are rich in vitamin C and are potent modifier of the healthy immune system. Vitamin D enhances the cellular resistance mainly by increasing the cytokines that the innate immune system causes. In addition, a combination of some herbs is also known to play a crucial role in the prevention COVID-19. People with low immunity are more prone to many acute and chronic diseases especially COVID-19. More research on immune boosting foods and herbs needs to be carried out significantly and to find their role in immunity-related issues. This may further minimise the immune-related diseases and complications most importantly in the COVID-19 aspects. More research is needed to find the role of foods employed for the prevention of many diseases and to know about the behaviour of coronavirus. In general, green foods and vegetables are a vital source against coronavirus pandemic by boosting the immunity of all aged groups. The health benefits of phytochemicals and the potential of these phytochemicals to be integrated in the food supplements as nutraceuticals have a tremendous effect on the food industry. Functional food

Table 13.1 Common major fruits and food product with immune boosting property

Fruits and food products	Main constituents/ compounds	Immune boosting property	Reference
Prebiotics	Fructo-oligosaccharides (FOS), galacto-oligosaccharides (GOS), trans-galacto-oligosaccharides and peptidoglycan	Stimulate the growth of bifidobacteria, conferring benefits upon host health, stimulates innate immune system against the pathogenic microorganisms	Davani-Davari et al. (2019); Guarner (2013)
Probiotics	*Lactobacillus*, Bifidobacteria	Modulates the immune system, downregulates hypersensitivity reactions and alleviates intestinal information	Plaza-Diaz et al. (2014)
Grapes	Anthocyanin, flavanols, flavonols, resveratrol	Immunomodulatory, cardioprotective, anticancer, anti-inflammation, antiaging, antimicrobial properties	González-Gallego et al. (2014); Xia et al. (2010)
Kiwi	Vitamin C, K, E, folate, Mg, antioxidants, fibres	*Strong antioxidant activities supporting immune boosting potential, health promotional activities* like treatment of hair fall, asthma, eye sight and depression	Soquetta et al. (2016)
Avocado	Antimicrobial peptides	Anti-infective, immunomodulatory therapeutic values, antimicrobials	Segovia et al. (2016); Tremocoldi et al. (2018)
Dragon fruit	Betanin, isobetanin, betacyanin polyphenols, flavonoids, oligosaccharides and vitamin C	Antioxidant activity, oligosaccharide with potent prebiotic and probiotic stimulation and immune boosting properties, increasing plasma IgA and IgG levels	Hanifa et al. (2016)
Goji berries	Polysaccharides, carotenoids, betaine, cerebroside, flavonoids, amino acids, minerals	Food supplements with medicinal properties like stimulation of liver and kidney functions, antioxidant and immunomodulatory properties, improved immune responses against vaccine and increase humoral immunity	Forino et al. (2016)
Apple	Polyphenols (proanthocyanidins), dihydrochalcones, flavanols	Potent anti-inflammatory derivatives, health promoting effects,	Mendoza-Wilson et al. (2016); Pádua et al. (2014)

(continued)

Table 13.1 (continued)

Fruits and food products	Main constituents/ compounds	Immune boosting property	Reference
		antioxidant properties, as therapeutic anticancer agent	
Berries, onions, tea	Flavonoids	Modulating cytokines and transcription factors such as nuclear factor kappa B (NF-κB), modulate the immune system, antioxidant and anti-inflammation	Ioannone et al. (2013)
Apricots, asparagus, carrots, tangerines, tomatoes	Carotenoids	Potential antioxidant, regulates immune system, reduces the toxic effects of ROS	Chew and Park (2004); Hughes (1999)
Rosemary	Carnosol, rosmarinic acid, hesperidin	Boosts immune function against allergic reactions, increases blood circulation and potent hepatoprotective, antifungal, insecticide, antioxidant and antibacterial properties	Ahmed and Babakir-Mina (2020); Nieto et al. (2018)
Andrographis paniculata	Andrographolide	Exhibits potent effects on cellular and humoral immune functions	Mukherjee et al. (2014)
Broccoli	Carotenoids, phenolic compounds, vitamin C, and glucosinolates	Boosts immunity, slows down hardening of the arteries, strong antibacterial activity	Singh (2008); Somasundaram et al. (2018)
Garlic	Allicin, diallyl disulphide, S-allylcysteine and diallyl trisulphide	Contains antioxidants and boosts immune response, possess anti- cancer, anti-bacterial, anti-viral properties	Mikaili et al. (2013); Thomson and Ali (2003)
Selenium		Enhanced proliferation of activated T cells and is an effective immunostimulant	Avery and Hoffmann (2018)
Vitamin A		Neutralizes ROS, prevents damage to the immune cells, improves immune function and increases resistance against infections	Stephensen (2001)
Vitamin B6		Benefits immunity of the organism, regulates immune responses associated with inflammation	Qian et al. (2017)

(continued)

Table 13.1 (continued)

Fruits and food products	Main constituents/ compounds	Immune boosting property	Reference
Vitamin C		Anti-inflammatory action and improved antibody production	Carr and Maggini (2017)
Vitamin E		Modulates host immune functions, modulates T cell function, impacts T cell membrane integrity	Lewis et al. (2019)
Magnesium		Strengthens immune system, regulates cardiovascular physiology	Faryadi (2012)
Zinc		Maintains immune function, important in cell division, cell growth and wound healing	John et al. (2010)

concept is a challenge for the food and health sector. The food industry can play a major and significant role in strengthening the nutritional density by the use of designer foods with abilities to provide traditional nutrients (proteins, fats and carbohydrates). Moreover, consumption of fruit- and vegetable-rich diet in combination with other medical therapies may be regarded as a novel treatment approach as many phytochemicals regulate the similar set of genes and pathways that are targeted by the use of drugs.

Conflict of Interest Authors declare no conflict of interest exists with this research work.

References

Ahmed EA, Omar HM, Ragb SM, Nasser AY (2011) The antioxidant activity of vitamin C, DPPD and L-cysteine against cisplatin-induced testicular oxidative damage in rats. Food Chem Toxicol 49(5):1115–1121

Ahmed HM, Babakir-Mina M (2020) Investigation of rosemary herbal extracts (Rosmarinus officinalis) and their potential effects on immunity. Phytother Res 34(8):1829–1837

Alpert PT (2017) The role of vitamins and minerals on the immune system. Home Health Care Manag Pract 29(3):199–202

Andre CM, Larsen L, Burgess EJ, Jensen DJ, Cooney JM, Evers D, Zhang J, Perry NB, Laing WA (2013) Unusual immuno-modulatory triterpene-caffeates in the skins of Russeted varieties of apples and pears. J Agric Food Chem 61(11):2773–2779

Arreola R, Quintero-Fabián S, López-Roa RI, Flores-Gutiérrez EO, Reyes-Grajeda JP, Carrera-Quintanar L, Ortuño-Sahagún D (2015) Immunomodulation and anti-inflammatory effects of garlic compounds. J Immunol Res 2015:PMC4417560

Ashaolu TJ (2020) Immune boosting functional foods and their mechanisms: a critical evaluation of probiotics and prebiotics. Biomed Pharmacother 130:110625

Aslani BA, Ghobadi S (2016) Studies on oxidants and antioxidants with a brief glance at their relevance to the immune system. Life Sci 146:163–173

Avery JC, Hoffmann PR (2018) Selenium, selenoproteins, and immunity. Nutrients 10(9):1203

Baidya B, Sethy P (2020) Importance of fruits and vegetables in boosting our immune system amid the COVID19. Food Sci Rep 1(7):50–55
Balch PA (2006) Prescription for nutritional healing. Penguin
Bankson DD, Kestin M, Rifai N (1993) Role of free radicals in cancer and atherosclerosis. Clin Lab Med 13(2):463–480
Behl T, Kumar K, Brisc C, Rus M, Nistor-Cseppento DC, Bustea C, Aron RAC, Pantis C, Zengin G, Sehgal A (2020) Exploring the multifocal role of phytochemicals as immunomodulators. Biomed Pharmacother 133:110959
Bendich A, Machlin L, Scandurra O, Burton G, Wayner D (1986) The antioxidant role of vitamin C. Adv Free Radical Biol Med 2(2):419–444
Bhuyan DJ, Alsherbiny MA, Perera S, Low M, Basu A, Devi OA, Barooah MS, Li CG, Papoutsis K (2019) The odyssey of bioactive compounds in avocado (Persea americana) and their health benefits. Antioxidants 8:10
Bocco A, Cuvelier M-E, Richard H, Berset C (1998) Antioxidant activity and phenolic composition of citrus peel and seed extracts. J Agric Food Chem 46(6):2123–2129
Boyer J, Liu RH (2004) Apple phytochemicals and their health benefits. Nutr J 3(1):1–15
Brindha P (2016) Role of phytochemicals as immunomodulatory agents: a review. Int J Green Pharm (IJGP) 10:1
Calder PC, Kew S (2002) The immune system: a target for functional foods? Br J Nutr 88(S2): S165–S176
Cameron E, Pauling LC (1979) Cancer and vitamin C: a discussion of the nature, causes, prevention, and treatment of cancer with special reference to the value of vitamin C. Linus Pauling Institute of Science and Medicine
Caminero A, Meisel M, Jabri B, Verdu EF (2019) Mechanisms by which gut microorganisms influence food sensitivities. Nat Rev Gastroenterol Hepatol 16(1):7–18
Carr AC, Maggini S (2017) Vitamin C and immune function. Nutrients 9(11):1211
Ceylan E, Fung DY (2004) Antimicrobial activity of spices 1. J Rapid Methods Autom Microbiol 12(1):1–55
Chandra RK (1997) Nutrition and the immune system: an introduction. Am J Clin Nutr 66(2):460S–463S
Chen H, Liu RH (2018) Potential mechanisms of action of dietary phytochemicals for cancer prevention by targeting cellular signaling transduction pathways. J Agric Food Chem 66(13): 3260–3276
Chew BP, Park JS (2004) Carotenoid action on the immune response. J Nutr 134(1):257S–261S
Cohen MM (2014) Tulsi-Ocimum sanctum: a herb for all reasons. J Ayurveda Integr Med 5(4):251
Cowan MM (1999) Plant products as antimicrobial agents. Clin Microbiol Rev 12(4):564–582
Davani-Davari D, Negahdaripour M, Karimzadeh I, Seifan M, Mohkam M, Masoumi SJ, Berenjian A, Ghasemi Y (2019) Prebiotics: definition, types, sources, mechanisms, and clinical applications. Foods 8(3):92
De Visser KE, Eichten A, Coussens LM (2006) Paradoxical roles of the immune system during cancer development. Nat Rev Cancer 6(1):24–37
Dey P (2019) Gut microbiota in phytopharmacology: a comprehensive overview of concepts, reciprocal interactions, biotransformations and mode of actions. Pharmacol Res 147:104367
Dhama K, Sachan S, Khandia R, Munjal A, Iqbal HMN, Latheef SK, Karthik K, Samad HA, Tiwari R, Dadar M (2016) Medicinal and beneficial health applications of Tinospora cordifolia (Guduchi): a miraculous herb countering various diseases/disorders and its immunomodulatory effects. Recent Pat Endocri Metab Immune Drug Discov 10(2):96–111
Di Renzo L, Gualtieri P, Romano L, Marrone G, Noce A, Pujia A, Perrone MA, Aiello V, Colica C, De Lorenzo A (2019) Role of personalized nutrition in chronic-degenerative diseases. Nutrients 11(8):1707
Drummond L (2013) Chapter three—the composition and nutritional value of kiwifruit. In: Boland M, Moughan PJ (eds) Advances in food and nutrition research. Academic Press, pp 33–57

Everitt AV, Hilmer SN, Brand-Miller JC, Jamieson HA, Truswell AS, Sharma AP, Mason RS, Morris BJ, Le Couteur DG (2006) Dietary approaches that delay age-related diseases. Clin Interv Aging 1(1):11

Faryadi Q (2012) The magnificent effect of magnesium to human health: a critical review. Int J Appl 2:3

Ferreira PS, Victorelli FD, Fonseca-Santos B, Chorilli M (2019) A review of analytical methods for p-coumaric acid in plant-based products, beverages, and biological matrices. Crit Rev Anal Chem 49(1):21–31

Fontham ET, Pickle LW, Haenszel W, Correa P, Lin Y, Falk RT (1988) Dietary vitamins a and C and lung cancer risk in Louisiana. Cancer 62(10):2267–2273

Fooks LJ, Fuller R, Gibson GR (1999) Prebiotics, probiotics and human gut microbiology. Int Dairy J 9(1):53–61

Forino M, Tartaglione L, Dell'Aversano C, Ciminiello P (2016) NMR-based identification of the phenolic profile of fruits of Lycium barbarum (goji berries). Isolation and structural determination of a novel N-feruloyl tyramine dimer as the most abundant antioxidant polyphenol of goji berries. Food Chem 194:1254–1259

Frei B (1998) Natural antioxidants in human health and disease. Nutrition 14(1):85–87

Frikke-Schmidt H, Lykkesfeldt J (2009) Role of marginal vitamin C deficiency in atherogenesis: in vivo models and clinical studies. Basic Clin Pharmacol Toxicol 104(6):419–433

Fürst R, Zündorf I (2014) Plant-derived anti-inflammatory compounds: hopes and disappointments regarding the translation of preclinical knowledge into clinical progress. Mediators of inflammation 2014

Gallaher CM, Gallaher DD (2008) Dried plums (prunes) reduce atherosclerosis lesion area in apolipoprotein E-deficient mice. Br J Nutr 101(2):233–239

Galluzzi L, Senovilla L, Zitvogel L, Kroemer G (2012) The secret ally: immunostimulation by anticancer drugs. Nat Rev Drug Discov 11(3):215–233

Ganesan K, Jayachandran M, Xu B (2020) Diet-derived phytochemicals targeting colon cancer stem cells and microbiota in colorectal cancer. Int J Mol Sci 21(11):3976

Gombart AF, Pierre A, Maggini S (2020) A review of micronutrients and the immune system–working in harmony to reduce the risk of infection. Nutrients 12(1):236

González-Gallego J, García-Mediavilla MV, Sánchez-Campos S, Tuñón MJ (2014) Anti-inflammatory and immunomodulatory properties of dietary flavonoids. Polyphenols Human Health Dis:435–452

Govind P (2011) Some important anticancer herbs: a review. Int Res J Pharm 2:45–53

Gregory-Mercado KY (2004) Predictors of fruit and vegetable consumption in older mostly Hispanic women in Arizona

Guarner F (2013) 11 – impacts of prebiotics on the immune system and inflammation. In: Calder PC, Yaqoob P (eds) Diet, immunity and inflammation. Woodhead Publishing, pp 292–312

Hanifa NI, Rumiyati S, Fakhrudin N (2016) Cytoprotective and antioxidant effects of ethanolic extract of red dragon fruit (Hylocereus polyrhizus) and carrot (Daucus carota L.). AIP Conf Proc 1755(1):030004

Hannum SM (2004) Potential impact of strawberries on human health: a review of the science. Crit Rev Food Sci Nutr 44(1):1–17

Harnly JM, Doherty RF, Beecher GR, Holden JM, Haytowitz DB, Bhagwat S, Gebhardt S (2006) Flavonoid content of US fruits, vegetables, and nuts. J Agric Food Chem 54(26):9966–9977

Hedges L, Lister C (2007) Nutritional attributes of spinach, silver beet and eggplant. Crop Food Res Confidential Rep 1928

Hemilä H, Chalker E (2013) Vitamin C for preventing and treating the common cold. Cochrane Database Syst Rev 1:CD000980

Hisano M, Bruschini H, Nicodemo AC, Srougi M (2012) Cranberries and lower urinary tract infection prevention. Clinics 67(6):661–668

Hosseinzadeh S, Jafarikukhdan A, Hosseini A, Armand R (2015) The application of medicinal plants in traditional and modern medicine: a review of Thymus vulgaris. Int J Clin Med 6(09): 635

Hughes DA (1999) Effects of carotenoids on human immune function. Proc Nutr Soc 58(3): 713–718

Hughes DA (2000) Dietary antioxidants and human immune function. Nutr Bull 25(1):35–41

Hussain F, Rana Z, Shafique H, Malik A, Hussain Z (2017) Phytopharmacological potential of different species of Morus alba and their bioactive phytochemicals: a review. Asian Pac J Trop Biomed 7(10):950–956

Hyson DA (2011) A comprehensive review of apples and apple components and their relationship to human health. Adv Nutr 2(5):408–420

Iddir M, Brito A, Dingeo G, Fernandez Del Campo SS, Samouda H, La Frano MR, Bohn T (2020) Strengthening the immune system and reducing inflammation and oxidative stress through diet and nutrition: considerations during the COVID-19 crisis. Nutrients 12(6):1562

Ioannone F, Miglio C, Raguzzini A, Serafini M (2013) 15 – flavonoids and immune function. In: Calder PC, Yaqoob P (eds) Diet, immunity and inflammation. Woodhead Publishing, pp 379–415

Isolauri E, Sütas Y, Kankaanpää P, Arvilommi H, Salminen S (2001) Probiotics: effects on immunity. Am J Clin Nutr 73(2):444s–450s

Jantan I, Ahmad W, Bukhari SNA (2015) Plant-derived immunomodulators: an insight on their preclinical evaluation and clinical trials. Front Plant Sci 6:655

Jepson RG, Craig JC (2007) A systematic review of the evidence for cranberries and blueberries in UTI prevention. Mol Nutr Food Res 51(6):738–745

Jiang Q, Ames BN (2003) γ-Tocopherol, but not α-tocopherol, decreases proinflammatory eicosanoids and inflammation damage in rats. FASEB J 17(8):816–822

John E, Laskow TC, Buchser WJ, Pitt BR, Basse PH, Butterfield LH, Kalinski P, Lotze MT (2010) Zinc in innate and adaptive tumor immunity. J Transl Med 8(1):1–16

Kalantar-Zadeh K, Berean KJ, Burgell RE, Muir JG, Gibson PR (2019) Intestinal gases: influence on gut disorders and the role of dietary manipulations. Nat Rev Gastroenterol Hepatol 16(12): 733–747

Kau AL, Ahern PP, Griffin NW, Goodman AL, Gordon JI (2011) Human nutrition, the gut microbiome and the immune system. Nature 474(7351):327–336

Kaur C, Kapoor HC (2001) Antioxidants in fruits and vegetables–the millennium's health. Int J Food Sci Technol 36(7):703–725

Kaur J, Singh BP, Chaudhary V, Elshaghabee FM, Singh J, Singh A, Rokana N, Panwar H (2021) Probiotics as live bio-therapeutics: prospects and perspectives, advances in probiotics for sustainable food and medicine. Springer, pp 83–120

Khanna K, Kohli SK, Kaur R, Bhardwaj A, Bhardwaj V, Ohri P, Sharma A, Ahmad A, Bhardwaj R, Ahmad P (2020) Herbal immune-boosters: substantial warriors of pandemic Covid-19 battle. Phytomedicine 85, 153361

Khoo H-E, Prasad KN, Kong K-W, Jiang Y, Ismail A (2011) Carotenoids and their isomers: color pigments in fruits and vegetables. Molecules 16(2):1710–1738

Konczak I, Roulle P (2011) Nutritional properties of commercially grown native Australian fruits: lipophilic antioxidants and minerals. Food Res Int 44(7):2339–2344

Lewis ED, Meydani SN, Wu D (2019) Regulatory role of vitamin E in the immune system and inflammation. IUBMB Life 71(4):487–494

Li W, Guo Y, Zhang C, Wu R, Yang AY, Gaspar J, Kong A-NT (2016) Dietary phytochemicals and cancer chemoprevention: a perspective on oxidative stress, inflammation, and epigenetics. Chem Res Toxicol 29(12):2071–2095

Lippman SM, Klein EA, Goodman PJ, Lucia MS, Thompson IM, Ford LG, Parnes HL, Minasian LM, Gaziano JM, Hartline JA (2009) Effect of selenium and vitamin E on risk of prostate cancer and other cancers: the selenium and vitamin E cancer prevention trial (SELECT). JAMA 301(1): 39–51

Lobo V, Patil A, Phatak A, Chandra N (2010) Free radicals, antioxidants and functional foods: impact on human health. Pharmacogn Rev 4(8):118

Ma ZF, Zhang H, Teh SS, Wang CW, Zhang Y, Hayford F, Wang L, Ma T, Dong Z, Zhang Y, Zhu Y (2019) Goji berries as a potential natural antioxidant medicine: an insight into their molecular mechanisms of action. Oxidative Med Cell Longev 2019:2437397

Madhuri S, Pandey G (2009) Some anticancer medicinal plants of foreign origin. Curr Sci 96:779–783

Matsuzaki T, Chin J (2000) Modulating immune responses with probiotic bacteria. Immunol Cell Biol 78(1):67–73

Mattila P, Hellström J, Törrönen R (2006) Phenolic acids in berries, fruits, and beverages. J Agric Food Chem 54(19):7193–7199

Mehrotra S, Agnihotri G, Singh S, Jamal F (2013) Immunomodulatory potential of Curcuma longa: a review. South Asian J Exp Biol 3(6):299–307

Mendoza-Wilson AM, Castro-Arredondo SI, Espinosa-Plascencia A, Robles-Burgueño MDR, Balandrán-Quintana RR, Bermúdez-Almada MDC (2016) Chemical composition and antioxidant-prooxidant potential of a polyphenolic extract and a proanthocyanidin-rich fraction of apple skin. Heliyon 2(2):e00073

Meydani SN, Han SN, Wu D (2005) Vitamin E and immune response in the aged: molecular mechanisms and clinical implications. Immunol Rev 205(1):269–284

Mikaili P, Maadirad S, Moloudizargari M, Aghajanshakeri S, Sarahroodi S (2013) Therapeutic uses and pharmacological properties of garlic, shallot, and their biologically active compounds. Iran J Basic Med Sci 16(10):1031–1048

Mitsuoka T (2014) Development of functional foods. Biosci Microbiota Food Health

Mukherjee PK, Nema NK, Bhadra S, Mukherjee D, Braga FC, Matsabisa MG (2014) Immunomodulatory leads from medicinal plants. Indian J Tradit Knowl 13(2):235–256

Neri-Numa IA, Arruda HS, Geraldi MV, Júnior MRM, Pastore GM (2020) Natural prebiotic carbohydrates, carotenoids and flavonoids as ingredients in food systems. Curr Opin Food Sci 33:98–107

Nieto G, Ros G, Castillo J (2018) Antioxidant and antimicrobial properties of rosemary (Rosmarinus officinalis, L.): a review. Medicines (Basel) 5(3):98

Nile SH, Park SW (2014) Edible berries: bioactive components and their effect on human health. Nutrition 30(2):134–144

Pádua TA, de Abreu BSSC, Costa TEMM, Nakamura MJ, Valente LMM, Henriques MDG, Siani AC, Rosas EC (2014) Anti-inflammatory effects of methyl ursolate obtained from a chemically derived crude extract of apple peels: potential use in rheumatoid arthritis. Arch Pharm Res 37(11):1487–1495

Pansai N, Chakree K, Takahashi Yupanqui C, Raungrut P, Yanyiam N, Wichienchot S (2020) Gut microbiota modulation and immune boosting properties of prebiotic dragon fruit oligosaccharides. Int J Food Sci Technol 55(1):55–64

Pasupuleti M, Schmidtchen A, Malmsten M (2012) Antimicrobial peptides: key components of the innate immune system. Crit Rev Biotechnol 32(2):143–171

Pengelly A (2020) The constituents of medicinal plants: an introduction to the chemistry and therapeutics of herbal medicine. Routledge

Pickard JM, Zeng MY, Caruso R, Núñez G (2017) Gut microbiota: role in pathogen colonization, immune responses, and inflammatory disease. Immunol Rev 279(1):70–89

Plaza-Diaz J, Gomez-Llorente C, Fontana L, Gil A (2014) Modulation of immunity and inflammatory gene expression in the gut, in inflammatory diseases of the gut and in the liver by probiotics. World J Gastroenterol: WJG 20(42):15632

Pratap K, Taki AC, Johnston EB, Lopata AL, Kamath SD (2020) A comprehensive review on natural bioactive compounds and probiotics as potential therapeutics in food allergy treatment. Front Immunol 11:996

Qian B, Shen S, Zhang J, Jing P (2017) Effects of vitamin B6 deficiency on the composition and functional potential of T cell populations. J Immunol Res 2017:2197975

Rabeta M, Chan S, Neda G, Lam K, Ong M (2013) Anticancer effect of underutilized fruits. Int Food Res J 20(2):551
Ran L, Zhao W, Wang J, Wang H, Zhao Y, Tseng Y, Bu H (2018) Extra dose of vitamin C based on a daily supplementation shortens the common cold: A meta-analysis of 9 randomized controlled trials. BioMed Res Int 2018:1837634
Richardson DP, Ansell J, Drummond LN (2018) The nutritional and health attributes of kiwifruit: a review. Eur J Nutr 57(8):2659–2676
Santos-Buelga C, Scalbert A (2000) Proanthocyanidins and tannin-like compounds–nature, occurrence, dietary intake and effects on nutrition and health. J Sci Food Agric 80(7):1094–1117
Schols J, De Groot C, Van Der Cammen T, Rikkert MO (2009) Preventing and treating dehydration in the elderly during periods of illness and warm weather. J Nutr Health Aging 13(2):150–157
Segovia FJ, Corral-Pérez JJ, Almajano MP (2016) Avocado seed: Modeling extraction of bioactive compounds. Ind Crop Prod 85:213–220
Seyed MA (2019) A comprehensive review on Phyllanthus derived natural products as potential chemotherapeutic and immunomodulators for a wide range of human diseases. Biocatal Agric Biotechnol 17:529–537
Shan B, Cai YZ, Sun M, Corke H (2005) Antioxidant capacity of 26 spice extracts and characterization of their phenolic constituents. J Agric Food Chem 53(20):7749–7759
Sharma K, Tayade A, Singh J, Walia S (2020) Bioavailability of nutrients and safety measurements. Funct Foods Nutraceuticals Springer:543–593
Sies H, Stahl W, Sundquist AR (1992) Antioxidant functions of vitamins: vitamins E and C, Beta-carotene, and other carotenoids a. Ann N Y Acad Sci 669(1):7–20
Singh G (2008) How to boost your immune system naturally? Lulu. com
Singh U, Devaraj S, Jialal I (2005) Vitamin E, oxidative stress, and inflammation. Annu Rev Nutr 25:151–174
Sinha R, Anderson D, McDonald S, Greenwald P (2003) Cancer risk and diet in India. J Postgrad Med 49(3):222
Škrovánková S, Mišurcová L, Machů L (2012) Antioxidant activity and protecting health effects of common medicinal plants. Adv Food Nutr Res 67:75–139
Somasundaram J, Sivagurunathan Moni S, Makeen H, Intakhabalam M, Pancholi S, Siddiqui R, Eltyepelmobark M (2018) Antibacterial potential of ethanolic extract of broccoli (Brassica oleracea var. italica) against human pathogenic bacteria. Int J Pharm Res 10:143–146
Soquetta MB, Stefanello FS, Huerta KDM, Monteiro SS, da Rosa CS, Terra NN (2016) Characterization of physiochemical and microbiological properties, and bioactive compounds, of flour made from the skin and bagasse of kiwi fruit (Actinidia deliciosa). Food Chem 199:471–478
Stephensen CB (2001) Vitamin a, infection, and immune function. Annu Rev Nutr 21(1):167–192
Su C-X, Duan X-G, Liang L-J, Feng W, Zheng J, Fu X-Y, Yan Y-M, Ling H, Wang N-P (2014) Lycium barbarum polysaccharides as an adjuvant for recombinant vaccine through enhancement of humoral immunity by activating Tfh cells. Vet Immunol Immunopathol 158(1):98–104
Szajdek A, Borowska E (2008) Bioactive compounds and health-promoting properties of berry fruits: a review. Plant Foods Hum Nutr 63(4):147–156
Szakiel A, Pączkowski C, Pensec F, Bertsch C (2012) Fruit cuticular waxes as a source of biologically active triterpenoids. Phytochem Rev 11(2–3):263–284
Tan BL, Norhaizan ME (2018) Plant-derived compounds in cancer therapy: traditions of past and drugs of future, anticancer plants: properties and application. Springer, pp 91–127
Thomson M, Ali M (2003) Garlic [Allium sativum]: a review of its potential use as an anti-cancer agent. Curr Cancer Drug Targets 3(1):67–81
Tiwari R, Latheef SK, Ahmed I, Iqbal H, Bule MH, Dhama K, Samad HA, Karthik K, Alagawany M, El-Hack ME (2018) Herbal immunomodulators-a remedial panacea for designing and developing effective drugs and medicines: current scenario and future prospects. Curr Drug Metab 19(3):264–301

Tremocoldi MA, Rosalen PL, Franchin M, Massarioli AP, Denny C, Daiuto ÉR, Paschoal JAR, Melo PS, Alencar SMD (2018) Exploration of avocado by-products as natural sources of bioactive compounds. PLoS One 13(2):e0192577

van't Veer P, Jansen MC, Klerk M, Kok FJ (2000) Fruits and vegetables in the prevention of cancer and cardiovascular disease. Public Health Nutr 3(1):103–107

Venkateshappa S, Sreenath K (2013) Potential medicinal plants of Lamiaceae. Am Int J Res Formal Appl Nat Sci 1(3):82–87

Verlangieri A, Kapeghian J, El-Dean S, Bush M (1985) Fruit and vegetable consumption and cardiovascular mortality. Med Hypotheses 16(1):7–15

Vinson JA, Bose P, Proch J, Al Kharrat H, Samman N (2008) Cranberries and cranberry products: powerful in vitro, ex vivo, and in vivo sources of antioxidants. J Agric Food Chem 56(14): 5884–5891

Wang X, Ouyang Y, Liu J, Zhu M, Zhao G, Bao W, Hu FB (2014) Fruit and vegetable consumption and mortality from all causes, cardiovascular disease, and cancer: systematic review and dose-response meta-analysis of prospective cohort studies. BMJ 349:g4490

Wang X, Quinn PJ (2000) The location and function of vitamin E in membranes. Mol Membr Biol 17(3):143–156

Wargovich MJ (2000) Anticancer properties of fruits and vegetables. HortScience 35(4):573–575

Wichienchot S, Jatupornpipat M, Rastall RA (2010) Oligosaccharides of pitaya (dragon fruit) flesh and their prebiotic properties. Food Chem 120(3):850–857

Wintergerst ES, Maggini S, Hornig DH (2007) Contribution of selected vitamins and trace elements to immune function. Ann Nutr Metab 51(4):301–323

Xia E-Q, Deng G-F, Guo Y-J, Li H-B (2010) Biological activities of polyphenols from grapes. Int J Mol Sci 11(2):622–646

Xiao J (2017) Dietary flavonoid aglycones and their glycosides: which show better biological significance? Crit Rev Food Sci Nutr 57(9):1874–1905

Xu L, Zhang Y, Wang L (2016) Structure characteristics of a water-soluble polysaccharide purified from dragon fruit (Hylocereus undatus) pulp. Carbohydr Polym 146:224–230

Xu Z, Knight R (2015) Dietary effects on human gut microbiome diversity. Br J Nutr 113(S1):S1–S5

Yahfoufi N, Alsadi N, Jambi M, Matar C (2018) The immunomodulatory and anti-inflammatory role of polyphenols. Nutrients 10(11):1618

Yamauchi R (2007) Addition products of α-tocopherol with lipid-derived free radicals. Vitamins Hormones 76:309–327

Yin G, Jeney G, Racz T, Xu P, Jun X, Jeney Z (2006) Effect of two Chinese herbs (Astragalus radix and Scutellaria radix) on non-specific immune response of tilapia. Oreochromis niloticus Aquacult 253(1–4):39–47

Zhang L, Virgous C, Si H (2019) Synergistic anti-inflammatory effects and mechanisms of combined phytochemicals. J Nutr Biochem 69:19–30

Zhu W, Zhou S, Liu J, McLean RJC, Chu W (2020) Prebiotic, immuno-stimulating and gut microbiota-modulating effects of Lycium barbarum polysaccharide. Biomed Pharmacother 121:109591

Solanaceous Plants for Immunomodulation 14

Neha Pandey, Anupam Tiwari, Ritesh Kumar Yadav, and Neelam S. Sangwan

Abstract

In the era of COVID pandemic, it is highly desirable to improve body's immune responses. Many of the diseases including COVID impact the immune system and make it defenseless. There is a growing interest of herbal medicine around the world owing to effective pharmacological actions attributed by its unique phytocompounds. Majority of the immunomodulatory drugs including both immunosuppressor and immuno-stimulators are synthetic organic compounds having side effects. Therefore, there is a growing interest to use plant-based products to regulate immune responses. The present chapter deals with the recognition of members of the family Solanaceae for their use as potential immunomodulators. There are quite a few reports of active phytocompounds isolated from different solanaceous plants that have a myriad of medicinal properties. Various phytochemicals such as phenolic compounds, flavonoids, alkaloids, terpenes, lactones, and glycosides have been shown to have various therapeutic effects on human body owing to their ability to influence body's immune system. The present chapter will give a comprehensive overview of plant-based therapeutics of solanaceous plants including withanolides isolated

N. Pandey
Department of Botany, CMP PG College, University of Allahabad, Prayagraj, India

A. Tiwari
Department of Botany, Dayalbagh Educational Institute, Agra, Uttar Pradesh, India

R. K. Yadav
National Institute of Plant Genome Research, New Delhi, India

N. S. Sangwan (✉)
School of Interdisciplinary and Applied Sciences, Department of Biochemistry, Central University of Haryana, Mahendergarh, Haryana, India
e-mail: nsangwan@cuh.ac.in

N. S. Sangwan et al. (eds.), *Plants and Phytomolecules for Immunomodulation*,
https://doi.org/10.1007/978-981-16-8117-2_14

from *Withania somnifera* and *Physalis* sps, capsaicinoids isolated from *Capsicum*, and lycopene from tomato. Solanaceae is one of the superfamily having huge repertoire of plants belonging to major food crops utilized throughout the world (tomato, potato, capsicum etc.) as well as in medicine category. Besides several members of the Solanaceae family such as *Physalis* sps, *Atropa* sps, *Datura stramonium*, and *Solanum xanthocarpum* have potential phytomolecules belonging to alkaloids, terpenoids, and steroid category, which are recognized as medicines. This chapter will also give a critical overview of the possibilities, facts, and prospects of the use of members of Solanaceae family as potential future immunomodulatory drugs.

Keywords

Solanaceae · *Atropa* sps · *Hyoscyamus* sps · *Physalis* sps · *Withania somnifera* · Capsicum · Withanolides · Capsaicinoids

14.1 Introduction

Immunity refers to the body's defense against a number of infectious agents including viruses, bacteria, fungal spores, and other pollutants and pathogens. The immune system of the body works in a highly complex and regulated way. One key characteristic of the immune system is its ability to differentiate between foreign and body's own entities. Identification of any foreign particle/cells immediately triggers a coordinated response against it. This collective response can be specific or non-specific and involves a variety of cells, tissues, and organs. Pathogens generally express pathogen-associated molecular patterns (PAMPs) which are recognized by evolutionarily conserved host sensor pattern recognition receptors (PRRs). The interaction between PAMPs and PRRs leads to an array of immune responses quickly triggered via induction of different type I interferons, chemokines, and cytokines. A PRR family includes different types of receptors such as DNA receptors, NOD-like receptors, RIG-I-like receptors, and toll-like receptors (Parkin and Cohen 2001).

Depending upon the type of response, body's immunity has been categorized into innate and adaptive immunity. Innate immune system provides non-specific, short-term immunity against any foreign entity. Physical and chemical barriers are sometimes included as part of innate immunity. Innate immunity system produces different kinds of cells such as mast cell, basophil cells, NK cells, macrophage, eosinophil, neutrophil, NK T cells, and dendrite cells (Fig. 14.1) which play pivotal roles in antibody-dependent cell-mediated cytotoxicity, secretion of cytokines, nitric oxide (NO) production and antigen presentation, processing, and phagocytosis. Phagocytosis includes the role of basophils, eosinophils, and neutrophils. These cells also release inflammatory mediators like cytokines and histamines.

Adaptive immune system provides specific immunity against a number of foreign particles/pollutants/pathogens and involves the role of antigen-specific T and B lymphocytes. Adaptive immunity begins to respond generally after innate immunity

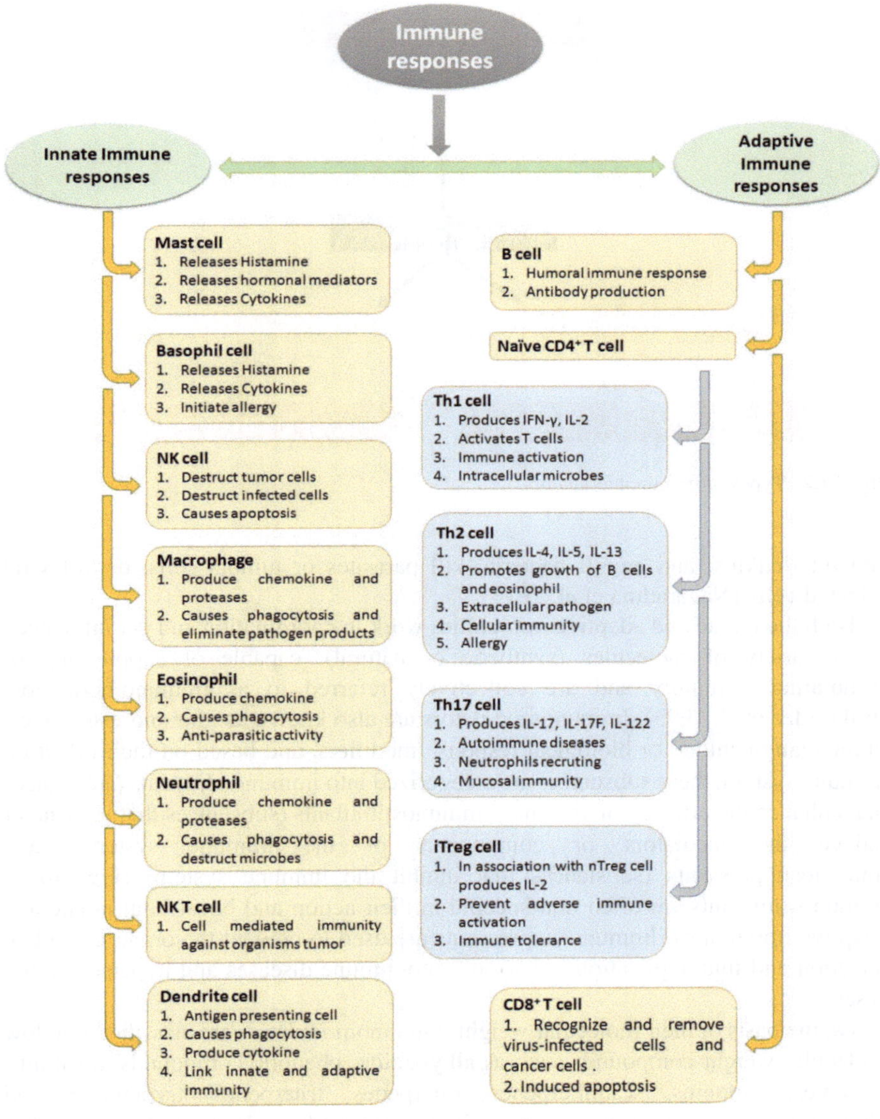

Fig. 14.1 Innate and adaptive immune systems and their roles

seems insufficient to target the antigen (pathogen or pollutant). Adaptive immunity has a long-lasting memory and often provides active immunity against a number of pathogens for lifetime (Jantan et al. 2015). The cells of the immune system take the help of T helper cells; activate signals through secretion of cytokines, lymphokines,

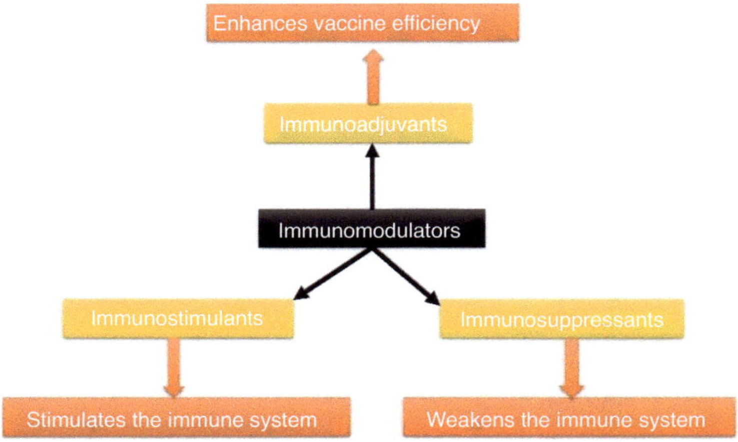

Fig. 14.2 Types of immunomodulators

and interleukins; and engulf bacteria, kill parasites or tumors cells, or kill viral infected cells (Nagarathna et al. 2013).

Both the innate and adaptive immunities work in coordination and get influenced by a variety of molecules (synthetic or natural), capable of suppressing or stimulating immunity and are collectively referred to as immunomodulators (Bakuridze et al. 1993). Immunomodulators are also known as immunorestoratives, immunoaugmentors, or biological response modifiers, and based on their effect on immune system, these substances are categorized into immunoadjuvants (substances that enhance the efficacy of vaccine), immunostimulants (substances that activate or induce the mediators or components of the immune system), and immunosuppressants (substances that inhibit the immune system) (Fig. 14.2). Immunostimulants are often non-specific in their action and boost both innate and adaptive immunities. Immunosuppressants are used to prevent the organ transplant rejection and find application in various autoimmune diseases and hypersensitivity cases.

On the basis of their molecular weight, immunomodulators are classified into low molecular weight compounds such as alkylamide, phenolic compounds, alkaloids, quinones, saponins, sesquiterpenes, diterpenes, triterpenes, tryptamine, and phytoestrogens and high molecular weight compounds such as proteins, peptides, polysaccharides, glycolipids, and glycoproteins (Wagner 1999).

A number of synthetic molecules including monoclonal antibodies are used as immunostimulants and immunosuppressants however with many adverse effects to the body. Glucocorticoids are frequently used as immunosuppressants as they reduce the expression of pro-inflammatory cytokines; however, they pose serious side effects by retarding the growth in children. They are also known to cause hypertension and hyperglycemia (Golan 2008; Jain 2008). There are many cytokine inhibitors such as etanercept and anakinra, which are used as immunosuppressants;

however, these cytokine inhibitors are reported to cause psoriasis and relapse of tuberculosis (Baudouin et al. 2003; Keane et al. 2001). Many immunostimulants including interferon alpha and gamma reportedly pose adverse effects on the body such as hypotension, gastro-intestinal distress, anorexia, and arrhythmias (Strayer and Carter 2012).

Immunomodulators with lesser side effects and greater safety are required, and natural compounds are probably a better alternative. Immunomodulation using plant-based natural products, plant extract, and their active moieties is a preferred alternative to the conventional chemotherapy (Vattem and Shetty 2005). Indian medicinal plants are the rich source of substances which can induce para-immunity, the non-specific immunomodulation of especially granulocytes, macrophages, natural killer cells, and competent functions. Immunomodulatory actions exhibited by plant extract or phytomolecule play an important role in the prevention and cure of infection (Jandú et al. 2017), inflammation (Schulze-Koops et al. 1999), and also immunodeficiencies (Ziauddin et al. 1996) by influencing the various cell types via cytokines and interleukins. These plant-based compounds may provide additional safety because plants and plant products have been in traditional use for medicinal purposes since pre-historic times.

A number of plant-derived compounds have immunomodulatory properties. The groups of compounds with such potential include alkaloids, monoterpenes, diterpenes, sesquiterpenes, polysaccharides, flavonoids, and many components of essential oils.

The Solanaceae family includes a number of plants with immense medicinal properties owing to their diverse groups of secondary metabolites. Many genera such as *Withania*, *Datura*, *Capsicum*, *Solanum*, and others have well-documented medicinal uses since ancient times.

14.2 Genus *Withania*

The genus *Withania* is an important member of the nightshade family, Solanaceae. At least twenty-three species of the genus *Withania* have been reported, out of which species *W. somnifera* and *W. coagulans* are recognized as medicinally and economically important. Other species include *W. japonica*, *W. begoniifolia*, *W. qaraitica*, *W. chevalieri*, and many others with lesser medicinal and economic importance (Tuli et al. 2009). *W. somnifera* finds its uses since pre-historic times in Ayurveda and many other traditional systems of medicine. In Indian ayurvedic literature, this plant has been reported to have "vata" soothing properties (Tuli et al. 2009). The plant is also called as "ashwagandha," the literal meaning of which is "smell of horse." The roots of the plant smell like horse, and therefore the name "ashwagandha" has been given in ancient times. The roots of *W. somnifera* are tuberous and approximately 20 to 30 cm long. Roots are brown to yellowish in color and have a bitter taste (John 2014). The plants bear berry fruits which turn bright orange to red at maturity. The plant produces fairly diverse population of secondary metabolites including few unique ones such as withanolides (Sangwan et al. 2017). The plant is known for a huge population of secondary metabolites which include many types of steroidal lactone (commonly called as withanolides), alkaloids, glycosides, nitrogen-containing compounds, and bioactive flavonoids. Important bioactive compounds of *W. somnifera* have been enlisted in Table 14.1. Withaferin A is the first withanolide discovered from the tissues of the plant.

Withania coagulans also contains several distinct steroidal phytomolecules; however, their medicinal roles are reported differently from that of *W. somnifera* (Kushwaha et al. 2013). *W. coagulans* berries contain milk coagulating properties. In addition, *W. coagulans* and their bioactive compounds are useful in many health-related disorders due to antihyperglycemic, anti-inflammatory, hepatoprotective, hypolipidemic, antimicrobial, anti-tumor, and immunomodulatory activities (Maurya 2010).

In addition to withanolides, *W. coagulans* contain structurally different bioactive compounds, namely, coagulin H, which is a derivative of withanolides and has coagulating properties. Few withanolides and coagulin H isolated from *W. coagulans* are also known for their immunosuppressive activities. More precisely, coagulin H has been reported to inhibit proliferation of lymphocytes and production of Th-1 cytokines (Mesaik et al. 2006). Further, Huang et al. (2009) have isolated six new withanolides, withacoagulin A, withacoagulin B, withacoagulin C, withacoagulin D, withacoagulin E, and withacoagulin F, from *W. coagulans* which have been shown to have immunosuppressive activities.

Anti-inflammatory activities of many withanolides from *W. coagulans* have been reported through many studies. For example, 3-β-hydroxy-2,3-dihydrowithanolide F has been shown to exhibit strong anti-inflammatory activity in rats (Budhiraja et al. 1984). This compound (3-β-hydroxy-2,3-dihydrowithanolide F) has also been shown to act as depressant of the central nervous system.

14.2.1 Repurposing of Withanolides for Immunomodulation during Treatment of SARS-CoV-19

The causative agent of COVID-19 is SARS-CoV-2 and has caused a major pandemic affecting the whole world. Though there is rapid scientific growth toward development of specific treatment line and vaccine, the disease is still far from control. This has compelled scientist's world over to investigate plant bioactive compounds with known immunomodulatory properties for their repurposing. In the year 2019–2020, *W. somnifera* has been investigated by many workers around the world, and many bioactive compounds of *W. somnifera* have been shown to be effective in treatment of COVID-19 patients. Many bioactive compounds such as withaferin A, withanolide D, withanone, caffeic acid, withanoside V, withanoside X, quercetin glucoside, and others have been reported to have anti-viral (against SARS-CoV-19) and immunomodulatory activities (Khanal et al. 2021).

After infection, SARS-CoV-19 enters inside the cell through ACE2 (angiotensin-converting enzyme 2) receptors and, through uncoating, releases its genetic material, which is in the form of single-stranded RNA. With the help of host machinery, the

Table 14.1 Bioactive compounds of *Withania somnifera*

Class of compound	Bioactive compound	Properties	References
Steroidal lactone	Withaferin-A 5,6-De-epoxy 5-en-7-one-17-hydroxy withaferin A	Anti-microbial, anti-tumor, anti-inflammatory, anti-angiogenic, immunomodulatory properties Cytotoxic	Jayaprakasam and Nair (2003), Jayaprakasam et al. (2003), Zhao et al. (2002), Sabir et al. (2008), Khedgikar et al. (2013), Khedgikar et al. (2014) and patent Sidique et al. (2014)
	Withanolide A	Immunomodulatory activities, anti-inflammatory, and cholinesterase inhibiting activities	Malik et al. (2007) Choudhary et al. (2004) Ahmad et al. (2016a, 2016b)
	Withanolide D	Immunomodulatory activities, anti-tumor, anti-metastatic activities	Leyon and Kuttan (2004)
	Withanolide E	Immunomodulatory activity (more specifically immunosuppressive)	Shohat et al. (1978)
	Withanolide F	Immunomodulatory activity, anticancer activity	Glotter et al. (1977) Seth et al. (2016)
	Withanolide G	NR	Glotter et al. (1973)
	Withanolide H	NR	Glotter et al. (1973)
	Withanolide I	NR	Glotter et al. (1973)
	Withanolide J	NR	Glotter et al. (1973)
	Withanolide K	NR	Glotter et al. (1973)
	Withanolide L	NR	Glotter et al. (1973)
	Withanolide M	NR	Glotter et al. (1973)
	Withanone	Anti-inflammatory, immunomodulatory, anti-arthritic	Kirson et al. (1971)
	Tubocapsenolide F	Anticancerous property	Chang et al. (2007)
	Viscosalactone B	Anticancerous property	Jayaprakasam et al. (2003)
Alkaloids	Withanine	Bradycardic action, prolonged hypotensive and respiratory-stimulant action	Malhotra et al. (1981)
	Withaninine		
	Somniferine		
	Tropeltigloate		
	Somniferinine		
	Somninine		
	Nicotine		
	Withasomine		
	Anaferine		
	Tropeltigloate		
	Withanol	NR	Bharti et al. (2016)

(continued)

Table 14.1 (continued)

Class of compound	Bioactive compound	Properties	References
Nitrogen-containing compounds	Somnisol	NR	
	Somnitol	NR	
Glycosides	Sitoindosides VII and VII	Anti-inflammatory, anti-stress properties	Bhattacharya and Muruganandam (2003)
Flavonoids	Quercetin	Anti-inflammatory, immunomodulatory	Nadkarni (1954)
	Kaempferol	Anti-inflammatory, immunomodulatory	Nadkarni (1954)

RNA molecule replicates and translates inside the cell. Various bioactive compounds from *W. somnifera* have been shown to block coronavirus infection at various stages of the virus life cycle. Withaferin A has been shown to bind to the spike protein of coronavirus, thereby disabling the coronavirus to enter the host cell through ACE2 receptor (Straughn and Kakar 2020). Besides this, withanolide D has also been shown to block the ACE2 receptor, through which SARS-CoV-19 enters inside the cell (Patel et al. 2020). In a report by Kumar et al. (2020), withaferin A and withanone have been predicted to interact with the SARS-CoV-2 protease and inhibit its function. Protease inhibition potential has also been reported for withanoside V by Tripathi et al. (2020). Inhibition of helicases and replicases (RNA-directed RNA polymerase) is also among the potential targets of drugs for treatment of COVID-19. Interestingly, withanoside X has been found to inhibit the replication of viral genome by blocking helicases and replicases coded by the viral genome (Chikhale et al. 2020).

Withanolides isolated from *W. somnifera* may have dual impact on patients suffering from coronavirus infection. Many withanolides have been reported to block or inhibit actions of all-important steps of viral life cycle, i.e., spike protein attachment and entry through ACE2 receptor, protease function, and replication. In addition to the anti-viral effects, withanolides (more precisely *W. somnifera* herbal preparations containing withanolides) are known for their immunomodulating properties since ancient times. In response to virus infection, a complex and coordinated host response is activated which includes production of pro-inflammatory cytokines and interferons. Overproduction of cytokines and interferons may cause cytokine storm which is fatal for the individual. Withanolides are known to manage the cytokine storm and help patients to manage the immune response in a better way (Agarwal et al. 1999; Davis and Kuttan 2000). Mechanism of withanolide's action against SARS-CoV-2 and immunomodulatory effects has been graphically represented in Fig. 14.3.

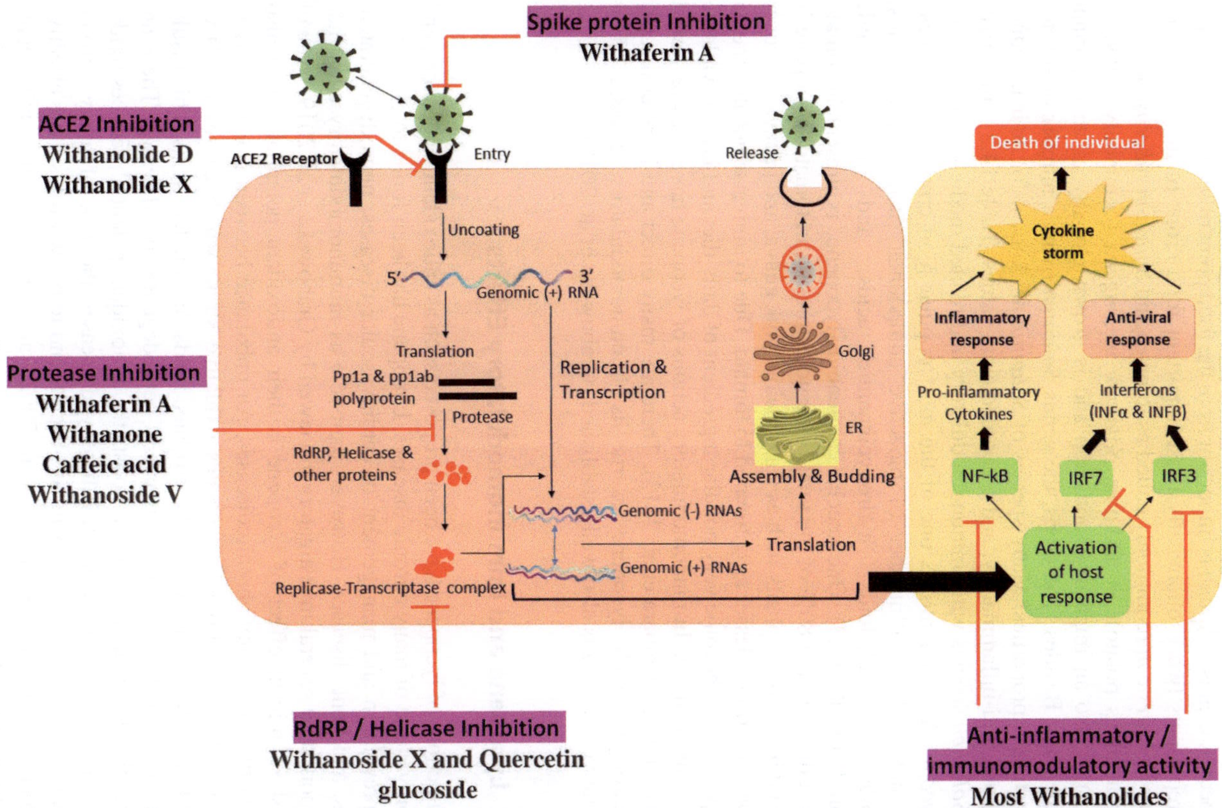

Fig. 14.3 Graphical representation of actions of bioactive compounds of *Withania somnifera* as anti-viral (against SARS-CoV-2) and immunomodulatory drugs

14.3 Genus *Solanum*

Solanum is one of the largest genera of flowering plants and roughly includes around 2000 species. Genus *Solanum* is important from the food point of view as well, as the genus represents two economically most important food crops of the world, i.e., *Solanum lycopersicon* (commonly called as tomato) and *Solanum tuberosum* (commonly called as potato). In addition, *Solanum melongena* (commonly called as eggplant) is also an important food crop and is grown globally as an important vegetable crop. Besides food crops, the genus *Solanum* owns a number of species with medicinal properties. Many species of *Solanum* are known for their immunomodulatory, anti-inflammatory, and other medicinal values. Table 14.2 summarizes list of *Solanum* species with immunomodulatory and related medicinal properties.

S. lycopersicon (tomato) is one of the most popular food crops of Solanaceae family and contains a number of bioactive compounds such as carotenoids, flavonoids, glycoalkaloids, hydroxycinnamic acids, and many volatile compounds (Table 14.3). One special carotenoid, lycopene, is present in tomato and constitutes around 80% of the total carotenoids present in tomato. Lycopene is also found in few other plants such as watermelon, pink guava, and papaya; however their concentration is lesser as compared to tomato. Due to immense health benefits of lycopene, it has become very popular ingredient of daily diet and is also available as health supplements. Immunomodulatory actions of lycopene have been reviewed by many workers (Camara et al. 2013). There are many direct and indirect mechanism through which lycopene exerts its immunomodulatory effects. Their antioxidative capacity is also well established (Grabowska et al. 2019).

14.3.1 Lycopene and Immunomodulatory Effects

Many functions of lycopene have been proposed such as antioxidant, anticancer, and in the prevention of many chronic diseases. Lycopene has been found to play a role in the prevention and management of cardiovascular diseases, neurodegenerative diseases, and bone diseases. However, their role as immunomodulatory agent has been reported by several investigators (reviewed by Grabowska et al. 2019). One of the most significant effect of lycopene is their antioxidant potential which also indirectly prevents many chronic diseases. An imbalance between reactive oxygen species (ROS) and antioxidants creates oxidative stress which causes oxidative damage to all vital biomolecules like proteins, lipids, and nucleic acids. This results in an increased risk of cardiovascular and neurodegenerative diseases. There are evidences that lycopene traps ROS, boosts the antioxidant potential, reduces oxidative damage to biomolecules, and thereby decreases the risk of many chronic diseases (Lee et al. 2000). Lycopene has been found to be two-folds more effective as compared to β-carotene (Böhm et al. 2002), and among various antioxidants present in the body, it has been found to possess greatest singlet oxygen-quenching potential. In addition, the conjugated double bonds in the structure of lycopene with ROS capturing potential make it work efficiently as antioxidant even at low oxygen

Table 14.2 List of important *Solanum* species with immunomodulatory and related medicinal properties

Genus and species	Bioactivities	Reference
S. capsicoides	Anti-inflammatory, immunomodulatory, anticancer properties	Chen et al. (2015), Petreanu et al. (2016)
S. aculeastrum	Anti-inflammatory, anti-microbial, anticancer, wound-healing properties	Wanyonyi et al. (2003), Koduru et al. (2007)
S. aethiopicum	Anti-inflammatory, anti-ulcer	Nwanna et al. (2013), Anosike et al. (2012)
S. americanum	Anti-microbial, anti-viral	Kaunda and Zhang (2019)
S. cathayanum	Anti-neurodegenerative, anti-inflammatory, anti-tumor, anti-inflammatory properties	Zhou et al. (2016)
S. chrysotrichum	Anti-inflammatory, anti-mycotic properties	Zamilpa et al. (2002), Alvarez et al. (2001)
S. dulcamara	Anti-inflammatory property	Tunon et al. (1995)
S. elaeagnifolium	Anti-inflammatory, anticancer, corticosteroid drug, hepatoprotective, analgesic, antispasmodic, antiepileptic properties	Feki et al. (2013)
S. erianthum	Analgesic, anti-inflammatory, anti-tumor properties	Chou et al. (2012), Peng et al. (2017), Priyadharshini and Sujatha (2013)
S. incanum	Anti-inflammatory, anticancer properties	Sundar and Pillaiy (2016)
S. indicum (Indian nightshade)	Immunomodulatory, anti-inflammatory, anti-tumor, appetite booster properties	Zhuang et al. (2018), Chiang et al. (1991)
S. khasianum	Anti-inflammatory, anticancer properties	Rosangkima and Jagetia (2015)
S. lycocarpum (wolf apple)	Anti-inflammatory, antihistamine, anticancer properties	Munari et al. (2014)
S. lycopersicum (tomato)	Antimicrobial, anti-inflammatory, anticancer properties	Munari et al. (2014), Miranda et al. (2012), Moreira et al. (2013)
S. lyratum	Anticancer, anti-inflammatory properties	Ren et al. (2009), Yao et al. (2013)
S. melongena (eggplant)	Antioxidant, anticancer, anti-inflammatory, antidiabetic, analgesic properties	Singh et al. (2009), Zhao et al. (2014), Yoshikawa et al. (1996)
S. muricatum	Anticancer, anti-inflammatory	Herraiz et al. (2016), Chang et al. (2016)
S. nigrum	Analgesic, anti-allergic, anti-histamine, anti-inflammatory, anti-tumor properties	Lin et al. (2008)
S. paniculatum	Anti-inflammatory, anti-ulcer, anti-viral, anti-bacterial properties	Valadares et al. (2009)
S. sessiliforum	Antioxidants, anti-microbial properties	Kaunda and Zhang (2019)
S. sisymbriifolium	Analgesic, anti-microbial, antioxidant properties	Kaunda and Zhang (2019)

(continued)

Table 14.2 (continued)

Genus and species	Bioactivities	Reference
S. surattense	Anti-inflammatory, anti-tumor, anti-ulcer, antimicrobial, wound healing properties	Kaunda and Zhang (2019)
S. torvum	Anti-platelet aggregation, anti-viral, antibacterial, analgesic, anti-inflammatory, wound healing properties	Kaunda and Zhang (2019)
S. trilobatum	Antioxidants, anti-inflammatory, antibacterial, antidiabetic, anticancer properties	Shahjahan et al. (2005)
S. tuberosum (potato)	Antioxidant, anti-microbial properties	Chandrashekara and Dharmesh (2014)
S. umbelliferum	Anticancer, anti-inflammatory properties	Kim et al. (1996)
S. violaceum	Anti-inflammatory, anticancer, antimicrobial, antioxidant properties	Raju et al. (2013)
S. Xanthocarpum	Immunomodulatory, anti-microbial, anti-inflammatory, antioxidant, hepatoprotective activities	Govindan et al. (1999)

pressure (Khachik et al. 1998, Di Mascio et al. 1989). Zhao et al. (2017) have shown that lycopene treatment causes overexpression of antioxidative enzymes. Lycopene's immunomodulatory effects are majorly due to inhibition of NF-kappaB (NF-kB) transcriptional pathway, thereby reducing production of pro-inflammatory cytokines, interleukins, and TNF-α. This suggests lycopene's anti-inflammatory role in various health disorders. The lycopene mode of action as immunomodulatory and antioxidative molecule has been graphically represented in Fig. 14.4.

In addition to NF-kB pathway, lycopene has recently shown to activate Nrf-2/HO-1 pathway for antioxidative functions. In aging chickens, lycopene has been reported to enhance the cell proliferation while reducing cell apoptosis. Besides, lycopene also minimizes oxidative damage in mitochondria, suggesting their role as positive regulator of aging (Liu et al. 2018).

14.4 Genus *Capsicum*

Capsicum is another most important solanaceous crop comprising of around 38 species. Only five (Table 14.4), *C. chinense*, *C. annuum*, *C. pubescens*, *C. baccatum*, and *C. frutescens*, are popularly cultivated for its pungent (hot pepper) and non-pungent (sweet pepper or bell pepper) fruits (USDA-ARS 2011). Chili fruits are popular as "wonder spice" as they add flavor, aroma, texture, pungency, color, and spice to the food. The spice of the chili peppers is mainly due to a group of secondary metabolites collectively known as capsaicinoids. Capsaicinoids

Table 14.3 Bioactive compounds of tomato (Tohge et al. 2014)

Class of compound	Subclass	Bioactive compound
Carotenoid	–	Lycopene
		β-carotene (pro-vitamin A)
Flavonoids	–	Chlorogenic acid
		Naringenin
		Quercetin
		Rutin
		Kaempferol
		Genistein
Glycoalkaloids	–	Dehydrotomatine
		α-Tomatine
		Tomatidine
Hydroxycinnamic acids	–	Caffeic acid
		Chlorogenic acid
		p-Coumaric acid
		Ferulic acid
		Resveratrol
Volatile organic compounds	Carotenoid-derived volatiles	β-Ionone
		6-methyl-5-hepten-2-one
		Geranylacetone
		β-Damascenone
	Fatty acid-derived volatiles	cis-3-Hexanal
		trans-2-Heptenal
		cis-2-Penten-1-ol
		trans-2-pentenal
		2-isobutylthiazole
	Terpenoid-derived volatiles	Lipophilic monoterpenoids, sesquiterpenoids, and diterpenoids
	Amino acid-derived volatiles	3-Methylbutanal
		Isobutylthiazole
		Isovaleronitrile
		Isobutyl acetate
		Guiacol
		Eugenol
		Methyl salicylate

(vanillylamine linked with a branched-chain fatty acid) are the most characteristic phenolic derivatives found in pepper fruits which are responsible for the pungency of peppers (de Jesús Ornelas-Paz et al. 2010).

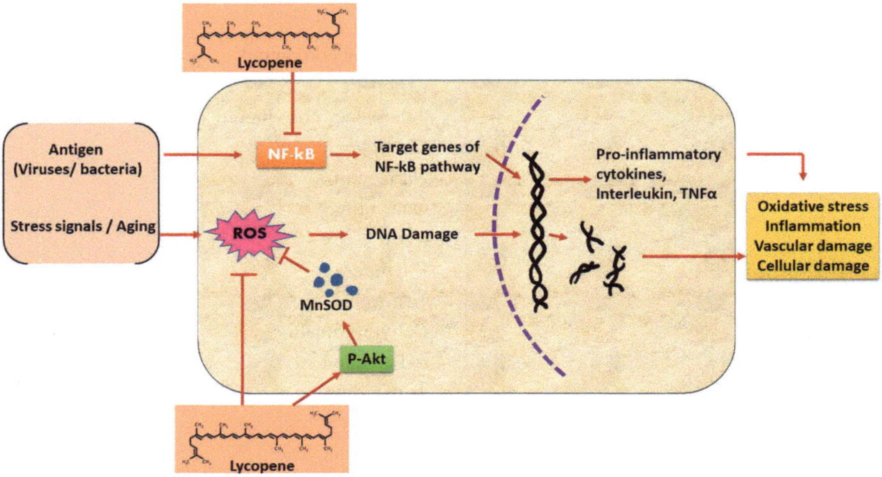

Fig. 14.4 Mode of action of lycopene as immunomodulatory and antioxidative molecule

14.5 Capsaicinoids and Capsinoids

Capsaicinoids are a group of chemically related compounds naturally biosynthesized by the genus *Capsicum* and produce pungency in chili pepper. Capsaicinoids are composed of a benzene ring of vanillyl group bonded with the branched fatty acids (alkyl moieties). The alkyl chain is responsible for the molecular diversity within capsaicinoids (Fig. 14.5). Different capsaicinoids identified from chili are capsaicin, dihydrocapsaicin, nordihydrocapsaicin, homocapsaicin I and II, homodihydrocapsaicin I and II, nornorcapsaicin/dinorcapsaicin, nornordihydrocapsaicin, ω-hydroxycapsaicin, and nonivamide. Quantitatively capsaicin, dihydrocapsaicin, and nordihydrocapsaicin together account for more than 90% of capsaicinoids in chili pepper fruit.

Capsinoids are a group of related compounds identified from non-pungent type of red pepper or CH-19 sweet pepper fruits and differ from capsaicinoids as they possess ester bond to join vanillyl moiety bonded with the fatty acid moiety instead of an amide moiety. The change of amide bond with ester bond causes disappearance of capsinoid pungency. Major capsinoids are capsiate ($C_{18}H_{26}O_4$), dihydrocapsiate ($C_{18}H_{28}O_4$), and nordihydrocapsiate ($C_{17}H_{26}O_4$) (Fig. 14.5).

Regular consumption of chili pepper is associated with several health benefits (Fig. 14.2) such as digestive (Prakash and Srinivasan 2012), antimicrobial (Cristian et al. 2013), anti-ulcer (Baruah et al. 2014), antioxidant (Morales-Soto et al. 2013), anticancer (Bessler and Djaldetti 2017), antidiabetic (Yuan et al. 2016), thermogenesis and weight reducing (Sanati et al. 2018), and anti-inflammatory activities (Srinivasan 2013). Besides, capsicum also possesses immunological effects on human health (Beltran et al. 2007) (Fig. 14.6).

Table 14.4 List of important species and varieties of the genus *Capsicum*

Genus and species	Important taxonomic varieties	Features
C. annuum	Anaheim (New Mexico chili pepper)	Mature fruits are red in color and are then called as Colorado
	Bird's eye chili	Fruits are thin and ends are pointed. Also called as Thia chili
	Black heart	An Australian variety with purple flowers and heart-shaped fruits which are red in color
	Cascabel	A Mexican variety with rounded, dark colored fruits with a nutty flavor
	Cayenne	Widely used in India and China as dried fruits grounded in powder form. Fruits are red with curved tip
	Chiltepin	Fruits are very small and oval in shape. Believed to be one of the ancient varieties of chili pepper.
	Friggitelli	Italian chili pepper with a sweet taste
	Jalapeno	Popular variety from United States and 5–10 cm long with round tip.
C. baccatum	Christmas bell	Fruits have a distinct 3D shape similar to hat of a bishop
	Lemon drop pepper	Fruits are green to yellow with a citrus flavor
C. chinensis	Aji dulce	Fruits are yellowish in color with milder taste
	Bhut jolokia	An interspecific hybrid and one of the hottest chili peppers, cultivated in North-East India
	Carolina Reaper	World's hottest chili pepper fruits are bright red with pointed small tip.
	Hainan yellow lantern chili	Mainly cultivated in China; fruits are bright yellow
C. frutescens	Siling labuyo	Fruits are small, red in color, triangular or cone shaped
	Kambuzi	Fruits are small, oval in shape
	African birdseye	Fruits are very hot and red in color
C. pubescens	Canario	Fruits are bigger and deep yellow in color
	Rocoto	Fruits are round and bright red in color

14.6 Capsaicin and Immunomodulatory Effect

The most popular bioactive compound isolated from *Capsicum* fruits is capsaicin. Capsaicin is the main ingredient of capsicum fruit and has potential immunomodulatory effect (Schnekenburger et al. 2019). Capsicum extract as well as capsaicin administration is reported to regulate the T cell immune responses and their immunomodulatory effects (Beltran et al. 2007). These immunomodulatory effects are either transient receptor potential channel vanilloid subfamily member 1 (TRPV1; a Ca2+ permeable ion channel) dependent or independent as close relationship has been demonstrated between the peripheral nervous system and immunological diseases (Stüve and Zettl 2014). TRPV family members exhibit innate and adaptive immune cell-mediated pro-inflammatory as well as anti-inflammatory properties

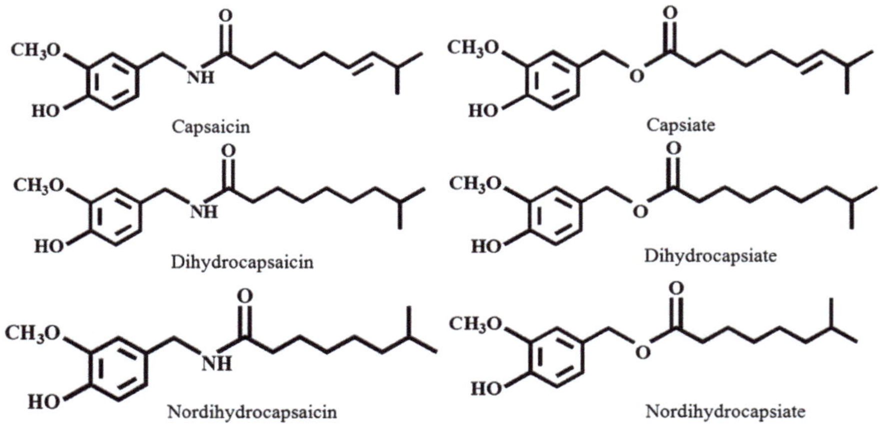

Fig. 14.5 Major capsaicinoids and capsinoids characterized from *Capsicum* species

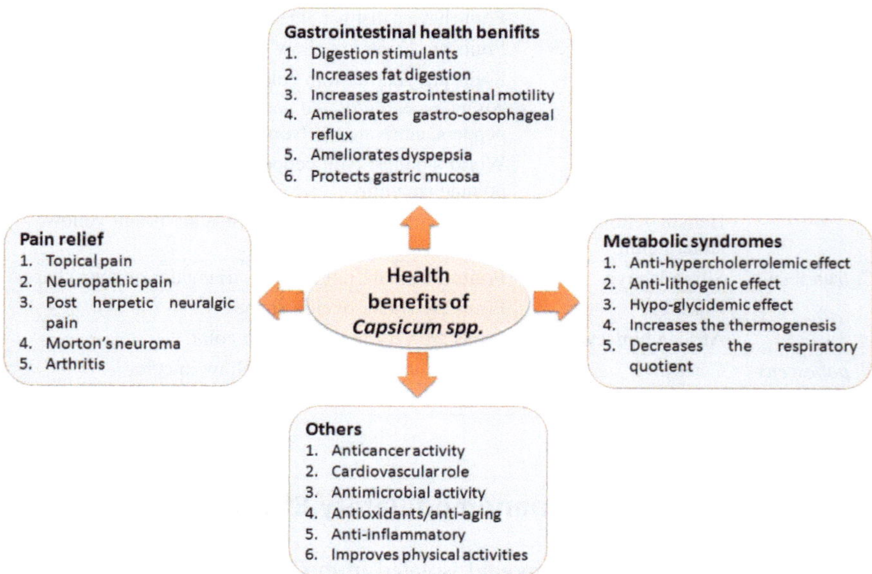

Fig. 14.6 Health benefits of regular consumption of *Capsicum* spp.

(Majhi et al. 2015) such as induction of dendritic cells (Basu and Srivastava 2005), neurogenic inflammation through the activation of mast cells (Biró et al. 1998), and suppression of T cell-mediated tumorigenesis. Activation of TRPV1 releases neuropeptides (involved in neurogenic inflammation) including pro-inflammatory peptides such as substance P, neurokinin A, and calcitonin gene-related peptide 1 (CGRP1), as well as the anti-inflammatory peptide somatostatin (Yoo et al. 2014).

Inactivation of capsaicin-sensitive sensory neurons in a mouse model of auto-antibody-induced arthritis resulted in severe arthritis and led to increased swelling of the joints, increased activity of matrix metalloproteinases and neutrophil-derived myeloperoxidase, increased ROS production and inflammatory cell accumulation, as well as histopathological alterations and decreased hyperalgesia in the late phase of arthritis (Borbély et al. 2015). Capsaicin influences inflammation by suppressing T cell activation and the progress of cell division to S phase (Sancho et al. 2002). In an in vivo animal experiment, Spiller et al. (2008) have reported that pre-treatment with red pepper juice adversely affected WBC migration, exudate volume, and lactate dehydrogenase (LDH) level in pleural fluid. It is assumed that anti-inflammatory effect of capsaicin is narrowly connected with production of inflammatory cytokines (Sancho et al. 2002). Capsicum extract (1 and 10 μg/ml) and capsaicin (3 and 30 μM) administration suppress the production of interferon (IFN)-gamma, interleukin (IL)-2, IL-4, and IL-5 in cultured murine Peyer's patch (PP) cells in vitro, while in an ex vivo experiment, IL-2, IFN gamma, and IL-5 were found induced in response to concanavalin A (Con A) on removed Peyer's patch (PP) cells from the mice after oral administration of capsicum extract (10 mg/kg/day) for 4 consecutive days. Dendritic cells have the receptor for capsaicin leading to powerful immune consequences (Oh et al. 2013). Capsaicin does inhibit the lipopolysaccharide (LPS)-induced proinflammatory cytokine production such as interleukin (IL)-1b, IL-6, and tumor necrosis factor (TNF)-a (Tang et al. 2015). Small doses of capsaicin cause reduction in the levels of pro-inflammatory cytokines TNF-a and IL-6 and induction of the anti-inflammatory IL-10 in blood plasma and attenuate the inflammatory responses (Demirbilek et al. 2004). Oral administration of capsaicin protected mice from developing type 1 diabetes mellitus by enhancing a discrete population of macrophages and attenuating the proliferation and activation of autoreactive T cells in pancreatic lymph nodes (Nevius et al. 2012). Consumption of capsaicin might attenuate immune hyperactivity. Capsaicin exerts an immunomodulatory and anti-oxidative effect on Schwann cells (SCs) by increasing the calcitonin gene-related peptide (CGRP) expression after capsaicin treatment and resulted in anti-inflammatory effect (Grüter et al. 2020). Capsaicin reduced major histocompatibility complex class II (MHC II) presentation after INF-γ stimulation and downregulated toll-like receptor 4 (TLR4) and intercellular adhesion molecule 1 (ICAM-1) expression (Grüter et al. 2020). ICAM-1 is a membrane glycoprotein initiating migration of leukocytes during inflammation.

It is a universally accepted fact that stress alters the immune system, causes inflammation, and increases the chance for infectious diseases. Chronic stress induces the inflammation as both the inflammatory markers, C-reactive protein (CRP) and myeloperoxidase (MPO), have showed a significant increase while the changes were not significant in capsaicin-treated groups. Exposure of chronic stress decreases the percentage of CD4 and CD8 cell level in blood (Marquardt et al. 2015). Chronic stress causes suppression of the natural killer cell activity and proliferation of mitogen-stimulated lymphocyte. The treatment of capsaicin in chronic stress-induced groups increases the concentration of CD4 and CD8 cells. Free radicals induce all the transcription factors related to inflammation such as nuclear factor kappa B, activator protein 1, NF-E2 related factor, etc. Capsaicin suppresses the TPA-induced activation of NF-kB (Manjunatha and Srinivasan 2006). Capsaicin inhibits the synthesis of inflammatory markers such as COX, LOX, and iNOS through regulating the signaling pathways such as JAK-STAT and MAP kinase (Savitha et al. 2015). The inflammation induced by carrageenan and iron-induced hepatotoxicity was found lowered by capsaicin administration (Saliha and Avadhany 2016).

Capsaicin influences tumor growth, through a number of pathways including stimulated apoptosis and increased number of reactive oxygen species, inhibited cell migration, intracellular calcium increase, and upregulation of P53 (Clark et al. 2015; Dıaz-Laviada and Rodrıguez-Henche 2014). The anticancer activity of capsaicin was expressed by increased TNF-α, IL-6, cyclooxygenase-2, and nuclear factor-kappa B in mice with benzo(a)pyrene-induced lung tumors (Anandakumar et al. 2012). Capsaicin caused a dose-dependent manner inhibition of TNF-α production by LPS-treated murine macrophages showing its anti-inflammatory properties (Park et al. 2004). Capsaicin causes inhibition of colon cancer cell proliferation in a dose-dependent manner but had no effect on peripheral blood mononuclear cell (PBMC) viability. Lower concentrations (10–100 μM) of capsaicin causes the secretion of TNF-α, IL-1β, IFN-γ, and IL-10, except IL-1ra. It also stimulated IL-6 (Bessler and Djaldetti 2017). Thus capsaicin in a concentration-dependent manner causes alteration of the immune balance between peripheral blood mononuclear cells (PBMC) and colon carcinoma cells (Bessler and Djaldetti 2017).

Capsaicin could target tumor-associated macrophages (TAMs) which leads to tumor-specific T cell-dependent apoptosis and destruction of tumor-associated stromal cells, causing the depletion of immunosuppressive host cell population at the tumor site and changing the cytokine profile (Ghosh and Basu 2012). The administrations of capsaicin causes increase of the CD8+ cytotoxic T cell (CTL) population and decrease of the numbers of Tregs and prevent the tumors escape from the body's immune system (Deng et al. 2020).

14.6.1 Other Genera of Family Solanaceae

Many other genera of the family *Solanaceae* also have many health benefits and are used either as nutritive supplements or as drugs for various medical conditions. Many species of the genus *Physalis* are well known for their tremendous health benefits due to being rich in many vitamins and minerals. Presence of sufficient quantities of potassium and vitamin C makes it a good supplement for the immune system. Similar to genus *Withania*, *Physalis* species are also known to biosynthesize withanolides. These withanolides from *Physalis* are known to possess anti-inflammatory and immunomodulatory activities. In addition, they have been reported to show significant anticancer potential (Shenstone et al. 2020).

Another species of family Solanaceae, *Datura stramonium*, has great medicinal values. The major bioactive compounds of *D. stramonium* majorly include tropane alkaloids such as hyoscyamine, scopolamine, tigloidine, and apoatropine and many minor alkaloids (Soni et al. 2012). To relieve pain of gout and rheumatism, *D. stramonium* leaf infusions are used. Due to their analgesic properties, they are also used to relieve pain of wounds.

Family Solanaceae is one of the most economically important families and includes a number of medicinal plants. Most of them are loaded with immunomodulatory properties. The bioactive compounds such as withanolides, capsaicinoids, or carotenoids are well-known immunomodulators and are in use since ancient times. Nowadays, these are used as purified compounds during treatment of many autoimmune, inflammatory, or cancer diseases. However, their use in daily meal is still popular and recommended for improvement of general health and boosting the immune system of the body.

Although these solanaceous plants are used since ancient times, their roles and exact mechanism of action are still not fully understandable. There is a need to get complete mechanism of action of withanolides, carotenoids, and capsaicinoids in order to make better use of these compounds as drugs.

Acknowledgements NSS gratefully acknowledge the financial support by DBT, DST, UGC and NMITLI-CSIR (Council of Scientific and Industrial Research), New Delhi, for grants of the projects at her lab, and fellowships to the young researchers who have contributed in making progress in projects.

References

Agarwal R, Diwanay S, Patki P, Patwardhan B (1999) Studies on immunomodulatory activity of *Withania somnifera* (Ashwagandha) extracts in experimental immune inflammation. J Ethnopharmacol 67(1):27–35

Ahmad H, Arya A, Agrawal S, Samuel SS, Singh SK, Valicherla GR, Sangwan N, Mitra K, Gayen JR, Paliwal S, Shukla R, Dwivedi AK (2016a) Phospholipid complexation of NMITLI118RT+: way to a prudent therapeutic approach for beneficial outcomes in ischemic stroke in rats. Drug Deliv 23(9):3606–3618. https://doi.org/10.1080/10717544.2016.1212950

Ahmad H, Samuel SS, Khandelwal K, Arya A, Tripathi S, Agrawal S, Sangwan NS, Shukla R, Dwivedi AK (2016b) Enduring protection provided by NMITLI118RT+ and its preparation NMITLI118RT+ CFM against ischemia/reperfusion injury in rats. RSC Adv 6:42827–42835

Alvarez L, del Carmen Pérez M, González JL, Navarro V, Villarreal ML, Olson JO (2001) SC-1, an antimycotic spirostan saponin from *Solanum chrysotrichum*. Planta Médica 67(04):372–374

Anandakumar P, Kamaraj S, Jagan S, Ramakrishnan G, Asok Kumar S et al (2012) Capsaicin inhibits benzo(a)pyrene induced lung carcinogenesis in an in vivo mouse model. Inflamm Res 61:1169–1175

Anosike CA, Obidoa O, Ezeanyika LU (2012) Membrane stabilization as a mechanism of the anti-inflammatory activity of methanol extract of garden egg (Solanum aethiopicum). DARU J Pharm Sci 20(1):76

Bakuridze AD, Kurtsikidze S, Pisarev VM, Makharadze RV, Berashvili DT (1993) Immunomodulators of plant origin. Pharm Chem J 27:589–595

Baruah S, Md ZK, Rajbongshi P, Das S (2014) A review on recent researches on *Bhut jolokia* and pharmacological activity of capsaicin. Int J Pharm Sci Rev Res 24:89–94

Basu S, Srivastava P (2005) Immunological role of neuronal receptor vanilloid receptor 1 expressed on dendritic cells. Proc Natl Acad Sci U S A 102:5120–5125

Baudouin V, Crusiaux A, Haddad E, Schandene L, Goldman M, Loirat C, Abramowicz D (2003) Anaphylactic shock caused by immunoglobulin E sensitization after retreatment with the chimeric anti–interleukin-2 receptor monoclonal antibody basiliximab. Transplantation 76(3): 459–463

Beltran J, Amiya K, Ghosh, Basu S (2007) Immunotherapy of Tumors with Neuroimmune ligand capsaicin. J Immunol 178:3260–3264

Bessler H, Djaldetti M (2017) Capsaicin modulates the immune cross talk between human Mononuclears and cells from two colon carcinoma lines. Nutr Cancer 69:14–20

Bharti VK, Malik JK, Gupta RC (2016) Ashwagandha: multiple health benefits. Nutraceuticals 52: 717–733

Bhattacharya SK, Muruganandam AV (2003) Adaptogenic activity of Withania somnifera: an experimental study using a rat model of chronic stress. Pharmacol Biochem Behavior 75(3): 547–555. https://doi.org/10.1016/S0091-3057(03)00110-2

Biró T et al (1998) Characterization of functional vanilloid receptors expressed by mast cells. Blood 91:1332–1340

Böhm V, Puspitasari-Nienaber NL, Ferruzzi MG, Schwartz SJ (2002) Trolox equivalent antioxidant capacity of different geometrical isomers of α-carotene, β-carotene, lycopene, and zeaxanthin. J Agric Food Chem 50(1):221–226

Borbély E et al (2015) Capsaicin-sensitive sensory nerves exert complex regulatory functions in the serum transfer mouse model of autoimmune arthritis. Brain Behav Immun 45:50–59

Budhiraja RD, Sudhir S, Garg KN (1984) Antiinflammatory activity of 3 β-Hydroxy-2, 3-dihydro-withanolide F. Planta Med 50(02):134–136

Camara M, Cortes M, Mata S, Ruiz VF, Camara RA, Manzoor S et al (2013) Lycopene: a review of chemical and biological activity related to beneficial health effects. Stud Nat Chem 40:383–423

Chandrashekara KD, Dharmesh SM (2014) Effect of various cooking processes on antioxidants of potato (Solanum tuberosum). J Pharm Res 8:1148–1157

Chang HC, Chang FR, Wang YC et al (2007) A bioactive withanolide Tubocapsanolide A inhibits proliferation of human lung cancer cells via repressing Skp2 expression. Mol Cancer Ther 6(5): 1572–1578

Chang VHS, Chiu TH, Fu SC (2016) In vitro anti-inflammatory properties of fermented pepino (Solanum muricatum) milk by γ-aminobutyric acid-producing Lactobacillus brevis and an in vivo animal model for evaluating its effects on hypertension. J Sci Food Agric 96(1):192–198

Chen BW, Chen YY, Lin YC, Huang CY, Uvarani C, Hwang TL, Sheu JH (2015) Capsisteroids A–F, withanolides from the leaves of Solanum capsicoides. RSC Adv 5(108):88841–88847

Chiang HC, Tseng TH, Wang CJ, Chen CF, Kan WS (1991) Apoptosis induced by dioscin in hela cells. Anticancer Res 11(5):1911–1917

Chikhale RV, Gurav SS, Patil RB, Sinha SK, Prasad SK, Shakya A, Prasad RS (2020) Sars-cov-2 host entry and replication inhibitors from Indian ginseng: an in-silico approach. J Biomol Struct Dyn:1–12

Chou SC, Huang TJ, Lin EH, Huang CH, Chou CH (2012) Antihepatitis B virus constituents of Solanum erianthum. Nat Prod Commun 7(2):1934578X1200700205

Choudhary MI, Yousuf S, Nawaz SA et al (2004) Cholinesterase inhibiting withanolides from Withania somnifera. Chem Pharm Bull 52(11):1358–1361

Clark R, Lee J, Lee SH (2015) Synergistic anticancer activity of capsaicin and 3,3-′-diindolylmethane in human colorectal cancer. J Agric Food Chem 63:4297–4304

Cristian D, Gigi C, Mihaela C, Petru A, Stefan D (2013) Antioxidant and antibacterial properties of capsaicine microemulsions. The annals of the University Dunarea de Jos of Galati fascicle VI. Food Technol 37:39–49

Davis L, Kuttan G (2000) Immunomodulatory activity of Withania somnifera. J Ethnopharmacol 71(1–2):193–200

de Jesús Ornelas-Paz J, Martínez-Burrola JM, Ruiz-Cruz S, Santana-Rodríguez V, Ibarra-Junquera-V, Olivas GI, Pérez-Martínez JD (2010) Effect of cooking on the capsaicinoids and phenolics contents of Mexican peppers. Food Chem 119:1619–1625

Demirbilek S, Ersoy MO, Demirbilek S, Karaman A, Gurbuz N et al (2004) Small-dose capsaicin reduces systemic inflammatory responses in septic rats. Anesth Analg 99:1501–1507

Deng L-J, Qi M, Li N, Yu-He L, Zhang D-M, Chen J-X (2020) Natural products and their derivatives: promising modulators of tumor immunotherapy. J Leukoc Biol 108:493–508

Di Mascio P, Wefers H, Do-Thi H-P, Lafleur MVM, Sies H (1989) Singlet molecular oxygen causes loss of biological activity in plasmid and bacteriophage DNA and induces single strand breaks. Biochim Biophys Acta 1007:151–157

Dıaz-Laviada I, Rodrıguez-Henche N (2014) The potential antitumor effects of capsaicin. Prog Drug Res 68:181–208

Feki H, Koubaa I, Jaber H, Makni J, Damak M (2013) Characteristics and chemical composition of Solanum elaeagnifolium seed oil. J Eng Appl Sci (Asian Res Publ Netw) 8(9):708–712

Ghosh AK, Basu S (2012) Tumor macrophages as a target for capsaicin mediated immunotherapy. Cancer Lett 324:91–97

Glotter E, Abraham A, Guenzberg IK (1977) Naturally occurring steroidal lactones with 17 a-oriented side chain. Structure of Withanolide E & related compounds. J Chem Soc Perkins Trans 1:341–343

Glotter E, Kirson I, Abraham A et al (1973) Constituents of Withania somnifera (Dunal) XIII—the withanolides of chemotype III. Tetrahedron 29:1353–1364

Golan DE (2008) Principles of pharmacology. In: Pennsylvania PA (ed) The pathophysiologic basic of drug therapy, 2nd edn. Lippincott Williams & Wilkins, pp 795–809

Govindan S, Viswanathan S, Vijayasekaran V, Alagappan R (1999) A pilot study on the clinical efficacy of solanum xanthocarpum and Solanum trilobatum in bronchial asthma. J Ethnopharmacol 66(2):205–210

Grabowska M, Wawrzyniak D, Rolle K, Chomczyński P, Oziewicz S, Jurga S, Barciszewski J (2019) Let food be your medicine: nutraceutical properties of lycopene. Food Funct 10(6): 3090–3102. https://doi.org/10.1039/c9fo00580c

Grüter T, Blusch A, Motte J et al (2020) Immunomodulatory and anti-oxidative effect of the direct TRPV1 receptor agonist capsaicin on Schwann cells. J Neuroinflammation 17:145

Herraiz FJ, Villaño D, Plazas M, Vilanova S, Ferreres F, Prohens J, Moreno DA (2016) Phenolic profile and biological activities of the Pepino (Solanum muricatum) fruit and its wild relative S. caripense. Int J Mol Sci 17(3):394

Huang CF, Ma L, Sun LJ, Ali M, Arfan M, Liu JW, Hu LH (2009) Immunosuppressive withanolides from Withania coagulans. Chem Biodivers 6(9):1415–1426

Jain S (2008) Handbook of pharmacology. Pars Publication, London

Jandú JJ, Moraes Neto RN, Zagmignan A, de Sousa EM, Brelaz-de-Castro MC, dos Santos Correia MT et al (2017) Targeting the immune system with plant lectins to combat microbial infections. Front Pharmacol 8:671

Jantan I, Ahmad W, Bukhari SNA (2015) Plant-derived immunomodulators: an insight on their preclinical evaluation and clinical trials. Front Plant Sci 6:655

Jayaprakasam B, Nair MG (2003) Cyclooxygenase-2 enzyme inhibitory withanolides from Withania somnifera leaves. Tetrahedron 59(6):841–849

Jayaprakasam B, Zhang Y, Seeram NP et al (2003) Growth inhibition of human tumor cell lines by withanolides from Withania somnifera leaves. Life Sci 74(1):125–132

John J (2014) Therapeutic potential of Withania somnifera: a report on phyto-pharmacological properties. IJPSR 5(6):2131

Kaunda JS, Zhang YJ (2019) The genus solanum: an ethnopharmacological, phytochemical and biological properties review. Nat Prod Bioprospect 9(2):77–137

Keane J, Gershon S, Wise RP, Mirabile-Levens E, Kasznica J, Schwieterman WD, Braun MM (2001) Tuberculosis associated with infliximab, a tumor necrosis factor α–neutralizing agent. N Engl J Med 345(15):1098–1104

Khachik F, Pfander H, Traber B (1998) Proposed mechanisms for the formation of synthetic and naturally occurring metabolites of lycopene in tomato products and human serum. J Agric Food Chem 46(12):4885–4890

Khanal P, Chikhale R, Dey YN, Pasha I, Chand S, Gurav N, Ayyanar M, Patil BM, Gurav S (2021) Withanolides from *Withania somnifera* as an immunity booster and their therapeutic options against COVID-19. J Biomol Struct Dyn 18:1–14. https://doi.org/10.1080/07391102.2020.1869588

Khedgikar V, Ahmad N, Kushwaha P, Gautam J, Nagar GK, Singh D, Trivedi PK, Mishra PR, Sangwan NS, Trivedi R (2014) Preventive effects of Withaferin A isolated from the leaves of an Indian Medicinal Plant *Withania somnifera* (L.): comparisons with 17-β-Estradiol and alendronate. Nutrition. https://doi.org/10.1016/j.nut.2014.05.010

Khedgikar V, Kushwaha P, Gautam J, Verma A, Changkija B, Kumar A, Sharma S, Nagar GK, Singh D, Trivedi PK, Sangwan NS, Mishra PK, Trivedi R (2013) Withaferin a: a proteasomal inhibitor promotes healing after injury and exerts anabolic effect on osteoporotic bone. Cell Death Dis 4:e778

Kim YC, Che QM, Gunatilaka AL, Kingston DG (1996) Bioactive steroidal alkaloids from Solanum umbelliferum. J Nat Prod 59(3):283–285

Kirson I, Glotter E, Lavis D et al (1971) Constituents of Withania somnifera Dunal XII. The withanolides of an Indian Chemotype. J Chem Soc (org) 52:2032–2044

Koduru S, Jimoh FO, Grierson DS, Afolayan AJ (2007) Antioxidant activity of two steroid alkaloids extracted from Solanum aculeastrum. J Pharmacol Toxicol 2:160–167

Kumar V, Dhanjal JK, Kaul SC, Wadhwa R, Sundar D (2020) Withanone and caffeic acid phenethyl ester are predicted to interact with main protease (Mpro) of SARS-CoV-2 and inhibit its activity. J Biomol Struct Dyn 39:1–17

Kushwaha AK, Sangwan NS, Tripathi S, Sangwan RS (2013) Molecular cloning and catalytic characterization of a recombinant tropine biosynthetic tropinone reductase from Withania coagulans leaf. Gene 516(2):238–247. https://doi.org/10.1016/j.gene.2012.11.091

Lee A, Thurnham DI, Chopra M (2000) Consumption of tomato products with olive oil but not sunflower oil increases the antioxidant activity of plasma. Free Radic Biol Med 29(10): 1051–1055

Leyon PV, Kuttan G (2004) Effect of Withania somnifera on B16F-10 melanoma induced metastasis in mice. Phytother Res 18(2):118–122

Lin HM, Tseng HC, Wang CJ, Lin JJ, Lo CW, Chou FP (2008) Hepatoprotective effects of Solanum nigrum Linn extract against CCl4-iduced oxidative damage in rats. Chem Biol Interact 171(3):283–293

Liu X, Lin X, Zhang S, Guo C, Li J, Mi Y, Zhang C (2018) Lycopene ameliorates oxidative stress in the aging chicken ovary via activation of Nrf2/HO-1 pathway. Aging (Albany NY) 10(8):2016

Majhi RK et al (2015) Functional expression of TRPV channels in T cells and their implications in immune regulation. FEBS J 282:2661–2681

Malik F, Singh J, Khajuria A et al (2007) A standardized root extract of Withania somnifera and its major constituent withanolide-A elicit humoral and cell-mediated immune responses by up regulation of Th1-dominant polarization in BALB/c mice. Life Sci 80(16):1525–1538

Manjunatha H, Srinivasan K (2006) Protective effect of dietary curcumin and capsaicin on induced oxidation of low-density lipoprotein, iron-induced hepatotoxicity and carrageenan induced inflammation in experimental rats. FEBS J 273:4528–4537

Marquardt JU, Gomez-Quiroz L, Camacho LO, Pinna F, Lee YH, Kitade M et al (2015) Curcumin effectively inhibits oncogenic NF-κB signaling and restrains stemness features in liver cancer. J Hepatol 63:661–669

Maurya R (2010) Chemistry and pharmacology of Withania coagulans: an Ayurvedic remedy. J Pharm Pharmacol 62(2):153–160. https://doi.org/10.1211/jpp.62.02.0001

Mesaik MA, Ul-Haq Z, Murad S, Ismail Z, Abdullah NR, Gill HK, Atta-ur-Rahman, Yousaf M, Siddiqui RA, Ahmad A, Choudhary MI (2006) Mol Immunol 43:1855

Miranda MA, Magalhães LG, Tiossi RFJ, Kuehn CC, Oliveira LGR, Rodrigues V et al (2012) Evaluation of the schistosomicidal activity of the steroidal alkaloids from Solanum lycocarpum fruits. Parasitol Res 111(1):257–262

Morales-Soto A, Gómez-Caravaca AM, García-Salas P, Segura-Carretero A, Fernández-Gutiérrez A (2013) High-performance liquid chromatography coupled to diode array and electrospray time-of-flight mass spectrometry detectors for a comprehensive characterization of phenolic and other polar compounds in three pepper (*Capsicum annuum* L.) samples. Food Res Int 51:977–984

Moreira RR, Martins GZ, Magalhaes NO, Almeida AE, Pietro RC, Silva FA, Cicarelli R (2013) In vitro trypanocidal activity of solamargine and extracts from Solanum palinacanthum and Solanum lycocarpum of Brazilian Cerrado. An Acad Bras Cienc 85(3):903–907

Munari CC, de Oliveira PF, Campos JCL, Martins SDPL, Da Costa JC, Bastos JK, Tavares DC (2014) Antiproliferative activity of solanum lycocarpumalkaloidic extract and their constituents, solamargine and solasonine, in tumor cell lines. J Nat Med 68(1):236–241

Nadkarni AK (1954) Indian materia medica, 3rd edn. Bombay Popular Prakashan, Mumbai, pp 1284–1286

Nagarathna PKM, Reena K, Sriram R, Wesley J (2013) Review on immunomodulation and immunomodulatory activity of some herbal plants. Int J Pharm Sci Rev Res 22(1):223–230

Nevius E, Srivastava PK, Basu S (2012) Oral ingestion of capsaicin, the pungent component of chili pepper, enhances a discreet population of macrophages and confers protection from autoimmune diabetes. Mucosal Immunol 5:76–86

Nwanna EE, Ibukun EO, Oboh G (2013) Inhibitory effects of methanolic extracts of two eggplant species from South-Western Nigeria on starch hydrolysing enzymes linked to type-2 diabetes. Afr J Pharm Pharmacol 7(23):1575–1584

Oh J, Hristov AN, Lee C, Cassidy T, Heyler K, Varga GA et al (2013) Immune and production responses of dairy cows to postruminal supplementation with phytonutrients. J Dairy Sci 96(12):7830–7843

Park JY, Kawada T, .Han IS, Kim BS, Goto T, et al.: Capsaicin inhibits the production of tumor necrosis factor alpha by LPS-stimulated murine macrophages, RAW 264.7: a PPARgamma ligand-like action as a novel mechanism. FEBS Lett 572, 266–270, 2004

Parkin J, Cohen B (2001) An overview of the immune system. Lancet 357:1777–1789. https://doi.org/10.1016/S0140-6736(00)04904-7

Patel CN, Kumar SP, Pandya HA, Rawal RM (2020) Identification of potential inhibitors of coronavirus hemagglutinin-esterase using molecular docking, molecular dynamics simulation and binding free energy calculation. Mol Divers 29:1–13. https://doi.org/10.1007/s11030-020-10135-w

Peng SY, Li H, Yang DP, Bai B, Zhu LP, Liu Q et al (2017) Solanerioside a, an unusual 14, 15-dinor-cyclophytane glucoside from the leaves of Solanum erianthum. Nat Prod Res 31(7):810–816

Petreanu M, Guimarães ÁAA, Broering MF, Ferreira EK, Machado ID, Gois ALT, Santin JR (2016) Antiproliferative and toxicological properties of methanolic extract obtained from Solanum capsicoides all. Seeds and carpesterol. Naunyn Schmiedeberg's Arch Pharmacol 389(10):1123–1131

Prakash UNS, Srinivasan K (2012) Fat digestion and absorption in spice pretreated rats. J Sci Food Agric 92:503–510

Priyadharshini SD, Sujatha V (2013) Antioxidant profile and GC-MS analysis of Solanum erianthum leaves and stem-A comparison. Int J Pharm Pharm Sci 5(3):652–658

Raju GS, Moghal MR, Dewan SMR, Amin MN, Billah M (2013) Characterization of phytoconstituents and evaluation of total phenolic content, anthelmintic, and antimicrobial activities of Solanum violaceum Ortega. Avicenna J Phytomed 3(4):313

Ren Y, Shen L, Zhang DW, Dai SJ (2009) Two new sesquiterpenoids from Solanum lyratum with cytotoxic activities. Chem Pharm Bull 57(4):408–410

Rosangkima G, Jagetia GC (2015) In vitro anticancer screening of medicinal plants of Mizoram state, India, against Dalton's Lymphoma, MCF-7 and HeLa cells. Int J Recent Sci Res 6(8): 5648–5653

Sabir F, Sangwan RS, Chaurasiya ND, Misra LN, Sangwan NS (2008) In vitro withanolides production by Withania somnifera Dunal cultures. Z fur Naturfor 63c:409–412
Saliha CK, Avadhany ST (2016) Comparative study of antioxidants: curcumin and capsaicin on stress induced rats and its effects on liver. Am J Pharmtech Res 6:1–19
Sanati S, Razavi BM, Hosseinzadeh H (2018) A review of the effects of *Capsicum annuum* L. and its constituent, capsaicin, in metabolic syndrome. Iran J Basic Med Sci 21:439–448
Sancho R, Lucena C, Macho A, Calzado MA, BlancoMolina M et al (2002) Immunosuppressive activity of capsaicinoids: capsiate derived from sweet peppers inhibits NFkappaB activation and is a potent antiinflammatory compound in vivo. Eur J Immunol 32:1753–1763
Sangwan NS, Tripathi S, Srivastava Y, Mishra B, Pandey N (2017) Phytochemical genomics of ashwagandha. In: Science of ashwagandha: preventive and therapeutic potentials. Edi: A Kaul, Wadhwa R, Springer, Cham, pp 3–36
Savitha D, Mani I, Ravikumar G, Avadhany ST (2015) Effect of curcumin in experimental peritonitis. Indian J Surg 77:502–507
Schnekenburger M, Dicato M, Diederich MF (2019) Anticancer potential of naturally occurring immunoepigenetic modulators: a promising avenue. Cancer-Am Cancer Soc 125:1612–1628
Schulze-Koops H, Burkhardt H, Kalden JR (1999) What we have learned from trials of immunomodulatory agents in rheumatoid arthritis: future directions. Drugs Today (Barc) 35:327351
Seth C, Mas C, Conod A et al (2016) Long-lasting WNT-TCF response blocking and epigenetic modifying activities of Withanolide F in human cancer cells. PLoS One 11(12):e0168170
Shahjahan M, Vani G, Shyamaladevi CS (2005) Effect of Solanum trilobatum on the antioxidant status during diethyl nitrosamine induced and phenobarbital promoted hepatocarcinogenesis in rat. Chem Biol Interact 156(2–3):113–123
Shenstone E, Lippman Z, & Van Eck J (2020) A review of nutritional properties and health benefits of Physalis species. *Plant foods for human nutrition (Dordrecht, Netherlands)*
Shohat B, Kirson I, Lavie D (1978) Immunosuppressive activity of two plant steroidal lactones withaferin A and withanolide E. Biomedicine 28(1):18–24
Sidique AA, Joshi P, Misra LN, Sangwan Neelam S, Darokar MP (2014) 5,6-De-epoxy 5-en-7-one-17-hydroxy withaferin a, a new cytotoxic steroid from *Withania somnifera* L. Dunal leaves. Nat Prod Res. https://doi.org/10.1080/14786419.2013.871545
Singh AP, Luthria D, Wilson T, Vorsa N, Singh V, Banuelos GS, Pasakdee S (2009) Polyphenols content and antioxidant capacity of eggplant pulp. Food Chem 114(3):955–961
Soni P, Siddiqui AA, Dwivedi J, Soni V (2012) Pharmacological properties of Datura stramonium L. as a potential medicinal tree: an overview. Asian Pac J Trop Biomed 2(12):1002–1008
Spiller F, Alves MK, Vieira SM, Carvalho TA, Leite CE et al (2008) Anti-inflammatory effects of red pepper (*Capsicum baccatum*) on carrageenan- and antigen-induced inflammation. J Pharm Pharmacol 60:473–478
Srinivasan K (2013) Biological activities of pepper alkaloids. In: Natural products. Springer, Berlin, pp 1397–1437
Straughn AR, Kakar SS (2020) Withaferin A: a potential therapeutic agent against COVID-19 infection. J Ovarian Res 13(1):1–5
Strayer DR, Carter WA (2012) Recombinant and natural human interferons: analysis of the incidence and clinical impact of neutralizing antibodies. J Interf Cytokine Res 32(3):95–102
Stüve O, Zettl U (2014) Neuroinflammation of the central and peripheral nervous system: an update. Clin Exp Immunol 175:333–335
Sundar S, Pillaiy JK (2016) Phytochemical screening and gas chromatograph-mass spectrometer profiling in the leaves of Solanum incanum. L As J Pharm Clin Res 3:179–188
Tang J, Luo K, Li Y, Chen Q, Tang D et al (2015) Capsaicin attenuates LPS-induced inflammatory cytokine production by upregulation of LXRa. Int Immunopharmacol 28:264–269
Tohge T, Alseekh S, Fernie AR (2014) On the regulation andfunction of secondary metabolism during fruit development andripening. J Exp Bot 65:4599–4611

Tripathi MK, Singh P, Sharma S, Singh TP, Ethayathulla AS, Kaur P (2020) Identification of bioactive molecule from Withania somnifera (Ashwagandha) as SARS-CoV-2 main protease inhibitor. J Biomol Struct Dyn 39:1–14

Tuli R, Sangwan RS, Kumar S, Bhattacharya S, Misra LN, Mandal C, Raghubir R, Trivedi PK, Tewari SK, Mishra P, Chaturvedi P, Sangwan Neelam S, et al (2009) Ashwagandha (*Withania somnifera*) A model Indian Medicinal Plant. Eds R Tuli and RS Sangwan, Publisher NMITLI-CSIR, New Delhi. Monograph Published by Council of Scientific and Industrial Research (CSIR), Govt. of India, ISBN No.978–93–80235-29-5

Tunon H, Olavsdotter C, Bohlin L (1995) Evaluation of anti-inflammatory activity of some Swedish medicinal plants. Inhibition of prostaglandin biosynthesis and PAF-induced exocytosis. J Ethnopharmacol 48(2):61–76

USDA-ARS (2011) Grin species records of capsicum. National 2063 Germplasm Resources Laboratory, Beltsville, Maryland

Valadares YM, Brandao'a GC, Kroon EG, Filho JDS, Oliveira AB, Braga FC (2009) J Biosci 64: 813–818

Vattem DA, Shetty K (2005) Biological function of ellagic acid: a review. J Food Biochem 29:234–266

Wagner H (1999) Immunomodulatory Agents from Plants. Springer Science & Business Media, 13 Chapters with 365 pages. ISBN 978–3–0348-8763-2

Wanyonyi AW, Chhabra SC, Mkoji G, Njue W, Tarus PK (2003) Molluscicidal and antimicrobial activity of Solanum aculeastrum. Fitoterapia 74(3):298–301

Yao F, Song QL, Zhang L, Li GS, Dai SJ (2013) Three new cytotoxic sesquiterpenoids from Solanum lyratum. Phytochem Lett 6(3):453–456

Yoo S, Lim JY, Hwang SW (2014) Sensory TRP channel interactions with endogenous lipids and their biological outcomes. Molecules 19:4708–4744

Yoshikawa K, Inagaki K, Terashita T, Shishiyama J, Kuo S, Shankel DM (1996) Antimutagenic activity of extracts from Japanese eggplant. Mutat Res/Genet Toxicol 371(1–2):65–71

Yuan LJ, Qin Y, Wang L, Zeng Y, Chang H, Wang J et al (2016) Capsaicin-containing chili improved postprandial hyperglycemia, hyperinsulinemia, and fasting lipid disorders in women with gestational diabetes mellitus and lowered the incidence of large-for-gestational-age newborns. Clin Nutr 35(2):388–393

Zamilpa A, Tortoriello J, Navarro V, Delgado G, Alvarez L (2002) Five new Steroidal Saponins from Solanum chrysotrichum leaves and their antimycotic activity. J Nat Prod 65(12): 1815–1819

Zhao B, Ren B, Guo R, Zhang W, Ma S, Yao Y, Liu X (2017) Supplementation of lycopene attenuates oxidative stress induced neuroinflammation and cognitive impairment via Nrf2/NF-κ B transcriptional pathway. Food Chem Toxicol 109:505–516

Zhao B, Sakurai Y, Shibata K, Kikkawa F, Tomoda Y, Mizukami H (2014) Cytotoxic fatty acid ketodienes from eggplants. Jpn J Food Chem Saf 21(1):42–47

Zhao J, Nakamura N, Hattori M et al (2002) Withanolide derivatives from the roots of Withania somnifera and their neurite outgrowth activities. Chem Pharm Bull 50(6):760–765

Zhou Y, Deng ZS, Cheng F, Dong WJ, Guo ZY, Wang JZ, Zou K (2016) A new Wutaifuranol derivative from solanum cathayanum. Chem Nat Compd 52(5):920–921

Zhuang YW, Wu CE, Zhou JY, Zhao ZM, Liu CL, Shen JY, Liu SL (2018) Solasodine reverses stemness and epithelial-mesenchymal transition in human colorectal cancer. Biochem Biophys Res Commun 505(2):485–491

Ziauddin M, Phansalkar N, Patki P, Diwanay S, Patwardhan B (1996) Studies on the immunomodulatory effects of Ashwagandha. J Ethnopharmacol 50(2):6976

Food and Vegetables as Source of Phytoactives for Immunomodulation

15

Jyoti Singh Jadaun, Manisha Chownk, Subir Kumar Bose, Swati Kumari, and Neelam S. Sangwan

Abstract

Immunomodulation deals with the alteration in immune responses under certain circumstances, and this is led by key molecules known as immunomodulators. Immunomodulators may be immunostimulators (involved in increase in response of immune system) or immunosuppressive (decreases the response of the immune system). It is of utmost importance that the function of the immune system is regulated to protect the human body against diverse ailments such as viral, cardiovascular, cancer and chronic inflammatory diseases. Recent studies reveal that certain foods have immunomodulatory properties that aid in the prevention of infection, and therefore, addition of diverse food in regular diet provides a wide spectrum of immunity. The most active constituents of food are vitamins,

Jyoti Singh Jadaun and Manisha Chownk contributed equally to this work

J. S. Jadaun
Department of Botany, Dayanand Girls Postgraduate College, Kanpur, Uttar Pradesh, India

M. Chownk
Department of Biotechnology, Panjab University, Chandigarh, India

S. K. Bose
Agriculture Department, Himalayan Garhwal University, Pauri, Garhwal, Uttarakhand, India

S. Kumari
School of Applied Sciences and Biotechnology, Shoolini University of Biotechnology and Management Sciences, Solan, HP, India

N. S. Sangwan (✉)
School of Interdisciplinary and Applied Sciences, Department of Biochemistry, Central University of Haryana, Mahendergarh, Haryana, India
e-mail: nsangwan@cuh.ac.in

N. S. Sangwan et al. (eds.), *Plants and Phytomolecules for Immunomodulation*,
https://doi.org/10.1007/978-981-16-8117-2_15

minerals, carotenes, flavonoids, etc. which provide health-promoting effects. Several vegetables such as spinach, celery, carrots, broccoli, etc. have been studied intensively for their immunomodulatory activity. Similarly, parsley, thyme, oregano and coriander, which are used as garnishing and as flavouring enhancer, are reported to control cytokine and chemokine responses. Recently, prebiotics, probiotics and various oligosaccharides and some essential oils have also been reported to have incredible immunomodulatory activities associated with their phenolic components. Proteins and protein hydrolysate derived from different foods (soybean, milk, fish, egg, rice, pea, spirulina, etc.) are also being investigated by researchers for their immunomodulatory effects. Here, in this chapter, we provide a detailed analysis of the immunomodulatory effects of different nutritional components of food and vegetables, with the probable mechanism of action.

Keywords

Cytokine · Immunity · Immunomodulation · Foods · Vegetables

15.1 Introduction

Immune response acts as first line of defence against any infection and disease, and it has long been known that abundant factors, including nutrition, stress and sleep, affect status of the immune system (Patel et al. 2012; Song et al. 2019; Gombart et al. 2020). A person with strong immunity is less prone to any disease outbreak, and now it has been proven that nutritional components of food have a major role as immunity booster (Marcos et al. 2003). Therefore, it is mandatory to take healthy diet which is full of nutrimental elements to aid and support the immune system to fight against diverse infections (Beck and Levander 2000). Currently, the whole world is facing a big challenge in the form of coronavirus disease (COVID-19). It is just a viral infection, but the pace of its transmission and virulence forced each country to take extreme precautions like nationwide lockdown and curfew to stop the spreading of this virus. Outbreak of COVID-19 is the perfect example establishing the relationship between the strength of immune system and severity of infection as the virus causes extreme illness in people with low immunity. As of now, without vaccine, a healthy immune system is the only way to fight the COVID-19 infection. Also, the WHO guidance on staying safe from the infection clearly suggests the importance of a healthy diet, which helps to build up strong enabled immune system of the body responding to fight at the time of any disease. A healthy diet provides several types of vitamins and minerals which have the ability to act as immunomodulators. Although a human body itself produces a number of immunomodulators for maintaining the homeostasis, a continuous supply from nutrient-rich food sources builds a stronger immune system. An immunomodulator is a term used for any chemical which has the ability to boost or supress the immune response (Siqueiros-Cendón et al. 2014). These are classified into three categories: immunosuppressant, immunostimulant and immunoadjuvants (Jantan et al. 2015). In clinical practices, importance and utilisation of immunomodulators are increasing rapidly due to their direct involvement in disease prevention. Market value of these immunomodulators has expanded tremendously in the past few years. Global market of immunomodulators was reported to be USD 165.21 billion in 2019, and it is estimated to exceed up to USD 251.69 billion by the year 2027 at 5.4% rate of common annual growth rate (www.reportsanddata.com/report). Most of the immunomodulators are synthetic or semisynthetic which are routinely employed in health and wellness industries, although there are several natural sources which have a very good measure of these compounds and their addition in regular diet assists the immune system to fight against any microbial attack. Fruits and vegetables are one of the most common sources of these compounds, and intake of a wide variety of these food groups is necessary to build a strong immune system. They are rich in antioxidants which provide protection against the cell damage caused by free radicals or assist in T cell development (Lobo et al. 2010; Cohen et al. 2017). Other than fruits and vegetables, dairy products also have an eminent role as potent immunomodulators, and milk is considered one of the healthiest and nutrient-rich foods. Human population is consuming milk and milk products since historic times due to their health-promoting effects, and now it has been proven that nutritional components of milk have immunomodulatory properties (Cross and Gill 2000; Zuurveld et al. 2020) although these properties vary with the source of milk (Hernell 2011). Cow milk has similar composition as breast milk; therefore, using human milk as reference and cow milk as protein source, several infant formulas have been developed (Dipasquale et al. 2020). But it has been reported that formula-fed and breast-fed babies show difference in the cognitive development, where breast-fed babies fared better and hence breast milk is considered an ideal meal for infants (Boquien 2018). Milk has different types of bioactive molecules such as lactenin, iron-binding protein lactoferrin and enzymes such as lysozyme and lacto-peroxidase which play an important role in the defensive action against a multitude of infections (Clare et al. 2003).

Since last decade, use of prebiotic and probiotics has become part of our diet due to their health-promoting effects. Although we are using the probiotics in our diet since ancient human civilisation like curd, pickles, kimchi, kefir, etc., now it has been proven experimentally that addition of probiotic or prebiotic or synbiotics (a combination of both) results in a healthy gut microbiota which provide protection against several infections (Singh et al. 2017; Narnoliya and Jadaun 2018). The immunomodulatory effects of pro- and prebiotics have been credited to the release of cytokines and chemokines, induced by the gut commensal bacteria which may result in the prevention of certain infections (O′ Flaherty et al. 2010; Azad et al. 2018).

Strong immunity is necessary to avoid uncountable health problems which can be maintained by the intake of a variety of fruits, vegetables, dairy and dairy products as a part of a well-balanced diet (Sarma and Khosa 1994). There have been numerous studies conducted by researchers all over the world reporting these effects, and during these dire times where the significance of a healthy diet has never looked so profound, we need to turn our focus to these natural immunomodulators. Therefore, in this chapter we describe in detail about the innumerable immunomodulatory properties of vegetables, dairy, dairy products, and pre- and probiotics and mechanism of action of various nutrients like vitamins and minerals. In addition, we also discuss certain food allergies which are the most common side effects of these immunomodulators (Table 15.1).

15.2 Production of Phytoactives and Immunomodulation by Vegetables

Plant-derived phytoactives and immunomodulators have been used by mankind for therapeutic purposes for many centuries to now. The immunomodulators consisting of immunostimulators and immunosuppressors maintain a balance in cellular system and play a crucial role in maintaining health. Researchers have been studying plant phytochemicals and their immunomodulatory properties extensively and found various compounds like steroids, terpenoids, phenolics, pigments, flavonoids and alkaloids modulating the immune response (Jantan et al. 2015). There are six plant-based compounds which have garnered researcher's attention in the past few years due to their great clinical potential: curcumin (*Curcuma longa*), colchicine (*Colchicum autumnale*), resveratrol [various fruit skin, blueberries (*Vaccinium corymbosum*), raspberries (*Rubus idaeus L.*)], capsaicin (*Capsicum annuum*), epigallocatechin-3-gallate (*Camellia sinensis*) and quercetin (*Allium cepa*). The phytochemicals occurring in common fruits and vegetables are excellent source of immunomodulators. There have been many studies conducted which demonstrate that the suppressed immune system can be alleviated by consuming certain food and food products and in turn aid in eliminating an infection from the body (Childs et al. 2019). It has been now proven that cancer patients have lower NK cell activity than those of healthy individuals. Also, using food as a medicine has been an integral part of traditional Indian medicine since many centuries (Kumar et al. 2012). Below we discuss the phytochemicals and immunomodulators found in various common fruits and vegetable families.

15.2.1 The Umbelliferae Family

The Umbelliferae family also known as Apiaceae includes spices (anise, caraway, coriander, cumin and fennel seeds), herbs (coriander, parsley leaves) as well as vegetables (carrots, parsley, parsnip and celery) and consists of over 3000 species of plants. Many of the essential oils found in the many varieties of plants from this family have been used for medicinal purposes, as aroma and flavours by many generations. The total phenolic content of this family's plants is very high and is usually found in leaves of herbs such as dill, parsley and coriander (Acimovic et al.

Table 15.1 Plants and their parts used for immunomodulatory properties

Serial no	Food/plant (botanical name)	Part used	Immunomodulatory properties	References
1.	*Apium graveolens*	Stem and leaf	Enhanced secretion of γ-interferon	Cherng et al. (2008)
2.	*Apium graveolens* L. var. dulce	Stems	Enhanced lymphocyte activation and secretion of IFN-γ	-do-
3.	*Coriandrum sativum*	Whole plant	Proliferation of human PBMC stimulation and IFN-γ secretion	-do-
4.	*Foeniculum vulgare*	Aerial parts and root	Micronutrient for immunity enhancement	Chandra (1991)
5.	*Petroselinum crispum*	Whole plant	Immunomodulatory nutrients such as vitamins (A, B2, C, E) and minerals (copper, zinc, iron, selenium)	Bhaskaram (2002), Chandra (1991)
6.	*Withania somnifera*	Powdered root extract	Increase in WBC count, bone marrow cellularity, increase in phagocytic activity; other pharmacological properties	Ahmad et al. (2016), Sangwan et al. (2017), Davis and Kuttan (2000)
7.	*Morus alba* Linn. (mulberry)	Fruit	Increase in phagocytic index, protection against cyclophosphamide-induced neutropenia and increased the adhesion of neutrophils. Increase in humoral and cell-mediated immunity	Bharani et al. (2010)
8.	*Sophora subprosrate*	Roots	Stimulated proliferation and IFN-gamma secretion of murine splenic lymphocytes and also increased the levels of interleukin-6 and tumour necrosis factor-alpha	Shuai et al. (2010)
9.	*Acacia catechu*	Extract	It increases both cell-mediated and humoral immunity	Ismail and Asad (2009)
10.	*Jatropha curcas* L.	80% aqueous methanol extract	Increases of the antibody titres, lymphocyte and macrophage cells	Abd-Alla et al. (2009)

(continued)

Table 15.1 (continued)

Serial no	Food/plant (botanical name)	Part used	Immunomodulatory properties	References
11.	*Achillea wilhelmsii*	Aqueous extract	Significant increase in the DTH response, stimulatory effect on both humoral and cellular immune functions	Sharififar et al. (2009)
12.	*Picrorhiza scrophulariiflora*	One glycoside (scrocaffeside A,) from the methanol extract	Enhanced proliferation of splenocytes and their response to polyclonal T cell; production of CD4/CD8 population and interleukin of splenocytes	An et al. (2009)
13.	*Plantago asiatica* L.	Seeds	Maturation of dendritic cells and significant role in primary immune system	Huang et al. (2009)
14.	*Panax ginseng*	Ginsenosides present in root extract	Anti-inflammatory effects and TNF-alpha stimulation, inhibition of TNF-alpha-induced CXCL-10 expression	Lee et al. (2009)
15.	*Caesalpinia bonducella*	Seed extract	Increase in percent neutrophil adhesion	Shukla et al. 2009
16.	*Allium sativum*	Raw garlic	Mitogenic activity towards human peripheral blood lymphocytes, murine splenocytes and thymocytes	Clement et al. (2010)
17.	*Cynodon dactylon*	Fresh juice of the grass	Evaluation of the immunomodulatory and DNA protective activities	Mangathayaru et al. (2009)
18.	*Terminalia arjuna*	Bark powder	Anti-inflammatory and immunomodulatory activity and also has antinociceptive action probably mediated via central opioid receptors	Halder et al. (2009)
19.	*Schisandra arisanensis*	Extract of the fruits	Enhanced immunomodulation	Cheng et al. (2009)
20.	*Pteridium aquilinum* (bracken fern)		Regulation of immune response both delayed-type hypersensitivity (DTH) analysis and evaluation of IFN gamma production by NK cells	Latorre et al. (2009)

(continued)

Table 15.1 (continued)

Serial no	Food/plant (botanical name)	Part used	Immunomodulatory properties	References
21.	*Actinidia eriantha*	Roots	Antitumor potentials of the polysaccharides; immunological responses on the growth of tumour transplanted in mice and the immune response in tumour-bearing mice	Xu et al. (2009)
22.	*Boerhavia diffusa*	Root extract	Enhanced proliferation of splenocytes, thymocytes and bone marrow cells; reduction of the LPS-induced elevated levels of proinflammatory cytokines, viz. TNF-alpha, IL-1-beta and IL-6 in mice	Manu and Kuttan (2009)
23.	*Andrographis paniculata*	Leaf extract and whole plant extract	Increased in plaque-forming cells in the spleen cells; stimulated phagocytosis, significant increase in total WBC count	Naik and Hule (2009)
24.	*Dioscorea japonica*	Storage protein dioscorin	Stimulated phagocytosis; proliferation of CD4 (+), CD8(+), Tim3(+) (Th1) cells in spleen and CD19(+) cells in spleen and thymus	Lin et al. (2009)
25.	*Curcuma longa*		Immunomodulation of granulomatous inflammation and liver pathology in acute schistosomiasis mansoni	Allam (2009)
26.	*Tinospora cordifolia* (guduchi)	Stem/leaf,	Lymphocyte proliferation and macrophage activation	Dahanukar and Thatte (1997)

2015). The polyacetylenes found in the roots of vegetables such as carrot, parsley and fennel bulbs not only contribute to their bitter taste but also act as antifungal and antibacterial, anti-inflammatory and anticancer compounds. The terpenoids found in the seeds of caraway, coriander, anise and caraway are responsible for their characteristic aroma and taste. These terpenoids make the major constituents of essential oils and show antimicrobial and antimycotic activities. Besides they also have antioxidant properties which provide benefits in cases of several disease conditions. The essential oils are frequently used in the ailment of degenerative diseases caused by aging (Cherng et al. 2008; Kaur and Arora 2010: Acimovic et al. 2015).

The coumarins and flavonoids obtained from vegetables including coriander, carrots, fennel, celery and parsley were found to be immunomodulatory in the human peripheral blood mononuclear cells (PBMC). These compounds directly increased the IFN-ɤ and PBMC activity and increased the number of $CD8^{+}T$ cells (Cherng et al. 2008). There are many studies focussed on the phytochemicals found in coriander (*Coriandrum sativum L.*) and suggest their strong antioxidant, anthelmintic and antimicrobial activity. The leaves and seeds of coriander were found to have high antioxidant activity which was quantified using different assays, such as scavenging of the diphenylpicrylhydrazyl (DPPH) radical method, inhibition of 15-lipoxygenase (15-LO) and inhibition of $Fe2^{+}$-induced porcine brain phospholipid peroxidation. The activity was found to be higher in seeds rather than in leaves, and the total phenolic content was directly related to the antioxidant activity of the ethyl acetate extract (Wangensteen et al. 2004). Similar studies demonstrated the antioxidant activity of coriander as well. One study suggested the antioxidant activity of the ether extract which was then fractionised using column chromatography into five fractions, β-carotene, β-cryptoxanthin epoxide, lutein-5,6-epoxide, violaxanthin and neoxanth (Guerra et al. 2005).

Another study demonstrated the extraction and identification of the phenolic compounds from the coriander leaf, shoot and root aqueous extract. The aqueous extract was subjected to column chromatography and gas and mass spectrometry which led to the identification of caffeic acid, protocatechinic acid and glycitin as the major phenolic compounds responsible for high antioxidant activity of the aqueous coriander extract (de Almeida Melo et al. 2005). In another study, the effect of coriander seeds on lipid parameters in 1,2-dimethyl hydrazine (DMH)-induced colon cancer in rats was studied, and it was found that while in the control group, the concentrations of cholesterol and cholesterol to phospholipid ratio decreased while the level of phospholipid increased, it remained unaffected in the case of spice administered group (Chithra and Leelamma 2000). The antioxidant potency of the coriander seed powder was found to inhibit the H_2O_2-induced oxidative stress in human lymphocyte cells. The H_2O_2 significantly reduced the antioxidant activity of enzymes such as superoxide dismutase, catalase, glutathione peroxidase, glutathione reductase and glutathione-S-transferase and decreased the amount of glutathione and increased the thiobarbituric acid-reacting substance (TBARS) content; all of these effects were reversed by polyphenolic compounds extracted from the coriander seed powder (Hashim et al. 2005). In addition, the crude and hydro-alcoholic extracts from coriander seeds were also studied for their in vitro and in vivo anthelmintic activity on the egg and adult nematode parasite *Haemonchus contortus*. The *in vivo* activity of the aqueous extract of coriander seeds was also tested on the parasite-infected sheep. A concentration less than 0.5 mg/ml. ED50 of aqueous extract effectively inhibited the egg hatchings completely. Crude aqueous extract of *Coriandrum sativum* at 0.45 and 0.9 g/kg dose levels successfully increased the faecal egg count reduction (FECR) and total worm count reduction (TWCR) as compared to the untreated groups (Eguale et al. 2007). Another protective effect of coriander antioxidant was demonstrated by pre-treatment of coriander (250 and 500 mg/kg, body weight) to the gastric mucosal-related injuries in rats. The gastric mucosal injuries were induced by NaCl, NaOH, ethanol, indomethacin and pylorus ligation accumulated gastric acid secretions. The coriander exposure protected the gastric mucosa not only by decreasing the ulcerogenic effects of necrotic agents but also by preventing the ethanol-induced lesions and decreased gastric acid secretions. It was concluded that these effects are mainly due to the antioxidant constituents like linalool, flavonoids, coumarins, catechins, etc. found in coriander (Al-Mofleh et al. 2006).

It has also been proved that the coriander extracts prevent the lead-induced immunotoxicity and lead deposition. In one study, the Nile tilapia (*Oreochromis niloticus L.*) were exposed to metal lead (20.2 mg/L) and then fed with coriander powder (20 mg/kg and 30 mg/kg) and extract (20 mg/kg and 30 mg/kg). It was found that coriander supplementation enhanced relative expression of interleukin-1β (IL-1β), serum lysozyme, nitric oxide and bactericidal activities in the fish as compared with the control group. In addition, the lead-induced fish mortality was greatly reduced (Ahmed et al. 2020). The preventive efficacy of coriander on the lead deposition was studied by Aga et al. (2001) by administrating coriander to mice by gastric intubation for 25 days. The coriander administration was started at the seventh day of lead exposure (1000 ppm), which was achieved by mixing lead acetate trihydrate in drinking water. It was found that the lead reached its highest concentration in the femur of the animal, and *meso*-2,3-dimercaptosuccinic acid (DMSA) was administered to mice group serving as control. The study demonstrated the chelating activity of DMSA and also that the coriander decreased the lead deposition in the femur and lead-associated injury to the kidneys. The lead deposition-linked enzyme inhibition activity of delta-aminolevulinic acid dehydratase (ALAD) was also reduced by MeOH extract of coriander in vitro.

In addition to the antioxidant activity of fresh coriander extracts, it was found that the freeze-dried and irradiated parsley (*Petroselinum crispum*) and coriander methanol extracts exhibited significant antioxidant and antimicrobial activities. The antioxidant activity was assessed by using an iron-induced linoleic acid oxidation model system, and the antimicrobial activity by studying the cell damage done by the herb extracts on *Bacillus subtilis* and *Escherichia coli* cells (Wong and Kitts 2006) (Table 15.2).

The study of immunostimulant activity of coriander incorporated diet on the fish *Catla catla* infected with *Aeromonas hydrophila* was done by Innocent et al. (2011), and it was found that the total erythrocyte count (TEC) and total leucocyte count (TLC) of the fish fed with coriander infused diet were significantly increased as compared to the control group. There was a noteworthy rise in the Hb content from 6.9 to 7.39% and serum protein level from 0.56% to 0.58%.

Not only the herbs like coriander possess antioxidant and immunomodulatory activities, but they can also be used as an antibiotic replacement in bird feed as well. A study found that 2% coriander seed powder mixed with the poultry feed lowered total cholesterol while blood urea was increased. In addition, the *Lactobacillus*

Table 15.2 Immunomodulatory properties of essential oils

Sr. no	Plant	Essential oil from part	Cell line	Immunomodulatory properties	References
1.	*Kunzea ericoides* and *Leptospermum scoparium*	Use of the bark, leaves, sap and seed capsules of these plants' essential oil	THP-1 human Monocyte/macrophage cell line	EOs have no major toxic side effects on THP-1 cells. Eos reduced the LPS-induced TNF-secretion but have no effect on IL-4 secretion	Chen et al. (2016a)
2.	*Litsea cubeba L.*	Aerial tissues	C57BL/6 mouse bone marrow-derived dendritic cells (DCs)	A slight cytotoxic effect was observed at 5–104-fold diluted EO. Release of TNF- and IL-12 by LPS-induced DCs was inhibited by EO in a dose-dependent fashion	Chen et al. (2016b)
3.	*Artemisia argyi*	Leaf essential oil	Murine macrophage RAW 264.7 cells	In LPS-induced cells, the EOs inhibited the release of NO, PGE2 and ROS and TNF, IL-6, IFN and MCP-1	Chen et al. (2017)
4.	*Origanum vulgare*; *Thymus vulgaris*; *Salvia sclarea*; *Eugenia caryophyllata*; *Lavandula angustifolia*; and *Thuja plicata*	From aerial part of the plant	Human embryo Lung hEL12469 cells	Essential oil presents toxic side effects at higher concentrations. Treatment with EOs did not induce any significant increase in DNA strand breaks; only *Thuja plicata* EO (0.2 -L/mL) showed a negative effect on DNA single-strand breaks in hEL 1269 cells	Puskarova et al. (2017)
5.	*Origanum heracleticum* L.	EOs obtained from flowers and younger leaves	Murine macrophage RAW264.7 cells	In LPS-stimulated RAW264.7 cells, all EOs from *Origanun heracleticum* L. showed anti-inflammatory activity by means of its capacity to decrease the NO production	Marrelli et al. (2018)

6.	*Origanum vulgare* L.	Air-dried herb of *Origanum vulgare* L.	Murine macrophage RAW264.7 cell	Low dose of EOs (1.25–20 g/mL) did not produce any toxicity. In LPS-induced RAW264.7 cells, pretreatment with the EOs reduced the expression and secretion of IL-1, IL-6 and TNF-. Inhibition of LPS-induced MAPK, PKB and NF-B was also observed. The EOs also inhibited the LPS-induced elevation of NADPH oxidase and oxidative stress	Cheng et al. (2018)
7.	*Pituranthostortuosus*	EOs from the aerial parts	Splenocyte suspension from Balb/c mice; murine melanoma B16F10 cell line	EO treatment was able to promote LPS-stimulated splenocyte proliferation. In addition, EO treatment was also able to increase the number of apoptotic cells	Krifa et al. (2015)
8.	*Cirsium japonicum* DC	Aerial parts essential oil	L02 cell line; human lung adenocarcinoma A549 cell line; murine macrophage	EOs have no major toxic side effects on L02 cells and even promoted cell proliferation. In the A549 cell line, EOs promote the proliferation of cancer cells. NO production was inhibited in LPS-induced RAW264.7 cells treated with EOs at 50 and 100 g/mL	Ma et al. (2019)

(continued)

Table 15.2 (continued)

Sr. no	Plant	Essential oil from part	Cell line	Immunomodulatory properties	References
9.	*Ferula iliensis*	EOs collected from dried samples obtained from flowers, umbels + seeds, leaves and stems	Human blood isolated neutrophils from healthy donors; bone marrow leukocytes isolated from Balb/c mice	EOs activated human neutrophil Ca2+ flux and also activated SOD-inhibitable ROS production in both human neutrophils and mouse bone marrow phagocytes	Ouzek et al. (2017)
10.	*Chamaecyparis obtusa*	Leaves essential oil	Murine macrophage RAW264.7 cells	In LPS-stimulated cells, EO treatment reduced nitric oxide, TNF and IL-6 production and inhibited iNOS and COX-2 expression	Park et al. (2016)
11.	*Pistacia vera L.*	4-Carene, −pinene and − 3-carene	Human blood isolated lymphocytes	EOs did not show any cytotoxic effects. In tert-butyl hydroperoxide-treated lymphocytes, incubation with EOs (20–12.5 g/mL) significantly increased cell viability	Smeriglio et al. (2017)
12.	*Allium roseum* L.	Methyl methane thiosulfinate, 3-vinyl-1,2 dithiacyclohex-5-ene and diallyl trisulfide were the major compounds	Cytotoxicity assay, 5 h; proliferation assay, 72 h	EOs did not show cytotoxic effects. Antiproliferative assay depicted that the number of cells was reduced by the incubation of hT29-D4 and Caco-2 cells with EOs in a dose-dependent fashion	Touihri et al. (2015)
13.	*Heracleum pyrenaicum* subsp. orsinii	Pinene, (Z)-ocimene and -pinene were the major compounds	Human cervix hela cell; human colon carcinoma LS174 cell; non-small cell lung carcinoma A549; human normal foetal lung fibroblast MRC-5 cell	The cytotoxic effect of EOs was prominent against heLa, LS174 and A549 cell lines. Eos did not show toxicity side effects against normal MRC-5 cell (IC50 > 200 g/mL)	Usjak et al. (2017)

14.	*Trachydium roylei*	Murine macrophage RAW264.7 cells	Phellandrene, myristicin and elemicine were the major compounds	In LPS-stimulated RAW264.7 cells, only a high concentration of EOs (40 mg/mL) showed a negative effect on cell viability. In addition, incubation with EOs inhibited the production of TNF, IL-1 and IL-6, whereas it increased the release of IL-10	Wang et al. (2016)

population was unaffected, but *E. coli* population was higher in the chicks not exposed to coriander. Also, the antibody titer was much higher in the coriander water fed chicks against the Newcastle, infectious bronchitis, and infectious bursal disease (Hosseinzadeh et al. 2014). In a similar study, it was found that the addition of coriander seed powder (3% and 5%) in rat feed resulted in the final body weight gain. There was also an increase in the Interlukin-6 concentration level in the experiment group as compared to the control group (El-Sayed and Ahmed 2017).

Recently, coriander extract when tested in male white mice was demonstrated to increase the phagocytic activity and the total leukocyte cell counts. The coriander extract in the doses of 100 mg/kg, 140 mg/kg and 200 mg/kg was administered orally for 7 days, and then the mice were infected with *Staphylococcus aureus* to help their immune system to express. The results showed that the phagocytic activity of macrophages increased by 44.6%, 54.2% and 60.2% in the mice exposed to 100, 140 and 200 mg/kg of coriander extract, respectively. The highest leukocyte cell increase was found at a dose of 200 mg/kg, which was 7310/μL of blood as compared to control group fed with 0.5% Na CMC (sodium carboxymethyl cellulose) suspension (Dillasamola et al. 2019).

Although coriander remains one of the most well-studied plants for its immunomodulatory activities, from the Umbelliferae family, others like fennel, caraway and carom seeds also have been researched extensively. Recently, a study provided insight into the in vitro antimicrobial activity of essential oil on the gram-positive and gram-negative bacteria isolated from oral cavities of patients suffering from periodontitis. It was found that the essential oils from *Anethum graveolens*, *Salvia officinalis* and *Satureja hortensis* exhibited significant antibiofilm and bactericidal activities at a concentration of 0.08–1.36 mg/mL which are due to their ability to permeate the cell membrane and inhibit efflux pump activities. In addition, it was found that summer savoury essential oil also induced cytokines of the THP-1 cells (Popa et al. 2020).

Similarly, in the case of rat colon, fennel essential oil has been shown to inhibit the growth of pathogens such as *Enterococcus* and *Clostridium perfringens* (Renjie et al. 2010) and has a hepatoprotective effect on the CCl4-induced liver damage in rats (Özbek et al. 2003). It has also been reported that fennel extracts inhibit the *Candida albicans* growth and immunomodulate peritoneal macrophages of mice and stimulate production of nitric oxide and ROS in mice (Naeini et al. 2009). The polyphenols and antioxidant activity of fennel hydro-alcoholic extracts demonstrated increased red and white blood cells and reduction of negative effects of free radicals on blood cells (Mansouri et al. 2015).

The immunomodulatory activity of different *Trachyspermum ammi* (carom) seeds was tested by measuring the delayed-type hypersensitivity (DTH) assay, and the skin thickness in rats after the treatment was assessed. It was found that the methanolic seed extract (500 mg/kg dose) stimulated the cell-mediated immunity and also resulted in the increased skin thickness in rats (Siddiqui et al. 2019). The antifungal activity of carom seeds and leaf extract was recently demonstrated by Khan and Jameel (2018).

The antioxidant, anticolitis, anti-inflammatory and immunomodulatory activity of caraway seeds has also been studied in detail. It was found that an anti-inflammatory agent, carvone, is responsible for the decrease in prostaglandins and leucotriene biosynthesis by inhibition of 5-lipoxygenase and cyclooxygenase (Keshavarz et al. 2013). Also, the presence of carvacrol provides the seeds with great antioxidant and antibacterial properties (Samojlik et al. 2010).

The Umbelliferae family plants have been used as a nutraceutical for many generations now, but it is only in the last 20 years that researchers have focused their attention on the possibility of utilising their extracts to replace conventional drugs.

15.2.2 Lamiaceae Family

The Lamiaceae family is known for its distinct aroma herbs such as rosemary, mints, thyme, oregano, basil, savoury, sage and lemon balm. Plants from this family are often used for pain management and are extensively studied for their immunomodulatory properties (Uritu et al. 2018). Herbs like holy basil and mint have been used for their medicinal benefits like antimicrobial, anti-inflammatory, antistressor, antimalarial, antiallergic and antidiabetic activities, in Ayurveda for many generations (Raja 2012).

Oregano essential oil (OEO) has been reported as demonstrating anti-inflammatory, immunomodulatory and anticancer activity in human skin disease model. It was found that the essential oil induced the antiproliferative effects and significantly reduced various inflammatory biomarkers, namely, monocyte chemoattractant protein 1, vascular cell adhesion molecule 1, intracellular cell adhesion molecule, interferon gamma-induced protein 10, etc. In addition, the essential oil also strongly inhibited macrophage colony-stimulating factor (M-CSF) and modulated other signalling pathways which are involved in skin inflammation, tissue remodelling and cancer (Han and Parker 2017). In addition, oregano has been recognised as an effective anti-inflammatory, antifungal and antibacterial agent. Another study on oregano methanol extract provided a detailed account of its cytoprotective and in vitro immunomodulatory activity by treating spleen mononuclear cells with oregano extract for a total of 48 h. It was found that the treatment did not alter concanavalin A-triggered proliferation or IFN-γ and IL-4 secretion but did decrease IL-17 secretion and downregulated production of nitric oxide (NO) in peritoneal cells (PC). It also saved pancreatic islets from cytotoxic effect of cytokines and fully preserved beta cell function (Vujicic et al. 2015).

Thyme has been used extensively as a fresh or dried herb in many cultures, but its immunomodulatory activity has gained interest quite recently. In 2016, a report suggested immune function enhancement and resistance to disease in sharptooth catfish fed regularly with thyme diet (1%). The catfish, after being fed thyme for 30 days, was studied for immune function, and it was found that the total serum protein, glutathione peroxidase (Gper) and catalase (CAT) increased significantly. The fish was also subjected to infection with *Aeromonas hydrophila* and weighed for any changes in its weight gain. Although the weight gain was lower than the control group, the cumulative mortalities were 40% as compared to 66.7% in the control group (Emeish and El-Deen 2016). It has been widely believed that that the immunomodulatory effect of *Thymus vulgaris* extract can be utilised to develop a therapeutic agent against multiple sclerosis, and it was supported by a study on the experimental autoimmune encephalomyelitis (EAE) in mice, exposed to thyme extract (100 mg/kg of body weight, every other day). The splenocytes of the MOG35–55 peptide-induced EAE model were studied for clinical symptoms and histopathological scores, production of IFN-γ and IL-6, which were found to be lower in the thyme-treated group as compared to the control group (Mahmoodi et al. 2019).

Therapeutic potential of *Ocimum basilicum*, the basil, has been the focus of many research studies, and it has been traditionally used in various treatments related to respiratory tract infections. Basil extract decreased the IL-4, IgE, PLA2 and TP levels, but increased IFN-γ/IL-4 ratio compared to untreated sensitised rats (Eftekhar et al. 2019). Moreover, it was concluded that the therapeutic potential of the plant in response to asthma was better than dexamethasone because of the significantly higher improvement effects as compared to control and rats administered with dexamethasone. Besides this, many other findings like stimulatory activity of aqueous extract of basil leaves on the DNA synthesis of human peripheral blood mononuclear cells (Tsai et al. 2011); increased RBC, WBC, serum protein and globulin in *Clarias batrachus* (walking catfish) after being fed aqueous extract of basil leaves (Nahak and Sahu 2014); increased production of TNF-α, IL-2, IFN- γ and IL-4; and improved haemoglobin concentration in spleen cells of myelosuppressed mice model (Hemalatha et al. 2012) suggest the significance of basil in maintaining optimal health.

15.2.3 Solanaceae Family

Solanaceae family plants have been used for ages, in the traditional medicine and human nutrition due to the presence of phytoactive secondary metabolites. They also contain high amounts of antimicrobial peptides and have garnered great interest recently to combat the growing concerns of antimicrobial resistance (Afroz et al. 2020). Food crops such as pepper, eggplant, tomato and potato are most widely popular crops of the family Solanaceae in terms of consumption, and other crops like *Atropa*, *Hyoscyamus*, *Withania* and *Nicotiana* have been used in the plant-based drug discovery for many years now (Chowanski et al. 2016). There have been many instances of evidences regarding the utilisation of Solanaceae family plants in the traditional Unani, Ayurvedic, Chinese and homeopathy medicines (Shah et al. 2013).

One of the most expansively studied Solanaceae members is *Solanum trilobatum* which reportedly has hepatoprotective, antimicrobial, antioxidant, cytotoxic, haemolytic, protective, immunomodulatory and anti-inflammatory properties, and due to that it has been used in the treatment of tuberculosis, respiratory problems and bronchial asthma (Sahu et al. 2013). The solvent extracts (ethanol, acetone and ethyl acetate) of *S. trilobatum* have been shown to incite an antibacterial response against gram-negative and gram-positive bacteria like *S. aureus*, *K. pneumoniae* and *P. aeruginosa*, and the effects were even more profound than common antibiotics (Kannabiran et al. 2008). The antihyperglycaemic and antihyperlipidaemic activity has been demonstrated by administering the albino rats with the ethanolic extracts (400 mg/kg) of the leaves of the plant. It was evident from the data that the underweight rats gained weight and showed the signs of reversal of the conditions and the results were equivalent to the results shown by glibenclamide. The antioxidant and anti-inflammatory activity of the leaf extracts of *S. trilobatum* was demonstrated by various studies where the effects of the extracts were either comparable to or significant than the conventional antioxidant and anti-inflammatory agents (Pandurangan et al. 2010; Sini and Devi 2004). The immunomodulatory activity of *S. trilobatum* was studied by administering a dose of 100, 200 and 400 mg/kg to experimental rats. The results showed increase in percentage of neutrophil adhesion, increase in haemagglutination antibody titer and significant increase in the phagocytic activity of reticuloendothelial system (Raja et al. 2009).

A recent study by Anandakumar et al. (2020) showed that the phytopharmaceutical constituents of *Solanum trilobatum* L. might have significant potential against COVID-19 [severe acute respiratory syndrome coronavirus 2 (SARS-CoV-2)]. They performed the molecular docking of the phytochemical constituents solanidine, solasodine and solanine against the main protease (Mpro) of SARS-CoV-2 and found that these bound to the active cavity site on the M^{pro}. This study provides a great basis to further study the plant extracts as the potential drug candidate through various biochemical and cell-based assays.

Another member of the Solanaceae family, *Solanum torvum*, has also been studied for its immunomodulatory activities such as delayed-type hypersensitivity (DTH) response and haemagglutinating antibody (HA) titre. The aqueous extracts of *S. torvum* fruits resulted in the significantly enhanced DTH response, increased HA titre and WBC count in the phenylhydrazine (PHZ)-induced anaemic rats. The 24-day courses of the extract also were able to reverse PHZ-induced anaemia and increase the RBC and Hb concentration (Koffuor et al. 2011). In addition, *S. torvum* has been studied extensively to demonstrate its cardio- and nephro-protection, anti-hypertensive, analgesic, anti-inflammatory, anti-ulcer and antimicrobial activities (Darkwah et al. 2020).

There are numerous documented health benefits of capsicum (*Capsicum annuum*), and it has been used in wide varities of dishes mainly due to its distinctive aroma and taste. There are studies which show that it possesses several pharmacological bioactivities including immunomodulatory properties. Although overindulgence of the plant can lead to irritation to mucous membrane, and some inflammation, the list of its benefits outweighs its toxicity (Parvez 2017).

A recent study shows the antimalarial activity of crude fruit extract of *Capsicum frutescens* var. minima against *Plasmodium berghei*-infected mice. Different doses of the fruit extracts (100, 200 and 400 mg/kg) were given to the infected mice, and it was found that all dosage exhibited significant inhibition of the parasite, although the maximum inhibition was found in the highest dose administered groups as compared to the control groups. It should also be noted that the extract did not cause any mortality and revealed no obvious sign of toxicity was reported (Habte and Assefa 2020).

Most of our country's population is starting to rely on the diet of high sugar, salt, trans fats, polyunsaturated fatty acids and other processed food items. This has led to unforeseen increase in the metabolic-related disorders in the last few decades which include heart-related illnesses like atherosclerosis, IBD, rheumatoid arthritis, type 2 diabetes and many different forms of cancers (Sebastia et al. 2013). All these debilitating illnesses can be easily avoided by replacing the diet with higher amounts of greens, vegetables and fruits. A study reported that an increased diet of cruciferous vegetables led to decreased serum levels of the proinflammatory cytokines IL-1β, TNFα and IL-6 (Jiang et al. 2014). There are many other reports available which all describe the anti-inflammatory and immunomodulatory activities of cruciferous vegetables. In fact, there are many aryl hydrocarbon ligands that have been identified in cruciferous vegetables like broccoli, which binds to aryl hydrocarbon receptor (a transcription factor expressed by immune cells, epithelial cells and some tumour cells), which in turn regulates many genes that control immunity and inflammation (Tilg 2015).

There are numerous documented health benefits of phytochemicals, glucosinolates, sulforaphane, polyphenols and antioxidants present in broccoli. Consuming steam cooked or blanched broccoli has reported gastroprotective, antimicrobial, antioxidant, anticancer, hepatoprotective, cardioprotective, anti-obesity, antidiabetic and anti-inflammatory effects (Owis 2015). The effect of sulforaphane on the immune system was studied by Thejass and Kuttan (2007) and described that it increases total WBC count, bone marrow cellularity, number of alpha-esterase-positive cells, circulating antibody titre and the number of plaque forming cells (PFC) in the spleen and phagocytic activity of peritoneal macrophages of BALB/c mice. It also led to decrease in the elevated level of TNF-alpha production by LPS-stimulated macrophages.

Water distillate of broccoli sprouts and freeze-dried broccoli sprouts have shown high antioxidant activities which are owed to the presence of 5-methylthiopentylnitrile, 4-methylthiobutylisothiocyanate, 4-methylthiobutylnitrile, 3-methylthiopropylisothiocyanate and 4-methylpentylisothiocyanate and phenolic antioxidants like 4-(1-methylpropyl) phenol, 4-methylphenol and 2-methoxy-4-vinylphenol (Jang et al. 2015). Broccoli biofortified with selenium was assessed for its antioxidant and antiproliferative activities. It was found that the seedlings of broccoli had higher phenolic compounds and antioxidant and antiproliferative activity (Bachiega et al. 2016).

Helicobacter pylori, a common gastric infection, is prevalent in regions where gastric cancer incidences are also higher. The sulforaphane [(−)-1-isothiocyanato-(4R)-(methylsulfinyl) butane present in certain broccoli varities and its sprouts can be bacteriostatic against 3 reference strains and 45 clinical isolates of *H. pylori.* Not only this, this isothiocyanate can also block gastric tumour formation. This dual activity can effectively reduce the usage of antibiotics and anti-cancerous drugs and could provide diet-based solution to the ailments (Fahey et al. 2002).

Momordica charantia L. (*M. charantia*) (bitter gourd, karela) is one of the most common vegetables consumed in Asia, Africa and the Caribbean. It originated in India and is widely popular for its numerous health benefits (anti-ulcer, anti-inflammatory, anti-leukemic, antimicrobial, anti-diabetic and anti-tumour). Although the bioactivity and mode of action of its extracts are not very well understood, its immunosuppressant as well as immunostimulant activities are well documented (Goo et al. 2016; Mahamat et al. 2020). A recent study was done on the methanol and diethyl ether leaf extracts of *Momordica charantia* on the *Salmonella typhi*-infected mouse models. Significantly higher antibody titre and mobilisation of leukocytes were found against the infection, and increased production of superoxide anion and nitric oxide and that of lysosomal acid phosphatase by macrophages and neutrophils were also reported (Mahamat et al. 2020).

Crude polysaccharide of *M. charantia* (MCP) fruit was prepared by hot water extraction and purified by DEAE-52 cellulose anion-exchange chromatography which was used to explore the immunomodulatory effects both in vivo and in vitro in a cyclophosphamide (Cy)-induced immunosuppressed mice. The extraction increased carbolic particle clearance index, serum haemolysin production, spleen index, thymus index and NK cell cytotoxicity to normal control levels and also stirred normal and concanavalin A-induced splenic lymphocyte proliferation in vitro at various dosages. This study led to a conclusion that the *M. charantia* extracts can be used as immunotherapeutic adjuvants (Deng et al. 2014).

The anti-cancerous activity of bitter melon extract was demonstrated by Bhattacharya et al. (2017) where the mouse head and neck cancer (SCCVII) cells injected syngeneic mice were administered with bitter melon extract. It was found that the extract fed mouse showed low expression of proliferating cell nuclear antigen (PCNA) and c-Myc, which is directly linked to inhibition of cell proliferation. Also, the extract reduced infiltrating regulatory T (Treg) cells by inhibiting Forkhead box (FoxP3+) protein, which is responsible of low tumour immunity. In addition, the treatment reduced Th17 cell population in the tumour. The study effectively provided evidence of effectiveness of bitter melon extract in the inhibition of head and neck tumour growth.

15.3 Immunomodulation by Milk and Fermented Milk (Probiotic Products)

Milk and milk products are the inherent part of our diet since the initiation of domestication of livestock. Milk or colostrum is a complex mixture of numerous propitious substances which have multidimensional role in gastrointestinal track (GIT) health of host. Milk is the only source of nutrition for neonate of mammals as it is essential for providing immune homeostasis in infants which aid in the prevention of several infectious diseases (Ebringer et al. 2008). Milk is considered a complete and healthy food due to its nutritional components. Beyond the nutritional value, milk also contains several bioactive components such as immunoglobulins, antimicrobial peptides, enzymes (lysozyme and lactoperoxidase), hormones, immunocompetent cells (T and B lymphocytes, granulocytes and neutrophils), etc. Colostrum and milk from other dairy species are the sources of bioactive metabolites which provide strength to the immune system. Since the past few years, researches are being conducted to identify the components of milk which have functional, nutritional or biological activities. Studies are done in in vivo and in vitro conditions to see the effects of identified components. Milk proteins constitute the most important and physiologically active fraction of milk. These proteins may be already in active form, or they may generate various bioactive peptides after digestion with enzymes (Park and Nam 2015). Bovine milk or colostrum has a number of functional peptides which remain in latent phase until these get digested in the gastrointestinal track and fermented by lactic acid bacteria (Meisel 2004). These proteins have a large number of physiological activities, and one of them is immunomodulatory as this has the ability to modulate the response of immune cells (Reyes Diaz et al. 2018). Naturally, these bioactive peptides are also present in fermented dairy products like yogurt and cheeses (Vinderola et al. 2007). Although structure and function of these bioactive peptides are not known in a comprehensive manner, their physiological activities motivate researchers to carry out more research in this area. Available literature reveals that functional role of some of the bioactive peptides has led them to be formulated in the product form for use as a food supplement. Different technologies are in practice to manufacture such products, and these are also supported by the people concerned about the problem of milk allergy in certain population (Clare and Swaisgood 2000). Occasionally, few infants show allergic response to cow milk, and it was observed in a study that it can be avoided by supplying the milk hydrolysates which are generated through enzymatic hydrolysis of milk. These are also able to boost the immunity by triggering the function of immune cells such as increased level of transforming growth factor-β, raising the level of haemolysin in serum and upgrading the phagocytosis of macrophages (Pan et al. 2013). Major proteins present in milk are immunoglobulins, lactoferrin, β-lactoglobulin, α-lactalbumin and serum albumin. They perform different kinds of physiological roles, but here we describe only their immunomodulatory functions. Colostrum is the first food which is given to a neonate because it has different kinds of immunoglobulins (IgG1, IgG2, IgM and secretory IgA) which provide passive immunity to newborn against microbial infections. More than half of the milk protein content is constituted by α-lactoglobulin and β-lactoglobulin proteins. Many immunomodulatory peptides are released from these proteins after hydrolysis (Tai et al. 2016). Another key milk protein is lactoferrin which is an iron-binding glycoprotein. It shows different beneficial activities in the body such as immunomodulatory, antimicrobial, anticancer as well as anti-inflammatory activities (Rascon-Cruz et al. 2021). Other than immunoglobulins many antimicrobial factors are reported in fresh milk which have bacteriostatic action, and these factors include lactoperoxidase, xanthene oxidoreductase and lysozyme (Ebringer et al. 2008). Lactoperoxidase provides protection to the mammary glands against microbial infections, and when it passes to the infant's body through mother's milk, it also helps in avoiding several other diseases (Silva et al. 2020).

A number of microbes are associated with fermented milk, and these are beneficial to support the healthy gut microbiota as they provide immunity from several infections and may also help in nutrient absorption and maintaining healthy nervous system functioning (Rinninella et al. 2019). Milk fermented with *Lactobacillus casei* CRL 431 exhibited the antitumor activity by dairy products which are known to have a significant abundance of lactic acid bacteria (LAB). These LABs are reported to exhibit anti-tumour effect via immunomodulatory action. A study conducted by Moineau and Goulet (1991) showed the positive effect of including yogurt in diet, on proliferation of macrophages. An animal study on mice conducted by this group described an enhancement in the phagocytic activity of macrophages in mice fed with milk constituting LAB compared to the control set of mice. This study was supported by another report, mentioning the induction in the production of IFN-alpha and IFN-beta by *L. acidophilus* in cell culture of murine peritoneal macrophage (Kitazawa et al. 1992). The lactic acid bacteria (LAB) administered orally can translocate to lymphatic organs present in the gut, resulting the production of sIgA (mucosal antibody) from PP (Payer's patch) cells. This antibody inhibits the infectious pathogens from colonisation in the gut. This effect of LAB on sIgA antibody is also studied on mice in a dose-dependent manner (Perdigón et al. 2001). It was proposed that the increase in sIgA may be due to proliferation of PP cells leading to the change in ratio of CD4+ T lymphocytes to CD8+ T lymphocytes. De Simone et al. reported the effect of LAB for survival against infectious *S. typhimurium*. It was described that the mice fed with yogurt diet had more survival rate against infection compared to control group. M. Medici et al. 2004, concluded a study on mice fed with probiotic cheese and concluded that it activates the peritoneal macrophages. An increase in the phagocytic activity was observed at the seventh day of feeding, which could be a result of reduced IgA cells in PFC group. Cousin et al. (2011), reported immunomodulatory effect of fermented milk with *Propionibacterium freudenreichii* on piglets. They observed an improved growth and diet intake of piglets fed with fermented milk. The immunomodulatory action of fermented milk led to the decreased secretion of interleukin-8 and TNF-α, and the microbial diversity and gut structure remained unaltered. This group suggested it as a potential therapy for inflammatory bowel diseases. Thus, these studies provide support to the idea of positive immunomodulatory effects of dairy on animals.

15.4 Mechanism of Immunomodulation of some Nutrients

It is believed that micronutrients work collectively to promote an optimal immune system. Copper, zinc, folate, selenium, iron and vitamins (A, B_{12}, B_6, C, D, E) are essential for boosting immune response (Gombart et al. 2020). The role of each of these nutrients is explained as follows:

15.4.1 Vitamins

Vitamins are vital for developing innate and adaptive immunity in the body. Vitamins are important components of our diet and have been known to affect the immune system for a long time. There are various significant roles of vitamins in modulating a broad range of immune processes, such as regulation of the immune response, activation and proliferation of lymphocyte, tissue-specific lymphocyte homing, T helper cell differentiation and the production of specific antibody isotypes. In recent years, vitamins A and D have been shown to have an unexpected and crucial effect on the immune response. The therapeutic ability of the metabolites of vitamin A and D is shown to modulate tissue-specific immune responses and to prevent and/or treat inflammation and autoimmunity (Mora et al. 2008).

15.4.2 Selenium

Selenium is a major component of the antioxidant system of the body, protecting the body from oxidative stress, a natural by-product of the metabolism of the body. There is now considerable evidence that selenium plays a crucial role in the immune system functioning. In 2008, Hoffmann and Berry related its role in nearly all tissues and cell types in regulating oxidative stress, redox responses and other cellular processes, including those involved in innate and adaptive immune responses. Interestingly, few other studies revealed a correlation between viral infections and selenium status (Broome et al. 2004 and Guillin et al. 2019). Selenium deficiency was correlated with upregulation of pro-inflammatory responses leading towards cancer and heart diseases (Arthur et al. 2003).

15.4.3 Zinc

Zinc is an essential nutrient for growth and development. In a wide range of reactions and a large number of enzymes, zinc has a key role as a catalyst. Shankar and Prasad showed in 1998 that zinc affects many elements of the immune system, from the skin barrier to gene regulation within lymphocytes. Maywald and colleagues reported that zinc insufficiency leads to changes in the numbers and activities of immune cell, resulting in increased vulnerability to infections and the development of inflammatory diseases (Maywald et al. 2017).

15.4.4 Iron

Iron is also a cofactor to enzymes in oxidation-reduction reactions. These reactions are important for the energy metabolism of cells. Research indicates that low levels of iron affect our body's ability to deliver a proper immune response (Soyano and Gomez 1999). Iron is essential for immune cell production and growth, particularly lymphocytes, which are related to the initiation of specific responses to infection (Soyano and Gomez 1999). Iron sequestration is an important innate host defence mechanism because many pathogens depend on this essential element. Dallman et al., in 1987, concluded that "abnormalities in cell-mediated immunity and ability of neutrophils to kill several types of bacteria" are usually seen in iron-deficient patients.

15.4.5 Copper

After iron and zinc, copper is the most abundant dietary trace mineral. Copper plays a vital role in innate immune response. It is a component of many enzymes and is needed to produce red and white blood cells (Bonham et al. 2002). Copper-dependent enzymes transport iron and load it into haemoglobin, a protein that carries oxygen through the blood. The deficiency of copper is defined as neutropenia, i.e., decrease in quantity of circulating neutrophils. Also, a loss in bactericidal activity is observed on impaired availability of copper. Thus it can make hosts more susceptible to bacterial infections, leading to severe health problems.

15.4.6 Folate

Folic acid has a major participation in cell-mediated immunity. It is not only crucial for the synthesis of DNA and protein but also is vital in cell proliferation mechanism. Its deficiency alters the functioning of thymus and also decreases the T lymphocyte response for several mitogens (Courtemanche et al. 2004). A decrease in immunity towards various bacterial infections due to less adequate folate in human system is well documented (Adhikari et al. 2016).

15.5 Conclusion

Intake of food and vegetables in diet is necessary to maintain the homeostasis of the body. Earlier these items are added in diet for their taste and flavours, but now research revealed that these are useful for their immunomodulatory effects. Green vegetables and herbs are rich source of different type of vitamins, macro- and micronutrients, other than these different plants, and include a lot of compounds which seem as potential molecules for therapy of several diseases. These special types of compounds are called as secondary metabolites, which are studied in detail for cure of severe diseases like diabetes, cancer, Alzheimer's, etc. Different bioactive peptides have been isolated from milk and fermented dairy products. Therefore, these are gaining more attention due to their beneficial effects on immune system by improving the quality of gut microbiota. Recent trend is focusing on the use of pre- or probiotic in any form in diet to boost the immune system as reports favour that their addition in meal provides protection against viral or bacterial infections.

References

Abd-Alla HI, Moharram FA, Gaara AH et al (2009) Phytoconstituents of *Jatropha curcas* L. leaves and their immunomodulatory activity on humoral and cell-mediated immune response in chicks. Z Naturforsch C 64(7–8):495–501

Acimovic MG, Kostadinovic LM, Popovic SJ et al (2015) Apiaceae seeds as functional food. J Agric Sci (Belgrade) 60(3):237–246

Adhikari PM, Chowta MN, Ramapuram JT (2016) Effect of vitamin B12 and folic acid supplementation on neuropsychiatric symptoms and immune response in HIV-positive patients. J Neurosci Rural Pract 7(3):362

Afroz M, Akter S, Ahmed A et al (2020) Ethnobotany and antimicrobial peptides from plants of the Solanaceae family: An update and future prospects. Front Pharmacol 11:565

Aga M, Iwaki K, Ueda Y (2001) Preventive effect of *Coriandrum sativum* (Chinese parsley) on localized lead deposition in ICR mice. J Ethnopharmacol 77(2–3):203–208

Ahmad H, Samuel SS, Khandelwal K, Arya A, Tripathi S, Agrawal S, Sangwan NS, Shukla R, Dwivedi AK (2016) Enduring protection provided by NMITLI118RT+ and its preparation NMITLI118RT+ CFM against ischemia/reperfusion injury in rats. RSC Adv 6:42827–42835

Ahmed SA, Reda RM, ElHady M (2020) Immunomodulation by *Coriandrum sativum* seeds (coriander) and its ameliorative effect on lead induced immunotoxicity in Nile tilapia (*Oreochromis niloticus* L.). Aquac Res 51(3):1077–1088

Allam G (2009) Immunomodulatory effects of curcumin treatment on murine schistosomiasis mansoni. Immunobiology 214(8):712–727

Al-Mofleh IA, Alhaider AA, Mossa JS et al (2006) Protection of gastric mucosal damage by *Coriandrum sativum* L. pre-treatment in Wistar albino rats. Environ Toxicol Pharmacol 22(1): 64–69

An N, Wang D, Zhu T et al (2009) Effects of scrocaffeside A from Picrorhiza scrophulariiflora on immunocyte function in vitro. Immunopharmacol Immunotoxicol 31(3):451–458

Anandakumar S, Kannan D, Wilson E, Narayanan KB, Suresh G, Kanakavalli K, Manoharan MT (2020) Potential phytopharmaceutical constituents of *Solanum trilobatum* L. as significant inhibitors against COVID-19: Robust-binding mode of inhibition by molecular docking, PASS-aid bioactivity and ADMET investigations.doi.org/https://doi.org/10.26434/chemrxiv.12781754.v1

Arthur JR, McKenzie RC, Beckett GJ (2003) Selenium in the immune system. J Nutr 133(5): 1457S–1459S

Azad M, Kalam A, Sarker M et al (2018) Immunomodulatory effects of probiotics on cytokine profiles. Bio Med Res Int 2018:1–10

Bachiega P, Salgado JM, de Carvalho JE et al (2016) Antioxidant and antiproliferative activities in different maturation stages of broccoli (*Brassica oleracea* Italica) biofortified with selenium. Food Chem 190:771–776

Beck MA, Levander OA (2000) Host nutritional status and its effect on a viral pathogen. J Infect Dis 182(Supplement_1):S93–S96

Bharani SE, Asad M, Dhamanigi SS et al (2010) Immunomodulatory activity of methanolic extract of *Morus alba* Linn (mulberry) leaves. Pak J Pharm Sci 23(1):63–68

Bhaskaram P (2002) Micronutrient malnutrition, infection and immunity: an overview. Nutr Rev 60(5):40–45

Bhattacharya S, Muhammad N, Steele R et al (2017) Bitter melon enhances natural killer–mediated toxicity against head and neck cancer cells. Cancer Prev Res 10(6):337–344

Bonham M, O'Connor JM, Hannigan BM (2002) The immune system as a physiological indicator of marginal copper status? Br J Nutr 87(5):393–403

Boquien CY (2018) Human milk: An ideal food for nutrition of preterm newborn. Front Pediatr 6: 295

Broome CS, McArdle F, Kyle JA (2004) An increase in selenium intake improves immune function and poliovirus handling in adults with marginal selenium status. Am J Clin Nutr 80(1):154–162

Chandra RK (1991) Nutrition and immunity: lessons from the past and insights into the future. Am J Clin Nutr 53(5):1087–1101

Chen CC, Yan SH, Yen MY et al (2016a) Investigations of kanuka and manuka essential oils for in vitro treatment of disease and cellular inflammation caused by infectious microorganisms. J Microbiol Immunol Infect 49:104–111

Chen HC, Chang WT, Hseu YC et al (2016b) Immunosuppressive effect of *Litsea cubeba* L. essential oil on dendritic cell and contact hypersensitivity responses. Int J Mol Sci 17:1319

Chen LL, Zhang HJ, Chao J et al (2017) Essential oil of *Artemisia argyi* suppresses inflammatory responses by inhibiting JAK/STATs activation. J Ethnopharmacol 204:107–117

Cheng C, Zou Y, Peng J (2018) Oregano essential oil attenuates RAW264.7 cells from lipopolysaccharide induced inflammatory response through regulating NADPH oxidase activation-driven oxidative stress. Molecules 23:1857

Cheng YB, Chang MT, Lo YW et al (2009) Oxygenated lignans from the fruits of *Schisandra arisanensis*. J Nat Prod 72(9):1663–1668

Cherng JM, Chiang W, Chiang LC (2008) Immunomodulatory activities of common vegetables and spices of Umbelliferae and its related coumarins and flavonoids. Food Chem 106(3):944–950

Childs CE, Calder PC, Miles EA (2019) Diet and immune function. Nutrients 11:1–9

Chithra V, Leelamma S (2000) *Coriandrum sativum* effect on lipid metabolism in 1, 2-dimethyl hydrazine induced colon cancer. J Ethnopharmacol 71(3):457–463

Chowanski S, Adamski Z, Marciniak P et al (2016) A review of bioinsecticidal activity of Solanaceae alkaloids. Toxins 8(3):60

Clare DA, Catignani GL, Swaisgood HE (2003) Biodefense properties of milk: the role of antimicrobial proteins and peptides. Curr Pharm Des 9(16):1239–1255

Clare DA, Swaisgood HE (2000) Bioactive milk peptides: a prospectus. J Dairy Sci 83(6): 1187–1195

Clement F, Pramod SN, Venkatesh YP (2010) Identity of the immunomodulatory proteins from garlic (*Allium sativum*) with the major garlic lectins or agglutinins. Int Immunopharmacol 10(3): 316–324

Cohen S, Danzaki K, MacIver NJ (2017) Nutritional effects on T-cell immunometabolism. Eur J Immunol 47(2):225–235

Courtemanche C, Elson-Schwab I, Mashiyama ST (2004) Folate deficiency inhibits the proliferation of primary human CD8+ T lymphocytes in vitro. J Immunol 173(5):3186–3192

Cousin FJ, Mater DD, Foligné B et al (2011) Dairy propionibacteria as human probiotics: a review of recent evidence. Dairy Sci Technol 91(1):1–26

Cross ML, Gill HS (2000) Immunomodulatory properties of milk. Br J Nutr 84(S1):81–89

Dahanukar SA, Thatte UM (1997) Current status of Ayurveda in phytomedicine. Phytomed 4:359–368

Dallman PR (1987) Iron deficiency and the immune response. Am J Clin Nutr 46(2):329–334

Darkwah WK, Koomson DA, Miwornunyuie N et al (2020) Phytochemistry and medicinal properties of *Solanum torvum* fruits. All Life 13(1):498–506

Davis L, Kuttan G (2000) Immunomodulatory activity of *Withania somnifera*. J Ethnopharmacol 71:193–200

de Almeida Melo E, Mancini Filho J, Guerra NB (2005) Characterization of antioxidant compounds in aqueous coriander extract (*Coriandrum sativum* L.). LWT-food. Sci Technol 38(1):15–19

Deng YY, Yi Y, Zhang LF et al (2014) Immunomodulatory activity and partial characterisation of polysaccharides from *Momordica charantia*. Molecules 19(9):13432–13447

Dillasamola D, Aldi Y, Kolobinti M (2019) The effect of coriander ethanol extract (*Coriandrum sativum* L.) against phagocytosis activity and capacity of the macrophage cells and the percentage of leukocyte cells in white male mice. Pharm J 11(6):1290–1298

Dipasquale V, Serra G, Corsello G, Romano C (2020) Standard and specialized infant formulas in Europe: making, marketing, and health outcomes. Nutr Clin Pract 35(2):273–281

Ebringer L, Ferencík M, Krajcovic J (2008) Beneficial health effects of milk and fermented dairy products- review. Folia Microbiol 53(5):378–394

Eftekhar N, Moghimi A, Roshan NM, Saadat S, Boskabady MH (2019) Immunomodulatory and anti-inflammatory effects of hydro-ethanolic extract of *Ocimum basilicum* leaves and its effect on lung pathological changes in an ovalbumin-induced rat model of asthma. BMC Compl Altern Med 19(1):349

Eguale T, Tilahun G, Debella A et al (2007) *In vitro* and *in vivo* anthelmintic activity of crude extracts of *Coriandrum sativum* against *Haemonchus contortus*. J Ethnopharmacol 110(3): 428–433

El-Sayed SAEG, Ahmed SY (2017) Effects of coriander seeds powder (*Coriandrum sativum*) as feed supplements on growth performance parameters and immune response in albino rats. Int J Livest Res 7(2):191–200

Emeish WFA, El-Deen AGS (2016) Immunomodulatory effects of thyme and fenugreek in sharp tooth catfish. *Clarias gariepinus*. Assiut Vet Med J 62(150):1–7

Fahey JW, Haristoy X, Dolan PM et al (2002) Sulforaphane inhibits extracellular, intracellular, and antibiotic-resistant strains of *Helicobacter pylori* and prevents benzo [a] pyrene-induced stomach tumors. Proc Natl Acad Sci 99(11):7610–7615

Flaherty SO, Saulnier D, Pot B, Versalovic J (2010) How can probiotics and prebiotics impact mucosal immunity? Gutmicrobes 1(5):293–300

Gombart AF, Pierre A, Maggini S (2020) A review of micronutrients and the immune system–working in harmony to reduce the risk of infection. Nutrients 12(1):236

Goo KS, Ashari S, Basuki N et al (2016) The bitter gourd *Momordica charantia* L.: morphological aspects, charantin and vitamin C contents. J Agric Vet Sci 9(10):76–81

Guerra NB, de Almeida ME, Mancini Filho J (2005) Antioxidant compounds from coriander (*Coriandrum sativum* L.) etheric extract. J Food Compos Anal 18(2–3):193–199

Guillin OM, Vindry C, Ohlmann T (2019) Selenium, selenoproteins and viral infection. Nutrients 11(9):2101

Habte G, Assefa S (2020) *In vivo* antimalarial activity of crude fruit extract of *Capsicum frutescens* var. minima (Solanaceae) against *Plasmodium berghei* infected mice. Biomed Res Int 2020:1–7

Halder S, Bharal N, Mediratta PK et al (2009) Anti-inflammatory, immunomodulatory and antinociceptive activity of *Terminalia arjuna* Roxb bark powder in mice and rats. Indian J Exp Biol 47(7):577–583

Han X, Parker TL (2017) Anti-inflammatory, tissue remodeling, immunomodulatory, and anticancer activities of oregano (*Origanum vulgare*) essential oil in a human skin disease model. Biochimie Open 4:73–77

Hashim MS, Lincy S, Remya V et al (2005) Effect of polyphenolic compounds from *Coriandrum sativum* on H2O2-induced oxidative stress in human lymphocytes. Food Chem 92(4):653–660

Hemalatha R, Karthik M, Babu KN, Kumar BD (2012) Immunomodulatory activity of *Triticum aestivum* and its effects on Th1/Th2 cytokines and NF κ BP 65 response. Am J Biochem Mol Biol 2(1):19–25

Hernell O (2011) Human milk vs. cow's milk and the evolution of infant formulas. In Milk and milk products in human nutrition 67:17–28 Karger Publishers

Hoffmann PR, Berry MJ (2008) The influence of selenium on immune responses Mol. Nutr Food Res 52(11):1273–1280

Hosseinzadeh H, Alaw Qotbi AA, Seidavi A et al (2014) Effects of different levels of coriander (*Coriandrum sativum*) seed powder and extract on serum biochemical parameters, microbiota, and immunity in broiler chicks. Sci World J 2014:1–11

Huang DF, Xie MY, Yin JY et al (2009) Immunomodulatory activity of the seeds of *Plantago asiatica* L. J Ethnopharmacol 124(3):493–498

Innocent BX, Fathima SAM, Sivarajani S (2011) Immune response of *Catla catla* fed with an oral immunostimulant *Plumbago rosea* and post challenged with *Aeromonas hydrophila*. Int J Appl Biol Pharm Technol 2(4):447–454

Ismail S, Asad M (2009) Immunomodulatory activity of *Acacia catechu*. Indian J Physiol Pharmacol 53(1):25–33

Jang HW, Moon JK, Shibamoto T (2015) Analysis and antioxidant activity of extracts from broccoli (*Brassica oleracea* L.) sprouts. J Agric Food Chem 63(4):1169–1174

Jantan I, Ahmad W, Bukhari SNA (2015) Plant-derived immunomodulators: an insight on their preclinical evaluation and clinical trials. Front Plant Sci 6:655

Jiang Y, Wu SH, Shu XO et al (2014) Cruciferous vegetable intake is inversely correlated with circulating levels of proinflammatory markers in women. J Acad Nutr Diet 114(5):700–708

Kannabiran K, Thanigairassu RR, Khanna VG (2008) Antibacterial activity of saponin isolated from the leaves of *Solanum trilobatum* Linn. J Appl Biol Sci 2(3):109–112

Kaur GJ, Arora DS (2010) Bioactive potential of *Anethum graveolens*, *Foeniculum vulgare* and *Trachyspermum ammi* belonging to the family Umbelliferae-current status. J Med Plant Res 4(2):087–094

Keshavarz A, Minaiyan M, Ghannadi A et al (2013) Effects of *Carum carvi* L.(caraway) extract and essential oil on TNBS-induced colitis in rats. Res Pharm Sci 8(1):1–8

Khan NT, Jameel N (2018) Antifungal activity of Ajawain seeds (*Trachyspermum ammi*). J Biomol Res Ther 7(2):1–2

Kitazawa H, Matsumura K, Itoh T et al (1992) Interferon induction in murine peritoneal macrophage by stimulation with Lactobacillus acidophilus. Microbiol Immunol 36(3):311–315

Koffuor GA, Amoateng P, Andey TA (2011) Immunomodulatory and erythropoietic effects of aqueous extract of the fruits of *Solanum torvum* Swartz (Solanaceae). Pharmacogn Res 3(2):130

Krifa M, El Mekdad H, Bentouati N et al (2015) Immunomodulatory and anticancer effects of *Pituranthostortuosus* essential oil. Tumour Biol 36:5165–5170

Kumar D, Arya V, Kaur R et al (2012) A review of immunomodulators in the Indian traditional health care system. J Microbiol Immunolog Infect 45(3):165–184

Latorre AO, Furlan MS, Sakai M et al (2009) Immunomodulatory effects of *Pteridium aquilinum* on natural killer cell activity and select aspects of the cellular immune response of mice. J Immunotoxicol 6(2):104–114

Lee DC, Yang CL, Chik SC (2009) Bioactivity-guided identification and cell signaling technology to delineate the immunomodulatory effects of *Panax ginseng* on human promonocytic U937 cells. J Transl Med 14(7):34

Lin PL, Lin KW, Weng CF et al (2009) Yam storage protein dioscorins from *Dioscorea alata* and *Dioscorea japonica* exhibit distinct immunomodulatory activities in mice. J Agric Food Chem 57(11):4606–4613

Lobo V, Patil A, Phatak A et al (2010) Free radicals, antioxidants and functional foods: impact on human health. Pharmacogn Rev 4(8):118

Ma Q, Jiang JG, Yuan X et al (2019) Comparative antitumor and anti-inflammatory effects of flavonoids, saponins, polysaccharides, essential oil, coumarin and alkaloids from *Cirsium japonicum* DC. Food Chem Toxicol 125:422–429

Mahamat O, Flora H, Tume C, Kamanyi A (2020) Immunomodulatory activity of *Momordica charantia* L.(Cucurbitaceae) leaf diethyl ether and methanol extracts on salmonella typhi-Infected mice and LPS-induced phagocytic activities of macrophages and neutrophils. Evidence-Based Complementary and Alternative Medicine 2020

Mahmoodi M, Ayoobi F, Aghaei A et al (2019) Beneficial effects of *Thymus vulgaris* extract in experimental autoimmune encephalomyelitis: clinical, histological and cytokine alterations. Biomed Pharmacother 109:2100–2108

Mangathayaru K, Umadevi M, Reddy CU (2009) Evaluation of the immunomodulatory and DNA protective activities of the shoots of *Cynodon dactylon*. J Ethnopharmacol 123(1):181–184

Mansouri E, Kooti W, Bazvand M et al (2015) The effect of hydro-alcoholic extract of *Foeniculum vulgare* mill on leukocytes and hematological tests in male rats. Jundishapur J Nat Pharm Prod 10(1):e18396

Manu KA, Kuttan G (2009) Immunomodulatory activities of punarnavine, an alkaloid from *Boerhaavia diffusa*. Immunopharmacol Immunotoxicol 31(3):377–387

Marcos A, Nova E, Montero A (2003) Changes in the immune system are conditioned by nutrition. Eur J Clin Nutr 57(1):S66–S69

Marrelli M, Araniti F, Abenavoli MR, Statti G, Conforti F (2018) Potential health benefits of Origanum heracleoticum essential oil: phytochemical and biological variability among *different Calabrian* populations. Nat Prod Commun 13

Maywald M, Wessels I, Rink L (2017) Zinc signals and immunity. Int J Mol Sci 18(10):2222

Medici M, Vinderola CG, Perdigón G (2004) Gut mucosal immunomodulation by probiotic fresh cheese. Int Dairy J 14(7):611–618

Meisel H (2004) Multifunctional peptides encrypted in milk proteins. Biofactors 21(14):55–61

Moineau S, Goulet J (1991) Effect of fermented milks on humoral immune response in mice. Int Dairy J 1(4):231–239

Mora JR, Iwata M, Von Andrian UH (2008) Vitamin effects on the immune system: vitamins A and D take centre stage. Nat Rev Immunol 8(9):685-98

Naeini A, Khosravi AR, Chitsaz M et al (2009) Anti-*Candida albicans* activity of some Iranian plants used in traditional medicine. J Mycol Med 19(3):168–172

Nahak G, Sahu R (2014) Immunostimulatory effects of *Ocimum sanctum* Linn. Leaf extracts in *Clarias batrachus* Linn. Asian J Pharm Clin Res 7(3):157–163

Naik SR, Hule A (2009) Evaluation of immunomodulatory activity of an extract of andrographolides from *Andrographis paniculata*. Planta Med 75(8):785–791

Narnoliya LK, Jadaun JS (2018) Synbiotics: necessity of today's meal. J Bioprocess Biotech 8:332. https://doi.org/10.4172/2155-9821.1000332

Ouzek G, Schepetkin IA, Utegenova GA et al (2017) Chemical composition and phagocyte immunomodulatory activity of *Ferula iliensis* essential oils. J Leukoc Biol 101:1361–1371

Owis AI (2015) Broccoli; the green beauty: a review. J Pharm Sci Res 7(9):696

Özbek H, Uğraş S, Dülger H, Bayram I, Tuncer I, Öztürk G, Öztürk A (2003) Hepatoprotective effect of Foeniculum vulgare essential oil. Fitoterapia 74(3):317–319

Pan DD, Wu Z, Liu J et al (2013) Immunomodulatory and hypoallergenic properties of milk protein hydrolysates in ICR mice. J Dairy Sci 96(8):4958–4964

Pandurangan A, Khosa RL, Hemalatha S (2010) Antinociceptive activity of steroid alkaloids isolated from *Solanum trilobatum* Linn. J Asian Nat Prod Res 12(8):691–695

Park Y, Yoo SA, Kim WU et al (2016) Anti-inflammatory effects of essential oils extracted from *Chamaecyparis obtusa* on murine models of inflammation and RAW264.7 cells. Mol Med Rep 13:3335–3341

Park YW, Nam MS (2015) Bioactive peptides in milk and dairy products: a review. Korean J Food Sci Anim Resour 35(6):831

Parvez GM (2017) Current advances in pharmacological activity and toxic effects of various capsicum species. Int J Pharm Sci Res 8(5):1900–1912

Patel SR, Malhotra A, Gao X (2012) A prospective study of sleep duration and pneumonia risk in women. Sleep 35(1):97–101

Perdigón G, Fuller R, Raya R (2001) Lactic acid bacteria and their effect on the immune system. Curr Issues Intest Microbiol 2(1):27–42

Popa M, Maruțescu L, Oprea E et al (2020) *In vitro* evaluation of the antimicrobial and immunomodulatory activity of culinary herb essential oils as potential perioceutics. Antibiotics 9(7):428. 1-14

Puskarova A, Buckova M, Krakova L et al (2017) The antibacterial and antifungal activity of six essential oils and their cyto/genotoxicity to human hEL 12469. Cells Sci Rep 7:8211

Raja L, Pooshan GV, Venkatachalam VV et al (2009) Immunomodulatory activity of leaves of *Solanum trilobatum* in experimental rats. Pharmacologyonline 3:758–765

Raja RR (2012) Medicinally potential plants of Labiatae (Lamiaceae) family: an overview. Res J Med Plant 6(3):203–213

Rascon-Cruz Q, Espinoza-Sanchez EA, Siqueiros-Cendon TS et al (2021) Lactoferrin: a glycoprotein involved in immunomodulation, anticancer, and antimicrobial processes. Molecules 26(1): 205

Renjie L, Zhenhong L, Shidi S (2010) GC-MS analysis of fennel essential oil and its effect on microbiology growth in rats' intestine. Afr J Microbiol Res 4(12):1319–1323

Reyes Diaz A, Gonzalez Cordova AF, Hernandez Mendoza A (2018) Immunomodulation by hydrolysates and peptides derived from milk proteins. Int J Dairy Technol 71(1):1–9

Rinninella E, Cintoni M, Raoul P et al (2019) Food components and dietary habits: keys for a healthy gut microbiota composition. Nutrients 11(10):2393

Sahu J, Rathi B, Koul S et al (2013) *Solanum trilobatum* (Solanaceae)-an overview. J Nat Remedies 13(2):76–80

Samojlik N, Lakic N, Mimica-Dukic N et al (2010) Antioxidant and hepatoprotective potential of essential oils of coriander (*Coriandrum sativum* L.) and caraway (*Carum carvi* L.) (Apiaceae). J Agric Food Chem 58(15):8848–8853

Sangwan NS, Tripathi S, Srivastava Y, Mishra B, Pandey N (2017) Phytochemical genomics of ashwagandha. In: Science of ashwagandha: preventive and therapeutic potentials. Edi: A Kaul, Wadhwa R, Springer, Cham, pp 3–36

Sarma DN, Khosa RL (1994) Immunomodulators of plant origin a review. Anc Sci Life 13(3–4): 326

Sebastia B, Balagopal P, Misra R (2013) Diet-related diseases: issues and solutions to nutrition transition and food programme policies in India. 1–17

Shah VV, Shah ND, Patrekar PV (2013) Medicinal plants from Solanaceae family. Res J Pharm Technol 6(2):143–151

Shankar AH, Prasad AS (1998) Zinc and immune function: the biological basis of altered resistance to infection. Am J Clin Nutr 68(2):447S–463S

Sharififar F, Pournourmohammadi S, Arabnejad MC (2009) Immunomodulatory activity of aqueous extract of *Achillea wilhelmsii* C. Koch in mice. Ind J Exp Biol 47:668–671

Shuai XH, Hu TJ, Liu HL et al (2010) Immunomodulatory effect of a *Sophora subprosrate* polysaccharide in mice. Int J Biol Macromol 46(1):79–84

Shukla S, Mehta A, John J et al (2009) Immunomodulatory activities of the ethanolic extract of *Caesalpinia bonducella* seeds. J Ethnopharmacol 125(2):252–256

Siddiqui MJ, Aslam A, Khan T (2019) Comparison and evaluation of different seed extracts of *Trachyspermum ammi* for immunomodulatory effect on cell-mediated immunity through delayed-type hypersensitivity assay skin thickness method. J Pharm Bioallied Sci 11(1):43

Silva E, Oliveira J, Silva Y et al (2020) Lactoperoxidase system in the dairy industry: challenges and opportunities. Czech J Food Sci 38(6):337–346

Singh SP, Jadaun JS, Narnoliya LK et al (2017) Prebiotic oligosaccharides: special focus on fructo-oligosaccharides, its biosynthesis and bioactivity. Appl Biochem Biotechnol 183(2):613–635

Sini H, Devi KS (2004) Antioxidant activities of the chloroform extract of *Solanum trilobatum*. Pharm Biol 42(6):462–466

Siqueiros-Cendón T, Arévalo-Gallegos S et al (2014) Immunomodulatory effects of lactoferrin. Acta Pharmacol Sin 35(5):557–566

Smeriglio A, Denaro M, Barreca D et al (2017) In vitro evaluation of the antioxidant, cytoprotective, and antimicrobial properties of essential oil from *Pistacia vera* L. variety Bronte hull. Int J Mol Sci 18:1212

Song H, Fall K, Fang F et al (2019) Stress related disorders and subsequent risk of life threatening infections: population based sibling controlled cohort study. BMJ 367:l5784

Soyano A, Gomez M (1999) Role of iron in immunity and its relation with infections. Arch Latinoam Nutr 49(3 Suppl 2):40S–46S

Tai CS, Chen YY, Chen WL (2016) β-Lactoglobulin influences human immunity and promotes cell proliferation. Biomed Res Int 2016:1–12

Thejass P, Kuttan G (2007) Immunomodulatory activity of sulforaphane, a naturally occurring isothiocyanate from broccoli (*Brassica oleracea*). Phytomedicine 14(7–8):538–545

Tilg H (2015) Cruciferous vegetables: prototypic anti-inflammatory food components. Clin Phytosci 1(1):1–6

Touihri I, Boukhris M, Marrakchi N et al (2015) Chemical composition and biological activities of *Allium roseum* L. var. grandiflorum Briq essential oil. J Oleo Sci 64:869–879

Tsai KD, Lin BR, Perng DS et al (2011) Immunomodulatory effects of aqueous extract of *Ocimum basilicum* (Linn.) and some of its constituents on human immune cells. J Med Plant Res 5(10): 1873–1883

Uritu CM, Mihai CT, Stanciu GD et al (2018) Medicinal plants of the family Lamiaceae in pain therapy: a review. Pain Res Manag 2018:1–44

Usjak L, Petrovic S, Drobac M et al (2017) Edible wild plant *Heracleum pyrenaicum* subsp. orsinii as a potential new source of bioactive essential oils. J Food Sci Technol 54:2193–2202

Vinderola G, Matar C, Palacios J et al (2007) Mucosal immunomodulation by the non-bacterial fraction of milk fermented by lactobacillus helveticus R389. Int J Food Microbiol 115(2): 180–186

Vujicic M, Nikolic I, Kontogianni VG et al (2015) Methanolic extract of *Origanum vulgare* ameliorates type 1diabetes through antioxidant, anti-inflammatory and anti-apoptotic activity. Br J Nutr 113(5):770–782

Wang YT, Zhu L, Zeng D et al (2016) Chemical composition and anti-inflammatory activities of essential oil from *Trachydium roylei*. J Food Drug Anal 24:602–609

Wangensteen H, Samuelsen AB, Malterud KE (2004) Antioxidant activity in extracts from coriander. Food Chem 88(2):293–297

Wong PY, Kitts DD (2006) Studies on the dual antioxidant and antibacterial properties of parsley (*Petroselinum crispum*) and cilantro (*Coriandrum sativum*) extracts. Food Chem 97(3):505–515

Xu HS, Wu YW, Xu SF et al (2009) Antitumor and immunomodulatory activity of polysaccharides from the roots of *Actinidia eriantha*. J Ethnopharmacol 125(2):310–317

Zuurveld M, van Witzenburg NP, Garssen J et al (2020) Immunomodulation by human milk oligosaccharides: the potential role in prevention of allergic diseases. Front Immunol 11:1–17

Immunomodulation Potential of Woody Plants

16

Francisco Geraldo Barbosa, Marcos Carlos de Mattos,
Fátima Miranda Nunes, Jair Mafezoli,
and Maria Conceição Ferreira Oliveira

Abstract

It is suggested that a myriad of plant-derived natural products may present immunomodulation properties, being an important source for future development of immunomodulators to be used in immunotherapies. Indeed, various phytomedicines were proved to modulate components of inflammatory pathways and are used to combat cell aging, atherosclerosis, cancer, diabetes, rheumatoid arthritis, autoimmune disease, and Parkinson's disease, among others. These plants show signs of operating through immunostimulant, immunoadjuvant, and immunosuppressant activities or by affecting the effector arm of the immune response. The immunomodulation potential of three important medicinal woody plants, *Moringa oleifera* Lam. (Moringaceae), *Azadirachta indica* A. Juss (Meliaceae), and *Ginkgo biloba* L. (Ginkgoceae), is discussed in this chapter, which were selected based on their economic importance worldwide. This chapter does not cover the vast literature available for these plant species and intends to arouse the reader's interest in them.

Keywords

Woody plants · Immunomodulation · Moringa oleifera · Azadirachta indica · Ginkgo biloba

F. G. Barbosa · M. C. de Mattos · F. M. Nunes · J. Mafezoli · M. C. F. Oliveira (✉)
Department of Organic and Inorganic Chemistry, Science Centre, Federal University of Ceará, Campus do Pici, Fortaleza, Ceará State, Brazil
e-mail: mcfo@ufc.br

N. S. Sangwan et al. (eds.), *Plants and Phytomolecules for Immunomodulation*,
https://doi.org/10.1007/978-981-16-8117-2_16

16.1 Introduction

The immune system (IS) is a unique and complex network that governs important pathways associated with inflammation, microbial recognition, microbial clearance, cell and tissue damage and death, and wound healing processes. Therefore, it is our IS that protects us from immune-associated diseases, such as microbial infection, allergy, and cancers, besides various chronic diseases. In this context, any substrate able to induce immunomodulatory activities (either altering or affecting immune cell systems) to cause an immune response (via dynamic adjustment of the target immune systems) is referred as an immunomodulator. When the substrate stimulates the immune systems by causing the activation or raising the activity of immune system components, it is referred as immunostimulator or immunostimulant (Wen et al. 2012).

It is suggested that a myriad of plant-derived natural products may present immunomodulation properties, being an important source for future development of immunomodulators to be used in immunotherapies. Indeed, various phytomedicines were proved to modulate components of inflammatory pathways. In general, phytomedicines (extracts, fractions, or isolated compounds) are trusted to interact with various targets to promote pharmacological or physiological effects at different degrees (cellular, tissue, or organ) (Wen et al. 2012).

Plants have been used as medication since ancient times, probably over 4000 years ago, and traditional herbal medicine was originated in China and India. Traditional medicine is found in several civilizations and cultures, including India (Ayurvedic), Egyptian, Western, Chinese, Japan (Kampo), and South Asia (Greco-Arab or Unani-Tibb) (Kumar et al. 2012). Various medicinal plants used in Rasayana (focusing the increasing of body's resistance), the Indian traditional system, are known to possess immunomodulatory activity. These medicinal herbs are mainly used to combat cell aging, atherosclerosis, cancer, diabetes, rheumatoid arthritis, autoimmune disease, and neurological diseases, among others. These herbs show signs of operating via immunomodulation activities (immunostimulant, immunoadjuvant, and immunosuppressant) or by causing some effect to the effector arm of the immune response (Kamath and Yadav 2020).

This chapter discusses the immunomodulation potential of three important medicinal woody plants, *Moringa oleifera* Lam. (Moringaceae), *Azadirachta indica* A. Juss (Meliaceae), and *Ginkgo biloba* L. (Ginkgoceae). The reason for choosing these particular species lays in the fact of their economic importance in the world scenario. As proof of their outstanding potential, *M. oleifera* was nominated as "Botany of the Year in 2007" (Gupta et al. 2018); *A. indica* was mentioned as "one of the most promising of all plants" and "a tree for solving global problems" (National Research Council 1992); *G. biloba* is considered an "archaic living fossil," living on earth for 270 million years. Billions of doses of standard extracts from leaves of *G. biloba* are sold worldwide, being ranked as one of the top five selling herbal supplements (Isah 2015). It is noteworthy that this chapter has no intention to cover the vast literature available for these plant species, but to arouse the reader's interest in them.

16.2 *Moringa oleifera* Lam. (Moringaceae)

16.2.1 Botanic Aspects and Occurrence

M. oleifera (syn. *M. pterygosperma* Gaertn.) (Fig. 16.1a) is related to the monogeneric family Moringaceae constituted of the single genus *Moringa* with about 13 species of shrubs and trees (Gupta et al. 2018; Jacques et al. 2020). This plant is native from Northwest India. As a plant that tolerates dryness, it grows in semi-arid, tropical, and subtropical regions, occurring in the Americas, Africa, Europe, Asia, and Oceania. The plant is broadly dispersed in India, Egypt, Philippines, Ceylon, Thailand, Malaysia, Burma, Pakistan, Singapore, West Indies, Brazil, Cuba, Jamaica, and Nigeria (Ramachandran et al. 1980; Foidl et al. 2001; Bosch 2017; Brilhante et al. 2017; Jacques et al. 2020) (Fig. 16.1b). In India, *M. oleifera* is called "drumstick" and "tree of life," while in some parts of the world, it is called "moringa" (Ramachandran et al. 1980; Gupta et al. 2018; Jacques et al. 2020). Due to its importance as a medicinal and nutraceutical plant, *M. oleifera* is economically valued and was honored by the National Institute of Health (NIH) as "Botany of the Year in 2007" (Gupta et al. 2018).

Botanically, *M. oleifera* can be described as a fast-growing, perennial tree with small to medium size (7–12 m in height); stem brittle; bark corky; whitish gray, soft, fissured, glabrous, drooping branches; and tuberous roots (Ramachandran et al. 1980; Foidl et al. 2001; Pandey et al. 2011; Bosch 2017). The plant leaves are spirally arranged (25–45 cm long), crowded at the end of the branches, pale green and bipinnate or more often tripinnate with opposite and ovoid leaves (Ramachandran et al. 1980; Pandey et al. 2011). Flowers (Fig. 16.1a) are white or cream, fragrant, bisexual, oblique, stalked, united into erect, axillary, many-flowered panicles, flowers with five free petals, oblong-spatulate (1–2 cm long), unequal, and finely veined (Ramachandran et al. 1980; Pandey et al. 2011). Fruits (Fig. 16.1a) are pendulous and elongated capsules (10–50 cm long), obtusely trigonous, ribbed, three-valved, valves spongy and thick, brown in color, and becoming dark brown when ripe. When dry, fruits form pods with three lobes, which open into three parts. Pods contain numerous (12–35 seeds *per* pod) seeds (Fig. 16.1a). The seeds are

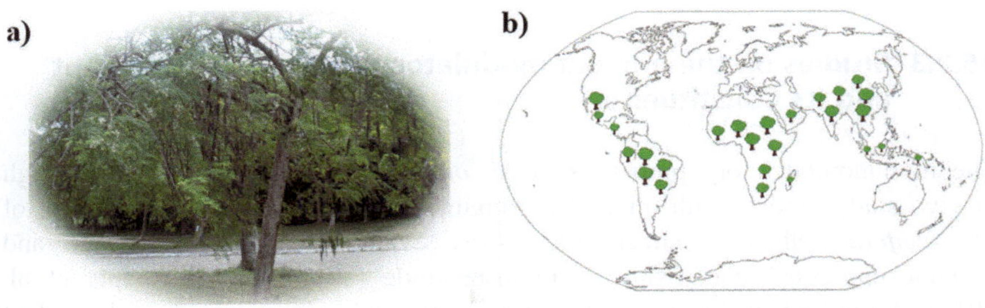

Fig. 16.1 (**a**) Picture of a specimen of *M. oleifera*. Source: F. G. Barbosa; (**b**) Representative image of the global distribution of the species

globular (1.0–1.5 cm diameter) and three-winged; thin wings are produced at the base and the apex (Ramachandran et al. 1980; Foidl et al. 2001; Bosch 2017). Moreover, *M. oleifera* exudes a gum via its lysigenous traumatic gum ducts found in the bark. The gum exudates have variable shapes and color, the latter varying from yellow to reddish brown or black (Bhandari 1995; Gupta et al. 2018).

16.2.2 Ethnopharmacology

The ethnopharmacological properties of *M. oleifera* have been known for a long time in the Indian traditional medicine. It has been mentioned by some authors as Chakradatta (A.D. 1050) and in the Sanskrit work on medicine *Bhavaprakasa* (Ramachandran et al. 1980). This tree is a medicinal plant reported as a natural source of anthelmintic, mild antibiotic, and detoxifier, being used in several countries to treat malnutrition and malaria (Nantachit 2006; Jacques et al. 2020). In addition, it is used as antiulcer, antibacterial, antifungal, antispasmodic, and diuretic, in wound healing, and to treat eye diseases (Caceres et al. 1991; Caceres et al. 1992; Rathi et al. 2006; Singh and Navneet. 2018). In Africa, *M. oleifera* is called "miraculous tree" and "never die" due to its potential to act toward more than 300 illnesses, including rheumatism and articular pain (N'diaye et al. 2002; Gupta et al. 2018). The tree has also been used to purify water in some poor regions due to its coagulant, flocculant, and antimicrobial properties (Gupta et al. 2018).

Studies indicate *M. oleifera* as an exceptional immune builder, being considered the cheapest and trustworthy alternative to provide good nutrition and healing and preventing many disorders (Dhakad et al. 2019). It is noteworthy that all parts of the plant, including leaves, flowers, immature pods, mature seeds, and roots, are consumed by humans (Gupta et al. 2018). The high nutritional value of this plant is due to the presence of various phenolic compounds, besides proteins and vitamins (Anwar et al. 2007; Manguro and Lemmen 2007; Brilhante et al. 2017). Thus, both the high nutritional value of *M. oleifera* and its medicinal properties, such as anti-inflammatory (Caceres et al. 1992), antioxidant, and anticarcinogenesis (Bharali et al. 2003), contribute to the immunomodulatory potential of this plant.

16.2.3 Studies on the Immunomodulatory Potential of the Plant and its Constituents

The immunomodulatory properties of *M. oleifera* have been evaluated through in vivo study models with mice. The immunomodulatory effect of the leaves of *M. oleifera* (ethanol extract 50%) was evaluated in both normal and immunosuppressed mice with cyclophosphamide (30 mg kg^{-1}) (Gupta et al. 2010). According to this study, the extract presented significant dose-dependent increase in leukocytes, percent neutrophils, and weight of thymus and spleen, along with phagocytic index in normal and immunosuppressed mice. These results suggested a significant reduction of cyclophosphamide-induced immunosuppression by stimulating both cellular and humoral immunities (Gupta et al. 2010).

Another study involving the immunomodulatory action of methanol extract from the plant leaves was reported (Sudha et al. 2010). In the study, the cellular immunity was assayed using neutrophil adhesion test, neutropenia induced by cyclophosphamide (200 mg kg^{-1}), and carbon clearance test. Additionally, the humoral immunity was evaluated through mice lethality test, serum immunoglobulin estimation, and indirect hemagglutination. The extract (250 and 750 mg kg^{-1}) significantly increased the levels of serum immunoglobulins and prevented the mortality induced by bovine *Pasteurella multocida* in mice. In addition, it was observed significant increment in both the circulating antibody titer in indirect hemagglutination test and adhesion of neutrophils. In carbon clearance assay, it was also observed the attenuation of the cyclophosphamide-induced neutropenia and the increment in phagocytic index. The study suggested that the plant acts by stimulating the immune system through cellular and humoral immunity. It is worth mentioning that the lowest dose of extract (250 mg/kg) was more effective than the highest dose (750 mg kg^{-1}).

The immunomodulatory potential of organic extracts (petroleum ether, chloroform, and methanol) from leaves of *M. oleifera* was assayed in Wistar albino rats (100, 200, and 400 mg kg^{-1} as tested doses) (Gaikwad et al. 2011). The immunomodulation assessment was carried out by delayed-type hypersensitivity (DTH), humoral antibody titer, cyclophosphamide-induced myelosuppression (30 mg kg^{-1}), and T cell population tests. Levamisole (50 mg kg^{-1}), a standard immunomodulatory drug, was used as positive control. As a result, petroleum ether (dose of 400 mg kg^{-1}) and methanol (all doses) extracts potentiated the DTH response ($p < 0.01$) after 24 h. The methanol extract showed dose dependence with the augmentation of antibody titer ($p < 0.05$). In addition, restoration of hematological parameters ($p < 0.01$) was suggested through the observed counteraction of cyclophosphamide-induced immunosuppression. Thus, the study suggested that the methanol extract from the plant leaves can stimulate both cellular and humoral immunities (Gaikwad et al. 2011).

The effects of the ethanol and ethanol:water (1:1) extracts from the leaves of *M. oleifera* were evaluated in vivo using Wistar rats (50, 100, and 200 mg kg^{-1}) (Banji et al. 2012). The hydroalcoholic extract (100 and 200 mg kg^{-1}) substantially enhanced cellular and humoral immune responses, neutrophil index, and phagocytic activity, while the ethanol extract was only efficient in the highest dose (200 mg kg^{-1}). The authors associated the enhancing of immunity due to the occurrence of flavonoids, vitamins, and minerals in the extracts (Banji et al. 2012). It is worth mentioning that similar results were found when the immunomodulatory potential of the methanol extract from plant leaves was investigated in Wistar albino rats. The observed results were attributed to the presence of macro- and micronutrients in the plant, besides the phytochemicals (Nfambi et al. 2015).

The in vitro immunomodulatory activity of aqueous extract from the plant leaf to $CD4^+$, $CD8^+$, and $B220^+$ cells in *Mus musculus* was determined. According to the study, the low-dose extract (0.1 μg mL^{-1}) increased the number of $CD4^+$ and $CD8^+$ cells, while the high-dose extract (10 μg mL^{-1}) increased the number of $B220^+$ cells, confirming an immunomodulatory action. The authors attributed the observed activity of the extract to the occurrence of flavonoids and saponins, which act as immunostimulant on $CD4^+$ (T helper cell) and $CD4^+$ (T cytotoxic cell) and $B220^+$ as well (Rachmawati and Rifa'i 2014).

The immunomodulatory potential of *M. oleifera* has also been exploited in the feeding supplementation of chickens. The effects of the powder plant as supplement (1.5%) were evaluated in the immunology, biochemical parameters, and carcass quality of broiler chicks (Mousa et al. 2017). According to the study, *M. oleifera* was a good feed additive, with antioxidant potential, increasing immunity and productivity of broiler chicks. In another study with dietary *M. oleifera* leaf supplementation in broilers, it was observed quadratically increase of total anti-oxidative capacity (T-AOC), the activity of total superoxide dismutase (T-SOD), and glutathione peroxidase (GSH-PX), besides decreased malondialdehyde (MDA) contents in plasma. These results corroborated the potential antioxidant properties of the plant (Cui et al. 2018). Dietetic supplementation of broiler chicks under heat stress conditions with *M. oleifera* modulated their immune response. In this case, there was a regulation of mRNA expression levels of the innate immune response mediators (interleukins IL2 and IL6), with decreasing the degenerative changes of the live tissue following heat stress conditions (El-Deep et al., 2019). Other studies confirmed the incorporation of the plant in chicken foods to improve their immune response and growth performance (Oghenebrorhie and Oghenesuvwe 2016; Khan et al. 2017).

When the immune system is attacked, the energy metabolism and protein requirements are enhanced to allow the process of cell replication, besides antibody and cytokine formation (Cunningham-Rundles 1982). In this context, the immune function of the plant is due to its high nutritional value, acting on the lymphoid system or indirectly affecting the cellular material or other organ systems that act as immune regulators. In addition, micronutrients (vitamins and minerals) present in *M. oleifera* can regulate metabolic and physiologic processes (Banji et al. 2012). Therefore, calcium mineral present in large quantity in this tree could be accountable for enhancing immune function and interleukin (IL-2) production (Gearing et al. 1985; Banji et al. 2012). The complete list of macro- and micronutrients from leaves, pods, and seeds of *M. oleifera* is reported in literature (Brilhante et al. 2017).

M. oleifera is rich in essential and non-essential amino acids and vitamins. Among these, L-arginine, an essential amino acid, increases cell immunity and enhances the proliferation of both lymphocyte and macrophage (Ferreria et al. 2008; Banji et al. 2012). Vitamin A acts on the immune system facilitating lymphocyte proliferation and increasing antibody production (Banji et al. 2012). In addition, the high levels of vitamins C and E can promote an important antioxidant action. Concerning the active phytochemicals present in the plant, flavonoids contribute to its immunological potential, since these compounds are known to activate lymphocytes (Cherng et al. 2008; Banji et al. 2012).

The most common and abundant flavonoids in the leaves of *M. oleifera* are kaempferol, quercetin, myricetin, isorhamnetin, and apigenin. These compounds are also reported to be glycosylated with a multitude of glycosyl moieties

Fig. 16.2 Chemical structures of major bioactive flavonoids and isothiocyanates found in *M. oleifera*

(Fig. 16.2; Bennett et al. 2003; Leone et al. 2015; Makita et al. 2016). It is known that some flavonoids, in addition to their antioxidant properties, possess antiviral activity and can contribute to improving the immune system. Among these, quercetin shows virucidal activity against enveloped viruses such as parainfluenza type 3, mengovirus, pseudorabies, respiratory syncytial, Sindbis, and herpes simplex (Choi et al. 2009). Quercetin, kaempferol, and myricetin, besides some of their glycosides, are among polyphenols that inhibited various steps during SARS-CoV-type virus entry and replication (Chinsembu 2020; Yang et al. 2020).

In addition to the importance of the antioxidant properties for the immunomodulatory action of *M. oleifera*, the anti-inflammatory and anti-tumor potential of the plant may support its therapeutic use in immunosuppression cases. The anti-inflammatory properties of this tree plant contribute to the modulation of the inflammatory process, which involves a protective immunovascular response with the participation of immune cells, blood vessels, and molecular mediators (Leone et al. 2015).

The ethyl acetate fraction of *M. oleifera* extract (water-methanol 50%) inhibits human macrophage cytokine production induced by cigarette smoke. According to this study, the tested fraction, containing high levels of phenolic compounds, was able to inhibit both the human macrophage cytokine production (TNF-α, IL-6, and IL-8) and the expression of gene *RelA*, implicated in the NF-κB p65 signaling in inflammation (Kooltheat et al. 2014).

The potential of a concentrated extract from *M. oleifera* leaves to relieve low-grade inflammation related to chronic diseases was proposed in literature (Waterman et al. 2014). Both the concentrate extract (1.66% isothiocyanates and 3.82% total polyphenols) and two isolated isothiocyanates, 4-[(α-L-rhamnosyloxy) benzyl]isothiocyanate and 4-[(4'-*O*-acetyl-α-L-rhamnosyloxy)benzyl]isothiocyanate (Fig. 16.2), significantly decreased gene expression and production of inflammatory markers in RAW macrophages, attenuating expression of inducible nitric oxide synthase (iNOS) and IL-1β and production of NO and TNF-α (Waterman et al. 2014).

Dried ethanol extract from *M. oleifera* leaves, besides quercetin (Fig. 16.2), inhibited the translocation of p65 subunit of NF-κB in mice fed with high-fat diet. It was also observed that the expressions of iNOS, IFN-γ, and C-reactive protein (CRP) were downregulated. In addition, liberation of serum inflammatory cytokines TNF-α and IL-6 was potently diminished (Das et al. 2013).

The boiled extracts from *M. oleifera* pods presented anti-inflammatory activity, which was verified through pro-inflammatory mediator expression in the lipopolysaccharide-induced murine RAW264.7, suggesting that the bioactive compounds present in the plant pods can contribute to make better the pathogenesis of inflammatory-associated chronic diseases (Muangnoi et al. 2012).

The hydroethanol extract from flowers of *M. oleifera* inhibited the production of NO, PGE2, pro-inflammatory cytokines, and inflammatory mediator in LPS-stimulated macrophages by preventing degradation of IκB-α in the NF-κB signaling pathway. These results suggested the anti-inflammatory action of the plant flowers (Tan et al. 2015).

16.2.4 Final Remarks

Besides its medicinal properties, *M. oleifera* is a relatively safe food plant. Numerous in vitro and in vivo studies confirmed the diverse ethnopharmacological and pharmacological properties of different parts of the plant. Leaves are the most used part of the plant, being rich in macro- and micronutrients (vitamins and minerals), besides bioactive compounds, including isothiocyanates, flavonoids, and polyphenols. The immunomodulatory properties of this tree plant and some of its secondary metabolites have been demonstrated through immunomodulation, anti-inflammatory, and antioxidant assays.

16.3 *Azadirachta indica* A. Juss (Meliaceae)

16.3.1 Botanic Aspects and Occurrence

The Swedish botanist Linnaeus was the first to describe the plant, as *Melia azadirachta*, in 1753. Almost 80 years later, the French botanist A. de Jussieu (1830) separated *Azadirachta* from *Melia*, and the plant was reclassified as *A. indica* A. Juss. Therefore, *Melia azadirachta* L. and *M. indica* (A. Juss.) Brand are reported as synonyms (van der Nat et al. 1991).

A. indica belongs to Meliaceae (Rutales order), a family of flowering plants (Angiosperms), and has only *A. excelsa* (Jack) Jacobs as accepted congener in The Plant List database (http://www.theplantlist.org). It is a perennial, fast-growing tree that reaches up to 30 m tall and 2.5 m in girth, with straight trunk. Whenever possible, its roots penetrate deeply the soil, especially when injured (National Research Council 1992). The plant leaves are comprised of 5–15 leaflets each, and its white flowers present honey smell and measure 0.42–0.51 cm (1/5–1/6 inch) long

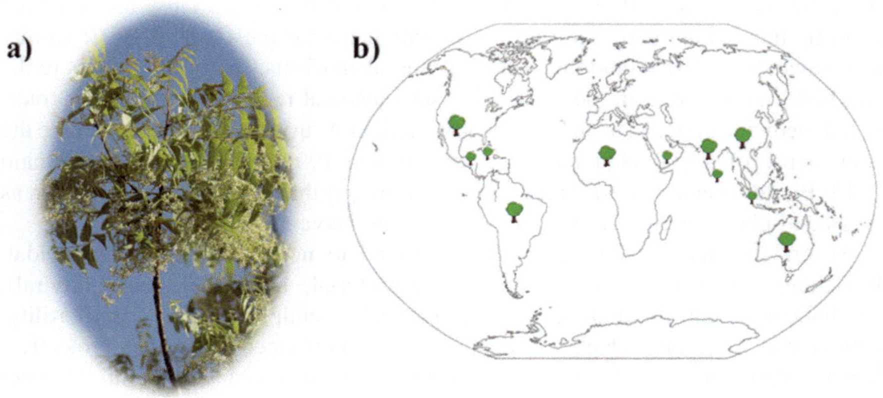

Fig. 16.3 (**a**) Picture of a specimen of *A. indica* A. Juss. Source: "*Azadirachta indica* leaves and flowers" by World Agroforestry is licensed with CC BY-NC-SA 2.0. To view a copy of this license, visit https://creativecommons.org/licenses/by-nc-sa/2.0/; (**b**) representative image of the global distribution of the species

(Yatish et al. 2019). Fructifying period is observed when the plant is about 3–5 years old, reaching its fully productive stage at the age of 10 years old. It is easily propagated, and its plantation can be done using seeds, seedlings, saplings, root suckers, or tissue (Fig. 16.3a; National Research Council 1992).

According to The Invasive Species Compendium (ISC), *A. indica* is native to India, Indonesia, Malaysia, Myanmar, Pakistan, Senegal, Sri Lanka, and Thailand. Later, it was introduced as an exotic plant in 87 other countries from African, American, Australian, and Asian continents (Fig. 16.3b) (Orwa et al. 2009). Because of its wide distribution, the species has received a variety of common names according to the country native language. Some examples are nim or nimb (Hindi); neem or Indian lilac (English); nimba, nimbou, or arishtha (Sanskrit); mindi (Indonesian); and veppu or aruveppu (Malayalam) (National Research Council 1992; Yatish et al. 2019). Among these common names, neem is the most used, and, therefore, it will be adopted in this chapter.

As shown in Fig. 16.3b, neem grows predominantly in the lowland tropics, adapting better in land areas with annual rainfalls of 400–1200 mm, and it grows under hot conditions where shade temperatures may reach over 50 °C. Nevertheless, the plant does not survive in extreme cold weathers. It is worth mentioning that neem is adaptable to acid soils due to its fallen leaves, which are slightly alkaline (pH 8.2) and, therefore, neutralize the soil acidity (National Research Council 1992).

16.3.2 Ethnopharmacology

Neem was considered as "one of the most promising of all plants" in the book entitled *Neem: A Tree for Solving Global Problems*, published by The National

Academies Press in 1992 (National Research Council 1992). Its use as medicinal plant by humankind dates ancient times (National Research Council 1992; Kumar and Navaratnam 2013), and the benefits of the plant (fruits, seeds, oil, leaves, roots, and bark) are reported in the earliest Sanskrit medical reports. Additionally, traditional medicine systems, such as the Siddha, Indian Ayurveda, and Unan, include the ancient use of neem (National Research Council 1992; Kumar and Navaratnam 2013; Bijauliya et al. 2018). Thus, it is not surprising that the first report on neem as medicinal plant dates ca. 4500 years (Kumar and Navaratnam 2013).

Several therapeutic applications are reported to neem, including insecticidal, larvicidal, anthelmintic, antimicrobial (antibacterial, antifungal, and antiviral), antidiabetic, anti-inflammatory, antimalarial, antiplasmodial, antifertility, antidiabetic, antioxidant, analgesic, anticancer, hepatoprotective, hypercholesteremic, and immunomodulatory activities (Atawodi and Atawodi 2009; Alzohairy 2016; Bijauliya et al. 2018; Fernandes et al. 2019; Yatish et al. 2019; Islas et al. 2020; Shukla et al. 2020).

16.3.3 Studies on the Immunomodulatory Potential of the Plant

The ethnopharmacological uses of neem suggest the involvement of the human immune system in the therapeutic actions of the plant and motivated experimental studies to corroborate the immunomodulatory potential of the plant. One of the first studies was performed by Nat and coworkers (van der Nat et al. 1987), who studied the in vitro immunomodulatory property of the lyophilized aqueous extract from neem bark collected in Sri Lanka. Experiments involved the C-depleting activity of the plant extract, besides its impact on chemiluminescence of human polymorphonuclear leukocytes (PMNs) induced by luminol and the generation of migration inhibition factor (MIF) by human lymphocytes. The extract resulted strong dose- and time-dependent anticomplementary effects, being more pronounced in the classical complement pathway assay. Additionally, a dose-dependent decreasing and a dose-dependent increasing were observed in the chemiluminescence of polymorphonuclear leukocytes and in the production of migration inhibition factor by lymphocytes, respectively.

A comprehensive review article on the immunomodulatory potential of Indian medicinal plants, including neem, was published by Agarwal and Singh in 1999 (Agarwal and Singh 1999). Therein, some results from literature, which corroborate the immunomodulatory properties of *A. indica* (extracts and oil), are summarized. Results from in vitro studies suggested that neem stimulates IL-1, IFN-γ, and TNF-α, and enhances prolative response of spleen cells to Con A and tetanus toxoid. As results from in vivo experiments, it was reported the following properties for the plant: activation of the immune system; enhancement of macrophage phagocytosis; expression of MHC II antigen, IgM, IgC, and anti-ovalbumin antibody; inhibition of macrophage migration; and DTH in patients with psoriasis. Clinical studies with psoriasis patients orally treated with neem extract resulted in erythema reduction, desquamation, and lesion infiltration (Agarwal and Singh 1999).

Immunomodulation activity of neem was investigated in BALB/c-mice via spleen and thymus enlargement test as well as by thymocyte counting (Beuth et al. 2006). The study was performed with the aqueous extract from the plant leaves, besides a fraction resulted from the chromatographic purification of the crude extract on Sephadex. Although only moderate increase in thymus weight and thymocyte counts (no statistical significance) was observed in response to the administration of neem samples, a strong immunomodulatory effect was obtained, resulting in a significant spleen weight gain.

Studies on the search for immunomodulators from neem leaf preparations resulted in the identification of a glycoprotein, named Neem Leaf Glycoprotein (NLGP), which presented sustained tumor growth restriction, besides angiogenic normalization chiefly by activating CD8+ T cells (Sarkar et al. 2007; Bose et al. 2009; Barik et al. 2013; Ghosh et al. 2016). More recently, the straight function of this glycoprotein as a regulator of tumor microenvironmental hypoxia and associated vascular endothelial growth factor (VEGF) production was investigated by the same research group (Saha et al. 2020). The study was performed with murine melanoma cells (B16Mel) and Lewis lung carcinoma (LL/2 LLC1). Considerable decrease in VEGF levels was observed in murine tumor (in vivo) and cancer cells (B16Mel, LLC; in vitro), besides macrophages treated with NLGP, suggesting that the neem leaves glycoprotein mediated the angiogenesis inhibition through VEGF suppression.

Elderly people present high risk of infection due to their compromised immune system. Thus, the immunomodulator effect of *A. indica* bark (capsules with powder plant) in immunocompromised elderly individuals ($\geq$ 60 years old; any gender) was investigated (Akmal et al. 2016). Total leukocyte count (TLC), differential leukocyte count (DLC), erythrocyte sedimentation rate (ESR), and T cell count (CD4 and CD8) were assessed before and after treatment. Plant administration to test group increased CD4 count (60th day; $p < 0.05$) but not CD8 count. Additionally, TLC was unchanged, while significant reduction in the control group was observed. The results suggested the therapeutic usefulness of neem bark in cases where immunomodulation is required.

Multiple sclerosis (MS) is a multifactorial neuroinflammatory and immunological disease of the central nervous system. Studies involving experimental animal model of MS (mouse hepatitis virus (MHV)-induced murine neuroinflammation model) were performed with neem bark extract (extracted with methanol) (Sarkar et al. 2020). RSA59 (recombinant demyelinating MHV strain) was preincubated with neem extract, and the effect of the plant was evaluated on viral replication and transcription, cytokine expression, and pathogenicity. It was observed reduction of viral entrance and replication, besides cell-to-cell fusion. In addition, RSA59 preincubated with the plant extract into C57BL/6 male mouse brain showed considerable reduction of acute hepatitis, meningoencephalomyelitis, and chronic progressive demyelination. The study suggested that the extract from neem bark binds directly to the virus-host attachment Spike glycoprotein (responsible for virus-cell fusogenicity) and suppresses MHV-induced neuroinflammation and neuropathogenesis.

All aforementioned examples of the immunomodulatory potential of *A. indica* involved studies with extracts from the plant, mainly leaves and barks. Nevertheless, more than 135 secondary metabolites from neem are reported in the literature, which can be classified into isoprenoid and non-isoprenoid groups. The most known compounds are azadirachtin, nimbosterol, and quercetin (Shukla et al. 2020). Despite of this, studies on the immunomodulatory activity of neem compounds are limited.

As studies on the immunomodulation potential of neem extracts are somehow associated to inflammatory process, herein, it is worth noting the anti-inflammatory activity of nimbidin, described as a mixture of secondary metabolites (tetranortriterpenes) isolated from neem seed oil (National Research Council 1992; Kaur et al. 2004; Kumar et al. 2018). These compounds are the major bitter principles in the oil and present several medicinal properties (Pillai and Laping 1984). Through in vivo and in vitro studies, it was proved that this compound mixture inhibits some of the functions of macrophages and neutrophils (Kaur et al. 2004).

16.3.4 Final Remarks

A. indica (neem) is a medicinal plant distributed in more than 90 countries. The use of this species by humankind dates ancient times, being included in traditional medicine systems. The ethnopharmacological data suggest the participation of the human immune system in the therapeutic role of the plant, motivating experimental studies to corroborate the immunomodulatory potential of the plant. Most of the examples of the immunomodulatory potential of *A. indica* reported in literature involve studies with extracts from the plant. Despite the high number of secondary metabolites isolated from neem, studies on the immunomodulatory activity of these compounds are limited, suggesting a field to be explored for better understanding the mechanisms involved in the plant immunomodulatory process.

16.4 *Ginkgo biloba* L.

16.4.1 Botanic Aspects and Occurrence

G. biloba has been a valuable plant for mankind for over 2000 years, being included in the list of the oldest living species on this planet. This woody plant (Fig. 16.4a) has existed for 270 million years, when the dinosaurs roamed the Earth, being considered an "archaic living fossil." Since there is no closely related species in the plant kingdom, it is believed that this species is the "missing link" between gymnosperms and angiosperms. Thus, *G. biloba* is extremely important for understanding the great variety of extinct seed plants that flourished ca. 100–200 million years ago (Isah 2015; Crane 2019).

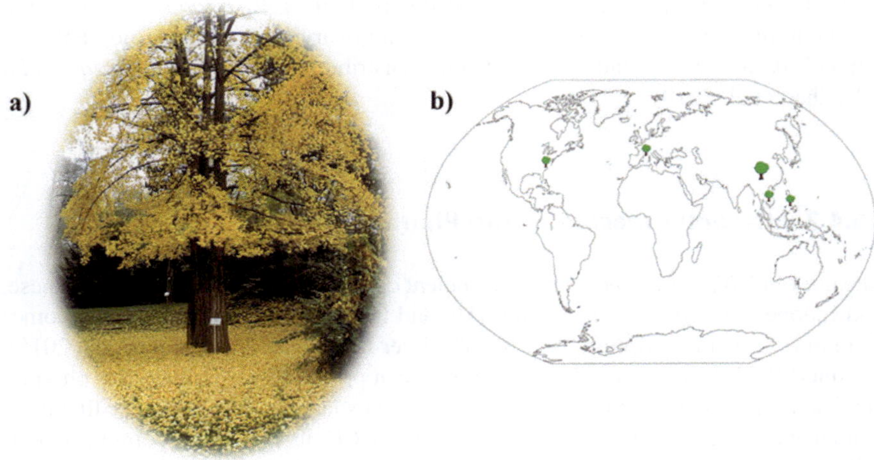

Fig. 16.4 (**a**) Picture of a specimen of *G. biloba* tree. Source: Marija Gajić, CC BY-SA 4.0 < https://creativecommons.org/licenses/by-sa/4.0>, via Wikimedia Commons; (**b**) representative image of the global distribution of the species

G. biloba is classified in a separate taxon as it is an exclusive plant: Ginkgophyta (division), Ginkgoales (order), and Ginkgoceae (family) (Isah 2015). It was in 1771 that the Swedish botanist Carl Linnaeus received a specimen of the plant and named it as *G. biloba* (Crane 2019). Botanically, this tree plant presents male and female reproductive organs in separate trees (about 1:1 ratio), being therefore considered a dioecious species. It is a beautiful tree, highly resistant to viral and bacterial infections, air pollution, and insects. One of the characteristics of this tree is its large trunk with a circumference that can reach 7 m and its height up to 30 m. The trees exhibit branched dimorphism with leaves that appear in golden-yellow clusters in autumn during senescence. The bilobed shape of its leaves is the origin of the *biloba* denomination of the species (Singh et al. 2012). The tree growing process is extremely slow, and its reproduction occurs from the age of 20, when the plant develops fruits containing naked seeds (Isah 2015; Fang et al. 2020). The seed-coat consists of a soft, fleshy outer layer (sarcotesta) that is normally referred as the *Ginkgo* nut (Fang et al. 2020). The evolutionary history and distribution of *G. biloba* are limited (Zhao et al. 2010). It is believed that the plant is originated from mountain valleys of Zhejiang, in eastern China. Later, the plant was spread to Japan and Korea and therefore being qualified as endemic to the three countries (Singh et al. 2008). There is evidence that during the last two millennia, *G. biloba* was introduced into Japan from China either via Korea or directly from China by human-mediated dispersal. The species was introduced to Europe in the eighteenth century, being widely cultivated as both ornamental and medicinal tree (Singh et al. 2008; Fang et al. 2020). Currently, *G. biloba* is found in almost all continents of the planet, especially in Asia, Europe, and North America (Fig. 16.4b) (Zhao et al. 2010).

To better understand the history and distribution of *G. biloba*, the following excellent literatures are suggested: "An evolutionary and cultural biography of ginkgo" (Crane 2019) and "Out of China: Distribution history of *Ginkgo biloba L.*" (Zhao et al. 2010).

16.4.2 Ethnopharmacology and Pharmacological Properties

The use of *G. biloba* can be seen in all ancient cultures, mainly in Chinese, Japanese, and Indonesian traditional medicines (Ho and Lai 2004). The name Ginkgo comes from the wrong transcription "Yin-Kwo" (silver fruit), a Japanese name (Isah 2015).

Since 1965, leaf extract of the plant has been produced in Germany for pharmaceutical purposes. Nevertheless, it was only 9 years later (in 1974) that the first plant extract was commercialized (brand name EGb761) in France, becoming a best-selling worldwide. Currently, billions of doses of standard extracts are sold worldwide by pharmaceutical companies, being ranked as one of the top five best-selling herbal supplements (Isah 2015). Several therapeutic actions of the commercialized supplement are reported, including regulation of cerebral blood flow, free radical protection, and retarding both the dementia and diabetes progress (Isah 2015).

Besides the plant leaf (either fresh or dried), *G. biloba* seeds are also used for medicinal purpose (Singh et al. 2008). Although extracts of *G. biloba* are recommended to treat several pathologies, including asthma, bronchitis, stomach pain, anxiety, hearing loss, tinnitus, diabetes mellitus, hypertension (EH), neurodegenerative diseases, tuberculosis, osteoarthritis (OA), geriatric complaints like vertigo, cataracts, age-related macular degeneration, cognitive dysfunction, arteriosclerosis, myocardial lesion, hippocampus neuronal lesions, liver damage, ischemic heart disease, morphometry testicular changes, cognitive dysfunction, among others (Bairy 2002; Mahmoud et al. 2000; Louajri et al. 2001; Koji 2005; Ramassamy et al. 2007; Dai et al. 2018; Ali et al. 2018; Zhang and Yao 2013; Almeida 2009; von Boetticher 2011; Dippolito 2005; Zhao et al. 2015; Ho et al. 2013; Fang et al. 2020; Souza et al. 2020; Xiong et al. 2014; Li et al. 2020), these diseases generally fall into one of three categories: peripheral vascular, cerebrovascular, or mitigation of tissue damage (Abad et al. 2010).

The composition referring to the constituent of the *G. biloba* leaf undergoes variations that depend on the origin and time of harvest. For this reason, the commercialization of the extract requires a standardization that involves the cultivation, the harvest, and the control of its known active compounds (Smith and Luo 2004). In a general way, the leaves are collected during the summer, being dried, and tested to detect pollutants and toxic compounds, such as aflatoxins and heavy metals. Quantification of flavonoids and terpenoids is used for extract standardization (Bilia 2002).

16.4.3 Studies on the Immunomodulatory Potential of the Plant and its Constituents Ginkgolides and Bilobalide (BB)

Elderly is committed by a common joint disorder named osteoarthritis. It contains plasma proteins in the synovial fluid, which prompt the production of inflammatory cytokines. In this case, interleukin IL-1A is a cytokine associated to the immunopathogenesis that acts by inducing the matrix metalloproteinases (MMPs) and proteases and which is responsible for the cartilage damage. In this context, Ho and coworkers investigated the protective effect of *G. biloba* leaf extract in proinflammatory cytokine-stimulated human chondrocytes in vitro (Ho et al. 2013). The authors found that the plant extract induced an immunomodulatory effect via inhibition of IL-1-induced JNK activation and degrading c-Jun protein in an ubiquitination-dependent way in human chondrocytes.

Immune responses of plant extracts can be assessed through lymphocyte transformation assay that explores the mitogenic activity of the plant on T lymphocyte proliferation. In this case, immunosuppression is a result of anti-proliferative activity on T lymphocyte culture. In contrast, immunostimulation occurs when proliferation is observed. In this sense, the modulatory effect of leaf extract of *G. biloba* on proliferation of T lymphocytes in vitro was investigated by Badria and coworkers (Badria et al. 2016). Crude methanol extract of the plant, besides fractions obtained through partition procedure with solvents (petrol ether, dichloromethane, ethyl acetate, methanol, and water), was assayed. Crude extract and water fraction were the assayed samples that showed the highest percentage inductions of lymphocyte transformation at 50 μg mL^{-1} (109.3 and 76.5%, respectively) in comparison with the standard mitogen (positive control) concanavalin A.

Despite the wide range of applications of *G. biloba* (EGb) extracts, the most popular use since the 1990s has been to treat neurological diseases (DeFeudis and Drieu 2000). Starting and developing of neurological diseases, particularly neuroimmune and/or neurodegenerative and ischemic cerebrovascular, may be associated with an excessive inflammatory immune response (IIR). This latter can be defined as a physiological or excessive systemic response, induced by inflammatory immune cells according to changes in the internal and external environments. The regulation of IIR has been established as a novel therapeutic target for neurological diseases (Han et al. 2018; Ritzel et al. 2018; Voet et al. 2019).

Recently, Li and coworkers published an elegant review paper on the role of ginkgolides in the inflammatory immune response of neurological diseases (Li et al. 2020). It is noteworthy that ginkgolides are one of the most relevant bioactive compounds found in *G. biloba*, presenting antiplatelet-activating, antiapoptotic, antioxidative, neurotrophic, and neuroimmunomodulatory effects (Huang et al. 2014). These compounds are terpene trilactones, displaying tertiary butyl functional groups in their structures (Li et al. 2020). Besides the flavonoids quercetin, kaempferol, and myricetin, the leaf extract of the plant contains ginkgolides A (GA), B (GB), C (GC), M (GM), J (GJ), K (GK), and L (GL) and bilobalide (BB) (Fig. 16.5) (Chávez-Morales et al. 2017).

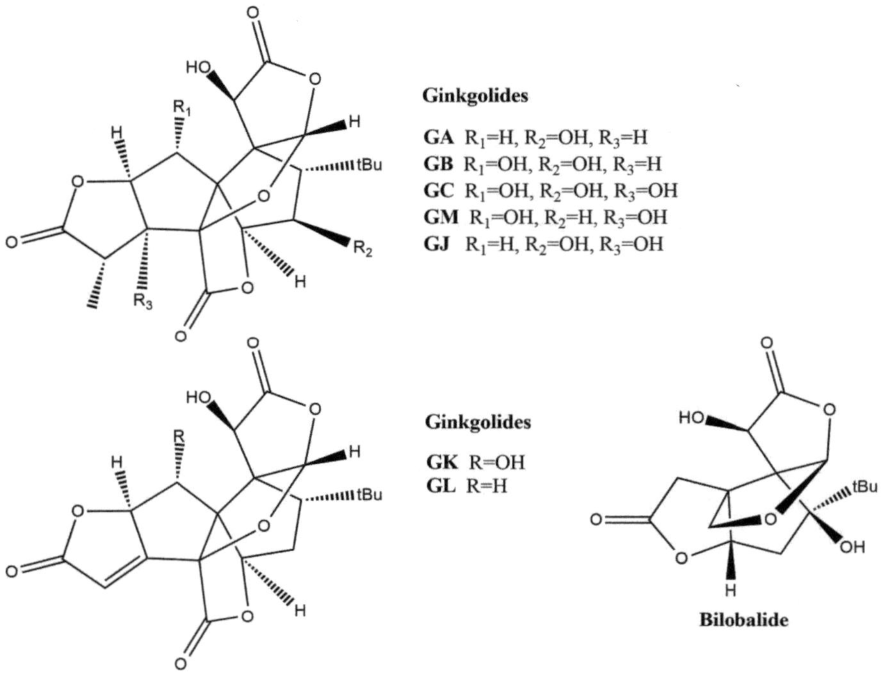

Fig. 16.5 Chemical structures of ginkgolides (G) A–C and J–M and bilobalide

In general, ginkgolides (G) and bilobalide (BB) regulate the inflammatory immune response (IIR), being considered potential compounds in the treatment of neurological diseases, such as ischemia stroke, Guillain-Barré syndrome (GBS), multiple sclerosis (MS), Parkinson's disease (PD), and Alzheimer's disease (AD). The immunomodulation responses of these natural products to some neurological diseases are summarized in Table 16.1.

Besides the immunomodulation response of ginkgolide B (GB) associated to some neurological diseases, the potential of this natural product to suppress T cell activation in peripheral blood of asthma patients was also investigated as initial approach for increasing the therapeutic property of antihistamines drugs (Mahmoud et al. 2012). The study used a model of T cell activation relying on lectin-mediated induction of the activation antigens CD25 (the IL-2 receptor) and the human leukocyte antigen-DR (HLA-DR). Cetirizine dihydrochloride (CTZ) and azelastine (AZE) were used as antihistamine drugs. Cultures stimulated with GB 10^{-6} M + CTZ 10^{-6} M showed the best suppression of activated cells (SICD3 + CD25+, $p = 0.024$, SICD3 + HLADR+, $p = 0.019$), suggesting that the natural product interacts with the antihistamine CTZ and potentiate its ability to downregulate the pathological immune activation.

It is worth noting that ginkgolide B (GB) has also been suggested in osteoarthritis therapy, since this compound showed anti-inflammatory and chondroprotective

Table 16.1 Roles of ginkgolides in immunomodulation responses associated to neurological diseases

Disease	Compound	Immunomodulation response
Ischemia stroke	GB	Downregulated TLR4 and NF-kB/reduced microglial activation in transient middle cerebral artery occlusion-induced cerebral I/R injury mice; decreased infarct size; serum levels of pro-inflammatory, such as TNF-α, IL-6, and IL-1β; and expressions of intercellular adhesion molecule-2 and E-selectin (Gu et al. 2012; Fan et al. 2020); promoted microglia/macrophage transferring from inflammatory M1 phenotype to a protective, anti-inflammatory M2 phenotype in vivo or in vitro (Shu et al. 2016); suppression of ERK/MAPK pathway, inhibition of Akt phosphorylation, and downregulation of p-TAK1, p-IkBa, and p-IKKb (Nabavi et al. 2015; Fan et al. 2020)
PD	BB	NF-kB activation contributes to the 6-OHDA-induced loss of dopaminergic neurons and the inhibition of the NF-kB pathway in rat model of PD (Li et al. 2008)
	GB	Attenuate the increase of the NF-kB DNA-binding activity induced by 1-methyl-4-phenyl-1,2,3,6-tetrahydropyridine through an NF-kB-dependent mechanism (Kim et al. 2013)
GBS	Ginkgolides	Regulation of MyD88/NF-kB via regulating the TLR/MyD88/NF-kB signaling pathway (Li et al. 2020)
MS	Ginkgolides	Regulation of TLR/MyD88/NF-kB signaling pathway and attenuation of the inflammatory responses to inhibit the productions of inflammatory factors mediated by OGD/R in microglial cells (Zhou et al. 2016); PAFR antagonist, reducing the increase of nuclear translocation of NF- kB p65 induced by PCP (Tran et al. 2018)
	GB	Inhibition of expressions of TLR4 and MyD88 induced by high glucose, then alleviation of TLR4-mediated inflammatory responses in endothelial cells (Chen et al. 2017); PAFR antagonist and effective prevention of synaptic damage in hippocampus of EAE mice without affecting microglial activation (Chen et al. 2017)
	GK	Autoimmune neuritis (EAN) through possible cellular and molecular mechanisms, especially as a peripheral immunomodulatory (Yu et al. 2019)
AD	GB	Inhibition of neurotoxicity induced by β-amyloid (Li et al. 2013; Gill et al. 2017)

Diseases, Guillain-Barré syndrome (GBS), multiple sclerosis (MS), Parkinson's disease (PD), and Alzheimer's disease (AD); compounds, ginkgolides (G) and bilobalide (BB)

effects in LPS-induced chondrocytes via inhibition of LPS-induced MAPK pathway activation (Hu et al. 2018).

16.4.4 Final Remarks

G. biloba is one of the oldest plants living on Earth. Although natural from Asian countries (China, Korea, and Japan), this plant is currently distributed in practically all continents. The standardized extract of the leaves of *G. biloba* is one of the most consumed herbal supplements and presents beneficial effects in the treatment of several pathologies. Finally, the immunomodulatory potential of the plant and some of its constituents was presented through literature studies involving important pathologies, such as osteoarthritis, asthma, and, especially, neurological diseases (ischemia stroke, Guillain-Barré syndrome, multiple sclerosis, Parkinson's disease, and Alzheimer's disease).

Acknowledgements The authors are grateful to the Fundação Cearense de Apoio ao Desenvolvimento Científico e Tecnológico (FUNCAP) for financial support (Process # DEP-0164-00236.02.00/19; SPU # 1910216698). Additionally, M. C. de Mattos (Process: 306043/2018-1) and M. C. F. de Oliveira (Process: 310881/2020-0) thank Conselho Nacional de Desenvolvimento Científico e Tecnológico (CNPq) for providing their research fellowship (PQ level 2).

References

Abad MJ, Bedoya LM, Bermejo P (2010) An update on drug interactions with the herbal medicine *Ginkgo biloba*. Curr Drug Metab 11(2):171–181. https://doi.org/10.2174/138920010791110818

Agarwal SS, Singh VK (1999) Immunomodulators: a review of studies on Indian medicinal plants and synthetic peptides. Part I: medicinal plants. PINSA-B: Proceedings of the Indian National Science Academy, Part B: Biological Sciences 65(3&4):179–204

Akmal M, Zarnigar, Rashid B (2016) Evaluation of Chale neem (*Azadirachta Indica*) as an Immunomodulator in elderly persons: A single blind randomized placebo controlled study. Res Rev J Pharmacol 6(3):48–52

Ali M, Khan T, Fatima K, Ali QUA, Ovais M, Khalil AT, Ullah I, Raza A, Shinwari ZK, Idrees M (2018) Selected hepatoprotective herbal medicines: evidence from ethnomedicinal applications, animal models, and possible mechanism of actions. Phytother Res 32(2):199–215. https://doi.org/10.1002/ptr.5957

Almeida ER (2009). Plantas adaptógenas e com ação no sistema nervoso central. São Paulo: Biblioteca 24 horas, 2009

Alzohairy MA (2016) Therapeutics role of Azadirachta indica (neem) and their active constituents in diseases prevention and treatment. Evid Based Complement Alternat Med 2016:7382506. https://doi.org/10.1155/2016/7382506

Anwar F, Latif S, Ashraf M, Gilani AH (2007) *Moringa oleifera*: A food plant with multiple medicinal uses. Phytother Res 21(1):17–25. https://doi.org/10.1002/ptr.2023

Atawodi SE, Atawodi JC (2009) Azadirachta indica (neem): a plant of multiple biological and pharmacological activities. Phytochem Rev 8(3):601–620. https://doi.org/10.1007/s11101-009-9144-6

Badria FAA, Bar FMARA, Elimam DMA, Zaghloul MHE (2016) Modulatory effects of plant extracts on proliferation of T-lymphocytes in vitro stimulated with concanavalin A. World J Pharm Pharm Sci 4(3):299–308

Bairy KL (2002) Wound healing potentials of plant products. J Nat Remedies 2(1):11–20. https://doi.org/10.18311/jnr/2002/355

Banji OJF, Banji D, Kavitha R (2012) Immunomodulatory effects of alcoholic and hydroalcoholic extracts of *Moringa oleifera* lam leaves. Indian J Exp Biol 50(4):270–276

Barik S, Banerjee S, Mallick A, Goswami KK, Roy S, Bose A, Baral R (2013) Normalization of tumor microenvironment by neem leaf glycoprotein potentiates effector T cell functions and therapeutically intervenes in the growth of mouse sarcoma. PLoS One 8:e66501. https://doi.org/10.1371/journal.pone.0066501

Bennett RN, Mellon FA, Foidl N, Pratt JH, Dupont MS, Perkins L, Kroon PA (2003) Profiling glucosinolates and phenolics in vegetative and reproductive tissues of the multi-purpose trees *Moringa oleifera* L. (horseradish tree) and *Moringa stenopetala* L. J Agric Food Chem 51(12): 3546–3553. https://doi.org/10.1021/jf0211480

Beuth J, Schneider H, Ko HL (2006) Enhancement of immune responses to neem leaf extract (*Azadirachta indica*) correlates with antineoplastic activity in BALB/c-mice. In Vivo 20(2): 247–252

Bhandari MM (1995) Flora of the Indian Desert, 2nd edn. MPS Repros, Jodhpur, India

Bharali R, Tabassum J, Azad MRH (2003) Chemomodulatory effect of Moringa oleifera, Lam, on hepatic carcinogen metabolizing enzymes, antioxidant parameters and skin papillomagenesis in mice. Asian Pac J Cancer Prev 4(2):131–139

Bijauliya RK, Alok S, Chanchal DK, Sabharwal M, Yadav RD (2018) An updated review of pharmacological studies on *Azadirachta indica* (neem). Int J Pharm Sci Res 9(7):2645–2655. https://doi.org/10.13040/IJPSR.0975-8232.9(7).2645-55

Bilia AR (2002) *Ginkgo biloba* L. Fitoterapia 73(3):276–279

Bosch CH. *Moringa oleifera* (PROTA), PlantUseFrançais. 2017 Consulté le 13 mai 2018 à 20:21 sur https://uses.plantnetproject.org/f/index.php?title=Moringa_oleifera(PROTA)&oldid=265013

Bose A, Chakraborty K, Sarkar K, Goswami S, Chakraborty T, Pal S, Baral R (2009) Neem leaf glycoprotein induces perforin-mediated tumor cell killing by T and NK cells through differential regulation of IFNg signaling. J Immunother 32(1):42–53. https://doi.org/10.1097/CJI.0b013e31818e997d

Brilhante RSN, Sales JA, Pereira VS, Castelo-Branco DSCM, Cordeiro RA, Sampaio CMS, Paiva MAN, Dos Santos JBF, Sidrim JJC, Rocha MFG (2017) Research advances on the multiple uses of *Moringa oleifera*: A sustainable alternative for socially neglected population. Asian Pac J Trop Med 10(7):621–630. https://doi.org/10.1016/j.apjtm.2017.07.002

Caceres A, Cabrera O, Morales O, Mollinedo P, Mendia P (1991) Pharmacological properties of *Moringa oleifera*. 1: preliminary screening for antimicrobial activity. J Ethnopharmacol 33(3): 213–216. https://doi.org/10.1016/0378-8741(91)90078-R

Caceres A, Saravia A, Rizzo S, Zabala L, De Leon E, Nave F (1992) Pharmacological properties of *Moringa oleifera*. 2: screening for antispasmodic, antiinflammatory and diuretic activity. J Ethnopharmacol 36(3):233–237. https://doi.org/10.1016/0378-8741(92)90049-W

Chávez-Morales RM, Jaramillo-Juárez F, Rodríguez-Vázquez ML, Martínez-Saldaña MC, Del Río FAP, Garfias-López JA (2017) The *Ginkgo biloba*, extract (GbE) protects the kidney from damage produced by a single and low dose of carbon tetrachloride in adult male rats. Exp Toxicol Pathol 69(7):430–434. https://doi.org/10.1016/j.etp.2017.04.003

Chen K, Sun W, Jiang Y, Chen B, Zhao Y, Sun J (2017) Ginkgolide B suppresses TLR4-mediated inflammatory response by inhibiting the phosphorylation of JAK2/STAT3 and p38 MAPK in high glucose-treated HUVECs. Oxidative Med Cell Longev 2017:9371602. https://doi.org/10.1155/2017/9371602

Cherng JM, Chiang W, Chiang LC (2008) Immunomodulatory activities of common vegetables and spices of Umbelliferae and its related coumarins and flavanoids. Food Chem 106(3):944–950. https://doi.org/10.1016/j.foodchem.2007.07.005

Chinsembu KC (2020) Coronaviruses and nature's pharmacy for the relief of coronavirus disease 2019. Rev Bras 30:603–621. https://doi.org/10.1007/s43450-020-00104-7

Choi HJ, Kim JH, Lee CH, Ahn YJ, Song JH, Baek SH, Kwon DH (2009) Antiviral activity of quercetin 7-rhamnoside against porcine epidemic diarrhea virus. Antivir Res 81(1):77–81. https://doi.org/10.1016/j.antiviral.2008.10.002

Crane PR (2019) An evolutionary and cultural biography of ginkgo. Plants People Planet 1(1): 32–37. https://doi.org/10.1002/ppp3.7

Cui YM, Wang JLW, Zhang HJ, Wu SG, Qi GH (2018) Effect of dietary supplementation with *Moringa Oleifera* leaf on performance, meat quality, and oxidative stability of meat in broilers. Poult Sci 97(8):2836–2844. https://doi.org/10.3382/ps/pey122

Cunningham-Rundles S (1982) Effects of nutritional status on immunological function. Am J Clin Nutr 35(5):1202–1210. https://doi.org/10.1093/ajcn/35.5.1202

Dai CX, Hu CC, Shang YS, Xie J (2018) Role of Ginkgo biloba extract as an adjunctive treatment of elderly patients with depression and on the expression of sérum S100B. Medicine 97(39): e12421. https://doi.org/10.1097/MD.0000000000012421

Das N, Sikder K, Bhattacharjee S, Majumdar SB, Ghosh S, Majumdar S, Dey S (2013) Quercetin alleviates inflammation after short-term treatment in high-fat-fed mice. Food Funct 4(6): 889–898. https://doi.org/10.1039/c3fo30241e

DeFeudis FV, Drieu K (2000) Ginkgo biloba extract (EGb761) and CNS functions: basic studies and clinical applications. Curr Drug Targets 1(1):25–58. https://doi.org/10.2174/1389450003349380

Dhakad AK, Ikram M, Sharma S, Khan S, Pandey VV, Singh A (2019) Biological, nutritional, and therapeutic significance of *Moringa oleifera lam.* Phytother Res 33(11):2870–2903. https://doi.org/10.1002/ptr.6475

Dippolito JAC (2005) Fitoterapia magistral: um guia prático para a manipulação de fitoterápicos. Editora Anfarmag, São Paulo

El-Deep MH, Dawood MAO, Assar MH, Ijiri D, Ohtsuka A (2019) Dietary Moringa oleifera improves growth performance, oxidative status, and immune related gene expression in broilers under normal and high temperature conditions. J Therm Biol 82:157–163. https://doi.org/10.1016/j.jtherbio.2019.04.016

Fan Q, Zhou J, Wang Y, Xi T, Ma H, Wang Z (2020) Chip-based serum proteomics approach to reveal the potential protein markers in the sub-acute stroke patients receiving the treatment of ginkgo Diterpene lactone meglumine injection. J Ethnopharmacol 260(5):112964. https://doi.org/10.1016/j.jep.2020.112964

Fang J, Wang Z, Wang P, Wang M (2020) Extraction, structure and bioactivities of the polysaccharides from Ginkgo biloba: A review. Int J Biol Macromol 162:1897–1905. https://doi.org/10.1016/j.ijbiomac.2020.08.141

Fernandes SR, Barreiros L, Oliveira RF, Cruz A, Prudêncio C, Oliveira AI, Pinho C, Santos N, Morgado J (2019) Chemistry, bioactivities, extraction and analysis of azadirachtin: state-of-the-art. Fitoterapia 134:141–150. https://doi.org/10.1016/j.fitote.2019.02.006

Ferreria PMP, Farias DF, Oliveira JTA, Carvalho AFU (2008) *Moringa oleifera*: bioactive compounds and nutritional potential. Rev Nutr 21(4):431–437. https://doi.org/10.1590/S1415-52732008000400007

Foidl N, Makkar HPS, Becker K (2001) The potential of *Moringa oleifera* for agricultural and industrial uses. In: Fuglie LJ (ed) The miracle tree/the multiple attributes of moringa. CTA, USA, pp 45–76

Gaikwad SB, Mohan DGK, Reddy KJ (2011) *Moringa oleifera* leaves: immunomodulation in Wistar albino rats. *International.* J Pharm Pharm Sci 3(Suppl 5):426–430

Gearing AJH, Wadhwa M, Perris AD (1985) Interleukin 2 stimulates T cell proliferation using a calcium flux. Immunol Lett 10(5):297–302. https://doi.org/10.1016/0165-2478(85)90105-1

Ghosh S, Sarkar M, Ghosh T, Guha I, Bhuniya A, Saha A, Dasgupta S, Barik S, Bose A, Baral R (2016) Neem leaf glycoprotein promotes dual generation of central and effector memory CD8+T cells against sarcoma antigen vaccine to induce protective anti-tumor immunity. Mol Immunol 71:42–53. https://doi.org/10.1016/j.molimm.2016.01.00720

Gill I, Kaur S, Kaur N, Dhiman M, Mantha AK (2017) Phytochemical Ginkgolide B attenuates amyloid-b1-42 induced oxidative damage and altered cellular responses in human neuroblastoma SH-SY5Y cells. J Alzheimers Dis 60(s1):S25–S40. https://doi.org/10.3233/JAD-161086

Gu JH, Ge JB, Li M, Wu F, Zhang W, Qin ZH (2012) Inhibition of NF-kB activation is associated with anti-inflammatory and anti-apoptotic effects of Ginkgolide B in a mouse model of cerebral ischemia/reperfusion injury. Eur J Pharm Sci 47(4):652–660. https://doi.org/10.1111/j.1462-5822.2011.01720.x

Gupta A, Gautam MK, Singh RK, Kumar MV, Rao CHV, Goel RK, Anupurba S (2010) Immunomodulatory effect of *Moringa oleifera* Lam. extract on cyclophosphamide induced toxicity in mice. Indian J Exp Biol 48(11):1157–1160

Gupta S, Jain R, Kachhwaha S, Kothari SL (2018) Nutritional and medicinal applications of *Moringa oleifera* Lam.- review of current status and future possibilities. J Herbal Med 11:1–11. https://doi.org/10.1016/j.hermed.2017.07.003

Han C, Li Y, Wang Y, Cui D, Luo T, Zhang Y, Ma Y, Wei W (2018) Development of inflammatory immune response-related drugs based on G protein-coupled receptor kinase 2. Cell Physiol Biochem 51(2):729–745. https://doi.org/10.1159/000495329

Ho LJ, Hung LF, Liu FC, Hou TY, Lin LC, Huang CY, Lai JH (2013) Ginkgo biloba extract individually inhibits JNK activation and induces c-Jun degradation in human chondrocytes: potential therapeutics for osteoarthritis. PLoS One 8(12):e82033. https://doi.org/10.1371/journal.pone.0082033

Ho LJ, Lai JH (2004) Chinese herbs as immunomodulators and potential disease-modifying Antirheumatic drugs in autoimmune disorders. Curr Drug Metab 5(4):181–192. https://doi.org/10.2174/1389200043489081

Hu H, Li Y, Xin Z, Zhanga X (2018) Ginkgolide B exerts anti-inflammatory and chondroprotective activity in LPS-induced chondrocytes. Adv Clin Exp Med 27(7):913–920. https://doi.org/10.17219/acem/70414

Huang P, Zhang L, Chai C, Qian XC, Li W, Li JS, Di LQ, Cai BC (2014) Effects of food and gender on the pharmacokinetics of ginkgolides A, B, C and bilobalide in rats after oral dosing with ginkgo terpene lactones extract. J Pharm Biomed Anal 100:138–144. https://doi.org/10.1016/j.jpba.2014.07.030

Isah T (2015) Rethinking *Ginkgo biloba* L.: medicinal uses and conservation. Pharmacogn Rev 9(18):140–148. https://doi.org/10.4103/0973-7847.162137

Islas JF, Acosta E, Buentello ZG, Delgado-Gallegos JL, Moreno-Treviño MG, Escalante B, Moreno-Cuevas JE (2020) An overview of neem (*Azadirachta indica*) and its potential impact on health. J Funct Foods 74:104171. https://doi.org/10.1016/j.jff.2020.104171

Jacques AS, Arnaud SSS, Fréjus OOH, Jacques DT (2020) Review on biological and immunomodulatory properties of *Moringa oleifera* in animal and human nutrition. J Pharmacogn Phytother 12(1):1–9. https://doi.org/10.5897/JPP2019.0551

Kamath G, Yadav S (2020) Review: antiviral and immunomodulatory properties of nutraceuticals and herbs. Int J Adv Res 8(05):797–813. https://doi.org/10.21474/IJAR01/10988

Kaur G, Alam MS, Athar M (2004) Nimbidin suppresses functions of macrophages and neutrophils: relevance to its antiinflammatory mechanisms. Phytother Res 18(5):419–424. https://doi.org/10.1002/ptr.1474

Khan I, Zaneb H, Masood S, Yousaf MS, Rehman HF, Rehman H (2017) Effect of *Moringa oleifera* leaf powder supplementation on growth performance and intestinal morphology in broiler chickens. J Anim Physiol Anim Nutr 101(S1):114–121. https://doi.org/10.1111/jpn.12634

Kim BK, Shin EJ, Kim HC, Chung YH, Dang DK, Jung BD (2013) Platelet-activating factor receptor knockout mice are protected from MPTP-induced dopaminergic degeneration. Neurochem Int 63(3):121–132. https://doi.org/10.1016/j.neuint.2013.05.010

Koji N (2005) Terpene trilactones from Gingko biloba: from ancient times to the 21st century. Bioorg Med Chem 13(17):4987–5000. https://doi.org/10.1016/j.bmc.2005.06.014

Kooltheat N, Sranujit RP, Chumark P, Potup P, Laytragoon-Lewin N, Usuwanthim K (2014) An ethyl acetate fraction of *Moringa oleifera* Lam. Inhibits human macrophage cytokine production induced by cigarette smoke. Nutrients 6(2):697–710. https://doi.org/10.3390/nu6020697

Kumar D, Arya V, Kaur R, Bhat ZA, Gupta VK, Kumar V (2012) A review of immunomodulators in the Indian traditional health care system. J Microbiol Immunol Infect 45(3):165–184. https://doi.org/10.1016/j.jmii.2011.09.030

Kumar R, Mehta S, Pathak SR. Chapter 4—bioactive constituents of neem. *In*: Tewari A, Tiwari, S. (org.). Synthesis of medicinal agents from plants. Elsevier, 2018, p. 75–103. doi: https://doi.org/10.1016/B978-0-08-102071-5.00004-0

Kumar VS, Navaratnam V (2013) Neem (*Azadirachta indica*): prehistory to contemporary medicinal uses to humankind. Asian Pac J Trop Biomed 3(7):505–514. https://doi.org/10.1016/S2221-1691(13)60105-7

Leone A, Spada A, Battezzati A, Schiraldi A, Aristil J, Bertoli S (2015) Cultivation, genetic, ethnopharmacology, phytochemistry and pharmacology of *Moringa oleifera* leaves: an overview. Int J Mol Sci 16(6):12791–12835. https://doi.org/10.3390/ijms160612791

Li C, Liu K, Liu S, Aerqin Q, Wu X (2020) Role of ginkgolides in the inflammatory immune response of neurological diseases: a review of current literatures. Front Syst Neurosci 14:45. https://doi.org/10.3389/fnsys.2020.00045

Li L, Zhang QG, Lai LY, Wen XJ, Zheng T, Cheung CW, Zhou SQ, Xu SY (2013) Neuroprotective effect of ginkgolide B on bupivacaine-induced apoptosis in SH-SY5Y. Oxidative Med Cell Longev 2013:1–11. https://doi.org/10.1155/2013/159864

Li LY, Zhao XL, Fei XF, Gu ZL, Qin ZH, Liang ZQ (2008) Bilobalide inhibits 6-OHDA-induced activation of NF-kB and loss of dopaminergic neurons in rat substantia nigra. Acta Pharmacol Sin 29(5):539–547. https://doi.org/10.1111/j.1745-7254.2008.00787.x

Louajri A, Harraga S, Godot V, Toubin G, Kantelip JP (2001) The effect of ginkgo biloba extract on free radical production in hypoxic rats. Biol Pharm Bull 24(6):710–712. https://doi.org/10.1248/bpb.24.710

Mahmoud F, Abul H, Onadeko B, Khadadah M, Hainea D, Morgan G (2000) In vitro effects of ginkgolide B on lymphocyte activation in atopic asthma: comparison with cyclosporin A. Jpn J Pharmacol 83(3):241–245

Mahmoud FF, Haines D, Al-Awadhi R, Arifhodzic N, Abal A, Azeamouzi C, Al-Sharah S, Tosaki A (2012) *In vitro* suppression of lymphocyte activation in patients with seasonal allergic rhinitis and pollen-related asthma by cetirizine or azelastine in combination with ginkgolide B or astaxanthin. Acta Physiol Hung 99(2):173–184. https://doi.org/10.1556/APhysiol.99.2012.2.11

Makita C, Chimuka L, Steenkamp P, Cukrowska E, Madala E (2016) Comparative analyses of flavonoid content in *Moringa oleifera* and *Moringa ovalifolia* with the aid of UHPLC-qTOF-MS fingerprinting. S Afr J Bot 105:116–122. https://doi.org/10.1016/j.sajb.2015.12.007

Manguro LO, Lemmen P (2007) Phenolics of *Moringa oleifera* leaves. Nat Prod Res 21(1):56–68. https://doi.org/10.1080/14786410601035811

Mousa MA, Osman AS, Hady HAMA (2017) Performance, immunology and biochemical parameters of *Moringa oleifera* and/or Cichorium intybus addition to broiler chicken ration. J Vet Med Anim Health 9(10):255–263. https://doi.org/10.5897/JVMAH2017.0611

Muangnoi C, Chingsuwanrote P, Praengamthanachoti P, Svasti S, Tuntipopipat S (2012) *Moringa oleifera* pod inhibits inflammatory mediator production by lipopolysaccharide-stimulated RAW 264.7 murine macrophage cell lines. Inflammation 35(2):445–455. https://doi.org/10.1007/s10753-011-9334-4

N'diaye M, Dieye AM, Mariko F, Tall A, Sall Diallo A, Faye B (2002) Contribution to the study of the anti-inflammatory activity of *Moringa oleifera* (Moringaceae). Dakar Med 47(2):210–212

Nabavi SM, Habtemariam S, Daglia M, Braidy N, Loizzo MR, Tundis R, Nabavi SF (2015) Neuroprotective effects of ginkgolide B against ischemic stroke: a review of current literature. Curr Top Med Chem 15(21):2222–2232. https://doi.org/10.2174/1568026615666150610142647

Nantachit K (2006) Antibacterial activity of the capsules of *Moringa oleifera* Lam. (Moringaceae). Chiang Mai Univ J 5(3):365–368

National Research Council (1992) Neem: a tree for solving global problems. The National Academies Press, Washington, DC. https://doi.org/10.17226/1924

Nfambi J, Bbosa GS, Sembajwe LF, Gakunga J, Kasolo JN (2015) Immunomodulatory activity of methanolic leaf extract of *Moringa oleifera* in Wistar albino rats. J Basic Clin Physiol Pharmacol 26(6):603–611. https://doi.org/10.1515/jbcpp-2014-0104

Oghenebrorhie O, Oghenesuvwe O (2016) Performance and haematological characteristics of broiler finisher fed *Moringa oleifera* leaf meal diets. J Northeast Agric Univ 23(1):28–34. https://doi.org/10.1016/S1006-8104(16)30029-0

Orwa C, Mutua A, Kindt R, Jamnadass R, Anthony S (2009) Agroforestree Database: a tree reference and selection guide version 4.0 (http://www.worldagroforestry.org/sites/treedbs/treedatabases.asp)

Pandey A, Pradheep K, Gupta R (2011) "Drumstick tree" (*Moringa oleifera* lam.): a multipurpose potential species in India. Genet Resour Crop Evol 58(3):453–460. https://doi.org/10.1007/s10722-010-9629-6

Pillai NR, Laping GSJ (1984) Some pharmacological actions of 'Nimbidin'- A bitter principle of *Azadirachta indica*-A Juss (neem). Ancient Sci Life IV(2):88–95

Rachmawati I, Rifa'i M (2014) *In vitro* immunomodulatory activity of aqueous extract of *Moringa oleifera* lam. Leaf to the $CD4^+$, $CD8^+$ and $B220^+$ cells in *Mus musculus*. J Exp Life Sci 4(1):15–20

Ramachandran C, Peter KV, Gopalakrishnan PK (1980) Drumstick (*Moringa oleifera*): a multipurpose Indian vegetable. Econ Bot 34(3):276–283. https://doi.org/10.1007/BF02858648

Ramassamy C, Longpre F, Christen Y (2007) *Ginkgo biloba* extract (EGb 761) in Alzheimer's disease: is there any evidence? Curr Alzheimer Res 4(3):253–262. https://doi.org/10.2174/156720507781077304

Rathi BS, Bodhankar SL, Bahetyi AM (2006) Evaluation of aqueous extract of *Moringa oleifera* Linn for wound healing in albino rats. Indian J Exp Biol 44(11):898–901

Ritzel RM, Lai YJ, Crapser JD, Patel AR, Schrecengost A, Grenier JM, Mancini NS, Patrizz A, Jellison ER, Morales-Scheihing D, Venna VR, Kofler JK, Liu F, Verma R, McCullough LD (2018) Aging alters the immunological response to ischemic stroke. *Acta Neuropathol* 136(1):89–110. https://doi.org/10.1007/s00401-018-1859-2

Saha A, Nandi P, Dasgupta S, Bhuniya A, Ganguly N, Ghosh T, Guha I, Banerjee S, Baral R, Bose A (2020) Neem leaf glycoprotein restrains VEGF production by direct modulation of HIF1-α-linked upstream and downstream cascades. Front Oncol 10:260. https://doi.org/10.3389/fonc.2020.00260

Sarkar K, Bose A, Laskar S, Choudhuri SK, Dey S, Roychowdhury PK, Baral R (2007) Antibody response against neem leaf preparation recognizes carcinoembryonic antigen. Int Immunopharmacol 7(3):306–312. https://doi.org/10.1016/j.intimp.2006.10.014

Sarkar L, Putchala RK, Safiriyu AA, Das SJ, Azadirachta indica A. (2020) Juss ameliorates mouse hepatitis virus-induced neuroinflammatory demyelination by modulating cell-to-cell fusion in an experimental animal model of multiple sclerosis. Front Cell Neurosci 14:116. https://doi.org/10.3389/fncel.2020.00116

Shu ZM, Shu XD, Li HQ, Sun Y, Shan H, Sun XY, Du RH, Lu M, Xiao M, Ding JH, Hu G (2016) Ginkgolide B protects against ischemic stroke via modulating micróglia polarization in mice. CNS Neurosci Ther 22(9):729–739. https://doi.org/10.1111/cns.12577

Shukla V, Khurshid MDD, Kumar B, Amandeep (2020) A review on Phytochemistry and pharmacological activity of *Azadirachta indica* (neem). Int J Pharm Biol Sci-IJPBSTM 10(3):172–180. https://doi.org/10.21276/ijpbs.2020.10.3.1

Singh A, Navneet. (2018) Ethnomedicinal, pharmacological and antimicrobial aspects of *Moringa oleifera* Lam.: a review. J Phytopharmacol 7(1):45–50

Singh B, Kaur P, Gopichand SRD, Ahuja PS (2008) Biology and chemistry of Ginkgo biloba. Fitoterapia 79(6):401–418. https://doi.org/10.1016/j.fitote.2008.05.007

Singh M, Mathur G, Jain CK, Mathur A (2012) Phyto-pharmacological potential of *Ginkgo biloba*: a review. J Pharm Res 5(10):5028–5030

Smith JV, Luo Y (2004) Studies on molecular mechanisms of *Ginkgo biloba* extract. Appl Microbiol Biotechnol 64(4):465–472. https://doi.org/10.1007/s00253-003-1527-9

Souza GA, Marqui SV, Matias JN, Guiguer EL, Barbalho SM (2020) Effects of Ginkgo biloba on diseases related to oxidative stress. Planta Med 86(6):376–386. https://doi.org/10.1055/a-1109-3405

Sudha P, Asdaq SMB, Dhamingi SS, Chandrakala GK (2010) Immunomodulatory activity of methanolic leaf extract of *Moringa oleifera* in animals. Indian J Physiol Pharmacol 54(2): 133–140

Tan WS, Arulselvan P, Karthivashan G, Fakurazi S (2015) *Moringa oleifera* flower extract suppresses the activation of inflammatory mediators in lipopolysaccharide-stimulated RAW 264.7 macrophages via NF-κB pathway. *Mediat Inflamm* 2015:720171. https://doi.org/10.1155/2015/720171

Tran TV, Park SJ, Shin EJ, Tran HQ, Jeong JH, Jang CG, Lee YJ, Nah SY, Nabeshima T, Kim HC (2018) Blockade of platelet-activating factor receptor attenuates abnormal behaviors induced by phencyclidine in mice through down-regulation of NF-kB. Brain Res Bull 137:71–78. https://doi.org/10.1016/j.brainresbull.2017.11.004

van der Nat JM, Klerx JPAM, van Dijk H, de Silva KTD, Labadie RP (1987) Immunomodulatory activity of an aqueous extract of *Azadirachta indica* stem bark. J Ethnopharmacol 19(2): 125–131. https://doi.org/10.1016/0378-8741(87)90036-5

van der Nat JM, van der Sluis WG, de Silva KTD, Labadie RP (1991) Ethnopharmacognostical survey of *Azadirachita indica* A. Juss. (Meliaceae). J Ethnopharmacol 35(1):1-24

Voet S, Srinivasan S, Lamkanfi M, van Loo G (2019) Inflammasomes in neuroinflammatory and neurodegenerative diseases. EMBO Mol Med 11(6):e10248. https://doi.org/10.15252/emmm.201810248

von Boetticher A (2011) Ginkgo biloba extract in the treatment of tinnitus: a systematic review. Neuropsychiatr Dis Treat 7:441–447

Waterman C, Cheng DM, Rojas-Silva P, Poulev A, Dreifus J, Lila MA, Raskin I (2014) Stable, water extractable isothiocyanates from *Moringa oleifera* leaves attenuate inflammation *in vitro*. Phytochemistry 103:114–122. https://doi.org/10.1016/j.phytochem.2014.03.028

Wen CC, Chen HM, Yang NS (2012) Developing phytocompounds from medicinal plants as Immunomodulators. Adv Bot Res 62:197–272. https://doi.org/10.1016/B978-0-12-394591-4.00004-0

Xiong XJ, Liu W, Yang XC, Feng B, Zhang YQ, Li YQ, Li XK, Wang J (2014) Ginkgo biloba extract for essential hypertension: A systemic review. Phytomedicine 21(10):1131–1136. https://doi.org/10.1016/j.phymed.2014.04.024

Yang Y, Islam MS, Wang J, Li Y, Chen X (2020) Traditional Chinese medicine in the treatment of patients infected with 2019-new coronavirus (SARS-CoV-2): a review and perspective. Int J Biol Sci 16(10):1708–1717. https://doi.org/10.7150/ijbs.45538

Yatish MR, Mitti JG, Joshi H (2019) Critical review on Nimba (*Azadirachta indica*). World J Pharm Pharm Sci 8(10):417–424

Yu WB, Wang Q, Chen S, Cao L, Tang J, Ma CG, Xiao W, Xiao BG (2019) The therapeutic potential of ginkgolide K in experimental autoimmune encephalomyelitis via peripheral immunomodulation. Int Immunopharmacol 70:284–294. https://doi.org/10.1016/j.intimp.2019.02.035

Zhang XY, Yao JK (2013) Oxidative stress and therapeutic implications in psychiatric disorders. Prog Neuro-Psychopharmacol Biol Psychiatry 46:197–199. https://doi.org/10.1016/j.pnpbp.2013.03.003

Zhao Q, Gao C, Cui Z (2015) Ginkgolide A reduces inflammatory response in high- glucose-stimulated human umbilical vein endothelial cells through STAT3-mediated pathway. Int Immunopharmacol 25(2):242–248. https://doi.org/10.1016/j.intimp.2015.02.001

Zhao Y, Paule J, Fu C, Koch MA (2010) Out of China: distribution history of *Ginkgo biloba* L. Taxon 59(2):495–504. https://doi.org/10.1002/tax.592014

Zhou JM, Gu SS, Mei WH, Zhou J, Wang ZZ, Xiao W (2016) Ginkgolides and bilobalide protect BV2 microglia cells against OGD/reoxygenation injury by inhibiting TLR2/4 signaling pathways. Cell Stress Chaperones 21(6):1037–1053. https://doi.org/10.1007/s12192-016-0728-y

Biotechnological Approaches for the Production of Immunomodulating Phytomolecules

17

Farzana Sabir

Abstract

Immunomodulating molecules can alter the immune response either by immunostimulation or immunosuppression. Manipulation of the immune system is commonly practiced in the case of destructive diseases mediated by uncontrolled immune responses. The use of chemical drugs as immunomodulators has several limitations, such as a higher risk of opportunistic infection and an adverse effect on the overall immune system. To address this issue, plenty of plant-derived molecules, like alkaloids, flavonoids, saponins, terpenoids, and steroid glycosides, have been utilized to manipulate the human immune system. Many traditional medicinal systems, including Ayurveda in India, have been employing phytomolecules as immunomodulators to control immunological disorders and strengthen the host immune system against many infectious diseases. Recent advances in biotechnological techniques offer tools for the genetic manipulation and metabolic engineering of biosynthetic pathways not only in plants but also in microorganisms. Therefore, it is also overviewed the use of yeasts for enhancing the production of phytomolecules in an easy-to-handle system. This chapter highlights the available information on the development of biotechnological tools to improve the accumulation of the desired natural compounds in controlled conditions. The major issues faced during the development of these biotechniques for the production of valuable phytomolecules are also addressed in this chapter.

F. Sabir (✉)
Linking Landscape, Environment, Agriculture and Food (LEAF), Instituto Superior de Agronomia, Universidade de Lisboa, Lisbon, Portugal

Research Institute for Medicines (iMed.ULisboa), Faculty of Pharmacy, Universidade de Lisboa, Lisbon, Portugal
e-mail: fsabir@isa.ulisboa.pt

N. S. Sangwan et al. (eds.), *Plants and Phytomolecules for Immunomodulation*,
https://doi.org/10.1007/978-981-16-8117-2_17

Keywords

Immune system · Natural compounds · Secondary metabolites · Medicinal plants · In vitro culture

17.1 The Immune System

The human immune system is a highly sophisticated defense system, which efficiently distinguishes the "self" (autoantigen) and "non-self" (alloantigen) antigens (Wilkes 2012). These antigens trigger the series of delicately regulated mechanisms to activate the immune system. The multidimensional architecture of the immune system protects the host at numerous levels by cooperative interaction of various types of cells, tissues, and organs. Additionally, particular chemical molecules are known as cytokines [classified in lymphokines, interleukins (IL), and chemokines] which also play a crucial role in the cross-talk between the cells of the immune system (Wilkes 2012).

The immune system is composed of two different but overlapping responses: (1) innate and (2) adaptive immune response (Parkin and Cohen 2001). The innate response involves the short-term mechanism as the first line of defense by the skin as a barrier and cytokines, complement system, phagocytes, natural killer (NK) cells, and gamma-delta T cells to provide the nonspecific response to the physical, biochemical, and cellular elements. On the other hand, the highly advanced adaptive immune response involves complex and diverse mechanisms generating the memory to the specific pathogen. It is composed of two types of responses: (a) humoral response mediated by B lymphocytes to secrete antibody and (b) cell-mediated response by T lymphocytes (Parkin and Cohen 2001).

In general, the cellular and molecular components of both systems coordinate to detect and eradicate the pathogens. The generic responses of the innate immune system are crucial for a healthy host; failing to this leads to secondary infections or death by sepsis (Parkin and Cohen 2001). Adaptive immunity, on the other hand, specifically aimed for a pathogen, which helps the improved response with subsequent exposures (Parkin and Cohen 2001).

17.2 Immunomodulation

The term immunomodulators is referred to synthetic or biological substances that can stimulate, enhance, or suppress any component or phase of the immune system, including both innate and adaptive systems (Mulcahy and Quinn 1986). Numerous specific and nonspecific methodologies can modulate the immune system (Mulcahy and Quinn 1986; Catanzaro et al. 2018). Nowadays, the concept of immunomodulation is receiving much attention to modulate the immune system for various disease managements. In fact, a newer branch of pharmacology named immunopharmacology has emerged, which is related to immunomodulators (Ballow

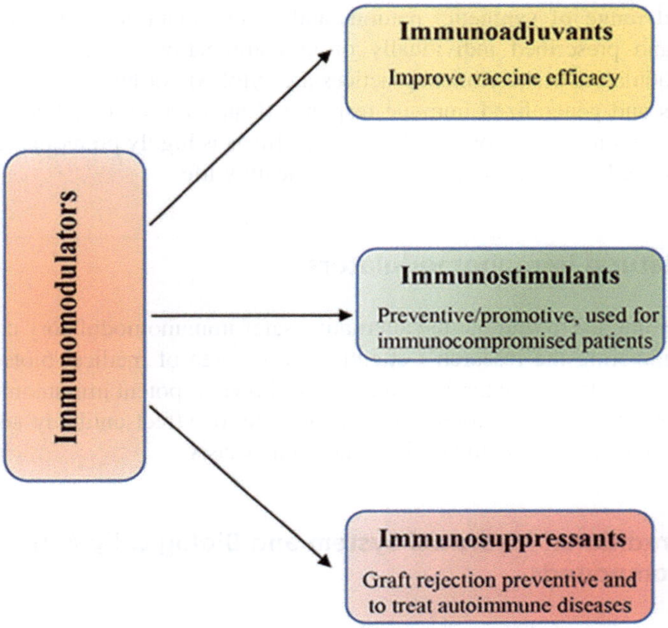

Fig. 17.1 Classification of immunomodulators and their clinical significance

and Nelson 1997). From clinical points of view, immunomodulators are classified into the following three categories (Fig. 17.1):

Immunoadjuvants: These substances are used to improve the effectiveness of vaccines, for example, Freund's adjuvant (Mulcahy and Quinn 1986). They enhance the immune response of the host when administered with an antigen by generating the danger signals. The alarmed immune system responds to the antigen as a real threat to the host. Immunoadjuvants are considered as the true modulators of immune response (Mulcahy and Quinn 1986; Parkin and Cohen 2001).

Immunostimulants: These substances nonspecifically improve the host's immunity by targeting primarily T or B lymphocytes or the complement system (Mulcahy and Quinn 1986; Parkin and Cohen 2001). Immunostimulants act as a preventive or promoting agent in a healthy immune system, whereas they are used as immunotherapeutic agents in the compromised immune system.

Immunosuppressants: This group of agents slows down the immune system in many clinical conditions, which need a suppressed immune response, for example, to prevent the rejection of organ transplants, graft versus host disease, and autoimmune diseases (Mulcahy and Quinn 1986; Parkin and Cohen 2001).

A broad range of synthetic, natural, and recombinant immunomodulators is available and prescribed individually or in combinations. The majority of the immunomodulators in the clinical practices are cytotoxic drugs, which exert severe side effects and generalized immune response (Catanzaro et al. 2018). Therefore, finding an immunomodulator with fewer side effects is highly prerequisite to obtain immunological homeostasis in the host for a healthy life.

17.3 Natural Immunomodulators

The current interest to find out the alternative safer immunomodulatory compounds has been gathering the research's attention in the area of medical biotechnology. Numerous plant-derived natural products have shown as potent immunomodulators, which either affect the functions of immune cells or affect antibody secretion to control the infection and maintain immune homeostasis.

17.3.1 Traditional Medicinal System and Biologically Active Compounds

Medicinal plants and their compounds have evidently been utilized since ancient times for the treatment of numerous diseases, including immunological disorders. Approximately 25% of modern-day medicines have been originated from natural resources, including medicinal plants. In fact, WHO mentioned that majority of the world population relies on plant-derived medicines, especially in developing countries (Kumar et al. 2012). Various traditional medicinal systems have been employing medicinal plants to improve immune responses. Indian traditional medicinal system, Ayurveda, is believed to be 5000 years old, and its branch Rasayana has the concept of utilizing the plants to promote health by strengthening host defenses against different diseases (Joshi and Bedekar 2017). According to the documented systems of medicine like Ayurveda, Unani, Siddha, and Homoeopathy (AYUSH), more than 7000 medicinal plant species are listed (Kumar et al. 2020). Indian medicinal plants are a rich source of natural compounds to induce nonspecific immunomodulation by manipulating the immune system at various levels like granulocytes, macrophages, natural killer cells, and complement function in mammalian models (Vrushali and Madhavi 2006).

Plants synthesize plenty of natural compounds, known as secondary metabolites, like alkaloids, terpenes, terpenoids, flavonoids, tannins, phenols, glycosides, and polysaccharides. These compounds are not strictly essential for the growth and reproduction of the plants but are believed to play some important role against external stimuli like biotic and abiotic factors and may accumulate in the specialized tissue/organ of the plants (Hussain et al. 2012). These natural compounds, possessing therapeutic potential, have always been of great interest to the researchers working on infectious diseases to improve the immune system. Potential natural immunomodulators can be divided into high- and low-molecular-weight compounds. Terpenoids, phenolic compounds, and alkaloids are low-molecular immunomodulatory compounds, whereas polysaccharides are the high-molecular-weight compounds (Varljen et al. 1989).

The most commonly known medicinal plants with immunomodulatory activities are *Aloe vera*, *Allium sativum*, *Andrographis paniculata*, *Artemisia annua*, *Azadirachta indica*, *Boerhaavia diffusa*, *Calendula officinalis*, *Cannabis sativa*, *Carica papaya*, *Centella asiatica*, *Chlorophytum borivilianum*, *Curcuma longa*, *Datura quercifolia*, *Emblica officinalis*, *Glycyrrhiza glabra*, *Hypericum perforatum*, *Matricaria chamomilla*, *Nelumbo nucifera*, *Nerium oleander*, *Nigella sativa*, *Ocimum sanctum*, *Panax ginseng*, *Silybum marianum*, *Tinospora cordifolia*, *Withania somnifera*, etc. [reviewed by Brindha (2016)].

17.3.2 Pharmacology of Natural Immunomodulators

For immunomodulatory studies, screening of the plant extracts, either separated as a pure phytomolecule or a single distilled fraction, extracted in various solvent systems, is employed. The pharmacological screening of natural compounds is determined by numerous in vitro and in vivo methods (Shantilal et al. 2018). Natural immunomodulators exert their effects through various mechanisms like changing white blood cell (WBC) counts, mast cells, and natural killer cells, cytokine modulation, activation of macrophages, stimulation of phagocytosis and lymphoid cells, and enhanced production of nonspecific immunity mediators and antigen-specific immunoglobulin. Additionally, reduced nitric oxide (NO) production and expression of co-stimulatory molecules (CD40, CD80, and CD86) by natural compounds also represent as the characteristics of immunomodulators (Kumar et al. 2012; Mukherjee et al. 2014).

17.4 Biotechnological Approaches

Identification and validation of the medicinal properties of a plant are generally a combination of both ethnobotanical information and systematic screening. Traditionally, the selected plants are collected, dried, ground up, and extracted in various organic solvents. This conventional method generally leads to over-exploitation of medicinal herbs, disturbing their natural fauna, and endangering them to extinction. This environment-unfriendly approach is time-consuming and labor-intensive. Additionally, the yield of phytomolecules depends on various factors like different biotic and abiotic stress, seed viability, and germination. On the other hand, in vitro chemical synthesis of these compounds at the industrial level is a difficult approach due to their complex chemical structures and complicated biosynthetic pathways. To meet the continuous market demand, development of novel approaches, ensuring sufficient supply of these molecules, is of great significance.

Biotechnology offers numerous cutting-edge tools for the continuous production of plant-based therapeutic compounds in an eco-friendlier way. Recent biotechnological advancements, including plant tissue culture methodologies, plant genetic engineering procedures, and metabolic and pathway engineering, offer a sustainable approach to conserve and produce the phytomolecules at a commercial scale. This section deals with the development of these techniques and their subsequent classic examples for the production of immunotherapeutic compounds. The biotechnological strategies for production of plant-derived immunomodulatory secondary metabolites of terpenoids and other chemical groups are summarized in Table 17.1 and Table 17.2, respectively.

17.4.1 Plant Tissue Culture

Plant tissue culture technology is an integral part of plant biotechnology, and it represents a potential renewable approach for the bioproduction of natural therapeutic compounds. This alternate strategy offers the production of important metabolites, bypassing the limitations of traditional methods like adverse geographical and climatic factors, difficult and lengthy cultivation period, genetic diversity, habitat degradation, species extinction, and low metabolite yield.

17.4.1.1 Types of Plant Tissue Culture Systems

In this technique, in vitro plant cell, tissue, and organs are cultivated under aseptic conditions at differentiated (shoots/roots) or undifferentiated (callus) level. Differentiated tissues like shoots and roots are induced at, respectively, higher and lower cytokinin/auxin ratio. They have been indicated to have a similar metabolic profile to native plants, and they biosynthesize the same types of phytochemicals under in vitro conditions (Chandran et al. 2020). On the other hand, the unorganized mass of cells known as callus is induced at higher auxin concentrations. Callus culture can be used to generate single-cell suspension cultures. Both callus and suspension cultures possess the ability to biosynthesize the phytoconstituents. It has been reported that callus cultures are used for the production of numerous natural immunomodulators of alkaloids, flavonoids, and terpenoids (Tables 17.1 and 17.2). Suspension cultures are a widely preferred system to produce bioactive phytomolecules either by batch or by continuous fermentation due to their relatively fast growth and possible extracellular secretion of phytochemicals, and they offer easier manipulation of the biosynthetic pathway at single-cell level (Chandran et al. 2020). However, the genetic instability of the fast-growing cells has always been a point of concern for researchers.

17.4.1.2 Hairy Root Cultures

Plant hairy root cultures present the advantage of genotypic and phenotypic stability; they are induced by the transformation of plants with *Agrobacterium rhizogenes*. Hairy root culture has been employed in the area of plant biotechnology for more than the last three decades, allowing the production of various pharmaceutical compounds by roots grown in bioreactors (Gutierrez-Valdes et al. 2020). They have established a beneficial biological system to study the biosynthesis of several

Table 17.1 Biotechnological intervention for the production of potential natural immunomodulatory agents of terpenoid group from some important medicinal plants (*full forms of abbreviations are in the footnote)

Plants (common name)	Natural compounds	Mode of actions	Biotechnological strategies
Artemisia annua (sweet wormwood)	Artemisinin (sesquiterpene lactone)	Inhibits NO production, promotes T cell function to accelerate the immune reconstitution (Yang et al. 1993; Konkimalla et al. 2008)	• Plant cell and tissue culture (Basile et al. 1993; Wetzstein et al. 2018) • Elicitation (Baldi and Dixit 2008) • Biotransformation (Sabir et al. 2010; Omar et al. 2015) • Heterologous expression in *S. cerevisiae* (Ro et al. 2006) • Metabolic engineering in *S. cerevisiae* (Westfall et al. 2012; Paddon et al. 2013) • Overexpression of transcription factors (Yu et al. 2012; Han et al. 2014)
Andrographis paniculata (green chiretta)	Andrographolides (labdane diterpenoid)	Increases IL-2 production, inhibits NO production (Rajagopal et al. 2003; Wang et al. 2010; Sudhakaran 2012)	• Plant cell and tissue culture (Praveen et al. 2009; Sharma and Jha 2012; Zaheer and Giri 2017; Das and Bandyopadhyay 2020) • Elicitation (Gandi et al. 2012; Vakil and Mendhulkar 2013; Sharma et al. 2015; Zaheer and Giri 2015; Zaheer and Giri 2017) • Biotransformation (De Sousa et al. 2018)
Azadirachta indica (neem tree)	Azadirachtin (triterpenoid-limonoid)	Increases IgM and IgG Inhibits NO production and degranulation of neutrophils (Vrushali and Madhavi 2006)	• Plant cell and hairy root culture (Singh and Chaturvedi 2013; Thakore and Srivastava 2017) • Elicitation (Satdive et al. 2007)

(continued)

Table 17.1 (continued)

Plants (common name)	Natural compounds	Mode of actions	Biotechnological strategies
Centella asiatica (gotu kola)	Asiaticoside (triterpene saponin)	Inhibits human PBMC mitogenesis and production of IL-2 and TNF-alpha (Punturee et al. 2005)	• Plant tissue and hairy root culture (Kim et al. 2007; Mercy et al. 2012; Singh et al. 2015a) • Elicitation (Kim et al. 2004; Kim et al. 2007; Gupta and Chaturvedi 2019) • Biotransformation (Alfarra and Omar 2014)
Glycyrrhiza glabra (licorice)	Glycyrrhizin (triterpenoid saponins)	Activates $CD4^+$ and $CD8^+$ population. Increases cytokines IL2, IL6, and IL 7 and decreases TNF-alpha levels (Ayeka et al. 2017)	• Plant cell and tissue culture (Hayashi et al. 1988; Shams-Ardakani et al. 2007) • Elicitation (Shabani et al. 2009) • Biotransformation (Hayashi et al. 1990; Hayashi et al. 1992) • Heterologous expression in *S. cerevisiae* (Wang et al. 2019)
Nigella sativa (black cumin)	Thymoquinone (monoterpene)	Regulates Th1/Th2 differentiation and NK cytotoxic activity (Majdalawieh and Fayyad 2015)	• Media optimization for plant tissue culture production (Rezaei et al. 2018; Kazmi et al. 2019; Golkar et al. 2020) • Elicitation (Scholz et al. 2009; Farag et al. 2015) • Nanoparticle-mediated stimulation of pathway genes (Kahila et al. 2018)
Panax ginseng (ginseng)	Ginsenosides (triterpenoid saponin)	Regulates immune cells, macrophages, natural killer cells, dendritic cells, T cells, and B cells (Kang and Min 2012)	• Plant tissue culture, hairy root culture, bioreactors, elicitation, biotransformation [reviewed by (Gantait et al. 2020)] • Heterologous expression in *S. cerevisiae*, metabolic engineering in *S. cerevisiae* [reviewed by (Chu et al. 2020)]

(continued)

Table 17.1 (continued)

Plants (common name)	Natural compounds	Mode of actions	Biotechnological strategies
Gymnema sylvestre (gymnema)	Gymnemic acid (triterpenoid saponin)	Restores the innate immune function by stimulating both myeloid and lymphoid components of the immune system (Singh et al. 2015b)	• Plant cell culture [reviewed by (Tiwari et al. 2014)] • Elicitation (Netala et al. 2016)
Withania somnifera (Ashwagandha)	Withanolides (steroidal lactone)	Increases total WBC count, circulating antibody titer, phagocytic activity of macrophages (Davis and Kuttan 2000)	• Plant cell and tissue and hairy root culture (Ray et al. 1996; Sabir et al. 2008; Nagella and Murthy 2010; Sivanandhan et al. 2013b; Sabir et al. 2012) • Elicitation (Sivanandhan et al. 2012; Sivanandhan et al. 2013a; Sivanandhan et al. 2014; Ahlawat et al. 2017) • Biotransformation (Sabir et al. 2011) • Overexpression of pathway key gene (Grover et al. 2013; Patel et al. 2015)

a*NO* nitric oxide, *IL* interleukin, *Ig* immunoglobulin, *NK* natural killer cells, *TNF* tumor necrosis factor, *NFκB* nuclear factor kappa-light-chain-enhancer of activated B cells, *PBMC* peripheral blood mononuclear cell, *Th1/Th2* T helper cell type 1/type 2

bioactive compounds like artemisinin, andrographolides, ginsenoside, withanolide, ginsenoside, and cannabinoids (Tables 17.1 and 17.2).

17.4.1.3 Optimization of Culture Conditions

In vitro biosynthesis rate of phytomolecules depends on various abiotic and biotic factors. Culture media composition and pH play a pivotal role in secondary metabolite production. The concentration of macro- and micronutrients, carbon source (sugar), and amino acid and optimized combination and concentration of plant growth regulators like cytokinins, auxins, and gibberellins influence the metabolite production. Moreover, culture physical conditions such as temperature, light, agitation, inoculum density, and aeration also affect the production of specialized metabolites. Alteration in the physical and nutritional components of the culture is the most fundamental way to manipulate the in vitro biosynthesis of the secondary metabolites. To optimize the culture parameters for improved phytomolecules

Table 17.2 Biotechnological strategies for the production of potential natural immunomodulatory compounds of other chemical groups (*full forms of abbreviations are in the footnote)

Plants (common name)	Natural compounds	Mode of actions	Biotechnological strategies
Boerhavia diffusa (punarnava)	Punarnavine (alkaloid)	Inhibits NK cell cytotoxicity, NO production, IL-2, and TNF-alpha (Mehrotra et al. 2002)	• Plant callus and shoot culture (Ahmad et al. 2020) • Naturally growing accessions (Shukla et al. 2003)
Curcuma longa (turmeric)	Curcumin (phenolics-diarylheptanoid)	Inhibits mitogen-stimulated lymphocyte proliferation, NFκB activation, and IL2 signaling (Ranjan et al. 2004)	• Plant shoot, callus, microrhizome culture (Pistelli et al. 2012; Wu et al. 2015; de Souza Ferrari et al. 2016; Gurav et al. 2020), • Immobilized cell culture (Chaturvedi et al. 2014) • Metabolic engineering in *E. Coli* (Katsuyama et al. 2008; Couto et al. 2017) • Biotransformation (Wang et al. 2013)
Hypericum perforatum (St. John's wort)	Hypericins (phenolics-naphthodianthrone)	Increases IL6 production (Froushani et al. 2015)	• *Agrobacterium-mediated* genetic transformation (Hou et al. 2016; Khan et al. 2020) • Plant cells and tissue culture (Santarém and Astarita 2003; Kirakosyan et al. 2004) • Nanoparticle-mediated stimulation (Ebadollahi et al. 2019)
Cannabis sativa (hemp)	Cannabinoids (prenylated polyketide)	Enhances the production of cytokines IL1, TNF alpha, and IL6 (Friedman et al. 1995)	• Plant cell and tissue culture (Schachtsiek et al. 2018) • Heterologous expression in tobacco (Schachtsiek et al. 2018) • De novo biosynthesis in *S. cerevisiae* (Zirpel et al. 2017; Luo et al. 2019) • Biotransformation and vacuolar compartmentalization of the product in yeast *Komagataella phaffii* and *S. cerevisiae* (Luo et al. 2019)

a*NO* nitric oxide, *IL* interleukin, *Ig* immunoglobulin, *NK* natural killer cells, *TNF* tumor necrosis factor, *NFκB* nuclear factor kappa-light-chain-enhancer of activated B cells, *PBMC* peripheral blood mononuclear cell, *Th1/Th2* T helper cell type 1/type 2 biosynthesis, a simulation-based modeling approach has been used for the development of hairy roots (Lenk et al. 2014).

17.4.1.4 Elicitors

In nature, plants produce secondary metabolites as a defense mechanism in the response of various physical stimuli or microbial infections (fungi, bacteria, viruses, yeast, etc.). The signal molecules generated from these stimuli are called elicitors (Chandran et al. 2020). Specific receptors present either on the plant cell surface or at the intracellular level recognize the elicitor, triggering the signal transduction pathways, which eventually stimulate the cells to activate the defense system of plants for synthesizing an array of metabolites involved in plant protection. Numerous secondary metabolic pathways have been shown to be triggered by utilizing elicitors, resulting in enhanced production of secondary metabolites, including immunomodulators (Tables 17.1 and Table 17.2). The selection of an elicitor varies according to the plant and to the relevant metabolite. Elicitors can be of biotic and abiotic origin. Physical factors like cold/heat, UV, water and osmotic stress, and chemical components like salts and heavy metals are abiotic elicitors (Halder et al. 2019), whereas biotic elicitors originated from the biological source are subclassified in exogenous and endogenous groups. The extracellular components like constituents of microbial cell walls (chitosan and chitin) or plant cell wall (polysaccharides or oligosaccharides) are considered as exogenous biotic elicitors. On the other hand, molecules produced by plants due to various stress and pathogen invasion are considered as endogenous biotic elicitors (Halder et al. 2019). For example, polysaccharides produced by degrading plant cell wall and intracellular proteins and phytohormones like salicylic acid or methyl jasmonate are in the category of endogenous biotic elicitors (Vanisree et al. 2004; Chandran et al. 2020). Utilization of various biotic elicitors like bacterial, fungal, and yeast extracts has been a point of interest of the researchers to produce enhanced phytomolecules like flavonoids, alkaloids, terpenoids, and phenylpropanoids (Table 17.1 and Table 17.2).

The combined use of various biotic and abiotic elicitors has also shown the synergistic effect to increase the production of secondary metabolites (Halder et al. 2019). Besides the selection of elicitor, their dosage and incubation time are also significant factors to consider for the optimized production of phytomolecules.

Additionally, nanoparticles (NPs) are also believed to trigger different cellular signal transduction pathways, modulating the plant secondary metabolism. NP-induced ROS may act as a signal to induce the plant secondary metabolism. The effects of some important metal oxide nanoparticles like titanium oxide and silicon oxide have been reported for improved secondary metabolite production in *H. perforatum* and *N. sativa* by stimulating the pathway key genes (Kahila et al. 2018; Ebadollahi et al. 2019).

17.4.1.5 Biotransformation and Cell Immobilization

Another powerful biotechnological tool to enhance the less abundant phytomolecules is through biotransformation by unique enzymes of plants and microbes. This technique not only improves the accumulation of a metabolite of interest, but it also has the potential to produce a novel compound, which is not present in the parent plants (Giri et al. 2001). This technique is useful to transform low-cost and abundant compounds, like pathway intermediates and industrial by-products, into high demand and expensive therapeutic compounds with higher bioactivity, bioavailability, and stability. Biotransformation can be achieved specifically targeting biosynthetic pathway intermediates or exogenously added foreign substrates to obtain the desired regio- and stereo-specific changes like hydroxylation, glycosylation, and epoxidation. Some examples of biotransformation for immunomodulator compounds are listed in Tables 17.1 and 17.2.

The rate of bioconversion by plant cells/organs depends on numerous factors such as enzyme activity and its localization, precursor solubility and presence of undesired by-products, and culture conditions like pH and elicitation (Giri et al. 2001). For efficient and cost-effective biotransformation, the source of biocatalyst needs to be stable and recyclable. Whole-cell culture is a stable system, and it has advantages over purified enzymes for reutilizing the essential cofactors and co-enzymes. Immobilization or entrapment of the plant cells avoids the damaging of the cells due to continuous shaking, which can be reutilized further. Moreover, the immobilization of single cells creates a microenvironment of cell clusters imitating the organized and differentiated tissues for the biosynthesis of secondary metabolites (Brodelius and Mosbach 1982). To immobilize the plant cells, gel entrapment through ion exchange, precipitation, and polymerization is the frequently used method. Enzyme immobilization by polyacrylamide matrix prevents the leakage of the enzymes, and microencapsulation forms a microsphere of polymeric membranes around the enzyme in solution.

Further, for large-scale biotransformation, membrane reactors can be employed, providing efficient controls of fluid dynamics and flow distribution. This system retains the plant cells and enzymes on the membrane while allowing the substrates, nutrients, and products to pass freely (Brodelius and Mosbach 1982; Giri et al. 2001).

17.4.2 Metabolic Engineering

Plants produce tremendous arrays of secondary metabolites, which are biosynthesized through one of the three major pathways: shikimate, polyketide, and terpenoid pathways (Chandran et al. 2020). Biosynthesis of flavonoids and alkaloids takes place through the shikimate pathway. The pathway intermediate p-coumaroyl-CoA initiates the biosynthesis of coumarins, flavonoids, lignans, stilbenoids, catechins, vanillin, gallic acid, and betalain compounds. Terpenoids are the most diverse group of plant secondary metabolites. More than one-third of the plant secondary metabolites are terpenoids, which are biosynthesized from the basic isoprene units, isopentenyl diphosphate (IDP) and dimethylallyl diphosphate (DMADP), originated from two independent pathways, mevalonic acid pathway (MVA) and methyl-D-erythritol (MEP) pathway. The MVA pathway synthesizes brassinosteroids, sesquiterpenoids, phytosterols, triterpenoids, and polyprenols (Scragg 2002). MEP pathway originates monoterpenoids, diterpenoids, hemiterpenoids, tocopherols, plastoquinone, and plant growth regulators like cytokinins, gibberellins, etc. The polyketide pathway drives the biosynthesis of fatty acids, originated from either acetyl CoA or malonyl CoA. Secondary metabolites established from the polyketide pathway include acetogenins, jasmonates, 6-methyl salicylic acid, plumbagin, coniine, and anthraquinones (Scragg 2002).

Recent advancements of biotechnological tools in the area of genomics, proteomics, and metabolomics have been allowed to elucidate the unknown endogenous secondary metabolite pathways and their regulation in numerous medicinal plants with therapeutic compounds (Scragg 2002). By utilizing the continuously developing biotechnological tools, secondary metabolite pathways of a selected phytomolecules can be upregulated, whereas the other competitive pathway can be downregulated to divert the endogenous carbon flux and common precursors and enzymes. Additionally, the whole or partial biosynthetic pathway can be reconstituted in heterologous hosts such as bacterial, plant, and yeast cells. The following sections explain these techniques in more detail:

17.4.2.1 Up- and Downregulation of the Pathways

The biosynthetic pathway of secondary metabolites is tightly regulated at various steps of key regulatory enzymes. The upregulation of the biosynthesis can be achieved by overexpression of genes of those key enzymes. Further, co-expression of multiple key genes of a metabolic pathway and overexpression of precursor biosynthetic genes have been shown to improve the production of natural immunomodulators (Tables 17.1 and 17.2). In a biosynthetic pathway, enzymes compete for the precursors. The silencing of the competitive pathway, which utilizes the common precursor and carbon flow, will redirect these components toward the biosynthesis of the metabolite of the interest (Tables 17.1 and 17.2).

The endogenous expression of the target gene is regulated by transcriptional factors. Overexpression of transcription factors is another approach for enhanced secondary metabolite production. Numerous transcription factors have exhibited the regulatory effect on the biosynthesis of secondary metabolites. For instance, transcription factors of WRKY family increased the production of secondary metabolites such as artemisinin in transgenic plants of *Artemisia annua* and *W. somnifera* (Han et al. 2014; Jadaun et al. 2020). AP2/ERF transcription factors are involved in plant response to biotic and abiotic stresses, as well as in the regulation of metabolism in various plant species. An improved accumulation of artemisinin and artemisinic acids was observed due to AP2/ERF overexpression (Yu et al. 2012; Tripathi et al. 2020).

17.4.2.2 Heterologous Expression

The pharmaceutical compounds obtained from the native plants are generally of low yield. Additionally, the purification of the desired metabolite from the complex mixture of other compounds produced by complex secondary metabolite pathways is tricky, leading to low final concentration. Genetic and metabolic engineering has, however, overcome this constraint. With the help of continuously developing molecular biology tools, genes of one organism can be heterologously expressed into another host organism, enabling to produce a higher amount of compound of interest (Scragg 2002; Chandran et al. 2020). Utilization of microbial cell factory for plant natural compound biosynthesis is gaining popularity due to their easy-to-handle system and limited secondary metabolite pathways to interfere with the heterologous pathway from plants. Plenty of plant secondary metabolite pathway genes, encoding pharmaceutically valuable compounds, have been cloned and expressed in microorganisms. *E. coli* was used as the host for the biosynthesis. But, yeast is currently widely preferred as a heterologous host for the production of plant secondary metabolites (Rahmat and Kang 2020). The budding yeast, *Saccharomyces cerevisiae*, is a key model organism for fundamental molecular biology research, and it was the first eukaryotic organism to have its genome completely sequenced. By combined utilization of synthetic biology and molecular biology tools, numerous plant secondary metabolite pathways have been constructed (Siddiqui et al. 2012), including natural immunomodulator compounds (Tables 17.1 and 17.2). The biosynthesis of the desired phytomolecule is achieved either by de novo biosynthesis from whole pathway construction or from partial pathway steps in the yeast cells with the addition of inexpensive pathway precursors. Overexpression of the pathway genes can be obtained by utilizing the vector with a strong constitutive promoter. Genetically engineered yeast cells with downregulated native ergosterol pathway have been utilized for redirection of pathway intermediate of terpenoid biosynthesis (Siddiqui et al. 2012). Under the control of a repressible promoter, the enhanced carbon flow also allows the higher accumulation of pathway precursors. The yeast strains expressing the pathway genes are selected at the suitable auxotrophic marker. Alternatively, combinatorial biosynthesis allows the combined utilization of genes from different plants to design a new set of gene cluster heterologous production of phytomolecules (Rahmat and Kang 2020).

Biosynthetic pathways of numerous crucial phytomolecules are not completely understood. In that case, another significant biotechnological tool, transcriptome analysis of the selected plant, provides an indication of the genes involved in the biosynthetic pathway of the phytomolecules. Differential regulation of metabolic pathways in a comparative transcriptomic manner identifies putative key genes of the pathway upregulation. The schematic presentation of the biotechnological strategies is depicted in Fig. 17.2.

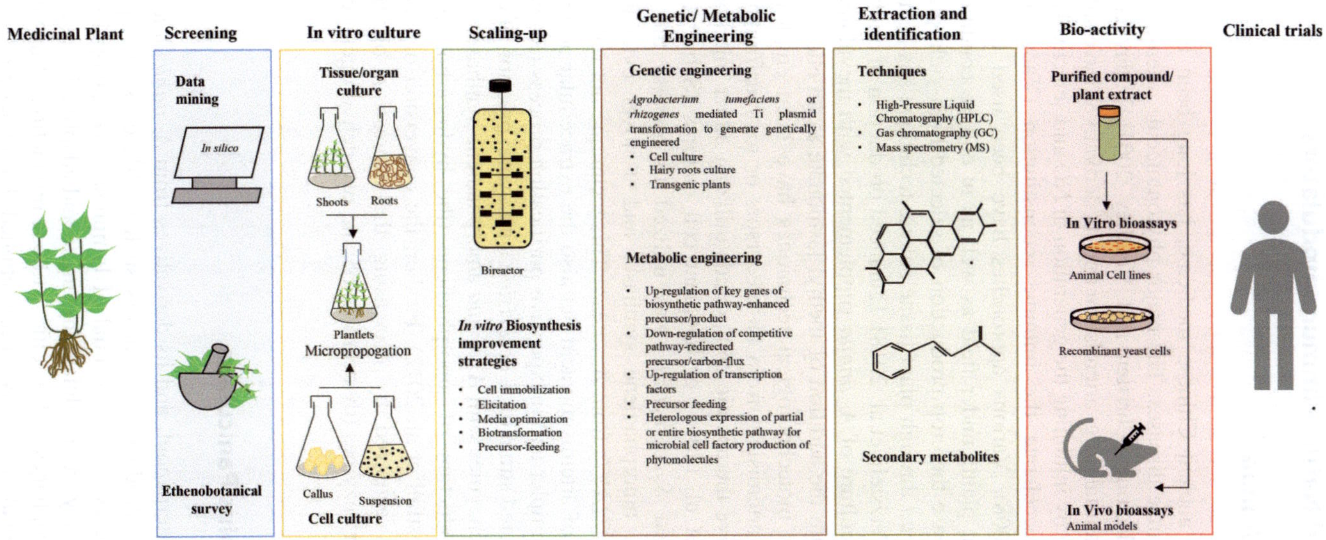

Fig. 17.2 Schematic representation of biotechnological strategies for in vitro production of plant-derived immunomodulators. The picture explains the various phases of screening of natural compounds from medicinal plants, their improved production by various in vitro biotechnological techniques, their isolation and purification techniques, and utilization as drugs to human

17.5 Some Case Studies of Biotechnological Production of Selected Natural Immunomodulators

17.5.1 Artemisia Annua

Artemisia annua, an ancient Chinese medicine, has long been utilized to treat malarial and autoimmune diseases. The major phytomolecule artemisinin is a sesquiterpene trioxane lactone (Christen and Veuthey 2001; Yadav et al. 2014). Besides, as an anti-malarial drug, artemisinin and its derivatives have shown immunomodulatory activity by inhibiting the generation of NO and selectively promoting T cell function to accelerate the immune reconstitution (Yang et al. 1993; Konkimalla et al. 2008). Various approaches have been used to enhance the artemisinin content in plants under field as well as in culture conditions (Yadav et al. 2014). To increase the crop production of selected high-yield plants, clonal propagation of *A. annua* through plant tissue culture technique has been attempted (Basile et al. 1993; Wetzstein et al. 2018). Enhanced production of artemisinin was observed in the cell culture of *A. annua* supplemented with mevalonic acid as a pathway precursor with the addition of methyl jasmonate as an elicitor (Baldi and Dixit 2008). Various biotechnological techniques have been employed for the biotransformation of artemisinin from its precursors or to transform it into less-toxic and high effective therapeutic compounds utilizing either microbial cultures reviewed by Omar et al. (2015) or by plant cell cultures (Sabir et al. 2010). Genetically engineered *S. cerevisiae* was utilized to express the engineered mevalonate pathway, amorphadiene synthase, and a novel cytochrome P450 monooxygenase (*CYP71AV1*) from *A. annua*. This technique not only leads to the improved production of artemisinic acid but also the extracellular accumulation of the compound for a simpler and inexpensive purification process (Ro et al. 2006). Moreover, the engineered strain of *S. cerevisiae* with overexpression of endogenous mevalonate pathway enzymes with *A. annua* amorphadiene synthase enzyme leads to the improved production of amorphadiene (the hydrocarbon precursor of artemisinic acid) (Westfall et al. 2012). Further, the conversion of amorphadiene artemisinic acid was achieved by decreasing the expression of *Aa*CPR and co-expression of other enzymes (cytochrome b5 and two dehydrogenases) (Paddon et al. 2013).

17.5.2 Andrographis Paniculata

Andrographolide, a diterpenoid, and its derivatives from *A. paniculata* have been shown to regulate the production of factors such as NK cells, IFN-γ, IL-2, and TNF-α, increasing the cytotoxic potential of lymphocyte (Rajagopal et al. 2003; Mishra et al. 2007). They can modulate the innate and adaptive immune responses by activating the macrophages and specific antibody production (Wang et al. 2010). Various biotechnological tools have been applied for enhanced production of andrographolide from the shoot, root, and cell cultures of *A. paniculata* supplemented without or with different elicitors like methyl jasmonate, salicylic acid, and various fungal elicitors like yeast extract, *Aspergillus niger*, and *Penicillium expansum* (Praveen et al. 2009; Gandi et al. 2012; Sharma and Jha 2012; Vakil and Mendhulkar 2013; Zaheer and Giri 2015; Zaheer and Giri 2017; Das and Bandyopadhyay 2020). In cell culture of *A. paniculata*, improved accumulation of andrographolide in the presence of methyl jasmonate was linked with the induction of biosynthetic pathway genes (Sharma et al. 2015). Various microbial source and purified enzymes have been utilized for the bioconversion of andrographolide and its derivatives with higher therapeutic value [reviewed by (De Sousa et al. 2018)].

17.5.3 Curcuma Longa

Curcumin, a diarylheptanoid compound, is obtained from the tuber of *Curcuma longa.* It has been used for various medicinal purposes, including anti-inflammatory and immunomodulatory activity, for centuries. Curcumin is one of the most extensively studied compounds for its immunomodulatory properties. It inhibits the NO production, nuclear factor-kappa B (NK-kB) activation, cyclooxygenase-2 (COX-2), IFN-γ, or TNF-α activated macrophages (Surh et al. 2001). A considerable amount of curcumin was obtained in shoot culture and in vitro microrhizomes (Pistelli et al. 2012). Various media optimizations have attempted for its efficient micropropagation (de Souza Ferrari et al. 2016) and phytomolecule accumulation in microrhizomes (Wu et al. 2015), callus (Gurav et al. 2020), and immobilized suspension culture (Chaturvedi et al. 2014). Commercial application of heterologous production of curcuminoids through the combinatorial biosynthesis approach has been employed in *E. coli* by utilizing pathway enzymes from different plants (Katsuyama et al. 2008). Further, two-step fermentation optimization and utilization of different *E. coli* strain enhanced the production of curcuminoids (Couto et al. 2017). Moreover, due to the addition of a 2-fluorol-phenylalanine precursor, three novel derivatives of curcuminoids were obtained (Wang et al. 2013).

17.5.4 Glycyrrhiza Glabra

Glycyrrhiza glabra is one of the most important medicinal plants known since ancient time. The triterpenoid saponins glycyrrhizin and glycyrrhetinic acid are the major pharmaceutical compounds, having various immunomodulatory activities targeting dendritic cells and upregulating the expression of CD40 and CD86 (Bordbar et al. 2012). In vivo bioactive molecule is glycyrrhetinic acid, which is derived from glycyrrhizin after oral administration. Glycyrrhizin accumulation in the cell culture of *G. glabra* was not detected, but other soyasaponins were produced by the in vitro culture (Hayashi et al. 1988). Cell cultures of *G. glabra* converted the externally supplied glycyrrhetinic acid to other derivatives of glycyrrhizin (Hayashi et al. 1990; Hayashi et al. 1992). Various pathway genes were regulated by elicitors methyl jasmonate and yeast extract in the cell culture (reviewed by (Hayashi 2009).

Additionally, the accumulation in the root culture was also improved due to the addition of elicitors methyl jasmonate and salicylic acid (Shabani et al. 2009). The entire biosynthetic pathway of glycyrrhetinic acid is reconstructed in the metabolically engineered *S. cerevisiae* by a combinatorial biosynthesis approach for improved production in fed-batch fermentation (Wang et al. 2019).

17.5.5 Withania Somnifera

Withania somnifera is one of the extensively studied medicinal plants of Ayurveda (Sangwan et al. 2017). The main phytoconstituent withanolide is a steroidal lactone, which has numerous medicinal properties, including immunomodulatory activities, by increasing the total WBC count and activating the phagocytic activity of macrophages (Davis and Kuttan 2000). Plenty of studies have been attempted for in vitro production of withanolides from a plant cell, shoot, adventitious, and hairy root cultures (Sabir et al. 2008; Nagella and Murthy 2010). Various culture optimization and elicitation have been utilized for enhanced production of withanolides (Sivanandhan et al. 2012; Sivanandhan et al. 2013a; Sivanandhan et al. 2014; Sivanandhan et al. 2015; Ahlawat et al. 2017). It has been suggested that the concentration and types of withanolide accumulation depend on the tissue system differentiation (Sharada et al. 2007; Sabir et al. 2013); however, enhanced de novo biosynthesis of withanolide A, which was not present in parental chemotype, in shoot cultures was reported by incorporation studies using [2-^{14}C] acetate as a precursor (Sangwan et al. 2007). The quantitative NMR-based study of radioactive ^{13}C labeled D-glucose incorporated in the plant's in vitro shoot cultures revealed that both DOXP and MVA pathways of isoprenogenesis contribute to withanolide biosynthesis (Chaurasiya et al. 2012). Furthermore, metabolic engineering of the biosynthetic pathway was attempted. Key enzyme squalene synthase genes either from *Withania* or from *Arabidopsis* were overexpressed in *Agrobacterium*-mediated genetically engineered cells and organs of plants for enhanced production of withanolides (Mirjalili et al. 2011; Grover et al. 2013; Patel et al. 2015; Mishra et al. 2016).

17.6 Major Concerns Related to Natural Immunomodulators

Despite the availability of a plethora of researches pointing out the significance of plant-derived compounds as natural immunomodulators, their clinical usefulness is limited. The major challenge to produce the plant-derived pharmaceutical compounds includes low yield, imprecise identification, and speciation of the compounds. The low concentration of the bioactive compound generally limits sufficient bioassays for the rapid screening of the newly identified phytomolecule. The in vivo testing on laboratory animals is restricted in some cases by ethical considerations. Insufficient pharmacological data, lack of practical therapy benefits, little knowledge about the plant drug interaction with modern medicine are foremost limiting factors.

Therefore, it is highly prerequisite to develop efficient and high-throughput techniques to integrate the separation, structure elucidation, and biological testing of therapeutic compounds in one continuum. Advanced biotechnological tools for efficient identification, purification, and bioassays of the natural compounds are being developed and employed. The coupling of the mass spectrometer with HPLC (high-performance liquid chromatography) and gas chromatographic techniques offers simultaneous separation and identification. To address the insufficient concentration for bioactivity, an alternate approach by Biacore Technology offers the coupling of separations to bioassays by utilizing a microchip with specific enzymes, receptors, and specific antibodies (Parker et al. 2010). The compound, which reacted with the microchip proteins, can be identified with their mode of action. Additionally, a recombinant yeast is widely utilized to screen the plant-based therapeutic compounds (Zhang et al. 2018). By reinforcing the target-based approach, a number of plan-based immunomodulators, such as curcumin, andrographolide, resveratrol, colchicine, capsaicin, etc., are undergoing clinical trials (Jantan et al. 2015).

17.7 Conclusion and Future Directions

The expanding knowledge of plant-derived compounds as therapeutic agents for the immune system, either for preventive or therapeutic purposes, can offer a powerful approach for the development of novel, safer, and affordable medicines. By applying the various biotechnological tools, it is possible to obtain the target compound under in vitro controlled conditions from plant cell, tissue, and hairy root cultures with faster growth cycles, without threatening the flora and biodiversity of the region. Moreover, the biotechnological approach offers to manipulate the biosynthesis of therapeutic compounds at a large scale (bioreactors) with strict quality control and simplified downstream chemical processing. However, the challenges encountered by utilizing the natural immunomodulators need to be addressed. Continuous advancements in the area of plant metabolic engineering hold a great promise for the improved production of phytoconstituents not only in the abovementioned tissue system but also in the microbial cell factories. The combinatorial approach for faster separation and bioactivity evaluation with lower quantity can be a game-changer for providing enough support for preclinical/clinical assessments. Employing nanotechnology can also serve as a high-end technique to deliver the natural immunomodulator nanoparticles for better bioavailability.

References

Ahlawat S, Saxena P, Ali A, Khan S, Abdin MZ (2017) Comparative study of withanolide production and the related transcriptional responses of biosynthetic genes in fungi elicited cell suspension culture of *Withania somnifera* in shake flask and bioreactor. Plant Physiol Biochem 114:19–28

Ahmad W, Husain I, Ahmad N, Amir M, Sarafroz M, Ansari MA, Zafar A, Ali A, Zafar R, Ashraf K (2020) Box–Behnken supported development and validation of robust HPTLC method: an application in estimation of punarnavine in leaf, stem, and their callus of *Boerhavia diffusa* Linn. 3 Biotech 10(4):1–10

Alfarra HY, Omar MN (2014) HPLC separation and isolation of Asiaticoside from *Centella asiatica* and its biotransformation by *A. niger*. Int J Pharma Med Biol Sci 3(3):1

Ayeka PA, Bian Y, Githaiga PM, Zhao Y (2017) The immunomodulatory activities of licorice polysaccharides (*Glycyrrhiza uralensis* Fisch.) in CT 26 tumor-bearing mice. BMC Complement Altern Med 17(1):1–9

Baldi A, Dixit VK (2008) Enhanced artemisinin production by cell cultures of *Artemisia annua*. Curr Trends Biotechnol Pharm 2(2):341–389

Ballow M, Nelson R (1997) Immunopharmacology: immunomodulation and immunotherapy. JAMA 278(22):2008–2017

Bansal S, Narnoliya LK, Mishra B, Chandra M, Yadav RK, Sangwan NS (2018) HMG-CoA reductase from camphor Tulsi (*Ocimum kilimandscharicum*) regulated MVA dependent biosynthesis of diverse terpenoids in homologous and heterologous plant systems. Sci Rep 8:3547. https://doi.org/10.1038/s41598-017-17153-z

Basile DV, Akhtari N, Durand Y, Nair MSR (1993) Toward the production of artemisinin through tissue culture: determining nutrient-hormone combinations suitable for cell suspension cultures. In Vitro Cell Dev Biol Plant 29(3):143–147

Bordbar N, Karimi MH, Amirghofran Z (2012) The effect of glycyrrhizin on maturation and T cell stimulating activity of dendritic cells. Cell Immunol 280(1):44–49

Brindha P (2016) Role of phytochemicals as immunomodulatory agents: a review. Int J Green Pharm (IJGP) 10:1

Brodelius P, Mosbach K (1982) Immobilized plant cells. In: Advances in applied microbiology. Elsevier, pp 1–26

Catanzaro M, Corsini E, Rosini M, Racchi M, Lanni C (2018) Immunomodulators inspired by nature: a review on curcumin and Echinacea. Molecules 23(11):2778

Chandran H, Meena M, Barupal T, Sharma K (2020) Plant tissue culture as a perpetual source for production of industrially important bioactive compounds. Biotechnol Rep:e00450

Chaturvedi P, Mehta S, Chatterjee P, Chowdhary A (2014) Media optimization in immobilized culture to enhance the content of curcumin in *Curcuma longa* (Zingiberaceae) and protein profile of treated samples in static culture. Nat Prod Chem Res

Chaurasiya ND, Sangwan NS, Sabir F, Misra L, Sangwan RS (2012) Withanolide biosynthesis recruits both mevalonate and DOXP pathways of isoprenogenesis in Ashwagandha *Withania somnifera* L.(Dunal). Plant Cell Rep 31(10):1889–1897

Christen P, Veuthey JL (2001) New trends in extraction, identification and quantification of artemisinin and its derivatives. Curr Med Chem 8(15):1827–1839

Chu LL, Montecillo JAV, Bae H (2020) Recent advances in the metabolic engineering of yeasts for ginsenoside biosynthesis. Frontiers in Bioengineering and Biotechnology 8

Couto MR, Rodrigues JL, Rodrigues LR (2017) Optimization of fermentation conditions for the production of curcumin by engineered *Escherichia coli*. J R Soc Interface 14(133):20170470

Das D, Bandyopadhyay M (2020) Novel approaches towards over-production of andrographolide in in vitro seedling cultures of *Andrographis paniculata*. S Afr J Bot 128:77–86

Davis L, Kuttan G (2000) Immunomodulatory activity of *Withania somnifera*. J Ethnopharmacol 71(1–2):193–200

De Sousa IP, Sousa Teixeira MV, Jacometti Cardoso Furtado NA (2018) An overview of biotransformation and toxicity of diterpenes. Molecules 23(6):1387

de Souza Ferrari MP, Antoniazzi D, Nascimento AB, Franz LF, Bezerra CS, Magalhães HM (2016) Evaluation of new protocols to *Curcuma longa* micropropagation: a medicinal and ornamental specie. J Med Plants Res 10(25):367–376

Ebadollahi R, Jafarirad S, Kosari-Nasab M, Mahjouri S (2019) Effect of explant source, perlite nanoparticles and TiO 2/perlite nanocomposites on phytochemical composition of metabolites in callus cultures of *Hypericum perforatum*. Sci Rep 9(1):1–15

Farag MA, El Sayed AM, El Banna A, Ruehmann S (2015) Metabolomics reveals distinct methylation reaction in MeJA elicited *Nigella sativa* callus via UPLC–MS and chemometrics. Plant Cell Tissue Organ Culture (PCTOC) 122(2):453–463

Friedman H, Klein TW, Newton C, Daaka Y (1995) Marijuana, receptors and immunomodulation. In: The brain immune axis and substance abuse. Springer, pp 103–113

Froushani SMA, Gouvarchin Galee HE, Khamisabadi M, Lotfallahzade B (2015) Immunomodulatory effects of hydroalcoholic extract of *Hypericum perforatum*. Avicenna J Phytomed 5(1):62

Gandi S, Rao K, Chodisetti B, Giri A (2012) Elicitation of andrographolide in the suspension cultures of *Andrographis paniculata*. Appl Biochem Biotechnol 168(7):1729–1738

Gantait S, Mitra M, Chen J-T (2020) Biotechnological interventions for ginsenosides production. Biomol Ther 10(4):538

Giri A, Dhingra V, Giri CC, Singh A, Ward OP, Narasu ML (2001) Biotransformations using plant cells, organ cultures and enzyme systems: current trends and future prospects. Biotechnol Adv 19(3):175–199

Golkar P, Bakhshi G, Vahabi MR (2020) Phytochemical, biochemical, and growth changes in response to salinity in callus cultures of *Nigella sativa* L. In Vitro Cell Dev Biol-Plant 56(2): 247–258

Grover A, Samuel G, Bisaria VS, Sundar D (2013) Enhanced withanolide production by overexpression of squalene synthase in *Withania somnifera*. J Biosci Bioeng 115(6):680–685

Gupta S, Chaturvedi P (2019) Enhancing secondary metabolite production in medicinal plants using endophytic elicitors: a case study of *Centella asiatica* (Apiaceae) and asiaticoside. endophytes for a growing. WORLD:310–323

Gurav SS, Gurav NS, Patil AT, Duragkar NJ (2020) Effect of explant source, culture media, and growth regulators on callogenesis and expression of secondary metabolites of *Curcuma longa*. J Herbs Spices Med Plants 26(2):172–190

Gutierrez-Valdes N, Häkkinen ST, Lemasson C, Guillet M, Oksman-Caldentey K-M, Ritala A, Cardon F (2020) Hairy root cultures—a versatile tool with multiple applications. Frontiers in Plant Science 11

Halder M, Sarkar S, Jha S (2019) Elicitation: a biotechnological tool for enhanced production of secondary metabolites in hairy root cultures. Eng Life Sci 19(12):880–895

Han J, Wang H, Lundgren A, Brodelius PE (2014) Effects of overexpression of AaWRKY1 on artemisinin biosynthesis in transgenic *Artemisia annua* plants. Phytochemistry 102:89–96

Hayashi H (2009) Molecular biology of secondary metabolism: case study for *Glycyrrhiza* plants. In: Recent advances in plant biotechnology. Springer, pp 89–103

Hayashi H, Fukui H, Tabata M (1988) Examination of triterpenoids produced by callus and cell suspension cultures of *Glycyrrhiza glabra*. Plant Cell Rep 7(7):508–511

Hayashi H, Fukui H, Tabata M (1990) Biotransformation of 18β-glycyrrhetinic acid by cell suspension cultures of *Glycyrrhiza glabra*. Phytochemistry 29(7):2149–2152

Hayashi H, Yamada K, Fukui H, Tabata M (1992) Metabolism of exogenous 18β-glycyrrhetinic acid in cultured cells of *Glycyrrhiza glabra*. Phytochemistry 31(8):2724–2733

Hou W, Shakya P, Franklin G (2016) A perspective on *Hypericum perforatum* genetic transformation. Front Plant Sci 7:879

Hussain MS, Fareed S, Saba Ansari M, Rahman A, Ahmad IZ, Saeed M (2012) Current approaches toward production of secondary plant metabolites. J Pharm Bioallied Sci 4(1):10

Jadaun JS, Kushwaha AK, Sangwan NS, Mishra S, Sangwan RS (2020) WRKY1-mediated regulation of tryptophan decarboxylase in tryptamine generation for withanamide production in *Withania somnifera* (Ashwagandha). Plant Cell Rep 39(11):1443–1465

Jantan I, Ahmad W, Bukhari SNA (2015) Plant-derived immunomodulators: an insight on their preclinical evaluation and clinical trials. Front Plant Sci 6:655

Joshi N, Bedekar SS (2017) Concept of Rasayana for a better health-a review. J Ayurveda Integr Med Sci. (ISSN 2456-3110) 2(1):209–212

Kahila MMH, Najy AM, Rahaie M, Mir-Derikvand M (2018) Effect of nanoparticle treatment on expression of a key gene involved in thymoquinone biosynthetic pathway in *Nigella sativa* L. Nat Prod Res 32(15):1858–1862

Kang S, Min H (2012) Ginseng, the 'immunity boost': the effects of *Panax ginseng* on immune system. J Ginseng Res 36(4):354

Katsuyama Y, Matsuzawa M, Funa N, Horinouchi S (2008) Production of curcuminoids by *Escherichia coli* carrying an artificial biosynthesis pathway. Microbiology 154(9):2620–2628

Kazmi A, Khan MA, Huma A (2019) Biotechnological approaches for production of bioactive secondary metabolites in *Nigella sativa*: an up-to-date review. Int J Secondary Metabolite 6(2): 172–195

Khan SA, Verma P, Parasharami VA (2020) Homo and heterologous expression of the HpPKS2 gene in *Hypericum perforatum* and *Bacopa monnieri*. Plant Cell Tissue Organ Culture (PCTOC). https://doi.org/10.1007/s11240-020-01965-5

Kim O-T, Bang K-H, Shin Y-S, Lee M-J, Jung S-J, Hyun D-Y, Kim Y-C, Seong N-S, Cha S-W, Hwang B (2007) Enhanced production of asiaticoside from hairy root cultures of *Centella asiatica* (L.) urban elicited by methyl jasmonate. Plant Cell Rep 26(11):1941–1949

Kim OT, Kim MY, Hong MH, Ahn JC, Hwang B (2004) Stimulation of asiaticoside accumulation in the whole plant cultures of *Centella asiatica* (L.) urban by elicitors. Plant Cell Rep 23(5): 339–344

Kirakosyan A, Sirvent TM, Gibson DM, Kaufman PB (2004) The production of hypericins and hyperforin by in vitro cultures of St. John's wort (*Hypericum perforatum*). Biotechnol Appl Biochem 39(1):71–81

Konkimalla VB, Blunder M, Korn B, Soomro SA, Jansen H, Chang W, Posner GH, Bauer R, Efferth T (2008) Effect of artemisinins and other endoperoxides on nitric oxide-related signaling pathway in RAW 264.7 mouse macrophage cells. Nitric Oxide 19(2):184–191

Kumar D, Arya V, Kaur R, Bhat ZA, Gupta VK, Kumar V (2012) A review of immunomodulators in the Indian traditional health care system. J Microbiol Immunol Infect 45(3):165–184

Kumar P, Shaunak I, Verma ML (2020) Biotechnological application of health promising bioactive molecules. In: Biotechnological production of bioactive compounds. Elsevier, pp 165–189

Lenk F, Sürmann A, Oberthür P, Schneider M, Steingroewer J, Bley T (2014) Modeling hairy root tissue growth in in vitro environments using an agent-based, structured growth model. Bioprocess Biosyst Eng 37(6):1173–1184

Luo X, Reiter MA, d'Espaux L, Wong J, Denby CM, Lechner A, Zhang Y, Grzybowski AT, Harth S, Lin W (2019) Complete biosynthesis of cannabinoids and their unnatural analogues in yeast. Nature 567(7746):123–126

Majdalawieh AF, Fayyad MW (2015) Immunomodulatory and anti-inflammatory action of *Nigella sativa* and thymoquinone: a comprehensive review. Int Immunopharmacol 28(1):295–304

Mehrotra S, Mishra KP, Maurya R, Srimal RC, Singh VK (2002) Immunomodulation by ethanolic extract of *Boerhaavia diffusa* roots. Int Immunopharmacol 2(7):987–996

Mercy S, Sangeetha N, Ganesh D (2012) In vitro production of adventitious roots containing asiaticoside from leaf tissues of *Centella asiatica* L. In Vitro Cell Dev Biol-Plant 48(2):200–207

Mirjalili MH, Moyano E, Bonfill M, Cusido RM, Palazon J (2011) Overexpression of the Arabidopsis thaliana squalene synthase gene in *Withania coagulans* hairy root cultures. Biol Plant 55(2):357–360

Mishra S, Bansal S, Mishra B, Sangwan RS, Jadaun JS, Sangwan NS (2016) RNAi and homologous over-expression based functional approaches reveal triterpenoid synthase gene-cycloartenol synthase is involved in downstream withanolide biosynthesis in *Withania somnifera*. PLoS One 11(2):e0149691

Mishra SK, Sangwan NS, Sangwan RS (2007) *Andrographis paniculata*, Kalmegh, a review. Pharm Rev 1(2):283–298

Mukherjee PK, Nema NK, Bhadra S, Mukherjee D, Braga FC, Matsabisa MG (2014) Immunomodulatory leads from medicinal plants. Indian J Tradit Knowl 13:235–256

Mulcahy G, Quinn PJ (1986) A review of immunomodulators and their application in veterinary medicine. J Vet Pharmacol Ther 9(2):119–139

Nagella P, Murthy HN (2010) Establishment of cell suspension cultures of *Withania somnifera* for the production of withanolide a. Bioresour Technol 101(17):6735–6739

Netala VR, Kotakadi VS, Gaddam SA, Ghosh SB, Tartte V (2016) Elicitation of gymnemic acid production in cell suspension cultures of *Gymnema sylvestre* R. Br through endophytic fungi. 3 Biotech 6(2):232

Omar MN, Hasali NHM, Khan NT, Moin SF, Alfarra HY (2015) Biotechnological transformation of artemisinin: toward an effective anti-malaria drug. Biomed Pharmacol J 5(1):19–24

Paddon CJ, Westfall PJ, Pitera DJ, Benjamin K, Fisher K, McPhee D, Leavell MD, Tai A, Main A, Eng D (2013) High-level semi-synthetic production of the potent antimalarial artemisinin. Nature 496(7446):528–532

Parker CE, Pearson TW, Anderson NL, Borchers CH (2010) Mass-spectrometry-based clinical proteomics–a review and prospective. Analyst 135(8):1830–1838

Parkin J, Cohen B (2001) An overview of the immune system. Lancet 357(9270):1777–1789

Patel N, Patel P, Kendurkar SV, Thulasiram HV, Khan BM (2015) Overexpression of squalene synthase in *Withania somnifera* leads to enhanced withanolide biosynthesis. Plant Cell Tissue Organ Culture (PCTOC) 122(2):409–420

Pistelli L, Bertoli A, Gelli F, Bedini L, Ruffoni B, Pistelli L (2012) Production of curcuminoids in different in vitro organs of *Curcuma longa*. Nat Prod Commun 7(8):1934578X1200700819

Praveen N, Manohar SH, Naik PM, Nayeem A, Jeong JH, Murthy HN (2009) Production of andrographolide from adventitious root cultures of *Andrographis paniculata*. Curr Sci 10: 694–697

Punturee K, Wild CP, Kasinrerk W, Vinitketkumnuen U (2005) Immunomodulatory activities of *Centella asiatica* and Rhinacanthus nasutus extracts. Asian Pac J Cancer Prev 6(3):396

Rahmat E, Kang Y (2020) Yeast metabolic engineering for the production of pharmaceutically important secondary metabolites. Appl Microbiol Biotechnol 104:4659–4674

Rajagopal S, Kumar RA, Deevi DS, Satyanarayana C, Rajagopalan R (2003) Andrographolide, a potential cancer therapeutic agent isolated from *Andrographis paniculata*. J Exp Ther Oncol 3(3):147–158

Ranjan D, Chen C, Johnston TD, Jeon H, Nagabhushan M (2004) Curcumin inhibits mitogen stimulated lymphocyte proliferation, NFκB activation, and IL-2 signaling. J Surg Res 121(2): 171–177

Ray S, Ghosh B, Sen S, Jha S (1996) Withanolide production by root cultures of *Withania somnifera* transformed with *Agrobacterium rhizogenes*. Planta Med 62(6):571–573

Rezaei F, Isik S, Kartal M, Erdem SA (2018) Effect of priming on thymoquinone content and in vitro plant regeneration with tissue culture of black cumin (*Nigella sativa* L.) seeds. J Chem Metrol 12(2):98

Ro D-K, Paradise EM, Ouellet M, Fisher KJ, Newman KL, Ndungu JM, Ho KA, Eachus RA, Ham TS, Kirby J (2006) Production of the antimalarial drug precursor artemisinic acid in engineered yeast. Nature 440(7086):940–943

Sabir F, Kumar A, Tiwari P, Pathak N, Sangwan RS, Bhakuni RS, Sangwan NS (2010) Bioconversion of artemisinin to its nonperoxidic derivative deoxyartemisinin through suspension cultures of *Withania somnifera* Dunal. Zeitschrift für Naturforschung C 65(9–10):607–612

Sabir F, Mishra S, Sangwan RS, Jadaun JS, Sangwan NS (2013) Qualitative and quantitative variations in withanolides and expression of some pathway genes during different stages of morphogenesis in *Withania somnifera* Dunal. Protoplasma 250(2):539–549

Sabir F, Sangwan NS, Chaurasiya ND, Misra LN, Sangwan RS (2008) In vitro withanolide production by *Withania somnifera* L. cultures. Zeitschrift für Naturforschung C 63(5–6): 409–412

Sabir F, Sangwan RS, Kumar R, Sangwan NS (2012) Salt stress induced responses in growth and metabolism in callus cultures and differentiating in vitro shoots of Indian ginseng (*Withania somnifera* Dunal). J Plant Growth Regul 31:537–548

Sabir F, Sangwan RS, Singh J, Misra LN, Pathak N, Sangwan NS (2011) Biotransformation of withanolides by cell suspension cultures of *Withania somnifera* (Dunal). Plant Biotechnol Rep 5(2):127–134

Sangwan NS, Tripathi S, Srivastava Y, Mishra B, Pandey N (2017) Phytochemical genomics of ashwagandha. InScience of Ashwagandha: preventive and therapeutic potentials. Springer, Cham, pp 3–36

Sangwan RS, Chaurasiya ND, Lal P, Misra L, Uniyal GC, Tuli R, Sangwan NS (2007) Withanolide a biogeneration in in vitro shoot cultures of ashwagandha (*Withania somnifera* D UNAL), a main medicinal plant in ayurveda. Chem Pharm Bull 55(9):1371–1375

Santarém ER, Astarita LV (2003) Multiple shoot formation in *Hypericum perforatum* L. and hypericin production. Braz J Plant Physiol 15(1):43–47

Satdive RK, Fulzele DP, Eapen S (2007) Enhanced production of azadirachtin by hairy root cultures of *Azadirachta indica* A. Juss by elicitation and media optimization. J Biotechnol 128(2): 281–289

Schachtsiek J, Warzecha H, Kayser O, Stehle F (2018) Current perspectives on biotechnological cannabinoid production in plants. Planta Med 84(04):214–220

Scholz M, Lipinski M, Leupold M, Luftmann H, Harig L, Ofir R, Fischer R, Prüfer D, Müller KJ (2009) Methyl jasmonate induced accumulation of kalopanaxsaponin I in *Nigella sativa*. Phytochemistry 70(4):517–522

Scragg A (2002) In: Verpoorte R, Alfermann AW (eds) Metabolic engineering of plant secondary metabolism. Kluwer Academic Publishers. 2000. 286, pp. ISBN 0 7923 6360 4. Plant Cell, Tissue and Organ Culture 68(2):211–212

Shabani L, Ehsanpour AA, Asghari G, Emami J (2009) Glycyrrhizin production by in vitro cultured *Glycyrrhiza glabra* elicited by methyl jasmonate and salicylic acid. Russ J Plant Physiol 56(5): 621–626

Shams-Ardakani M, Mohagheghzadeh A, Ghannadi A, Barati A (2007) Formation of glycyrrhizin by in vitro cultures of *Glycyrrhiza glabra*. Chem Nat Compd 43(3):353–354

Shantilal S, Vaghela JS, Sisodia SS (2018) Review on immunomodulation and immunomodulatory activity of some medicinal plant. Eur J Biomed 5(8):163–174

Sharada M, Ahuja A, Suri KA, Vij SP, Khajuria RK, Verma V, Kumar A (2007) Withanolide production by in vitro cultures of *Withania somnifera* and its association with differentiation. Biol Plant 51(1):161–164

Sharma SN, Jha Z (2012) Production of andrographolide from callus and cell suspension culture of *Andrographis paniculata*. J Cell Tissue Res 12(3):3423–3429

Sharma SN, Jha Z, Sinha RK, Geda AK (2015) Jasmonate-induced biosynthesis of andrographolide in *Andrographis paniculata*. Physiol Plant 153(2):221–229

Shukla N, Sangwan Neelam S, Misra HO, Sangwan RS (2003) Genetic diversity in *Boerhavia diffusa* L. of different geographic locations in India using RAPD markers. Genet Resour Crop Evol 50:587–601

Siddiqui MS, Thodey K, Trenchard I, Smolke CD (2012) Advancing secondary metabolite biosynthesis in yeast with synthetic biology tools. FEMS Yeast Res 12(2):144–170

Singh M, Chaturvedi R (2013) Sustainable production of azadirachtin from differentiated in vitro cell lines of neem (*Azadirachta indica*). Aob Plants 5:plt034

Singh J, Sabir F, Sangwan RS, Narnoliya LK, Saxena S, Sangwan NS (2015a) Enhanced secondary metabolite production and pathway gene expression by leaf explants-induced direct root morphotypes are regulated by combination of growth regulators and culture conditions in *Centella asiatica* (L.) urban. Plant Growth Regul 75:55–66

Singh VK, Dwivedi P, Chaudhary BR, Singh R (2015b) Immunomodulatory effect of *Gymnema sylvestre* (R. Br.) leaf extract: an in vitro study in rat model. PLoS One 10(10):e0139631

Sivanandhan G, Arun M, Mayavan S, Rajesh M, Mariashibu TS, Manickavasagam M, Selvaraj N, Ganapathi A (2012) Chitosan enhances withanolides production in adventitious root cultures of *Withania somnifera* (L.) Dunal. Ind Crop Prod 37(1):124–129

Sivanandhan G, Dev GK, Jeyaraj M, Rajesh M, Arjunan A, Muthuselvam M, Manickavasagam M, Selvaraj N, Ganapathi A (2013a) Increased production of withanolide a, withanone, and withaferin a in hairy root cultures of *Withania somnifera* (L.) Dunal elicited with methyl jasmonate and salicylic acid. Plant Cell Tissue Organ Culture (PCTOC) 114(1):121–129

Sivanandhan G, Dev GK, Jeyaraj M, Rajesh M, Muthuselvam M, Selvaraj N, Manickavasagam M, Ganapathi A (2013b) A promising approach on biomass accumulation and withanolides production in cell suspension culture of *Withania somnifera* (L.) Dunal. Protoplasma 250(4): 885–898

Sivanandhan G, Selvaraj N, Ganapathi A, Manickavasagam M (2014) Enhanced biosynthesis of withanolides by elicitation and precursor feeding in cell suspension culture of *Withania somnifera* (L.) Dunal in shake-flask culture and bioreactor. PLoS One 9(8):e104005

Sivanandhan G, Selvaraj N, Ganapathi A, Manickavasagam M (2015) Effect of nitrogen and carbon sources on in vitro shoot multiplication, root induction and withanolides content in *Withania somnifera* (L.) Dunal. Acta Physiol Plant 37(2):12

Sudhakaran MV (2012) Botanical pharmacognosy of *Andrographis paniculata* (Burm. F.) Wall. Ex. Nees. Pharm J 4(32):1–10

Surh Y-J, Chun K-S, Cha H-H, Han SS, Keum Y-S, Park K-K, Lee SS (2001) Molecular mechanisms underlying chemopreventive activities of anti-inflammatory phytochemicals: down-regulation of COX-2 and iNOS through suppression of NF-κB activation. Mutat Res Fundam Mol Mech Mutagen 480:243–268

Thakore D, Srivastava AK (2017) Production of biopesticide azadirachtin using plant cell and hairy root cultures. Eng Life Sci 17(9):997–1005

Tiwari P, Mishra BN, Sangwan NS (2014) Phytochemical and pharmacological properties of *Gymnema sylvestre*: an important medicinal plant. Biomed Res Int 2014:830285

Tripathi S, Srivastava Y, Sangwan RS, Sangwan NS (2020) *In silico* mining and functional analysis of AP2/ERF gene in *Withania somnifera*. Sci Rep 10:4877

Vakil MM, Mendhulkar VD (2013) Enhanced synthesis of andrographolide by Aspergillus niger and Penicillium expansum elicitors in cell suspension culture of *Andrographis paniculata* (Burm. f.) Nees. Bot Stud 54(1):1–8

Vanisree M, Lee C-Y, Lo S-F, Nalawade SM, Lin CY, Tsay H-S (2004) Studies on the production of some important secondary metabolites from medicinal plants by plant tissue cultures. Bot Bull Acad Sin 45(1):1–22

Varljen J, Lipták A, Wagner H (1989) Structural analysis of a rhamnoarabinogalactan and arabinogalactans with immuno-stimulating activity from *Calendula officinalis*. Phytochemistry 28(9):2379–2383

Vrushali D, Madhavi I (2006) Immunostimulatory activity of Amoora rohituka and *Azadirachta indica*. Adv Pharmacol Toxicol 7:5–12

Wang C, Su X, Sun M, Zhang M, Wu J, Xing J, Wang Y, Xue J, Liu X, Sun W (2019) Efficient production of glycyrrhetinic acid in metabolically engineered Saccharomyces cerevisiae via an integrated strategy. Microb Cell Factories 18(1):95

Wang S, Zhang S, Zhou T, Zeng J, Zhan J (2013) Design and application of an in vivo reporter assay for phenylalanine ammonia-lyase. Appl Microbiol Biotechnol 97(17):7877–7885

Wang W, Wang J, Dong S, Liu C, Italiani P, Sun S, Xu J, Boraschi D, Ma S, Qu D (2010) Immunomodulatory activity of andrographolide on macrophage activation and specific antibody response. Acta Pharmacol Sin 31(2):191–201

Westfall PJ, Pitera DJ, Lenihan JR, Eng D, Woolard FX, Regentin R, Horning T, Tsuruta H, Melis DJ, Owens A (2012) Production of amorphadiene in yeast, and its conversion to dihydroartemisinic acid, precursor to the antimalarial agent artemisinin. Proc Natl Acad Sci 109(3):E111–E118

Wetzstein HY, Porter JA, Janick J, Ferreira JF, Mutui TM (2018) Selection and clonal propagation of high artemisinin genotypes of *Artemisia annua*. Front Plant Sci 9:358

Wilkes DS (2012) Autoantibody formation in human and rat studies of chronic rejection and primary graft dysfunction. In: Seminars in immunology. Elsevier, pp 131–135

Wu K, Zhang X, Sun S, Wang X (2015) Factors affecting the accumulation of curcumin in microrhizomes of *Curcuma aromatica* Salisb. Biomed Res Int 2015

Yadav RK, Sangwan RS, Sabir F, Srivastava AK, Sangwan NS (2014) Effect of prolonged water stress on specialized secondary metabolites, peltate glandular trichomes, and pathway gene expression in *Artemisia annua* L. Plant Physiol Biochem 74:70–83

Yang S, Xie S, Gao H, Long Z (1993) Artemisinin and its derivatives enhance T lymphocyte-mediated immune responses in normal mice and accelerate immunoreconstitution of mice with syngeneic bone marrow transplantation. Clin Immunol Immunopathol 69(2):143–148

Yu Z-X, Li J-X, Yang C-Q, Hu W-L, Wang L-J, Chen X-Y (2012) The jasmonate-responsive AP2/ERF transcription factors AaERF1 and AaERF2 positively regulate artemisinin biosynthesis in *Artemisia annua* L. Mol Plant 5(2):353–365

Zaheer M, Giri CC (2015) Multiple shoot induction and jasmonic versus salicylic acid driven elicitation for enhanced andrographolide production in *Andrographis paniculata*. Plant Cell Tissue Organ Culture (PCTOC) 122(3):553–563

Zaheer M, Giri CC (2017) Enhanced diterpene lactone (andrographolide) production from elicited adventitious root cultures of *Andrographis paniculata*. Res Chem Intermed 43(4):2433–2444

Zhang Q, Hu S, Wang K, Cui M, Li X, Wang M, Hu X (2018) Engineering a yeast double-molecule carrier for drug screening. Artif Cells Nanomed Biotechnol 46(sup2):386–396

Zirpel B, Degenhardt F, Martin C, Kayser O, Stehle F (2017) Engineering yeasts as platform organisms for cannabinoid biosynthesis. J Biotechnol 259:204–212

Technological Mapping of Plant-Derived Immunomodulator Drugs: A Patent-Guided Overview about Species and its Main Compounds

18

Jose de Brito Vieira Neto, Maria Francilene Souza Silva, Lana Grasiela Alves Marques, Carlos Roberto Koscky Paier, Paulo Michel Pinheiro Ferreira, and Claudia Pessoa

Abstract

Research on natural products has investigated and employed many (bio)-technologies to find out plant active fractions and optimize extraction and isolation of molecules looking for innovative clinical therapies for several clinical conditions, as well as immune-related diseases. Indeed, the world incidence and prevalence of autoimmune diseases have increased over the last years, while immune therapy has arisen as a new promising tool for cancer treatment. Also, in the emergence of COVID-19 pandemic, immunomodulation has been proved effective to reduce the "cytokine storm" and avoid worsening the clinical condition of patients in the acute stage of respiratory syndrome. These health issues have also driven the search for new immunomodulatory compounds. In this context, prospective analysis is an important tool to identify the most relevant opportunities and demands in research and development (R&D) of pharmaceutical medicines, including substances able to modulate immune and inflammatory responses. In addition, prospection allows understanding the landscape of immunomodulatory plant-derived drugs and the associated technologies described in patents. This chapter employed the descriptors "Immunomodulator*" and "drug" and "plant" in search strategy to map the technological potential of

J. de Brito Vieira Neto · M. F. S. Silva · C. R. K. Paier · C. Pessoa (✉)
Department of Physiology and Pharmacology, Drug Research and Development Center, Federal University of Ceará, Fortaleza, Brazil
e-mail: cpessoa@ufc.br

L. G. A. Marques
Postgraduate in Biotechnology and Natural Resources, Federal University of Ceara, Fortaleza, Brazil

P. M. P. Ferreira
Department of Biophysics and Physiology, Federal University of Piauí, Teresina, Brazil

N. S. Sangwan et al. (eds.), *Plants and Phytomolecules for Immunomodulation*,
https://doi.org/10.1007/978-981-16-8117-2_18

immunomodulatory herbal drugs, using the software VantagePoint, Cortellis Competitive Intelligence, and Orbit Intelligence, respectively. The results provided quantitative data and technological indicators to drive future strategies for the development of these drugs. In conclusion, the most cited plant species in patents expressing molecules with immunomodulatory properties were *Curcuma longa*, *Moringa oleifera*, *Remirea maritima*, *Maytenus* ssp., *Angelica* sp., *Fagopyrum esculentum*, as well as plants from the families Leguminosae, Rosaceae, Iridaceae, Moraceae, and Amaranthaceae, among others. The main chemical classes implicated in immunomodulation were xanthones, coumarins, flavonoids, and (tri)terpenes. The technological mapping also showed that in the year 2002, there was an increased deposit of patents, reaching the highest peak in 2009 with 56 patents. Nowadays, the United States is the country with the biggest number of patents, followed by Canada, Australia, and Korea.

Keywords

Immunomodulation · Autoimmune diseases · Cytokines · Adjuvant treatment · Patent

18.1 Introduction

18.1.1 Immunomodulatory Drugs

The immune response is a highly controlled and specific process that involves different sorts of cells and molecules to fight pathogens, pollutants, and innocuous but exogenous particles, which might be dangerous for the organism's health (Catanzaro et al. 2018). This system aims to protect the skin, respiratory airways, intestinal tract, and other tissues/organs from cancer cells and foreign antigens, such as microbes (bacteria, fungi, and parasites), viruses, environmental proteins, and toxins. Based on the **receptor proteins** expressed on their **plasma membranes**, defense cells orchestrate immune responses through straightforward cell-to-cell interactions and cell-to-environment communications via the secretion of water-soluble signaling molecules **(e.g., cytokines)** (Nicholson 2016).

In this context, two "lines of defense" are considered: innate immunity and adaptive immunity. Innate immunity represents the first line of defense against a pathogen, a non-specific mechanism used by the host immediately or within hours. This response has no immunologic memory, so it is not able to recognize or "memorize" the same pathogen in a second contact (Turvey and Broide 2010). Adaptive immunity, on the other hand, is antigen-dependent and antigen-specific, involving a lag time between the exposure to the antigen and the maximal response. The adaptive immune response acquires memory to develop faster and more effective cellular and humoral responses following a second exposure to the antigen (Bonilla and Oettgen 2010). Both innate and adaptive immunities are complementary mechanisms whose defects result in host vulnerability or inappropriate

Table 18.1 Drugs approved as immunomodulatory agents

Class	Drug	Pharmacological effects
TNF-α inhibitors	Infliximab Adalimumab Certolizumab Golimumab	Inhibition of TNF-α membrane bound
Glucocorticoids	Hydrocortisone Prednisone Dexamethasone	Inhibition of immune system cells (B cell, T cell, and phagocytes)
Calcineurin inhibitors	Cyclosporine Tacrolimus	Inhibition of T cell activation

responses. So, in some situations, immunological processes fail to distinguish between the own organism components and foreign molecules eliciting some disorders, such as allergy, autoimmune diseases, and immunosuppression. Therefore, managing immunological-related threats depends on understanding how to maximize the potential of our sophisticated immune system in the service of human health (Nicholson 2016).

Immunomodulatory drugs (IDs) are a class of medications to treat diseases by modulating the immune system through its activation or inactivation to regulate/reduce/activate the inflammatory process and to combat infections, autoimmune aggressions, and cancer (Riccio and Lauritano 2020). Such drugs can be divided into two general groups: small molecules or biologic agents (Lin and Flynn 2012). The IDs acting as immunosuppressants may actuate through different mechanisms, such as inhibition of immune cell replication, suppression of cell-to-cell interaction, blockade of cell trafficking, cytokine targeting, receptor inhibition, or downstream cascade inactivation (Davis et al. 2020). In this perspective, there are chemically diverse classes of immunomodulatory agents (Table 18.1), including TNF-α inhibitors, glucocorticoids, and calcineurin inhibitors. It is important to note that most of immunomodulatory drugs approved by federal health agencies are synthetic substances as demonstrated in Table 18.1. However, it is worth mentioning that there is a myriad of plant-derived molecules with proven in vitro and in vivo immunomodulatory activity. For example, the alkaloid piperine found in the *Piper nigrum* and *Piper longum* (Piperaceae) is known for its immunomodulatory actions through different mechanisms, such as downregulating T helper cells in murine model of asthma, inhibition of LPS-induced endotoxin shock, and interference on functions of B cells (Bae et al. 2010; Kim and Lee 2009; Soutar et al. 2017). Moreover, a study conducted by Zheng and collaborators (2017) evaluated the effectivity of butein in vitro and in vivo to treat osteoarthritis. Butein is a natural product belonging to the chemical class of chalcones extracted from the stem bark of different species of plants. In such study, butein effectively reduced pro-inflammatory molecules [nitric oxide (NO), prostaglandin E_2 (PGE_2), tumoral necrosis factor-α (TNF-α), interleukin-6 (IL-6), cyclooxygenase-2 (COX-2), and inducible nitric oxide synthase (iNOS)] induced by IL-1β in human osteoarthritis chondrocytes. Besides that, the in vivo evaluation of butein in osteoarthritis mice model disclosed it was capable of reducing cartilage damage, therefore presenting a chondroprotective effect (Zheng et al. 2017).

Despite of the common idea of immunomodulators as immunosuppressive drugs, some IDs act stimulating the immune system. They comprise a diverse group of small molecules, such as the imide chemicals and the drug isoprinosine. The imide class is a group of small molecules that stand out for its immunomodulatory activity against a variety of hematological malignancies (Pessoa et al. 2010; Costa et al. 2015a, b; Gao et al. 2020a, b). The imide group is represented by thalidomide and its analogues lenalidomide and pomalidomide. All of them stimulate the immune system activity by inducing natural killer (NK) cells and T cells to produce cytokines (Gao et al. 2020a, b). Furthermore, these drugs provide a better uptake of tumor antigens by dendritic cells and a decrease in regulatory T cell production (Nadeem et al. 2020). In a similar way, isoprinosine enhances the immune system through the activation of T cells and cytokines to fight infections and other immune-related diseases (Silva et al. 2019). Drugs belonging to the class of methyl inosine monophosphate can also improve the immune system by increasing T cell differentiation and lymphocyte mitogen-induced proliferation (Kayser et al. 2003).

Apart from small molecules, thymosin alpha 1 ($T\alpha_1$) is a recombinant peptide composed of 28 amino acids whose development was based on the so-called lymphocytopoietic factor thymosin. Therefore, $T\alpha_1$ mechanism boosts up T cell development and differentiation into $CD4^+$ and $CD8^+$ subpopulations and modulates a diversity of chemokines (Li et al. 2010). Another important group of immunomodulatory recombinant proteins are the cytokines, interleukins, and interferon gamma (IFN-γ). Interleukin-12 is a well-recognized pro-inflammatory cytokine for inducing the differentiation of naïve T cells into T helper ones, as well as activate T cell and NK cells and increase the production of IFN-γ (Zundler and Neurath 2015). Interferon-γ has broad actions upon different immune cells and activates macrophages and NK cells and the interaction with antibodies to modulate B cells. Moreover, it is important for migration of leukocytes during the inflammatory process (Kak et al. 2018).

18.1.2 Natural Products and Inspiration for Drug Development

Nature has been a source of medicinal products for thousands of years, and also it has been used for many years in the screening of molecules with biological activities (Cragg and Newman 2013; Martel et al. 2017) Although synthetic drugs have largely replaced herbal medicine in the twentieth century, many cultures continue to use plant-derived medicines to prevent and treat disease, especially in some developing countries, where plants still represent the main alternative of treatment' (Pereira et al. 2015). The interest in herbal medicine and alternative options mainly derives from the consumers' interest in disease prevention and disappointment with allopathic medicine (Martel et al. 2017). Pharmaceutical companies continue to use plants as sources of drug candidates, and about half of the current pharmaceutical drugs have been derived, directly or indirectly, from natural sources (Cragg and Newman 2013; Mahady 2001).

Indeed, between 1981 and 2010, 1073 new chemical entities were approved as drugs by the Food and Drug Administration (FDA/USA), and 64% of them are natural, derived, or synthesized from natural products. In modern techniques of organic synthesis like combinatorial chemistry, natural products remain as an important source of new chemical scaffolds to the development of therapeutic agents against infections (fungal or bacterial), disease vectors, cancer, dyslipidemia, and immune system misregulation (Butler 2004; Newman and Cragg 2012; Ferreira et al. 2009; Farias et al. 2013; Araújo et al. 2015; Ferreira et al. 2019). This is explained by different reasons, including high chemical diversity, biochemical specificity, molecular flexibility, and other structural properties that make natural carbon skeletons more favorable to originate lead molecules for drug discovery. Moreover, the fraction of natural product structures with two or more "rule-of-five" violations is quite low ($\approx$ 10%) and like that of trade drugs (Lipinski et al. 2001; Feher and Schmidt 2003; Pascolutti and Quinn 2014; Sousa et al. 2017).

Traditional medicine in some parts of the world affirms the potential of herbal plants for strengthening the immune system (Samuel et al. 2010). Medicinal plants have been known as immunostimulants (Tan and Vanitha 2012; Catanzaro et al. 2018), and the modulation of immune functions using their products as an alternative therapeutic approach has become accepted in recent years (Catanzaro et al. 2018). An increase in prevalence and incidence of immunological diseases has been detected in the past decades, and great attention has been given to the development of molecules capable of modulating the immune response. Plant natural products can act as inhibitors or stimulators of the immune cells, depending on the experimental conditions they were submitted in. For example, the method of chemical extraction may determine the effects of vegetal products on the defense immune cells, possibly due to the solubility and different potencies of extracted stimulants and inhibitors. Aqueous extracts generally activate immune cells, while ethanol extracts inhibit them (Martel et al. 2017).

The main properties of medicinal plants are cell growth promotion, immune system enhancement, and antibacterial and antiviral activities (Van Hai 2015). Several classes of natural products have immunostimulatory potential, both of low (alkaloids, terpenoids, and phenols) and high (lectins, polysaccharides, and nucleotides) molecular weights (Wagner 1993). Polysaccharides from plants have received attention in recent years due to their broad-spectrum therapeutic potential combined with low toxicity (Schepetkin and Quinn 2006).

Another example is the phytol, a diterpene with an acyclic unsaturated long-chain alcohol, which has demonstrated anxiolytic, metabolism-modulating, cytotoxic, antioxidant, autophagy-promoting, apoptosis-inducing, antinociceptive, anti-inflammatory, immunomodulatory, and antimicrobial activities (Islam et al. 2018). Additionally, there are condensed tannins, which have been studied to produce nitric oxide, stimulate TNF-α secretion, and activate nuclear factor NF-κB in macrophages. Lignans strongly inhibit the production of TNF-α in lipopolysaccharide-stimulated macrophage-like murine cells (Cho et al. 1999; Yang et al. 2019).

In cancer immunotherapy, one of the outstanding current anticancer treatment, the immune system itself is exploited against tumors. The purpose of the immunotherapy is to overcome immunosuppression in the tumor microenvironment aiming the defense cells eliminate tumors without causing intolerable side effects (Pessoa et al. 2010; Banday et al. 2015; Ferreira and Pessoa 2017). Immunomodulatory properties of medicinal plants are also able to mitigate the growth of cancer cells, and such capacity is dependent on cytotoxic T cells, natural killer cells, and cytokines, as chemokines and tumor necrosis factor-α (Anderson 2014; Baraya et al. 2017). Additionally, other classes of plant polysaccharides showed in vitro and in vivo anticancer, immunomodulatory, and antipathogenic properties (Azike et al. 2015; Yoo et al. 2012; Zong et al. 2012).

18.1.3 The Use of Patents as a Source of Information about Immunomodulation

Technological progress has been necessary for the development of more sophisticated and specialized natural or synthetic products. The current scenario of research, technological development, and production of bioproducts is extremely dependent on scientific growth and innovation, which are found in patent documents. In this context, the search for scientific and technological information with health purposes has received increasing attention from governments and private agencies. This effort has recognized potential advances in scientific and technological fields applied to the health systems, to promote not only cost reduction but also the expansion of capacity for cure and treatment of diseases. On this issue, several reports have highlighted a great variety of technological prospecting methods, such as the competitive intelligence technology, data and text mining, and technological mapping (Oliveira et al. 2019; Quintella et al. 2019; Clark et al. 2011; Newman et al. 2003). These methods may be employed to create a systematic process of technological analysis, characteristics, development routes, and potential impacts on the future of research, development, and innovation (RD&I).

The prospective or prospecting analysis is a planning technique used to improve the information available to decision-making management (Calixto et al. 2016). It comes from the necessity to perform a general investigation work to determine the most important information to direct investigation steps. Although it was created for planning in economic sectors, the prospecting also presents possibilities in the context of management of productive chains in science and technology (S&T) (Castro 2001). According to Santos et al. (2004), analysis of future technologies is a systematic process to produce judgments about emerging technologies, development routes, and potential impacts in the future. Technological prospection incorporates a variety of technological prospecting methods and requests identification of key changes and/or technological guidelines, through pre-defined methods and approaches, to map the opportunities and perceive the risks involved in changing the paradigm of meeting social and economic demands (Antunes and Magalhães 2008).

Ethical technological prospection consists of steps of collection, data search, and use of public and published information available on trends, events, and stakeholders, outside the company's boundaries, to avoid the competitive internal environment of the company. To reach such level of analysis, the process takes into consideration filtered integrated ways (Calixto et al. 2016; Santos et al. 2004).

The technological prospection may obtain bibliographic and technical information from a patent document. Due to the abundance of information present in patent documents and articles, industries, companies, and researchers must closely follow the technological innovations and inventions of interest, especially when the panorama is characterized by the growing launch of new products related to research with medicinal plants (Castro 2001; Simões et al. 2015; Silva et al. 2016), which is considered one of the most relevant fields in megadiverse countries. Then, to do so while monitoring biodiversity loss and conservation efforts, it is crucial to understand the nation's resources to allow fauna and flora preservation, sustainable forest management, and wood certification (UN Environment Programme 2019). This chapter presents a global scenario about the relationship between immunomodulation and plant-based medicines and conducts a mapping of protected technologies under patent rights.

18.2 Methodology

For a complete and reliable review, the databases used for search (Table 18.2) were the European Patent Office (EPO), World Intellectual Property Organization (WIPO), United States Patent and Trademark Office (USPTO), and the National Institute of Industrial Property (INPI) using the descriptors (immunomodulat* AND drug* AND plant*), both in the title and in the abstract of the patents. Bibliometric analyses, data generation, and identification of patent indicators did not consider the date limit and included outcomes until the present day.

18.3 Results and Discussion

A total of 167 patents were identified in preliminary evaluations, 22 of which were excluded because they were not related to plant-derived immunomodulatory drugs. After reviewing, only three patents of the USPTO database were included due to their claims for the invention of a plant-derived immunomodulatory drug (Fig. 18.1). A total of 145 patents were identified after preliminary scrutiny, 14 of which were excluded once duplicity of documents was detected. Moreover, after reading the abstract and claims of the 131 patents, five of them were excluded considering that their contents were not the focus of this chapter. Lastly, 126 patents were selected for technological mapping. Figure 18.1 illustrates the guidelines used for the search in patent databases.

Table 18.2 Main information about patent databases

Databases	Country	Language	Search engines	URL
European Patent Office – EPO	European community (database provides documents from systems such as WIPO and Patent of Japan)	English interface	Boolean operators (AND, OR, NOT, AND NOT) and truncation character (*? #)	https://worldwide.espacenet.com
World Intellectual Property Organization – WIPO	Geneva (Switzerland). Body responsible for establishing common parameters and policies in the area of intellectual property in the international context	Interface is available in more 10 languages	Offers a wide range of operators that can be used to combine search terms, including Boolean operators, proximity operators, and range operators. Wildcard operators to search for variants of terms based on a common stem or root. The following are supported: AND, "+," OR, NOT, AND NOT, and "-"as Boolean operators	https://wipo.int/patentscope
United States Patent and Trademark Office – USPTO	United States	English interface	Boolean operators (AND, OR, NOT, AND NOT) and truncation character (* $)	http://www.uspto.gov
National Institute of Industrial Property – INPI	Brazil	The interface is entirely in Portuguese	Standard English connectors Boolean operators – (AND, OR, NOT, AND NOT) and truncation character * (asterisk)	https://www.gov.br/inpi/pt-br

18.3.1 Timeline Patent of Plant-Derived Immunomodulator Drugs

Firstly, it was noted a growing number of patents between 1985 and 2002, followed by a decrease in the next 2 years. The evolution of deposited patents since 1994 is clear, but a wave behavior was found, once the number of deposited patents has varied widely every 2–3-year interval until 2017–2020, when an augmentation was observed (Fig. 18.2).

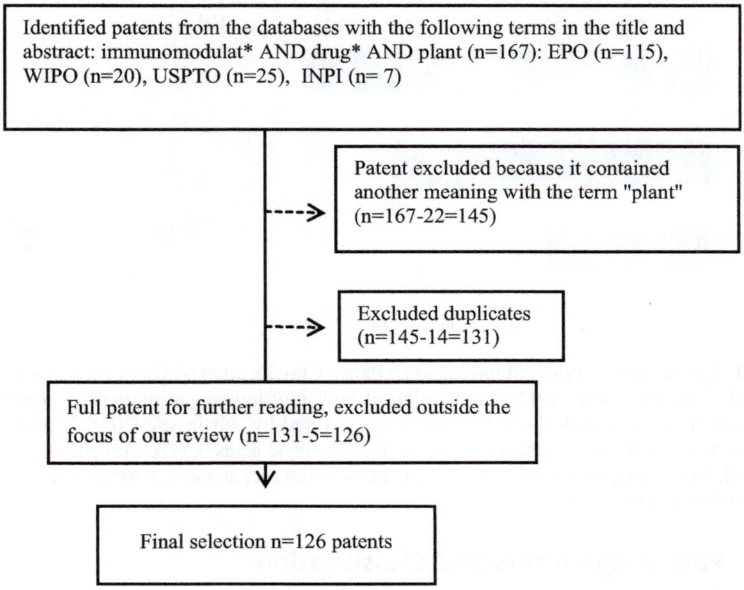

Fig. 18.1 General flowchart of the search process, screening, and eligibility of articles to compose the results and the discussion of the thematic proposal

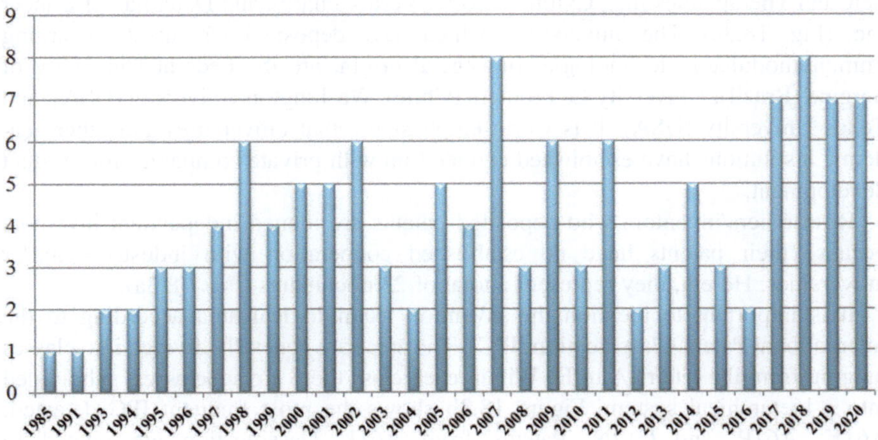

Fig. 18.2 Published patents by decade on plant-derived immunomodulator drugs. **Source**: Crude data from EPO, USPTO, WIPO, and INPI

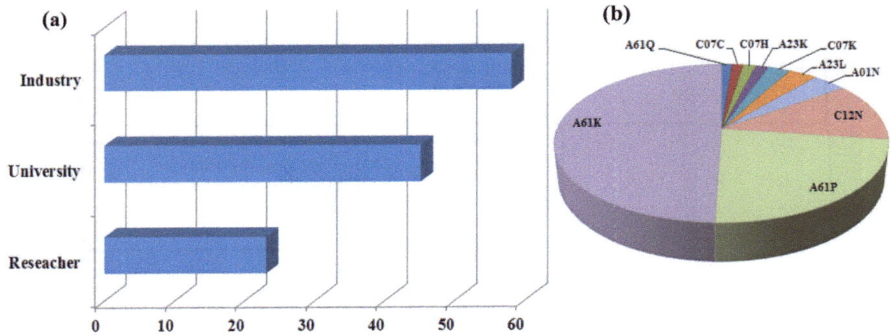

Fig. 18.3 Patent applicant (**a**) and International Patent Classification (IPC) for immunomodulation of plant-derived drugs (**b**). A01N Preservation of bodies of humans or animals or plants; A23L Foods, foodstuffs, or non-alcoholic beverages, not covered by subclasses; C07K Peptides; C07H Sugars; derivatives thereof; nucleosides; nucleotides; nucleic acids; CO7C Acyclic or carbocyclic compound; A61Q Specific use of cosmetics or similar toilet preparations. **Source**: Crude data from EPO, USPTO, WIPO, and INPI

18.3.2 Patent Applicants and Classification

The owners and/or inventors, who have the legitimacy to protect industrial property rights, are grouped into three main economic segments (industry, university, and independent inventors). The industries lead the ranking with 58 patent deposits, for example, Aché Lab. Pharm. S.A, Eurovita A/S, Bristol-Myers Squibb Company, SciCann Therapeutics Inc., Cytonics Corp., Vertex Pharm, and Dana Farber Cancer Inc. (Fig. 18.3a). The universities, which have deposited 45 patents regarding immunomodulatory technologies of vegetal origin, are the Federal University of Sergipe (Brazil), University La Frontera (Chile), Washington University (USA), and Texas University (USA). It is worth emphasizing that universities and other academic institutions have established cooperation with private companies for product development.

Researchers/inventors who deposited patents are named independent inventors because their patents have no established cooperation with industries and/or universities. Herein, they represent a total of 23 documents (Fig. 18.3a).

It is important to mention the invention technological area according to the International Patent Classification (IPC). The areas are generally divided into classes ranging from the letters A to H. Within each class there are subclasses, subdivided into a hierarchical system. Figure 18.3b shows the most frequent IPC detected: A61K, A61P, and C12N. Patents from A61K classification are related to "Preparations for medical, dental or hygienic purposes" (332 occurrences); A61P describes "Specific therapeutic activity of chemical compounds or medicinal preparations" (156 occurrences), and C12N refers to "Microorganisms or enzymes; their compositions; propagation, conservation, or maintenance of microorganisms; genetic or mutation engineering; culture media" (40 occurrences). A patent can comprise one or more classifications.

18.3.3 Most Common Chemicals under Patent Protection

Table 18.3 summarizes 13 plant-derived patents from extracts and compounds with auspicious immunomodulatory properties. Six of them stands out for their biological activity. Therefore, they received more attention.

Spilanthol (Fig. 18.4a) is a bioactive *N*-alkyl amide originated from secondary metabolism and found in different plants (Fig. 18.4c) such as *Heliopsis longipes*, a plant known as "chilcuague," and in many species from the *Acmella* genus (e.g., *A. oleracea*, known as "jambu") (Koren 2019). It presents lots of bioactivities, as antioxidant, antimutagenic, anticancer, antinociceptive, anti-inflammatory, antimicrobial, antibacterial, antifungal, and analgesic actions and also acts as an endocannabinoid agonist (Koren 2019).

"Cannabinoid" is a general term to describe compounds that interfere with the cannabinoid receptors (National Center for Biotechnology Information 2004). These compounds may be obtained from different origins, such as endogenous (Hernández-Cervantes et al. 2017), synthetic, and vegetal sources (*Cannabis* genus). The so-called phytocannabinoids (pCBs) belong to a class of organic molecules from the reaction of resorcinol and an isoprenoid (Citti et al. 2019). They can act as agonists or antagonists at the cannabinoid receptor (Hernández-Cervantes et al. 2017). There are seven well-studied pCBs in a molecular pharmacology point of view, namely, Δ9-tetrahydrocannabinol (Fig. 18.4b), Δ9-tetrahydrocannabivarin, cannabinol, cannabidiol, cannabidivarin, cannabigerol, and cannabichromene (Turner et al. 2017). They are known for their vast biological activity, such as immunomodulatory, anti-inflammatory, anticonvulsive, and analgesic properties (Oláh et al. 2017). For example, Δ9-tetrahydrocannabinol (THC) was neuroprotective under neuroinflammation experimental conditions. THC was interfered with different functions of $CD8^+$-T cells, reduced astrocyte activation, and modulated IFNγ and degranulation processes (Henriquez et al. 2020).

Limonin (Fig. 18.4c) is a limonoid, a tetracyclic triterpenoid found in different plant families (Rutaceae, Meliaceae, and Ranunculaceae) and extracted from distinct parts of plants, such as root, seeds, and fruits. Limonin presents biological activity documented in different studies, including anticancer, anti-inflammatory, analgesic, antiviral, and antibacterial ones (Rahman et al. 2015; Yang et al. 2014a, b; Vikram et al. 2012; Battinelli et al. 2003). Yang et al. (2014a, b) show a combination of berberine and limonin capable of diminishing both IgE and mononuclear cells of food-allergic patients, mainly through modulation of IκBα and upregulation of signal transducer, and berberine was the compound with the highest immunomodulatory activity. Moreover, another study tested a combination of curcumin and limonin in mice to evaluate its ability to suppress $CD4^+$ T cell. Mice fed with curcumin and limonin showed a remarkable suppression of NF-κB activation, resulting in a decrease of CD4+ T cell activation (Kim et al. 2009).

Solanine (Fig. 18.4d) belongs to the class of glycoalkaloids found in different species of plants, such as *Solanum tuberosum*, *S. lycopersicum*, *S. melongena*, and *Malus domestica* (Ordóñez-Vásquez et al. 2017). Solanine is a potential in vitro and in vivo anticancer compound against different histological types of cancers (Pan

Table 18.3 Details of patents about immunomodulatory properties of plant-derived compounds

Applicant / Ref	Year	Country	Species/family/ part of the plant	Invention	Dose or concentration	Mechanism	Assay	Claims products*	Chemical class/compounds
Univ La Frontera /Eulefi et al. (2018)	2018	WO Chile	Extract – *Blechnum chilense* (family)	Obtaining or preparation of medicine for preventing and treating autoimmune diseases such as lupus erythematosus, rheumatoid arthritis, multiple sclerosis, psoriasis, chronic inflammatory diseases, and inflammatory bowel diseases	0.1–100 μg/mL	Inhibiting of lymphocyte proliferation	In vitro/ in vivo$CD3^+$ T lymphocyte culture viability assay; $CD3^+$ T cell flow cytometry assay	Pharmaceutical composition: syrups, lozenges, tablets, solution for injection, eye drops, creams, or ointments	Polydine, butanediol, limonin, 5Cl-phenyltrimethylsilane, α –gallocatechin, and γ-Glutamylmethionine
Council of Scientific and Industrial Research / Mandalet et al. (2018)	2018	WO India	Leaves, stem, root, and flowers Extract – *Murraya koenigii* Extract – *Alpinia galanga, Rubia cordifolia, Camellia sinensis, Aegle marmelos, Cinnamomum tamala, Spinacia oleracea, Ocimum tenuiflorum, ipomoea aquatic*	Formulation comprising extracts of selected plants for the management of cancers and cancer stem-like cells via growth inhibitory activity, inhibition of autophagy, and immunomodulation	0–200 μg/mL	Inhibition of autophagy, migration, and epithelial-mesenchymal transition and inducing of anoikis cell death in cancer cells	In vitro/in vivo Acute toxicity studies in normal rodents; subacute toxicity studies	A kit for treatment and management of cancers comprising (i) formulation of extracts optionally with one or more pharmaceutically acceptable preservatives, lubricants, diluents, and other excipients; (ii) clinically approved anti-cancer drug;(iii) a pamphlet containing instruction of use	NS

SciCann Therapeutics Inc./Koren (2019)	2019	WO Canada	*Cannabis* sp., *Heliopsis longipes*, species from *Acmella* genus	An FDC drug, also known as combination drug, is a fixed-dose combination (cannabinoid and spilanthol), which includes two or more active pharmaceutical ingredients, combined in a manufactured single fixed dosage	1–20 mg	Synergistic, additive, or a mutual potentiating effect in the prevention and treatment	In vitro/in vivo/ clinical evaluation	The method and pharmaceutical composition with cannabidiol and spilanthol	Cannabinoid, spilanthol (fatty acid amide)
Bristol-Myers Squibb Company/Siegall et al. (1994)	1994	United States	*Bryonia dioica*	Isolation and characterization of a novel ribosome-inactivating protein from *B. dioica*. It also cites immunoconjugates comprising the new protein and antibodies immunologically specific for tumor-associated antigens	1–2000 mg/m. sup.2.	Inactivation of ribosomes via blockade of productive interactions with elongation factor-2	In vitro translation assay using rabbit reticulocyte lysate. In vivo toxicity of bryodin 2	Pharmaceutical composition comprising an immunotoxin with bryodin 2 and a ligand	Gelonin, saporin, trichosanthin, and bryodin; ricin and abrin
Merck Patent GmbH/ Eichmann and Modolell (1989)	1989	WO Denmark	Extracts from fresh and dried leaves of *Acanthospermum hispidum*	Usage of extract immunomodulation (influencing the immune state); for prophylaxis and therapy of viral and retroviral diseases and cancer therapy	Administration of extract over a period of 5 days	Cell destruction (cytopathic effect) or possible inhibition by interferon. Induce the proliferation of B lymphocytes	In vivo action on tumor models induced by methyl cholanthrene	Method for immunomodulation; therapy with an immunomodulatory extract	Melampolides, glycosides, acetylenes

(continued)

Table 18.3 (continued)

Applicant / Ref	Year	Country	Species/family/ part of the plant	Invention	Dose or concentration	Mechanism	Assay	Claims products*	Chemical class/compounds
Eurovita A/S/ Weidner et al. (1999)	1999	WO Denmark	Fresh or dried *Alpinia galanga* or parts thereof, preferably the rhizome	Pharmaceutical compound, a cosmetic or a supplement dietary containing the compositions and a pharmaceutically or cosmetically acceptable carrier	NS	Inhibitory activity on three enzymes, leukotriene C4 synthase, 5-lipoxygenase, and phosphodiesterases-IV	Phosphodiesterase-IV assay; leukotriene C4 synthase assay	Drug, cosmetic, or a supplement	Terpenoids (general), 1′-acetoxychavicol acetate, acetoxyeugenol acetate, (E)-*p*-coumaryl diacetate, coniferyl diacetate, 1′-hydroxychavicol acetate, 1′-hydroxychavicol, *p*-hydroxy-(E)-cinnamaldehyde, *p*-methoxy-(E)-cinnamylalcohol, and 3, 4-dimethoxy-(E)-cinnamylalcohol
Anahit et al. 2014.	2014	WO United States	*Solanum tuberosum*	Methods for tumor immunotherapy, for prevention and treatment of metastatic tumor growth by activating of dendritic cells, macrophages, and NK cells		To suppress activity of T regulatory cells or conversely to inhibit sensitivity of cells to the suppressive effects of T cells		A pharmaceutical kit for treatment of cancer	Acidic peptidoglycan
Federal University of Sergipe/Araujo et al. (2014)	2014	Brazil	*Remirea maritima*	Formulation with antitumor and immunomodulatory purposes containing *R. maritima*	25 and 2000 mg/kg		In vivo	Formulations with antitumor and immunomodulatory effects characterized by solid, liquid or topical dosage presentations.	Xanthones, coumarins, flavonoids, triterpenes

Lomapharm, Rudolf Lohmann GmbH KG, and Pharmazeutische Fabrik/Hildebert et al. (1986).	1986	WO Denmark	*Echinacea purpurea* and *E. angustifolia*	Vegetal polysaccharides acting on the immune system and a process for their isolation from plant cell cultures	NS	Activation of T and B lymphocytes	In vitro/in vivo Influence of polysaccharide fractions from cell culture on carbon clearance on mice	Biochemically pure polysaccharide fractions isolated from plant cell structures of *Echinacea purpurea* or *Echinacea angustifolia*. Treatment methods and pharmaceutical preparations using fractions with immunostimulating activity	Polysaccharides
Obreshkova and Petrov (1999)	1999	WO Bulgaria	*Clinopodium vulgare* L.	Used as a general recovery, immunomodulation, and biostimulation formulation to improve the metabolism, as well as for prevention and additional recovery therapy after tumor chemotherapies	NS	NS	NS	Dry drug or alcohol extract from the aerial part of *C. vulgare*	NS
Aché lab. Pharm. S.A/ Pianowski and Brandão (2003)	2003	Brazil	*Maytenus* spp.	Antiulcerogenic drug, comprised of an active fraction of the *Maytenus* spp. as an effective component in pharmaceutical composition to treat ulcers, as immunomodulator and immunostimulant	10–100 mg/kg	NS	In vivo Experimental model of indomethacin-induced ulcer	Process, composition, method, and use of extract from *Maytenus* spp.	NS

(continued)

Table 18.3 (continued)

Applicant / Ref	Year	Country	Species/family/ part of the plant	Invention	Dose or concentration	Mechanism	Assay	Claims products*	Chemical class/compounds
Gubarev and Enioutina (1998)	1998	WO/RU United States	Plant-derived alkaloids	To encompass steroid-like glycoalkaloids of potential innate immunity stimulant action	5–15 mg/kg	NS	In vivo	A plant-derived immunomodulator. The plant-derived immunomodulator contains glycoalkaloids. Characterized by the fact that the plant-derived immunomodulator is solanine and chaconine	Glycoalkaloids include solanine and chaconine
Lozoya et al. (2005)	2005	WO Mexico	*Mimosa tenuiflora*	Preparing plant extract with standardization of *M. tenuiflora* to develop useful phytomedicine. This plant extract has different effects, such as cicatrizant, antimicrobial, anti-inflammatory, and immunomodulatory	5–10 mg		In vitro	Method, use, and pharmaceutical formulations, i.e., gels and creams	Alkaloids, tannins, group of triterpenoid saponins

WO World Intellectual Property Organization
[a]Among other claims. *NS* not specified or not documented

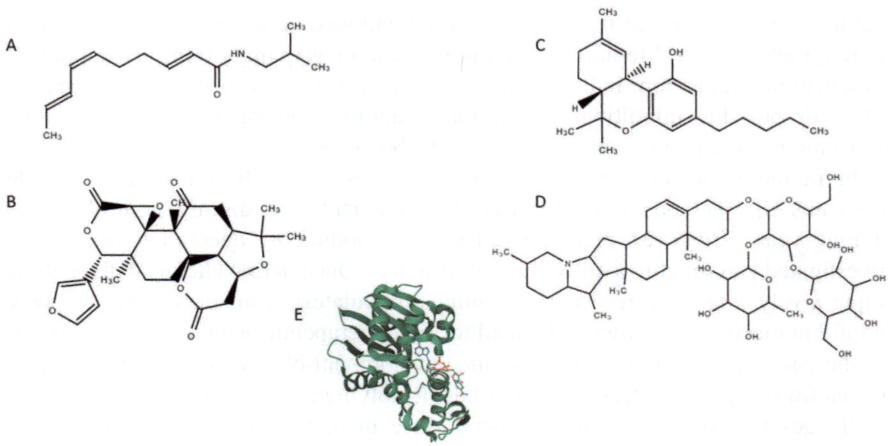

Fig. 18.4 Chemical structures or 3D representation of the five main compounds described in the patents: (**a**) spilanthol, (**b**) Δ9-tetrahydrocannabinol, (**c**) limonin, (**d**) solanine, (**e**) trichosanthin (retrieved from Protein Data Bank). *Chemical structures drawing using ChemSpider

et al. 2016). A study conducted by Gao et al. 2020a, b) showed that solanine was able to improve the antitumor immune response in a hepatoma mouse model. The compound's immunomodulatory activity diminished the number of CD4 + CD25+ Treg and induced the expression of transforming growth factor-β (TGF-β) (Gao et al. 2020a, b).

Trichosanthin (TCS) (Fig. 18.4e) was first described as an active protein extracted from the roots of *Trichosanthes kirilowii*. Furthermore, trichosanthin has multiple pharmacological activities, such as antitumor, antiviral, and immunomodulatory actions (Shi et al. 2018). Regarding immunomodulatory properties, trichosanthin could alter the humoral immune system through modulation of T cells as well as downregulation of IFN-γ, meanwhile increasing the expression of some interleukins (Shi et al. 2018).

18.4 Conclusion and Future Prospects

The technological mapping of patents encompassing plant-derived chemicals with immunomodulatory properties showed that in 2002 there was an increase in the number of deposits, reaching the highest peak in 2009 with 56 patents. The United States was the country that filed the largest number of patents, followed by Canada, Australia, and Korea. The most cited plant species in these patents were *Curcuma longa*, *Moringa oleifera*, *Remirea maritima*, *Maytenus* ssp., *Angelica* sp., *Fagopyrum esculentum*, as well as plants from the families Leguminosae, Rosaceae, Iridaceae, Moraceae, and Amaranthaceae, among others. The main chemical classes involved in immunomodulation properties were xanthones, coumarins, flavonoids, and triterpenes. Their mechanisms of action mentioned in patents include inhibition of B lymphocyte proliferation; inhibition of autophagy; migration and epithelial-mesenchymal transition; induction of anoikic cell death in cancer cells; cytopathic effect or possible inhibition by interferon, additive or mutual enhancer in the prevention or treatment; and inactivation of ribosomes.

Immunomodulators derived from natural sources, mainly from flora, are valuable adjuvants for therapeutic or prophylactic treatments for disease control in the coming years. However, plant-derived immunomodulatory agents are still poorly investigated and roughly available for clinical use. Once herbal chemical diversity is impressively huge, the research on immunomodulators from plants may be very helpful in meeting the growing demand for such therapeutic agents, as demonstrated by the patent prospection. Naturally, the development of new immune-modulating agents from vegetal sources may contribute to solving the current and future clinical challenges regarding immune disorders, since there has been an increase in the incidence of diseases with pathogenesis related to the imbalance of the immune system.

References

Anahit G et al (2014) Applicant. Method for the cancer treatment and prevention of metastatic disease. Patent WO US20170274069

Anderson C (2014) Application of central immunologic concepts to cancer: helping T cells and B cells become intolerant of tumors. Eur J Immunol 44(7):1921–1924. https://doi.org/10.1002/eji.201444826

Antunes MAS, Magalhães JL (2008) Patenteamento & prospecção tecnológica no setor farmacêutico. Interciência. UFRJ, Editora

Araujo AAS et al (2014) Inventors, Federal University of Sergipe, Applicant. Formulações com efeito antitumoral eimunomodular contendo Remirea marítima. Patent BR 102014018366-3

Araújo EJF, Oliveira GAL, Sousa LQ, Bolzani VS, Cavalheiro AJ, Tome AR, Peron AP, Santos AG, Cito AMGL, Pessoa C, Freita RM, Ferreira PMP (2015) Counteracting effects on free radicals and histological alterations induced by a fraction with casearins. An Acad Bras Cienc 87(3):1791–1807. https://doi.org/10.1590/0001-3765201520150149

Azike CG, Charpentier PA, Lui EM (2015) Stimulation and suppression of innate immune function by American ginseng polysaccharides: biological relevance and identification of bioactives. Pharm Res 32(3):876–897. https://doi.org/10.1007/s11095-014-1503-3

Bae GS et al (2010) Inhibition of lipopolysaccharide-induced inflammatory responses by piper-ine. Eur J Pharmacol 642(1–3):154–162. https://doi.org/10.1016/j.ejphar.2010.05.02

Banday AH, Jeelani S, Hruby VJ (2015) Adjuvantes de vacina contra o câncer—progresso clínico recente e perspectivasfuturas. Immunopharmacol Immunotoxicol 37:1–11. https://doi.org/10.3109/08923973.2014.971963

Baraya YS, Wong KK, Yaacob NS (2017) The immunomodulatory potential of selected bioactive plant-based compounds in breast cancer: a review. Anti Cancer Agents Med Chem 17(6): 770–783. https://doi.org/10.2174/1871520616666160817111242

Battinelli L, Mengoni F, Lichtner M, Mazzanti G, Saija A, Mastroianni CM, Vullo V (2003) Effect of limonin and nomilin on HIV-1 replication on infected human mononuclear cells. Planta Med 69(10):910–913. https://doi.org/10.1055/s-2003-45099

Bonilla FA, Oettgen HC (2010) Adaptive immunity. J Allergy Clin Immunol 125(2):S33–S40. https://doi.org/10.1016/j.jaci.2009.09.017

Butler MS (2004) The role of natural product chemistry in drug discovery. J Nat Prod 67(12): 2141–2153. https://doi.org/10.1021/np040106y

Calixto JB, Rafael CD, Campos MM, Santos ARS (2016) Medicinal plants in Brazil: pharmacological studies, drug discovery, challenges and perspectives. Pharmacol Res 112:4–29

Castro AMG (2001) Análise prospectiva de cadeias produtivas agropecuárias. Transinformação 13(2):55–72

Catanzaro M, Corsini E, Rosini M, Racchi M, Lanni C (2018) Immunomodulators inspired by nature: a review on curcumin and Echinacea. Molecules 23(11):2778–2795. https://doi.org/10.3390/molecules23112778

Cho JY, Kim AR, Yoo ES, Baik KU, Park MH (1999) Immunomodulatory effect of Arctigenin, a Lignan compound, on tumour necrosis factor-α and nitric oxide production, and lymphocyte proliferation. J Pharm Pharmacol 51:1267–1273. https://doi.org/10.1211/0022357991777001

Christie AF, Dent C, McIntyre P, Lachlan W, Studdert DM (2013) Patents associated with high-cost drugs in Australia. PLoS One 8(4):e60812. https://doi.org/10.1371/journal.pone.0060812

Citti C et al (2019) A novel phytocannabinoid isolated from Cannabis sativa L. with an in vivo cannabimimetic activity higher than Δ9-tetrahydrocannabinol: Δ9-Tetrahydrocannabiphorol. Sci Rep 9:20335. https://doi.org/10.1038/s41598-019-56785-1

Clark K, Cavicchi J, Jensen K, Fitzgerald R, Bennett A, Kowalski SP (2011) Patent data mining: a tool for accelerating HIV vaccine innovation. Vaccine 29(24):4086–4093. https://doi.org/10.1016/j.vaccine.2011.03.052

Costa NDJ, Oliveira SFC, Silva JN, Pacheco ACL, Abreu MC, Cavalcante AACM, Ferreira PMP (2015b) Technological and therapeutic potential of *Calotropis procera*. Revista GEINTEC: gestao, inovacao e tecnologias 5(3):2222–2236. https://doi.org/10.7198/S2237-0722201500030001

Costa PM, Costa MP, Carvalho AA, Cavalcanti SMT, Cardoso MVO, Bezerra Filho GO, Viana DA, Fechine-Jamacaru FV, Leite ACL, Moraes MO, Pessoa C, Ferreira PMP (2015a) Improvement of *in vivo* anticancer and antiangiogenic potential of thalidomide derivatives. Chem Biol Interact 239:174–183. https://doi.org/10.1016/j.cbi.2015.06.037

Cragg GM, Newman DJ (2013) Natural products: a continuing source of novel drug leads. Biochim Biophys Acta – General Subjects 1830(6):3670–3369. https://doi.org/10.1016/j.bbagen.2013.02.008

Davis J, Ferreira D, Paige E et al (2020) Infectious complications of biological and small molecule targeted immunomodulatory therapies. Clin Microbiol Rev 33(3):1–117. https://doi.org/10.1128/CMR.00035-19

Eichmann K, Modolell M (1989) Inventors Merck patent GmbH, Applicant. Extracts of the *Acanthospermum hispidum* plant. Patent US 5256416

Eulefi MEE et al (2018) Inventors Univ La Frontera (2018) Applicant Natural Blechnum chilense extract, with immunosuppressive and anti-inflammatory activity for inhibiting the lymphocyte activation associated with auto-immune and inflammatory diseases and the rejection of tissues in transplant patients. Patent WO2019234476

Farias DF, Souza TM, Viana MP, Soares BM, Cunha AP, Vasconcelos IM, Ricardo NMPS, Ferreira PMP, Melo VMM, Carvalho AFU (2013) Antibacterial, antioxidant, and anticholinesterase activities of plant seed extracts from Brazilian semiarid region. Biomed Res Int 3:1–9. https://doi.org/10.1155/2013/510736

Feher M, Schmidt JM (2003) Property distributions: differences between drugs, natural products, and molecules from combinatorial chemistry. J Chem Inf Comput 43(1):218–227. https://doi.org/10.1021/ci0200467

Ferreira PMP, Carvalho AFFU, Farias DF, Cariolano NG, Melo VMM, Queiroz MGR, Martins MAC, Machado-Neto JG (2009) Larvicidal activity of the water extract of *Moringa oleifera* seeds against Aedes aegypti and its toxicity upon laboratory animals. An Acad Bras Cienc 81(2):207–216. https://doi.org/10.1590/s0001-37652009000200007

Ferreira PMP, Lopes LAR, Carnib LPA, Sousa PVL, Lugo LMN, Nunes NMF, Silva JN, Araujo LS, Frota KMG (2019) Cruciferous vegetables as antioxidative, chemopreventive and antineoplasic functional foods: preclinical and clinical evidences of sulforaphane against prostate cancers. Curr Pharm Des 24(40):4779–4793. https://doi.org/10.2174/1381612825666190116124233

Ferreira PMP, Pessoa C (2017) Molecular biology of human epidermal receptors, signaling pathways and targeted therapy against cancers: new evidences and old challenges. Braz J Pharm Sci 53(2):e16076. https://doi.org/10.1590/s2175-97902017000216076

Gao J, Ying Y, Wang J, Cui Y (2020a) Solanine inhibits immune escape mediated by hepatoma Treg cells via the TGFβ/Smad signaling pathway. Biomed Res Int 2020:10. https://doi.org/10.1155/2020/9749631

Gao S, Wang S, Song Y (2020b) Novel immunomodulatory drugs and neo-substrates. Biomarker Res 8(1):1–8. https://doi.org/10.1186/s40364-020-0182-y

Gubarev MJ, Enioutina EY (1998) Method to enhance innate immunity defense mechanisms by treatment with plant-derived alkaloids. Patent WO/1998/058543

Henriquez J, Bach A, Matos-Fernandez K, Crawford R, Kaminski N (2020) Δ9-tetrahydrocannabinol (THC) impairs CD8+ T cell-mediated activation of astrocytes. J Neuroimmune Pharmacol 15:863–874. https://doi.org/10.1007/s11481-020-09912-z

Hernández-Cervantes R, Méndez-Díaz M, Prospéro-García Ó, Morales-Montor J (2017) Immunoregulatory role of cannabinoids during infectious disease. Neuroimmunomodulation. https://doi.org/10.1159/000481824

Hildebert W et al (1986) Inventor Pharmazeutische Fabrik (1986) applicant. Immunostimulating polysaccharides, method for using such, and pharmaceutical preparations containing them. Patent WO 4857512

Islam MT, Ali ES, Uddin SJ, Shaw S, Islam MA, Ahme MI, Shil MC, Karmakar UK, Yarla NS, Khan IN, Billah MM, Piecynska MD, Zengin G, Maliner C, Nicoletti F, Gulei D, Beridan-Neagoe I, Apostolov A, Banach M, Yeung AWK, El-Demerdash A, Xiao J, Dey P, Yele S, Jozwik A, Strzalkpwska N, Marchewka J, Rengasamy K, Horbanczuk J, Kamal MA, Mubarak M, Mishras K, Shilpi JA, Atanasov AG (2018) Phytol: a review of biomedical activities. Food Chem Toxicol 121:82–94. https://doi.org/10.1016/j.fct.2018.08.032

Kak G, Raza M, Tiwari B (2018) Interferon-gamma (IFN-γ): exploring its implications in infectious diseases. Biomol Concepts 9(1):64–79. https://doi.org/10.1515/bmc-2018-0007

Kayser O, Masihi KN, Kiderlen AF (2003) Natural products and synthetic compounds as immunomodulators. Expert Rev Anti-Infect Ther 1(2):319–335. https://doi.org/10.1586/14787210.1.2.319

Kim SH, Lee YC (2009) Piperine inhibits eosinophil infiltration and airway hyperresponsiveness by suppressing T cell activity and Th2 cytokine production in the ovalbumin-induced asthma model. J Pharm Pharmacol 61(3):353–359. https://doi.org/10.1211/jpp/61.03.0010

Kim W, Fan Y-Y, Smith R, Patil B, Jayaprakash G, McMurray D, Chapkin R (2009) Dietary curcumin and Limonin suppress CD4+ T-cell proliferation and Interleukin-2 production in mice. J Nutr 139(5):1042–1048. https://doi.org/10.3945/jn.108.102772

Koren Z (2019) SciCann Therapeutics Inc. Compositions comprising a cannabinoid and spilanthol. CA Toronto, Patent US010512614B2

Li L, Liu C, Wang F (2010) Thymosin alpha 1: biological activities, applications and genetic engineering production. Peptides 31(11):2151–2158. https://doi.org/10.1016/j.peptides.2010.07.026

Lin P, Flynn J (2012) Tuberculosis Research using Nonhuman Primates. Nonhuman Primates in Biomedical Research. https://doi.org/10.1016/B978-0-12-381366-4.00003-1

Lipinski CA, Lombardo F, Dominy BW, Feeney PJ (2001) Experimental and computational approaches to estimate solubility and permeability in drug discovery and development settings. Adv Drug Deliv Rev 46(1–3):3–26. https://doi.org/10.1016/s0169-409x(00)00129-0

Lozoya L, Xavier et al. (2005) Applicant. Improved *Mimosa tenuiflora* extracts, methods for its obtaining and use for the treatment of ulcers in mammals. Patent WO 2007/072233

Maciel AM, Sátiro DSP, Pereira JDS, Rossi CGF, Rocha HAO (2015) Veiculação da fração alcolóidica de Croton cajucara Benth para aplicação na terapêutica imunomoduladora. Patent BR102015014704-0

Mahady GB (2001) Global harmonization of herbal health claims. J Nutr 131(3):1120S–1123S. https://doi.org/10.1093/jn/131.3.1120s

Mandal C, Satyavarapu EM, Chandra BP (2018) Council of Scientific and Industrial Research, Applicant. Pharmaceutical formulations for management of cancers. Patent WO2019116391

Martel J, Yun-Fei K, David MO, Chia-Chen L, Chih-Jung C, Chuan-Sheng L, Hsin-Chih L, John DY (2017) Immunomodulatory properties of plants and mushrooms. Trends Pharmacol Sci 38: 11. https://doi.org/10.1016/j.tips.2017.07.006

Nadeem O, Tai YT, Anderson KC (2020) Immunotherapeutic and targeted approaches in multiple myeloma. Immunotargets Ther 9:201–215. https://doi.org/10.2147/ITT.S240886

National Center for Biotechnology Information (2004) PubChem Compound Summary for CID 9852188, Cannabinoids. https://pubchem.ncbi.nlm.nih.gov/compound/Cannabinoids. Accessed Dec. 16, 2020

Newman DJ, Cragg GM (2012) Natural products as sources of new drugs over the 30 years from 1981 to 2010. J Nat Prod 75(3):311–335. https://doi.org/10.1021/np200906s

Newman DJ, Cragg GM, Snader KM (2003) Natural products as a source of new drugs over the period 1981-2002. J Nat Prod 66(7):1022–1037. https://doi.org/10.1021/np030096l

Nicholson LB (2016) The immune system Essays in biochemistry. 60(3):275–301. https://doi.org/10.1042/EBC20160017

Obreshkova DP, Petrov AP (1999) Inventors. Prophylactic form – medicinal herb preparation of the *Clinopodium vulgare* L. plant. WO 03160. Disponível:https://patentscope.wipo.int/search/en/detail.jsf?docId=BG226518227&_cid=P20-KHC1FE-11504-1. Acesso: novembro 2020

Oláh A, Szekanecz Z, Bíró T (2017) Targeting cannabinoid signaling in the immune system: "high"-ly exciting questions, possibilities, and challenges. Front Immunol 8(1487):1–14. https://doi.org/10.3389/fimmu.2017.01487

Oliveira LB, Russo SL, Marques LG, Gomila JMV (2019) Technological productivity on control of *Boophilus microplus* tick: a patentometric study. Int J Adv Eng Res Sci 6(2):134–143. https://doi.org/10.22161/ijaers.6.2.17

Ordóñez-Vásquez A, Jaramillo-Gómez L, Duran-Correa C et al (2017) A reliable and reproducible model for assessing the effect of different concentrations of α-Solanine on rat bone marrow mesenchymal stem cells. Bone Marrow Res 2017:1–7. https://doi.org/10.1155/2017/2170306

Pan B, Zhong W, Deng Z, Lai C, Jiao G, Liu J, Zho Q (2016) Inhibition of prostate cancer growth by solanine requires the suppression of cell cycle proteins and activation of ROS/P38 signaling pathway. Q Cancer Med 5(911):3214–3222. https://doi.org/10.1002/cam4.916

Pascolutti M, Quinn RJ (2014) Natural products as lead structures: chemical transformations to create lead-like libraries. Drug Discov Today 19(3):215–221. https://doi.org/10.1016/j.drudis.2013.10.013

Pereira JBA, Rodrigues MM, Morais IR, Vieira CRS, Sampaio JPM, Moura MG, Damasceno MFM, Silva JN, Calou IBF, Deus FA, Peron AP, Abreu MC, Militão GCG, Ferreira PMP (2015) The therapeutic role of the program Farmácia viva and the medicinal plants in the center-south of Piauí. Revista Brasileira Plantas Medicinais 17:550–561. https://doi.org/10.1590/1983-084X/14_008

Pessoa C, Ferreira PMP, Costa-Lotufo LV, Moraes MO, Cavalcanti SMT, Coelho LCD, Hernandez MZ, Leite ACL, De Simone CA, Costa VMA, Souza VMO (2010) Discovery of phthalimides as immunomodulatory and antitumor drug prototypes. ChemMedChem 5(4):523–528. https://doi.org/10.1002/cmdc.200900525

Pianowski LF, Brandão DC (2003) Inventors Aché Lab. Pharm. SA, Applicant. Processo de obtenção de extrato potencializado de *Maytenus* spp., fração ou extrato, composição, unidade de dosagem, método para previnir, inibir ou suspender processo ulcerativo, método de tratamento de processo ulcerativo e uso. Patent PI 0300734-0

Quintella CM, Mata AMT, Lima LCP (2019) Overview of bioremediation with technology assessment and emphasis on fungal bioremediation of oil contaminated soils. J Environ Manag. 241156–166. https://doi.org/10.1016/j.jenvman.2019.04.019

Rahman A, Siddiqui SA, Jakhar R, Kang SC (2015) Growth inhibition of various human cancer cell lines by imperatorin and limonin from *Poncirus trifoliata* rafin. Seeds Anticancer Agents Med Chem 15(2):236–241. https://doi.org/10.2174/1871520614666140922122358

Riccio G, Lauritano C (2020) Microalgae with immunomodulatory activities. Mar Drugs 18:1. https://doi.org/10.3390/md18010002

Samuel AJSJ, Kalusalingam A, Chellappan DK, Gopinath R, Radhamanni S, Husain HA, Muruganandham V, Promwichit P (2010) Ethnomedical survey of plants used by the orang Asli in Kampung Bawong, Perak. West Malaysia J Ethnobiol Ethnomed 6(5):1–6. https://doi.org/10.1186/1746-4269-6-5

Santos MM, Massari G, Santos DM, Fellows L (2004) Prospecção de tecnologias de futuro: métodos, técnicas e abordagem. Parcerias Estratégicas 19(9):189–229

Schepetkin IA, Quinn MT (2006) Botanical polysaccharides: macrophage immunomodulation and therapeutic potential. Int Immunopharmacol. https://doi.org/10.1016/j.intimp.2005.10.005

Siegall CB, Gawlak SL, Marquartdt H (1994) Inventors Bristol-Myers Squibb Company, Applicant. Bryodin 2 a ribosomeinactivating protein isolated from the plant Bryonia dioica. Patent US 5597569

Shi WW, Wong KB, Shaw PC (2018) Structural and functional investigation and pharmacological mechanism of Trichosanthin, a type 1 ribosome-inactivating protein. Toxins (Basel) 10(8):335. https://doi.org/10.3390/toxins10080335

Silva J, Pantzartzi C, Votava M (2019) Inosine Pranobex: a key player in the game against a wide range of viral infections and non-infectious diseases. Adv Ther 36(8):1878–1905. https://doi.org/10.1007/s12325-019-00995-6

Silva JN, Drumond RR, Moncao NBN, Peron AP, Sousa JMC, Cito AMGL, Ferreira PMP (2016) Prospective study about antineoplasic properties of plants from Fabaceae family: emphasis in *Mimosa caesalpiniifolia* Benth. Revista GEINTEC: gestao, inovacao e tecnologias 6(3): 3304–3318. https://doi.org/10.7198/S2237-072220160003005

Simões ERB, Santos EA, Abreu MC, Silva JN, Nunes NMF, Costa MP, Pessoa ODL, Pessoa C, Ferreira PMP (2015) Biomedical properties and potentiality of *Lippia microphylla* Cham. And its essential oils. J Intercultural Ethnopharmacol 4(3):256–263. https://doi.org/10.5455/jice.20150610104841

Sousa LQ, Machado KC, Oliviera SFC, Araujo LS, Moncao-Filho ES, Cavalcante AACM, Vieira-Junior GM, Ferreira PMP (2017) Bufadienolides from amphibians: a promising source of anticancer prototypes for radical innovation, apoptosis triggering and Na+/K+-ATPase inhibition. Toxicon 127:63–76. https://doi.org/10.1016/j.toxicon.2017.01.004

Soutar D, Doucette C, Liwski R, Hoskin D (2017) Piperine, a pungent alkaloid from black pepper, inhibits B lymphocyte activation and effector functions. Phytother Res 31:466–474. https://doi.org/10.1002/ptr.5772

Tan B, Vanitha J (2012) Immunomodulatory and antimicrobial effects of some traditional Chinese medicinal herbs: a review. Curr Med, Chem. https://doi.org/10.2174/0929867043365161

Turner S, Williams CM, Iversen L, Whalley BJ (2017) Molecular pharmacology of phytocannabinoids. In: Kinghorn A, Falk H, Gibbons S, Kobayashi J (eds) Phytocannabinoids, vol 103. Springer, Cham, pp 61–101

Turvey SE, Broide DH (2010) Innate immunity. J Allergy Clin Immunol 125(2):S24–S32. https://doi.org/10.1016/j.jaci.2009.07.016

United Nations Environment Programme (UN Environment Programme). Megadiverse Brazil: giving biodiversity an online boost. 2019. https://www.unenvironment.org/news-and-stories/story/megadiverse-brazil-giving-biodiversity-online-boost . Accessed on: 12 dec. 2020

Van Hai N (2015) The use of medicinal plants as immunostimulants in aquaculture: a review. Aquaculture. https://doi.org/10.1016/j.aquaculture.2015.03.014

Vikram A, Jayaprakasha GK, Jesudhasan PR, Pillai SD, Patil BS (2012) Limonin 7-methoxime interferes with *Escherichia coli* biofilm formation and attachment in type 1 pili and antigen 43 dependent manner. Food Control 26(2):427–438. https://doi.org/10.1016/j.foodcont.2012.01.030

Wagner H (1993) Leading structures of plant origin for drug development. J Ethnopharmacol 38(2–3):105–112. https://doi.org/10.1016/0378-8741(93)90004-O

Weidner MS. et al. (1999) Inventors Eurovita A/S, Applicant. Composições sinergísticas contendo compostos aromáticos e terpenóides presentes em *Alpina galanga*. WO 1999/053935

Yaguang L (2003) Special preparation of anticancer drugs made by novel nanotechnology. Patent US20040192641

Yang N, Wang J, Liu C, Liu C, Song Y, Zhang S, Zi J, Zhan J, Masilamai M, Cox A, Nowak-Wegryn A, Sampson H, Ciu-Min L (2014a) Berberine and limonin suppress IgE production by human B cells and peripheral blood mononuclear cells from food-allergic patients. Ann Allergy Asthma Immunol 113:556–564. https://doi.org/10.1016/j.anai.2014.07.021

Yang Q, Huang M, Cai X, Jia L, Wang S (2019) Investigation on activation in RAW264.7 macrophage cells and protection in cyclophosphamide-treated mice of *Pseudostellaria heterophylla* protein hydrolysate. Food Chem Toxicol 134:110816–110820. https://doi.org/10.1016/j.fct.2019.110816

Yang Y, Wang X, Zhu Q, Gong G, Luo D, Jiang A, Yang L, Xu Y (2014b) Synthesis and pharmacological evaluation of novel limonin derivatives as anti-inflammatory and analgesic agents with high water solubility. Bioorg Med Chem Lett 24(7):1851–1855. https://doi.org/10.1016/j.bmcl.2014.02.003

Yoo DG, Kim MC, Park MK, Park KM, Quan FS, Song JM, Wee JJ, Wang BZ, Keol YK, Compan RW, Kang SM (2012) Protective effect of ginseng polysaccharides on influenza viral infection. PLoS One 7(3):e33678. https://doi.org/10.1371/journal.pone.0033678

Zheng W, Zhang H, Jin Y, Wang Q, Chen L, Feng Z, Chen H, Wu Y (2017) Butein inhibits IL-1β-induced inflammatory response in human osteoarthritis chondrocytes and slows the progression of osteoarthritis in mice. Int Immunopharmacol 42:1–10. https://doi.org/10.1016/j.intimp.2016.11.009

Zong A, Cao H, Wang F (2012) Anticancer polysaccharides from natural resources: A review of recent research. Carbohydr Polym 90(4):1395–1410. https://doi.org/10.1016/j.carbpol.2012.07.026

Zundler S, Neurath M (2015) Interleukin-12: functional activities and implications for disease. Cytokine Growth Factor Rev 26(5):559–568. https://doi.org/10.1016/j.cytogfr.2015.07.003

MIX
Papier aus verantwortungsvollen Quellen
Paper from responsible sources
FSC® C105338

If you have any concerns about our products,
you can contact us on
ProductSafety@springernature.com

In case Publisher is established outside the EU,
the EU authorized representative is:
Springer Nature Customer Service Center GmbH
Europaplatz 3, 69115 Heidelberg, Germany

Printed by Libri Plureos GmbH
in Hamburg, Germany